Edited by
Valentine Nenajdenko

Isocyanide Chemistry

Related Titles

Cordova, A. (ed.)

Catalytic Asymmetric Conjugate Reactions

2010
ISBN: 978-3-527-32411-8

Cossy, J., Arseniyadis, S., Meyer, C. (eds.)

Metathesis in Natural Product Synthesis

Strategies, Substrates and Catalysts

2010
ISBN: 978-3-527-32440-8

Blaser, H.-U., Federsel, H.-J. (eds.)

Asymmetric Catalysis on Industrial Scale

Challenges, Approaches and Solutions

2010
ISBN: 978-3-527-32489-7

Quin, L. D., Tyrell, J.

Fundamentals of Heterocyclic Chemistry

Importance in Nature and in the Synthesis of Pharmaceuticals

2010
E-Book
ISBN: 978-0-470-62653-5

Edited by Valentine Nenajdenko

Isocyanide Chemistry

Applications in Synthesis and
Material Science

WILEY-
VCH

WILEY-VCH Verlag GmbH & Co. KGaA

The Editor

Prof. Dr. Valentine G. Nenajdenko
Moscow State University
Leninskie Gory
119992 Moscow
Russia

All books published by **Wiley-VCH** are carefully produced. Nevertheless, authors, editors, and publisher do not warrant the information contained in these books, including this book, to be free of errors. Readers are advised to keep in mind that statements, data, illustrations, procedural details or other items may inadvertently be inaccurate.

Library of Congress Card No.: applied for

British Library Cataloguing-in-Publication Data
A catalogue record for this book is available from the British Library.

Bibliographic information published by the Deutsche Nationalbibliothek
The Deutsche Nationalbibliothek lists this publication in the Deutsche Nationalbibliografie; detailed bibliographic data are available on the Internet at <http://dnb.d-nb.de>.

© 2012 Wiley-VCH Verlag & Co. KGaA, Boschstr. 12, 69469 Weinheim, Germany

All rights reserved (including those of translation into other languages). No part of this book may be reproduced in any form – by photoprinting, microfilm, or any other means – nor transmitted or translated into a machine language without written permission from the publishers. Registered names, trademarks, etc. used in this book, even when not specifically marked as such, are not to be considered unprotected by law.

Composition   Toppan Best-set Premedia Limited, Hong Kong

Printing and Binding   Markono Print Media Pte Ltd, Singapore

Cover Design   Schulz Grafik-Design, Fußgönheim

Print ISBN: 978-3-527-33043-0
ePDF ISBN: 978-3-527-65256-3
ePub ISBN: 978-3-527-65255-6
mobi ISBN: 978-3-527-65254-9
oBook ISBN: 978-3-527-65253-2

Printed in Singapore
Printed on acid-free paper

Contents

Preface *XIII*
List of Contributors *XV*

1 **Chiral Nonracemic Isocyanides** *1*
Luca Banfi, Andrea Basso, and Renata Riva
1.1 Introduction *1*
1.2 Simple Unfunctionalized Isocyanides *1*
1.3 Isocyanides Containing Carboxylic, Sulfonyl, or Phosphonyl Groups *4*
1.3.1 α-Isocyano Esters *4*
1.3.2 α-Isocyano Amides *7*
1.3.3 Other Isocyano Esters or Amides *9*
1.3.4 Chiral Sulfonylmethyl or Phosphonylmethyl Isocyanides *10*
1.4 Isocyanides Containing Amino or Alcoholic Functionalities *11*
1.4.1 Chiral Amino or Azido Isocyanides *11*
1.4.2 Chiral Hydroxy Isocyanides *12*
1.5 Natural Isocyanides *16*
1.5.1 Isolation and Natural Sources *16*
1.5.2 Synthesis of Naturally Occurring Isocyanides *17*
1.6 Isocyanides Used in the Synthesis of Chiral Polyisocyanides *23*
1.6.1 Properties *24*
1.6.2 Synthesis *25*
1.6.3 Applications *26*
References *26*

2 **General Aspects of Isocyanide Reactivity** *35*
Maxim A. Mironov
2.1 Introduction *35*
2.2 Isocyanide–Cyanide Rearrangement *37*
2.3 Oxidation/Reduction of the Isocyano Group *41*
2.3.1 Oxidation of the Isocyano Group *41*
2.3.2 Reactions with Sulfur and Selenium *43*

2.3.3	Reduction of the Isocyano Group 45
2.4	Reactions of Isocyanides with Electrophiles 47
2.4.1	Reaction with Acids 49
2.4.2	Reactions with Halogens and Acyl Halides 52
2.4.3	Reactions with Activated Alkenes and Alkynes 55
2.4.4	Reactions with Carbonyl Compounds and Imines 58
2.4.5	Reactions with Activated Heterocumulens 60
2.5	Reactions of Isocyanides with Nucleophiles 62
2.5.1	Reactions with Organometallic Compounds 62
2.5.2	Reactions with Hydroxide, Alcohols, and Amines 64
2.6	Conclusions 66
	References 67
3	**α-Acidic Isocyanides in Multicomponent Chemistry** *75*
	Niels Elders, Eelco Ruijter, Valentine G. Nenajdenko, and Romano V.A. Orru
3.1	Introduction 75
3.2	Synthesis of α-Acidic Isocyanides 76
3.3	Reactivity of α-Acidic Isocyanides 78
3.4	MCRs Involving α-Acidic Isocyanides 80
3.4.1	van Leusen Imidazole MCR 81
3.4.2	2,6,7-Trisubstituted Quinoxaline MCR 82
3.4.3	4,5-Disubstituted Oxazole MCR 83
3.4.4	Nitropyrrole MCR 83
3.4.5	2,4,5-Trisubstituted Oxazole MCR 84
3.4.5.1	2,4,5-Trisubstituted Oxazoles 84
3.4.5.2	Variations on the 2,4,5-Trisubstituted Oxazole MCR 86
3.4.5.3	Oxazole MCR and *In-Situ* Domino Processes 88
3.4.6	2-Imidazoline MCR 91
3.4.6.1	2-Imidazoline MCR in the Union of MCRs 93
3.4.7	Dihydropyridone MCR 95
3.5	Conclusions 97
	References 98
4	**Synthetic Application of Isocyanoacetic Acid Derivatives** *109*
	Anton V. Gulevich, Alexander G. Zhdanko, Romano V.A. Orru, and Valentine G. Nenajdenko
4.1	Introduction 109
4.2	Synthesis of α-Isocyanoacetate Derivatives 109
4.3	Alkylation of Isocyanoacetic Acid Derivatives 113
4.4	α-Isocyanoacetates as Michael Donors 115
4.5	Reaction of Isocyanoacetic Acids with Alkynes: Synthesis of Pyrroles 119
4.6	Reaction of Isocyanoacetic Acid Derivatives with Carbonyl Compounds and Imines 121
4.6.1	Aldol-Type Reaction of Isocyanoacetic Acids with Aldehydes: Synthesis of Oxazolines 122

4.6.2	Transition Metal-Catalyzed Aldol-Type Reactions	*124*
4.6.3	Reaction of Isocyanoacetic Acids with Imines: Imidazoline Formation	*126*
4.7	Reaction with Acylating Agents	*129*
4.8	Multicomponent Reactions of Isocyanoacetic Acid Derivatives	*133*
4.9	Chemistry of Isocyanoacetates Bearing an Additional Functional Group	*134*
4.10	Reactions of Isocyanoacetic Acids with Sulfur Electrophiles	*138*
4.11	Miscellaneous Reactions	*139*
4.12	Concluding Remarks	*144*
4.13	Notes Added in Proof	*145*
	References	*145*
5	**Ugi and Passerini Reactions with Carboxylic Acid Surrogates** *159*	
	Laurent El Kaïm and Laurence Grimaud	
5.1	Introduction	*159*
5.2	Carboxylic Acid Surrogates	*160*
5.2.1	Thiocarboxylic Acids	*160*
5.2.2	Carbonic Acid and Derivatives	*163*
5.2.3	Selenide and Sulfide	*165*
5.2.4	Silanol	*165*
5.2.5	Isocyanic Acid and Derivatives	*166*
5.2.6	Hydrazoic Acid	*167*
5.2.7	Phenols and Derivatives	*171*
5.2.8	Cyanamide	*179*
5.3	Use of Mineral and Lewis Acids	*180*
5.3.1	Ugi and Passerini Reactions Triggered by Mineral Acids	*181*
5.3.2	Ugi and Passerini Reactions Triggered by Lewis Acids	*184*
5.4	Conclusions	*189*
	References	*189*
6	**Amine (Imine) Component Surrogates in the Ugi Reaction and Related Isocyanide-Based Multicomponent Reactions** *195*	
	Mikhail Krasavin	
6.1	Introduction	*195*
6.2	Hydroxylamine Components in the Ugi Reaction	*196*
6.3	Hydrazine Components in the Ugi Reaction	*200*
6.4	Miscellaneous Amine Surrogates for the Ugi Reaction	*218*
6.5	Activated Azines in Reactions with Isocyanides	*220*
6.6	Enamines, Masked Imines, and Cyclic Imines in the Ugi Reaction	*223*
6.7	Concluding Remarks	*227*
	Acknowledgments	*227*
	References	*227*

7	**Multiple Multicomponent Reactions with Isocyanides** 233
	Ludger A. Wessjohann, Ricardo A.W. Neves Filho, and Daniel G. Rivera
7.1	Introduction 233
7.2	One-Pot Multiple IMCRs 234
7.2.1	Synthesis of Multivalent Glycoconjugates 236
7.2.2	Synthesis of Hybrid Peptide–Peptoid Podands 237
7.2.3	Covalent Modification and Immobilization of Proteins 240
7.2.4	Assembly of Polysaccharide Networks as Synthetic Hydrogels 241
7.2.5	Synthesis of Macromolecules by Multicomponent Polymerization 243
7.3	Isocyanide-Based Multiple Multicomponent Macrocyclizations 243
7.3.1	Synthesis of Hybrid Macrocycles by Double Ugi-4CR-Based Macrocyclizations 244
7.3.2	Synthesis of Macrobicycles by Threefold Ugi-4CR-Based Macrocyclization 246
7.4	Sequential Isocyanide-Based MCRs 248
7.4.1	Sequential Approaches to Linear and Branched Scaffolds 248
7.4.2	Sequential Approaches to Macrocycles 254
7.4.3	Convergent Approach to Natural Product Mimics 256
7.5	Conclusions 257
	References 258

8	**Zwitterions and Zwitterion-Trapping Agents in Isocyanide Chemistry** 263
	Ahmad Shaabani, Afshin Sarvary, and Ali Maleki
8.1	Introduction 263
8.2	Generation of Zwitterionic Species by the Addition of Isocyanides to Alkynes 265
8.2.1	CH-Acids as Zwitterion-Trapping Agents 266
8.2.2	NH-Acids as Zwitterion-Trapping Agents 271
8.2.3	OH-Acids as Zwitterion-Trapping Agents 273
8.2.4	Carbonyl Compounds as Zwitterion-Trapping Agents 275
8.2.5	Imine Compounds as Zwitterion-Trapping Agents 278
8.2.6	Electron-Deficient Olefins as Zwitterion-Trapping Agents 279
8.2.7	Miscellaneous Compounds as Zwitterion-Trapping Agents 280
8.3	Generation of Zwitterionic Species by the Addition of Isocyanides to Arynes 283
8.4	Generation of Zwitterionic Species by the Addition of Isocyanides to Electron-Deficient Olefins 284
8.5	Miscellaneous Reports for the Generation of Zwitterionic Species 286
8.6	Isocyanides as Zwitterion-Trapping Agents 287
8.7	Conclusions 289
	Acknowledgments 289
	References 289

9	**Recent Progress in Nonclassical Isocyanide-Based MCRs** *299*	
	Rosario Ramón, Nicola Kielland, and Rodolfo Lavilla	
9.1	Introduction *299*	
9.2	Type I MCRs: Isocyanide Attack on Activated Species *300*	
9.3	Type II MCRs: Isocyanide Activation *308*	
9.4	Type III MCRs: Formal Isocyanide Insertion Processes *320*	
9.5	Conclusions *327*	
	Acknowledgments *327*	
	References *327*	
10	**Applications of Isocyanides in IMCRs for the Rapid Generation of Molecular Diversity** *335*	
	Muhammad Ayaz, Fabio De Moliner, Justin Dietrich, and Christopher Hulme	
10.1	Introduction *335*	
10.2	Ugi/Deprotect/Cyclize (UDC) Methodology *337*	
10.2.1	Ugi-4CC: One Internal Nucleophile *337*	
10.2.2	TMSN$_3$-Modified Ugi-4CC: One Internal Nucleophile *343*	
10.2.3	Ugi-4CC: Two Internal Nucleophiles *344*	
10.2.4	Ugi-4CC: Three Internal Nucleophiles *347*	
10.2.5	Ugi-5CC: One Internal Nucleophile *348*	
10.3	Secondary Reactions of Ugi Products *350*	
10.3.1	Nucleophilic Additions and Substitutions *351*	
10.3.1.1	Alkylations *351*	
10.3.1.2	Mitsunobu Reactions *352*	
10.3.1.3	Lactonization and Lactamization *354*	
10.3.2	Base- or Acid-Promoted Condensations *355*	
10.3.3	Nucleophilic Aromatic Substitutions *355*	
10.3.4	Palladium-Mediated Reactions *356*	
10.3.5	Ring-Closing Metatheses *358*	
10.3.6	Staudinger–aza-Wittig Reactions *358*	
10.3.7	Cycloadditions *359*	
10.4	The Bifunctional Approach (BIFA) *361*	
10.4.1	Applications of Amino Acids *363*	
10.4.2	Applications of Cyclic Imines *365*	
10.4.3	Applications of Tethered Aldehyde and Keto Acids *366*	
10.4.4	Heterocyclic Amidines as a Tethered Ugi Input *371*	
10.4.5	Combined Bifunctional and Post-Condensation Modifications *372*	
	Acknowledgments *375*	
	Abbreviations *375*	
	References *376*	

11 Synthesis of Pyrroles and Their Derivatives from Isocyanides *385*
Noboru Ono and Tetsuo Okujima
11.1 Introduction *385*
11.2 Synthesis of Pyrroles Using TosMIC *386*
11.3 Synthesis of Pyrroles Using Isocyanoacetates *391*
11.3.1 Synthesis from Nitroalkenes *391*
11.3.2 Synthesis from α,β-Unsaturated Sulfones *396*
11.3.3 Synthesis from Alkynes *401*
11.3.4 Synthesis from Aromatic Nitro Compounds: Isoindole Derivatives *402*
11.4 Synthesis of Porphyrins and Related Compounds *407*
11.4.1 Tetramerization *407*
11.4.2 Meso-Tetraarylporphyrins via the Lindsey Procedure *412*
11.4.3 [3+2] and [2+2] Methods *414*
11.4.4 Expanded, Contracted, and Isomeric Porphyrins *414*
11.4.5 Functional Dyes from Pyrroles *420*
11.5 Conclusion *423*
References *424*

12 Isocyanide-Based Multicomponent Reactions towards Benzodiazepines *431*
Yijun Huang and Alexander Dömling
12.1 Introduction *431*
12.2 1,4-Benzodiazepine Scaffolds Assembled via IMCR Chemistry *433*
12.2.1 Two-Ring Systems *433*
12.2.2 Fused-Ring Systems *440*
12.3 1,5-Benzodiazepine Scaffolds Assembled via IMCR Chemistry *443*
12.4 Outlook *446*
References *446*

13 Applications of Isocyanides in the Synthesis of Heterocycles *451*
Irini Akritopoulou-Zanze
13.1 Introduction *451*
13.2 Furans *451*
13.3 Pyrroles *453*
13.4 Oxazoles *459*
13.5 Isoxazoles *461*
13.6 Thiazoles *464*
13.7 Imidazoles *466*
13.8 Pyrazoles *466*
13.9 Oxadiazoles and Triazoles *470*
13.10 Tetrazoles *471*
13.11 Benzofurans and Benzimidazoles *473*
13.12 Indoles *473*
13.13 Quinolines *477*

13.14	Quinoxaline *479*	
	Abbreviations *480*	
	References *480*	
14	**Renaissance of Isocyanoarenes as Ligands in Low-Valent Organometallics** *493*	
	Mikhail V. Barybin, John J. Meyers, Jr, and Brad M. Neal	
14.1	Historical Perspective *493*	
14.2	Isocyanidemetalates and Related Low-Valent Complexes *497*	
14.2.1	Introduction *497*	
14.2.2	Four-Coordinate Isocyanidemetalates and Redox-Related Complexes *497*	
14.2.3	Five-Coordinate Isocyanidemetalates *502*	
14.2.4	Six-Coordinate Isocyanidemetalates and Redox-Related Complexes *504*	
14.3	Coordination and Surface Chemistry of Nonbenzenoid Isocyanoarenes *508*	
14.3.1	Isocyanoazulenes *508*	
14.3.2	Organometallic η^5-Isocyanocyclopentadienides *509*	
14.3.3	Homoleptic Complexes of Nonbenzenoid Isocyanoarenes *510*	
14.3.4	Bridging Nonbenzenoid Isocyanoarenes *514*	
14.3.5	Self-Assembled Monolayer Films of Nonbenzenoid Isocyano- and Diisocyanoarenes on Gold Surfaces *517*	
14.4	Conclusions and Outlook *521*	
	Acknowledgments *522*	
	References *523*	
15	**Carbene Complexes Derived from Metal-Bound Isocyanides: Recent Advances** *531*	
	Konstantin V. Luzyanin and Armando J.L. Pombeiro	
15.1	Introduction *531*	
15.2	Coupling of the Isocyanide Ligand with Simple Amines or Alcohols *532*	
15.3	Coupling of the Isocyanide Ligand with Functionalized Amines or Alcohols *537*	
15.4	Coupling of the Isocyanide Ligand with a Hydrazine or Hydrazone *537*	
15.5	Coupling of the Isocyanide Ligand with an Imine or Amidine *538*	
15.6	Intramolecular Cyclizations of Functionalized Isocyanide Ligands *540*	
15.7	Coupling of Isocyanides with Dipoles *543*	
15.8	Other Reactions *544*	
15.9	Final Remarks *546*	
	Acknowledgments *546*	
	References *547*	

16	**Polyisocyanides** *551*
	Niels Akeroyd, Roeland J.M. Nolte, and Alan E. Rowan
16.1	Introduction *551*
16.1.1	Chiral Polymers *551*
16.1.2	Polyisocyanides and Their Monomers *553*
16.2	The Polymerization Mechanism *553*
16.3	Conformation of the Polymeric Backbone *556*
16.4	Polyisocyanopeptides *561*
16.5	Polyisocyanides as Scaffolds for the Anchoring of Chromophoric Molecules *563*
16.6	Functional Polyisocyanides *570*
16.7	Conclusions and Outlook *575*
	References *576*

Index *587*

Preface

Isocyanide (isonitrile) chemistry began in 1859 when Lieke obtained the first compound of this type. In 1958, isocyanides became generally available by dehydration of formamides prepared from primary amines. This discovery and many other important inventions in the chemistry of isocyanides have been attributed to Ivar Ugi. He was probably the first person to understand the exceptional nature of isocyano functionality and its rich synthetic possibilities. Being stable carbenes, isonitriles are highly reactive compounds that can react with almost any type of reagents (electrophiles, nucleophiles and even radicals). Today isocyanide chemistry is a broad and important part of organic chemistry; however inorganic, coordination, polymeric, combinatorial and medicinal chemistry explore the rich reactivity of isonitriles as well. Multicomponent reactions with isocyanides are used for synthesis of broad varieties of peptides and peptide mimetics. The renaissance of isocyanide chemistry was at the end of the 20th century when thousands of new compounds libraries became highly desirable for diversity-oriented synthesis, high-throughput screening and drug discovery. Isocyanide-based multicomponent reactions are out of competition in terms of effectiveness and economy to synthesize drugs like compounds or natural compounds only in a single synthetic step.

In this book, an effort has been made to provide a comprehensive modern view of all the most significant branches of isocyanide chemistry, demonstrating how important are these compounds to date and how significant is their impact on chemistry. It should be pointed out that the book *Isonitrile Chemistry* was published by Ivar Ugi in 1971. Since then a number of excellent reviews and monograph chapters regarding isocyanides, in particular their multicomponent reactions, have been published. However, a book devoted to the chemistry of isocyanides has not been published for more than 40 years.

It is a great honor and pleasure for me to be the editor of this book. I would like to thank all the authors of the individual chapters for their excellent contributions. These outstanding scientists are known experts in the field of isocyanide chemistry. This book is a result of worldwide cooperation of contributors from many countries. I would like also to thank all my collaborators at Wiley-VCH for help to realize this project.

I also wish to use this opportunity to mention my personal love for isocyanide chemistry. Almost 25 years ago as a student, I read *Isonitrile Chemistry* by Ivar Ugi.

Such a beautiful and rich chemistry made me dream to do something important and interesting in this field. However, it was impossible at that time because I was still a student. Nevertheless, I synthesized my first isocyanide and had experience with specific odors of isonitriles. My next step to isocyanides was the conference in Yaroslavl, Russia, in 2001, where I met Ivar Ugi. We had a long and fruitful discussion, and this talk supported me significantly. Since then my laboratory has been involved in isocyanide chemistry. I would like to dedicate this book to the memory of an outstanding chemist and major pioneer of isocyanide chemistry, Ivar Ugi.

Moscow, 2012　　　　　　　　　　　　　　　　　　　　　　*Valentine Nenajdenko*

List of Contributors

Niels Akeroyd
Radboud University Nijmegen
Institute for Molecules and Materials
Heyendaalseweg 135
6525 AJ Nijmegen
The Netherlands

Irini Akritopoulou-Zanze
Abbott Laboratories
Scaffold-Oriented Synthesis
Abbott Park, IL 60064
USA

Muhammad Ayaz
The University of Arizona
College of Pharmacy
BIO5 Oro Valley
Tucson, AZ 85737
USA

Luca Banfi
Università a degli Studi di Genova
Dipartimento di Chimica e
Chimica Industriale
Via Dodecaneso 31
16146 Genova
Italy

Mikhail V. Barybin
The University of Kansas
Department of Chemistry
1251 Wescoe Hall Drive
Lawrence, KS 66045
USA

Andrea Basso
Università a degli Studi di Genova
Dipartimento di Chimica e
Chimica Industriale
Via Dodecaneso 31
16146 Genova
Italy

Fabio De Moliner
The University of Arizona
College of Pharmacy
BIO5 Oro Valley
Tucson, AZ 85737
USA

Justin Dietrich
The University of Arizona
College of Pharmacy
BIO5 Oro Valley
Tucson, AZ 85737
USA

List of Contributors

Alexander Dömling
University of Pittsburgh
School of Pharmacy
Department of Pharmaceutical
Sciences
Pittsburgh, PA 15261
USA

Laurent El Kaïm
Ecole Nationale Supérieure des
Techniques Avancées
Unité Chimie et Procédés
UMR 7652
CNRS-ENSTA-Polytechnique
32 Bd Victor
75012 Paris
France

Niels Elders
VU University Amsterdam
Department of Chemistry &
Pharmaceutical Sciences
De Boelelaan 1083
1081 HV Amsterdam
The Netherlands

Laurence Grimaud
Ecole Nationale Supérieure des
Techniques Avancées
Unité Chimie et Procédés
UMR 7652
CNRS-ENSTA-Polytechnique
32 Bd Victor
75012 Paris
France

Anton V. Gulevich
Moscow State University
Department of Chemistry
Leninskie Gory
Moscow 119991
Russia

Yijun Huang
University of Pittsburgh
School of Pharmacy
Department of Pharmaceutical
Sciences
Pittsburgh, PA 15261
USA

Christopher Hulme
The University of Arizona
College of Pharmacy
BIO5 Oro Valley
Tucson, AZ 85737
USA

Nicola Kielland
Barcelona Science Park
University of Barcelona
Baldiri Reixac 10-12
08028 Barcelona
Spain

Mikhail Krasavin
Griffith University
Eskitis Institute
Brisbane, QLD 4111
Australia

Rodolfo Lavilla
Barcelona Science Park
University of Barcelona
Baldiri Reixac 10-12
08028 Barcelona
Spain

Konstantin V. Luzyanin
Technical University of Lisbon
Centro de Química Estrutural
Instituto Superior Técnico
1049-001 Lisbon
Portugal

List of Contributors | XVII

Ali Maleki
Iran University of Science and
Technology
Department of Chemistry
Narmak
Tehran 16846-13114
Iran

John J. Meyers, Jr
The University of Kansas
Department of Chemistry
1251 Wescoe Hall Drive
Lawrence, KS 66045
USA

Maxim A. Mironov
Ural Federal University
Department of Technology
for Organic Synthesis
str. Mira, 19
620002 Ekaterinburg
Russia

Brad M. Neal
The University of Kansas
Department of Chemistry
1251 Wescoe Hall Drive
Lawrence, KS 66045
USA

Valentine G. Nenajdenko
Moscow State University
Department of Chemistry
Leninskie Gory
119991 Moscow
Russia

Ricardo A.W. Neves Filho
Leibniz Institute of Plant Biochemistry
Department of Bioorganic Chemistry
Weinberg 3
06120 Halle (Saale)
Germany

Roeland J.M. Nolte
Radboud University Nijmegen
Institute for Molecules and Materials
Heyendaalseweg 135
6525 AJ Nijmegen
The Netherlands

Tetsuo Okujima
Ehime University
Graduate School of Science and
Engineering
Department of Chemistry and Biology
2-5 Bunkyo-cho
Matsuyama 790-8577
Japan

Noboru Ono
Kyoto University
Institute for Integrated Cell-Material
Sciences (iCeMS)
Nishikyo-ku
Kyoto 615-8510
Japan

Romano V.A. Orru
Vrije Universiteit Amsterdam
Department of Chemistry and
Pharmaceutical Sciences
De Boelelaan 1083
1081 HV Amsterdam
The Netherlands

Armando J.L. Pombeiro
Technical University of Lisbon
Centro de Química Estrutural
Instituto Superior Técnico
1049-001 Lisbon
Portugal

Rosario Ramón
Barcelona Science Park
University of Barcelona
Baldiri Reixac 10-12
08028 Barcelona
Spain

Renata Riva
Università a degli Studi di Genova
Dipartimento di Chimica e
Chimica Industriale
Via Dodecaneso 31
16146 Genova
Italy

Daniel G. Rivera
Leibniz Institute of Plant Biochemistry
Department of Bioorganic Chemistry
Weinberg 3
06120 Halle (Saale)
Germany
and
Faculty of Chemistry
University of Havana
Center for Natural Products Study
Zapata y G
10400 La Habana
Cuba

Alan E. Rowan
Radboud University Nijmegen
Institute for Molecules and Materials
Heyendaalseweg 135
6525 AJ Nijmegen
The Netherlands

Eelco Ruijter
VU University Amsterdam
Department of Chemistry &
Pharmaceutical Sciences
De Boelelaan 1083
1081 HV Amsterdam
The Netherlands

Afshin Sarvary
Shahid Beheshti University
Department of Chemistry
G. C. P. O. Box 19396-4716
Tehran
Iran

Ahmad Shaabani
Shahid Beheshti University
Department of Chemistry
G. C. P. O. Box 19396-4716
Tehran
Iran

Ludger A. Wessjohann
Leibniz Institute of Plant Biochemistry
Department of Bioorganic Chemistry
Weinberg 3
06120 Halle (Saale)
Germany

Alexander G. Zhdanko
Moscow State University
Department of Chemistry
Leninskie Gory
Moscow 119991
Russia

1
Chiral Nonracemic Isocyanides
Luca Banfi, Andrea Basso, and Renata Riva

1.1
Introduction

Although isocyanides have proven to be very useful synthetic intermediates – especially in the field of multicomponent reactions – most research investigations performed to date on isocyanides have involved commercially available, unfunctionalized and achiral (or chiral racemic) compounds. Two reasons can be envisioned for the infrequent use of enantiomerically pure isocyanides: (i) the general lack of asymmetric induction produced by them; and (ii) the high tendency to lose stereochemical integrity in some particular classes of isonitriles. However, it is believed that when these drawbacks are overcome, the use of chiral nonracemic isocyanides in multicomponent reactions can be very precious, allowing a more thorough exploration of diversity (in particular stereochemical diversity) in the final products. Recently, several reports have been made describing the preparation and use of new classes of functionalized chiral isocyanides. In fact, several chiral isocyanides may be found in nature, and these will be briefly described in Section 1.5, with attention focused on their total syntheses. Another growing application of chiral isocyanides is in the synthesis of chiral helical polyisocyanides.

It is hoped that this review will encourage chemists first to synthesize a larger number of chiral isocyanides, and subsequently to exploit them in multicomponent reactions, in total synthesis, and in the material sciences.

1.2
Simple Unfunctionalized Isocyanides

The standard method used to prepare chiral isocyanides (whether functionalized, or not) begins from the corresponding amines, and employs a two-step sequence of formylation and dehydration (Scheme 1.1). Many enantiomerically pure amines are easily available from natural sources, classical resolution [1], or asymmetric synthesis. Formylation is commonly achieved via four general

Isocyanide Chemistry: Applications in Synthesis and Material Science, First Edition. Edited by Valentine Nenajdenko.
© 2012 Wiley-VCH Verlag GmbH & Co. KGaA. Published 2012 by Wiley-VCH Verlag GmbH & Co. KGaA.

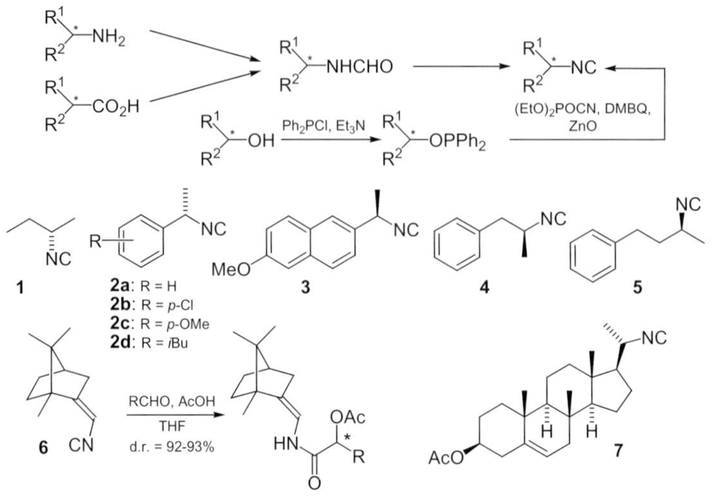

1 Chiral Nonracemic Isocyanides

Scheme 1.1

methods: (i) refluxing the amine in ethyl formate [2]; (ii) reacting the amine with the mixed formic–acetic anhydride [2]; (iii) reacting the amine with formic acid and DCC (dicyclohexylcarbodiimide) [3] or other carbodiimides [4]; and (iv) reacting the amine with an activated formic ester, such as cyanomethyl formate [5], p-nitrophenyl formate [6], or 2,4,5-trichlorophenyl formate [7]. For the dehydration step, several reagents are available, with the commonest and mildest methods involving $POCl_3$, diphosgene, or triphosgene at low temperatures in the presence of a tertiary amine [2]. Although less commonly used, Burgess reagent (methyl N-(triethylammoniumsulfonyl)carbamate) [8] and the $CCl_4/PPh_3/Et_3N$ system [7] have also been employed.

Alternatively, formamides can be obtained from chiral carboxylic acids, through a stereospecific Curtius rearrangement followed by reduction of the resulting isocyanate [9, 10].

Isocyanides may also be prepared from alcohols, by conversion of the alcohol into a sulfonate or halide, followed by S_N2 substitution with AgCN [11]; however, this method works well only with primary alcohols. In contrast, a series of chiral isocyanides have been synthesized from chiral secondary alcohols via a two-step protocol that involves conversion first into diphenylphosphinite, followed by a stereospecific substitution that proceeds with a complete inversion of configuration [12]. The substitution step is indeed an oxidation–substitution, that employs dimethyl-1,4-benzoquinone (DMBQ) as a stoichiometric oxidant and ZnO as an additive. Alternatively, primary or secondary alcohols can be converted into formamides through the corresponding alkyl azides and amines.

Some examples of simple chiral isocyanides are shown in Scheme 1.1. These materials have all been prepared in a traditional manner, starting from chiral amines; the exception here is **5**, which was synthesized from the secondary alcohol.

The compounds comprise fully aliphatic examples such as **1** [13], α-substituted benzyl isocyanides such as **2** [1, 2, 13, 14] and **3** [14, 15], and α-substituted phenethyl or phenylpropyl isocyanides such as **4** [2] and **5** [12].

Because of the great synthetic importance of isocyanide-based multicomponent reactions, these chiral isocyanides have been often used as inputs in these reactions. The use of enantiomerically pure isocyanides can, in principle, bring about two advantages: (i) the possibility to obtain a stereochemically diverse adduct, controlling the absolute configuration of the starting isonitrile; and (ii) the possibility to induce diastereoselection in the multicomponent reaction. With regards to the second of these benefits, the results have been often disappointing, most likely because of the relative unbulkiness of this functional group. For example, Seebach has screened a series of chiral isocyanides, including **2a** and **4** in the $TiCl_4$-mediated addition to aldehydes, but with no diastereoselection at all [2]. This behavior seems quite general also for the functionalized isocyanides described later, the only exception known to date being represented by the camphor-derived isocyanide **6** [16], which afforded good levels of diastereoselection in Passerini reactions. The same isonitrile gave no asymmetric induction in the corresponding Ugi reaction, however. Steroidal isocyanides have also been reported (i.e., **7**) [17, 18].

Apart from multicomponent reactions, and the synthesis of polyisocyanides (see Section 1.6), chiral unfunctionalized isocyanides have been used as intermediates in the synthesis of chiral nitriles, exploiting the stereospecific (retention) rearrangement of isocyanides into nitriles under flash vacuum pyrolysis conditions (FVP) [14, 19]. This methodology was used for the enantioselective synthesis of the anti-inflammatory drugs ibuprofen and naproxen, from **2d** and **3**, respectively. As isocyanides are usually prepared from amines, the overall sequence represents the homologation of an amine to a carboxylic derivative, and is therefore opposite to the Curtius rearrangement.

Another interesting application of **2a**, as a chiral auxiliary, was reported by Alcock *et al.* (Scheme 1.2). Here, the chiral isocyanide reacts with racemic vinylketene tricarbonyliron(0) complex **8** to produce two diastereomeric (vinylketeneimine) tricarbonyliron complexes **9** that can be separated. Subsequent reaction with an organolithium reagent, followed by an oxidative work-up, was found to be highly diastereoselective, forming only adduct **10**. This represents a useful method for

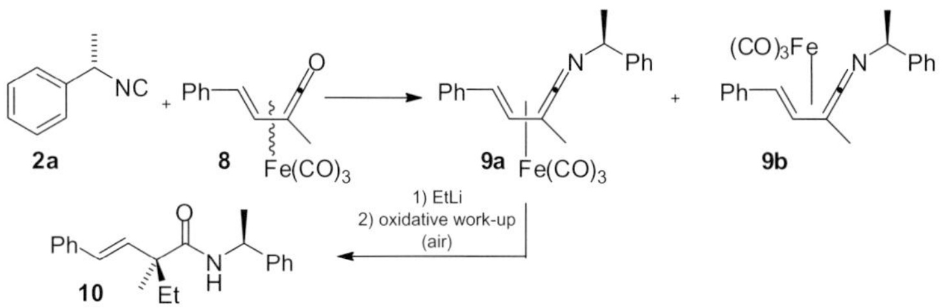

Scheme 1.2

accessing quaternary stereogenic centers, with the induction being clearly due to the tricarbonyliron group, while the isocyanide chirality serves only as a means of separating the two axial stereoisomers **9a** and **9b** [15].

1.3
Isocyanides Containing Carboxylic, Sulfonyl, or Phosphonyl Groups

As the reactivity of α-isocyano esters and amides is reviewed in Chapters 3 and 4 of this book, attention at this point will be focused only on stereochemical issues; reactions exploiting reactivity at the α position will not be described.

1.3.1
α-Isocyano Esters

Enantiomerically pure α-isocyano esters **12** can be prepared by the dehydration of formamides **11**, which in turn are synthesized in two steps from the corresponding α-amino acids [20, 21] (Scheme 1.3). The most critical step is dehydration, which has been demonstrated in some instances to be partly racemizing. The combination of diphosgene with N-methylmorpholine (NMM) at a low (<−25 °C) temperature has been reported in various studies to be able to avoid racemization and to be superior to the use of $POCl_3$ with more basic amines [2, 22–25]. In a recent extensive study, the use of triphosgene/NMM at −30 °C was suggested as the method of choice [26], although a direct comparison of triphosgene with diphosgene was not carried out.

These isocyanides would be very useful in multicomponent reactions, such as the Passerini and Ugi condensations, for the straightforward preparation of depsipeptides or peptides, although racemization may be a relevant issue. Under Passerini conditions, these compounds appear to be configurationally stable during reaction with various aldehydes [22, 27–29], and this approach has been used, for

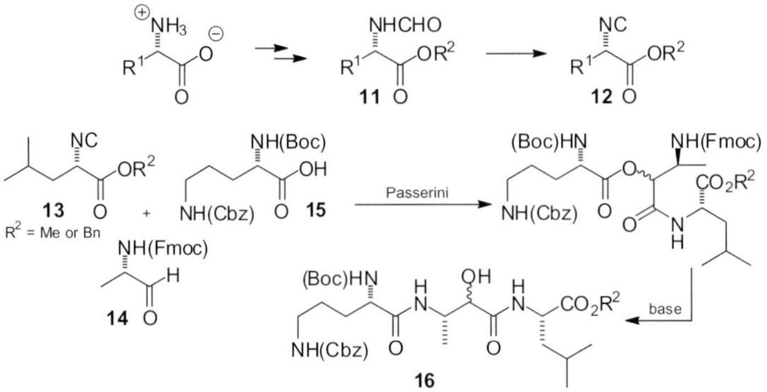

Scheme 1.3

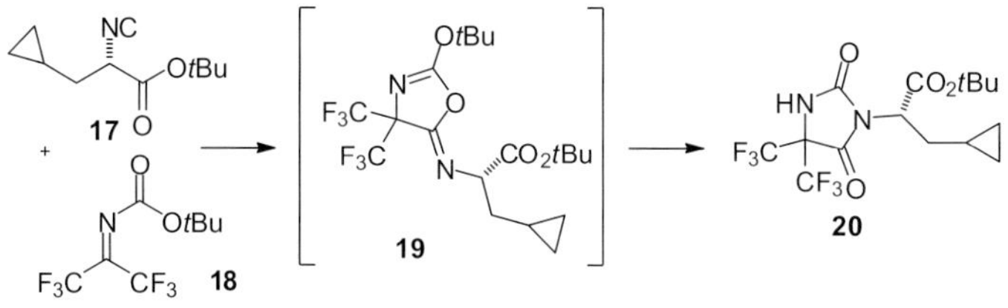

Scheme 1.4

example, in the total synthesis of eurystatin A [22] (Scheme 1.3). The quite complex tripeptide **16** has been assembled in just two steps by using a PADAM (Passerini–Amine Deprotection–Acyl Migration) strategy [30], starting from three enantiomerically pure substrates **13**, **14**, and **15**. Once again, none of the three chiral inputs was able to induce any diastereoselection, but at least three of the four stereogenic centers could be fully controlled by the appropriate substrate configurations. α-Isocyano esters are also configurationally stable during the $TiCl_4$-mediated condensation of isocyanides with aldehydes [2].

With ketones, the Passerini reaction is slower such that some degree of racemization may occur, depending on the carboxylic acid employed [31]. Chiral α-isocyano esters have been used also in the synthesis of optically active hydantoins such as **20** (Scheme 1.4) [5]; however, the enantiomeric purity was not precisely assessed, and it could not be ascertained if these conditions were racemizing, or not.

In contrast, the conditions of the Ugi reaction are often incompatible with the stereochemical integrity of chiral α-isocyano esters [20, 25, 32]. A careful study of reaction conditions has shown that – at least for the reaction with ketones – racemization can be almost completely suppressed by carrying out the reaction in CH_2Cl_2 with $BF_3 \cdot Et_2O$ as catalyst [25]. In this case, racemization is believed to be provoked by the free amine; in fact, α-isocyano esters will readily racemize when treated with amines at room temperature (r.t.) [2]. On this basis, the use of preformed imines would be expected to be capable of preventing racemization, although such success was stated only in few cases, that always involved preformed cyclic imines (Scheme 1.5). For example, Joullié has reported the formation of only two diastereomers **23** in the condensation of chiral imine **21** with chiral isocyano ester **22** [33]. Similarly, Sello has obtained only two diastereomers in the condensation of achiral imine **24** with chiral isocyanide **25** and Boc-proline. Interestingly, the two diastereomers have been obtained in 70:30 ratio [34]; this was unusual since, in most cases, chiral isocyanides and chiral carboxylic acids provide no stereochemical control in Ugi reactions. The absence of racemization in Ugi–Joullié reactions is not general. Rather, the present authors experienced the formation of four diastereomers in the reaction of chiral pyrroline **27** with leucine-derived isocyano ester **28** [24].

An ingenious approach to avoid these racemization issues was recently devised by Nenajdenko and coworkers [35], who employed orthoesters **30** as surrogates of

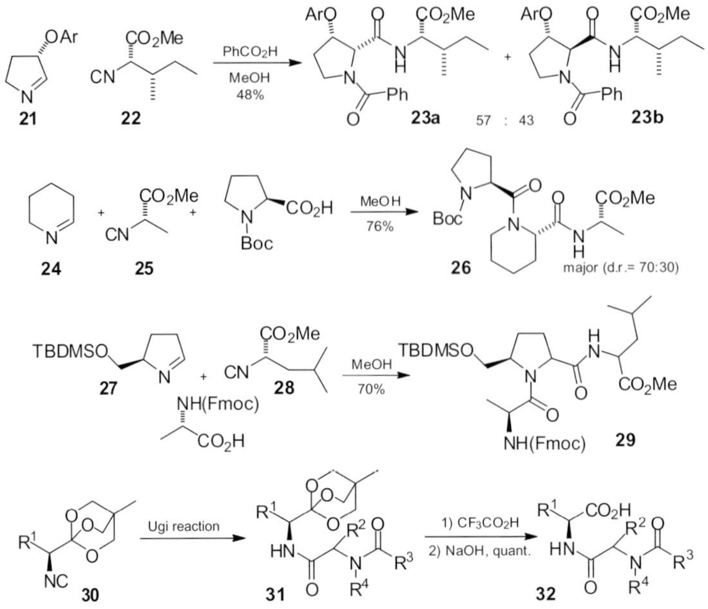

Scheme 1.5

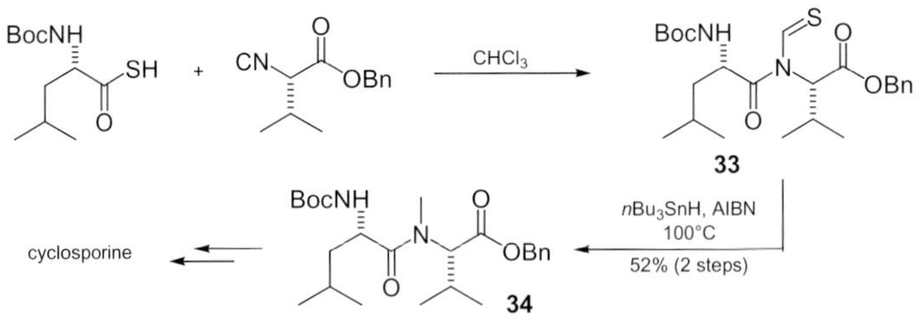

Scheme 1.6

α-isocyano esters. After the Ugi reaction, which proceeds with no racemization, the free carboxylic acids **32** could be obtained in quantitative yield via a two-step/one-pot methodology.

A non-multicomponent application of chiral α-isocyano esters was recently developed by Danishefsky, who created a general method for the synthesis of N-methylated peptides, a moiety which is present in many important natural substances, such as cyclosporine [36] (Scheme 1.6). The coupling of an isocyano ester with a thioacid produces a thioformyl amide that can be conveniently reduced by tributyltin hydride, with the overall sequence taking place without racemization.

1.3 Isocyanides Containing Carboxylic, Sulfonyl, or Phosphonyl Groups

α-Isocyano esters can provide a variety of reactions involving enolization at the α positions (these are reviewed in Chapters 3 and 4). Whilst deprotonation clearly brings about the loss of the stereogenic center, if chirality is present elsewhere (e.g., in the alcoholic counterpart of the ester), then asymmetric induction is, in principle, viable. To date, very few α-isocyanoacetates of chiral alcohols have been prepared [37, 38], and their efficient application in asymmetric synthesis has never been reported [21].

1.3.2
α-Isocyano Amides

Although, in principle, chiral α-isocyano amides can be prepared by the reaction of α-isocyano esters with amines, the easy racemization of the latter compounds under basic conditions makes this approach unfeasible. By using chiral enantiomerically pure amines, it is even possible to realize a dynamic kinetic resolution of racemic α-isocyano esters, obtaining α-isocyanoamides in good diastereomeric ratios, as in the case of compound **35** [2] (Scheme 1.7). Due to the lower α-acidity, the stereoconservative preparation of chiral α-isocyano amides from the corresponding formamides is less problematic than that of the corresponding esters, and combinations of $POCl_3$ with Et_3N may also be used. The only exception here is represented by penicillin- or cephalosporin-derived isocyanides [6]. Cephalosporin-derived isocyanide **39** can be obtained without epimerization, but only when weaker NMM is used as the base (with Et_3N, extensive epimerization takes place). On the other hand, with penicillin-derived formamide **36** a near to 1:1 epimeric mixture is obtained, even with NMM.

α-Isocyano amides are also less prone to racemize during multicomponent reactions, although in this case the yields may be impaired by concurrent oxazole formation [24, 39] (see Chapter 3).

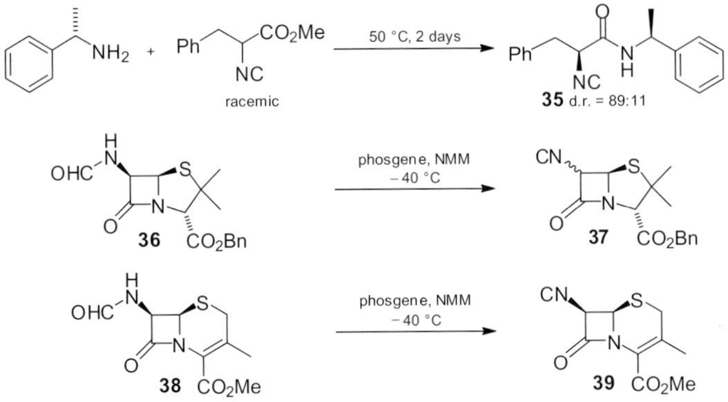

Scheme 1.7

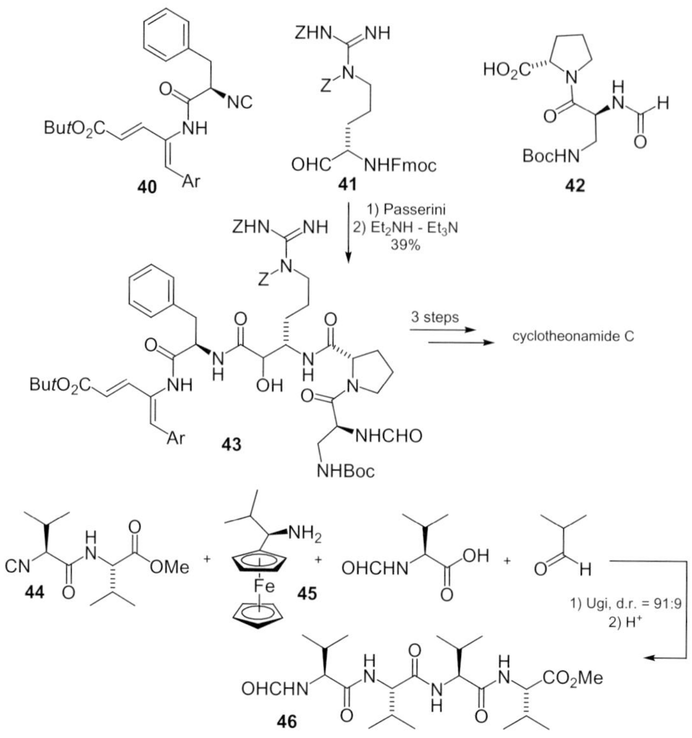

Scheme 1.8

Nonetheless, α-isocyanoamides have been employed successfully in both the Passerini [27, 40] and Ugi reactions [41], some representative examples of which are shown in Scheme 1.8. Aitken and Fauré have accomplished a highly convergent synthesis of cyclotheonamide C by exploiting the above-cited PADAM strategy [30], and using three polyfunctionalized substrates, namely α-isocyano amide **40**, protected α-amino aldehyde **41**, and protected amino acid **42** [40]. Despite none of these three chiral substrates being capable of affording any stereoselection, this is unimportant because the new stereogenic center is later lost by oxidation. Ugi et al. have demonstrated the applicability of their reaction in the straightforward synthesis of tetrapeptide **46** [41] although, in this case, two problems had first to be resolved: (i) the poor asymmetric induction provided by both the chiral isocyanide and the carboxylic acid; and (ii) the need for secondary amides (the use of ammonia is often inadequate in Ugi reactions). Ultimately, both issues were resolved by using the chiral ferrocenyl auxiliary **45**, which afforded a good stereoinduction and could easily be removed under acidic conditions.

The α-isocyano amides may also provide a wide variety of stereoselective reactions that involve enolization at the α positions, provided that chirality is present in the amine counterpart. Consequently, various chiral α-isocyano amides have

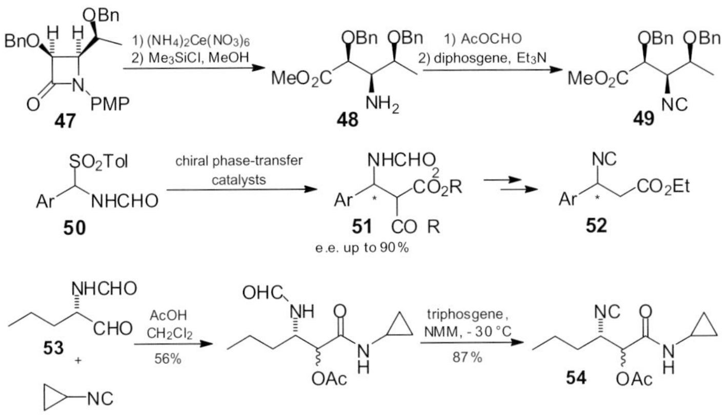

1.3 Isocyanides Containing Carboxylic, Sulfonyl, or Phosphonyl Groups

Scheme 1.9

been prepared [6, 42–46], some of which have provided good levels of diastereoselectivity [6, 42, 45, 46] (these reactions are reviewed in Chapters 3 and 4).

1.3.3
Other Isocyano Esters or Amides

The β-isocyano esters are not expected to suffer from the racemization issues of their α counterparts, and may be very valuable inputs for the multicomponent assembly of peptidomimetics. Somewhat surprisingly, however, very few reports have been made on this class of compound, most likely because of the limited availability of enantiomerically pure β-amino acids (Scheme 1.9). Previously, Palomo et al. [47] have successfully prepared β-isocyano esters such as **49** through an opening of the β-lactam **47**, which in turn was stereoselectively accessed by the Staudinger condensation of a lactaldehyde imine. Although this approach may represent a fairly general entry to these isocyanides, its potential has not been further exploited, and the isocyanide **49** has been used simply as an intermediate for deamination procedures.

Within the present authors' group, a general organocatalytic entry to β-isocyano esters of general formula **52** in both enantiomeric forms has recently been identified. While N-formyl imines have been demonstrated to be unstable, they can be generated in situ from sulfonyl derivatives **50** under phase-transfer conditions. Subsequently, the use of quinine- or quinidine-derived catalysts allowed, after careful optimization, malonates **51** to be obtained in both enantiomeric forms, with enantiomeric excess (e.e.) values ranging between 64% and 90%. Moreover, the yields were almost quantitative and the e.e.-values could be brought to 98% by crystallization. Subsequently, malonates **51** have been converted in high yields into isocyanides **52** by decarboxylation and dehydration (F. Morana, et al., unpublished results).

1 Chiral Nonracemic Isocyanides

During the total synthesis of the antiviral agent telaprevir, Orru and Ruijter have recently reported an interesting approach (that in principle may be general) for the synthesis of protected α-hydroxy-β-isocyano esters such as **54**, based on the Passerini reaction of a chiral α-formylamino aldehyde **53** [48]. The only drawback of this methodology is the low stereoselection of the Passerini reaction, though this is not influential if the targeted products are peptidomimetic and contain the α-keto-β-amino amide transition state mimic.

1.3.4
Chiral Sulfonylmethyl or Phosphonylmethyl Isocyanides

Sulfonylmethyl isocyanides are synthetic equivalents of formaldehyde mono- or di-anions, and have found several useful applications. Chiral derivatives can, in principle, be used for achieving asymmetric induction, with Van Leusen and colleagues having prepared a series of chiral analogues with either stereocenters in the group attached to sulfur (i.e., **55**) or with a stereogenic sulfur atom (**56**) (Scheme 1.10). These chiral *p*-toluenesulfonylmethyl isocyanide (TosMIC) analogues were tested in the synthesis of cyclobutanones [49] or oxazolines [50]. In the latter case, two *trans* diastereomers (**57a** and **57b**) were usually obtained, and the best results in terms of stereoselectivity were obtained with sulfonimide **56** (diastereomeric excess (d.e.) = 80%). The preparation of enantiomerically pure sulfonimide **56** is not trivial, however. Oxazolines **57** can be hydrolyzed to α-hydroxyaldehydes **58**.

Van Leusen has also prepared the chiral phosphonylmethyl isocyanide **61** (as well as its *trans* epimer), starting from enantiomerically pure dioxaphosphorinane **59** [51]. Here, the key step is an Arbuzov reaction of **59** with a *N*-methylformamide equipped with a good leaving group. It is worth noting that this represents an unconventional formamide synthesis, that does not proceed through a primary amine.

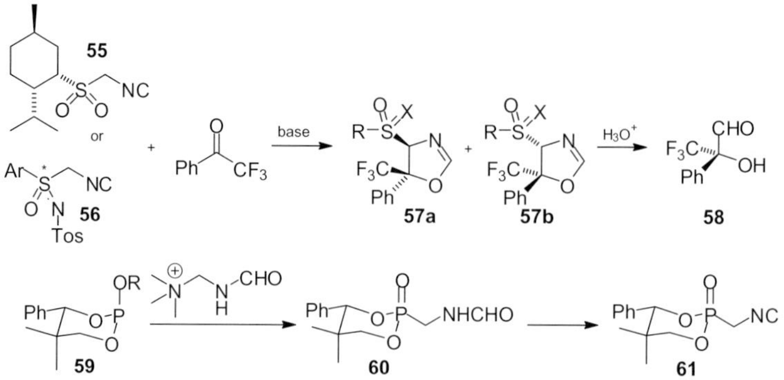

Scheme 1.10

1.4
Isocyanides Containing Amino or Alcoholic Functionalities

1.4.1
Chiral Amino or Azido Isocyanides

A series of protected chiral β-amino isocyanides of general formula **63** has been prepared in enantiomerically pure form via two general strategies (Scheme 1.11). The first strategy begins with protected α-amino acids and involves transformation into the nitriles, reduction, formylation, and dehydration [52]. The second strategy [10] is considerably shorter, but starts from less readily available β-amino acids that are converted in a one-pot reaction, via a Curtius rearrangement, into the same formamides **62**. These isocyanides have been used in the cycloaddition with trimethylsilyl azide to produce tetrazoles **65** [10], and also in the synthesis of isoselenocyanates **66** and selenoureas [52].

A few chiral γ-isocyano amines are also known [10]. For example, compounds **67** have been obtained by the reaction of chiral N-tosyl aziridines with α-lithiated benzyl isocyanides [53], though the reaction is poorly stereoselective and two separable diastereomers were obtained. Likewise, the complex nucleosidic γ-amino isocyanide **68** has been prepared as an advanced intermediate in the convergent total synthesis of muraymicyn D2, and used as input in an Ugi reaction with a chiral carboxylic acid and achiral aldehyde and amine [54]. No asymmetric induction was observed, however.

Whereas, the use of isocyanides in the Ugi reactions leads to peptide-like structures, the Huisgen cycloaddition of azides and alkynes produces triazoles, which are also deemed as peptide surrogates. Consequently, the incorporation of both an isocyanide and an azide into the same building block represents a valuable

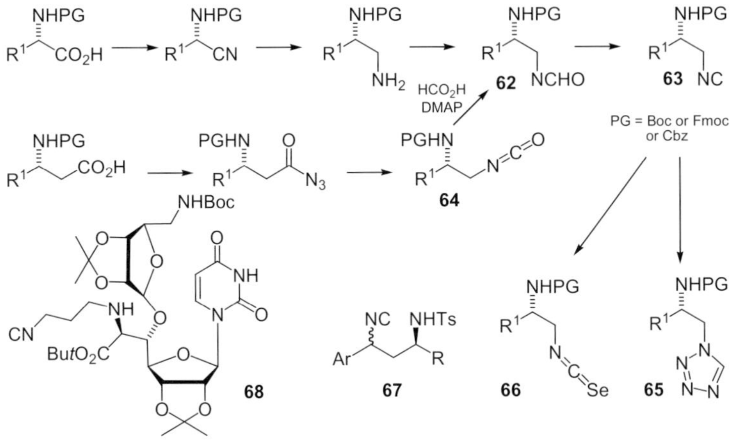

Scheme 1.11

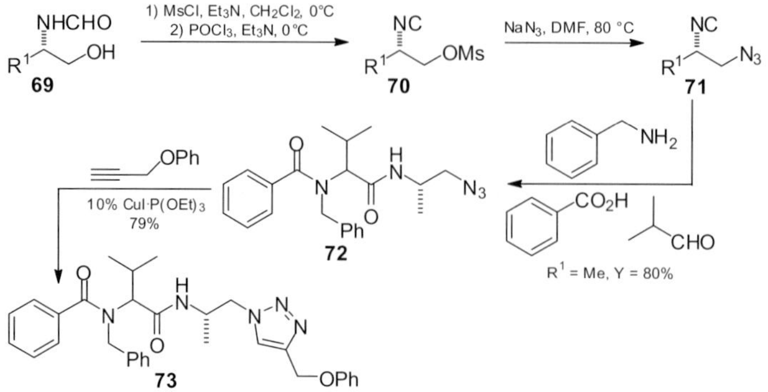

Scheme 1.12

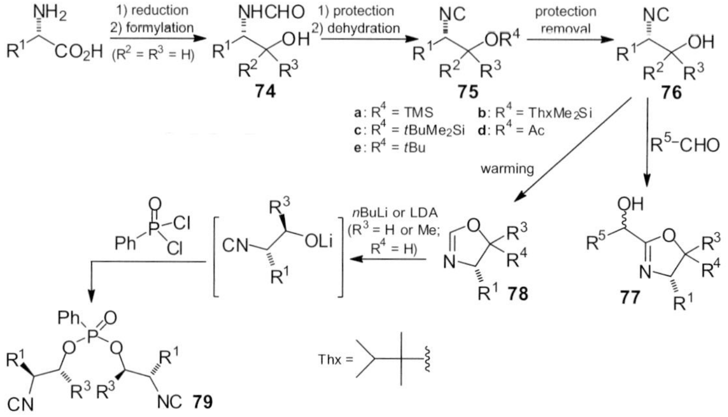

Scheme 1.13

strategy to build up peptidomimetics in a very convergent manner [55]. Compounds **71** (Scheme 1.12) have been prepared from β-formamido alcohols **69**, in turn obtained from α-amino acids; in this case, a one-pot mesylation–dehydration step provided the sulfonate **70**, which was then substituted by sodium azide. These isocyanides are configurationally stable under either Ugi or Passerini conditions. An example of an application featuring a tandem Ugi–Huisgen protocol is shown in Scheme 1.12 where, as usual, no asymmetric induction by the chiral isocyanide was noted in the Ugi step.

1.4.2
Chiral Hydroxy Isocyanides

β-Hydroxy isocyanides **76** or their protected derivatives **75** represent very useful synthons for the synthesis of peptidomimetic structures through multicomponent reactions (Scheme 1.13). Compounds **76** have also been prepared as potential anti-AIDS drugs (i.e., nucleoside mimics with reverse transcriptase inhibitory activity) [56]. For $R^2 = R^3 = H$, the alcoholic function can be later oxidized, making **75–76** synthetically equivalent to easily racemizable α-isocyano esters or the likely unstable α-isocyano aldehydes [57].

For $R^2 = R^3 = H$, formamides **74** can be easily obtained in two steps from α-amino acids, whereas α- or α,α′-substituted derivatives have been obtained through longer routes [58, 59]. As the direct conversion of **74** into isocyanides **76** was reported to be troublesome [59], the best route seems to involve a temporary protection of **74** to produce **75a** or **75b**, followed by dehydration and finally deprotection with $BF_3 \cdot Et_2O$ (**75a**) [59] or nBu_4NF (**75b**) [56]. However, when R^2 and R^3 are different from hydrogen [59], or if the Burgess reagent is employed [8], then dehydration to **76** can be carried out directly on **74**. One of the main uses of compounds **76** is the two-component synthesis of oxazolines **77** by reaction with aldehydes. This reaction displays no stereoselectivity, and consequently oxazolines **77** are obtained as a 1:1 separable mixture of diastereomers that have been used as bidentate chiral ligands in the asymmetric diethylzinc addition to aldehydes [60].

In contrast, the protected isocyano alcohols **75c** and **75d** have been employed in classical Passerini [61] and Ugi reactions [57, 62] although, again, no stereoselection was observed. Compounds **75d** have also been submitted to the isocyanide–cyanide rearrangement under FVP conditions to produce β-acetoxy nitriles [19]. Finally, **75e** has been employed in the synthesis of formamidines which have, in turn, been used as chiral auxiliaries [63].

Alcohols **76** are not very stable, and must be conserved at low temperatures or used immediately after their preparation, because they tend to be converted into unsubstituted oxazolines **78**. This cyclization may be reverted under strong basic conditions to produce the conjugate bases of **76**, that have been exploited for the preparation of *pseudo*-C_2-symmetric ligands **79** [58, 64]. These ligands have found various applications in organometal catalysis; for example, iron(II) complexes have recently been used in the asymmetric transfer hydrogenations of aromatic and heteroaromatic ketones [58].

A general approach to γ-isocyano alcohols is represented by the biocatalytic desymmetrization of 2-substituted 1,3-propanediols, followed by the substitution of one of the two hydroxy functions with the isocyanide, through the corresponding azides and formamides (Scheme 1.14). The synthetic equivalence of the two hydroxymethyl arms allows the enantiodivergent synthesis of both enantiomeric isocyanides, starting from the same monoacetate. The present authors' group has recently prepared both enantiomers of a series of isocyanides **80** and **81** by this strategy, and used them in stereoselective Ugi–Joullié coupling with chiral imines [65]. The nucleosidic γ-isocyano alcohol **83** has been prepared, again by reduction, formylation, and dehydration from azido alcohol **82**, in an attempt to identify potential anti-AIDS drugs, such as azidothymidine (AZT) analogues [7].

Previously, several isocyano sugars have been synthesized, with the isocyanide either being bound to the anomeric positions, or not. Glycosyl isocyanides, such as **86** (Scheme 1.15), may be prepared starting from fully benzylated glycosyl halides **84** by reaction with silver cyanide [66], with the β-anomer usually being

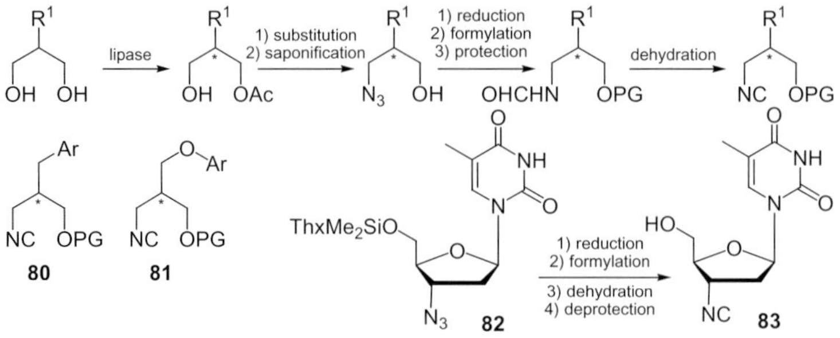

Scheme 1.14

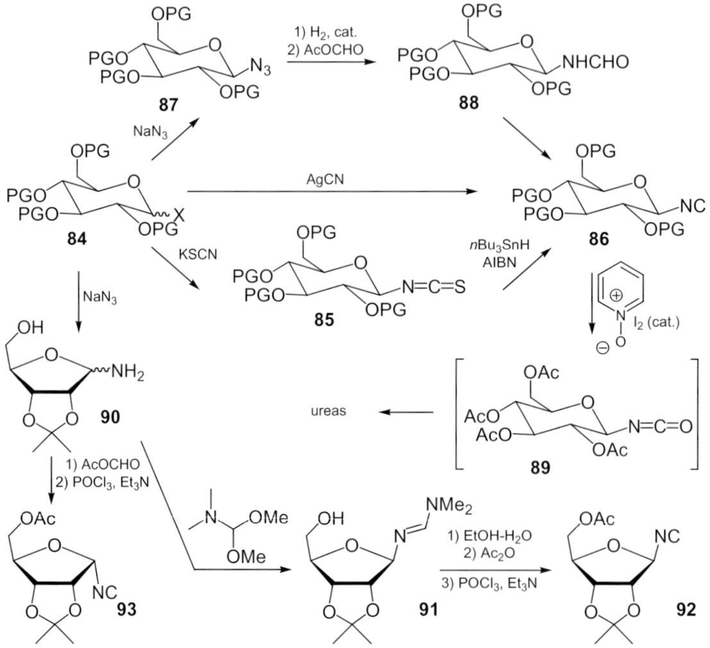

Scheme 1.15

favored. A more efficient methodology, that is also more compatible with fully acetylated glycosyl halides, involves the initial transformation into the isothiocyanate **85**, followed by a controlled radical reduction with $n\text{Bu}_3\text{SnH}$ initiated by azo-*bis*-isobutyronitrile (AIBN) [67].

Alternatively, glycosyl isocyanides may be prepared by a longer (but often more stereoselective) route that involves the formation of a glycosyl azide **87**, followed by reduction, formylation, and dehydration [68, 69]. A shorter route, which allows the preparation of both anomers, has been developed in the field of pentofuranoses

1.4 Isocyanides Containing Amino or Alcoholic Functionalities

(ribosyl isocyanides are shown as an example) [70]. The treatment of protected ribofuranosylamine **90** with the mixed acetic formic anhydride affords directly the α formamide, with concomitant acetylation of the 5-OH; dehydration then yields α isocyanide **93**. In contrast, when **90** is converted into the amidine **91**, only the β-anomer is formed. A careful hydrolysis produces a β-formamide that, upon acetylation and dehydration, leads to the β-isocyanide **92**.

The main application of glycosyl isocyanides relies on their transformation into isocyanates such as **89**, *en route* to ureido-linked disaccharides or sugar-amino acid conjugates [68, 71, 72]. These isocyanides have also been converted into amidines [73].

In contrast, 2-deoxy-2-isocyano sugars have been synthesized starting from the corresponding 2-deoxy-2-aminosugars (glucosamine, galactosamine) (Scheme 1.16) [74]. Thus, glucosamine **94** was formylated to produce (after peracetylation of the hydroxy groups) formamide **95**, which was then either directly dehydrated to **96** or first activated and coupled with various glycosyl acceptors, and later dehydrated, affording isocyano disaccharides [75, 76]. In each of these cases the isocyano group was finally reductively (nBu$_3$SnH) removed. Thus, its ultimate function was simply an elimination of the amino group in order to obtain 2-deoxysugars.

Surprisingly, despite the excellent chemistry that has been developed for their synthesis, these sugar isocyanides have to date been employed only very rarely in multicomponent reactions (some examples are depicted in Scheme 1.17). Ziegler

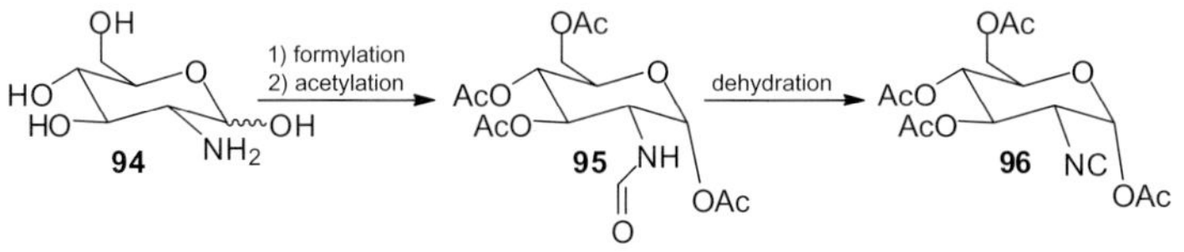

Scheme 1.16

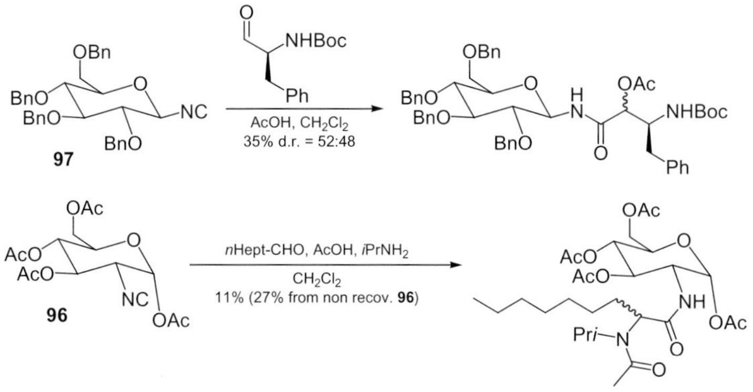

Scheme 1.17

has reported a series of Ugi and Passerini reactions of glycosyl isocyanides such as **97** [69], while Beau and again Ziegler have reported Ugi and Passerini reactions of 2-isocyano sugars such as **96** [77]. The yields of these reactions are not very high although, in general, glycosyl isocyanides behave better than 2-isocyano sugars. Among the latter compounds the bulkier α-anomers function worse, often giving rise to sluggish reactions. Finally, the Passerini condensations afford better yields than their Ugi counterparts, and in all cases – even when chiral aldehydes have been used – the stereoselectivity was very poor; that is, the diastereomeric ratio (d.r.) never was >60:40).

1.5
Natural Isocyanides

1.5.1
Isolation and Natural Sources

This topic has been widely examined, and information obtained up to mid 2003 has been included in three excellent reviews [78–80]. Interestingly, the first naturally occurring isocyanide, xanthocillin **98** (Figure 1.1), was isolated only in 1957 from a culture of *Penicillium notatum*, but this is not a chiral compound. The first enantiomerically pure isocyanide, axisonitrile-1 **99**, was isolated and characterized only in 1973 [81]. In general, these compounds have been identified in marine invertebrates, such as sponges and nudibranch mollusks, and less frequently in fungi or cyanobacteria (blue-green algae). Natural isocyanides are often accompanied by the corresponding isothiocyanates and formamides – compounds that have been shown as being biogenetically related. In natural chiral isonitriles, the isocyano group is in most cases attached directly to the stereogenic center which is, in turn, a tertiary or often a quaternary carbon.

Marine derivatives display almost exclusively a terpene-derived skeleton (sesqui- or diterpenes) and, in some cases, also interesting biological properties, such as anti-malarial activity, antibiotic properties, and cytotoxicity. On the contrary, com-

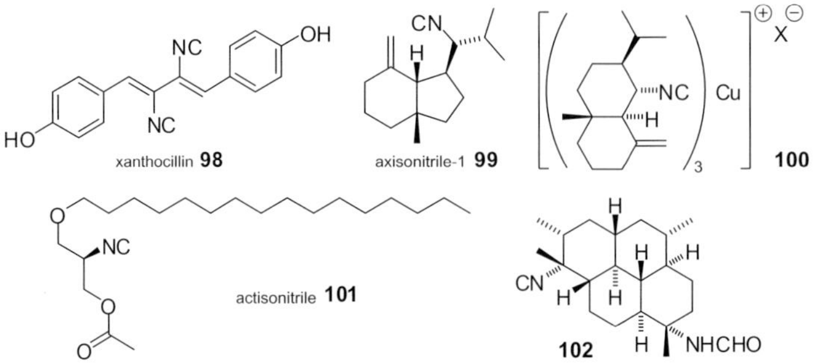

Figure 1.1 Examples of natural isocyanides.

pounds isolated from cyanobacteria have rather complex alkaloid structures, whereas to date very few examples of natural isocyanides in the field of macrolides or carbohydrates have been reported [82, 83].

During the past eight years, only a limited number of new isocyanides have been isolated, among which three compounds are worthy of mention: **100**, a copper(I) complex [84]; **101**, with a unique lipidic arrangement [85]; and **102**, one of the few examples of structures which bear simultaneously an isocyanide and a formamide function [86].

1.5.2
Synthesis of Naturally Occurring Isocyanides

Among the plethora of total syntheses that have been reported to date, to the best of the present authors' knowledge only a few dozen are related to the total synthesis of chiral natural isocyanides. These syntheses typically afford enantiomerically pure molecules, although in some instances they may be either enantiomers [87, 88] or epimers [11] of the actual natural compounds. In addition, efficient–but racemic–syntheses have been reported, including the preparation of racemic **99** [89, 90], **103** [91], **104** and its diastereoisomer [92], and **105** [93]. The anti-malarial β-lactam **106** was prepared via a semi-synthesis from another natural substance [94], whereas in the case of **107** only a synthetic approach to the racemic target was reported [95] (Figure 1.2). Those compounds which seem to have attracted the most synthetic efforts are the complex hapalindole alkaloids, for which an exhaustive collection of references is available [96]. In essentially all of the synthesized compounds, however, the isocyanide moiety is bound to a stereogenic center (often quaternary), with the exception of **105** [93] and **108–109** [11].

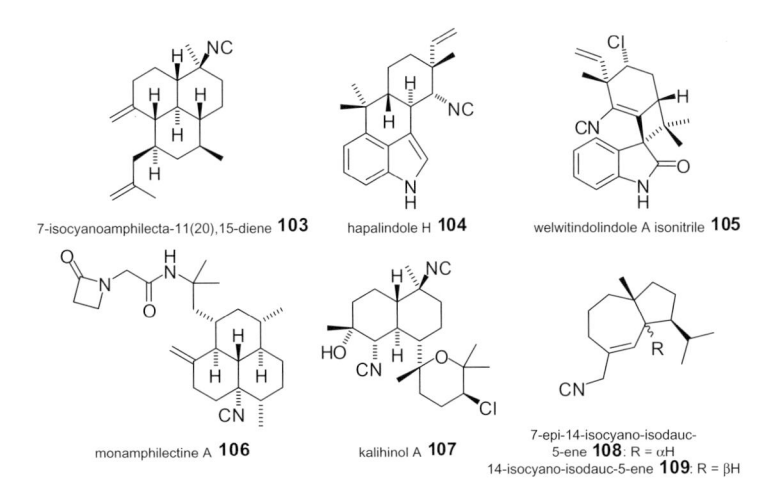

Figure 1.2 Examples of natural isocyanides that have been the targets of total syntheses.

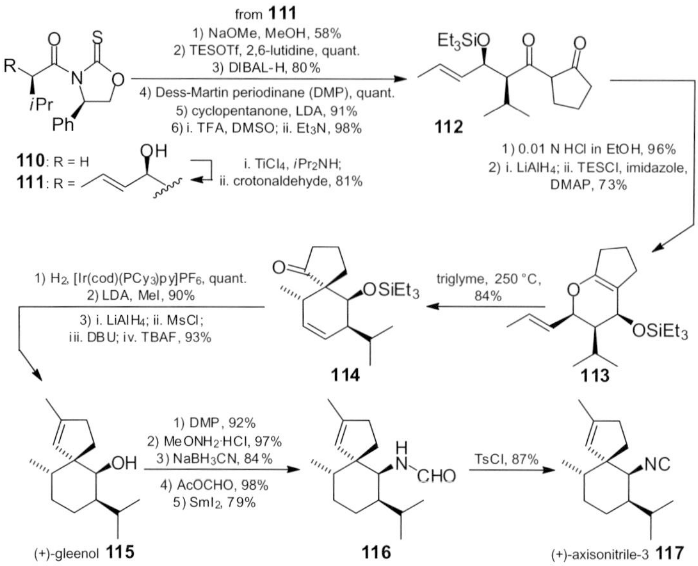

Scheme 1.18

Typically, the introduction of the isocyanide moiety has been performed as the final step, by dehydration of the corresponding formamide. The only exceptions to this are compounds **106**, in which an advanced intermediate already bearing the isocyanide group (**146**) was used [94], and **108** in which a nucleophilic substitution of an allylic iodide by means of AgCN has been performed at the end of the synthesis (see Schemes 1.22 and 1.21, respectively) [11]. Other exceptions are represented by the syntheses of hapaindole derivatives, where often the formation of an isocyanide group is not the last transformation [97]. Hence, on this basis, more attention will be dedicated to these compounds.

A brief survey of the syntheses of enantiomerically pure sesquiterpenes [11, 87, 88, 90, 98, 99] and diterpenoids [91, 94, 95] is reported in Schemes 1.18–1.22.

(+)-Axisonitrile-3 **117**, an anti-malarial compound isolated from the marine sponge *Axinella cannabina*, was synthesized from chiral oxazolidinethione **110** by exploiting a non-Evans *syn*-aldol reaction to produce **111** (Scheme 1.18) [99]. After several steps, the key spiranic intermediate (+)-gleenol **115** was obtained through a Claisen rearrangement of dihydropyran **113** to **114**, followed by methylation and other simple transformations. The stereoselective conversion of the hydroxy group of **115** into the isocyanide required the oxidation to a ketone, and the introduction of the nitrogen function through a highly diastereoselective reduction of the corresponding *O*-methyl oxime. The resulting hydroxylamine was formylated, while a reductive cleavage of the N–O bond produced formamide **116** which, eventually, was dehydrated.

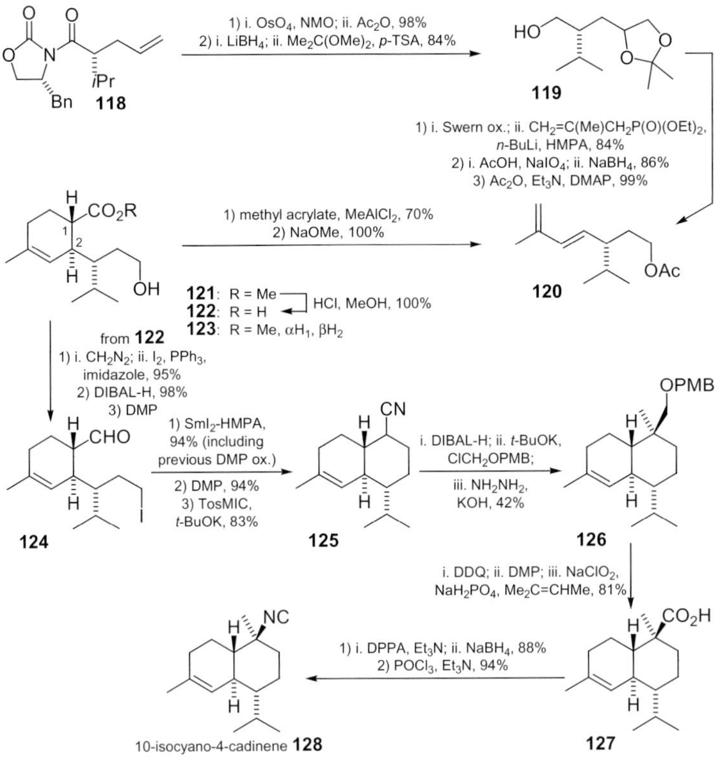

1.5 Natural Isocyanides

Scheme 1.19

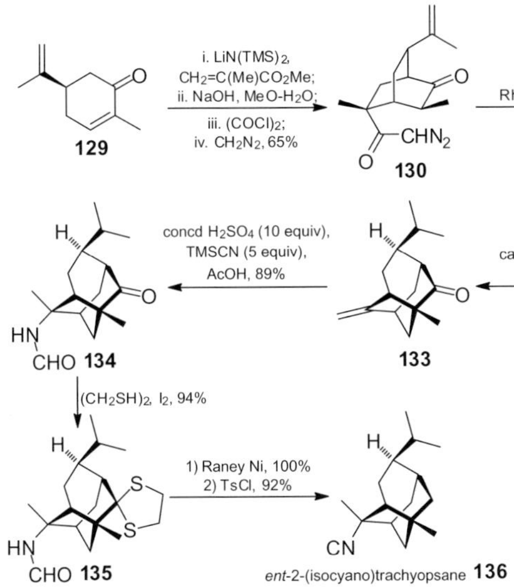

Scheme 1.20

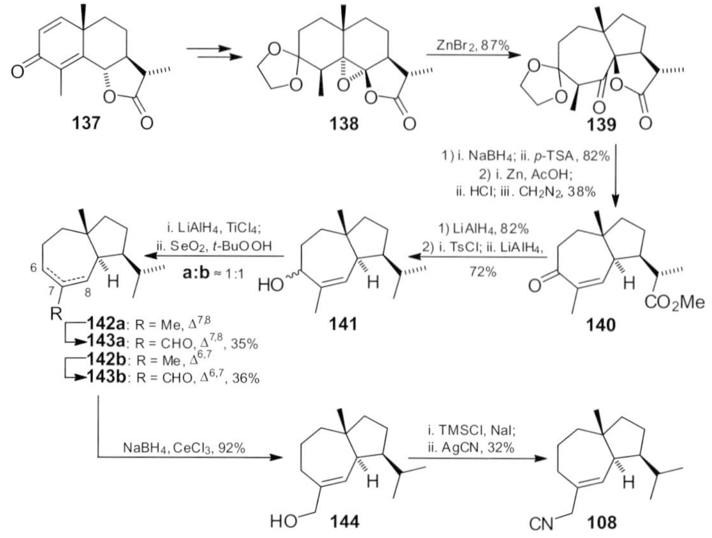

Scheme 1.21

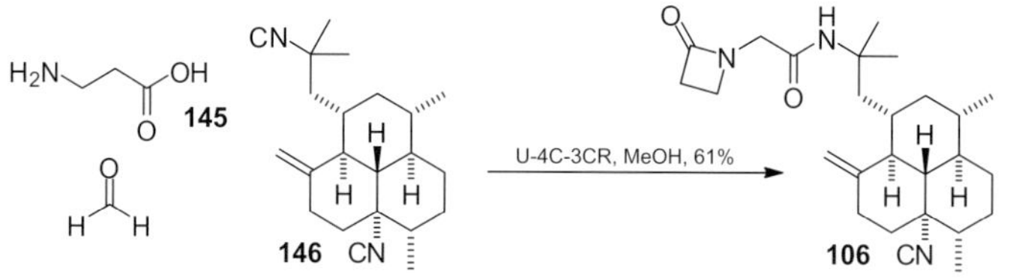

Scheme 1.22

10-Isocyano-4-cadinene **128**, a marine sesquiterpene with anti-fouling activity isolated from nudibranchs of the family *Phyllidiidae*, was prepared from a known allylated oxazolidinone **118**, which was the precursor of diene **120** (Scheme 1.19). Here, the key step is the formation of the cyclohexene moiety with a *trans* relationship of the 1,2-substituents. The Diels–Alder reaction of diene **120** with methyl acrylate afforded the expected products **121**, but as a mixture of four diastereomers. However, equilibration with NaOMe/MeOH gave (apart from cleavage of the acetate) only the *trans* ester **121** and its *trans* isomer **123** in a 2:1 ratio. The desired acid **122** was finally obtained in pure form by the slow addition of 1 M HCl, followed by selective precipitation; **122** was then transformed into decaline **125** and, via a rather lengthy route, into the carboxylic acid **127**. Finally, a Curtius rearrangement (see Section 1.2) allowed production of the formamide precursor of **128**, with the isocyanide on a quaternary carbon [98].

The biomimetic synthesis of *ent*-2-(isocyano)trachyopsane **136**, the enantiomer of a complex tricyclo[4.3.1.0^{3,8}]decane which was extracted from the nudibranch

Phyllidia varicosa and displayed anti-fouling activity, was achieved from carvone **129** (Scheme 1.20), which was first transformed into diazoketone **130** by a double Michael reaction, followed by functional group transformations. Isotwistane dione **131**, bearing a neopupukeanane skeleton, was obtained through a known regioselective C–H insertion of the corresponding Rh carbenoid. Regioselective and stereoselective reduction to **132**, followed by treatment with camphorsulfonic acid, promoted the rearrangement affording trachyopsane **133**, an advanced precursor of **136** [88]. The required nitrogen function was introduced stereoselectively through a Ritter reaction, using cyanotrimethylsilane and H_2SO_4 to yield **134**.

7-Epi-14-isocyano-isodauc-5-ene **108**, the epimer of natural **109** extracted from the marine sponge *Acantkella acuta*, was prepared from natural α-(–)-santonin **137**, which was converted into eudesmane derivative **138** by a reported procedure (Scheme 1.21). This intermediate was submitted to a $ZnBr_2$-mediated rearrangement to give the typical isodaucane skeleton of **139**, while subsequent functional group manipulation afforded **141**. The reductive removal of allylic hydroxy group afforded an inseparable mixture of the regioisomeric alkenes **142a,b**, after which selective oxidation of the allylic methyl group gave alcohols **143a,b**. Following separation of the two regioisomers, **143a** was finally converted (in moderate yield) into isocyanide **108** by substitution of the corresponding iodide with AgCN [11].

Monamphilectine A **106** is a diterpenoid β-lactam alkaloid that has recently been extracted from the marine sponge *Hymeniacidon* sp. and shows a potent antimalarial activity. The synthesis of **106** is the only one in which a multicomponent reaction was employed for a semi-synthetic approach (Scheme 1.22) [94], whereby the β-lactam moiety was introduced through a Ugi four-center, three-component reaction (U-4C-3CR) reacting together β-alanine **145**, formaldehyde, and *bis*isocyanide **146**. Interestingly, only one isocyanide group (probably the less-hindered) takes part in the multicomponent reaction.

Hapalindole-type natural compounds form a family of over 60 biogenetically related structures that have been isolated from blue-green algae (cyanobacteria) since 1984, and which are characterized by a broad range of biological activities. Typically, they have an indole (and in few cases a fragment derived from the oxidative degradation of the indole) with a monoterpene unit bonded to C_3. Most of these compounds have an isocyanide or an isothiocyanate bound (with few exceptions) to a stereogenic carbon that is part of a cyclohexane; moreover, they present in the vicinal position an all-carbon quaternary center (with methyl and vinyl substituents). The tricyclic framework of hapalindole may become either tetracyclic (some hapalindoles, fischerindoles, and some ambiguines) or even pentacyclic (more complex ambiguines). Welwitindolinones may be tetracyclic with a spirocyclic cyclobutanone centered around C_3, or they can be characterized by a [4.3.1] bicyclononanone moiety. Some representative examples of their structures are shown in Figure 1.2 (compounds **104** and **105**) and Figure 1.3 (compounds **147–151**) [96, 97, 100].

Apart from an early report on the synthesis of a hapalindole [101], the most impressive syntheses of these alkaloids in enantiomerically pure form are those reported by Baran's group [96, 97, 100]. Each of these syntheses is based on a very

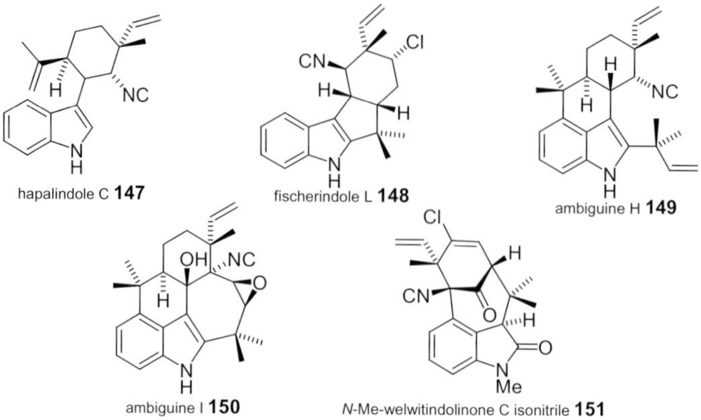

Figure 1.3 Structures of some hapalindoles.

simple principle: maximize "atom", "step," and "redox-economy", where the latter term indicates a minimization of the superfluous redox manipulations. In this way, the preparation of the target molecule avoids the use of protecting groups and also exploits the natural reactivity of functional groups, such that the basic skeleton can be built on the gram scale.

An example of this, the synthesis of ambiguine H **149** and of hapalindole U **156**, the precursor of **147** lacking the prenyl unit at C_2, is shown in Scheme 1.23 [97]. Here, the intermediate **152**, which is readily available from commercial *p*-menth-1-en-9-ol, was coupled with 4-bromoindole **153** to produce **154**, without any need to protect the NH group. The direct Friedel–Crafts annulation on the analogue of **154** (having H instead of Br) was unsuccessful because it was sitoselective on C_2 instead of on C_4; hence, a switch was made to **154**, which allowed the possibility of forcing a formation of the fourth ring in the appropriate position. The desired 6-*exo-trig* cyclization (reductive Heck) onto C_4 to afford **155** was effective when promoted by Hermann's catalyst. Subsequently, transformation into hapalindole U **156** was accomplished by a reductive amination under microwave heating, followed by conventional introduction of the isocyanide group.

Introduction of the prenyl unit, leading to **149**, was the most critical step because direct C–C bond formation was impossible, due to the unusual reactivity of the indole moiety, and to an incompatibility of the isocyanide with acids and transition metals. However, instead of resolving the problem by means of protective groups, the high reactivity of both the indole and the isocyanide was exploited simultaneously by the treatment of **156** with *t*BuOCl followed by prenyl 9-borabicyclononane (BBN). The electrophilic chlorination of the isocyanide is presumably followed by an addition of the chloronitrilium ion to C_3 of the indole, and by coupling of the borane to the indolenine nitrogen to produce **157**. Finally, B → C migration yields the crystalline chlorimidate **158**, with the *t*-prenyl group correctly bound at C_2. The following Norrish-type homolytic cleavage, promoted by irradiation,

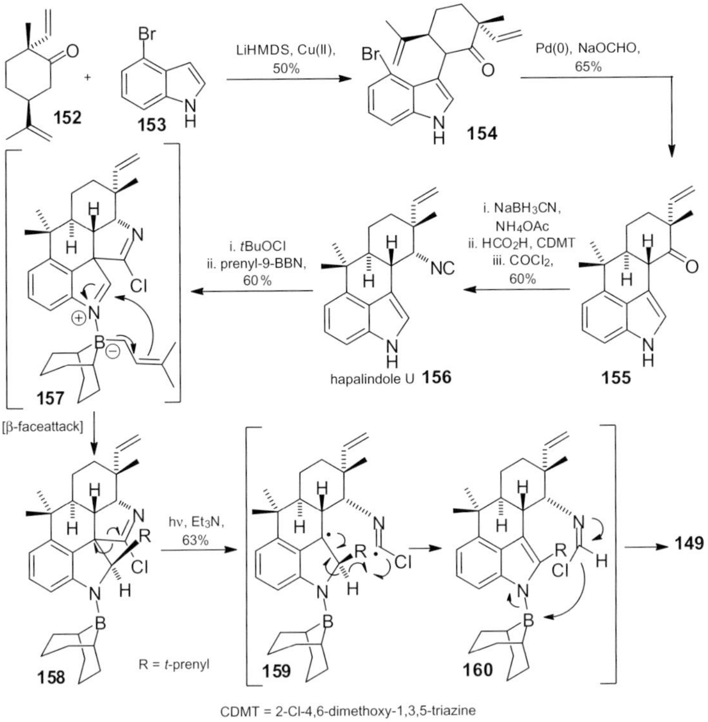

Scheme 1.23

furnished eventually ambiguine H **149**, most likely through the fragmentation cascade depicted in Scheme 1.23. Hence, the synthesis of **149** was achieved via a very rapid strategy which fits perfectly with the initial assumptions quoted above.

1.6
Isocyanides Used in the Synthesis of Chiral Polyisocyanides

Atropisomerism, a stereochemical property that is well known in organic chemistry but very rare in polymer chemistry, was first demonstrated in 1974 in polymers of isocyanides [102] (Figure 1.4a). In this case, it was shown that poly(*tert*-butyl isocyanide) could be resolved chromatographically into separate fractions that displayed positive and negative optical rotations. Subsequent investigations [13] revealed that the optical rotation was due to a helical configuration of the polymer backbone. Remarkably, these helices did not undergo racemization, even at elevated temperatures, due to the presence of bulky substituents which prevented the kinetically formed helical polymer from unfolding. (For reviews on the subject of polyisocyanides, see Refs [104–105] and Chapter 16.)

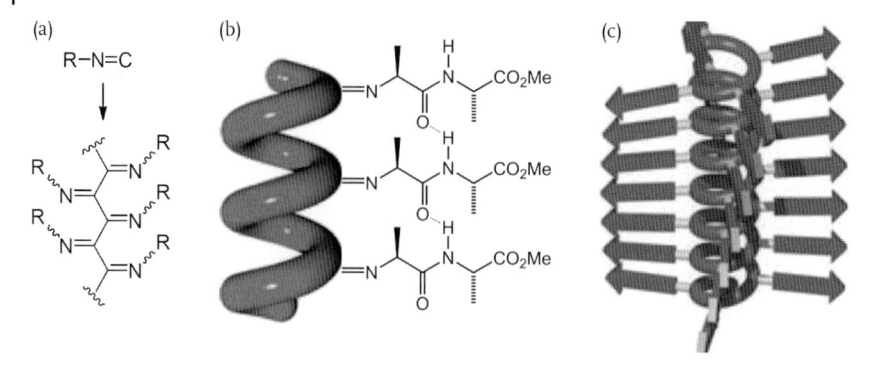

Figure 1.4 Structure of polyisocyanides. Reproduced in part, with permission from Prof. Dr R. J. M. Nolte, from Ref. [103]; © 2010, Wiley.

1.6.1
Properties

One special characteristic of polyisocyanides (also known as polyiminomethylenes or polycarbonimidoyls) is the fact that every carbon atom in the polymer backbone bears a substituent and, as a consequence, the side chains experience sufficient steric hindrance that causes the polymer to adopt a nonplanar conformation. The polymers fold into a 4_1-helical conformation containing four repeat units per turn and a helical pitch of 4.1–4.2 Å [106] (Figure 1.4b). Both, left- (M) and right-handed (P) helices can be distinguished, with absolute helic sense being determined by using circular dichroism (CD) spectroscopy or X-ray diffraction analyses on liquid crystalline phases [107].

Besides the helical conformation, a so-called "*syndio*" conformation was discussed by Clericuzio and Alagona [108] on the basis of *ab initio* calculations, while Green [109], relying on comprehensive measurements such as viscosity, light scattering and nuclear magnetic resonance (NMR) spectroscopy, suggested an alternative irregular conformation, in which the stereo irregularity is associated with a *syn–anti* isomerism about the imine bond.

The helical conformation of the polyisocyanide backbone can be effectively stabilized if a well-defined hydrogen bonded network is present between the side chains at positions n and $n + 4$. This additional stabilization increases the persistence length of these types of macromolecule, which were found to be even more rigid than DNA [110]. As in biomolecules, the disruption of these hydrogen-bonding arrays is a cooperative effect. When the pendant side chains are peptidic fragments, they are stacked above each other at a distance of approximately 4.6 Å, with a general structure as depicted in Figure 1.4b,c [103]. Analogously to the denaturation of proteins, such secondary structure can be irreversibly disrupted by treatment with strong acids or heating.

1.6.2
Synthesis

Polyisocyanides are prepared by the polymerization of isocyanides. Shortly after their discovery, it was realized that the isonitriles readily polymerize, the driving force being the conversion of a formally divalent carbon atom in the monomer to a tetravalent carbon in the polymer, which yields a heat of polymerization of around 81 kJ mol^{-1} [111]. The polymerization process may be either acid-mediated or catalyzed by transition metal complexes based on Ni(II), Rh(III), or Pd(II)/Pt(II) couples. The latter approach has been more widely used, with the Pd–Pt heteronuclear complex mainly being used for the polymerization of aryl isocyanides, while the Rh catalysts were effectively employed where the isonitrile monomer displayed bulky substituents [112, 113]. In contrast, the acid-mediated polymerizations can offer polymers with exceptionally long lengths and high stereospecificities [114].

Based on kinetic measurements and experiments with optically active isocyanides, a so-called "merry-go-round" mechanism has been proposed for the Ni(II) catalyzed polymerization and extensively reviewed by Cornelissen and Nolte [104]. The polymerization is initiated by a nucleophile (an amine or an alcohol), and the isocyanide monomers coordinate to the Ni center and are incorporated into the growing chain by a series of consecutive α-insertions. When achiral isocyanides are used, the intermediate formed after attack of the nucleophile on one of the four coordinated isocyanide molecules has no preference to attack either the left or the right neighboring isocyanide and, as a consequence, an equal amount of M and P helices are formed. A preferred helical handedness can be achieved when a chiral initiator or a chiral isocyanide is used; some of the chiral isocyanides that have been successfully employed are shown in Figure 1.5. Remarkably, isocyanides bearing a remote chiral group are also capable of chirality transfer onto the ongrowing polymer chain.

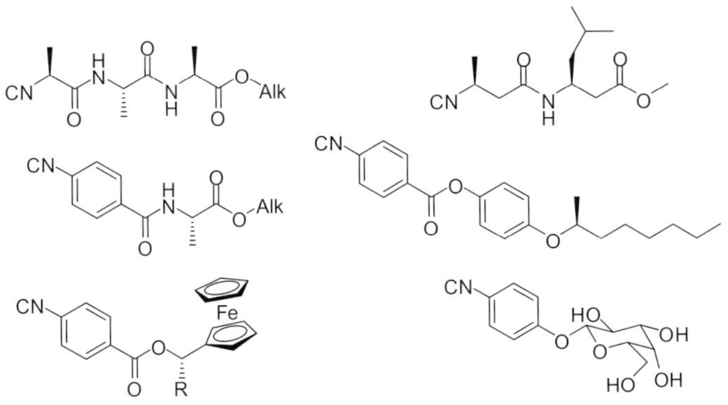

Figure 1.5 Chiral isocyanides used in the diastereoselective synthesis of polyisocyanides.

The many factors that can influence the helix-sense of the polymer may include not only the nature of the isocyanide, but also the solvent, the temperature, and the metal catalyst employed in the polymerization process. For example it has been found, as a consequence, that the same isocyanide can fold to a preferred P or M helix, depending on the specific reaction conditions [115]. The assignment of the preferred helical sense by CD spectroscopy is, however, often hampered by an overlap of the signals that arise from the polymer backbone and the side chains, thus rendering determination of the helical sense excess a difficult task. In fact, high-resolution atomic force microscopy (AFM) has often been the only method used to determine such helical sense [116].

1.6.3
Applications

Stable helical polymers with a preferred handedness are materials that offer intriguing characteristics which might be exploited in the fields of electronics, biosensing, and catalysis. For example, the group of Gomar-Nadal has developed polyisocyanides bearing chiral tetrathiafulvalene derivatives to be used as multistate redox-switchable organic materials in molecular devices [117]. Redox-active polyisocyanides have been reported also by Takahashi, by employing chiral ferrocenyl isocyanide monomers [118], sugar polyisocyanides have been synthesized to study the effect of saccharide arrays along the backbone on molecular recognition phenomena [119], and cholesterol-containing isocyanides have been polymerized to obtain liquid-crystalline polymers [120]. Of particular benefit have been the bioinspired polyisocyanides bearing peptidic appendages; for example, imidazole-containing polyisocyanides have been used in enantioselective ester hydrolysis [121], while cysteine-based polyisocyanides have been synthesized and preliminary experiments performed with the aim of exploiting thio-specific "click" reactions to obtain multichromophoric scaffolding, platforms for nucleation of proteins, or for the binding of metal ions [103]. Dipeptide-derived polyisocyanides have also been used as probes to prove the Davydov hypothesis, that energy-transport in proteins and enzymes occurs via a vibrational soliton mechanism [122]. Finally, giant vesicles (employed as simple model systems for living cells) have been electroformed from diblock copolymers of styrene and isocyanoalanine-2-(thiophen-3-ylethyl)amide [123].

References

1 Saegusa, T., Ito, Y., Kinoshita, H., and Tomita, S. (1971) Synthetic reactions by complex catalysts. XIX. Copper-catalyzed cycloaddition reactions of isocyanides. Novel synthesis of δ-1-pyrroline and δ-2-oxazoline. *J. Org. Chem.*, **36**, 3316–3323.

2 Seebach, D., Adam, G., Gees, T., Schiess, M., and Weigand, W. (1988) Scope and limitations of the $TiCl_4$-mediated additions of isocyanides to aldehydes and ketones with formation of α-hydroxycarboxylic acid amides. *Chem. Ber.*, **121**, 507–517.

3 Waki, M. and Meienhofer, J. (1977) Efficient preparation of N-α-formylamino acid tert-butyl esters. *J. Org. Chem.*, **42**, 2019–2020.

4 Zhang, X., Zou, X., and Xu, P. (2005) Template synthesis of peptidomimetics composed of aspartic acid moiety by Ugi four-component condensation reaction. *Synth. Commun.*, **35**, 1881–1888.

5 Wehner, V., Stilz, H.-U., Osipov, S.N., Golubev, A.S., Sieler, J., and Burger, K. (2004) Trifluoromethyl-substituted hydantoins, versatile building blocks for rational drug design. *Tetrahedron*, **60**, 4295–4302.

6 Bentley, P.H., Clayton, J.P., Boles, M.O., and Girven, R.J. (1979) Transformations using benzyl 6-isocyanopenicillanate. *J. Chem. Soc., Perkin Trans. 1*, 2455–2467.

7 Maillard, M., Faraj, A., Frappier, F., Florent, J.-C., Grierson, D.S., and Monneret, C. (1989) Synthesis of 3′-substituted-2′,3′-dideoxynucleoside analogs as potential anti-AIDS drugs. *Tetrahedron Lett.*, **30**, 1955–1958.

8 Creedon, S.M., Crowley, H.K., and McCarthy, D.G. (1998) Dehydration of formamides using the Burgess reagent: a new route to isocyanides. *J. Chem. Soc., Perkin Trans. 1*, 1015–1018.

9 Walborsky, H.M. and Niznik, G.E. (1972) Synthesis of isonitriles. *J. Org. Chem.*, **37**, 187–190.

10 Sureshbabu, V.V., Narendra, N., and Nagendra, G. (2009) Chiral N-Fmoc-β-amino alkyl isonitriles derived from amino acids: first synthesis and application in 1-substituted tetrazole synthesis. *J. Org. Chem.*, **74**, 153–157.

11 Li, D.R., Xia, W.J., Shi, L., and Tu, Y.Q. (2003) A general approach from eudesmane to isodaucane sesquiterpenes: synthesis of 7-epi-14-isocyanato-isodauc-5-ene from α-(−)-santonin. *Synthesis*, 41–44.

12 Masutani, K., Minowa, T., Hagiwara, Y., and Mukaiyama, T. (2006) Cyanation of alcohols with diethyl cyanophosphonate and 2,6-dimethyl-1,4-benzoquinone by a new type of oxidation-reduction condensation. *Bull. Chem. Soc. Jpn*, **79**, 1106–1117.

13 Kamer, P.C.J., Cleij, M.C., Nolte, R.J.M., Harada, T., Hezemans, A.M.F., and Drenth, W. (1988) Atropisomerism in polymers. Screw-sense selective polymerization of isocyanides by inhibiting the growth of one enantiomer of a racemic pair of helices. *J. Am. Chem. Soc.*, **110**, 1581–1587.

14 Wolber, E.K.A. and Rüchardt, C. (1991) Neue synthesen für Ibuprofen und Naproxen. *Chem. Ber.*, **124**, 1667–1672.

15 Alcock, N.W., Pike, G.A., Richards, C.J., and Thomas, S.E. (1990) Generation of homochiral quaternary carbon centres from (vinylketenimine)tricarbonyliron(0) complexes. *Tetrahedron: Asymmetry*, **1**, 531–534.

16 Bock, H. and Ugi, I. (1997) Multicomponent reactions 2. Stereoselective synthesis of 1(S)-camphor-2-cis-methylidene-isocyanide and its application in Passerini- and Ugi-reaction. *J. Prakt. Chem.*, **339**, 385–389.

17 Hertler, W., and Corey, E. (1958) Notes. A novel preparation of isonitriles. *J. Org. Chem.*, **23**, 1221–1222.

18 Stoelwinder, J., Van Zoest, W.J., and Van Leusen, A.M. (1992) Chemistry of N,P-acetals: application to the synthesis of 20-ketosteroids. *J. Org. Chem.*, **57**, 2249–2252.

19 Rüchardt, C., Meier, M., Haaf, K., Pakusch, J., Wolber, E.K.A., and Müller, B. (1991) The isocyanide–cyanide rearrangement; mechanism and preparative applications. *Angew. Chem. Int. Ed. Engl.*, **30**, 893–901.

20 Failli, A., Nelson, V., Immer, H., and Goetz, M. (1973) Model experiments directed towards the synthesis of N-aminopeptides. *Can. J. Chem.*, **51**, 2769–2775.

21 Gulevich, A.V., Zhdanko, A.G., Orru, R.V.A., and Nenajdenko, V.G. (2010) Isocyanoacetate derivatives: synthesis, reactivity, and application. *Chem. Rev.*, **110**, 5235–5331.

22 Owens, T.D., Araldi, G.-A., Nutt, R.F., and Semple, J.E. (2001) Concise total synthesis of the prolyl endopeptidase inhibitor eurystatin A via a novel Passerini reaction–deprotection–acyl migration strategy. *Tetrahedron Lett.*, **42**, 6271–6274.

23 Skorna, G. and Ugi, I. (1977) Isocyanide synthesis with diphosgene. *Angew. Chem. Int. Ed. Engl.*, **16**, 259–260.

24 Banfi, L., Basso, A., Guanti, G., Merlo, S., Repetto, C., and Riva, R. (2008) A convergent synthesis of enantiopure bicyclic scaffolds through multicomponent Ugi reaction. *Tetrahedron*, **64**, 1114–1134.

25 Berlozecki, S., Szymanski, W., and Ostaszewski, R. (2008) Application of isocyanides derived from α-amino acids as substrates for the Ugi reaction. *Synth. Commun.*, **38**, 2714–2721.

26 Zhu, J.L., Wu, X.Y., and Danishefsky, S.J. (2009) On the preparation of enantiomerically pure isonitriles from amino acid esters and peptides. *Tetrahedron Lett.*, **50**, 577–579.

27 Banfi, L., Guanti, G., Riva, R., Basso, A., and Calcagno, E. (2002) Short synthesis of protease inhibitors via modified Passerini condensation of N-Boc-α-aminoaldehydes. *Tetrahedron Lett.*, **43**, 4067–4069.

28 Basso, A., Banfi, L., Riva, R., Piaggio, P., and Guanti, G. (2003) Solid-phase synthesis of modified oligopeptides via Passerini multicomponent reaction. *Tetrahedron Lett.*, **44**, 2367–2370.

29 Oaksmith, J.M., Peters, U., and Ganem, B. (2004) Three-component condensation leading to β-amino acid diamides: convergent assembly of β-peptide analogues. *J. Am. Chem. Soc.*, **126**, 13606–13607.

30 Banfi, L., Guanti, G., and Riva, R. (2000) Passerini multicomponent reaction of protected α-aminoaldehydes as a tool for combinatorial synthesis of enzyme inhibitors. *Chem. Commun.*, 985–986.

31 Berlozecki, S., Szymanski, W., and Ostaszewski, R. (2008) α-Amino acids as acid components in the Passerini reaction: influence of N-protection on the yield and stereoselectivity. *Tetrahedron*, **64**, 9780–9783.

32 Immer, H., Nelson, V., Robinson, W., and Goetz, M. (1973) Anwendung der Ugi-reaktion zur synthese modifizierter eledoisin-teilsequenzen. *Liebigs Ann. Chem.*, 1789.

33 Bowers, M.M., Carroll, P., and Joullié, M.M. (1989) Model studies directed toward the total synthesis of 14-membered cyclopeptide alkaloids: synthesis of prolyl peptides via a four-component condensation. *J. Chem. Soc., Perkin Trans. 1*, 857–865.

34 Socha, A.M., Tan, N.Y., LaPlante, K.L., and Sello, J.K. (2010) Diversity-oriented synthesis of cyclic acyldepsipeptides leads to the discovery of a potent antibacterial agent. *Bioorg. Med. Chem.*, **18**, 7193–7202.

35 Zhdanko, A.G., and Nenajdenko, V.G. (2009) Nonracemizable isocyanoacetates for multicomponent reactions. *J. Org. Chem.*, **74**, 884–887.

36 Wu, X.Y., Stockdill, J.L., Wang, P., and Danishefsky, S.J. (2010) Total synthesis of cyclosporine: access to N-methylated peptides via isonitrile coupling reactions. *J. Am. Chem. Soc.*, **132**, 4098–4100.

37 Solladié-Cavallo, A. and Quazzotti, S. (1992) An efficient synthesis of (+)-8-phenylmenthyl isocyanoacetate. *Tetrahedron: Asymmetry*, **3**, 39–42.

38 Ilankumaran, P., Kisanga, P., and Verkade, J.G. (2001) Facile synthesis of a chiral auxiliary bearing the isocyanide group. *Heteroat. Chem.*, **12**, 561–562.

39 Bughin, C., Zhao, G., Bienaymé, H., and Zhu, J.P. (2006) 5-aminooxazole as an internal traceless activator of C-terminal carboxylic acid: rapid access to diversely functionalized cyclodepsipeptides. *Chem. Eur. J.*, **12**, 1174–1184.

40 Fauré, S., Hjelmgaard, T., Roche, S.P., and Aitken, D.J. (2009) Passerini reaction-amine deprotection-acyl migration peptide assembly: efficient formal synthesis of cyclotheonamide C. *Org. Lett.*, **11**, 1167–1170.

41 Urban, R., Eberle, G., Marquarding, D., Rehn, D., Rehn, H., and Ugi, I. (1976) Synthesis of an isomer-free tetravaline derivative by stereoselective four-component condensation and exponential multiplication of selectivity. *Angew. Chem. Int. Ed. Engl.*, **15**, 627–628.

42 Schoellkopf, U., Hausberg, H.-H., Segal, M., Reiter, U., Hoppe, I., Saenger, W., and Lindner, K. (1981) Asymmetric syntheses via heterocyclic intermediates, IV. Asymmetric synthesis of α-

methylphenylalanine and its analogues by alkylation of 1-chiral substituted 4-methyl-2-imidazoline-5-one. *Liebigs Ann. Chem.*, 439–458.

43 Yamamoto, Y., Kirihata, M., Ichimoto, I., and Ueda, H. (1985) Asymmetric synthesis of α-methylglutamic acid and α-methylornithine by a chiral isocyano amide reagent. *Agric. Biol. Chem.*, **49**, 1761–1766.

44 Tang, J.S. and Verkade, J.G. (1996) Chiral auxiliary-bearing isocyanides as synthons: synthesis of strongly fluorescent (+)-5-(3,4-dimethoxyphenyl)-4-[[N-[(4S)-2-oxo-4-(phenylmethyl)-2-oxazolidinyl]]carbonyl]oxazole and its enantiomer. *J. Org. Chem.*, **61**, 8750–8754.

45 Aratani, M., Hirai, H., Sawada, K., Yamada, A., and Hashimoto, M. (1985) Synthetic studies on carbapenem antibiotics from penicillins. III. Stereoselective radical reduction of a chiral 3-isocyanoazetidinone: a total synthesis of optically active carpetimycins. *Tetrahedron Lett.*, **26**, 223–226.

46 John, D.I., Tyrrell, N.D., and Thomas, E.J. (1983) Stereoselective preparation of 6β-substituted penicillanates: triorganotin hydride reduction of 6-isocyano-, 6-phenylselenenyl-, 6-halo-, and 6-isothiocyanato-penicillanates. *Tetrahedron*, **39**, 2477–2484.

47 Palomo, C., Aizpurua, J.M., Urchegui, R., and Garcia, J.M. (1993) A new entry to 1,3-polyols, 2-amino 1,3-polyols, and β-(1-hydroxyalkyl)isoserines using azetidinone frameworks as chiral templates via iterative asymmetric [2+2] cycloaddition reactions. *J. Org. Chem.*, **58**, 1646–1648.

48 Znabet, A., Polak, M.M., Janssen, E., de Kanter, F.J.J., Turner, N.J., Orru, R.V.A., and Ruijter, E. (2010) A highly efficient synthesis of telaprevir by strategic use of biocatalysis and multicomponent reactions. *Chem. Commun.*, **46**, 7918–7920.

49 van Leusen, D., Rouwette, P.H.F.H., and van Leusen, A.M. (1981) Synthesis of (+)-(neomenthylsulfonyl)methyl isocyanide. Synthesis and absolute configuration of (R)-(+)-2-methylcyclobutanone and (S)-(−)-2-methylcyclobutanone. *J. Org. Chem.*, **46**, 5159–5163.

50 Hundscheid, F.J.A., Tandon, V.K., Rouwette, P.H.F.M., and van Leusen, A.M. (1987) Synthesis of chiral sulfonylmethyl isocyanides, and comparison of their propensities in asymmetric induction reactions with acetophenones. *Tetrahedron*, **43**, 5073–5088.

51 Weener, J.-W., Versleijen, J.P.G., Meetsma, A., ten Hoeve, W., and van Leusen, A.M. (1998) Cis- and trans-2-(isocyanomethyl)-5,5-dimethyl-2-oxo-4-phenyl-1,3,2-dioxaphosphorinane – synthesis and structure of the first chiral isocyanomethylphosphonate synthons. *Eur. J. Org. Chem.*, **1998**, 1511–1516.

52 Chennakrishnareddy, G., Nagendra, G., Hemantha, H.P., Das, U., Guru Row, T.N., and Sureshbabu, V.V. (2010) Isoselenocyanates derived from Boc/Z-amino acids: synthesis, isolation, characterization, and application to the efficient synthesis of unsymmetrical selenoureas and selenoureidopeptidomimetics. *Tetrahedron*, **66**, 6718–6724.

53 Kaiser, A. and Balbi, M. (1999) Chiral 1,3-diamines from a lithiated isocyanide and chiral aziridines. *Tetrahedron: Asymmetry*, **10**, 1001–1014.

54 Tanino, T., Ichikawa, S., Shiro, M., and Matsuda, A. (2010) Total synthesis of (−)-muraymycin D2 and its epimer. *J. Org. Chem.*, **75**, 1366–1377.

55 Nenajdenko, V.G., Gulevich, A.V., Sokolova, N.V., Mironov, A.V., and Balenkova, E.S. (2010) Chiral isocyanoazides: efficient bifunctional reagents for bioconjugation. *Eur. J. Org. Chem.*, **2010**, 1445–1449.

56 Genevois-Borella, A., Florent, J.-C., Monneret, C., and Grierson, D.S. (1990) Synthesis of 1-(3-R-amino-4-hydroxy butyl)thymine acyclonucleoside analogs as potential anti-aids drugs. *Tetrahedron Lett.*, **31**, 4879–4882.

57 Mroczkiewicz, M. and Ostaszewski, R. (2009) A new and general method for the synthesis of tripeptide aldehydes

based on the multi-component Ugi reaction. *Tetrahedron*, **65**, 4025–4034.
58. Naik, A., Maji, T., and Reiser, O. (2010) Iron(II)-bis(isonitrile) complexes: novel catalysts in asymmetric transfer hydrogenations of aromatic and heteroaromatic ketones. *Chem. Commun.*, **46**, 4475–4477.
59. Bauer, M. and Kazmaier, U. (2009) Straightforward synthesis of chiral hydroxy isocyanides. *Eur. J. Org. Chem.*, **2009**, 2360–2366.
60. Bauer, M. and Kazmaier, U. (2006) A new, modular approach towards 2-(1-hydroxyalkyl)oxazolines, effective bidentate chiral ligands. *J. Organomet. Chem.*, **691**, 2155–2158.
61. Soeta, T., Kojima, Y., Ukaji, Y., and Inomata, K. (2010) O-Silylative Passerini reaction: a new one-pot synthesis of α-siloxyamides. *Org. Lett.*, **12**, 4341–4343.
62. Mroczkiewicz, M., Winkler, K., Nowis, D., Placha, G., Golab, J., and Ostaszewski, R. (2010) Studies of the synthesis of all stereoisomers of MG-132 proteasome inhibitors in the tumor targeting approach. *J. Med. Chem.*, **53**, 1509–1518.
63. Meyers, A.I. and Bailey, T.R. (1986) An asymmetric synthesis of (+)-morphinans in high enantiomeric purity. *J. Org. Chem.*, **51**, 872–875.
64. Naik, A., Meina, L., Zabel, M., and Reiser, O. (2010) Efficient aerobic Wacker oxidation of styrenes using palladium bis(isonitrile) catalysts. *Chem. Eur. J.*, **16**, 1624–1628.
65. Cerulli, V., Banfi, L., Basso, A., Rocca, V., and Riva, R. (2012) Diversity oriented and chemoenzymatic synthesis of densely functionalized pyrrolidines through a highly diastereoselective Ugi multicomponent reaction. *Org. Biomol. Chem.*, **10**, 1255–1274.
66. Boullanger, P., Marmet, D., and Descotes, G. (1979) Synthesis of glycosyl isocyanides. *Tetrahedron*, **35**, 163–167.
67. Witczak, Z.J. (1986) Desulfurization of glycosyl isothiocyanates with tributyltin hybrid. *Tetrahedron Lett.*, **27**, 155–158.
68. Ichikawa, Y., Ohara, F., Kotsuki, H., and Nakano, K. (2006) A new approach to the neoglycopeptides: synthesis of urea- and carbamate-tethered N-acetyl-D-glucosamine amino acid conjugates. *Org. Lett.*, **8**, 5009–5012.
69. Ziegler, T., Kaisers, H.-J., Schlömer, R., and Koch, C. (1999) Passerini and Ugi reactions of benzyl and acetyl protected isocyanoglucoses. *Tetrahedron*, **55**, 8397–8408.
70. Ewing, D.F., Hiebl, J., Humble, R.W., Mackenzie, G., Raynor, A., and Zbiral, E. (1993) Stereoselective synthesis of N-(α- and β-ribofuranosyl)-formamides and related glycosyl formamides – precursors for sugar isocyanides. *J. Carbohydr. Chem.*, **12**, 923–932.
71. Nishiyama, T., Kusumoto, Y., Okumura, K., Hara, K., Kusaba, S., Hirata, K., Kamiya, Y., Isobe, M., Nakano, K., Kotsuki, H., and Ichikawa, Y. (2010) Synthesis of glycocinnasperimicin D. *Chem. Eur. J.*, **16**, 600–610.
72. Prosperi, D., Ronchi, S., Lay, L., Rencurosi, A., and Russo, G. (2004) Efficient synthesis of unsymmetrical ureido-linked disaccharides. *Eur. J. Org. Chem.*, **2004**, 395–405.
73. Marmet, D., Boullanger, P., and Descotes, G. (1981) Reaction des glycosyl isonitriles avec les amines et les alcools. *Can. J. Chem.*, **59**, 373–378.
74. Barton, D.H.R., Bringmann, G., Lamotte, G., Motherwell, W.B., Motherwell, R.S.H., and Porter, A.E.A. (1980) Reactions of relevance to the chemistry of aminoglycoside antibiotics. Part 14. A useful radical-deamination reaction. *J. Chem. Soc., Perkin Trans. 1*, 2657–2664.
75. Tavecchia, P., Trumtel, M., Veyrières, A., and Sinaÿ, P. (1989) Glycosylations with N-formylamino sugars: a new approach to 2'-deoxy-β-disaccharides. *Tetrahedron Lett.*, **30**, 2533–2536.
76. Trumtel, M., Tavecchia, P., Veyrières, A., and Sinaÿ, P. (1989) The synthesis of 2'-deoxy-β-disaccharides: novel approaches. *Carbohydr. Res.*, **191**, 29–52.
77. Grenouillat, N., Vauzeilles, B., and Beau, J.-M. (2007) Lipid analogs of the nodulation factors using the Ugi/Passerini multicomponent reactions: preliminary studies on the carbohydrate monomer. *Heterocycles*, **73**, 891–901.

78 Garson, M.J. and Simpson, J.S. (2004) Marine isocyanides and related natural products – structure, biosynthesis and ecology. *Nat. Prod. Rep.*, **21**, 164–179.

79 Scheuer, P.J. (1992) Isocyanides and cyanides as natural products. *Acc. Chem. Res.*, **25**, 433–439.

80 Edenborough, M.S. and Herbert, R.B. (1988) Naturally occurring isocyanides. *Nat. Prod. Rep.*, **5**, 229–245.

81 Cafieri, F., Fattorusso, E., Magno, S., Santacroce, C., and Sica, D. (1973) Isolation and structure of axisonitrile-1 and axisothiocyanate-1 two unusual sesquiterpenoids from the marine sponge *Axinella cannabina*. *Tetrahedron*, **29**, 4259–4262.

82 Carmeli, S., Moore, R.E., Patterson, G.M.L., Mori, Y., and Suzuki, M. (1990) Isonitriles from the blue-green alga *Scytonema mirabile*. *J. Org. Chem.*, **55**, 4431–4438.

83 Gloer, J.B., Poch, G.K., Short, D.M., and McCloskey, D.V. (1988) Structure of brassicicolin A: a novel isocyanide antibiotic from the phylloplane fungus *Alternaria brassicicola*. *J. Org. Chem.*, **53**, 3758–3761.

84 Ishiyama, H., Kozawa, S., Aoyama, K., Mikami, Y., Fromont, J., and Kobayashi, J.-I. (2008) Halichonadin F and the Cu(I) complex of halichonadin C from the sponge *Halichondria* sp. *J. Nat. Prod.*, **71**, 1301–1303.

85 Manzo, E., Carbone, M., Mollo, E., Irace, C., Di Pascale, A., Li, Y., Ciavatta, M.L., Cimino, G., Guo, Y.-W., and Gavagnin, M. (2011) Structure and synthesis of a unique isonitrile lipid isolated from the marine mollusk *Actinocyclus papillatus*. *Org. Lett.*, **8**, 1897–1899.

86 Wright, A.D. and Lang-Unnasch, N. (2009) Diterpene formamides from the tropical marine sponge *Cymbastela hooperi* and their antimalarial activity in vitro. *J. Nat. Prod.*, **72**, 492–495.

87 Caine, D. and Deutsch, H. (1978) Total synthesis of (−)-axisonitrile-3. An application of the reductive ring opening of vinylcyclopropanes. *J. Am. Chem. Soc.*, **100**, 8030–8031.

88 Srikrishna, A., Ravi, G., and Venkata Subbaiah, D.R.C. (2009) Enantioselective first total syntheses of 2-(formylamino) trachyopsane and ent-2-(isocyano) trachyopsane via a biomimetic approach. *Synlett*, 32–34.

89 Piers, E., Yeung, B.W.A., and Rettig, S.J. (1987) Methylenecyclohexane annulation. Total synthesis of (±)-axamide-1, (±)-axisonitrile-1, and the corresponding C-10 epimers. *Tetrahedron*, **43**, 5521–5535.

90 Kuo, Y.-L., Dhanasekaran, M., and Sha, C.-K. (2009) Total syntheses of (±)-axamide-1 and (±)-axisonitrile-1 via 6-*exo-dig* radical cyclization. *J. Org. Chem.*, **74**, 2033–2038.

91 Miyaoka, H. and Okubo, Y. (2011) Total synthesis of amphilectane-type diterpenoid (±)-7-isocyanoamphilecta-11(20),15-diene. *Synlett*, 547–550.

92 Muratake, H., Kumagami, H., and Natsume, M. (1990) Synthetic studies of marine alkaloids hapalindoles. Part 3. Total synthesis of (±)-hapalindoles H and U. *Tetrahedron*, **46**, 6351–6360.

93 Reisman, S.E., Ready, J.M., Weiss, M.M., Hasuoka, A., Hirata, M., Tamaki, K., Ovaska, T.V., Smith, C.J., and Wood, J.L. (2008) Evolution of a synthetic strategy: total synthesis of (±)-welwitindolinone A isonitrile. *J. Am. Chem. Soc.*, **130**, 2087–2100.

94 Avíles, E. and Rodríguez, A.D. (2010) Monamphilectine A, a potent antimalarial β-lactam from marine sponge Hymeniacidon sp: isolation, structure, semisynthesis, and bioactivity. *Org. Lett.*, **12**, 5290–5293.

95 White, R.D. and Wood, J.L. (2001) Progress toward the total synthesis of kalihinane diterpenoids. *Org. Lett.*, **3**, 1825–1827.

96 Richter, J.M., Ishihara, Y., Masuda, T., Whitefield, B.W., Llamas, T., Pohjakallio, A., and Baran, P.S. (2008) Enantiospecific total synthesis of the hapalindoles, fischerindoles, and welwitindolinones via a redox economic approach. *J. Am. Chem. Soc.*, **130**, 17938–17954.

97 Baran, P.S., Maimone, T.J., and Richter, J.M. (2007) Total synthesis of marine natural products without using protecting groups. *Nature*, **446**, 404–408.

98 Nishikawa, K., Nakahara, H., Shirokura, Y., Nogata, Y., Yoshimura, E., Umezawa, T., Okino, T., and Matsuda, F. (2010) Total synthesis of 10-isocyano-4-cadinene and determination of its absolute configuration. *Org. Lett.*, **12**, 904–907.

99 Tamura, K., Nakazaki, A., and Kobayashi, S. (2009) Stereocontrolled total synthesis of antimalarial (+)-axisonitrile-3. *Synlett*, 2449–2452.

100 Baran, P.S. and Richter, J.M. (2005) Enantioselective total syntheses of welwitindolinone A and fischerindoles I and G. *J. Am. Chem. Soc.*, **127**, 15394–15396.

101 Fukuyama, T. and Chen, X. (1994) Stereocontrolled synthesis of (−)-hapalindole G. *J. Am. Chem. Soc.*, **116**, 3125–3126.

102 Nolte, R.J.M., van Beijnen, A.J.M., and Drenth, W. (1974) Chirality in polyisocyanides. *J. Am. Chem. Soc.*, **96**, 5932–5933.

103 Le Gac, S., Schwartz, E., Koepf, M., Cornelissen, J.J.L.M., Rowan, A.E., and Nolte, R.J.M. (2010) Cysteine-containing polyisocyanides as versatile nanoplatforms for chromophoric and bioscaffolding. *Chem. Eur. J.*, **16**, 6176–6186.

104 Cornelissen, J., Rowan, A.E., Nolte, R.J.M., and Sommerdijk, N. (2001) Chiral architectures from macromolecular building blocks. *Chem. Rev.*, **101**, 4039–4070.

105 Schwartz, E., Koepf, M., Kitto, H.J., Nolte, R.J.M., and Rowan, A.E. (2011) Helical poly(isocyanides): past, present and future. *Polym. Chem.*, **2**, 33–47.

106 Van Beijnen, A.J.M., Nolte, R.J.M., Drenth, W., and Hezemans, A.M.F. (1976) Screw sense of polyisocyanides. *Tetrahedron*, **32**, 2017–2019.

107 Hase, Y., Nagai, K., Iida, H., Maeda, K., Ochi, N., Sawabe, K., Sakajiri, K., Okoshi, K., and Yashima, E. (2009) Mechanism of helix induction in poly(4-carboxyphenyl isocyanide) with chiral amines and memory of the macromolecular helicity and its helical structures. *J. Am. Chem. Soc.*, **131**, 10719–10732.

108 Clericuzio, M., Alagona, G., Ghio, C., and Salvadori, P. (1997) Theoretical investigations on the structure of poly(iminomethylenes) with aliphatic side chains. Conformational studies and comparison with experimental spectroscopic data. *J. Am. Chem. Soc.*, **119**, 1059–1071.

109 Green, M.M., Gross, R.A., Schilling, F.C., Zero, K., and Crosby, C. (1988) Macromolecular stereochemistry–effect of pendant group-structure on the conformational properties of polyisocyanides. *Macromolecules*, **21**, 1839–1846.

110 Cornelissen, J., Donners, J., de Gelder, R., Graswinckel, W.S., Metselaar, G.A., Rowan, A.E., Sommerdijk, N., and Nolte, R.J.M. (2001) β-Helical polymers from isocyanopeptides. *Science*, **293**, 676–680.

111 Nolte, R.J.M. (1994) Helical poly(isocyanides). *Chem. Soc. Rev.*, **23**, 11–19.

112 Onitsuka, K., Mori, T., Yamamoto, M., Takei, F., and Takahashi, S. (2006) Helical sense selective polymerization of bulky aryl isocyanide possessing chiral ester or amide groups initiated by arylrhodium complexes. *Macromolecules*, **39**, 7224–7231.

113 Onitsuka, K., Yamamoto, M., Mori, T., Takei, F., and Takahashi, S. (2006) Living polymerization of bulky aryl isocyanide with arylrhodium complexes. *Organometallics*, **25**, 1270–1278.

114 Metselaar, G.A., Cornelissen, J., Rowan, A.E., and Nolte, R.J.M. (2005) Acid-initiated stereospecific polymerization of isocyanopeptides. *Angew. Chem. Int. Ed.*, **44**, 1990–1993.

115 Kajitani, T., Okoshi, K., and Yashima, E. (2008) Helix-sense-controlled polymerization of optically active phenyl isocyanides. *Macromolecules*, **41**, 1601–1611.

116 Onouchi, H., Okoshi, K., Kajitani, T., Sakurai, S.I., Nagai, K., Kumaki, J., Onitsuka, K., and Yashima, E. (2008) Two- and three-dimensional smectic ordering of single-handed helical polymers. *J. Am. Chem. Soc.*, **130**, 229–236.

117 Gomar-Nadal, E., Veciana, J., Rovira, C., and Amabilino, D.B. (2005) Chiral teleinduction in the formation of a macromolecular multistate chiroptical redox switch. *Adv. Mater.*, **17**, 2095–2098.

118 Hida, N., Takei, F., Onitsuka, K., Shiga, K., Asaoka, S., Iyoda, T., and Takahashi, S. (2003) Helical, chiral polyisocyanides bearing ferrocenyl groups as pendants: synthesis and properties. *Angew. Chem. Int. Ed.*, **42**, 4349–4352.

119 Hasegawa, T., Matsuura, K., Ariga, K., and Kobayashi, K. (2000) Multilayer adsorption and molecular organization of rigid cylindrical glycoconjugate poly(phenylisocyanide) on hydrophilic surfaces. *Macromolecules*, **33**, 2772–2775.

120 van Walree, C.A., van der Pol, J.F., and Zwikker, J.W. (1990) A liquid-crystalline poly(iminomethylene) with cholesterol-containing pendant groups. *Rec. Trav. Chem. Pays-Bas*, **109**, 561–565.

121 van der Eijk, J.M., Nolte, R.J.M., Richters, V.E.M., and Drenth, W. (1981) Activity and enantioselectivity in the hydrolysis of substituted phenyl esters catalysed by imidazole-containing poly(iminomethylenes). *Rec. Trav. Chem. Pays-Bas*, **100**, 222–226.

122 Schwartz, E., Bodis, P., Koepf, M., Cornelissen, J.J.L.M., Rowan, A.E., Woutersen, S., and Nolte, R.J.M. (2009) Self-trapped vibrational states in synthetic β-sheet helices. *Chem. Commun.*, 4675–4677.

123 Vriezema, D.M., Kros, A., de Gelder, R., Cornelissen, J.J.L.M., Rowan, A.E., and Nolte, R.J.M. (2004) Electroformed giant vesicles from thiophene-containing rod-coil diblock copolymers. *Macromolecules*, **37**, 4736–4739.

2
General Aspects of Isocyanide Reactivity

Maxim A. Mironov

2.1
Introduction

The isocyano group is an unusual functionality, which may react with electrophiles, nucleophiles, and radicals under various conditions, giving rise to different primary imine adducts [1]. Therefore, isocyanides may serve as versatile building blocks in organic synthesis and as interesting ligands to form complexes with metals. The chemistry of isocyanides is characterized by the great diversity of transformations that includes hundreds of multicomponent reactions and transition metal-catalyzed insertions, as well as very large numbers of oligomerizations and polymerizations. Many of these reactions have been employed as an initial step in numerous preparative methods, and especially for the syntheses of various heterocycles.

The aim of this chapter is to describe those reactions which can be regarded as a core of isocyanide reactivity; hence, two-component intermolecular reactions of isocyanides with electrophiles and nucleophiles, as well as the oxidation and reduction of isocyano groups and isocyanide–cyanide rearrangement are discussed. Despite these reactions having been discovered more than 100 years ago, their practical use in organic synthesis has been initiated only during the past two decades. It should be noted that transition metal catalyzed or organocatalyzed transformations, and also formal [4+1] cycloadditions, are beyond the intended scope of this chapter.

The electronic structure of isocyanides can be represented by the alternative formulas of **1a** or **1b** being responsible for their reactivity (Figure 2.1). According to these structures, an isocyanide should be not only a strong nucleophile but also a 1,1-dipolar partner in different types of cycloaddition, owing to the presence of a monocoordinated carbon atom having an electron lone pair. However, only a few examples of [2+1] cycloadditions or direct alkylations of isocyano group have been described to date. In contrast, isocyanides can react with nucleophiles, sometimes without catalysis; in that way, the formula of **1c** with a donor–acceptor triple bond and an electron lone pair at the carbon atom may be suitable to describe the isocyanide reactivity. The structural parameters of isocyano group, such as the

Isocyanide Chemistry: Applications in Synthesis and Material Science, First Edition. Edited by Valentine Nenajdenko.
© 2012 Wiley-VCH Verlag GmbH & Co. KGaA. Published 2012 by Wiley-VCH Verlag GmbH & Co. KGaA.

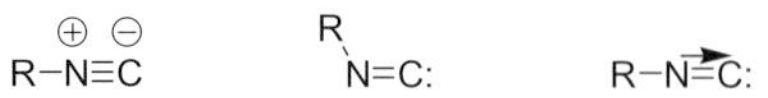

2 General Aspects of Isocyanide Reactivity

$\overset{\oplus}{R-N}\equiv\overset{\ominus}{C}$ $\underset{R}{N}=C:$ $R-N\overset{\rightarrow}{=}C:$

  1a   1b   1c

Figure 2.1 Electronic structure of isocyanides.

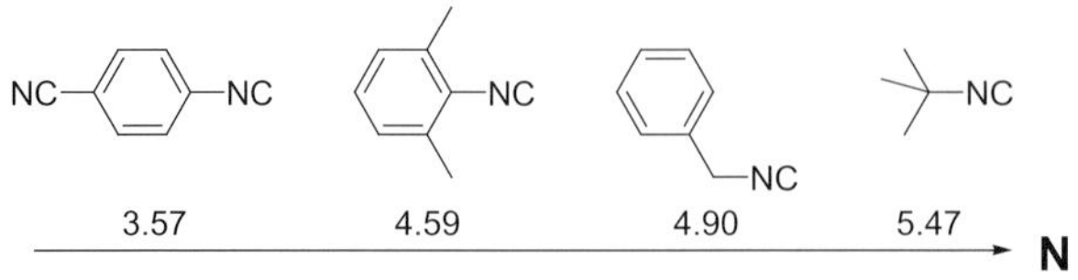

NC–⌬–NC	(2,6-dimethylphenyl)–NC	Ph–CH$_2$–NC	t-Bu–NC
3.57	4.59	4.90	5.47

$\longrightarrow$ N

Figure 2.2 Nucleophilicity parameter N for isocyanides.

C–N–C angle (close to 180°) and the C–N bond length (in the range of 0.116–0.117 nm), both of which are typical of compounds having a triple bond, confirm this suggestion [1].

Recently, the nucleophilicity of the isocyanides was determined and compared to that of other nucleophiles [2, 3]. As a consequence, it was shown that useful rankings of nucleophiles could be made on the basis of kinetic experiments, when only carbon electrophiles such as benzhydrylium ions were considered as the reaction partners. In this way comprehensive nucleophilicity scales, including π-, n-, and σ-nucleophiles, have been constructed by Mayr and Ofial [2]. According to these data, the nucleophilicities (N) of most alkyl and aryl isocyanides are affected only slightly by the substituents, and can be compared to those of furans, pyrroles, allylsilanes, and silyl enol ethers [3]. This ranking is in good agreement with the relative proton affinities of the isocyano group [4]. The nucleophilicity parameters, N, for four aliphatic and aromatic isocyanides are in the range of 3.5 to 5.5 N (Figure 2.2). Analogous kinetic investigations have yielded a nucleophilicity parameter of N = 16.3 on the logarithmic N-scale for free cyanide ion; this means that major changes proceeded during the attachment of an alkyl or aryl group to its N terminus.

Although aryl isocyanides bearing one electron-withdrawing group are only one to two orders of magnitude less nucleophilic than aliphatic isocyanides [3], the introduction of two and more acceptor substituents leads to isocyanides with an enhanced electrophilicity. These compounds can react with nucleophiles such as amines and water, without additional activation.

In summary, isocyanides may be regarded as a poor nucleophiles and electrophiles and, in general, are inert towards alkenes, alkynes, heterocumulenes, alkyl halides, amines, alcohols, water and many others reagents, under typical conditions. However, the ability of an isocyano group to add two (and more) electrophilic and nucleophilic educts in a sequential manner creates an indispensable functionality in organic synthesis.

2.2
Isocyanide–Cyanide Rearrangement

The thermal isocyanide–cyanide rearrangement discovered in 1873 by Weith [5] is formally a cationotropic 1,2-shift (Scheme 2.1). During the past two decades, this reaction has been widely employed as a convenient method for the stereoselective transformation of amines into corresponding nitriles. In addition, owing to their noncatalytic nature, rearrangements of this type have been proved to serve as excellent model reactions for testing the kinetic theories of unimolecular gas-phase reactions, thermal explosions, and vibrational energy transfer [6].

The majority of isocyanides isomerize quantitatively at 200–250 °C, with a half-life of a few hours, in diluted solutions of high-boiling solvents. This reaction also proceeds smoothly on the heating of pure isocyanides at 200 °C under reduced pressure. It should be noted that the structure of aliphatic and aromatic isocyanides has very little influence on the isocyanide–cyanide rearrangement [7], and therefore a general procedure may be adopted for all commercially available isocyanides.

Several research groups have reported side reactions occurring in this system that include an elimination of the cyano group and a reduction to hydrocarbons [8, 9]. For example, whilst the rearrangement of 2-adamantyl isocyanide without solvent produced 2-adamantyl nitrile, this was contaminated with considerable amounts of adamantan-2-one as a byproduct. Likewise, in diglyme, considerable quantities of adamantane were produced as a byproduct from 1- and 2-adamantyl isocyanides (Scheme 2.2) [9].

It has been revealed that a free-radical chain mechanism was competing with a synchronous pathway at higher isocyanide concentrations. In the presence of free-radical inhibitors (e.g., 1,1-diphenylethylene and 3-cyanopyridine; 0.1–0.15 mol l^{-1}),

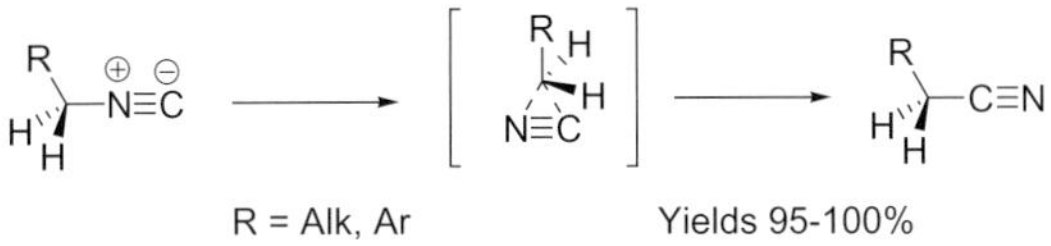

Scheme 2.1

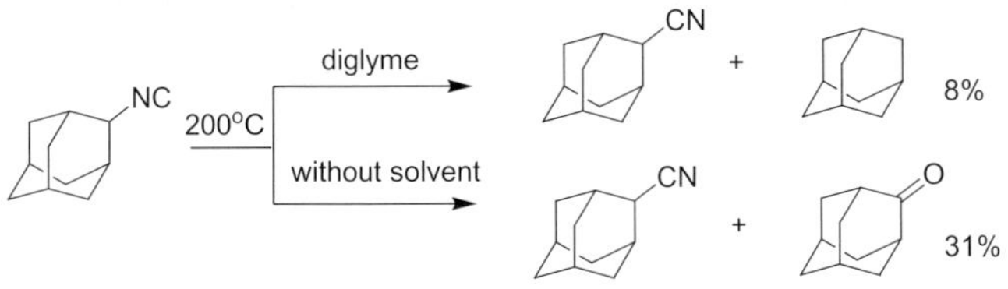

Scheme 2.2

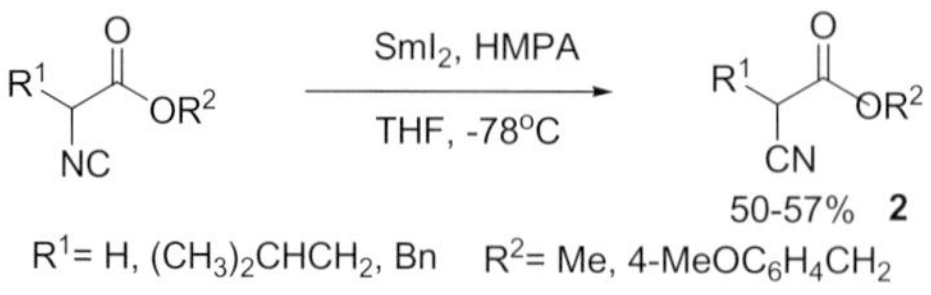

Scheme 2.3

R^1 C(=O)OR^2 with NC group → (SmI$_2$, HMPA, THF, −78 °C) → R^1 C(=O)OR^2 with CN group, 50–57%, **2**

R^1 = H, (CH$_3$)$_2$CHCH$_2$, Bn R^2 = Me, 4-MeOC$_6$H$_4$CH$_2$

the yields of nitriles are almost quantitative, and only minor amounts of byproduct can be detected [10].

Aliphatic nitriles have also been obtained from isocyanides by flash pyrolysis, in almost quantitative yields and with an almost complete retention of the stereochemistry. The reaction proceeded smoothly at 500–540 °C, 10^{-2} Torr and 5–30 min retention time [11], and similar conditions were used for the synthesis of aromatic and heterocyclic nitriles from the corresponding isocyanides [12].

Kang and Koh [13] reported the details of a samarium(II) iodide-promoted rearrangement of isocyanides to nitriles **2** that contrasted with the classical thermolysis (Scheme 2.3). Although this reaction occurs under very mild conditions (SmI$_2$, THF–HMPA, −78 °C), it requires the presence of an α-alkoxycarbonyl group at the carbon atom bearing the isocyano group. The samarium(II) iodide-promoted rearrangement tolerated various types of functionality, and provided final products in moderate yields. It was also shown that diastereoselectivity might be achieved if an appropriate chiral environment could be maintained; however, this reaction was not sufficiently effective for an asymmetric synthesis.

The mechanism of the isocyanide–cyanide rearrangement is of great interest in the theory of organic reactions, and many past investigations have been focused on this topic. For example, Casanova et al. [14] showed that cyclobutyl isocyanide would isomerize in the gas phase without any skeletal rearrangement of the migrating group, and that methyl, ethyl, isopropyl, and tert-butyl isocyanide would isomerize with very similar rates that were only slightly slower than those of phenyl, p-chlorophenyl, and p-methoxyphenyl isocyanides. This observation was in accordance with that expected for an entropy-controlled rather than for an enthalpy-controlled reaction. Optically active substrates – for example, (+)-sec-butyl isocyanide – undergo isomerization with a retention of configuration at the migrating carbon atom. The secondary α-deuterium kinetic isotope effect showed a k_H/k_D-value of 1.11 for the isomerization of benzyl isocyanide [15]. Taken together, these facts indicate that the bond-breaking and bond-making processes are essentially synchronous, and that little charge separation develops in the transition state. Consequently, the isocyanide–cyanide rearrangement should be regarded as a process that proceeds via an almost neutral, hypervalent, three-centered intermediate.

However, a free-radical chain mechanism can compete with a cationotropic 1,2-shift and, under certain conditions, predominate in the system. Unfortunately, the mechanism is usually accompanied by a poor reproducibility and complex mixtures, including reduction products. Thus, the photoinduced isomerization reaction from alkyl isocyanides to nitriles in benzene solution at room temperature was found to follow a free-radical chain mechanism [16]. The alkyl radical intermediates in the isomerization reaction were detected using electron spin resonance (ESR) spectroscopy, in combination with spin traps. The initiation of an isomerization in the condensed-phase can be explained by a homolytic bond breaking between the alkyl group and a nitrogen atom in the photoirradiated isocyanide derivatives.

When Meier and Rüchardt investigated the structure–reactivity relationship for isocyanide–cyanide rearrangement [7], the reproducible rates of isomerization of aliphatic isocyanides to nitriles in solution were determined (using gas-liquid chromatography or infrared spectroscopy) when free-radical inhibitors were added to suppress any competing radical chain reactions. As a consequence, the reactivities of 19 primary, secondary, tertiary, cyclic, bicyclic, bridgehead, benzyl, substituted-benzyl, α-carbomethoxymethyl, and triphenylmethyl isocyanides in this rearrangement reaction were found to be practically identical. One exception to this was 9-triptycyl isocyanide, which exhibited a slow rate of rearrangement due to steric hindrance by the three peri hydrogens. This essential independence of rate on large variations in structure confirmed that the bond angles and steric constraints are not changed between the ground state and the transition state.

Typically, aromatic isocyanides isomerize no more than 10-fold faster than aliphatic isocyanides, independent of any polar *para*-substituents and bulky *ortho*-substituents; for example, phenyl migrates only 3.5-fold faster than 1-octyl [7]. A hypervalent orthogonal transition state with retention of the aromatic sextet was proposed in contrast to the popular phenonium-type transition states for aryl migration in other 1,2-rearrangements (Figure 2.3). In this case, *ortho*-substituents will have only a minimal influence on the rate of the reaction.

Despite a faintly pronounced effect of substituents on this rearrangement, several groups have reported Hammett and Taft correlations for this situation. Thus, Kim *et al.* [15] obtained the absolute and relative rates of thermal rearrangements of substituted benzyl isocyanides at temperatures between 170 and 230 °C. Whilst the relative rates were independent of temperature and exhibited excellent Hammett correlations, the enthalpies of activation remained almost invariable (ca. 33.9 kcal mol^{-1}), so as to declare sparse substituent effects. In fact, the substituents were able to exercise only a minor influence on the enthalpies of activation, because the requirements for bond formation and breakage were in opposition. For example, an electron-withdrawing group would facilitate the bond-making process, but retard the bond-breaking process. In contrast, the entropies of

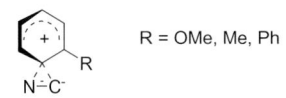

Figure 2.3 Transition state for aryl isocyanides.

activation could be closely related to the differences in number and character of the degrees of freedom between the transition state and the starting compounds. The cyclic structure of the transition state barely allows an internal freedom, whereas isocyanide enjoys the free rotations. Moreover, the electron-donating substituents tend to increase the extent of bond cleavage, leading to a relatively looser transition state structure. Together, these changes result in a decrease in the reaction time for the *p*-methoxy and *p*-methyl derivatives, which in turn means that the rates of the rearrangement may be delicately controlled through the entropic term derived from the degrees of the bond cleavages.

Sung [17] examined these substituent effects on the isomerization energies of isocyanides and nitriles by *ab initio* calculations, and also by correlating the isomerization energies with Taft's dual-substituent parameters. The isocyanide–cyanide rearrangement energies have exhibited a good correlation with Taft's dual-substituent parameters. Typically, isomerization was disfavored by the σ-donating substituents but favored by the σ-accepting substituents; in contrast, isomerization was favored by the π-donating substituents but disfavored by the π-accepting substituents.

The isocyanide–cyanide rearrangement was studied for heteroanalogues of the isocyanides, where the N-atom of the isocyano group is connected to O, S, N, P, and Si [18–22]. Thus, Wentrup and Winter [20] showed that cyanamides **3** would be formed in the gas phase during the pyrolysis of instable isocyanoamides **4** (Scheme 2.4). At lower temperatures (500 and 450 °C), most of the starting material was recovered unchanged, whereas pyrolyses conducted at higher temperatures (700 °C) resulted in an increased formation of cyanamides at the expense of isocyanoamides. The disubstituted amino group in isocyanoamides accelerates the isocyanide–cyanide rearrangement – an effect which is even more pronounced in the case of the diphenylamino group.

In organic isocyanides, the Si–NC bond is considerably more labile than the C–NC bond, with the silyl isocyanides **5** being reported as in equilibrium with silyl nitriles **6** at room temperature (Scheme 2.5) [22]. An intramolecular migration of triorganosilyl groups, from the nitrogen to the carbon of a cyanide group, was proposed as a possible mechanism for this, although some exchanges involving

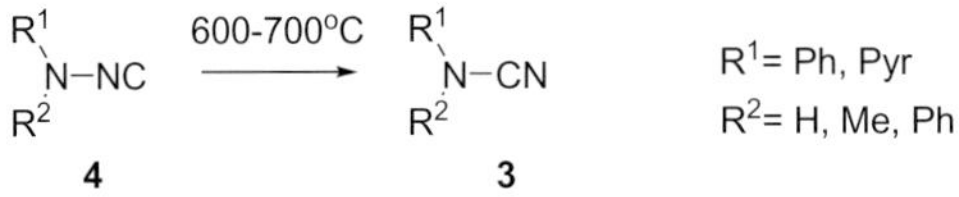

$$\underset{\underset{R^2}{R^1}}{N-NC} \xrightarrow{600-700\,°C} \underset{\underset{R^2}{R^1}}{N-CN} \qquad R^1 = Ph, Pyr$$
$$\qquad \textbf{4} \qquad\qquad\qquad \textbf{3} \qquad\qquad R^2 = H, Me, Ph$$

Scheme 2.4

$$\underset{\underset{R}{R}}{R-Si-CN} \rightleftharpoons \underset{\underset{R}{R}}{R-Si-NC} \qquad R = Me, Et, \textit{i}-Pr$$
$$\qquad \textbf{6} \qquad\qquad\qquad \textbf{5}$$

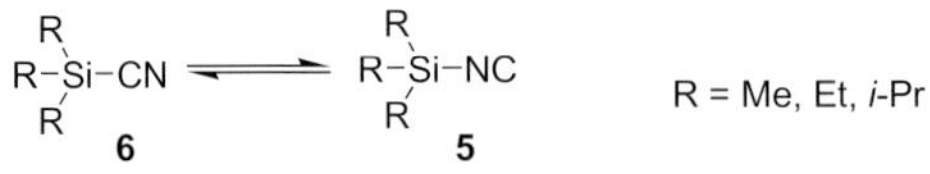

Scheme 2.5

silyl cyanides must proceed through an intermolecular transfer. All of the presently available evidence indicates that the relative position of the equilibrium will depend on: (i) the nature of the groups on silicon; (ii) the temperature; (iii) the phase conditions; and (iv) the solvation and surface effects. For example, the heating of Alk_3SiCN in 1-chloronaphthalene at 225 °C sharply enhanced the N-bonded form but slightly reduced the C-bonded form. Hence, electron-withdrawing groups tended to favor the C-bonded form, while steric effects were most likely secondary. Subsequent calculations using *ab initio* methods revealed that the SiNC isomer lay 1.5 kcal mol^{-1} above the SiCN species [23].

The preparative application of this rearrangement reaction requires the suppression of side reactions, and is best carried out using flash pyrolysis; indeed, excellent chemical and optical yields (>96% retention) of cyanides were achieved using this procedure. For example, *trans*-2-butenyl isocyanide was rearranged, without any concomitant allylic isomerization, to *trans*-2-butenyl cyanide [6], while optically active 1-(formyloxymethyl)-2-phenylethyl cyanide was obtained from optically active L-phenylalanine as a new type of chiral pool synthon [24]. As a consequence, new and economically interesting syntheses for the known nonsteroidal anti-inflammatory drugs ibuprofen and (S)-naproxen have been described. Glycosyl cyanides, which are important precursors in the synthesis of C-glycosides or C-nucleosides, may also be prepared via the thermal isomerization of glycosyl isocyanides [8].

These examples demonstrate the possible uses of isocyanide–cyanide rearrangement as a powerful tool in organic synthesis for the transformation of amines into corresponding nitriles.

2.3
Oxidation/Reduction of the Isocyano Group

2.3.1
Oxidation of the Isocyano Group

Investigations into the oxidation of isocyanides have been made since their initial discovery in 1867, with Gautier [25] showing that methyl isocyanide and ethyl isocyanide would each be oxidized by mercuric oxide so as to produce a complex mixture from which a small amount of ethyl isocyanate could be isolated.

Subsequently, a variety of oxidants and catalytic systems were tested for creating isocyanates from isocyanides, including the reaction of alkyl isocyanides with ozone [26]. In this case, the yields of isocyanates **7** ranged from 7% to 73%, with high-molecular-weight isocyanides providing the best conversions to isocyanates than their low-molecular-weight counterparts (Scheme 2.6).

Simple salts of nickel(II) were found to serve as catalysts for the reaction of isocyanides with oxygen [27]. For example, when oxygen was bubbled continuously through a solution of *tert*-butyl isocyanide and $NiCl_2$ in THF, an almost quantitative oxidation to the corresponding isocyanate was observed by Deming and Novak

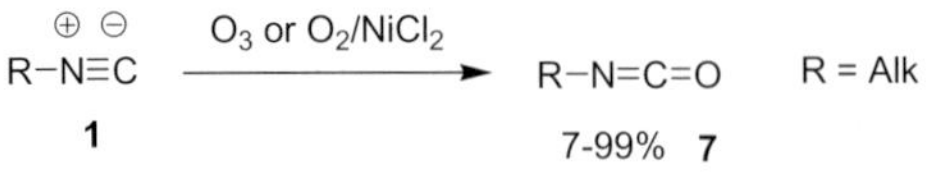

Scheme 2.6

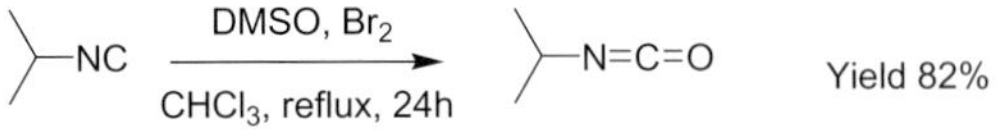

$$R-N\equiv C \xrightarrow{O_3 \text{ or } O_2/NiCl_2} R-N=C=O \quad R = Alk$$
$$\text{1} \qquad\qquad\qquad 7\text{-}99\% \quad \text{7}$$

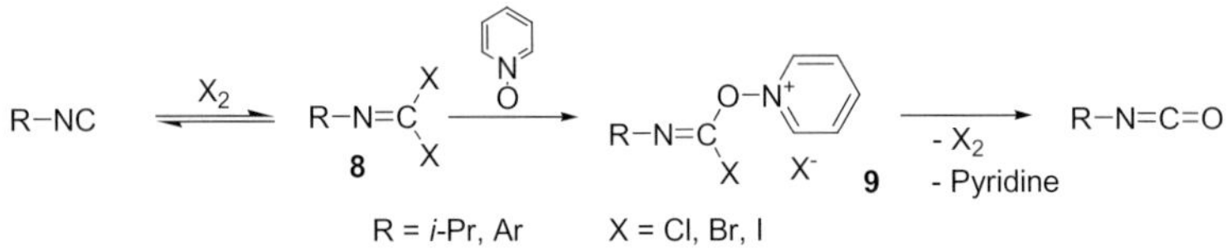

Scheme 2.7

$$\text{iPr-NC} \xrightarrow[\text{CHCl}_3,\text{ reflux, 24h}]{\text{DMSO, Br}_2} \text{iPr-N=C=O} \quad \text{Yield 82\%}$$

Scheme 2.8

$$R-NC \rightleftharpoons \underset{\text{8}}{R-N=CX_2} \xrightarrow{\text{pyridine N-oxide}} \underset{\text{9}}{R-N=C(X)(O-N^+\text{Py})\ X^-} \xrightarrow[-\text{Pyridine}]{-X_2} R-N=C=O$$

R = i-Pr, Ar   X = Cl, Br, I

(Scheme 2.6) [28]. In this case, the authors considered that nickel(II) had been reduced to nickel(I) by the isocyanide, and then reoxidized to nickel(II) by oxygen. In the presence of an excess of oxygen, oxidation of the isocyanide was observed, whereas a catalytic amount of oxygen favored polymerization.

Johnson and Daughetee [29] reported the oxidations of isopropylisocyanide and phenylisocyanide to the corresponding isocyanates with dimethyl sulfoxide (DMSO) in the presence of halogens. The addition of 5 mol% of halogens (bromine, iodine, or chlorine) to an equimolar mixture of isopropylisocyanide and DMSO in chloroform solvent at refluxing temperatures led to the smooth formation of isopropylisocyanate and dimethyl sulfide (Scheme 2.7). Phenylisocyanide also produced phenylisocyanate with bromine and DMSO.

The same groups proposed another oxidant, namely pyridine N-oxide in the presence of iodine [30]. The oxidation of isopropylisocyanide and aromatic isocyanides has been studied in detail. Since pyridine N-oxide had no effect on the isocyanide in the absence of a halogen, the formation was suggested of an unstable isocyanide diiodide **8**, which reacted with pyridine N-oxide (Scheme 2.8). A subsequent destruction of the corresponding salts **9** resulted in the formation of target isocyanates.

During recent years, the oxidation of isocyanides has played an increasing role in the chemistry of saccharides. An example is the oxidation of glycosyl isocyanides to generate highly reactive glycosyl isocyanates, which have in turn been used in the synthesis of glycoconjugates and ureas [31–36]. A typical example of the synthesis of α-glucopyranosyl urea is shown in Scheme 2.9 [31], where a pyridine

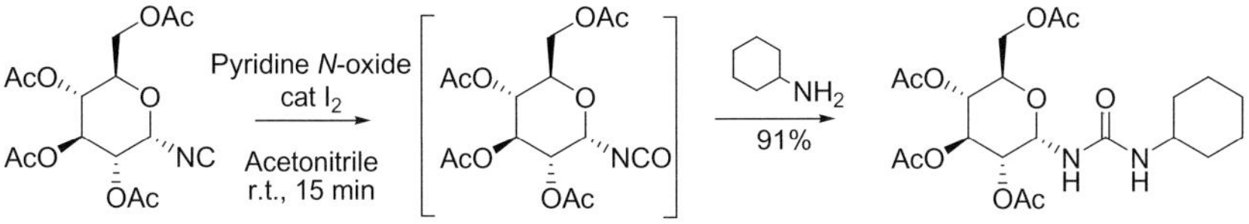

Scheme 2.9

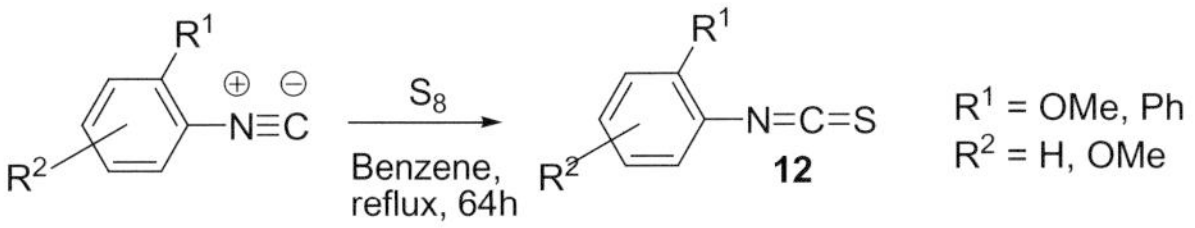

Scheme 2.10

N-oxide-mediated oxidation of the α-glucopyranosyl isocyanides **10**, catalyzed by iodine, afforded the α-glucopyranosyl isocyanates. The latter could then be treated successively with a variety of amines in a one-pot process to provide the glucopyranosyl ureas **11** in good yields. This method appeared useful for the preparation of a combinatorial library of glucopyranosyl urea derivatives which may ultimately find widespread use within the pharmaceutical industries.

Similar approaches have been applied to obtain α- and β-xylopyranosyl ureas [32], bicyclic pyranoid glycosyl ureas [33], and β-ribofuranosyl isocyanate [34]. Pyridine N-oxide (3.0 equiv.) and a catalytic amount of iodine (0.07 equiv.) in acetonitrile provided the most satisfactory oxidation conditions for these transformations [31–33]. Moreover, lead tetraacetate [34], iodobenzene bis-(trifluoroacetate) [35], and 2,4,6-trimethylbenzonitrile oxide [36] were each employed for the oxidation of glycosyl isocyanides.

2.3.2
Reactions with Sulfur and Selenium

Although, when heated with sulfur, the isocyanides were converted into isothiocyanates, no more than trace amounts of both ethyl and phenyl isothiocyanates were reported as being produced in the initial experiments conducted during the late nineteenth century [5, 37].

Subsequently, the first preparative synthesis of aryl isothiocyanates **12** from isocyanides was reported almost 100 years later (Scheme 2.10) [38], when the sulfurization of aromatic isocyanides bearing electron-donating groups by elemental sulfur (1.0–1.5 molar excess) was carried out in a solution of anhydrous benzene at reflux temperatures for 64 h. However, the need to include an *ortho*-substituent to achieve a consistently high yield proved to be a serious obstacle for the practical use of this method.

During the past two decades, several novel catalytic methods have been introduced for the transformation of isocyanides into isothiocyanates. As a result, a

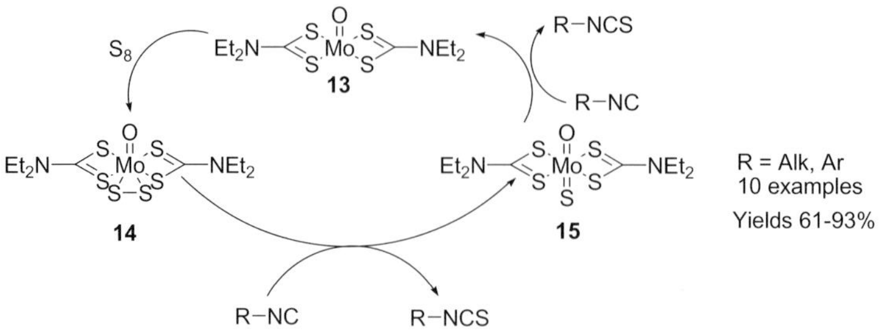

Scheme 2.11

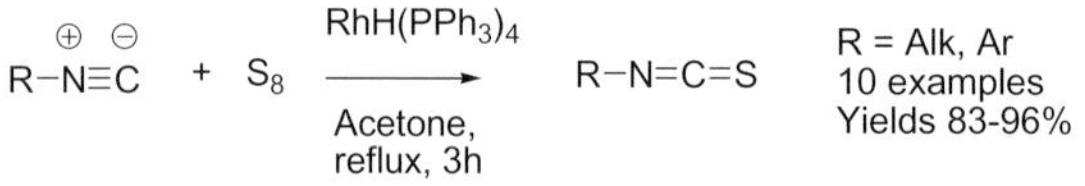

Scheme 2.12

range of aliphatic and aromatic isothiocyanates has been synthesized, in good to excellent yields, from corresponding isocyanides and elemental sulfur under mild conditions by using catalytic amounts of elemental selenium [39]. About a year later, the catalytic activity of tellurium to create isothiocyanates from isocyanides and sulfur was shown to be extremely high, and far superior to that of selenium [40]. Unfortunately, the toxicity of these chalcogens limits their practical use in organic syntheses, however.

The direct molybdenum-catalyzed sulfuration of a variety of isocyanides with elemental sulfur was described by Adam et al. [41]. Indeed, a catalytic cycle was suggested in which the molybdenum–oxo–disulfur complex operated as the active sulfur-transferring species. In this reaction, the molybdenum complex **13** reacts first with elemental sulfur to generate the disulfur complex **14**, which is then desulfurized by two isocyanide molecules back to complex **13** through intermediate **15** (Scheme 2.11). The yields of the isolated products were generally good to excellent, and the consumption of isocyanides usually was complete within 72 h in refluxing acetone under an inert atmosphere. The molybdenum catalyst has also been used for the direct synthesis of thioureas from isocyanides [42].

Recently, it was shown that rhodium complexes RhH(PPh$_3$)$_4$ and Rh(acac)(CH$_2$=CH$_2$)$_2$ could catalyze the reactions of aliphatic and aromatic isocyanides with sulfur, thus producing isothiocyanates in high yields [43]. This reaction was shown to be much faster than its molybdenum-based counterpart, and in most cases was complete within 3 h in refluxing acetone (Scheme 2.12). The reaction mechanism most likely involves the initial formation of a low-valent rhodium complex (RNC)$_m$RhL$_n$ from isocyanide and a rhodium species. An activated sulfur species generated from elemental sulfur undergoes oxidative addition to the complex, after which the transfer of a sulfur atom to isocyanide occurs to produce isothiocyanate.

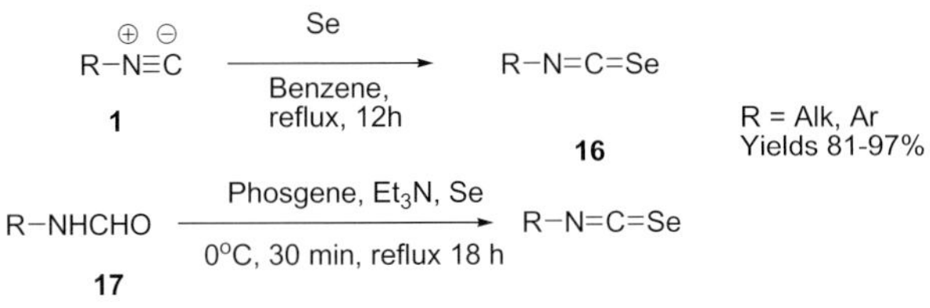

Scheme 2.13

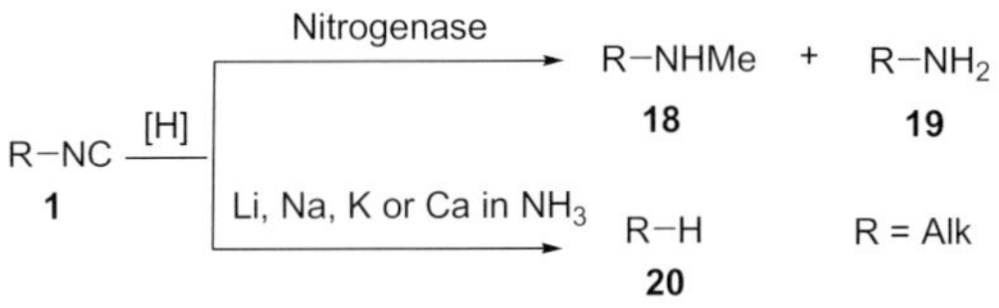

R−N≡C (1) →[Se, Benzene, reflux, 12h] R−N=C=Se (16) R = Alk, Ar; Yields 81–97%

R−NHCHO (17) →[Phosgene, Et₃N, Se, 0°C, 30 min, reflux 18 h] R−N=C=Se

Scheme 2.14

R−NC (1) →[H]:
- Nitrogenase → R−NHMe (18) + R−NH₂ (19)
- Li, Na, K or Ca in NH₃ → R−H (20), R = Alk

This mechanism is consistent with the observation that it is essential to add isocyanide to the rhodium complex prior to the addition of sulfur. A direct isothiocyanate synthesis corresponding to a biomimetic methodology has been documented in the sulfuration of isocyanides by a sulfur-containing metalloenzyme [44].

One of the classical methods employed for the synthesis of organic isoselenocyanates **16** involves the addition of elemental selenium to isocyanides [45, 46]. Previously, Barton et al. [47] reported a high-yielding, one-pot procedure for the preparation of isoselenocyanates from the corresponding formamides **17** (Scheme 2.13). In this case, various aromatic and aliphatic primary amines were employed in order to prepare the isoselenocyanates, so as to establish the generality of the procedure [48]. A similar approach was employed for the synthesis of bicyclic sugar selenoureas [49].

2.3.3
Reduction of the Isocyano Group

The reduction of isocyanides leads to three main products: N-methyl amines **18**; primary amines **19**; and hydrocarbons **20** (Scheme 2.14). From the viewpoint of organic synthesis, the formal deamination reaction via isocyanides is of major interest as a powerful method for functional groups transformation [50], while the creation of primary and secondary amines may be more widely exploited in the field of biochemistry [51].

When Ugi and Bodesheim [52] observed the near-quantitative reduction of isocyanides to their corresponding hydrocarbons by solutions of metals (lithium, sodium, potassium, calcium) in liquid ammonia (Scheme 2.14), they hypothesized that the isocyanide would accept two electrons – by either a one-step or a two-step process – followed by cleavage of the C–N bond.

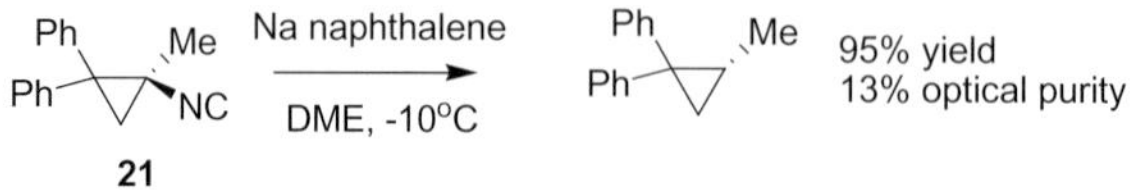

Scheme 2.15

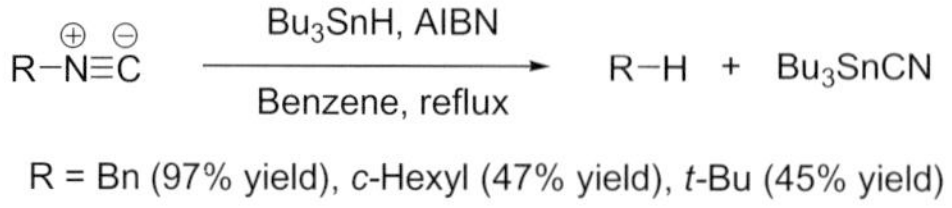

$$R-N\overset{\oplus}{\equiv}\overset{\ominus}{C} \xrightarrow[\text{Benzene, reflux}]{\text{Bu}_3\text{SnH, AIBN}} R-H + \text{Bu}_3\text{SnCN}$$

R = Bn (97% yield), c-Hexyl (47% yield), t-Bu (45% yield)

Scheme 2.16

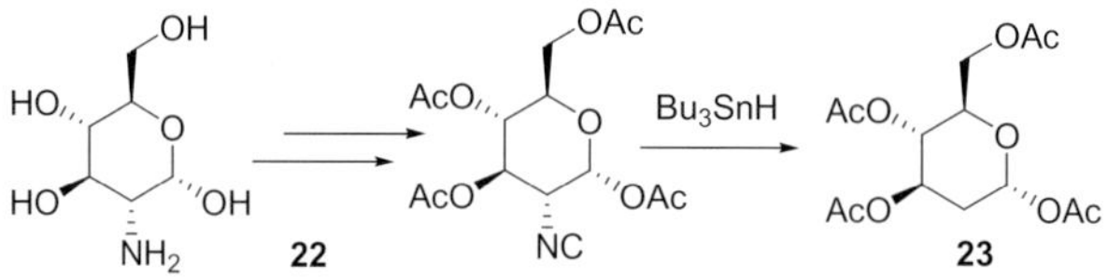

Scheme 2.17

Later, Buchner and Dufaux [53] found that the reductive cleavage of isocyanides also occurred when THF was employed as solvent, as well as in liquid ammonia. These authors postulated a two-step mechanism in which the addition of the first electron led to the formation of a radical anion intermediate that could accept a second electron; this then led to the formation of a carbanion intermediate and cyanide ions.

An alternative reducing medium – sodium naphthalene in 1,2-dimethoxyethane (DME) – was introduced by Niznik and Walborsky [54, 55]. From a synthetic viewpoint, this would be a method of choice as both aromatic and aliphatic primary amines can be converted to hydrocarbons via isocyanides, both conveniently and in high yield. However, all methods based on reduction with metal solvents usually lead to the formation of a racemic product from a chiral isocyanide. Although a low stereoselectivity was observed in the reduction of a chiral cyclopropyl isocyanide **21** (Scheme 2.15), in most cases the starting reagent was totally racemized [55].

Saegusa et al. [56] described the radical reduction of aliphatic isocyanides by tributyltin hydride, which can be used for the deamination of a great variety of amines (Scheme 2.16).

The marked differences in the rates of reduction of tertiary, secondary, and primary aliphatic isocyanides permit a selective reduction to be carried out (Scheme 2.16), as aromatic isocyanides are not reduced under these conditions. A representative example of the application of this reaction is shown in Scheme 2.17, which demonstrates the pathway from the readily available glucosamine **22** to 2-deoxyglucose **23** [50, 57].

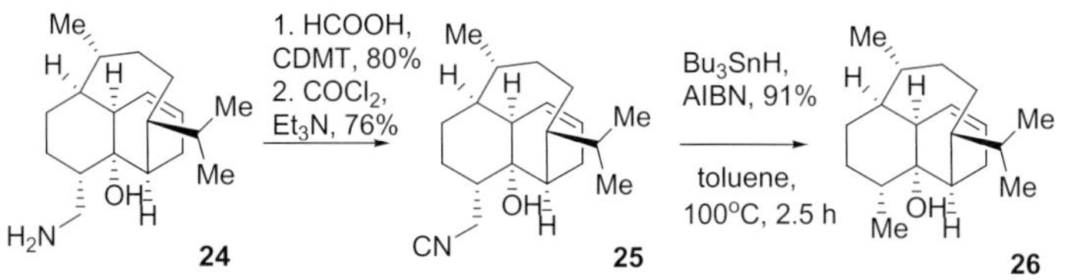

Scheme 2.18

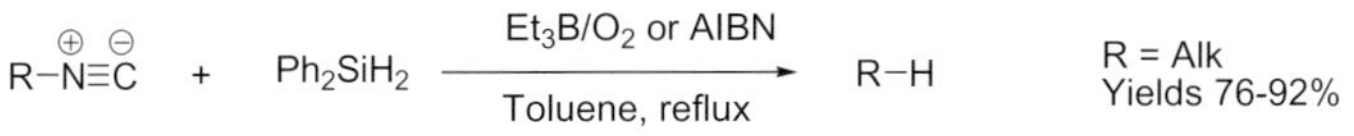

Scheme 2.19

Recently, Baran et al. [58] employed the Saegusa deamination as a key step in the total synthesis of the biologically significant diterpene, vinigrol. The extreme difficulty in preparing this molecule stems from its highly congested decahydro-1,5-butanonaphthalene ring system, which contains eight contiguous stereocenters. In this synthesis, the aminomethyl group was transformed into the desired methyl group via a three-step sequence: (i) immediate formylation of the crude amine **24**; (ii) dehydration to a primary isocyanide **25**; and (iii) treatment with Bu$_3$SnH in the presence of AIBN in 56% overall yield (Scheme 2.18). The robustness of this overall route is evident from the fact that over 5 g of the target product **26** was easily prepared, and that all of the steps leading to this key intermediate could be conducted on the gram scale.

Other reagents may also be used for the radical chain deamination. For example, isocyanides were deaminated to the corresponding hydrocarbons or olefins with diphenylsilane (Ph$_2$SiH$_2$) in good yield (Scheme 2.19) [59]. The reduction of aliphatic isocyanides (methyl isocyanide) can be catalyzed by the component proteins of nitrogenase (see Scheme 2.14) [51].

In summary, the oxidation and reduction of isocyanides play prominent roles in modern organic synthesis as convenient methods for the transformation of functional groups, and especially of amines into corresponding hydrocarbons, ureas, and thioureas.

2.4
Reactions of Isocyanides with Electrophiles

The distinctive feature of electrophilic addition to isocyano group is the formation of highly reactive species (O- and S-acyl isoimides **27**; dihalogenoisocyanides **28**; imidoyl halides **29**; and zwitterionic intermediates **30–32**) that can react further with various nucleophilic reagents, thereby opening a great multitude of synthesis pathways (Figure 2.4).

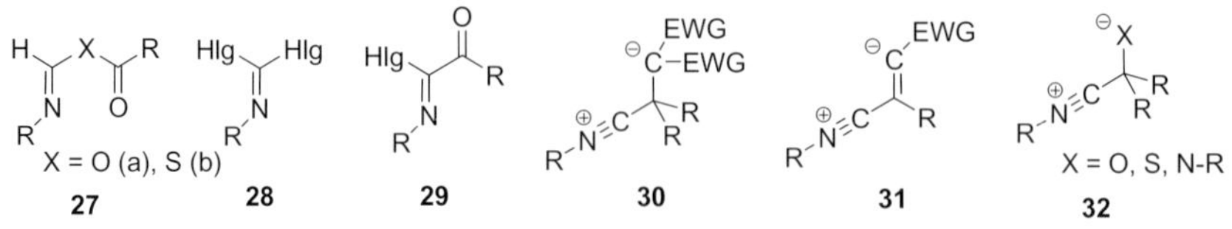

Figure 2.4 Intermediates in reactions of isocyanides with electrophiles.

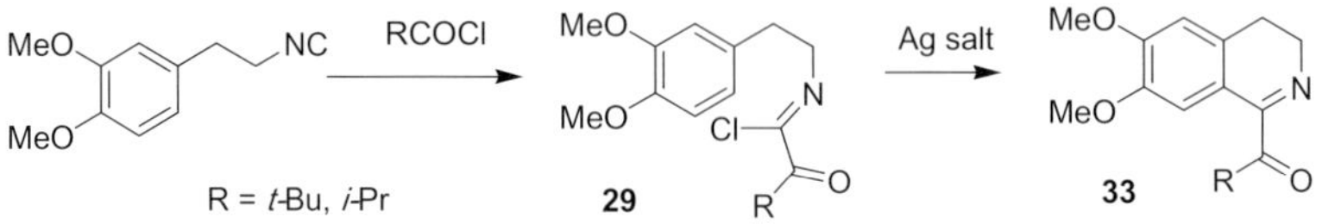

Scheme 2.20

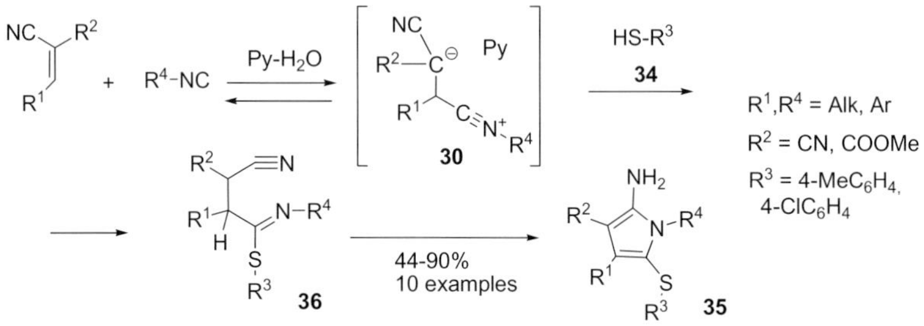

Scheme 2.21

These intermediates can be utilized in the synthesis of heterocycles in the case of intramolecular post-transformations, or for the design of novel multicomponent reactions in the case of intermolecular reactions. A representative example of the first strategy is the formation of 1-acyl-3,4-dihydroisoquinolines **33** in moderate to good yields as a result of the silver(I)-mediated cyclization of the corresponding imidoyl halides **29** [60]. This method is considered to be a useful supplement to the well-known Bischler–Napiralski synthesis of 3,4-dihydroisoquinolines (Scheme 2.20). Many other cyclizations of highly reactive intermediates in the reactions of isocyanides with electrophiles are capable of providing access to hundreds of heterocyclic scaffolds. Recent progress in this area has been reviewed by Lygin and de Meijere [61].

The second strategy may be illustrated by the reaction of isocyanides with gem-diactivated olefins and thiophenols [62]. In this case, the instable zwitterionic intermediate **30** was trapped via the addition of thiophenols **34** to the reaction mixture and, depending on the choice of isocyanides, substituted 2-aminopyrroles **35** or thioimidates **36** have been isolated in high yields (Scheme 2.21). This method provides the possibility to vary the four substituents in the pyrrole ring, while the presence of a primary amino group opens the route to further modifications of

2.4 Reactions of Isocyanides with Electrophiles

the compounds obtained. During the past decade, a variety of novel approaches has been introduced for the creation of multicomponent reactions, based on the replacement of either one ("single-reactant replacement" method) [63] or two ("reaction-operator" method) [64] educts in the reactions of isocyanides with electrophiles.

2.4.1
Reaction with Acids

The highly reactive O-acyl isoimides **27a** are key intermediates in all reactions of isocyanides with carboxylic acids [1]. Although the isolation of these mixed anhydrides has been claimed by several groups, a detailed analysis was subsequently provided by Danishefsky et al. [65] that showed most of these earlier reports to be incorrect [66]. Although, at present, pure O-acyl isoimides **27a** are unknown, Rebek et al. [67] have brought to bear impressive evidence as to their existence in special settings. Thus, cavitand **37** has been introduced which contains an "introverted" acid functionality on an anthracene skeleton; the subsequent reaction of **37** with small, aliphatic isocyanides bound inside the cavity provides an elusive, mixed anhydride. By using this approach, the labile O-acyl isoimide intermediate **38** was first observed using ^{1}H nuclear magnetic resonance (NMR) and infrared spectroscopies (Scheme 2.22).

The possibility of performing a bimolecular reaction between isocyanides and carboxylic acids in the cylindrical host **39** was also demonstrated by utilizing NMR methods [68]. The supramolecular structure can self-assemble only in the presence of a suitable guest (or combinations of different guests) that correctly fills the space (Scheme 2.23). Hence, a combination of p-tolylacetic acid and n-butyl isocyanide provided a single encapsulation complex with the host, in which the isocyanide and acid groups were located close to one another within the nanocapsule. The subsequent demonstration (using ^{1}H NMR spectroscopy) that the initial encapsulation complex had been replaced by the rearrangement product provided gave

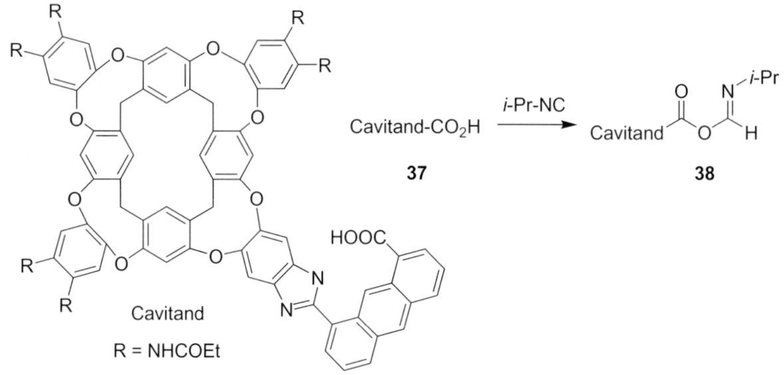

Scheme 2.22

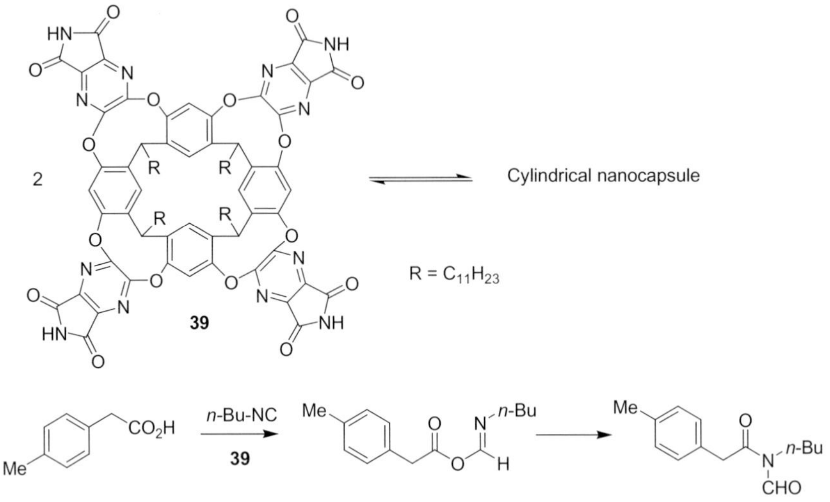

Scheme 2.23

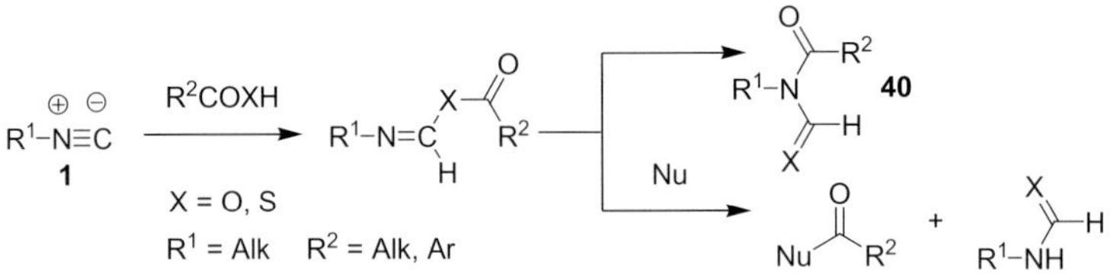

Scheme 2.24

strong evidence of O-acyl isoimide formation in the reaction of isocyanides with carboxylic acids.

Depending on the reaction conditions and the structure of an isocyanide, this reaction can result in the formation of various types of amide. Consequently, by using an excess of carboxylic acid it is possible to obtain a symmetrical anhydride of carboxylic acid and formamide [1]. Danishefsky et al. showed that, under microwave heating at 150 °C in CHCl$_3$, this reaction would lead to disubstituted amides **40** via acyl migration, and provided an efficient formation of amide bonds (Scheme 2.24) [69].

Moreover, it was discovered that – unlike the oxa version – the overall thio counterpart of this reaction occurs at room temperature [70]. In this case, the S-acyl isothioimide **27b** intermediates generated from the reactions of isocyanides with thioacids are much more reactive than their oxy analogues; consequently, the corresponding rearrangement is far more rapid and takes place under mild conditions.

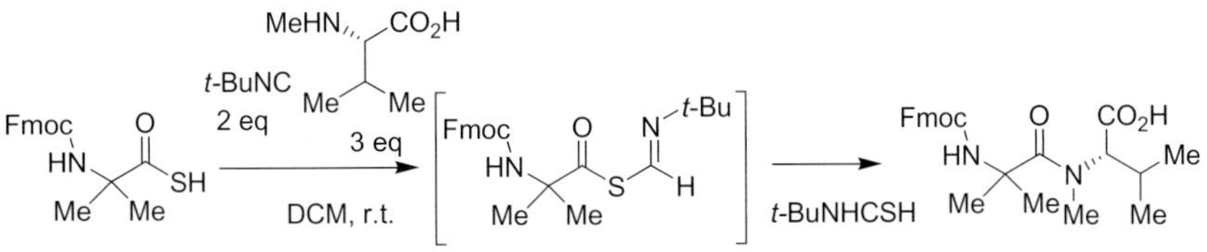

Scheme 2.25

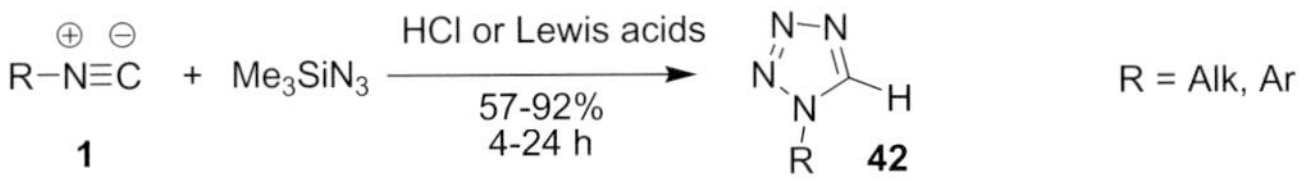

Scheme 2.26

Both, O- and S-intermediates **27** are able to react with alcohols, and also with primary and secondary amines (Scheme 2.24), to form various esters and amides in good yields [65, 69, 70]. Even more impressive was the success of the method in producing the highly hindered tertiary amide **41**. Currently, the reactions of S-acyl isothioimide with amines are conducted under neutral conditions at room temperature in dichloromethane [70, 71], in which case the isocyanides are not incorporated into the final structure but rather serve as an activating agent, such as N,N-dicyclohexylcarbodiimide (DCC) (Scheme 2.25). As a result, this method may be regarded as a notable advance in the construction of amide bonds.

The formation of amide bonds in the presence of isocyanides has also been applied to the fashioning of biologically important molecules. Early investigations in this area led to reports of peptide bond formation in the presence of isocyanides and, indeed, several short peptides were obtained using this approach [72]. Subsequently, Danishefsky et al. [71] explored various methods of generating tertiary amides in the context of a total synthesis of a challenging target system. One compound to be selected as a worthy goal was cyclosporine A, a reversible inhibitor of cytokines in T-helper cells that initially was isolated from the fungus *Tolypocladium inflatum gams*. The subsequent use of a novel approach to the formation of amide bonds with the participation of isocyanides led to a total synthesis of cyclosporine A, in a fashion that allowed for a more detailed mapping of the material's structure–activity relationship. These findings showed clearly that the chemistry of isocyanides could be applied to the construction of a variety of tertiary amides, including highly complex natural structures.

Isocyanides are able to interact with various acids, including hydrochloric, hydrazoic, and thiocyanic acids [1]. Among these materials, the reaction with hydrazoic acid has received the most attention due to its extensive use in the synthesis of 1-substituted tetrazoles **42** (Scheme 2.26) [73]. Tetrazoles are regarded as being biologically equivalent to the carboxylic acid group, and extensive studies on these compounds have been conducted in the field of medicinal chemistry. Although the reaction of isocyanides with hydrazoic acid has been well known for almost a century, the classical procedure requires the direct addition of a large excess of hydrazoic acid, which is both dangerous and potentially harmful. Hence, recent research has focused more on the development of a straightforward (but generally available) method to produce 1-substituted tetrazoles. In order to achieve this, Jin et al. [74] used trimethylsilylazide (1.5 equiv.) in methanol in the presence of various acid catalysts to generate hydrazoic acid. In the absence of an acid catalyst the reaction provided a low yield of 1-substituted tetrazoles. However, the addition of HCl as the catalyst HCl led to a very high yield (up to 90%), with Lewis acid catalysts such as $CeCl_3$ and $ZnCl_2$ also being effective.

Sureshbabu et al. [75] employed $ZnBr_2$ as a catalyst in dry methanol as solvent to produce a new class of unnatural amino acids that contained 1-substituted tetrazoles within the side chains, in almost 80% yields. The tetrazole-forming reaction proceeded well with aromatic isocyanides, irrespective of the position and electronic nature of the substituents on the aromatic ring, but longer reaction times were required for reaction of the alkyl isocyanides. Although, in general, the final product was acquired at high yields, the isocyanides were partially hydrolyzed to formamides in the presence of a trace amount of moisture in the reaction medium, and this led to a decreased yield of the tetrazoles. Hence, performing the reaction in completely dry conditions and under a nitrogen atmosphere would lead to a significant improvement in yield, by minimizing the degree of hydrolysis.

Other acids react with isocyanides in similar fashion, leading to a variety of heterocyclic compounds. For example, isocyanides may add two molecules of thiocyanic acid to produce triazines **43** [76], or two molecules of HCl to yield imidazolidines **44** [77] (Scheme 2.27).

To summarize, during the past decade the reaction of isocyanides with acids has become a useful tool for organic synthesis, most notably as a versatile method of constructing amide bonds and their equivalent.

2.4.2
Reactions with Halogens and Acyl Halides

Aromatic isocyano dichlorides (carbonimidyl chlorides) **28** were first prepared by Nef in 1892 [78] , by the treatment of aromatic isocyanides with chlorine in cold

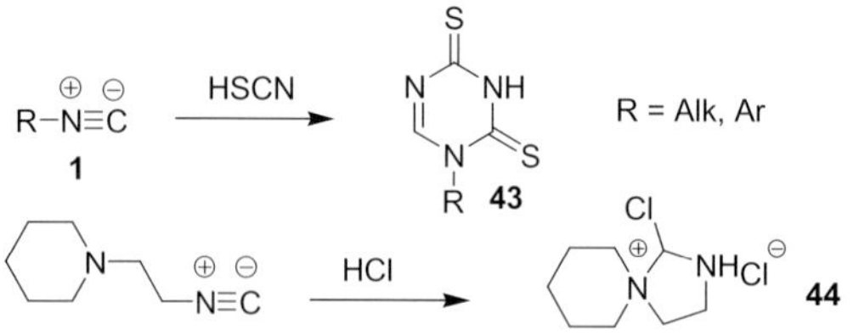

Scheme 2.27

chloroform solution. More recently, however, these compounds have been more conveniently produced via the chlorination of isothiocyanates, amides, thioamides, or carbamates. Although, for many years, the chlorination of isocyanides was considered to be an exotic reaction which had serious limitations for practical use in organic synthesis, the situation changed following the introduction of convenient methods to obtain the isocyanides. Both, the scope and the routes of the various processes developed to prepare isocyanide dihalides – and particularly of isocyanide dichlorides – have been critically reviewed [79].

Dihalogeno isocyanides **28** are relatively stable derivatives that may be isolated following the addition of chlorine or bromine to a solution of isocyanides. The reaction proceeds well with all types of isocyanide, whether aromatic, aliphatic, hindered, or even electrophilic [79]. The presence of two halogen atoms activates the C=N double bond towards a nucleophilic attack, with possible additions occurring in a sequential manner. Hence, these compounds may serve as a basis for either domino or tandem reactions in which several steps are performed as a one-stop process, without the isolation of any intermediates.

One good example of a three-component strategy starting from isocyanides is the straightforward synthesis of substituted tetrazoles **45**, as developed by El Kaïm et al. [80]. This new cascade involves three steps: (i) the bromination of isocyanides; (ii) the addition of an azide and cyclization; and (iii) a Suzuki coupling (Scheme 2.28). A similar approach which involved an isocyanide dihalogenide conversion to chloropyridines, followed by a palladium coupling, has been reported for the synthesis of variolin analogues [81].

Another reaction of isocyanides, named after Nef, represents the interaction of acyl halides and isocyanides [78]. Although, usually, this reaction provides the corresponding ketoimidoyl halides **29** in high yields, an excess of isocyanides may lead to iterative insertions and result in the formation of oligomerized products. The subsequent addition of nucleophilic reagents may lead to the formation of stable adducts such as pyruvamides **46a** (X = O) after water addition, or acylthioamides **46b** (X = S) following the addition of hydrogen sulfide [78, 82] (Scheme 2.29).

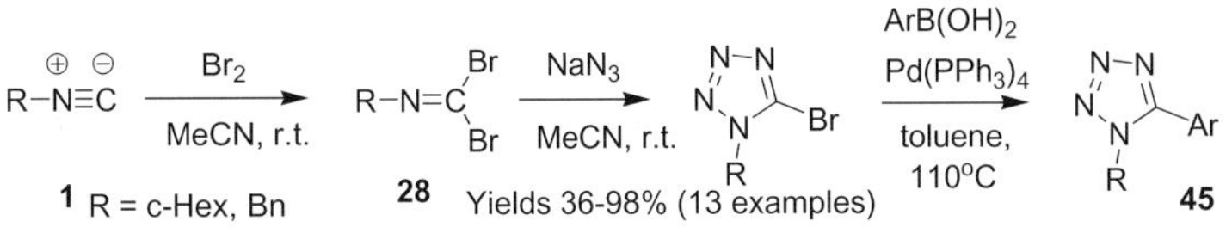

Scheme 2.28

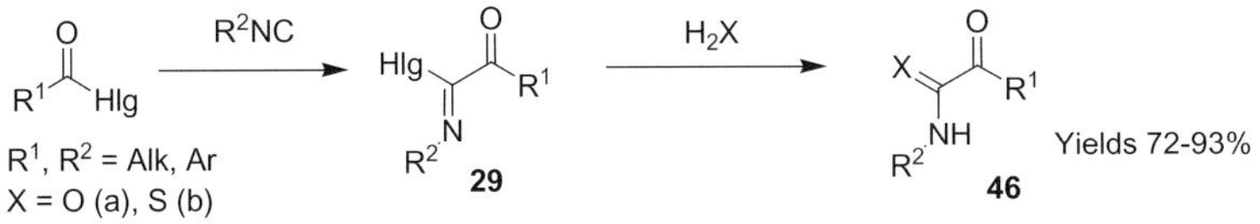

Scheme 2.29

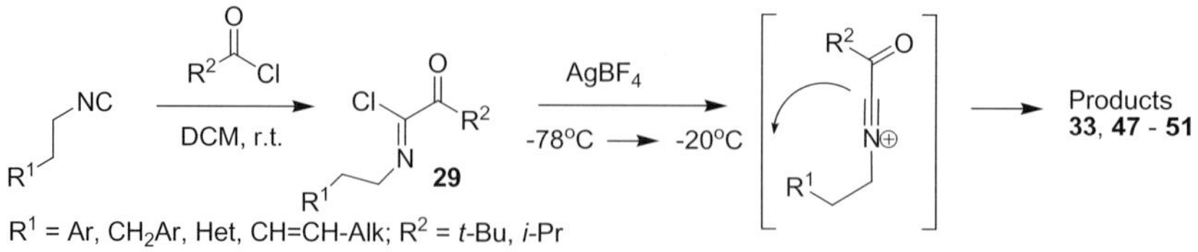

Scheme 2.30

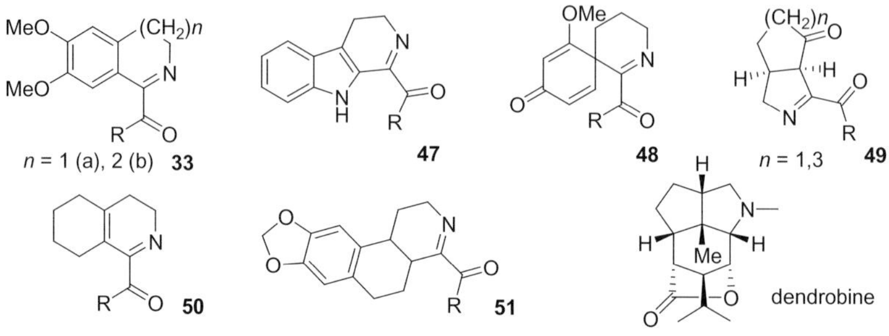

Figure 2.5 Products in reaction of isocyanides with acyl halides.

Livinghouse et al. have developed a versatile method for the synthesis of N-containing heterocycles, by using the reaction of isocyanides and acyl halides as a basis [60] (Scheme 2.30). In this case, the products of the reaction derived from 2-phenylethyl isocyanides were found to undergo a silver(I)-mediated cyclization to form 1-acyl-3,4-dihydroisoquinolines **33a**, in moderate to good yields [83]. The mechanism of such of a heterocyclization involves the formation of acylnitrilium cations as key intermediates, and a further intramolecular addition to the electron-rich aromatic system. Benzene rings bearing electron-donating groups can be replaced in this synthesis with furan, indole moieties, or an electron-rich double bond, such that furan- and indole-annelated dihydropyridines **47** have been synthesized in good yields. Seven-membered heterocycles such as 2-acylbenzazepines **33b**, as well as spiroannelated tetrahydropyridines **48** (Figure 2.5), have been conveniently prepared in a two-step, one-pot procedure by using this methodology [60].

Westling and Livinghouse also reported the details of a short pathway to 2-acylpyrrolines **49**, which was based on the addition of acylnitrilium ions to a silyl enol ether moiety and subsequent desilylation [84]. The 2-acylpyrrolines were obtained in moderate to good yields using a one-pot approach, and the method was employed for synthesis of the key precursor in a total synthesis of dendrobine (Figure 2.5) [85], the alkaloid isolated from *Dendrobinium nobile* L. , which is used as a component of traditional Chinese medicine. It was also revealed that the unactivated double bond and aromatic ring may be involved in a cascade cyclization, leading to the bicyclic and tricyclic compounds **50** and **51**, in 54% yield (Figure 2.5) [84].

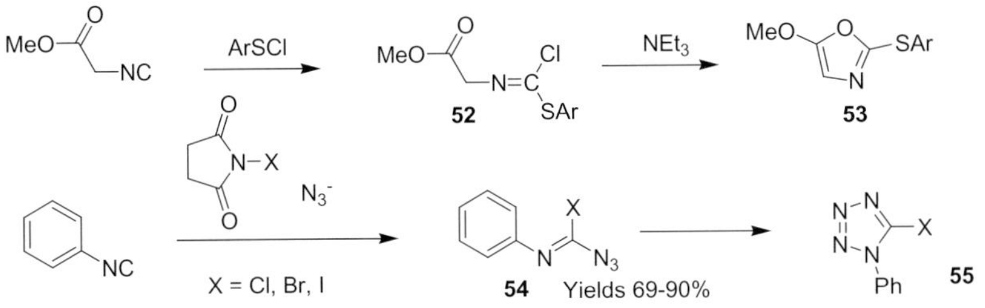

Scheme 2.31

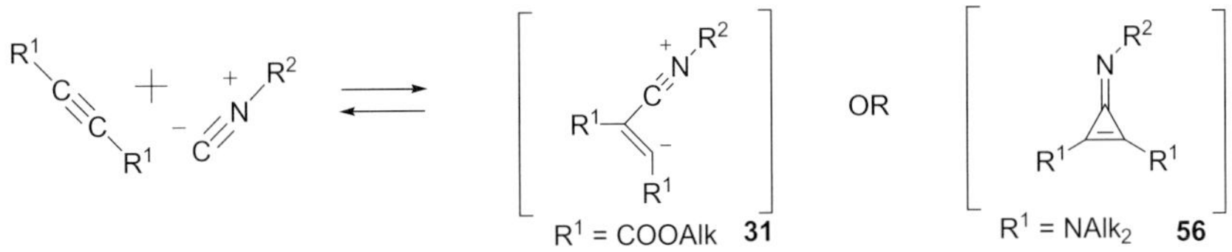

Figure 2.6 Intermediates in reactions of isocyanides with alkynes.

Similar to the acyl chlorides, other halogen-containing compounds react with isocyanides to produce unstable intermediates that, in most cases, undergo further cyclization. Many synthetic pathways have been based on this methodology, from which two representative examples have been selected here. First, the reaction of arylsulfenyl chlorides **52** with the ethyl ester of isocyanoacetic acid results in the formation of 2-arylthio-5-alkoxyoxazoles **53** (Scheme 2.26) [86]. Likewise, adducts of isocyanides with Hlg-N$_3$ **54** have been shown to undergo cyclizations to produce halogen-substituted tetrazoles **55** (Scheme 2.31) [87].

2.4.3
Reactions with Activated Alkenes and Alkynes

The presence of a monocoordinated carbon atom having an electron lone pair in the structure of the isocyano group makes possible its participation in different types of cycloaddition as a 1,1-dipolar partner. In the case of compounds with activated double and triple bonds, this cycloaddition may lead to cyclopropanes and cyclopropenes **56** (Figure 2.6). In fact, isocyanides have been known to react with various dipolarophiles such as disilenes, diphosphenes, and silenes [88], whereas aryl- and alkyl-substituted alkenes and alkynes exhibit an inertness towards isocyanides under heating [1]. At present, one report exists [89] in which cyclopropenimines **56** have been postulated as a product of the reaction between aryl isocyanides bearing electron-withdrawing groups and electron-rich alkynes (ynamines). According to these data, electron-withdrawing substituents in the aromatic ring tend to accelerate the cycloaddition, such that diamino-substituted alkynes will add isocyanides much faster than will monoamino-substituted alkanes. In addition, Krebs et al. [89] have identified an excellent Hammett correlation for aryl isocyanides and the quite small solvent effect on the addition rate. Consequently, this example is beyond the intended scope of the present section, which is devoted to cycloaddition reactions of isocyanides with electrophiles.

Theoretical studies employing density functional theory (DFT) have demonstrated the stepwise character of this reaction [88], in which the rate-determining step is the addition of an isocyano group to a carbon atom of the alkyne, giving rise to the zwitterionic intermediate **31**. The second step is a ring closure to yield the final product (cyclopropenimine). Although, in all cases the isocyano group was considered to be an electrophile, isocyanides are known also to react with electron-poor alkynes such as dimethyl acetylenedicarboxylate (DMAD); however, in this case the cyclopropene intermediates were neither isolated nor fixed. Instead of an intramolecular cyclization, the zwitterionic intermediates **31** undergo a multiple of intermolecular reactions with the excess of isocyanides or alkynes [90]. In this way, all experimental and theoretical studies can provide compelling arguments against the synchronous cycloaddition of isocyanides to form double and triple bounds.

The zwitterionic species **31** resulting from isocyanides and DMAD can be captured by an excess of educts, solvents, or suitable substrates and, after a series of transformations, provide a stable product [91, 92]. Depending on the reaction conditions and the ratio of educts, however, these reactions may lead to a variety of final products that include bicyclic systems with various proportions of components (1:2, 2:1, 3:2) [93, 94]. It should be noted here that, despite the unique flexibility of this reaction, no data are presently available relating to its practical use in organic synthesis, due mainly to the low yields of the products and the poor reproducibility of results. Nonetheless, the multicomponent reactions that have been discovered on the basis of this chemistry represent much more promising synthetic tools than the initial two-component reactions [90].

Other acceptor-substituted alkynes have shown similar reactivities towards the isocyanides. Notably, the products obtained from benzoyl- and trifluoromethyl-acetylene are cyclic 2:1 and bicyclic 1:2 (alkyne-isocyanide) adducts **57–59** [95, 96] (Scheme 2.32).

An alkyne having electron-withdrawing substituents may serve as an initiator of oligomerization reactions; consequently, the oligomerization of 4-bromo-2,6-dimethyl-phenylisocyanide in the presence of methyl 3-phenylpropiolate was observed by Takizawa et al. [97]. As a result, cyclopentane derivatives **60** were isolated in 41% yield (Scheme 2.33).

To date, very few examples have been reported of interactions between activated alkenes and isocyanides. Thus, electron-poor alkenes such as 1,1-diciano-2,2-bis-trifluoromethylethene **61** have been shown to react with two equivalents of t-butyl isocyanide to generate a diiminocyclobutane derivate **62** [98]. A similar reaction with benzoquinone resulted in the formation of bicyclic or tricyclic products **63** and **64**, depending on the excess of isocyanide [99]. It is interesting to note that

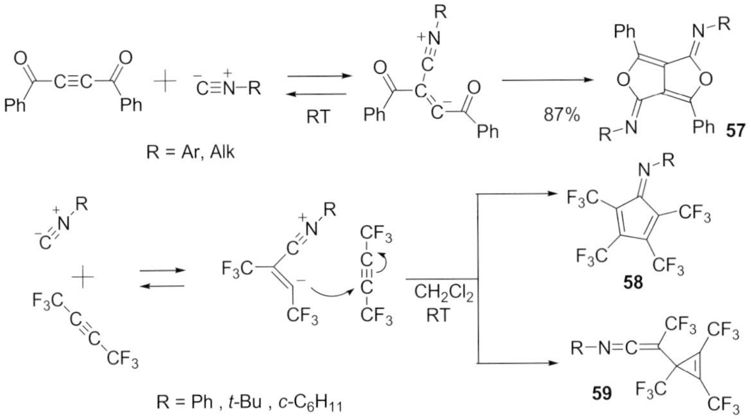

2.4 Reactions of Isocyanides with Electrophiles | 57

Scheme 2.32

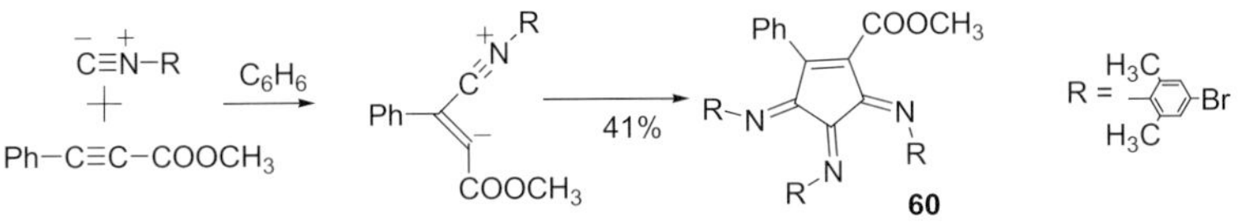

Scheme 2.33

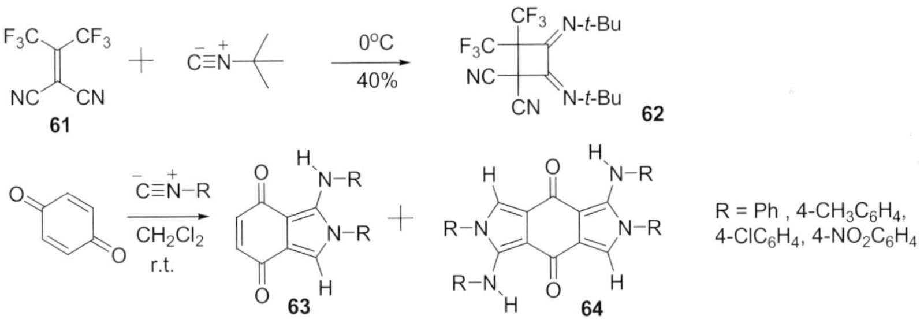

Scheme 2.34

the reaction proceeded well with both electron-rich and electron-poor aromatic isocyanides (Scheme 2.34).

One very interesting example of reactions with alkenes is the synthesis of methyl *trans*-iminocrotonates **65** from aliphatic isocyanides and methyl acrylate in *t*-butanol [100]. Whilst the mechanism of the reaction is unknown, it is possible to predict the formation of a zwitterionic intermediate and hydride ion transfer (Scheme 2.35).

Similar to alkynes, alkenes bearing electron-withdrawing groups may serve as initiators for the oligomerization of isocyanides, with high selectivity. In this type

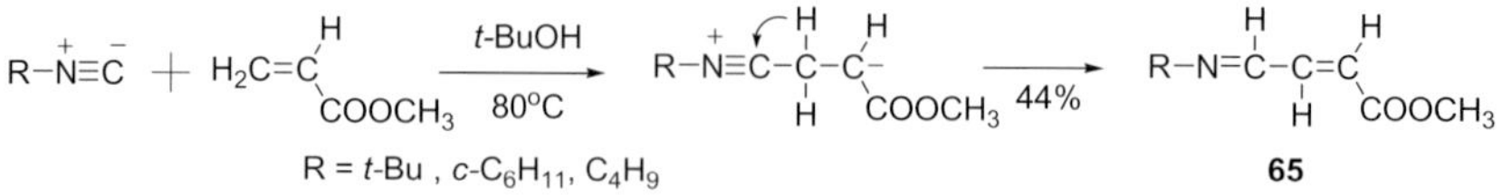

2 General Aspects of Isocyanide Reactivity

$$R-\overset{+}{N}\equiv\overset{-}{C} + H_2C=C\begin{smallmatrix}H\\COOCH_3\end{smallmatrix} \xrightarrow[80°C]{t\text{-BuOH}} R-\overset{+}{N}\equiv C-\overset{H}{\underset{H}{C}}-\overset{H}{\underset{COOCH_3}{C}}- \xrightarrow{44\%} R-\overset{H}{\underset{H}{N}}=\overset{}{\underset{}{C}}-C=\overset{H}{\underset{COOCH_3}{C}}$$

R = t-Bu, c-C$_6$H$_{11}$, C$_4$H$_9$

65

Scheme 2.35

$$3\,\overset{-}{C}\equiv\overset{+}{N}\text{-}t\text{-Bu} + \text{R}\underset{H}{\overset{}{\diagdown}}\!\!=\!\!\underset{CN}{\overset{CN}{\diagup}} \xrightarrow[\text{reflux}]{CH_3CN} \text{(cyclopentene derivative 66)}$$

R = 3-CH$_3$O, 4-Cl, 4-F, 4-NO$_2$

66

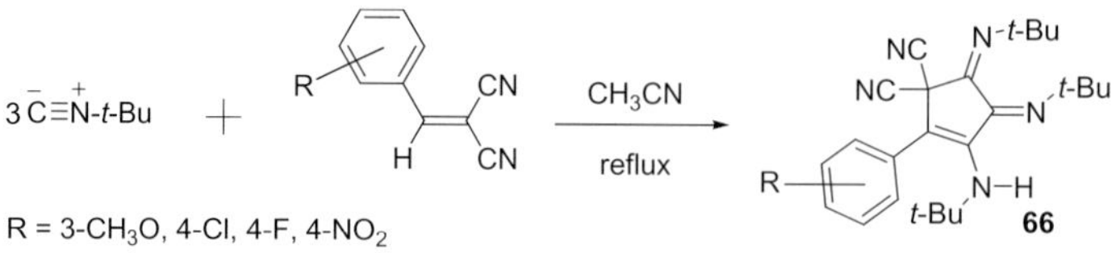

Scheme 2.36

of oligomerization, an initiator is included in a final structure, thus opening a way to the synthesis of products with different initiator-to-isocyanide ratios. A good example of such a transformation is a novel reaction of 3 equiv. of isocyanides with 1,1-dicyanoethylenes [101], leading to the derivatives of cyclopentene **66** in good yields (Scheme 2.36). A polar solvent such as acetonitrile will provide the stabilization of any zwitterionic intermediates, and in turn provides an explanation for the high selectivity. During this process four new C–C bonds are formed; moreover, by taking into account the easy access to activated olefins this method allows the creation of five C–C bonds, simultaneously. Significant disadvantages of the above-described method were a sensitivity to the starting materials and the presence of several electron-withdrawing substituents in the olefin.

In general, the reactions of isocyanides with alkenes are of minimal importance for planning strategies in organic synthesis, though they may represent a good source for identifying novel multicomponent reactions.

2.4.4
Reactions with Carbonyl Compounds and Imines

A variety of aliphatic and aromatic aldehydes and ketones are known to cyclize with isocyanides, in the presence of catalytic amounts of Lewis acids such as BF$_3$. These additions generally involve the interactions of three and more molecules with various proportions of carbonyl compounds and isocyanides. The most common type of these reactions is characterized by the addition of two isocyanide molecules to one C=O bond [102]. Iminooxetanes **67**, which are the primary reaction product, may be isolated in many cases, or they may be readily converted to other products under the same reaction conditions (Scheme 2.37). These highly reactive substrates undergo a ring-opening process, with the further addition of other intermediates present in the reaction mixture, as illustrated in Scheme 2.37. The possible pathways include a rearrangement to five-membered heterocycles,

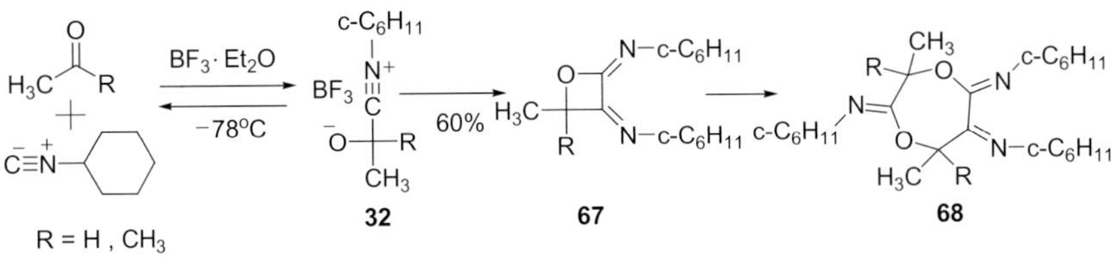

Scheme 2.37

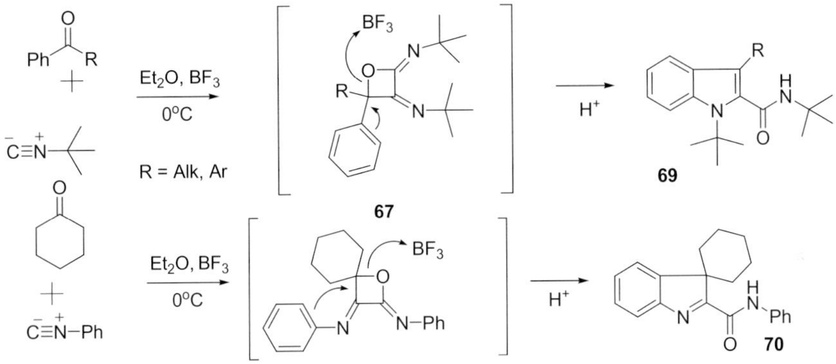

Scheme 2.38

the formation of larger rings (e.g., **68**), and aromatization [103]. The substituent effects and the reaction conditions each determine which synthesis pathway is followed. Typically, reactions of this type are carried out in an aprotic medium in completely dry conditions under a nitrogen atmosphere, because the isocyanides may be partially hydrolyzed to formamides if only trace amounts of moisture are present in the reaction mixture [104].

The preparative application of this chemistry requires the suppression of side reactions that, in most cases, is not possible. However, several groups have reported interesting syntheses for drug-like molecules and heterocyclic compounds, based on the reaction of isocyanides with carbonyl compounds. Thus, 2,3-bis-alkyliminooxetanes **67** derived from this reaction can serve as precursors for β-lactams and aryloxypropanolamines, which are analogues of β-adrenergic receptor blocking agents [105]. Various indoles **69** and 3H-indoles **70** were obtained in moderate to good yields when oxetanes underwent ring opening and recyclization [104]. Generally, two approaches exist to the formation of an indole ring: (i) an electrophilic attack of the activated C=N bond on the phenyl ring of an aromatic ketone; and (ii) a molecule of an aromatic isocyanide is used as a basis for formation of the indole ring (Scheme 2.38).

Similar to aldehydes and ketones, azomethines derived from *p*-nitroaniline undergo cycloaddition with two molecules of *t*-butyl isocyanide to produce the 2,3-bis-iminoazetidines **72** [106]. This reaction was carried out under relatively

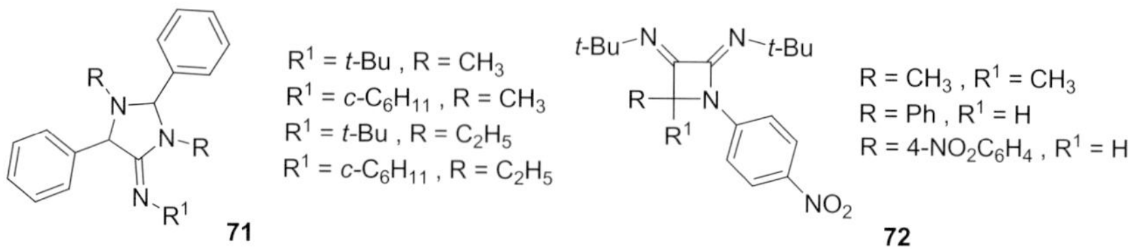

Figure 2.7 Products in reaction of isocyanides with azomethines.

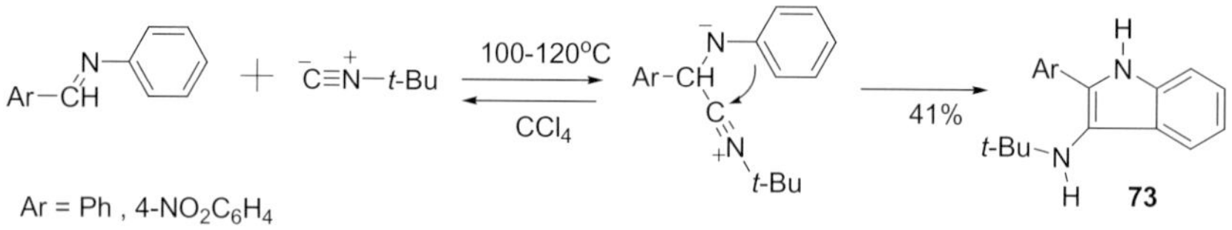

Scheme 2.39

harsh conditions (heating in sealed vial at 120 °C for 13 h in the presence of HCl), and consequently its practical application is limited. In the case of azomethines derived from aliphatic amines, substituted imidazolines **71** were obtained in moderate yields [107] (Figure 2.7).

In contrast to aldehydes and ketones, azomethines from aniline and electron-rich aromatic amines can form 1 : 1 adducts. Deyrup and Vestling have determined the structure of these adducts, which were found to be 3-aminoindoles **73** [106]. In this case, the nitrilium ion derived from isocyanides reacts selectively with the pendant electron-rich π-system, which in turn leads to a five-membered cycle, as shown in Scheme 2.39. This reaction was recently rediscovered by Sorensen et al. [108], who used triflyl phosphoramide as a catalyst; in this case, a novel practical method for the synthesis of both 3-aminoindoles **73** and substituted indoxyls was elaborated.

In summary, the reaction of isocyanides and carbonyl compounds/imines represents a versatile tool for the synthesis of four-membered heterocycles and indoles.

2.4.5
Reactions with Activated Heterocumulens

Heterocumulens are known to undergo a number of various transformations with isocyanides to produce four- and five-membered heterocycles. Thus, Ugi et al. [109] reported the formation of 2 : 1 adducts **74** in the reaction of diphenylketene with isocyanides (Scheme 2.40); notably, the high yields of the product **74** in reactions of this type were possible only in the case of ketenes with electron-withdrawing groups [110].

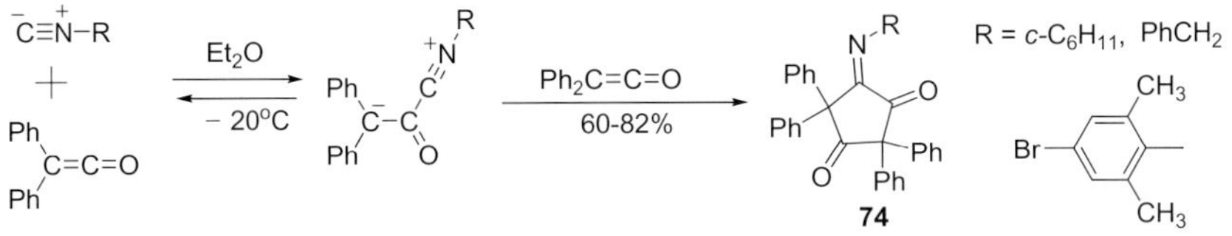

Scheme 2.40

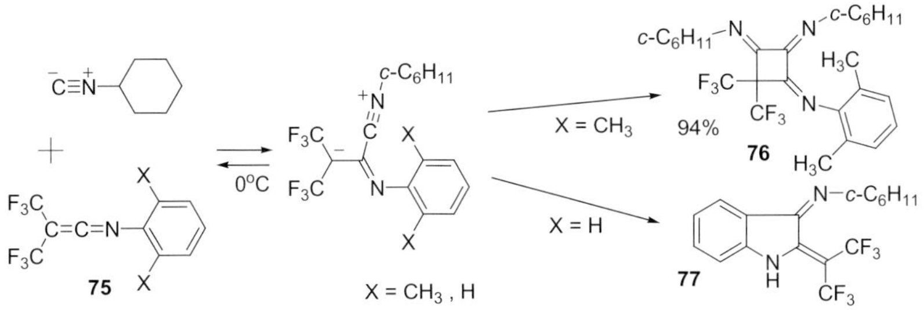

Scheme 2.41

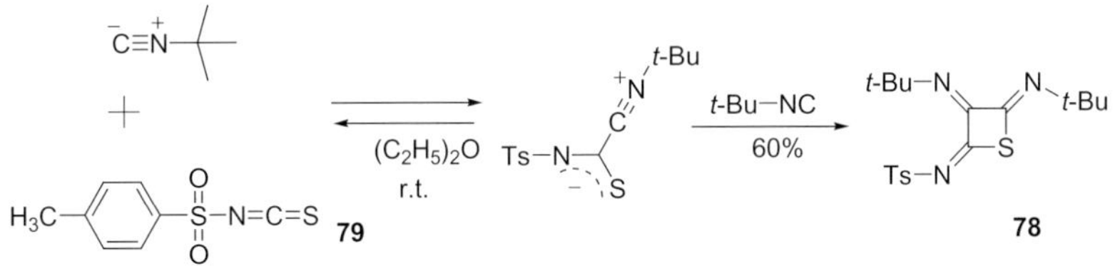

Scheme 2.42

A number of ketenimines have been also tested as a partners in reactions with isocyanides, but in most cases no products were detected. It appears from these experiments that, under the usual conditions, aliphatic ketenimines are inert towards isocyanides, and this fact has been confirmed by the successful syntheses of ketenimines from alkyl isocyanides. In contrast, activated aryl ketenimines are able to react with isocyanides to produce 1:1 and 1:2 adducts. Thus, ketenimines bearing two trifluoromethyl groups and an *ortho*-disubstituted phenyl moiety **75** produced the four-membered cycle **76** [111] although, if one of *ortho*-positions had been unoccupied, the reaction product would have been the 3H-iminoindoles **77** (Scheme 2.41). A similar cyclization was observed with dihaloketenimines generated from carbenes and 2,6-dimethylphenyl isocyanide [112].

The triiminothietanes **78** have been obtained starting from isothiocyanate **79** and *t*-butyl isocyanide [113, 114]. The proposed mechanism involves the formation of the externally stabilized 1,3-dipole as an intermediate, and a cycloaddition reaction with a second molecule of isocyanide (Scheme 2.42).

2.5
Reactions of Isocyanides with Nucleophiles

Although the addition of carbanions and other nucleophiles to the isocyanide functionality is an important tool in organic synthesis, only intramolecular transformations of this type have been widely used. Thus, α-metalated, γ-metalated *ortho*-methylphenyl and *ortho*-lithiophenyl isocyanides represent versatile precursors for certain types of nitrogen-containing heterocycle [61]. Although the transition metal-catalyzed reactions of isocyanides with nucleophiles have also been employed as a convenient method in heterocyclic chemistry [115], such reactions are beyond the scope of this chapter, as they proceed via an insertion of coordinated isocyanides to a ligand on the transition metal coordination sphere.

In contrast, the noncatalyzed intermolecular reactions with nucleophiles – including the successive insertions of isocyanides into a carbon–metal bond of typical metals such as Li, Mg, and Ca – have been not extensively studied. These reactions demonstrate several features that limit their practical use in organic synthesis:

- The isocyanides are very poor electrophiles, that can only react with very strong nucleophiles such organometallic compounds.

- The electron-withdrawing effect of the isocyano group enhances the acidity of the α-CH bond in aliphatic isocyanides, as well as the *ortho*-methyl group in aromatic isocyanides.

Thus, owing to the competing deprotonation, only limited types of isocyanides can be used in reactions with nucleophiles; these include *tert*-alkyl isocyanides, 2,6-disubstituted aryl isocyanides, 1,2-di-isocyanoarene, and isocyanides with an enhanced electrophilicity. Nonetheless, despite these limitations this area of chemistry has undergone rapid development during the past two decades.

2.5.1
Reactions with Organometallic Compounds

Various organometallic compounds of typical metals undergo an insertion reaction with isocyanide. As described by both Ugi [116] and Walborsky [117], Grignard reagents may be reacted with isocyanide to produce the corresponding magnesium aldimines in good yields. The synthetic use of these reagents has been limited, however, because they cannot compete with lithiated 1,3-dithianes and vinyl ethers as masked acyl anions. One exception to this is the reaction of Grignard reagents with 1,2-di-isocyanoarene **80**, which results in the formation of quinoxalines **81** and its oligomers after hydrolysis (Scheme 2.43) [118]. In this case, successive insertions of the two *ortho*-isocyano groups of 1,2-di-isocyanoarene into the carbon–magnesium bond have been observed.

Isocyanides react similarly with alkyllithiums to produce α-(*N*-substituted imino)alkyllithium compounds. 1,1,3,3-Tetramethylbutyl isocyanide (TMBI; Walborsky reagent) was introduced by Walbosky and coworkers as a powerful reagent

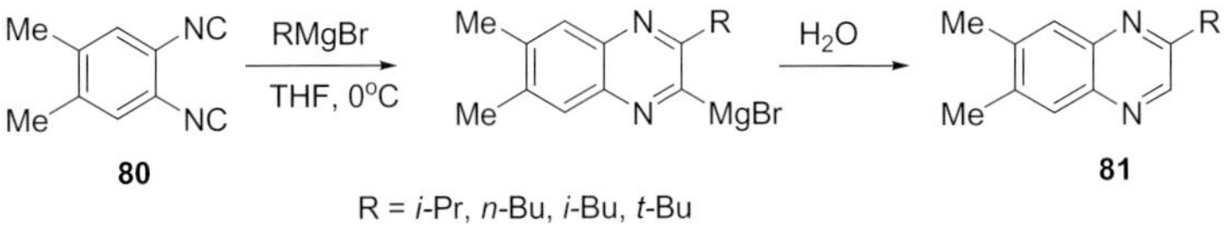

2.5 Reactions of Isocyanides with Nucleophiles

Scheme 2.43

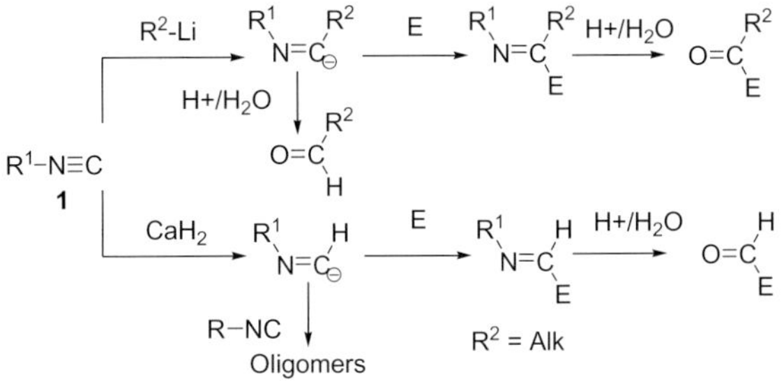

Scheme 2.44

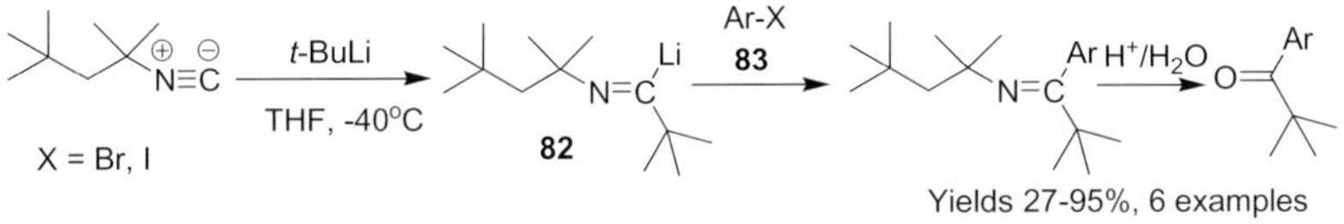

Scheme 2.45

for the synthesis of carbonyl compounds [119]. Lithium aldimines, resulting from the addition of alkyllithiums to this isocyanide, have been shown to serve as versatile acyl anion equivalents (Scheme 2.44), and were treated with primary alkyl, aryl, vinyl, and acetylenic halides, carbon dioxide, ethyl chloroformate, trimethysilyl chloride, nonenolizable aldehydes, propylene oxide, and water to produce – after hydrolysis – the corresponding carbonyl compound [120]. A good example of this was the reaction of lithium aldimines **82** derived from TMBI with arylhalides **83** such as bromo or iodobenzene (Scheme 2.45). It should be noted that only the alkyllithiums reacted with TMBI in high yield to produce lithium aldimine, although intramolecular versions of these transformations have been employed successfully in the synthesis of various heterocycles [61].

The addition of a metal hydride to an isocyanide functionality would likewise produce the synthetic equivalent of the formyl anion, which is an important building block in organic synthesis (Scheme 2.44). Thus, the reaction of the calcium hydride complex with TMBI resulted in formation of the calcium aldimine complex [121]. The addition of this complex to the free isocyanide produces a C–C-coupled intermediate that rearranges to a ketenimine, a reaction which is comparable to the first step in the polymerization of isocyanides.

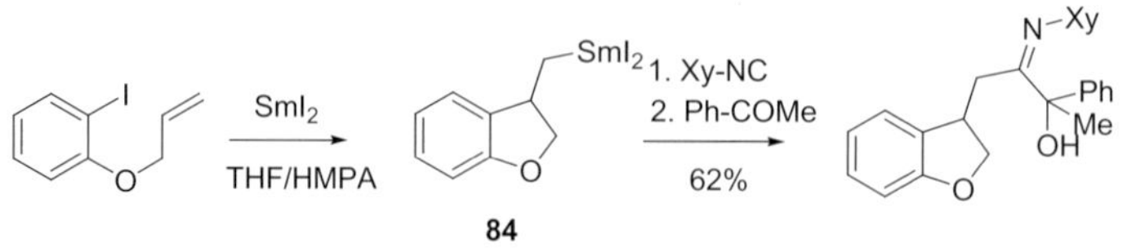

2 General Aspects of Isocyanide Reactivity

Scheme 2.46

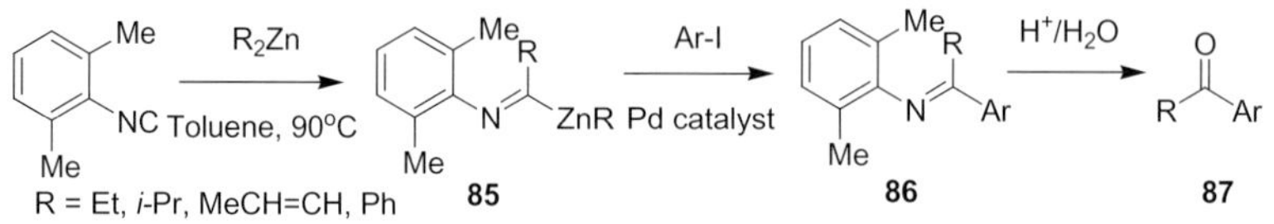

Scheme 2.47

The alkylsamarium reagents **84** obtained via a reduction of the corresponding radicals with samarium iodide are moderately stable in solution, and can react with isocyanides to yield highly reactive samarium aldimines. Ito and coworkers [122] have reported the formation of these reagents by the reduction of benzyl chloromethyl ether with samarium iodide in the presence of xylyl isocyanide. In this case, the samarium aldimines were trapped by aldehydes and ketones, without isolation. Subsequently, Curran et al. conducted a stepwise experiment that involved three stages: (i) a reduction of the alkyl iodides with SmI_2; (ii) the addition of xylyl isocyanide; and (iii) the addition of acetophenone [123]. These investigations confirmed the mechanism involved in the addition of alkylsamarium reagents to xylyl isocyanide (Scheme 2.46).

The insertion of *ortho*-disubstituted isocyanides into aliphatic organozinc compounds occurs only on heating, to yield α-(*N*-arylimino)alkylzincs **85** [124]. The coupling of these organometallic compounds, as prepared from 2,6-xylyl isocyanide with aromatic iodides, is catalyzed efficiently by a palladium catalyst to provide the corresponding *N*-aryl arylimine derivatives **86**, which can be readily hydrolyzed to the corresponding aromatic ketones **87** in high yields (Scheme 2.47). In contrast, alkyl isocyanides do not give the expected products [124].

In summary, the isocyanides may provide access to great variety of metalloaldimines that can be utilized as masked acyl or formyl anions.

2.5.2
Reactions with Hydroxide, Alcohols, and Amines

In general, isocyanides are inert towards water, hydroxide, and amines without the addition of catalysts such as acids and transition-metal complexes. However, aromatic isocyanides bearing electron-withdrawing groups, isocyanoazines and

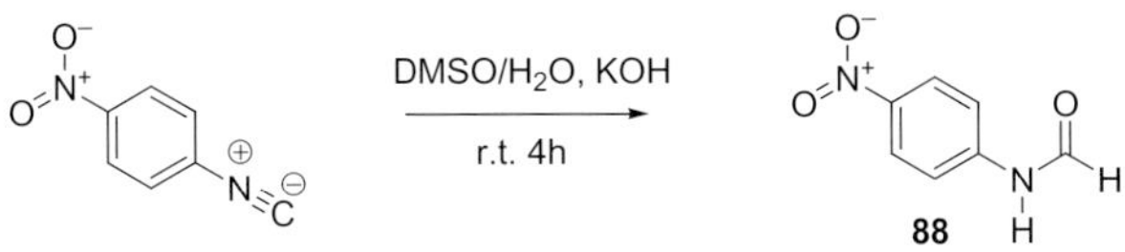

2.5 Reactions of Isocyanides with Nucleophiles | 65

Scheme 2.48

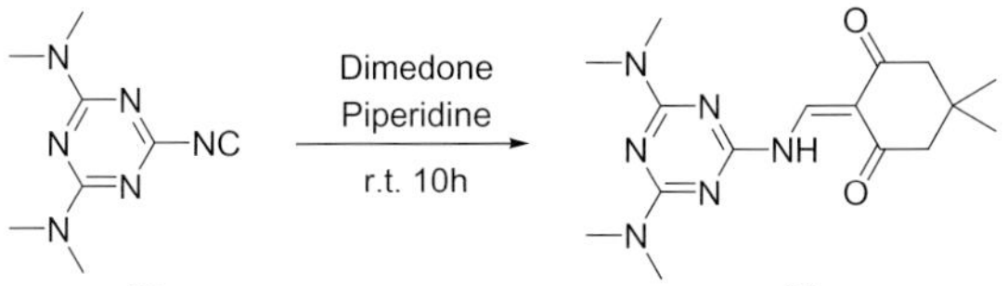

Scheme 2.49

related compounds undergo these transformations without any activation of the isocyano group. The choice of isocyanides with an enhanced electrophilicity is limited, because reagents of this type undergo spontaneous uncatalyzed polymerization, and consequently an optimal combination of high electrophilicity and stability should be identified for the practical use of these reagents [125]. The most popular isocyanides with enhanced electrophilicity are 4-nitro- and 2-cyano-isocyanobenzene, 2- and 3-isocyanopyridine, isocyanotriazines, and 1,2-di-isocyanobenzene.

Thus, in contrast to other aromatic isocyanides, 4-nitro-isocyanobenzene reacts easily with hydroxides in aqueous DMSO to produce the corresponding formamide **88** [126]. Although this reaction has no practical use, it may be considered as valid evidence of the noticeable changes in the reactivity of the isocyano group (Scheme 2.48). Another example of this is the reaction of isocyano-1,3,5-triazine **89** with dimedone in the presence of piperidine [127] (Scheme 2.49). In this case, the mechanism of formation of the final product **90** involves the formation of amidine from isocyanide and amine, and a further substitution of the amino function with dimedone.

Transformations of this type may be employed successfully as a convenient method for obtaining nitrogen-containing heterocycles. For example, Ito et al. reported a convenient and efficient synthesis of 2-diethylaminoquinazoline **91** via the cyclization of 2-cyano-isocyanobenzene with diethylamine [128]. In this case, a nucleophile attacks the isocyano group activated with an electron-withdrawing group in the *ortho*-position (Scheme 2.50). The resultant imidoyl anion then undergoes an electrocyclization to provide the final product **91**, after protonation of its valence tautomer. The similar cyclization of *ortho*-alkynylphenyl isocyanides **92** with alcohols, amines, and diethyl malonate leads to 2,3-disubstituted quinolines **93** (Scheme 2.50) [128].

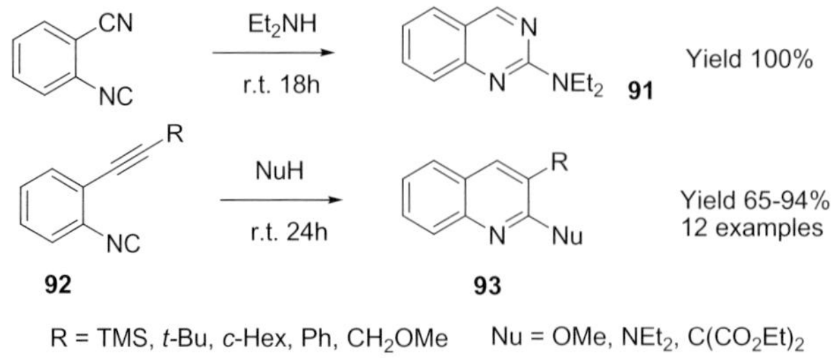

Scheme 2.50

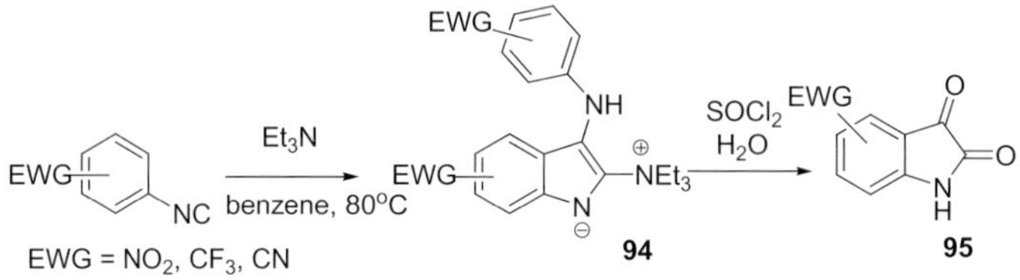

Scheme 2.51

The isocyanides bearing electron-withdrawing groups are able to add tertiary alkyl amines, thus opening the way to the indole derivatives **94** [129]. The treatment of indolates with thionylchloride and hydrolysis afforded isatines **95** in high yields (Scheme 2.51). The same mechanism also involves a reversible formation of the imidoyl anion, and further addition the free electrophilic isocyanide, followed by cyclization.

Hence, imidoyl anions derived from isocyanides and aliphatic amines represent interesting intermediate compounds, the potential of which has not yet been fully exploited.

2.6
Conclusions

With some of the reactions described in this chapter having been discovered more than 100 years ago, and well defined in classic books on organic chemistry, this area would seem to have been totally exploited and to show very little future promise for novel findings. Yet, during the past two decades this "old chemistry" has undergone a major renovation, with the discovery of novel methods for amide bond formation via reactions between isocyanides and thioacids [69], through conventional methods for the synthesis of isothiocyanates [40, 43] and tetrazoles [74] from isocyanides, the novel syntheses of quinolines [128], indoles [108] and isatines [129], and the samarium-mediated reactions of isocyanides [122, 123]. Moreover, the classical reactions of the isocyanides have more recently been applied to a variety of syntheses of natural products and drugs, including ibuprofen [24], vinigrol [58], cyclosporine A [71], variolin [81], and dendrobine [85]. Two-component reactions of isocyanides represent a source for the discovery of novel multicomponent reactions, which may serve as a very powerful tool in drug design [63, 64]. Clearly, the "old chemistry" of isocyanides can be regarded as an on-going area of development capable of providing access to many novel reactions and methods.

References

1 Ugi, I. (ed.) (1971) *Isonitrile Chemistry*, Academic Press, New York.
2 Mayr, H. and Ofial, A.R. (2008) Do general nucleophilicity scales exist? *J. Phys. Org. Chem.*, **21**, 584–595.
3 Tumanov, V.V., Tishkov, A.A., and Mayr, H. (2007) Nucleophilicity parameters for alkyl and aryl isocyanides. *Angew. Chem. Int. Ed.*, **46**, 3563–3566.
4 Meot-Ner (Mautner), M., Karpas, Z., and Deakyn, C.A. (1986) Ion chemistry of cyanides and isocyanides. 1. The carbon pair as proton acceptor: proton affinities of isocyanides. Alkyl cation affinities of N, O, and C lone-pair donors. *J. Am. Chem. Soc.*, **108**, 3913–3919.
5 Weith, W. (1873) Beziehungen zwischen aromatischen senfolen und cyanuren. *Chem. Ber.*, **6**, 210–214.
6 Rüchardt, C., Meier, M., Haaf, K., Pakusch, J., Wolber, E.K.A., and Müller, B. (1991) The isocyanide–cyanide rearrangement; mechanism and preparative applications. *Angew. Chem. Int. Ed. Engl.*, **30** (8), 893–901.
7 Meier, M., Muller, B., and Ruchardt, C. (1987) The isonitrile-nitrile rearrangement. A reaction without a structure–reactivity relationship. *J. Org. Chem.*, **52**, 648–652.
8 Boullanger, P., Marmet, D., and Descotes, G. (1979) Synthesis of glycosyl isocyanides. *Tetrahedron*, **35**, 163–167.
9 Sasaki, T., Eguchi, S., and Katada, T. (1974) Synthesis of adamantane derivatives. XXV. Synthesis and reactions of 1-and 2-adamantyl isocyanides. *J. Org. Chem.*, **39** (9), 1239–1242.

10 Meier, M. and Rüchardt, C. (1983) A free-radical chain mechanism for the isonitrile-nitrile rearrangement in solution and its inhibition. *Tetrahedron Lett.*, **24** (43), 4671–4674.
11 Meier, M. and Rüchardt, C. (1984) Flash pyrolysis of isonitriles, a stereospecific high yield reaction. *Tetrahedron Lett.*, **25** (32), 3441–3444.
12 Wentrup, C., Stutz, U., and Wollweber, H.-J. (1978) Synthese von aryl- und heteroaryl isocyaniden aus nitrosoverbindungen. *Angew. Chem.*, **90** (9), 731–732.
13 Kang, H.-Y., Pae, A.N., Cho, Y.S., Koh, H.Y., and Chung, B.Y. (1997) Isonitrile–nitrile rearrangement promoted by samarium(II) iodide. *Chem. Commun.*, 821–822.
14 Casanova, J., Jr, Werner, N.D., and Schuster, R.E.J. (1966) The isonitrile nitrile isomerization. *Org. Chem.*, **31**, 3473–3477.
15 Kim, S.S., Choi, W.J., Zhu, Y., and Kim, J.H. (1998) Thermal isomerizations of substituted benzyl isocyanides: relative rates controlled entirely by differences in entropies of activation. *J. Org. Chem.*, **63**, 1185–1189.
16 Sun, J., Qian, X., Chen, D., Liu, Y., Wang, D., and Chang, Q. (1999) Free radical mechanism for photoinduced isonitrile-nitrile isomerization in solution. *J. Photochem. Photobiol. A*, **126** (1–3), 23–26.
17 Sung, K. (1999) Substituent effects on stability and isomerization energies of isocyanides and nitriles. *J. Org. Chem.*, **64**, 8984–8989.

18 Wentrup, C., Gerecht, B., Laqua, D., Briehl, H., Winter, H.-W., Reisenauer, H.P., and Winnewisser, M. (1981) Organic fulminates, R-O-NC. *J. Org. Chem.*, **46**, 1046–1048.

19 Buschmann, J., Lentz, D., Luger, P., Perpetuo, G., Preugschat, D., Thrasher, J.S., Willner, H., and Wolk, H.-J. (2004) Crystal and molecular structures of the pentafluorosulfanyl compounds, SF_5X (X = -NC, -CN, -NCO, -NCS and -$NCCl_2$) by X-ray diffraction at low temperature. *Z. Anorg. Allg. Chem.*, **630**, 113–114.

20 Wentrup, C. and Winter, H.-W. (1981) Isocyanoamines, R-NH-NC. *J. Org. Chem.*, **46**, 1045–1046.

21 Kingston, J.V., Ellern, A., and Verkade, J.G. (2005) A stable structurally characterized phosphorus-bound isocyanide and its thermal and catalyzed isomerization to the corresponding cyanide. *Angew. Chem.*, **117**, 5040–5043.

22 Seckar, J.A. and Thayer, J.S. (1976) Normal-iso rearrangement in cyanotrialkylsilanes. *Inorg. Chem.*, **15** (3), 501–504.

23 Richardson, N.A., Yamaguchi, Y., and Schaefer, H.F. (2003) Isomerization of the interstellar molecule silicon cyanide to silicon isocyanide through two transition states. *J. Chem. Phys.*, **119** (24), 12946–12955.

24 Meier, M. and Rüchardt, C. (1987) The synthetic potential of the isocyanide-cyanide rearrangement. *Chem. Ber.*, **120** (1), 1–4.

25 Gautier, A. (1869) Ueber die producte der oxidation der carbylamine. *Liebigs Ann. Chem.*, **149** (3), 311–318.

26 Feuer, H., Rubinstein, H., and Nielsen, A.T. (1958) Reaction of alkyl isocyanides with ozone. A new isocyanate synthesis. *J. Org. Chem.*, **23**, 1107–1109.

27 Otsuka, S., Nakamura, A., and Tatsuno, Y. (1967) Catalytic oxidation of isocyanides: a nickel-oxygen complex. *Chem. Commun.*, **16**, 836.

28 Deming, T.J. and Novak, B.M. (1993) Mechanistic studies of the nickel-catalyzed polymerization of isocyanides. *J. Am. Chem. Soc.*, **115**, 9101–9111.

29 Johnson, H.W. and Daughhetee, P.H. (1964) A new conversion of isonitriles to isocyanates. *J. Org. Chem.*, **99**, 246–247.

30 Johnson, H.W. and Krutzsch, H. (1967) The halogen-catalyzed pyridine N-oxide oxidation of isonitriles to isocyanates. *J. Org. Chem.*, **32**, 1939–1941.

31 Ichikawa, Y., Nishiyama, T., and Isobe, M. (2001) Stereospecific synthesis of the α- and β-D-glucopyranosyl ureas. *J. Org. Chem.*, **66**, 4200–4205.

32 Ichikawa, Y., Watanabe, H., Kotsuki, H., and Nakano, K. (2010) Anomeric effect of the nitrogen atom in the isocyano and urea groups. *Eur. J. Org. Chem.*, 6331–6337.

33 Prosperi, D. (2006) Polyhydroxylated bicyclic ureas from glycosyl isocyanides. *Synlett*, (5), 786–788.

34 Hiebl, J. and Zbiral, E. (1988) Synthese von Glycofuranosylformamiden, -isocyaniden und -isocyanaten ausgehend von den entsprechenden Glycosylaziden. *Liebigs Ann. Chem.*, 765–774.

35 Ichikawa, Y., Nishiyama, T., and Isobe, M. (2000) A new synthetic method for α- and β-glycosyl ureas and its application to the synthesis of glycopeptide mimics with urea-glycosyl bonds. *Synlett*, 1253–1256.

36 Ichikawa, Y., Ohara, F., Kotsuki, H., and Nakano, K. (2006) A new approach to the neoglycopeptides: synthesis of urea-and carbamate-tethered N-acetyl-D-glucosamine amino acid conjugates. *Org. Lett.*, **8**, 5009–5012.

37 Nef, J.U. (1894) Ueber das zweiwertige kohlenstoffatom. *Liebigs Ann. Chem.*, **280** (2-3), 291–342.

38 Boyer, J.H. and Ramakrishnan, V.T. (1972) Sulfurization of isocyanides. *J. Org. Chem.*, **37** (9), 1360–1364.

39 Fujiwara, S., Shin-Ike, T., Sonoda, N., Aoki, M., Okada, K., Miyoshi, N., and Kambe, N. (1991) Novel selenium catalyzed synthesis of isothiocyanates from isocyanides and elemental sulfur. *Tetrahedron Lett.*, **32** (29), 3503–3506.

40 Fujiwara, S., Shin-Ike, T., Okada, K., Aoki, M., Kambe, N., and Sonoda, N. (1992) A marvelous catalysis of tellurium in the formation of isothiocyanates from isocyanides and sulfur. *Tetrahedron Lett.*, **33** (46), 7021–7024.

41 Adam, W., Bargon, R.M., Bosio, S.G., Schenk, W.A., and Stalke, D. (2002) Direct synthesis of isothiocyanates from isonitriles by molybdenum-catalyzed sulfur transfer with elemental sulfur. *J. Org. Chem.*, **67**, 7037–7041.

42 Byrne, J.J. and Vallee, Y.A. (1999) Dithioxo-bishydroxyaminomolybdenium complex as sulfur and nitrogen transfer reagent. Synthesis of thioureas from isonitriles. *Tetrahedron Lett.*, **40**, 489–490.

43 Arisawa, M., Ashikawa, M., Suwa, A., and Yamaguchi, M. (2005) Rhodium-catalyzed synthesis of isothiocyanate from isonitrile and sulfur. *Tetrahedron Lett.*, **46**, 1727–1729.

44 Simpson, J.S. and Garson, M.J. (2001) Advanced precursors in marine biosynthetic study. Part 2: the biosynthesis of isocyanides and isothiocyanates in the tropical marine sponge *Axinyssa* n. sp. *Tetrahedron Lett.*, **42**, 4267–4269.

45 Bulka, E., Ahlers, K.-D., and Tućek, E. (1967) Synthese und IR-spektren von aryl-isoselenocyanaten. *Chem. Ber.*, **100**, 1367–1372.

46 Garud, D.R., Koketsu, M., and Ishihara, H. (2007) Isoselenocyanates: a powerful tool for the synthesis of selenium-containing heterocycles. *Molecules*, **12**, 504–535.

47 Barton, D.H.R., Parekh, S.I., Tajbakhsh, M., Theodorakis, E.A., and Tse, C.-L. (1994) A convenient and high yielding procedure for the preparation of isoselenocyanates. Synthesis and reactivity of O-alkylselenocarbamates. *Tetrahedron*, **50**, 639–654.

48 Su, W.K. and Liang, X.R. (2003) An efficient and convenient route to some isoselenocyanates via reaction of formamides with bis(trichloromethyl) carbonate and selenium. *J. Indian Chem. Soc.*, **80**, 645–647.

49 Fernández-Bolaños, J.G., López, Ó., Ulgar, V., Maya, I., and Fuentes, J. (2004) Synthesis of O-unprotected glycosyl selenoureas. A new access to bicyclic sugar isoureas. *Tetrahedron Lett.*, **45**, 4081–4084.

50 Barton, D.H.R. and Motherwell, W.B. (1981) New and selective reactions and reagents in natural product chemistry. *Pure Appl. Chem.*, **53**, 1081–1099.

51 Rubinson, J.F., Corbin, J.L., and Burgess, B.K. (1983) Nitrogenase reactivity: methyl isocyanide as substrate and inhibitor. *Biochemistry*, **22**, 6260–6268.

52 Ugi, I. and Bodesheim, F. (1961) Notiz zur reduktion von isonitrilen mit alkali- und erdalkalimetallen in flüssigem ammoniak. *Chem. Ber.*, **94**, 1157–1158.

53 Buchner, W. and Dufaux, R. (1966) Die reaktion von isonitrilen und nitrilen mit alkali- und erdalkalimetallen. *Helv. Chim. Acta*, **49** (3), 1145–1150.

54 Walborsky, H.M. and Niznik, G.E. (1972) Synthesis of isonitriles. *J. Org. Chem.*, **37**, 187–190.

55 Walborsky, H.M. and Niznik, G.E. (1978) Isocyanide reductions. A convenient method for deamination. *J. Org. Chem.*, **43** (12), 2396–2399.

56 Saegusa, T., Kobayashi, S., Ito, Y., and Yasuda, N. (1968) Radical reaction of isocyanide with organotin hydride. *J. Am. Chem. Soc.*, **90** (15), 4182.

57 John, D.I., Thomas, E.J., and Tyrrell, N.D. (1979) Reduction of 6α-alkyl-6β-isocyanopenicillanates by tri-n-butyltin hydride. A stereoselective synthesis of 6β-alkylpenicillanates. *Chem. Commun.*, 345–347.

58 Maimone, T.J., Shi, J., Ashida, S., and Baran, P.S. (2009) Total synthesis of vinigrol. *J. Am. Chem. Soc.*, **131**, 17066–17067.

59 Barton, D.H.R., Jang, D.O., and Jaszberenyi, J.C. (1993) The invention of radical reactions. Part XXXI. Diphenylsilane: a reagent for deoxygenation of alcohols via their thiocarbonyl derivatives, deamination via isonitriles, and dehalogenation of bromo- and iodo- compounds by radical chain chemistry. *Tetrahedron*, **49** (33), 7193–7214.

60 Livinghouse, T. (1999) C-Acylnitrilium ion initiated cyclizations in heterocycle synthesis. *Tetrahedron*, **55** (33), 9947–9978.

61 Lygin, A.V. and de Meijere, A. (2010) Isocyanides in the synthesis of nitrogen

heterocycles. *Angew. Chem. Int. Ed.*, **49**, 9094–9124.

62 Kolontsova, A.N., Ivantsova, M.N., Tokareva, M.I., and Mironov, M.A. (2010) Reaction of isocyanides with thiophenols and gem-diactivated olefins: a one-pot synthesis of substituted 2-aminopyrroles. *Mol. Divers.*, **14** (3), 543–550.

63 Ganem, B. (2009) Strategies for innovation in multicomponent reaction design. *Acc. Chem. Res.*, **42** (3), 463–472.

64 Mironov, M.A. (2006) Design of multi component reactions: from libraries of compounds to libraries of reactions. *QSAR Comb. Sci.*, **25** (5–6), 423–432.

65 Li, X., Yuan, Y., Berkowitz, W.F., Todaro, L.J., and Danishefsky, S.J. (2008) On the two-component microwave-mediated reaction of isonitriles with carboxylic acids: regarding alleged formimidate carboxylate mixed anhydrides. *J. Am. Chem. Soc.*, **130**, 13222–13224.

66 Shaabani, A., Soleimani, E., and Rezayan, A.H. (2007) A novel approach for the synthesis of aryl amides. *Tetrahedron Lett.*, **48**, 6137–6141.

67 Restorp, P. and Rebek, J. (2008) Reaction of isonitriles with carboxylic acids in a cavitand: observation of elusive isoimide intermediates. *J. Am. Chem. Soc.*, **130**, 11850–11851.

68 Hou, J.-L., Ajami, D., and Rebek, J. (2008) Reaction of carboxylic acids and isonitriles in small spaces. *J. Am. Chem. Soc.*, **130**, 7810–7811.

69 Li, X. and Danishefsky, S.J. (2008) New chemistry with old functional groups: on the reaction of isonitriles with carboxylic acids – a route to various amide types. *J. Am. Chem. Soc.*, **130**, 5446–5448.

70 Rao, Y., Li, X., and Danishefsky, S.J. (2009) Thio FCMA intermediates as strong acyl donors: a general solution to the formation of complex amide bonds. *J. Am. Chem. Soc.*, **131**, 12924–12926.

71 Wu, X., Stockdill, J.L., Wang, P., and Danishefsky, S.J. (2010) Total synthesis of cyclosporine: access to N-methylated peptides via isonitrile coupling reactions. *J. Am. Chem. Soc.*, **132**, 4098–4100.

72 Yasuda, N., Ariyoshi, Y., and Toi, K. (1974) Formation of peptide bonds in the presence of isonitriles. US Patent, 3933783, filed December 9, 1974 and issued January 20, 1976.

73 Smith, P.A. and Kalenda, N.W. (1958) Investigation of some dialkylamino isocyanides. *J. Org. Chem.*, **23**, 1599–1603.

74 Jin, T., Kamijo, S., and Yamamoto, Y. (2004) Synthesis of 1-substituted tetrazoles via the acid-catalyzed [3+2] cycloaddition between isocyanides and trimethylsilyl azide. *Tetrahedron Lett.*, **45**, 9435–9437.

75 Sureshbabu, V.V., Narendra, N., and Nagendra, G. (2009) Chiral N-Fmoc-amino alkyl isonitriles derived from amino acids: first synthesis and application in 1-substituted tetrazole synthesis. *J. Org. Chem.*, **74**, 153–157.

76 Gördeler, J. and Weber, D. (1964) Monomere und dimere imidoyl-isothiocyanate. *Tetrahedron Lett.*, **5**, 799–800.

77 Polyakov, A.I., Baskakov, Y.A., Artamonova, O.S., and Baranova, S.S. (1983) Synthesis of imidazolinium salts by cyclization of amino isonitriles. *Chem. Heterocycl. Compd.*, **19** (6), 684.

78 Nef, J.U. (1892) Ueber das zweiwertige kohlenstoffatom. *Liebigs Ann. Chem.*, **270** (3), 267–335.

79 Kuhle, E., Anders, B., and Zumach, G. (1967) Syntheses of isocyanide dihalides. *Angew. Chem. Int. Ed. Engl.*, **6**, 649–665.

80 El Kaim, L., Grimaud, L., and Patil, P. (2011) Three-component strategy toward 5-membered heterocycles from isocyanide dibromides. *Org. Lett.*, **13** (5), 1261–1263.

81 Baeza, A., Burgos, C., Alvarez-Builla, J., and Vaquero, J.J. (2007) Selective palladium-catalyzed amination of the heterocyclic core of variolins. *Tetrahedron Lett.*, **48**, 2597–2601.

82 Walter, W. and Bode, K.D. (1966) Syntheses of thiocarboxamides. *Angew. Chem. Int. Ed. Engl.*, **5** (5), 447–461.

83 Westling, M. and Livinghouse, T. (1985) Intramolecular cyclizations of α-ketoimidoyl halides derived from organic isonitriles. An expedient approach to the synthesis of 1-acyl-3,4-dihydroisoquinolines. *Tetrahedron Lett.*, **26** (44), 5389–5392.

84 Westling, M. and Livinghouse, T. (1987) Acylnitrilium ion cyclizations in heterocycle synthesis. A convergent method for the preparation of 2-acylpyrrolines via the intramolecular acylation of silyloxyalkenes with α-keto imidoyl chlorides. *J. Am. Chem. Soc.*, **109** (2), 590–592.

85 Lee, C.H., Westling, M., Livinghouse, T., and Williams, A.C. (1992) Acylnitrilium ion initiated heteroannulations in alkaloid synthesis. An efficient, stereocontrolled, total synthesis of Orchidaceae alkaloid (±)-dendrobine. *J. Am. Chem. Soc.*, **114** (11), 4089–4095.

86 Bossio, R., Marcaccini, S., and Pepino, R. (1986) A novel synthetic route to oxazoles: one pot synthesis of 2-arylthio-5-alkoxyoxazoles. *Heterocycles*, **24**, 2003–2005.

87 Collibee, W.L., Nakajima, M., and Anselme, J.-P. (1995) 5-Halo-1-phenyltetrazoles. *J. Org. Chem.*, **60** (2), 468–469.

88 Nguyen, L.T., Le, T.N., De Proft, F., Chandra, A.K., Langenaeker, W., Nguyen, M.T., and Geerlings, P. (1999) Mechanism of [2 + 1] cycloadditions of hydrogen isocyanide to alkynes: molecular orbital and density functional theory study. *J. Am. Chem. Soc.*, **121**, 5992–6001.

89 Krebs, A. and Kimling, H. (1971) Investigations on strained cyclic acetylenes. 2. [1+2] cycloadditions of isocyanides to alkynes. *Angew. Chem. Int. Ed. Engl.*, **10** (6), 409–410.

90 Nair, V., Rajesh, C., Vinod, A.U., Bindu, S., Sreekanth, A.R., Mathen, J.S., and Balagopal, L. (2003) Strategies for heterocyclic construction via novel multicomponent reactions based on isocyanides and nucleophilic carbenes. *Acc. Chem. Res.*, **36**, 899–907.

91 Winterfeldt, E., Schumann, D., and Dillinger, H.J. (1969) Additionen an die Dreifachbindung–XI Struktur und Reaktionen des 2:1-Adducktes aus Acetylenedicarbonester und Isonitrilen. *Chem. Ber.*, **102**, 1656–1664.

92 Dillinger, H.J., Fengler, G., Schumann, D., and Winterfeldt, E. (1974) Additionen an die Dreifachbindung-XXII: das Thermodynamisch Kontrollierte Adukt aus *tert*-Butylisonitril und Acetylendicarbonester. *Tetrahedron*, **30**, 2561–2564.

93 Junjappa, H., Saxena, M.K., Ramaiah, D., Lohray, B.B., Rath, N.P., and George, M.V. (1998) Structure and thermal isomerization of the adducts formed in the reaction of cyclohexyl isocyanide with dimethyl acetylenedicarboxylate. *J. Org. Chem.*, **63**, 9801–9805.

94 Cheng, F.-Y., Sung, K., Lee, G.-H., and Wang, Y. (2000) An efficient synthetic method and single crystal structure of a 2:3 adduct of cyclohexyl isocyanide and dimethyl acetylenedicarboxylate. *J. Chin. Chem. Soc.*, **47**, 1295–1298.

95 Ott, W., Kollenz, G., Peters, K., Peters, E-M., Von Schnering, H.G., and Quast, H. (1983) Synthesis of 1*H*,4*H*-furo[3,4-c] furans via "criss-cross" cycloaddition reaction of isocyanides to 1,4-diphenylbutyne-1,4-dione. *Liebigs Ann. Chem.*, (4), 635–641.

96 Oakes, T.R., David, H.G., and Nagel, F.J. (1969) Small ring systems from isocyanides. I. The reaction of isocyanides with hexafluorobutyne-2. *J. Am. Chem. Soc.*, **91**, 4761.

97 Takizawa, T., Obata, N., Suzuki, Y., and Yanagida, T. (1969) A novel cycloaddition reaction of 4-bromo-2,6-dimethyl-phenylisonitrile with acetylene derivatives. *Tetrahedron Lett.*, **10** (39), 3407–3410.

98 Middleton, W.J. (1965) 1,1-dicyano-2,2-bis(trifluoromethyl)ethylene. *J. Org. Chem.*, **30**, 1402.

99 Walter, O., Formacek, V., and Seidenspinner, H.M. (1984) Reaction with isocyanides. Synthesis of dark-blue dyes from 1,4-quinones and aryl isocyanides. *Liebigs Ann. Chem.*, (5), 1003–1012.

100 Saegusa, T., Ito, Y., Tomita, S., Kinoshita, H., and Taka-ishi, N. (1971) Reaction of isocyanide with α,β-unsaturated carbonyl and nitrile compounds. *Tetrahedron*, **27** (1), 27–31.

101 Maltsev, S.S., Mironov, M.A., and Bakulev, V.A. (2006) Synthesis of cyclopentene derivatives by the cyclooligomerization of isocyanides with substituted benzylidenemalononitriles. *Mendeleev Commun.*, **16**, 201–202.

102 Moderhack, D. (1985) Four-membered rings from isocyanides. Recent advances. *Synthesis*, **17** (12), 1083–1096.
103 Saegusa, T., Taka-ishi, N., and Fujii, H. (1968) Reaction of carbonyl compound with isocyanide. *Tetrahedron*, **24**, 3795–3798.
104 Zeeh, B. (1969) Additions reaktionen zwischen Isocyaniden und Doppelbindungssystemen. *Synthesis*, **1**, 65–78.
105 Lumma, W.C. (1981) Modification of the Passerini reaction: facile synthesis of analogs of isoproterinol and (aryloxy) propanolamine β-adrenergic blocking agents. *J. Org. Chem.*, **46** (18), 3668–3671.
106 Deyrup, J.A., Vestling, M.M., Hagan, W.V., and Yun, H.Y. (1969) Reactions of imines with *t*-butyl isocyanide. *Tetrahedron*, **25**, 1467–1478.
107 Saegusa, T., Taka-ishi, N., Tamura, I., and Fujii, H. (1969) Acid- catalyzed reaction of isocyanide with a Schiff base. *J. Org. Chem.*, **34** (4), 1145–1147.
108 Schneekloth, J.S., Jr, Kim, J., and Sorensen, E.J. (2009) An interrupted Ugi reaction enables the preparation of substituted indoxyls and aminoindoles. *Tetrahedron*, **65** (16), 3096–3101.
109 Ugi, I. and Rosendahl, K. (1961) Umsetzungen von isonitrilen mit ketenen. *Chem. Ber.*, **94**, 2233–2238.
110 Moore, H.W. and Yu, C.C. (1981) Cycloaddition of *tert*-butylcyanoketene to isocyanides. *J. Org. Chem.*, **46** (24), 4935–4938.
111 Gambaryan, N.P., Avetisyan, E.A., Deltsova, D.P., and Safronova, S.V. (1981) Fluoro-containing heterocumulens XXI. *Armen. Chem. J.*, **34** (5), 380–388.
112 Obata, N., Mizuno, H., Koitabashi, T., and Takizawa, T. (1975) The cycloaddition reaction of isonitrile to ketenimine formed by the reaction of isonitrile with carbene. *Bull. Chem. Soc. Jpn*, **48** (8), 2287–2293.
113 L'abbe, G., Huybrechts, I., Toppet, S., Deelereq, J.P., Germain, G., and Van Meerssehe, M. (1978) Cycloaddition reactions of tris(imino)thietanes. *Bull. Soc. Chim. Belg.*, **87** (11–12), 893–901.
114 L'abbe, G., Huybrechts, I., Deelereq, J.P., Germain, G., and Van Meerssehe, M. (1979) Synthesis (imino)thietanes; X-ray crystal structure of 2,3-bis-(*tert*-butylimino)-4-*p*-tolylsulfonyliminothitane. *Chem. Commun.*, 160–161.
115 Saegusa, T. and Ito, Y. (1975) Synthesis of cyclic compounds via copper-isonitrile complexes. *Synthesis*, **7** (5), 291–300.
116 Ugi, I. and Fetzer, U. (1961) Die umsetzung von cyclohexylisocyanide mit phenylmagnesium- bromiden. *Chem. Ber.*, **94** (10), 2239–2243.
117 Niznik, G.E., Morrison, W.H., and Walborsky, H.M. (1974) Metallo aldimines. Masked acyl carbanion. *J. Org. Chem.*, **39** (5), 600–604.
118 Ito, Y., Ihara, E., Hirai, M., Ohsaki, H., Ohnishi, A., and Murakami, M. (1990) Aromatizing oligomerization of 1,2-di-isocyanoarene to quinoxaline oligomers. *Chem. Commun.*, 403–405.
119 Walborsky, H.M., Morrison, W., and Niznik, G.E. (1970) Metallo aldimines. A versatile synthetic intermediate. *J. Am. Chem. Soc.*, **92** (2), 6675–6676.
120 Marks, M.J. and Walborsky, H.M. (1982) Metallo aldimines. 3. Coupling of lithium aldimines with aryl, vinyl, and acetylenic halides. *J. Org. Chem.*, **47** (1), 52–56.
121 Spielmann, J. and Harder, S. (2007) Hydrocarbon-soluble calcium hydride: a "worker-bee" in calcium chemistry. *Chem. Eur. J.*, **13**, 8928–8938.
122 Murakami, M., Kawano, T., and Ito, Y. (1990) [2-Benzyloxy)-1-(*N*-2,6-xylylimino) ethyl]samarium as a synthetic equivalent to α-hydroxyacetyl anion. *J. Am. Chem. Soc.*, **112** (6), 2437–2439.
123 Curran, D.P. and Totleben, M.J. (1992) The samarium Grignard reaction. *In situ* formation and reactions of primary and secondary alkylsamarium(III) reagents. *J. Am. Chem. Soc.*, **114**, 6050–6058.
124 Ito, Y. (1990) New metallation and synthetic applications of isonitriles. *Pure Appl. Chem.*, **62** (4), 583–588.
125 Barton, D.H.R., Ozbalik, N., and Vacher, B. (1988) The invention of radical reactions. Part XVIII. A convenient solution of the 1-carbon problem. *Tetrahedron*, **44** (12), 3501–3522.

126 Cannighem, I.D., Buist, G., and Arkl, S. (1991) Chemistry of isocyanides Part 2. Nucleophile addition of hydroxide to aromatic isocyanides in aqueous dimethyl sulfoxide. *J. Chem. Soc., Perkin Trans. 2*, (5), 589–593.

127 Hashida, Y., Imai, A., and Sakigushi, S. (1989) Preparation and reactions of isocyano-1,3,4-triazines. *J. Heterocycl. Chem.*, **26** (11), 901–905.

128 Suginome, M., Fukuda, T., and Ito, Y. (1999) New access to 2,3-disubstituted quinolines through cyclization of o-alkynylisocyanobenzenes. *Org. Lett.*, **1** (12), 1977–1979.

129 Mironov, M.A. and Mokrushin, V.S. (1998) Reaction of aromatic isocyanides with triethylamine: a new method for the synthesis of indole betaines. *Mendeleev Commun.*, **8** (3), 242–243.

3
α-Acidic Isocyanides in Multicomponent Chemistry
Niels Elders, Eelco Ruijter, Valentine G. Nenajdenko, and Romano V.A. Orru

3.1
Introduction

During the past century-and-a-half, multicomponent reactions (MCRs) [1–12] have evolved from random academic oddities to key methods in combinatorial chemistry, diversity-oriented synthesis (DOS), and natural product synthesis. Today, MCRs are defined as one-pot processes in which at least three compounds react to form a single product that contains essentially all the atoms of the starting materials. Early examples of MCRs include the Strecker [13, 14], Hantzsch [15], Biginelli [16, 17], and Mannich [18] reactions (Scheme 3.1); however, the discovery of the Passerini [19, 20] and Ugi [21, 22] reactions—both of which are isocyanide-based MCRs (IMCRs) [11]—marked the dawn of truly versatile multicomponent chemistry.

The widespread application of the Passerini and Ugi reactions underlines the importance of isocyanides in multicomponent chemistry. The exceptional reactivity of isocyanides became evident shortly after their first isolation by Lieke in 1859 [23]. For example, their ability to readily form radicals and to react both as a nucleophile and electrophile at the same atom (α-addition) is not encountered in any other functional groups. The α-addition reaction of isocyanides plays a key role in classical IMCRs such as the Ugi and Passerini reactions (see Scheme 3.2).

Although smaller in number, the IMCRs are considered more versatile and diverse than the other classes of MCR. A third property determining the reactivity of the isocyanides is their α-acidity. Isocyanides which carry an electron-withdrawing group (EWG) at the α-position have α-protons which are considerably acidic. Consequently, after α-proton abstraction the α-acidic isocyanides possess – in addition to the already unusual reactivity of the isocyanide functional group – a second nucleophilic center. This alternative reactivity offers ample possibility for new MCRs, leading to a range of interesting products that include N-heterocycles. Among the various classes of α-acidic isocyanides, the isocyano esters (**1**) [24], isocyanoamides (**2**) [25], and arylsulfonylmethyl isocyanides (**3**) [26] have been studied most intensely; consequently, attention in this chapter will be mainly focused on MCRs that involve these compounds.

Isocyanide Chemistry: Applications in Synthesis and Material Science, First Edition. Edited by Valentine Nenajdenko.
© 2012 Wiley-VCH Verlag GmbH & Co. KGaA. Published 2012 by Wiley-VCH Verlag GmbH & Co. KGaA.

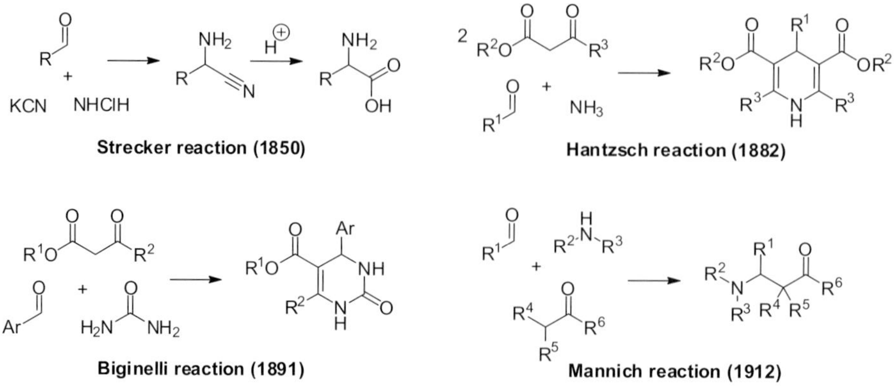

Scheme 3.1 Early examples of multicomponent reactions.

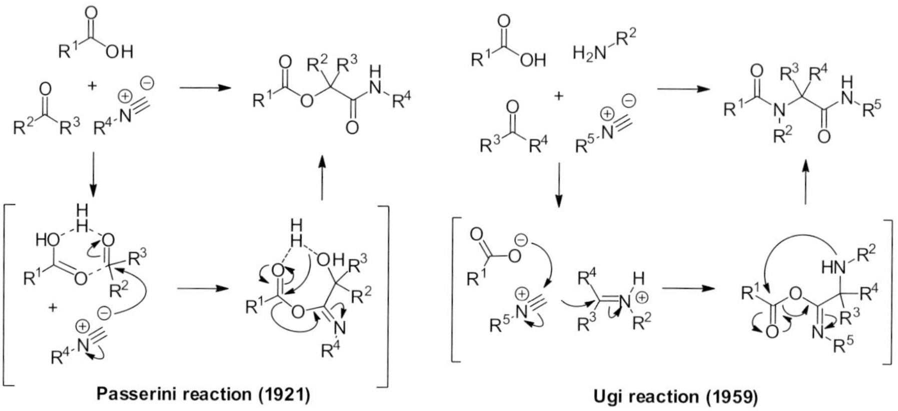

Scheme 3.2 The Passerini and Ugi reactions – the classical isocyanide-based MCRs.

3.2
Synthesis of α-Acidic Isocyanides

Several approaches to the synthesis of α-acidic isocyanides **1–3** have been reported, the most general of which are depicted in Scheme 3.3. For example, due to the acidity of the α-protons, α-substituted α-acidic isocyanides can be synthesized from the corresponding α-unsubstituted compounds (**4**) by deprotonation and subsequent alkylation (Scheme 3.3, pathway **A**). Dialkylation (a problem which is often observed for all three classes of α-acidic isocyanides [27–29]) can be suppressed by employing methods that lead to the selective formation of monoalkylated derivatives. For the selective monoalkylation of arylsulfonylmethyl isocyanides with primary halides, phase-transfer catalysis (employing nBu$_4$X, NaOH, CH$_2$Cl$_2$) is often the method of choice [30].

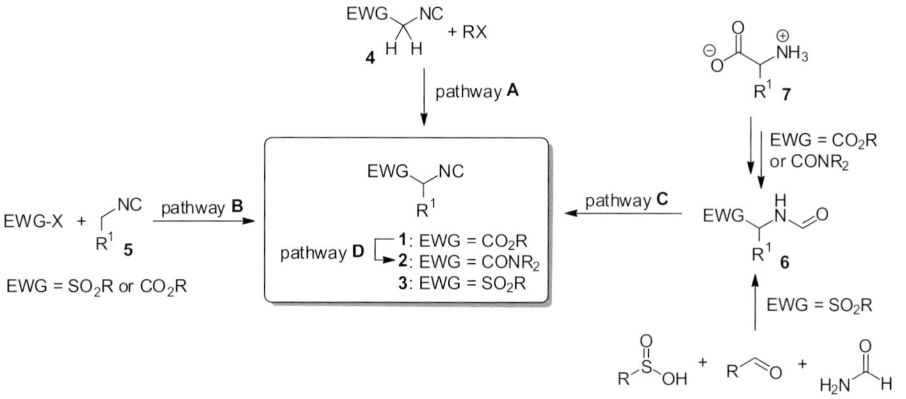

3.2 Synthesis of α-Acidic Isocyanides

Scheme 3.3 General synthetic routes to α-acidic isocyanides.

Ito et al. have described the monoalkylation of isocyanoamides by a Michael reaction with α,β-unsaturated ketones in the presence of a catalytic amount of tetra-n-butylammonium fluoride (TBAF) in tetrahydrofuran (THF) [31]. The selective monoalkylation of isocyanoamides using CsOH•H$_2$O (1.5 equiv.) in MeCN at 0 °C was recently reported by Zhu [32]. A slight excess of alkylating agent (1.2 equiv.) can be used to obtain the monoalkylated products with a broad range of primary halides (X = Br or I).

Schöllkopf reported the monoalkylation of α-isocyano esters using *tert*-butyl isocyano acetate (R^1 = *t*Bu) [28, 33]. Besides primary halides, 2-iodopropane can also be used to produce the α-alkylated product (**1**) by this method (KO*t*Bu in THF). Ito reported several methods for the monoalkylation of isocyano esters, including a TBAF-catalyzed Michael addition (see above) [31], a Claisen rearrangement [34], and Pd-catalyzed asymmetric allylation [35]. Finally, Zhu recently reported the synthesis of methyl α-isocyano *p*-nitrophenylacetate by nucleophilic aromatic substitution (CsOH•H$_2$O, MeCN, 0 °C) [36].

Arylsulfonyl methyl isocyanides (**3**) and isocyano esters (**1**) can also be prepared by reaction with tosyl fluoride and dialkyl carbonates or ethyl chloroformate, respectively, after deprotonation of alkyl isocyanides (**5**) by a strong base (*n*BuLi or NaH) (Scheme 3.3, pathway **B**) [27, 37]. However, the use of low-molecular-weight, foul-smelling alkyl isocyanides makes this route less attractive.

As with all isocyanides, α-acidic isocyanides can be synthesized from the corresponding formamides (**6**) by dehydration (Scheme 3.3, pathway **C**, for example, using (tri)phosgene or POCl$_3$ under basic conditions). In the case of arylsulfonyl-methyl isocyanides, the corresponding formamides can be generated using a three-component reaction (3CR) between sulfinates, aldehydes and formamide, as reported in 1972 by van Leusen [38]. By using this procedure, a wide range of (optically pure) sulfinates can be converted to the corresponding arylsulfonyl-methyl isocyanides (**3**) [38–40]. The scope with respect to the aldehyde input remained limited to formaldehyde and benzaldehyde until Sisko reported a modification (+TMSCl, solvent = toluene:MeCN (1:1), 50°C, 5 h) which allowed the use of a broad range of aldehydes to obtain the formamides on the multikilogram scale [41].

The formamido esters and amides can be generated in two steps, starting from (natural) α-amino acids (7). Owing to the wide availability of optically pure formamides, the synthesis of optically pure α-substituted α-acidic isocyanoamides (2) and esters (1) has been envisioned using the dehydration route. Although the synthesis of optically active α-chiral α-acidic isocyano esters by dehydration of the corresponding formamides using $POCl_3$ [42], phosgene [43, 44], diphosgene [45–49], or triphosgene [50] has been reported, the isolation of such compounds is far from trivial. Although limited information regarding the enantiomeric excess (e.e.) of the adjacent stereocenter is generally provided, racemization during the dehydration step is a well-known complication [51]. In 2009, Danishefsky et al. compared several procedures and confirmed that the use of triphosgene and N-methylmorpholine (NMM) at −78 to −30°C led to the conservation of optical purity, whereas $POCl_3/Et_3N$ provided full racemization [52]. By using the former protocol, however, several chiral α-isocyano esters could be synthesized in high yield with e.e.-values of up to >98%.

Chiral α-isocyanoamides can also be obtained by using various dehydration methods [46, 52–65]. Because of the lower acidity of the α-proton compared to the corresponding α-isocyano esters, α-isocyanoamides can be obtained in high e.e., using relatively harsh dehydration methods ($POCl_3/Et_3N$, −20°C) [61], although the triphosgene/NMM protocol remains the method of choice [52].

Finally, α-isocyano esters can be converted to the corresponding amides by amidation [32, 42, 46, 66–70]. Although this procedure (Scheme 3.3, pathway **D**) is compatible with α,α-unsubstituted isocyano esters and a wide variety of primary and secondary amines, steric bulk in either of the reactants should be avoided.

3.3
Reactivity of α-Acidic Isocyanides

α-Acidic isocyanides of types **1**, **2**, and **3** [71] have found widespread application in the construction of heterocyclic compounds (Schemes 3.4 and 3.5). Most of the examples reported to date have involved [3+2] cycloadditions with carbon–heteroatom double bonds under (highly) basic conditions. For example, the cycloaddition of deprotonated isocyano esters [72] with C=O, C=N, or C=S double bonds yields oxazolines (Scheme 3.4, **A**) [73, 74], imidazolines (**C**) [75–78], and thiazolines (**F**) [79], respectively. The corresponding aromatic heterocycles – that is, **B** (oxazoles) [80–84] and **D** (imidazoles) [85–87] – can be obtained when the carbon–heteroatom double bonds are substituted with a leaving group. In addition, imidazoles **E** can be obtained by cycloaddition of the anion derived from **1** with other isocyanides [88–90], although the number of diversity points is reduced to two in this case. Thiazoles (**G**) can be obtained by the reaction of **1** with isothiocyanates under basic conditions [91–93].

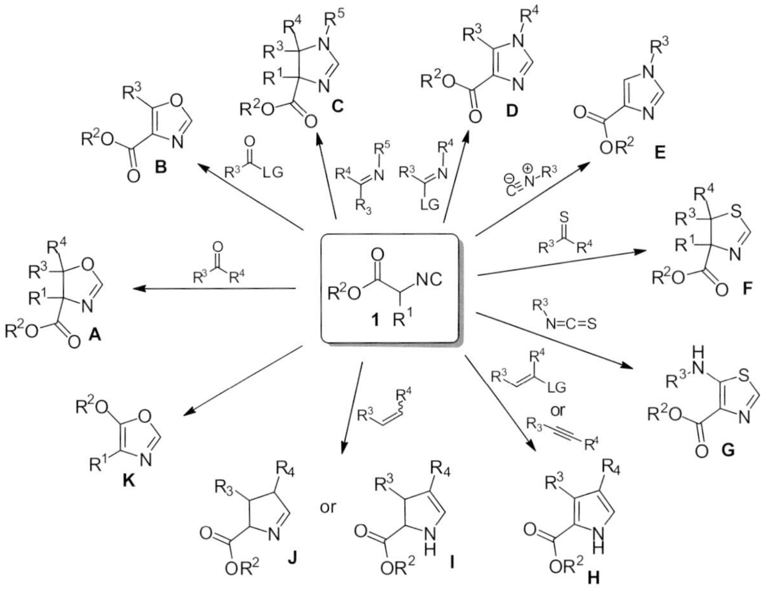

Scheme 3.4 Selected examples for the formation of heterocycles, starting from isocyano esters.

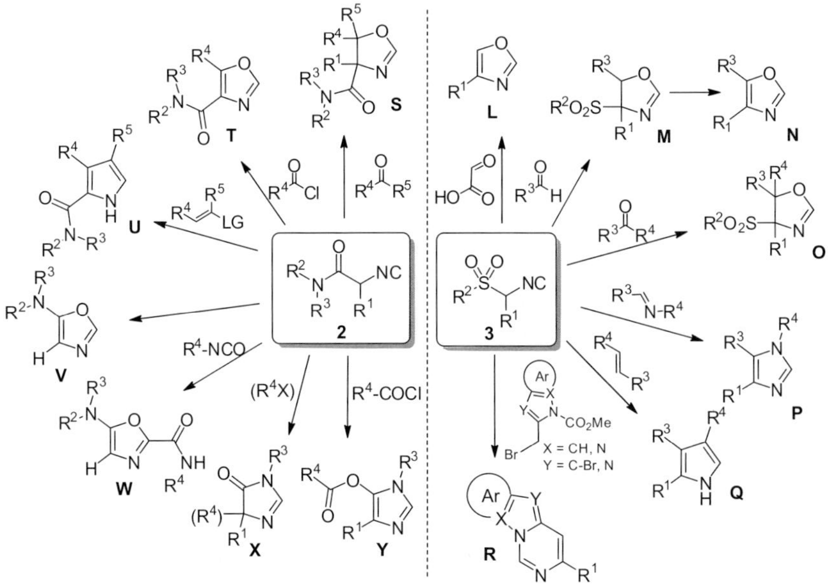

Scheme 3.5 Selected examples for the formation of heterocycles, starting from either isocyanoamides (**2**) or sulfonylmethyl isocyanides (**3**).

The cycloaddition of deprotonated **1** with electron-deficient alkenes as dipolarophiles to give pyrroles (Scheme 3.4, **H**) has been achieved using isocyanoalkenes [94–96], nitroalkenes [97–101], α,β-unsaturated sulfones [102–105], and α,β-unsaturated nitriles [106]; alkynes [107–109] have also been used to access pyrroles (**H**). Similarly, the cycloaddition of **1** with activated olefins yielding pyrrolines (**I** and **J**) has been accomplished using Cu_2O catalysis [73], AgOAc catalysis [88, 110], or organocatalysis [110], while the uncatalyzed reaction using [60]fullerene has also been reported [111].

Finally, the cyclization of α-acidic isocyano esters (**1**) under basic conditions has been reported to afford 5-alkoxyoxazoles (Scheme 3.4, **K**) [112, 113].

Similarly, isocyanoamides (**2**) and arylsulfonylmethyl isocyanides (**3**) have also found widespread application in the construction of heterocyclic compounds (Scheme 3.5). For example, [3+2] cycloadditions of arylsulfonylmethyl isocyanides (**3**) with aldehydes yielding 4,5-disubstituted oxazoles (**N**) have been reported [114]. The cycloaddition takes place in the same manner as for α-isocyano esters (**A**, $R^3 = H$), but the reaction typically does not stop at the trisubstituted oxazoline (**M**) stage due to the facile elimination of RSO_2H [114, 115]. With the use of ketones, the elimination of RSO_2H is not possible and the corresponding oxazolines (**O**) can be isolated [39, 116]. Monosubstituted oxazoles of type **L** can be obtained using glyoxylic acid [117], while imidazoles (**P**) and pyrroles (**Q**) can be obtained by the reaction of **3** with aldimines [118] and alkenes [114, 119, 120], respectively. Finally, azopyrimidines (**R**) can be accessed by the reaction of arylsulfonylmethyl isocyanides with N-protected bromomethylazoles under phase-transfer conditions [121].

The isocyanoamides (**2**) have been reported to undergo [3+2] cycloadditions with carbonyl compounds to give oxazolines (**S**) [122, 123] and oxazoles (**T**) [124]. Furthermore, pyrroles (**U**) can be produced by the reaction of **2** with nitroalkenes [97, 99, 125] or isocyanoalkenes [96].

Intramolecular cyclization reactions of α-acidic isocyanoamides (**2**) have also been described. For example, α-unsubstituted isocyanoamides (**2**, $R^1 = H$) readily cyclize via the amide oxygen to give $2H$-5-aminooxazoles (**V**) [126], while the addition of an isocyanate affords the corresponding 2-amido analogues **W** [127]. Cyclization via the amide nitrogen ($R^2 = H$) furnishes imidazolin-5-ones (**X**) [128, 129], in which an additional substituent (R^4) can be introduced by C4-alkylation. Alternatively, the addition of an acid chloride provides **Y** [129].

3.4
MCRs Involving α-Acidic Isocyanides

In addition to the numerous applications of α-acidic isocyanides in classical isocyanide-based MCRs, such as the Ugi and Passerini reactions, the presence of an acidic α-proton allows the straightforward generation of a formal 1,3 dipolar species with a unique reactivity that is intrinsically different from the typical α-addition reactivity of normal isocyanides, and this has opened new avenues towards the development of novel MCRs. In the following section, attention will be focused on novel MCRs of α-acidic isonitriles, as reported since 1998.

3.4.1 van Leusen Imidazole MCR

The first MCR involving the explicit use of α-acidic isonitriles was reported in 1998 by Sisko [130]. The reaction involved the formal cycloaddition of TosMIC derivative **8** to an *in situ*-generated imine (**10**), followed by the elimination of *p*-toluenesulfinic acid (TsH), as described previously in 1977 by van Leusen for preformed imines (Scheme 3.5, **P**) [118]. Although several potential pitfalls for the conversion of the classical van Leusen [3+2] cycloaddition to a one-pot protocol (the van Leusen three-component reaction, vL-3CR) were expected,[1)] a straightforward *in situ* preformation of the imine, followed by addition of the TosMIC derivative **8** and a base, afforded the corresponding imidazoles (**9**) in high yield [130] (Scheme 3.6).

The removal of water from the reaction mixture by adding $MgSO_4$ or molecular sieves did not have any major effect on the reaction outcome. Moreover, further optimization studies indicated that the reaction proceeded well in a wide range of common organic solvents (e.g., EtOAc, THF, MeCN, DMF, DCM, MeOH), and only a mild base such as piperazine, morpholine or K_2CO_3 would be required to promote the reaction. A wide variety of functional groups, including alcohols, esters, alkenes and even additional aldehydes, proved compatible with the vL-3CR. In addition to the broad scope with respect to the amine and carbonyl inputs, various α-aryl (and biaryl) -substituted TosMICs (**8**) could be used in the vL-3CR. Remarkably, despite their successful application in the stepwise van Leusen imidazole synthesis [118], α-alkyl-substituted TosMICs (**8**, R^1 = alkyl) and TosMIC itself (**8a**, R^1 = H) proved to be unreactive in the multicomponent approach [131], most likely due to the increased pK_a of the α-proton.

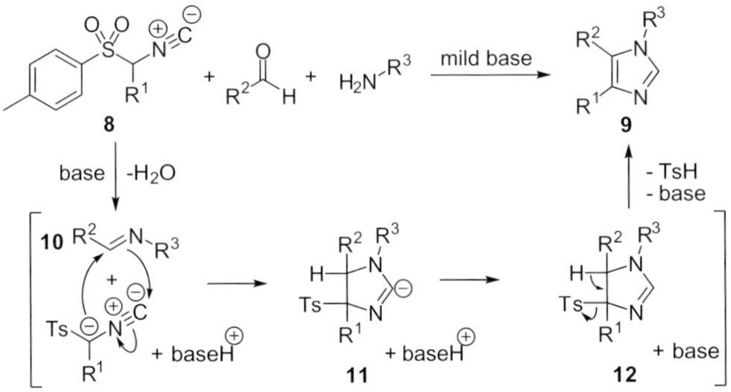

Scheme 3.6 Reaction mechanism for the van Leusen 3CR towards imidazoles.

1) Initially, the presence of water and competition between the aldehyde and imine were expected to cause problems in the reaction set-up.

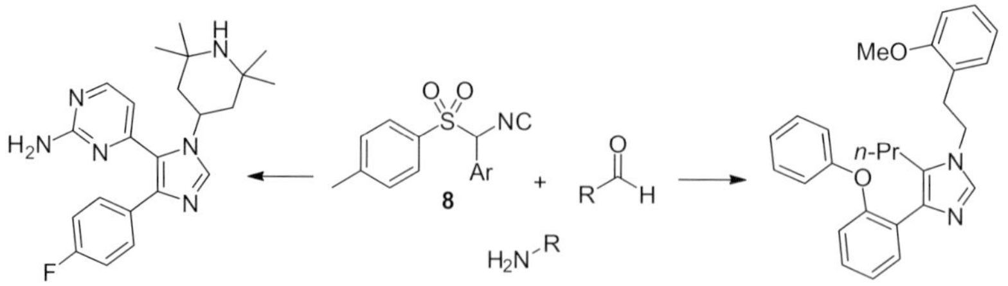

9a
Potent P38 MAP
kinase inhibitor

9b
Inhibitor of protein-
protein
interaction between Bcl-
w and Bak-BH3

Scheme 3.7 Two representative examples of the application of the van Leusen 3CR in the synthesis of biologically active compounds.

The broad substrate scope for the vL-3CR for the synthesis of imidazoles (**9**), in combination with the straightforward reaction set-up, has led to the reaction becoming a popular MCR for the preparation of biologically active compounds. For example, Sisko initially developed the vL-3CR for the synthesis of **9a** (Scheme 3.7) [130]; this is an inhibitor of p38 MAP kinase, and appears to be involved in an inflammation regulatory pathway [132]. Furthermore, the construction of a focused library of about 60 bioactive trisubstituted imidazoles by Dömling resulted in the identification of novel α-helix mimetics that could be synthesized in a single reaction step [133]. Initial screening results indicated a disruption of the interaction between Bcl-w and Bak-BH3 peptide, constituting an effective and selective possibility for the treatment of cancer, and making this novel class of lead molecules well suited to further optimization (Scheme 3.7).

3.4.2
2,6,7-Trisubstituted Quinoxaline MCR

In 2009, the three-component condensation of *o*-phenylenediamines (**13**), aldehydes, and TosMIC (**8a**) to produce 2,6,7-trisubstituted quinoxalines (**14**) was reported [134]. The reaction most likely proceeds via an initial attack of the deprotonated TosMIC on the *in situ*-generated imine to yield intermediate **16** (Scheme 3.8). Apparently, an attack of the secondary amine in **15** on the terminal carbon of the isocyanide to produce the corresponding imidazole (vL-3CR; see Scheme 3.6) is not favored in this case. Rather, 6-*endo-tet* cyclization by the displacement of Ts$^-$ by the primary aromatic amine, followed by HCN elimination and oxidation, afforded product **14**. Unfortunately, only reaction entries using aromatic aldehydes and the parent TosMIC input were reported. However, the straightforward preparation of otherwise difficult-to-access quinoxalines in moderate to high yield (46–91%), in combination with the unusual reactivity of TosMIC (contributing only CH to the product), makes this MCR appealing for further investigations.

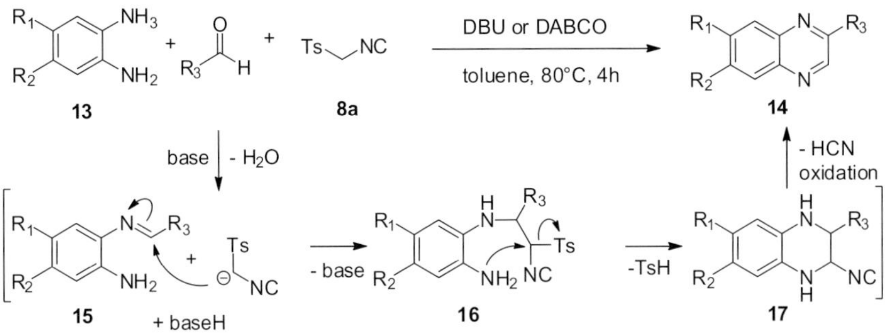

Scheme 3.8 Formation of 2,6,7-trisubstituted quinozalines (**14**) from the MCR between o-phenylenediamines (**13**), aldehydes, and TosMIC (**8a**).

3.4.3
4,5-Disubstituted Oxazole MCR

In 2009, an elegant multicomponent combination of two reactions originally developed by van Leusen leading to 4,5-disubstituted oxazoles (**19**) was reported [135]. The MCR involves the selective monoalkylation of TosMIC (**8a**) followed by the formal cycloaddition with an aldehyde (Scheme 3.9). Although dialkylation is a problem often observed with TosMIC, the use of K_2CO_3 as the base in an ionic liquid smoothly afforded the monoalkylated TosMIC derivative within a 12 h period. However, because similar basic reaction conditions are required for the van Leusen oxazole synthesis, the reaction could be carried out in the same pot to furnish 4,5-disubstituted oxazoles (**19**) in ≥75% yield. The scope with respect to the aldehyde input proved very broad, as the included aliphatic, aromatic and heteroaromatic aldehydes were all compatible with the reaction. Both, primary and secondary aliphatic halides (X = Cl, Br, I) could be used, although since an S_N2-type reaction is involved, the tertiary halides were clearly unsuccessful. Anhydrous conditions were not required, which further simplified the experimental procedure, while the ionic liquid could be reused at least five times without any significant loss of yield (Scheme 3.9).

3.4.4
Nitropyrrole MCR

The reactivity of TosMIC and α-isocyano esters towards nitroalkenes, as investigated by van Leusen [119, 120] and Zard [97, 99], respectively, has also been modified to a 3CR (Scheme 3.10) [136]. In this MCR approach towards nitropyrroles (**23**), TosMIC is treated with 2.0 equiv. of nBuLi[2] at −78 °C, followed by the addition of a chloroformate (**21**) to generate α-deprotonated 2-isocyano-2-tosylacetate (**24**). The reaction of **24** with the nitroalkene **22** (room temperature, 2–4 days) resulted

---
2) 2.0 equiv. nBuLi seems important to suppress the cyclodimerization of TosMIC; see: Ref. [125].

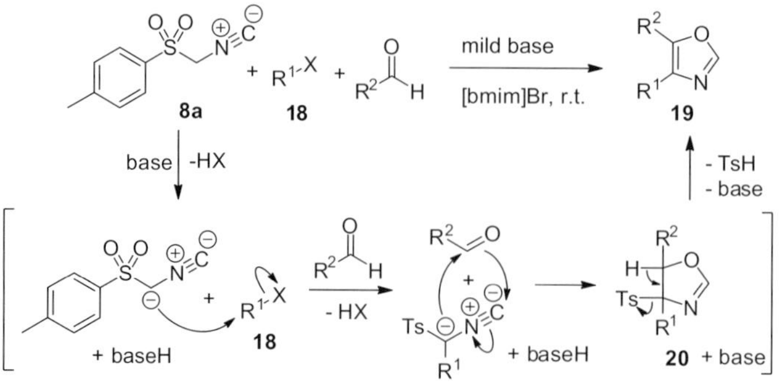

Scheme 3.9 MCR between TosMIC (**8a**), aliphatic halides (**18**), and aldehydes yielding 4,5-disubstituted oxazoles (**19**).

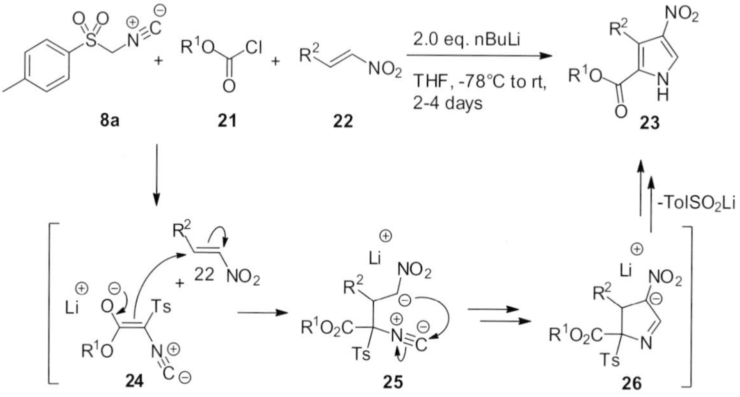

Scheme 3.10 MCR approach toward nitropyrroles (**23**).

in the formation of the corresponding nitropyrroles **23** via the intermediate adduct **25**, cyclization, and TsLi elimination. Although, unfortunately, only ethyl chloroformate (**21a**, R^1 = Et) and 2-nitrostyrene derivatives (R^2 = Ar) were investigated, the corresponding nitropyrroles could be isolated in good yields (63–88%) by using a polymer-assisted catch-and-release work-up.

3.4.5
2,4,5-Trisubstituted Oxazole MCR

3.4.5.1 2,4,5-Trisubstituted Oxazoles

The 3CR towards 2,4,5-trisubstituted oxazoles (**27**; Scheme 3.11) from α-isocyanoamides (**2**), amines, and aldehydes or ketones was first reported by Zhu in 2001 [137], and the reaction has since been used as the basis of various multi-

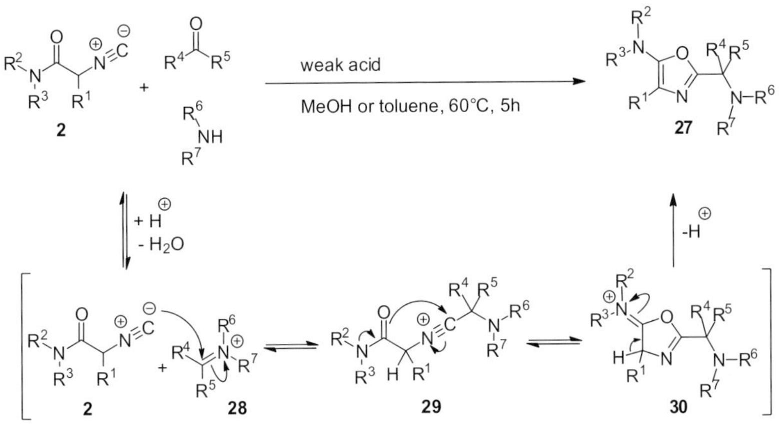

Scheme 3.11 The trisubstituted oxazole MCR.

component approaches towards diverse heterocyclic scaffolds [137–147]. Initially, the reaction employed tertiary isocyanoamides, where $-NR^1R^2$ was typically a morpholino group, although other dialkylamino groups can also be used. The mechanism of this MCR most likely involves an attack of the terminal isocyanide carbon on the (pre-formed) imine to produce intermediate **29**, which subsequently undergoes intramolecular cyclization (see also Ref. [148] and Scheme 3.11). The isocyanide R^3 substituent may be both aromatic and aliphatic, while aldehydes as well as ketones can be used as the carbonyl input. This MCR is compatible with both primary and secondary amines, although the isolated yields are typically lower using primary amines (40–75% compared to 51–96% for secondary amines). In contrast to most other MCRs involving α-acidic isocyanides, the reaction is catalyzed by weak Brønsted acids such as NH_4Cl, Py•HCl or Et_3N•HCl by the faster formation of iminium ion **28** (with the use of secondary amines) or activation of the aldimine (in the case of primary amines). Both, toluene and MeOH can be used as the solvent, and the inputs can be used in equimolar amounts to simplify the work-up procedure.

Interestingly, the use of primary isocyanoamides (**31**) under otherwise identical conditions did not afford the expected oxazoles (**27**, $R^2 = R^3 = H$), but rather the corresponding N-(cyanomethyl)amides (**32**) [149]. The unexpected formation of **32** could be rationalized by an attack of the isocyanide terminal carbon on the *in situ*-generated iminium ion (**28**), followed by cyclization and proton abstraction to form the 5-iminooxazoline intermediate **33** (Scheme 3.12); this could, in principle, undergo tautomerization to form the aromatic primary 5-aminooxazole. Instead, however, proton abstraction at the exocyclic imine moiety followed by ring opening occurs to afford **32**.

Zhu *et al.* reported the first application of an isocyanoester in the MCR towards 2,4,5-trisubstituted oxazoles in 2007. Stirring the highly acidic methyl α-isocyano *p*-nitrophenyl acetate (**34a**, $R^1 = p\text{-}NO_2$) in the presence of an amine and carbonyl

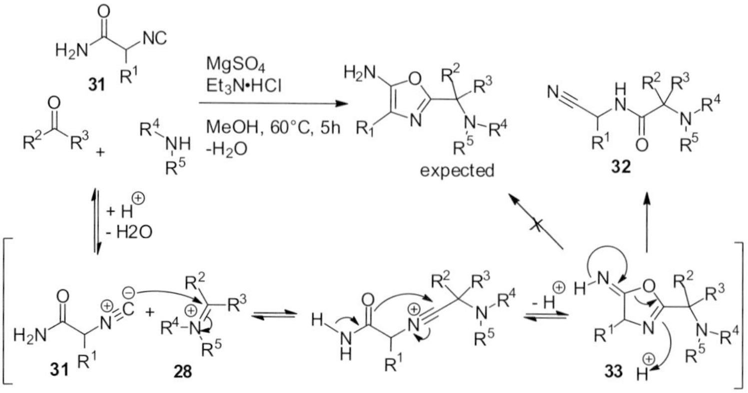

Scheme 3.12 Formation of N-(cyanomethyl)amides in the MCR between primary isocyanoamides, aldehydes/ketones, and amines.

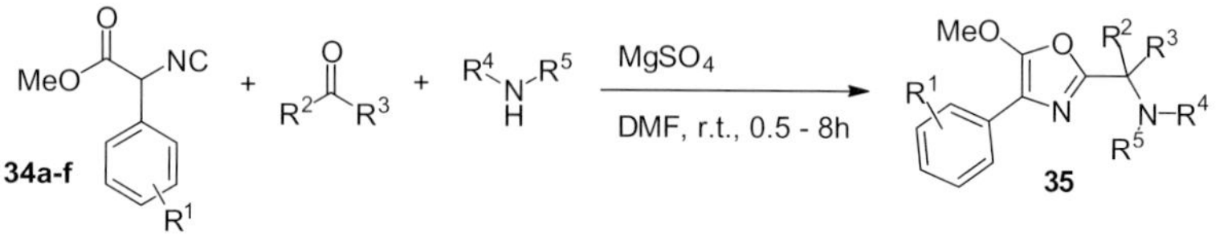

Scheme 3.13 Scope of isocyano esters in the MCR towards trisubstituted oxazoles.

34a: R^1 = p-NO$_2$, 34b: R^1 = H, 34c: R^1 = o-Cl, 34d: R^1 = m-Cl, 34e: R^1 = p-Cl, 34f: R^1 = p-OMe

component led to the production of 5-methoxyoxazole **35** (Scheme 3.13) [36]. Again, a wide variety of aldehydes and amines could be combined to produce the corresponding oxazoles (**35**, R^1 = p-NO$_2$) in moderate to excellent yield (23–97%). Later studies conducted by the present authors extended the range of compatible isocyanide inputs to less-acidic α-aryl α-isocyano esters (**34b–f**), by switching to DMF as the solvent (Scheme 3.13) [150]. In addition, ketones were shown to be compatible inputs and preformation of the imine was found to be unnecessary. Unfortunately, when the pK_a of the α-proton was too high (as in **34g**, R^1 = o,p-dimethoxyphenyl), no reaction was observed [150]. Furthermore, the MCR towards 5-methoxyoxazoles appeared to be limited to the use of α-aryl isocyano esters, as the reaction of α-alkyl isocyano esters, amines and aldehydes or ketones typically led to the corresponding 2-imidazolines (see Section 3.4.6) [151].

3.4.5.2 Variations on the 2,4,5-Trisubstituted Oxazole MCR

In 2003, Ganem et al. applied their single reactant replacement (SRR) strategy to the 5-aminooxazole MCR to develop variations of this reaction [152]. Substitution of the amine input by R$_3$SiCl in the reaction of α-unsubstituted isocyanoamides (**36**), with Zn(OTf)$_2$-mediated catalysis, led to the formation of 2,5-disubstituted

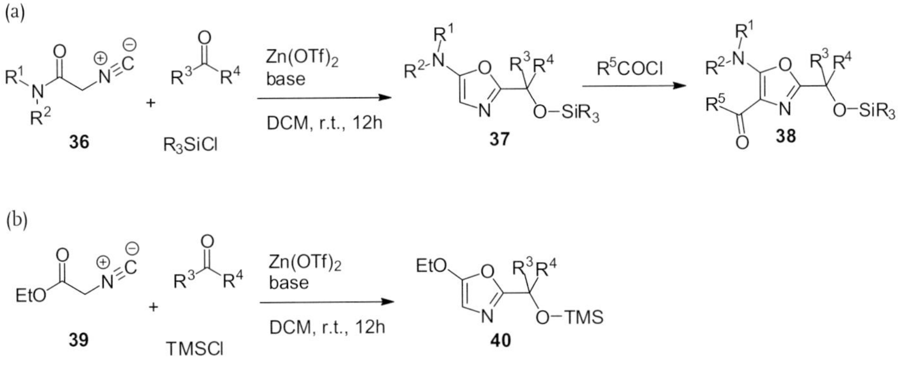

Scheme 3.14 Variations of the trisubstituted oxazole MCR, using the single reactant replacement approach.

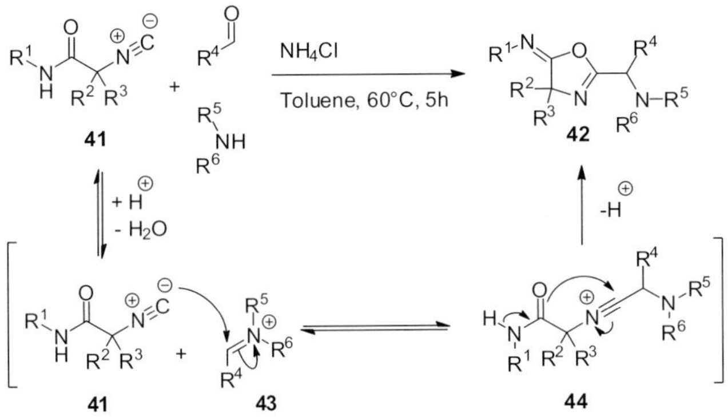

Scheme 3.15 Formation of 5-iminooxazolines (**42**) by the MCR between α,α-disubstituted isocyanoamides (**41**), aldehydes, and amines.

oxazole **37** (Scheme 3.14a) [142]. The dimensionality of the approach was increased by the addition of an acid chloride after complete formation of **37** to produce 2,4,5-trisubstituted oxazoles **38** in a four-component, one-pot process [143]. A similar variation in the MCR, using ethyl isocyanoacetate (**39**) and leading to **40**, had previously been described by Ganem and coworkers (Scheme 3.14b) [153].

In 2007, Tron and Zhu reported the multicomponent synthesis of 5-iminooxazolines (**42**) starting from α,α-disubstituted secondary isocyanoamides (**41**), amines, and aldehydes (Scheme 3.15) [154]. This reaction presumably follows a similar pathway as the 2,4,5-trisubstituted oxazole MCR (see Scheme 3.11). However, due to an absence of α-protons at the isocyanoamide **41**, the aromatization cannot occur. As in the 2,4,5-trisubstituted oxazole MCR, toluene was found to be the optimal solvent, while a weak Brønsted acid promoted the reaction. A

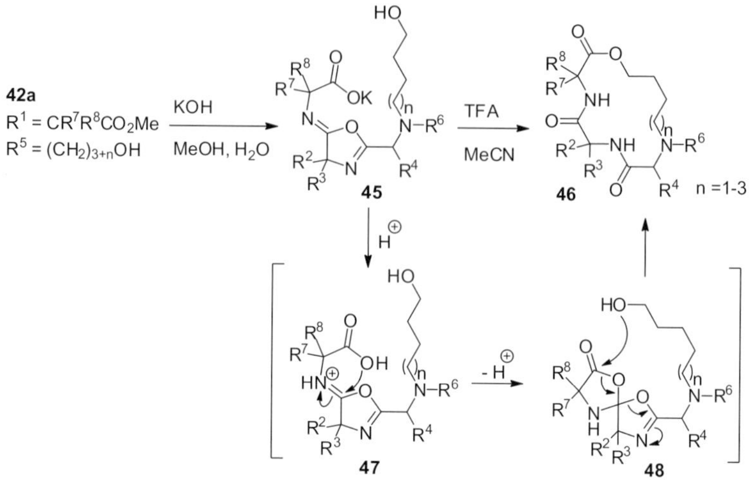

Scheme 3.16 Synthesis of macrocyclodepsipeptides from 5-iminooxazolines.

range of aldehydes and secondary amines was shown to be suitable inputs for the reaction, while several functional groups – including alcohols, carbamates, and esters – were found to be compatible with the reaction. Although NMR analysis of the crude products revealed a near-quantitative formation of the desired products, the isolated yields were moderate (50–68%), though this may have been the result of partial degradation occurring during the purification stages. Unfortunately, the MCR appears to be limited to the use of secondary α,α-disubstituted α-isocyanoacetamides (**44**, R^2 and $R^3 \neq H$), as complex reaction mixtures were obtained when α-monosubstituted isocyanoamides were used.

By incorporating functional groups in the substituents ($R^1 = CR^7R^8CO_2Me$; $R^5 = (CH_2)_{3+n}OH$), the resulting 5-iminooxazolines **42a** could be converted into macrocyclodepsipeptides of type **46** (Scheme 3.16). Hydrolysis of the methyl ester (KOH, MeOH/H$_2$O), followed by protonation of the exocyclic imine and attack of the neighboring carboxyl oxygen, then led to the formation of spirolactone **48**. Subsequent attack of the tethered OH and fragmentation then generated the desired 14-, 15-, and 16-membered macrocycles in 27–54% yield [154].

3.4.5.3 Oxazole MCR and *In-Situ* Domino Processes

The MCR between tertiary isocyanoamides, aldehydes, and amines can be combined with *in-situ* intramolecular domino processes to obtain various heterocyclic scaffolds of high molecular diversity and complexity (Scheme 3.17). Most of these domino processes are based on the reactivity of the oxazole ring towards C=C or C≡C bonds in hetero Diels–Alder reactions, followed by ring-opening reactions to produce highly variable complex heterocycles.

For example, pyrrolo[3,4-b]pyridine-5-ones (**50**) can be obtained via a four-component domino process which combines the original 3CR with acylation by

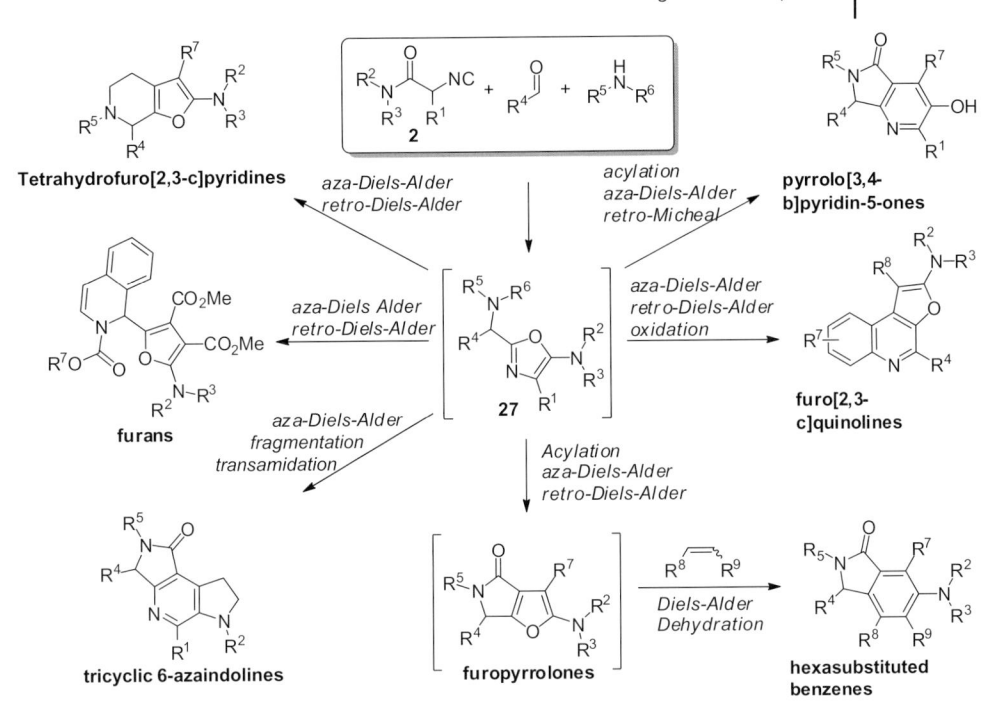

Scheme 3.17 Selected examples for scaffold diversity obtained by the trisubstituted oxazole MCR, followed by intramolecular tandem processes.

an acryloyl chloride derivative (49). After complete formation of the intermediate oxazole 27, the acylating agent 49 is added at 0 °C to produce intermediate 51. Subsequent heating at 110 °C for 12 h triggers an intramolecular *aza*-Diels–Alder reaction to afford 52 (Scheme 3.18). A subsequent base-induced *retro*-Michael reaction and amine elimination then leads to intermediate 53, which immediately tautomerizes to the pyrrolo[3,4-b]pyridine-5-ones 50 [137, 139, 146].

The use of propiolyl chlorides (54, or the corresponding pentafluorophenyl esters) as acylating agent under similar reaction conditions leads to furopyrrolones 55 (Scheme 3.19). Acylation in this case leads to 58, which undergoes an *aza*-Diels–Alder reaction to yield intermediate 59, which in turn undergoes a *retro*-Diels–Alder reaction to afford the furopyrrolone (55). Although the furopyrrolones obtained via this triple domino sequence can be isolated (one example 55a: NR^2R^3 = morpholine, R^4 = cHex, R^5 = nBu, R^6 = Ph; >95%), a second domino sequence towards hexasubstituted benzene derivatives (57) could be initiated by the addition of an alkene (Diels–Alder reaction, dehydration). Because of the similar reaction conditions required for the formation of 55 and 57, both reactions can be combined in a one-pot, double five-component domino process towards hexasubstituted benzenes 57 [141].

The introduction of the *aza*-dienophile on the isocyanoacetamide component 2 provides access to 6-azaindolines 61 (Scheme 3.20) [155]. After completion of the

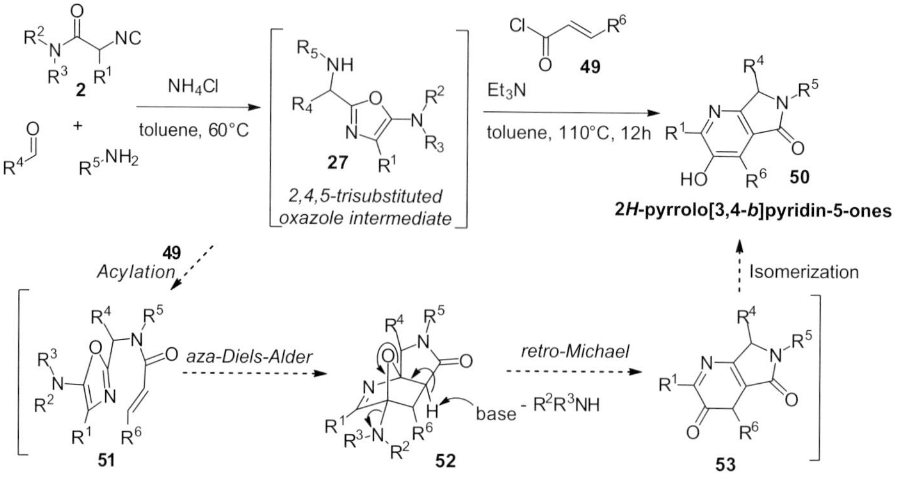

Scheme 3.18 Proposed reaction mechanism for the construction of pyrrolo[3,4-b]pyridine-5-ones (**50**) from the MCR between isocyanoamides, aldehydes, amines, and acryloyl chloride derivatives.

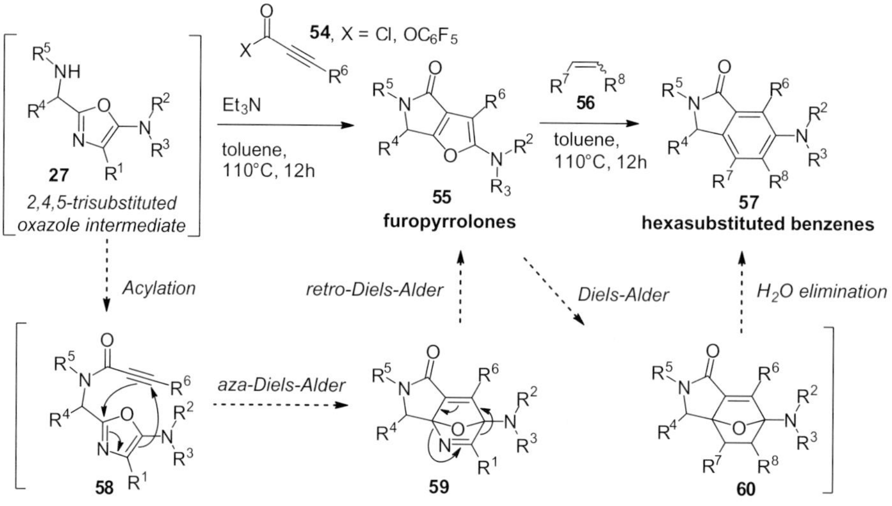

Scheme 3.19 Suggested reaction mechanism for the construction of furopyrrolones (**55**) and hexasubstituted benzenes (**57**), based on the trisubstituted oxazole MCR.

initial 3CR, oxazole intermediate **27** undergoes an intramolecular *aza*-Diels–Alder reaction affording the *oxa*-bridged intermediate **63**; subsequent fragmentation with a loss of water finally furnishes the 6-azaindolines **61**. The use of primary amines (R^6 = H) in combination with R^3 = CO_2Me adds one more step to the domino sequence by facile intramolecular transamidation, affording tricyclic 6-azaindolines (**62**).

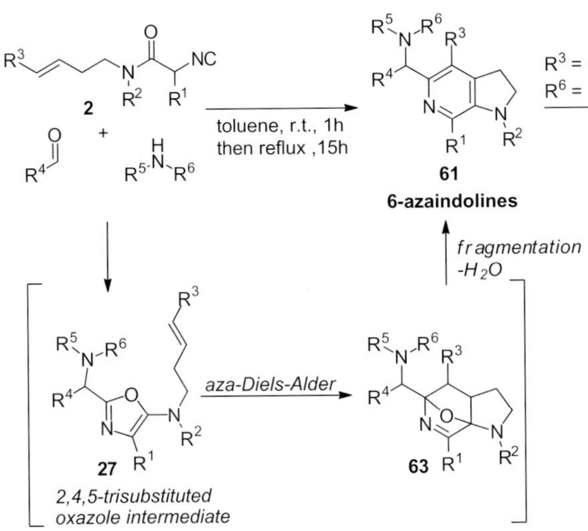

Scheme 3.20 Suggested reaction mechanism for the construction of (tricyclic) 6-azaindolines (**62**), based on the trisubstituted oxazole MCR.

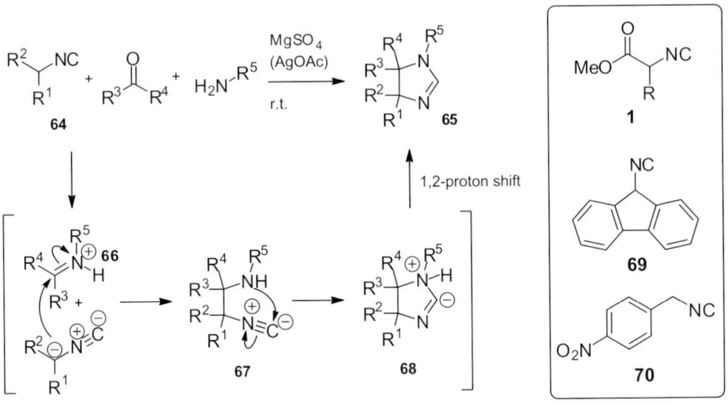

Scheme 3.21 Proposed reaction mechanism for the 2-imidazoline MCR.

3.4.6
2-Imidazoline MCR

In 2003, a three-component reaction of α-acidic isocyanides (**64**), primary amines, and aldehydes or ketones to produce highly substituted 2-imidazolines (**65**) was reported by the present authors [156], based on pioneering studies conducted by Schöllkopf during the 1970s [75]. The mechanism for this MCR most likely involves a Mannich-type addition of α-deprotonated isocyanide to (protonated) imine **66** and a subsequent cyclization/1,2-proton shift (Scheme 3.21). However, a concerted cycloaddition of **66** and deprotonated **64** to produce **65** cannot be excluded.

The reaction proceeds smoothly with methyl α-phenyl isocyanoacetate (**1**, R = Ph) and 9-fluorenyl isocyanide (**69**), in combination with a wide range of aldehydes and amines in CH_2Cl_2 [156]. Further reaction optimization [149, 150, 157, 158] indicated that ketones are also reactive inputs, and that preformation of the imine is not required. Furthermore, the 3CR could be performed in a wide range of solvents, while many functional groups in the inputs—including terminal alkenes, alcohols, esters, and even aliphatic isocyanides—would be tolerated. Moreover, the scope with respect to the isocyanide input was expanded to include *p*-nitrobenzyl isocyanide (**70**) [157] and other (less-acidic) α-substituted isocyano esters (**1**, R = H, Me, *i*Bu, *i*Pr) [158]. It was also shown that the reaction could be catalyzed by silver(I) salts, such as AgOAc. In this case, the Ag^+ most likely coordinates to the terminal NC carbon atom, leading to an increased α-acidity and electrophilicity of the isocyanide terminal carbon (Scheme 3.22). Remarkably, unlike most other reactions of α-acidic isocyanides, no additional base or acid is required for this 3CR; rather, the imine intermediate is probably sufficiently basic to deprotonate the isocyanide.

The scope of isocyanide inputs was even further increased to include also tertiary isocyanoamides (**2**) and primary isocyanoamides (**31**; see Scheme 3.22). As dis-

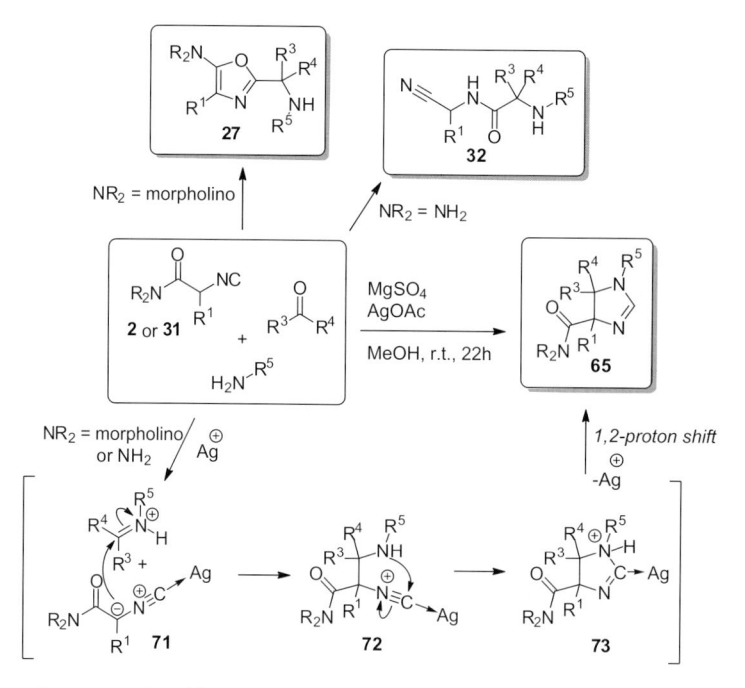

Scheme 3.22 Possible reaction paths for the reaction between α-acidic isocyanoamides (**2** or **31**), primary amines, and carbonyl components, including the Ag^+-catalyzed formation of 2-imidazolines (**65**).

cussed above, these classes of α-acidic isocyanides typically react with imines to produce 2,4,5-trisubstituted oxazoles (**27**; Scheme 3.11) [137–147] and *N*-(cyanomethyl)amides (**32**; Scheme 3.12) [149], respectively. However, the addition of only 2 mol% AgOAc (or CuI), in combination with a slow addition of the isocyanide to the imine, produced the corresponding 2-imidazolines in moderate to very good yields (39–91%; Scheme 3.22). This dramatic change in the reaction outcome to favor 2-imidazoline formation under these reaction conditions can be explained by formation of the (silver/copper-coordinated) isocyanide anion **71** (reactive species). Furthermore, AgOAc may block the formation of oxazoles (**27**) and *N*-(cyanomethyl)amides (**32**) by coordination of the isocyanide carbon to Ag$^+$, thus reducing its nucleophilicity. α-Alkyl-substituted isocyanoamides (**2** or **31**, R^1 = alkyl), however, could not be converted to the corresponding 2-imidazolines, most likely due to the increased pK_a of the α-proton. Thus, the MCR between α-acidic isocyanoamides, aldehydes or ketones, and primary amines could be directed towards three alternative reaction paths by careful selection of the substrates and experimental procedures. This strategy provides a valuable approach to the rapid generation of molecular diversity, allowing the variation of at least four diversity points in three distinct scaffolds.

The MCR toward 2-imidazolines (**65**) has found application in the construction of *N*-heterocyclic carbene (NHC) complexes (**74**). N3-Alkylation, followed by deprotonation at C2 with a strong base (NaH or KO*t*Bu), resulted in the formation of the free carbene species which could be trapped and isolated as the corresponding Ir or Rh complexes [159]. The corresponding *in situ*-generated Ru complexes were shown to serve as highly active and selective catalysts for the transfer hydrogenation of furfural to furfurol, using *i*PrOH as a hydrogen source [160].

Although the 3CR affords highly substituted 2-imidazolines (**65**, R^1–R^5), variation at C2 is not possible; however, in order to overcome this limitation, a further elaboration of **65** towards C2-arylated 2-imidazolines (**75**) was also developed. Here, the key steps involved oxidation (or thionation) at C2 and Liebeskind–Srogl crosscoupling of the resulting cyclic thioureas with arylboronic acids. The final products resembled the promising anticancer agents known as Nutlins [161], and could be prepared in good overall yields (23–35%). The high overall yields and appendage diversity of the products make this procedure amenable to the library synthesis of potential p53–HDM2 interaction inhibitors [162, 163] (Scheme 3.23).

3.4.6.1 2-Imidazoline MCR in the Union of MCRs

By virtue of the high functional group tolerance and high conversions generally observed (>90%), the 2-imidazoline MCR could be combined with other MCRs in the same pot to generate higher-order MCRs (referred to as the "union of MCRs" [164, 165]). In addition, the broad solvent compatibility observed for the 2-imidazoline MCR allows selection of the optimal solvent for the follow-up MCR. The most straightforward approach to providing such combinations of MCRs is the incorporation of a functional group in one of the inputs of the primary MCR that does not participate in the reaction, but does react as one of the components in the secondary MCR. For example, the use of an amino acid as the amine input

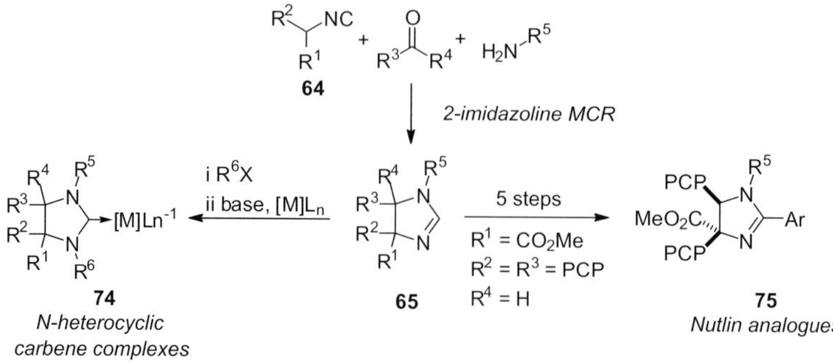

Scheme 3.23 Application of the 2-imidazoline MCR in the synthesis of N-heterocyclic carbene complexes (**74**) and Nutlin analogues (**75**).

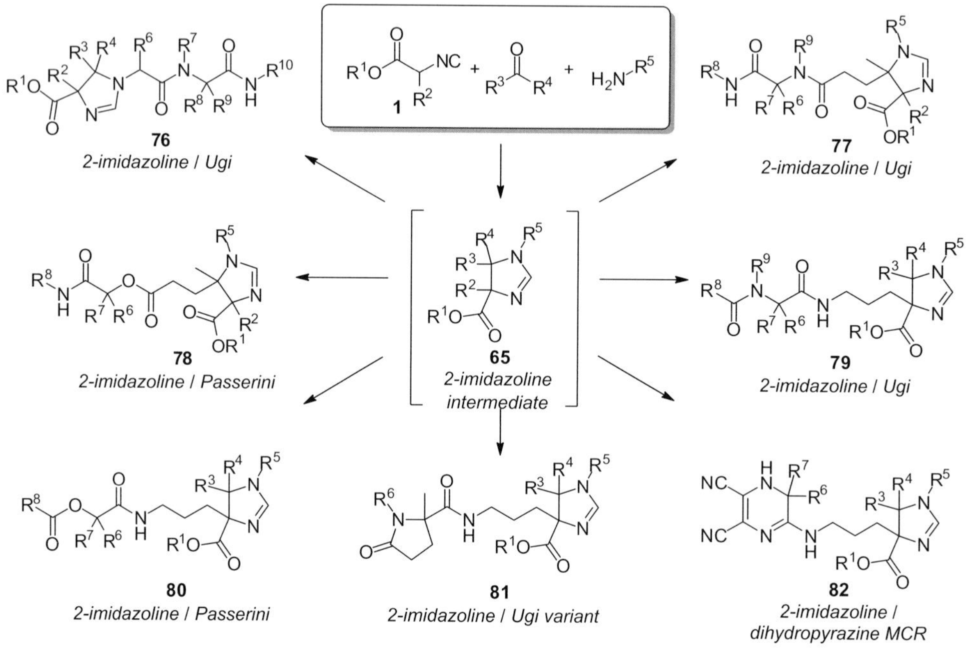

Scheme 3.24 Generation of higher-order MCRs, based on the 2-imidazoline MCR.

(R^5 = CHR^6CO_2H) under basic conditions in the 2-imidazoline MCR results in the formation of an imidazoline carboxylate (**65**, R^5 = $CHR^6CO_2^-$) as an intermediate, which could subsequently be used as the carboxylic acid input in an Ugi 4CR after protonation [166]. The united 2-imidazoline/Ugi products (**76**) generated by this formal 6CR were isolated in 38–62% yield, which was excellent considering the number of bond formations (93% yield per bond formation). Moreover, this approach allows variation on no less than nine positions, in a single operation (Scheme 3.24).

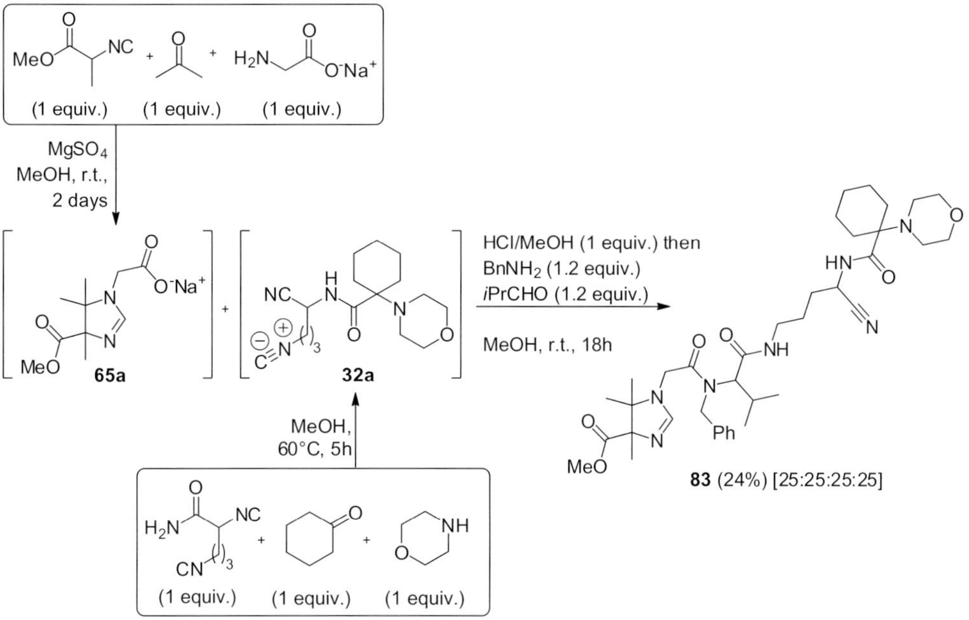

Scheme 3.25 One-pot 8CR based on three sequential MCRs.

A similar approach using levulinic acid as the carbonyl component allowed the intermediate 2-imidazoline (**65**, R^3 = Me, R^4 = $(CH_2)_2CO_2H$) to react as the carboxylic acid component in an Ugi 4CR (to produce **77**) or a Passerini 3CR (to produce **78**) (32–58% yield). The introduction of an aliphatic isocyanide group by using a diisocyanide (R^2 = $(CH_2)_3NC$) allowed the union of the 2-imidazoline 3CR with Ugi, Passerini, and other isocyanide-based MCRs at the C4 position of the 2-imidazoline (**79–82**; 41–78% yield). Similarly, the use of a diisocyanide (**31**, R^1 = $(CH_2)_3NC$) also allowed union of the N-(cyanomethyl)amide MCR with various isocyanide-based MCRs [149]. Finally, the first example of a triple MCR process was developed which resulted in an 8CR affording **83**, based on 2-imidazoline (**65**) and N-(cyanomethyl)amide (**32**) MCRs united with the Ugi-4CR (Scheme 3.25).

3.4.7
Dihydropyridone MCR

In 2006, the present authors' group reported a novel MCR based on the unusual reactivity of α-isocyano esters (**1**) towards 1-azadienes (**84**) generated *in situ* from phosphonates, nitriles, and aldehydes [167]. Remarkably, the resulting dihydropyridone products (**85**) retained the isocyanide functionality at C3. The formation of the 3-isocyano-3,4-dihydro-2-pyridone scaffold can be explained by the Michael

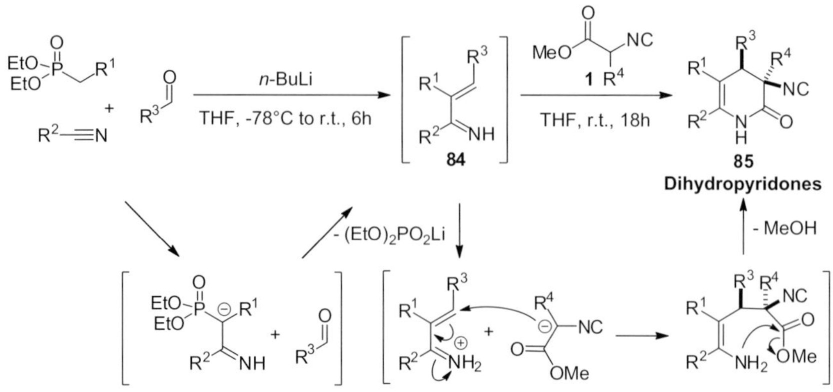

Scheme 3.26 Proposed reaction mechanism for the dihydropyridone 4CR.

addition of the α-deprotonated isonitrile (**1**) to the *in situ*-generated (protonated) 1-azadiene (**84**), followed by lactamization via an attack of the intermediate enamine on the ester function. Although the isocyano functionality is, in principle, not required for the formation of these dihydropyridones (**85**), all attempts to apply other α-acidic esters (e.g., malonates, α-nitro esters and α-cyano esters) resulted in either no conversion or low yields of the corresponding dihydropyridone derivatives [168] (Scheme 3.26).

The scope of the aldehyde inputs includes a wide range of (hetero)aromatic aldehydes, as well as α,β-unsaturated aldehydes. With the use of aliphatic aldehydes, side products due to aldol condensations were observed. The allowed R^4 substituents included Ph, H, Me, Et, Bn, *i*Bu, and *i*Pr, while the scope with respect to the nitrile input included aromatic, heteroaromatic and (hindered) aliphatic nitriles. However, the use of primary aliphatic nitriles should be avoided.

Because of the retained isocyanide functionality, the dihydropyridone MCR product **85** can be used in various follow-up (multicomponent) reactions. For example, the Passerini reaction between **85**, a carboxylic acid, and an aldehyde or ketone produced a number of dihydropyridone-based conformationally constrained depsipeptides **86** [169]. The subsequent Passerini reaction could also be performed in the same pot, resulting in a novel 6CR towards these complex products containing up to seven points of variation. The reaction of **85** with an aliphatic aldehyde or ketone and a primary or secondary amine resulted in the isolation of dihydrooxazolopyridines (DHOPs, **87**) [170], in analogy to the formation of iminooxazolines (**42**; Scheme 3.15) as reported by Tron and Zhu (see Scheme 3.15) [154]. These DHOPs (**87**), which represent a new synthetic scaffold, could be isolated in good to excellent yield (62–100%) for a wide variety of reaction inputs. In addition, owing to their structural similarities, the DHOPs can be considered as promising alternatives for the bioactive oxazolopyridines.

Remarkably, attempts to the direct Ugi reaction of **85** led predominantly to the formation of DHOPs (**87**) [170], although by applying an MCR/alkylation/MCR

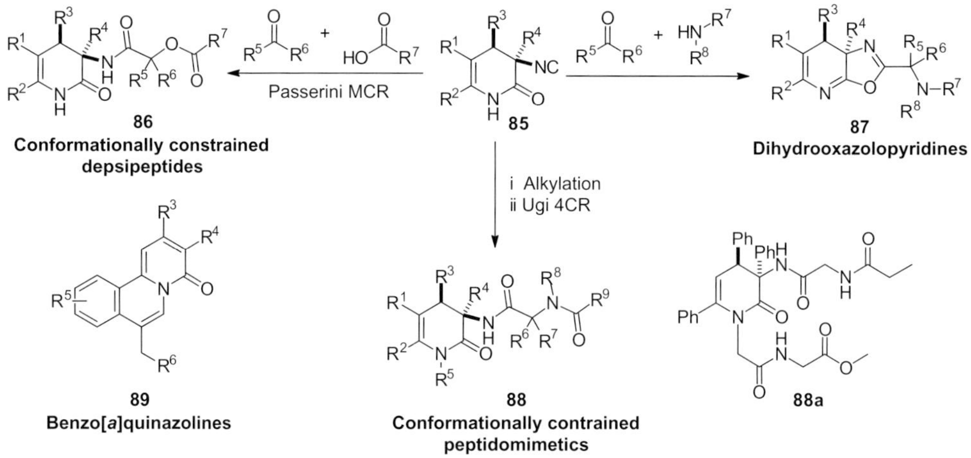

Scheme 3.27 Application of 3-isocyano-3,4-dihydro-2-pyridones (**85**) in further (multicomponent) approaches to increase the molecular complexity and diversity, resulting in conformationally constrained depsipeptides (**86**), dihydrooxazolopyridines (**87**), and conformationally constrained peptidomimetics (**88**), and benzo[a]quinazolines (**89**).

strategy the Ugi products (**88**) could be isolated in high yields [171]. The conformationally constrained peptidomimetics obtained allowed an unprecedented diversification, and ultimately were evaluated for their turn-inducing properties, indicating an open-turn structure for **88a**.

Finally, it was shown that the isocyanide readily undergoes β-elimination upon treatment with a strong base (NaH), which can be combined in one pot with N1-alkylation. When R_1 was an o-bromoaryl group and the N_1 substituent was allylic, a subsequent intramolecular Heck reaction afforded highly substituted benzo[a]quinazolines of type **89**, which resembled naturally occurring alkaloids of the protoberberine class and displayed interesting fluorescence properties [172] (Scheme 3.27).

3.5
Conclusions

The unique reactivity of isocyanides has led to the development of a variety of synthetically useful novel reactions, including many MCRs. α-Acidic isocyanides also display a unique and versatile reactivity that is distinctly different from the classical α-addition reactivity of "normal" isocyanides. Indeed, their use has led to the new class of versatile isocyanide-based MCRs that have been described in this chapter. From only three basic classes of α-acidic isocyanides, a wide range of novel reaction types have been discovered during the past decade, inspired by the early studies of Schöllkopf and van Leusen. Consequently, many highly substituted and/or complex heterocyclic scaffolds can be synthesized in straightforward fashion. Moreover, these usually very robust reactions can easily be combined into very short reaction sequences with other complexity-generating reactions, such as cycloadditions or even additional MCRs. This approach has led to yet other classes of heterocyclic motifs, offering exciting opportunities for the library design of highly functionalized heterocyclic scaffolds. Indeed, it is envisioned that these MCRs, employing α-acidic isocyanides, will find widespread application not only in medicinal chemistry and chemical biology but also in homogeneous catalysis.

References

1 (a) Dömling, A. and Ugi, I. (2000) Multikomponentenreaktionen mit Isocyaniden. *Angew. Chem.*, **112** (18), 3300–3344; (b) Dömling, A. and Ugi, I. (2000) Multicomponent reactions with isocyanides. *Angew. Chem. Int. Ed.*, **39** (18), 3168–3210.

2 Hulme, C. and Gore, V. (2003) Multi-component reactions: emerging chemistry in drug discovery. *Curr. Med. Chem.*, **10** (1), 51–80.

3 Orru, R.V.A. and De Greef, M. (2003) Recent advances in solution-phase multicomponent methodology for the synthesis of heterocyclic compounds. *Synthesis*, (10), 1471–1499.

4 Jacobi von Wangelin, A., Neumann, H., Gördes, D., Klaus, S., Strübing, D., and Beller, M. (2003) Multicomponent coupling reactions for organic synthesis: chemoselective reactions with amide–aldehyde mixtures. *Chem. Eur. J.*, **9** (18), 4286–4294.

5 Nair, V., Rajesh, C., Vinod, A.U., Bindu, S., Sreekanth, A.R., Mathen, J.S., and Balagopal, L. (2003) Strategies for heterocyclic construction via novel multicomponent reactions based on isocyanides and nucleophilic carbenes. *Acc. Chem. Res.*, **36** (12), 899–907.

6 Zhu, J. (2003) Recent developments in the isonitrile-based multicomponent synthesis of heterocycles. *Eur. J. Org. Chem.*, (7), 1133–1144.

7 Balme, G., Bossharth, E., and Monteiro, N. (2003) Pd-assisted multicomponent synthesis of heterocycles. *Eur. J. Org. Chem.*, (21), 4101–4111.

8 Simon, C., Constantieux, T., and Rodriguez, J. (2004) Utilisation of 1,3-dicarbonyl derivatives in multicomponent reactions. *Eur. J. Org. Chem.*, (24), 4957–4980.

9 (a) Ramón, D.J. and Yus, M. (2005) Neue Entwicklungen in der asymmetrischen mehrkomponentenreaktion. *Angew. Chem.*, **117** (11), 1628–1661; (b) Ramón, D.J. and Yus, M. (2005) Asymmetric multicomponent reactions (AMCRs): the new frontier. *Angew. Chem. Int. Ed.*, **44** (11), 1602–1634.

10 Tempest, P. (2005) Recent advances in heterocycle generation using the efficient Ugi multiple-component condensation reaction. *Curr. Opin. Drug. Discov. Dev.*, **8** (6), 776–788.

11 Dömling, A. (2006) Recent developments in isocyanide based multicomponent reactions in applied chemistry. *Chem. Rev.*, **106** (1), 17–89.

12 Godineau, E. and Landais, Y. (2009) Radical and radical–ionic multicomponent processes. *Chem. Eur. J.*, **15** (13), 3044–3055.

13 Strecker, A. (1850) Ueber die künstliche Bildung der Milchsäure und einen neuen, dem Glycocoll homologen Körper. *Liebigs Ann. Chem.*, **75** (1), 27–45.

14 Strecker, A. (1854) Ueber einen neuen aus Aldehyd–Ammoniak und Blausäure entstehenden Körper. *Ann. Chem. Pharm.*, **91** (3), 349–351.

15 Hantzsch, A. (1882) Condensationsprodukte aus

Aldehydammoniak und Ketoniartigen Verbindungen. *Chem. Ber.*, **14** (2), 1637–1638.
16 Biginelli, P. (1891) Ueber Aldehyduramide des Acetessigäthers. *Ber. Dtsch Chem. Ges.*, **24** (1), 1317–1319.
17 Biginelli, P. (1891) Ueber Aldehyduramide des Acetessigäthers. II. *Ber. Dtsch Chem. Ges.*, **24** (2), 2962–2967.
18 Mannich, C. and Krosche, W. (1912) Ueber ein Kondensationsprodukt aus Formaldehyd, Ammoniak und Antipyrin. *Arch. Pharm.*, **250** (1), 647–667.
19 Passerini, M. and Simone, L. (1922) *Gazz. Chim. Ital.*, **52**, 126–129.
20 Passerini, M. and Simone, L. (1922) *Gazz. Chim. Ital.*, **52**, 181–189.
21 Ugi, I. and Meyr, R. (1958) Neue Darstellungsmethode für Isonitrile. *Angew. Chem.*, **70** (23), 702–703.
22 Ugi, I., Meyr, R., and Steinbrückner, C. (1959) Versuche mit Isonitrilen. *Angew. Chem.*, **71** (11), 386.
23 Lieke, W. (1859) Über das Cyanallyl. *Liebigs Ann. Chem.*, **112** (3), 316–321.
24 For the first isolated example of an α-isocyano ester, see: Ugi, I., Betz, W. (1960) DE1177146; see also: *Chem. Abstr.*, **61**, 14536e.
25 For the first isolated example of an α-acidic isocyano amide, see: Beecham Group Ltd (1974) FR2260577; see also: *Chem. Abstr.*, **83**, 206251.
26 For the first report of TosMIC, see: van Leusen, A.M., Strating, J. (1970) *Q. Rep. Sulfur Chem.*, **5**, 67.
27 van Leusen, A.M., Boerma, G.J.M., Helmholdt, R.B., Siderius, H., and Strating, J. (1972) Chemistry of sulfonylmethylisocyanides. Simple synthetic approaches to a new versatile chemical building block. *Tetrahedron Lett.*, **13** (23), 2367–2378.
28 Schöllkopf, U., Hoppe, D., and Jentsch, R. (1975) Synthesen mit α-metallierten Isocyaniden, XXIX. Höhere Aminosäuren durch Alkylieren von α-metallierten α-Isocyan-propionsäure- und -essigsäureestern. *Chem. Ber.*, **108** (6), 1580–1592.
29 Matsumoto, K., Suzuki, M., Yoneda, N., and Miyoshi, M. (1977) Alkylation of α-isocyanoacetamides; Synthesis of 1,4,4-trisubstituted 5-oxo-4,5-dihydroimidazoles. *Synthesis*, (4), 249–250.
30 van Leusen, A.M., Bourma, R.J., and Possel, O. (1975) Phase-transfer mono-alkylation of tosylmethylisocyanide. *Tetrahedron Lett.*, **16** (40), 3487–3488.
31 Murakami, M., Hasegawa, N., Tomita, I., Inouye, M., and Ito, Y. (1989) Fluoride catalyzed Michael reaction of α-isocyanoesters with α,β-unsaturated carbonyl compounds. *Tetrahedron Lett.*, **30** (10), 1257–1260.
32 Housseman, C. and Zhu, J. (2006) Mono alkylation of α-Isocyano acetamide to its higher homologues. *Synlett*, (11), 1777–1779.
33 (a) Schöllkopf, U., Hoppe, D., and Jentsch, R. (1971) Höhere Aminosäuren durch Alkylieren von α-metallierten Isocyan-essig- oder -propionsäureestern. *Angew. Chem.*, **83** (5), 357–358; (b) Schöllkopf, U., Hoppe, D., and Jentsch, R. (1971) Higher amino acids by alkylation of α-metalated isocyano-acetic or -propionic esters. *Angew. Chem. Int. Ed.*, **10** (5), 331–333.
34 Ito, Y., Higuchi, N., and Murakami, M. (1988) Claisen rearrangement of allylic α-isocyano-esters – regioselective allylation of α-isocyanoesters at the α-carbon. *Tetrahedron Lett.*, **29** (40), 5151–5154.
35 Ito, Y., Sawamura, M., Matsuoka, M., Matsumoto, Y., and Hayashi, T. (1987) Palladium-catalyzed allylation of α-isocyanocarboxylates. *Tetrahedron Lett.*, **28** (41), 4849–4852.
36 (a) Bonne, D., Dekhane, M., and Zhu, J. (2007) Modulating the reactivity of α-isocyanoacetates: multicomponent synthesis of 5-methoxyoxazoles and furopyrrolones. *Angew. Chem.*, **119** (14), 2537–2540; (b) Bonne, D., Dekhane, M., and Zhu, J. (2007) Modulating the reactivity of α-isocyanoacetates: multicomponent synthesis of 5-methoxyoxazoles and furopyrrolones. *Angew. Chem. Int. Ed.*, **46** (14), 2485–2488.
37 Matsumoto, K., Suzuki, M., and Miyoshi, M. (1973) Synthesis of amino

acids and related compounds. 4. New synthesis of α-amino acids. *J. Org. Chem.*, **38** (11), 2094–2096.

38 Olijnsma, T., Engberts, J.B.F.N., and Strating, J. (1972) *Rec. Trav. Chim. Pays-Bas*, **91**, 209.

39 Hundscheid, F.J.A., Tandon, V.K., Rouwette, P.H.F.M., and van Leusen, A.M. (1987) Synthesis of chiral sulfonylmethyl isocyanides, and comparison of their propensities in asymmetric induction reactions with acetophenones. *Tetrahedron*, **43** (21), 5073–5088.

40 Barrett, A.G.M., Cramp, S.M., Hennessy, A.J., Procopiou, P.A., and Roberts, R.S. (2001) Oxazole synthesis with minimal purification: synthesis and application of a ROMPgel tosmic reagent. *Org. Lett.*, **3** (2), 271–274.

41 Sisko, J., Mellinger, M., Sheldrake, P.W., and Baine, N.H. (1996) An efficient method for the synthesis of substituted TosMIC precursors. *Tetrahedron Lett.*, **37** (45), 8113–8116.

42 Wright, J.J.K., Cooper, A.B., McPhail, A.T., Merrill, Y., Nagabhushan, T.L., and Puar, M.S. (1982) X-Ray crystal structure determination and synthesis of the new isonitrile-containing antibiotics, hazimycin factors 5 and 6. *J. Chem. Soc. Chem. Commun.*, (20), 1188–1190.

43 Urban, R., Marquarding, D., Seidel, P., Ugi, I., and Weinelt, A. (1977) Notiz zur Synthese optisch aktiver α-Isocyancarbonsäure-Derivate für Peptidsynthesen mittels Vier-Komponenten-Kondensation (4 CC). *Chem. Ber.*, **110** (5), 2012–2015.

44 Waki, M. and Meienhofer, J. (1977) Efficient preparation of N^α-formylamino acid *tert*-butyl esters. *J. Org. Chem.*, **42** (11), 2019–2020.

45 Kamer, P.C.J., Cleij, M.C., Nolte, R.J.M., Harada, T., Hezemans, A.M.F., and Drenth, W. (1988) Atropisomerism in polymers. Screw-sense selective polymerization of isocyanides by inhibiting the growth of one enantiomer of a racemic pair of helices. *J. Am. Chem. Soc.*, **110** (5), 1581–1587.

46 Seebach, D., Adam, G., Gees, T., Schiess, M., and Weigand, W. (1988) Scope and limitations of the TiCl$_4$-mediated additions of isocyanides to aldehydes and ketones with formation of α-hydroxycarboxylic acid amides. *Chem. Ber.*, **121** (3), 507–518.

47 Bowers, M.M., Carroll, P., and Joullié, M.M. (1989) Model studies directed toward the total synthesis of 14-membered cyclopeptide alkaloids: synthesis of prolyl peptides via a four-component condensation. *J. Chem. Soc., Perkin Trans. 1*, (5), 857–865.

48 Owens, T.D., Araldi, G.-L., Nutt, R.F., and Semple, J.E. (2001) Concise total synthesis of the prolyl endopeptidase inhibitor eurystatin A via a novel Passerini reaction–deprotection–acyl migration strategy. *Tetrahedron Lett.*, **42** (36), 6271–6274.

49 Wehner, V., Stilz, H.-U., Osipov, S.N., Golubev, A.S., Sieler, J., and Burger, K. (2004) Trifluoromethyl-substituted hydantoins, versatile building blocks for rational drug design. *Tetrahedron*, **60** (19), 4295–4302.

50 Horwell, D.C., Nichols, P.D., and Roberts, E. (1994) Conformationally constrained amino acids: synthesis of novel 3,4-cyclised tryptophans. *Tetrahedron Lett.*, **35** (6), 939–940.

51 Obrecht, R., Herrmann, R., and Ugi, I. (1985)) Isocyanide synthesis with phosphoryl chloride and diisopropylamine. *Synthesis*, (4), 400–402.

52 Zhu, J., Wu, X., and Danishefsky, S.J. (2009) On the preparation of enantiomerically pure isonitriles from amino acid esters and peptides. *Tetrahedron Lett.*, **50** (5), 577–579.

53 Pifferi, G., LiBassi, G., and Broccali, G. (1976) Semisynthetic β-lactam antibiotics. I. α-Isocyanobenzyl-penicillins and cephalosporins. *Heterocycles*, **4** (4), 759–765.

54 (a) Skorna, G. and Ugi, I. (1977) Isocyanid-Synthese mit Diphosgen. *Angew. Chem.*, **89** (4), 267–268; (b) Skorna, G. and Ugi, I. (1977) Isocyanide synthesis with diphosgene. *Angew. Chem. Int. Ed. Engl.*, **16** (4), 259–260.

55 van der Eijk, J.M., Nolte, R.J.M., Richters, V.E.M., and Drenth, W. (1981) *J. R. Neth. Chem. Soc.*, **100**, 222–226.

56 Nishimura, S., Yasuda, N., Sasaki, H., Matsumoto, Y., Kanimura, T., Sakane, K., and Takaya, T. (1990) Synthesis and β-lactamase inhibitory activity of 3-cyano-3-cephem derivatives. *Bull. Chem. Soc. Jpn*, **63** (2), 412–416.

57 Ikota, N. (1990) Synthetic studies on optically active β-lactams. II. Asymmetric synthesis of β-lactams by [2+2]cyclocondensation using heterocyclic compounds derived from L-(+)-tartaric acid, (S)- or (R)-glutamic acid, and (S)-serine as chiral auxiliaries. *Chem. Pharm. Bull.*, **38** (6), 1601–1608.

58 Creedon, S.M., Crowley, H.K., and McCarthy, D.G. (1998) Dehydration of formamides using the Burgess reagent: a new route to isocyanides. *J. Chem. Soc., Perkin Trans. 1*, (6), 1015–1018.

59 Banfi, L., Guanti, G., Riva, R., Basso, A., and Calcagno, E. (2002) Short synthesis of protease inhibitors via modified Passerini condensation of N-Boc-α-aminoaldehydes. *Tetrahedron Lett.*, **43** (22), 4067–4070.

60 de Witte, P.A.J., Castriciano, M., Cornelissen, J.J.L.M., Scolaro, L.M., Nolte, R.J.M., and Rowan, A.E. (2003) Helical polymer-anchored porphyrin nanorods. *Chem. Eur. J.*, **9** (8), 1775–1781.

61 Zhao, G., Bughin, C., Bienaymé, H., and Zhu, J. (2003) Synthesis of isocyano peptides by dehydration of the N-formyl derivatives. *Synlett*, (8), 1153–1154.

62 Vriezema, D.M., Kros, A., de Gelder, R., Cornelissen, J.J.L.M., Rowan, A.E., and Nolte, R.J.M. (2004) Electroformed giant vesicles from thiophene-containing rod–coil diblock copolymers. *Macromolecules*, **37** (12), 4736–4739.

63 Bughin, C., Zhao, G., Bienaymé, H., and Zhu, J. (2006) 5-aminooxazole as an internal traceless activator of C-terminal carboxylic acid: rapid access to diversely functionalized cyclodepsipeptides. *Chem. Eur. J.*, **12** (4), 1174–1184.

64 Metselaar, G.A., Adams, P.J.H.M., Nolte, R.J.M., Cornelissen, J.J.L.M., and Rowan, A.E. (2007) Polyisocyanides derived from tripeptides of alanine. *Chem. Eur. J.*, **13** (3), 950–960.

65 Schwartz, E., Kitto, H.J., de Gelder, R., Nolte, R.J.M., Rowan, A.E., and Cornelissen, J.J.L.M. (2007) Synthesis, characterisation and chiroptical properties of "click"able polyisocyanopeptides. *J. Mater. Chem.*, **17** (19), 1876–1884.

66 Nunami, K.-I., Suzuki, M., Hayashi, K., Matsumoto, K., Yoneda, N., and Takiguchi, K. (1985) Synthesis and herbicidal activities of α-isocyanocycloalkylideneacetamides. *Agric. Biol. Chem.*, **49** (10), 3023–3028.

67 Nunami, K.-I., Suzuki, M., Matsumoto, K., Yoneda, N., and Takiguchi, K. (1984) Syntheses and biological activities of isonitrile dipeptides. *Agric. Biol. Chem.*, **48** (4), 1073–1076.

68 Takiguchi, K., Yamada, K., Suzuki, M., Nunami, K.-I., Hayashi, K., and Matsumoto, K. (1989) Antifungal activities of α-isocyano-β-phenylpropionamides. *Agric. Biol. Chem.*, **53** (1), 77–82.

69 Soloshonok, V.A. and Hayashi, T. (1994) Gold(I)-catalyzed asymmetric aldol reactions of fluorinated benzaldehydes with an α-isocyanoacetamide. *Tetrahedron: Asymmetry*, **5** (6), 1091–1094.

70 Beck, B., Larbig, G., Mejat, B., Magnin-Lachaux, M., Picard, A., Herdtweck, E., and Dömling, A. (2003) Short and diverse route toward complex natural product-like macrocycles. *Org. Lett.*, **5** (7), 1047–1050.

71 For an excellent review on the synthetic use of tosylmethyl isocyanide (TosMIC) see: van Leusen, D., van Leusen, A.M. (2001) Synthetic uses of tosylmethyl isocyanide (TosMIC). *Org. React.*, **57** (3), 417–679.

72 (a) For a review on applications of α-metalated isocyanides see: Schöllkopf, U. (1977) Neuere Anwendungen α-metallierter Isocyanide in der organischen Synthese. *Angew. Chem.*, **89** (6), 351–360; (b) Schöllkopf, U. (1977) Recent applications of α-metalated isocyanides in organic synthesis. *Angew. Chem. Int. Ed.*, **16** (6), 339–348.

73 Seagusa, T., Ito, Y., Kinoshita, H., and Tomita, S. (1971) Synthetic reactions by complex catalysts. XIX. Copper-catalyzed cycloaddition reactions of isocyanides.

Novel synthesis of Δ¹-pyrroline and Δ²-oxazoline. *J. Org. Chem.*, **36** (22), 3316–3323.

74 Hoppe, D. and Schöllkopf, U. (1972) Synthesen mit α-metallierten Isocyaniden, XV: 4-Äthoxycarbonyl-2-oxazoline und ihre Hydrolyse zu N-Formyl-β-hydroxy-α-aminosäureäthylestern. *Liebigs Ann. Chem.*, **763** (1), 1–16.

75 Meyer, R., Schöllkopf, U., and Böhme, P. (1977) Synthesen mit α-metallierten Isocyaniden, XXXIX. 2-Imidazoline aus α-metallierten Isocyaniden und Schiff-Basen; 1,2-Diamine und 2,3-Diaminoalkansäuren. *Liebigs Ann. Chem.*, (7), 1183–1193.

76 Hayashi, T., Kishi, E., Soloshonok, V.A., and Uozumi, Y. (1996) Erythro-selective aldol-type reaction of N-sulfonylaldimines with methyl isocyanoacetate catalyzed by gold(I). *Tetrahedron Lett.*, **37** (28), 4969–4972.

77 Lin, Y.-R., Zhou, X.-T., Dai, L.-X., and Sun, J. (1997) Ruthenium complex-catalyzed reaction of isocyanoacetate and N-sulfonylimines: stereoselective synthesis of N-sulfonyl-2-imidazolines. *J. Org. Chem.*, **62** (6), 1799–1803.

78 Benito-Garagorri, D., Bocokíc, V., and Kirchner, K. (2006) Copper(I)-catalyzed diastereoselective formation of oxazolines and N-sulfonyl-2-imidazolines. *Tetrahedron Lett.*, **47** (49), 8641–8644.

79 Meyer, R., Schöllkopf, U., Madawinata, K., and Stafforst, D. (1978) Totalsynthese von (±)-2,2-Diethyl-3-methyl- und (±)-2,2-Diethyl-3,6-dimethyl-6-(phenoxyacetamido)penam-3-carbonsäure. *Liebigs Ann. Chem.*, (12), 1982–1989.

80 Suzuki, M., Iwasaki, T., Matsumoto, K., and Okumura, K. (1972) Convenient syntheses of aroylamino acids and α-amino ketones. *Communication*, **2** (4), 237.

81 Suzuki, M., Iwasaki, T., Miyoshi, M., Okumura, K., and Matsumoto, K. (1973) Synthesis of amino acids and related compounds. 6. New convenient synthesis of α-C-acylamino acids and α-amino ketones. *J. Org. Chem.*, **38** (20), 3571–3575.

82 Matsumoto, K., Suzuki, M., Yoneda, N., and Miyoshi, M. (1976) A new, convenient synthesis of 3-amino-4-hydroxy-2-oxo-1,2-dihydroquinoline derivatives via 5-(2-acylaminophenyl)-4-methoxycarbonyl-1,3-oxazoles. *Synthesis*, (12), 805–807.

83 Nunami, K.-I., Suzuki, M., Matsumoto, K., Miyoshi, M., and Yoneda, N. (1979) Reaction of phthalic anhydrides with methyl isocyanoacetate: a useful synthesis of 1,2-dihydro-1-oxoisoquinolines. *Chem. Pharm. Bull.*, **27** (6), 1373–1377.

84 Henneke, K.-W., Schöllkopf, U., and Neudecker, T. (1979) Synthesen mit α-metallierten Isocyaniden, XLII. Bi-, Ter- und Quateroxazole aus α-anionisierten Isocyaniden und Acylierungsmitteln; α-Aminoketone und α,α-Diaminoketone. *Liebigs Ann. Chem.*, (9), 1370–1387.

85 Gerecke, M., Kyburz, E., Borer, R., and Gassner, W. (1994) New tetracyclic derivatives of imidazo[1,5-a][1,4]benzodiazepines and of imidazo[1,5-a]-thieno[3,2-f][1,4]-diazepines. *Heterocycles*, **39** (2), 693–722.

86 Matecka, D., Wong, G., Gu, Z.-Q., Skolnick, P., and Rice, K.C. (1995) *Med. Chem. Res.*, **5** (1), 63–76.

87 Jacobsen, E.J., Stelzer, L.S., TenBrink, R.E., Belonga, K.L., Carter, D.B., Im, H.K., Bin Im, W., Sethy, V.H., Tang, A.H., VonVoigtlander, P.F., Petke, J.D., Zhong, W.-Z., and Mickelson, J.W. (1999) Piperazine imidazo[1,5-a]quinoxaline ureas as high-affinity GABA$_A$ ligands of dual functionality. *J. Med. Chem.*, **42** (7), 1123–1144.

88 Grigg, R., Lansdell, M.I., and Thornton-Pett, M. (1999) Silver acetate catalysed cycloadditions of isocyanoacetates. *Tetrahedron*, **55** (7), 2025–2044.

89 Kanazawa, C., Kamijo, S., and Yamamoto, Y. (2006) Synthesis of imidazoles through the copper-catalyzed cross-cycloaddition between two different isocyanides. *J. Am. Chem. Soc.*, **128** (33), 10662–10663.

90 Bonin, M.-A., Giguere, D., and Roy, R. (2007) N-Arylimidazole synthesis by cross-cycloaddition of isocyanides using

a novel catalytic system. *Tetrahedron*, **63** (23), 4912–4917.

91 Suzuki, M., Moriya, T., Matsumoto, K., and Miyoshi, M. (1982) A new convenient synthesis of 5-amino-1,3-thiazole-4-carboxylic acids. *Synthesis*, (10), 874–875.

92 Solomon, D.M., Rizvi, R.K., and Kaminski, J.J. (1987) Observations on the reactions of isocyanoacetane esters with isothiocyanates and isocyanates. *Heterocycles*, **26** (3), 651–674.

93 Boros, E.E., Johns, B.A., Garvey, E.P., Koble, C.S., and Miller, W.H. (2006) Synthesis and HIV-integrase strand transfer inhibition activity of 7-hydroxy[1,3]thiazolo[5,4-*b*]pyridin-5(4H)-ones. *Bioorg. Med. Chem. Lett.*, **16** (21), 5668–5672.

94 Suzuki, M., Miyoshi, M., and Matsumoto, K. (1974) Synthesis of amino acids and related compounds. 7. Convenient synthesis of 3-substituted pyrrole-2,4-dicarboxylic acid esters. *J. Org. Chem.*, **39** (13), 1980.

95 Matsumoto, K., Suzuki, M., Ozaki, Y., and Miyoshi, M. (1976) The synthesis of 3-substituted pyrrole-2,4-dicarboxylic acid esters: reaction of methyl isocyanoacetate with aldehydes. *Agric. Biol. Chem.*, **40** (11), 2271–2274.

96 Sakai, K., Suzuki, M., Nunami, K.-I., Yoneda, N., Onoda, Y., and Iwasawa, Y. (1980) Syntheses of 3-substituted pyrrole derivatives with antiinflammatory activity. *Chem. Pharm. Bull.*, **28** (8), 2384–2393.

97 Barton, D.H.R. and Zard, S.Z. (1985) A new synthesis of pyrroles from nitroalkenes. *J. Chem. Soc. Chem. Commun.*, 1098–1100.

98 Ono, N., Kawamura, H., and Maruyama, K. (1989) A convenient synthesis of trifluoromethylated pyrroles and porphyrins. *Bull. Chem. Soc. Jpn*, **62** (10), 3386–3388.

99 Barton, D.H.R., Kervagoret, J., and Zard, S.Z. (1990) A useful synthesis of pyrroles from nitroolefins. *Tetrahedron*, **46** (21), 7587–7598.

100 Lash, T.D., Bellettini, J.R., Bastian, J.A., and Couch, K.B. (1994) Synthesis of pyrroles from benzyl isocyanoacetate. *Synthesis*, (2), 170–172.

101 Ono, N., Katayama, H., Nisyiyama, S., and Ogawa, T. (1994) Regioselective synthesis of 5-unsubstituted benzyl pyrrole-2-carboxylates from benzyl isocyanoacetate. *J. Heterocycl. Chem.*, **31** (4), 707–710.

102 Uno, H., Sakamoto, K., Tominaga, T., and Ono, N. (1994) A new approach to β-fluoropyrroles based on the Michael addition of isocyanomethylide anions to α-fluoroalkenyl sulfones and sulfoxides. *Bull. Chem. Soc. Jpn*, **67** (5), 1441–1448.

103 Vicente, M.G.H., Tome, A.C., Walter, A., and Cavaleiro, J.A.S. (1997) Synthesis and cycloaddition reactions of pyrrole-fused 3-sulfolenes: a new versatile route to tetrabenzoporphyrins. *Tetrahedron Lett.*, **38** (20), 3639–3642.

104 Schmidt, W. and Monforts, F.-P. (1997) Synthesis of a novel enantiomerically pure chlorin as a potential subunit for an artificial photosynthetic reaction center. *Synlett*, (8), 903–904.

105 Abel, Y., Haake, E., Haake, G., Schmidt, W., Struve, D., Walter, A., and Monforts, F. (1998) Eine einfache und flexible Synthese von Pyrrolen aus α,β-ungesättigten Sulfonen. *Helv. Chim. Acta*, **81** (11), 1978–1996.

106 Bullington, J.L., Wolff, R.R., and Jackson, P.F. (2002) Regioselective preparation of 2-substituted 3,4-diaryl pyrroles: a concise total synthesis of Ningalin B. *J. Org. Chem.*, **67** (26), 9439–9442.

107 Kamijo, S., Kanazawa, C., and Yamamoto, Y. (2005) Phosphine-catalyzed regioselective heteroaromatization between activated alkynes and isocyanides leading to pyrroles. *Tetrahedron Lett.*, **46** (15), 2563–2566.

108 Kamijo, S., Kanazawa, C., and Yamamoto, Y. (2005) Copper- or phosphine-catalyzed reaction of alkynes with isocyanides. Regioselective synthesis of substituted pyrroles controlled by the catalyst. *J. Am. Chem. Soc.*, **127** (25), 9260–9266.

109 (a) Larionov, O.V. and de Meijere, A. (2005) Vielseitige direkte Synthese von oligosubstituierten Pyrrolen durch Cycloaddition von α-metallierten Isocyaniden an Acetylene. *Angew. Chem.*,

117 (35), 5809–5813; (b) Larionov, O.V. and de Meijere, A. (2005) Versatile direct synthesis of oligosubstituted pyrroles by cycloaddition of α-metalated isocyanides to acetylenes. *Angew. Chem. Int. Ed.*, **44** (35), 5664–5667.

110 (a) Guo, C., Xue, M.-X., Zhu, M.-K., and Gong, L.-Z. (2008) Organocatalytic asymmetric formal [3+2] cycloaddition reaction of isocyanoesters to nitroolefins leading to highly optically active dihydropyrroles. *Angew. Chem.*, **120** (18), 3462–3465; (b) Guo, C., Xue, M.-X., Zhu, M.-K., and Gong, L.-Z. (2008) Organocatalytic asymmetric formal [3+2] cycloaddition reaction of isocyanoesters to nitroolefins leading to highly optically active dihydropyrroles. *Angew. Chem. Int. Ed.*, **47** (18), 3414–3417.

111 Tsunenishi, Y., Ishida, H., Itoh, K., and Ohno, M. (2000) Heterocyclization of [60]fullerene with isocyanides. *Synlett*, (9), 1318–1320.

112 Schöllkopf, U. and Porsch, P.-H. (1973) Synthesen mit α-metallierten Isocyaniden, XXV. 4-Cyan-2-isocyanalkansäure-äthylester, 4-Cyan-2-(formylamino)alkansäure-äthylester und 4-Cyan-5(4)-pyrrolin-2-carbonsäure-äthylester aus α-metallierten Isocyanalkansäure-äthylestern und Acrylnitrilen. *Chem. Ber.*, **106** (10), 3382–3390.

113 Schöllkopf, U. and Hantke, K. (1973) Synthesen mit α-metallierten Isocyaniden, XXIV. α-Isocyanglutarsäure-diäthylester, Glutaminsäure-Derivate und Pyrrolincarbonsäure-äthylester aus α-metalliertem Isocyanessigsäure- oder -propionsäure-äthylester und Acrylsäureestern. *Liebigs Ann. Chem.*, (9), 1571–1582.

114 Possel, O. and van Leusen, A.M. (1977) Synthesis of oxazoles, imidazoles and pyrroles with the use of mono-substituted tosylmethyl isocyanides. *Heterocycles*, **7** (1), 77–80.

115 Horne, D.A., Yakushijin, K., and Buechi, G. (1994) A two-step synthesis of imidazoles from aldehydes via 4-tosyloxazolines. *Heterocycles*, **39** (1), 139–154.

116 Hundscheid, F.J.A., Tandon, V.K., Rouwette, P.H.F.M., and van Leusen, A.M. (1987) *Rec. Trav. Chim. Pays-Bas*, **106**, 159–160.

117 Atkins, J.M. and Vedejs, E. (2005) A two-stage iterative process for the synthesis of poly-oxazoles. *Org. Lett.*, **7** (15), 3351–3354.

118 van Leusen, A.M., Wildeman, J., and Oldenziel, O.T. (1977) Chemistry of sulfonylmethyl isocyanides. 12. Base-induced cycloaddition of sulfonylmethyl isocyanides to C,N double bonds. Synthesis of 1,5-disubstituted and 1,4,5-trisubstituted imidazoles from aldimines and imidoyl chlorides. *J. Org. Chem.*, **42** (7), 1153–1159.

119 van Leusen, A.M., Siderius, H., Hoogenboom, B.E., and van Leusen, D. (1972) A new and simple synthesis of the pyrrole ring system from Michael acceptors and tosylmethylisocyanides. *Tetrahedron Lett.*, **13** (52), 5337–5340.

120 ten Have, R., Leusink, F.R., and van Leusen, A.M. (1996) An efficient synthesis of substituted 3(4)-nitropyrroles from nitroalkenes and tosylmethyl isocyanides. *Synthesis*, (7), 871–876.

121 Beaza, A., Mendiola, J., Burgos, C., Alvarez-Builla, J., and Vaquero, J.J. (2005) Heterocyclizations with tosylmethyl isocyanide derivatives. A new approach to substituted azolopyrimidines. *J. Org. Chem.*, **70** (12), 4879–4882.

122 Hoppe, I. and Schöllkopf, U. (1976) Cycloaddition von "Isocyanketen" an Benzophenonanil und Thioimidsäureester zu β-Lactamen. *Chem. Ber.*, **109** (2), 482–487.

123 Ito, Y., Sawamura, M., Hamashima, H., Emura, T., and Hayashi, T. (1989) Asymmetric aldol reaction of α-ketoesters with isocyanoacetate and isocyanoacetamide catalyzed by a chiral ferrocenylphosphine-gold(I) complex. *Tetrahedron Lett.*, **30** (35), 4681–4684.

124 Ozaki, Y., Maeda, S., Iwasaki, T., Matsumoto, K., Odawara, A., Sasaki, Y., and Morita, T. (1983) Syntheses of 5-substituted oxazole-4-carboxylic acid derivatives with inhibitory activity on blood platelet aggregation. *Chem. Pharm. Bull.*, **31** (12), 4417–4424.

125 For examples using Weinreb amides see: Coffin, A.R., Roussell, M.A., Tserlin, E., and Pelkey, E.T. (2006) Regiocontrolled synthesis of pyrrole-2-carboxaldehydes and 3-pyrrolin-2-ones from pyrrole Weinreb amides. *J. Org. Chem.*, **71** (17), 6678–6681.

126 Chupp, J.P. and Leschinsky, K.L. (1980) Heterocycles from substituted amides. VII. Oxazoles from 2-isocyanoacetamides. *J. Heterocycl. Chem.*, **17** (4), 705–709.

127 Chupp, J.P. and Leschinsky, K.L. (1980) Heterocycles from substituted amides. VIII. Oxazole derivatives from reaction of isocyanates with 2-isocyanoacetamides. *J. Heterocycl. Chem.*, **17** (4), 711–715.

128 (a) Schöllkopf, U., Hausberg, H.-H., Hoppe, I., Segal, M., and Reiter, U. (1978) Asymmetrische Synthese von α-Alkyl-α-aminocarbonsäuren durch Alkylierung von 1-chiral-substituierten 2-Imidazolin-5-onen. *Angew. Chem.*, **90** (2), 136–138; (b) Schöllkopf, U., Hausberg, H.-H., Hoppe, I., Segal, M., and Reiter, U. (1978) Asymmetric synthesis of α-alkyl-α-aminocarboxylic acids by alkylation of 1-chiral-substituted 2-imidazolin-5-ones. *Angew. Chem. Int. Ed.*, **17** (2), 117–119.

129 Schöllkopf, U., Hausberg, H.-H., Segal, M., Reiter, U., Hoppe, I., Saenger, W., and Lindner, K. (1981) Asymmetrische Synthesen über heterocyclische Zwischenstufen, IV. Asymmetrische Synthese von α-Methylphenylalanin und seinen Analoga durch Alkylieren von 1-chiral substituiertem 4-Methyl-2-imidazolin-5-on. *Liebigs Ann. Chem.*, (3), 439–458.

130 Sisko, J. (1998) A one-pot synthesis of 1-(2,2,6,6-tetramethyl-4-piperidinyl)-4-(4-fluorophenyl)-5-(2-amino-4-pyrimidinyl)- imidazole: a potent inhibitor of P38 MAP kinase. *J. Org. Chem.*, **63** (13), 4529–4531.

131 In the original [3+2] cycloaddition, NaH as a base was found to essential (see Ref. [123]). Recently, allyl TosMIC was found to give the desired oxazole, although the yield was moderate (32%); see: Gracias, V., Gasiecki, A.F., and Djuric, S.W. (2005) Synthesis of fused bicyclic imidazoles by sequential van Leusen/ring-closing metathesis reactions. *Org. Lett.*, **7** (15), 3183–3186.

132 Lee, J.C., Laydon, J.T., McDonnell, P.C., Gallagher, T.F., Kumar, S., Green, D., McNulty, D., Blumenthal, M.J., Keys, J.R., Landvatter, S.W., Strickler, J.E., McLaughlin, M.M., Siemens, I.R., Fisher, S.M., Livi, G.P., White, J.R., Adams, J.L., and Young, P.R. (1994) A protein kinase involved in the regulation of inflammatory cytokine biosynthesis. *Nature*, **372** (6508), 739–746.

133 Antuch, W., Menon, S., Chen, Q.-Z., Lu, Y., Sakamuri, S., Beck, B., Schauer-Vukašinović, V., Agarwal, S., Hess, S., and Dömling, A. (2006) Design and modular parallel synthesis of a MCR derived α-helix mimetic protein–protein interaction inhibitor scaffold. *Bioorg. Med. Chem. Lett.*, **16** (6), 1740–1743.

134 Neochoritis, C., Stephanidou-Stephanatou, J., and Tsoleridis, C.A. (2009) Heterocyclizations via TosMIC-based multicomponent reactions: a new approach to one-pot facile synthesis of substituted quinoxaline derivatives. *Synlett*, (2), 302–305.

135 Wu, B., Wen, J., Zhang, J., Li, J., Xiang, Y.-Z., and Yu, X.-Q. (2009) One-pot van Leusen synthesis of 4,5-disubstituted oxazoles in ionic liquids. *Synlett*, (3), 500–504.

136 Baxendale, I.R., Buckle, C.D., Ley, S.V., and Tamborini, L. (2009) A base-catalysed one-pot three-component coupling reaction leading to nitrosubstituted pyrroles. *Synthesis*, (9), 1485–1493.

137 Sun, X., Janvier, P., Zhao, G., Bienaymé, H., and Zhu, J. (2001) A novel multicomponent synthesis of polysubstituted 5-aminooxazole and its new scaffold-generating reaction to pyrrolo[3,4-*b*]pyridine. *Org. Lett.*, **3** (6), 877–880.

138 Lobrégat, V., Alcaraz, G., Bienaymé, H., and Vaultier, M. (2001) Application of the "resin-capture–release" methodology to macrocyclisation via intramolecular Suzuki–Miyaura coupling. *Chem. Commun.*, (9), 817–818.

139 Janvier, P., Sun, X., Bienaymé, H., and Zhu, J. (2002) Ammonium chloride-promoted four-component synthesis of pyrrolo[3,4-b]pyridin-5-one. *J. Am. Chem. Soc.*, **124** (11), 2560–2567.

140 (a) Fayol, A. and Zhu, J. (2002) Synthesis of furoquinolines by a multicomponent domino process. *Angew. Chem.*, **114** (19), 3785–3787; (b) Fayol, A. and Zhu, J. (2002) Synthesis of furoquinolines by a multicomponent domino process. *Angew. Chem. Int. Ed.*, **41** (19), 3633–3635.

141 (a) Janvier, P., Bienaymé, H., and Zhu, J. (2002) A five-component synthesis of hexasubstituted benzene. *Angew. Chem.*, **114** (22), 4467–4470; (b) Janvier, P., Bienaymé, H., and Zhu, J. (2002) A five-component synthesis of hexasubstituted benzene. *Angew. Chem. Int. Ed.*, **41** (22), 4291–4294.

142 Wang, Q., Xia, Q., and Ganem, B. (2003) A general synthesis of 2-substituted-5-aminooxazoles: building blocks for multifunctional heterocycles. *Tetrahedron Lett.*, **44** (36), 6825–6827.

143 Wang, Q. and Ganem, B. (2003) New four-component condensations leading to 2,4,5-trisubstituted oxazoles. *Tetrahedron Lett.*, **44** (36), 6829–6832.

144 Fayol, A. and Zhu, J. (2004) Synthesis of polysubstituted 4,5,6,7-tetrahydrofuro[2,3-c]pyridines by a novel multicomponent reaction. *Org. Lett.*, **6** (1), 115–118.

145 Cuny, G., Gámez-Montano, R., and Zhu, J. (2004) Truncated diastereoselective Passerini reaction, a rapid construction of polysubstituted oxazole and peptides having an α-hydroxy-β-amino acid component. *Tetrahedron*, **60** (22), 4879–4885.

146 Fayol, A., Housseman, C., Sun, X., Janvier, P., Bienaymé, H., and Zhu, J. (2005)) Synthesis of α-isocyano-α-alkyl(aryl)acetamides and their use in the multicomponent synthesis of 5-aminooxazole, pyrrolo[3,4-b]pyridin-5-one and 4,5,6,7-tetrahydrofuro[2,3-c]pyridine. *Synthesis*, (1), 161–165.

147 Pirali, T., Tron, G.C., and Zhu, J. (2006) One-pot synthesis of macrocycles by a tandem three-component reaction and intramolecular [3+2] cycloaddition. *Org. Lett.*, **8** (18), 4145–4148.

148 The base-promoted ring closure has been reported for the formation of 2H-oxazoles see: Mishchenco, V.V., Itov, Z.I., L'vova, S.D., Shostakovskaya, G.K., and Gunar, V.I. (1988) Synthesis of 4-methyl-5-propoxy-oxazoles from propyl α-isocyanopropionate. *Pharm. Chem. J.*, **22** (7), 568–572.

149 Elders, N., Ruijter, E., de Kanter, F.J.J., Janssen, E., Lutz, M., Spek, A.L., and Orru, R.V.A. (2009) A multicomponent reaction towards N-(cyanomethyl) amides. *Chem. Eur. J.*, **15** (25), 6096–6099.

150 Elders, N., Ruijter, E., de Kanter, F.J.J., Groen, M.B., and Orru, R.V.A. (2008) Selective formation of 2-imidazolines and 2-substituted oxazoles by using a three-component reaction. *Chem. Eur. J.*, **14** (16), 4961–4973.

151 Elders, N. (2010) Multicomponent Approaches to Molecular Diversity & Complexity. PhD Thesis, VU University Amsterdam, the Netherlands. For details of the MCR towards 2-imidazolines, see Section 4.6.

152 For an overview of strategies for innovation in multicomponent reactions see: (a) Ganem, B. (2009) Strategies for innovation in multicomponent reaction design. *Acc. Chem. Res.*, **42** (3), 463–471; (b) Ruijter, E., Scheffelaar, R., and Orru, R.V.A. (2011) Mehrkomponentenreaktionen als Weg zu molekularer Komplexität und Diversität. *Angew. Chem.*, **123** (28), 6358–6371; (c) Ruijter, E., Scheffelaar, R., and Orru, R.V.A. (2011) Multicomponent reaction design in the quest for molecular diversity and complexity. *Angew. Chem. Int. Ed.*, **50** (28), 6324–6346.

153 Xia, Q. and Ganem, B. (2002) Metal-promoted variants of the Passerini reaction leading to functionalized heterocycles. *Org. Lett.*, **4** (9), 1631–1634.

154 Pirali, T., Tron, G.C., Masson, G., and Zhu, J. (2007) Ammonium chloride promoted three-component synthesis of 5-iminooxazoline and its subsequent transformation to

macrocyclodepsipeptide. *Org. Lett.*, **9** (25), 5275–5278.

155 Fayol, A. and Zhu, J. (2005) Three-component synthesis of polysubstituted 6-azaindolines and its tricyclic derivatives. *Org. Lett.*, **7** (2), 239–242.

156 Bon, R.S., Hong, C., Bouma, M.J., Schmitz, R.F., de Kanter, F.J.J., Lutz, M., Spek, A.L., and Orru, R.V.A. (2003) Novel multicomponent reaction for the combinatorial synthesis of 2-imidazolines. *Org. Lett.*, **5** (20), 3759–3762.

157 Bon, R.S., van Vliet, B., Sprenkels, N.E., Schmitz, R.F., de Kanter, F.J.J., Stevens, C.V., Swart, M., Bickelhaupt, F.M., Groen, M.B., and Orru, R.V.A. (2005) Multicomponent synthesis of 2-imidazolines. *J. Org. Chem.*, **70** (9), 3542–3553.

158 Elders, N., Schmitz, R.F., de Kanter, F.J.J., Ruijter, E., Groen, M.B., and Orru, R.V.A. (2007) A resource-efficient and highly flexible procedure for a three-component synthesis of 2-imidazolines. *J. Org. Chem.*, **72** (16), 6135–6142.

159 Bon, R.S., de Kanter, F.J.J., Lutz, M., Spek, A.L., Jahnke, M.C., Hahn, F.E., Groen, M.B., and Orru, R.V.A. (2007) Multicomponent synthesis of N-heterocyclic carbene complexes. *Organometallics*, **26** (15), 3639–3650.

160 (a) Strassberger, Z., Mooijman, M., Ruijter, E., Alberts, A.H., de Graaff, C., Orru, R.V.A., and Rothenberg, G. (2010) A facile route to ruthenium-carbene complexes and their application in furfural hydrogenation. *Appl. Organomet. Chem.*, **24** (2), 142–146; (b) Strassberger, Z., Mooijman, M., Ruijter, E., Alberts, A.H., Maldonado, A.G., Orru, R.V.A., and Rothenberg, G. (2010) Finding furfural hydrogenation catalysts via predictive modelling. *Adv. Synth. Catal.*, **352** (13), 2201–2210.

161 Vassilev, L.T., Vu, B.T., Graves, B., Carvajal, D., Podlaski, F., Filipovic, Z., Kong, N., Kammlott, U., Lukacs, C., Klein, C., Fotouhi, N., and Liu, E.A. (2004) In vivo activation of the p53 pathway by small-molecule antagonists of MDM2. *Science*, **303** (5659), 844–848.

162 Bon, R.S., Sprenkels, N.E., Koningstein, M.M., Schmitz, R.F., de Kanter, F.J.J., Dömling, A., Groen, M.B., and Orru, R.V.A. (2008) Efficient C2 functionalisation of 2H-2-imidazolines. *Org. Biomol. Chem.*, **6** (1), 130–137.

163 Srivastava, S., Beck, B., Wang, W., Czarna, A., Holak, T.A., and Dömling, A. (2009) Rapid and efficient hydrophilicity tuning of p53/mdm2 antagonists. *J. Comb. Chem.*, **11** (4), 631–639.

164 Combination of MCRs in a single pot performed either as a one-pot process (true novel MCR) or a short sequence (tandem); see: Mironov, M.A. (2006) Design of multi-component reactions: from libraries of compounds to libraries of reactions. *QSAR Comb. Sci.*, **25** (5–6), 423–431, and references cited therein.

165 For the first examples of union of MCRs see: (a) Dömling, A. and Ugi, I. (1993) Die siebenkomponentenreaktion. *Angew. Chem.*, **105** (4), 634–635; (b) Dömling, A. and Ugi, I. (1993) The seven-component reaction. *Angew. Chem. Int. Ed.*, **32** (4), 563–564; (c) Dömling, A., Herdtweck, E., and Ugi, I. (1998) MCR V: the seven-component reaction. *Acta Chem. Scand.*, **52** (1), 107–113.

166 (a) Elders, N., van der Born, D., Hendrickx, L.J.D., Timmer, B.J.J., Krause, A., Janssen, E., de Kanter, F.J.J., Ruijter, E., and Orru, R.V.A. (2009) The efficient one-pot reaction of up to eight components by the union of multicomponent reactions. *Angew. Chem.*, **121** (32), 5970–5973; (b) Elders, N., van der Born, D., Hendrickx, L.J.D., Timmer, B.J.J., Krause, A., Janssen, E., de Kanter, F.J.J., Ruijter, E., and Orru, R.V.A. (2009) *Angew. Chem. Int. Ed.*, **48** (32), 5856–5859.

167 Paravidino, M., Bon, R.S., Scheffelaar, R., Vugts, D.J., Znabet, A., Schmitz, R.F., de Kanter, F.J.J., Lutz, M., Spek, A.L., Groen, M.B., and Orru, R.V.A. (2006) Diastereoselective multicomponent synthesis of dihydropyridones with an isocyanide functionality. *Org. Lett.*, **8** (23), 5369–5372.

168 Scheffelaar, R., Paravidino, M., Znabet, A., Schmitz, R.F., de Kanter, F.J.J., Lutz, M., Spek, A.L., Fonseca Guerra, C., Bickelhaupt, F.M., Groen, M.B., Ruijter, E., and Orru, R.V.A. (2010) Scope and limitations of an efficient four-component reaction for dihydropyridin-2-ones. *J. Org. Chem.*, **75** (5), 1723–1732.

169 Paravidino, M., Scheffelaar, R., Schmitz, R.F., de Kanter, F.J.J., Groen, M.B., Ruijter, E., and Orru, R.V.A. (2007) A flexible six-component reaction to access constrained depsipeptides based on a dihydropyridinone core. *J. Org. Chem.*, **72** (26), 10239–10242.

170 Scheffelaar, R., Paravidino, M., Muilwijk, D., Lutz, M., Spek, A.L., de Kanter, F.J.J., Orru, R.V.A., and Ruijter, E. (2009) A novel three-component reaction toward dihydrooxazolopyridines. *Org. Lett.*, **11** (1), 125–128.

171 Scheffelaar, R., Klein Nijenhuis, R.A., Paravidino, M., Lutz, M., Spek, A.L., Ehlers, A.W., de Kanter, F.J.J., Groen, M.B., Orru, R.V.A., and Ruijter, E. (2009) Synthesis of conformationally constrained peptidomimetics using multicomponent reactions. *J. Org. Chem.*, **74** (2), 660–668.

172 den Heeten, R., van der Boon, L., Broere, D.L.J., Janssen, E., de Kanter, F.J.J., Ruijter, E., and Orru, R.V.A. (2011) Concise synthesis of highly substituted benzo[a]quinolizines by a multicomponent reaction/allylation/heck reaction sequence. *Eur. J. Org. Chem.*, 275–280.

4
Synthetic Application of Isocyanoacetic Acid Derivatives

Anton V. Gulevich, Alexander G. Zhdanko, Romano V.A. Orru, and Valentine G. Nenajdenko

4.1
Introduction

Since their first preparation in 1961, α-isocyanoacetic acid derivatives have received much attention [1]. Each molecule of isocyanoacetate contains four reaction centers – an isocyanide group, an active CH fragment, a substituent R (which can also handle functional groups), and a carboxylic acid with a protecting group (Figure 4.1). Moreover, isocyanoacetic acid derivatives can also be obtained in an optically active form, from natural amino acids. The combination of these potential four reaction centers within the molecule results in the isocyanoacetic acids having a diverse reactivity, which in turn offers numerous applications in the different branches of organic, inorganic, coordination, polymeric, combinatorial, and medicinal chemistries.

Among other isocyanides, isocyanoacetate derivatives are especially attractive objects for investigation, due to their interesting structure and exceptional synthetic application. As the chemistry of isocyanoacetic acid derivatives was recently and comprehensively reviewed by Nenajdenko *et al.* [2], attention in this chapter will be focused on the most important synthetic applications of isocyanoacetic acid derivatives, and especially on the most recent investigations.

4.2
Synthesis of α-Isocyanoacetate Derivatives

Three general synthetic routes to isocyanoacetates **5** (esters of α-isocyanoacetic acid) have been identified, including: (i) the dehydration of formamides (**1**); (ii) the carboxylation of acidic isocyanides (**2** or **4**); and (iii) the esterification of isocyanoacetic acid salts (**3**) (Scheme 4.1).

Thus, alkyl isocyanoacetates **5** can be obtained very easily via the dehydration of primary formamides **1**, and the process accomplished using a variety of reagents such as phosgene [3], diphosgene [4], triphosgene [5], phosphoryl chloride [6, 7],

Isocyanide Chemistry: Applications in Synthesis and Material Science, First Edition. Edited by Valentine Nenajdenko.
© 2012 Wiley-VCH Verlag GmbH & Co. KGaA. Published 2012 by Wiley-VCH Verlag GmbH & Co. KGaA.

4 Synthetic Application of Isocyanoacetic Acid Derivatives

Carboxylic group: Acylation etc.

Isocyanide group: Multicomponent Reactions, Nucleophilic Addition, Metal Coordination etc.

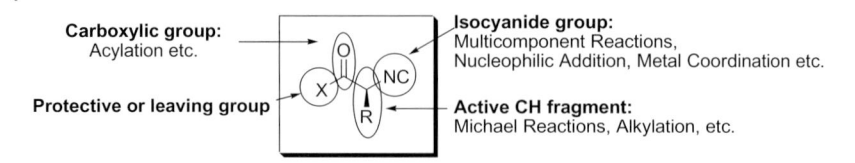

Protective or leaving group

Active CH fragment: Michael Reactions, Alkylation, etc.

Figure 4.1 Reactivity profiles of α-isocyanoacetate derivatives.

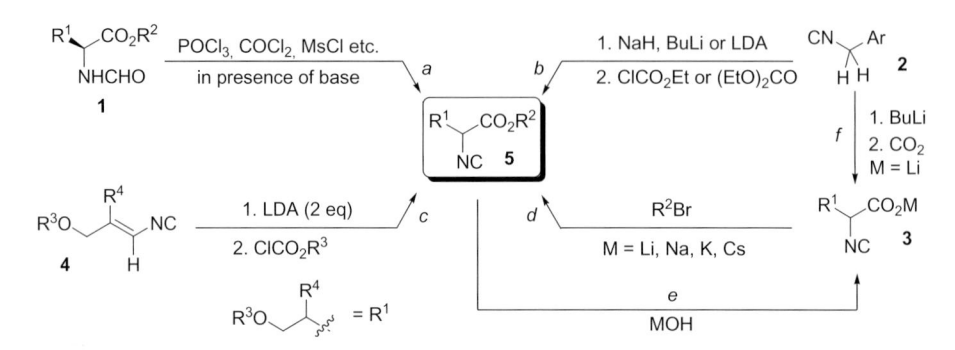

Scheme 4.1

mesyl chloride, oxalyl chloride [8], generally in combination with bases – in most cases, tertiary amines (route a in Scheme 4.1). An alternative method is based on the insertion of a carboxylic acid group to metalated isocyanides. Thus, benzyl isocyanides **2** can be deprotonated at the α-position and will react smoothly with dialkyl carbonate [9] and methyl chloroformate [10] to afford isocyanoacetates **5** in good yields (route b). Similarly, α-allylic isocyanoacetate derivatives can be obtained by deprotonation/migration of the double bond/alkoxycarbonylation of α,β-unsaturated isocyanides **4** with chloroformates (route c) [11]. It is notable, that salts of isocyanoacetic acids **3** can be used for the synthesis of isocyanoacetic esters via alkylation (route d) [12–14]; typically, salts **3** are easily obtained from isocyanoacetates **5** by saponification with ethanolic alkali hydroxides (route e) [15–17]. Yet another approach is based on the deprotonation of benzyl isocyanides **2** with BuLi, with subsequent carboxylation using CO_2 (route f) [18]. Clearly, racemic isocyanoacetates are easily accessible from the corresponding amino acids via well established and convenient protocols.

Although free isocyanoacetic acids are relatively unstable compounds [16], they can be obtained as complexes with transition metals (e.g., Re [19, 20], Cr, and W [21, 22]), coordinated to the isocyanide group.

Such chiral isocyanoacetates are configurationally unstable under basic conditions due to the relatively high acidity of the α-hydrogen which is induced by the two strong electron-withdrawing groups, NC and CO_2R. Thus, their preparation in the presence of basic amines such as triethylamine, diisopropylamine, or even N-methylmorpholine (NMM) at elevated temperatures (above −30 °C) leads to a partial or complete racemization [23]. The application of the NMM/diphosgene

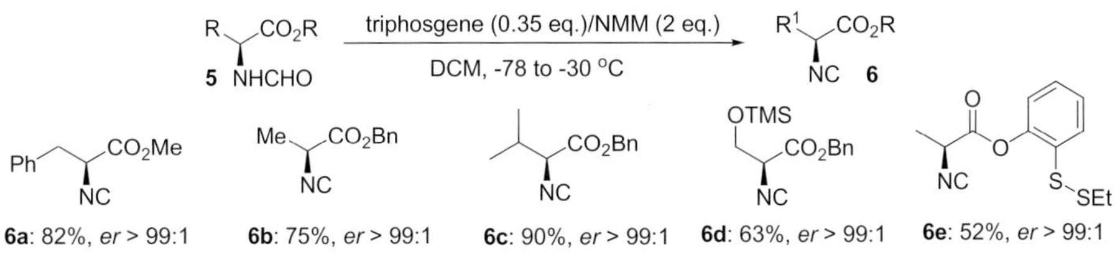

4.2 Synthesis of α-Isocyanoacetate Derivatives

Scheme 4.2

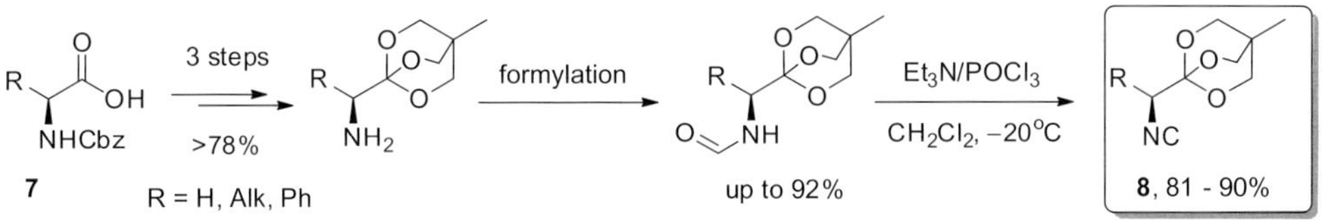

Reaction: R-CH(NHCHO)-CO₂R (**5**) → R¹-CH(NC)-CO₂R (**6**) using triphosgene (0.35 eq.)/NMM (2 eq.), DCM, −78 to −30 °C.

- **6a**: Ph-CH(NC)-CO₂Me; 82%, er > 99:1
- **6b**: Me-CH(NC)-CO₂Bn; 75%, er > 99:1
- **6c**: iPr-CH(NC)-CO₂Bn; 90%, er > 99:1
- **6d**: (OTMS)CH₂-CH(NC)-CO₂Bn; 63%, er > 99:1
- **6e**: CH₃-CH(NC)-C(O)O-C₆H₄-S-SEt; 52%, er > 99:1

Scheme 4.3

R-CH(NHCbz)-C(O)OH (**7**) → [3 steps, >78%] → R-CH(NH₂)-OBO → [formylation, up to 92%] → R-CH(NHCHO)-OBO → [Et₃N/POCl₃, CH₂Cl₂, −20 °C] → R-CH(NC)-OBO (**8**, 81–90%)

R = H, Alk, Ph

---

system at lower temperatures proved to be more effective, and has been employed for syntheses of many other optically pure isocyanoacetates derived from L-valine, L-alanine, L-isoleucine [24], and L-phenylalanine [25].

The most recent data reported by Danishefsky et al. have shown that the NMM/triphosgene system, at low temperatures, provides the desired isocyanides **6** in high optical purity (Scheme 4.2) [23]. Alternative racemization-free syntheses were conducted by using a non-basic dehydrating reagent 1,1′-carbonyldiimidazole (CDI·2MeSO₃H) at 0–5 °C [26], and cyanuric chloride/Py under microwave heating [27].

Consequently, optically active chiral α-isocyanoalkyl esters are easily available from natural or artificial amino acids, under mild weakly basic conditions and at low temperatures. However, it is known that chiral isocyanoacetates could be configurationally unstable under the condition of the Ugi reaction (see Chapter 4.13). In order to overcome racemization issue, Nenajdenko et al. suggested the use of OBO (4-methyl-2,6,7-trioxabiciclo [2.2.2]octyl derivatives) esters of isocyanoacetic acid **8** as a configurationally stable isocyanoacetate equivalent. It was shown that OBO esters of isocyanoacetic acids **8** are stable towards racemization under relatively harsh basic conditions, therefore undergoing the Ugi reaction without racemization [28]. The OBO-protected isocyanoacetates are easily available from the corresponding Cbz-protected α-amino acids **7** in high total yields, by using simple and scalable procedures (Scheme 4.3). More recently, isocyanides **8** were also employed by Wessjohann et al. in the synthesis of tubulysin analogues, to introduce a chiral amino acid residue [29].

Isocyanoacetamides **9** may be easily prepared from the corresponding isocyanoesters **10** (Scheme 4.4, route a), from the related formamides **11** (route b), or from their salts **12** (route c). In this case, the ester group in isocyanoacetates **10** is activated towards a nucleophilic attack due to the electron-withdrawing effect of the isocyano group; consequently, the aminolysis of isocyanoacetates may occur

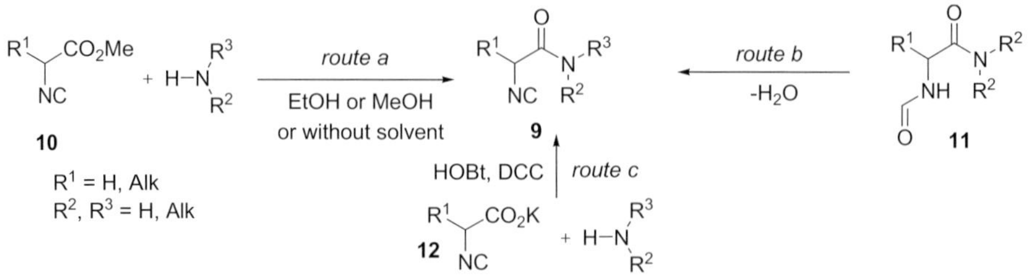

Scheme 4.4

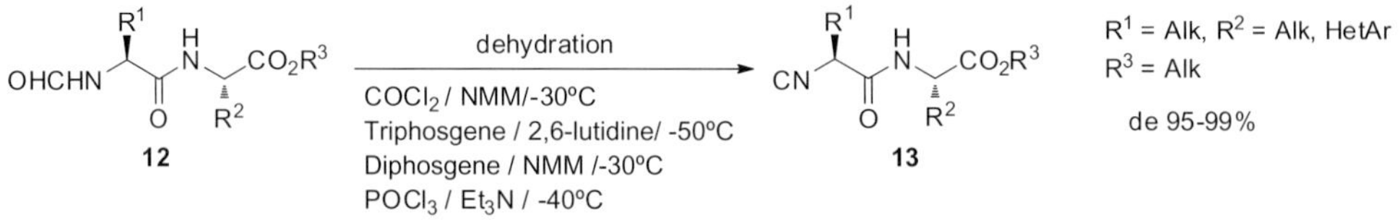

Scheme 4.5

under rather mild conditions (route *a*). In general, the methyl isocyanoacetates **10** react readily with either ammonia [30] or primary and secondary aliphatic amines [31–33] in ethanol or methanol at room temperature, whereas anilines are poor substrates. Not unexpectedly, the methyl esters are more reactive than ethyl esters [34], and the reaction is also efficient in solvent-free conditions [34–36]. Only racemic isocyanoacetates were involved in the amination reaction since, apparently, chiral isocyanoacetates would be configurationally unstable in the presence of amines.

Isocyanoamides may also be synthesized via the dehydration of formamides **11**, using a standard $POCl_3/NEt_3$ system (Scheme 4.4, route *b*) [37–40], with a variety of isocyanoacetamides [41–43] – including the α-isocyanoacetanilides [35] and Weinreb amides – being accessible via this methodology [44, 45].

Finally, isocyanoacetamides are available via the amination of potassium isocyanoacetates **12** with the corresponding amine in the presence of activating agents, such as HOBt/DCC (*N,N'*-dicyclohexylcarbodiimide) [32], EDCI (1-ethyl-3-(3-dimethylaminopropyl)carbodiimide) [33], and others (route *c*) [46].

It should be noted that isocyanoacetamides **9** are much more configurationally stable in comparison with the esters of isocyanoacetic acid **6**. Thus, optically active isocyanopeptide **13** can be obtained from the corresponding formamides **12** by using a variety of reagents such as the phosgene/NMM system [47], triphosgene/2,6-lutidine [23], and diphosgene/NMM [48–51]. Even stronger basic conditions (e.g., $POCl_3/NEt_3$) may be applied to prepare chiral isocyanoacetamides (Scheme 4.5) [52, 53].

Unsubstituted isocyanoacetonitrile **14** is a highly unstable compound that decomposes rapidly above −5 °C. The first such isocyanoacetonitrile **14** was syn-

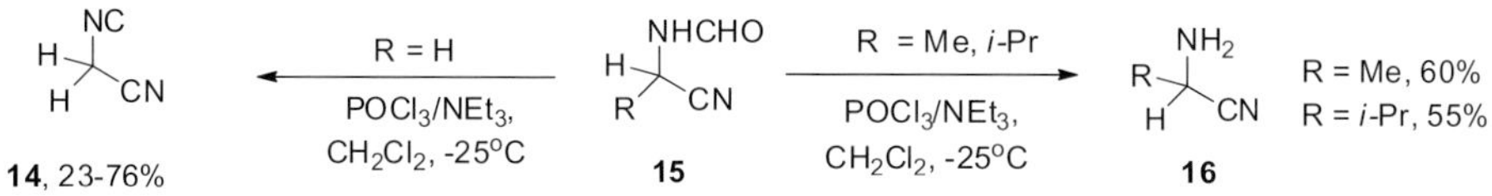

Scheme 4.6

thesized in 1975 by Schöllkopf et al., via dehydration of the corresponding formamide **15**, in 23% yield. Subsequently, the yield was increased to 76% by Lentz et al., by using a low-temperature, high-vacuum condensation [54]. In contrast, substituted isocyanoacetonitriles **16** appeared to be bench-stable and could be easily isolated using regular distillation (Scheme 4.6) [55].

In conclusion, a broad variety of isocyanoacetic acid derivatives can be prepared by creating a carboxylic or isocyano function in a prefunctionalized molecule. However, the most common approach is based on the transformation of a primary amine group to the isocyanide, via a standard formylation/dehydration protocol.

4.3
Alkylation of Isocyanoacetic Acid Derivatives

α-Isocyanoacetic acid derivatives contain two powerful electron-withdrawing substituents at the α-position, and may be completely or reversibly deprotonated by a variety of bases that includes potassium *tert*-butoxide, sodium hydride, sodium ethoxide, sodium cyanide, or even tertiary amines [Et$_3$N or DBU (1,8-diazabicyclo[5.4.0]undec-7-ene)] [56]. Usually, the alkylation of metalated unsubstituted methyl or ethyl isocyanoacetates leads to dialkylation products, even with only 1 equiv. of alkylating agent (in several cases, intermediate monoalkylated products may be isolated in 10–20% yields) [57]. However, the monoalkylation of isocyanoacetate can be accomplished with *tert*-butyl ester **17**, since here the bulky ester group prevents attack of the α-anion to a second alkyl halide molecule (Scheme 4.7). When 2 equiv. of alkylating agent (*n*-alkyl or benzyl halide) are used

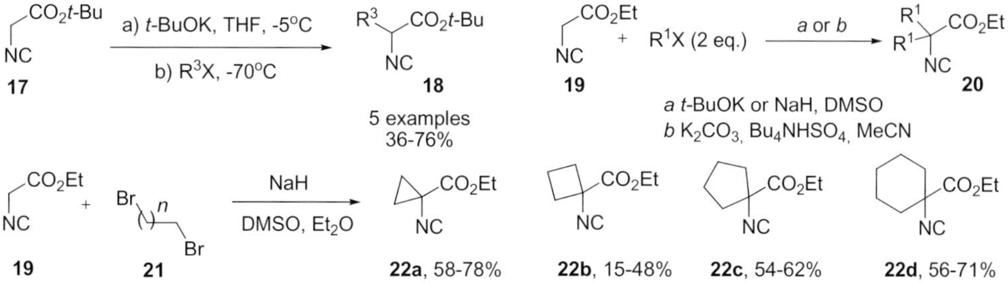

Scheme 4.7

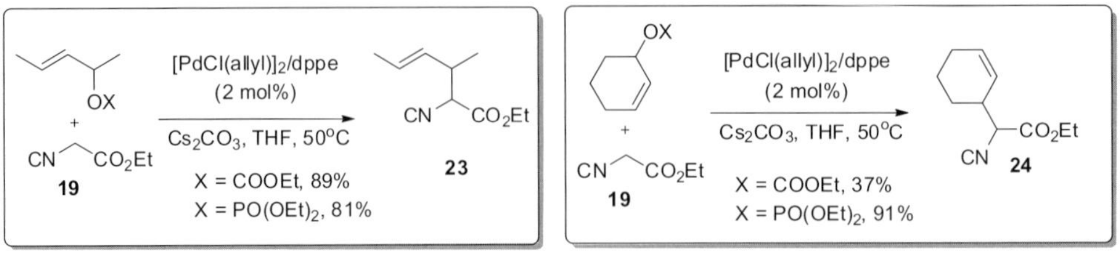

Scheme 4.8

in the combination with methyl- or ethyl-isocyanoacetates, however, the α,α-dialkylated products **20** are formed in yields of up to 90% (Scheme 4.7) [58–61]. α-Monosubstituted isocyanoacetates may undergo alkylation by alkyl halides, using t-BuOK or NaH as a base, to form α,α-disubstituted isocyanoacetates in good yields [58, 59].

In 1997, Kotha et al. adopted phase-transfer conditions for the dialkylation of isocyanoacetates, using tertiary ammonium salts as a phase-transfer agent. Thus, propargyl bromide [62], allyl bromide [63], a broad variety of substituted benzyl bromides, and even thienylmethyl bromide, were shown to be efficient alkylating reagents for this reaction (Scheme 4.7) [64]. The use of dihalogen compounds **21** opens up broad possibilities for the synthesis of cyclic compounds, such as **22a–d** (Scheme 4.7) [58, 65–67]. The alkylation of isocyanoacetates bearing chiral (−)-menthyl, (+)-bornyl [68], and (−)-8-phenylmenthyl moieties yielded the desired products, though only in low selectivity and with a maximum diastereomeric excess (d.e.) of up to 48% [69].

In 1987, Ito, Hayashi and coworkers demonstrated the Pd-catalyzed allylation of α-isocyanoacetate esters with allyl acetates [70], while in 2003 Kazmaier and Ackerman found carbonates and phosphates to be more efficient leaving groups. For example, the isocyanoacetates **22** and **24** were obtained from the corresponding allylic carbonates and phosphates in good to excellent yields (Scheme 4.8) [71].

The alkylation and allylation of α-isocyanoacetic acid derivatives is of major interest, since its overall effect is the production of longer-chain or branched amino acids. In this case, α-isocyanoacetates are considered to serve as an efficient glycine anion equivalent. Although the flexibility of this method for the syntheses of diversely substituted amino acid derivatives was well demonstrated by Kotha [72, 73], it should be noted that selective monoalkylation and the development of a stereoselective alkylation process remain an ongoing challenge.

Since α-isocyanoacetamides are generally less acidic then α-isocyanoacetates, a selective monoalkylation can be achieved under mild basic or acidic conditions. Consequently, Zhu et al. developed an efficient protocol for the monoalkylation of tertiary isocyanoacetamides **25** in the presence of CsOH, to afford α-substituted isocyanoacetamides **26** in high yields (Scheme 4.9) [33].

In contrast, it was shown by Matsumoto et al. that secondary isocyanoacetamides would undergo both alkylation and cyclization under strong basic conditions,

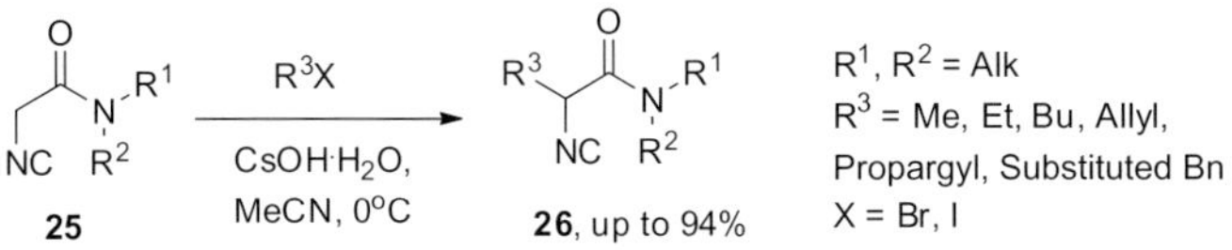

Scheme 4.9

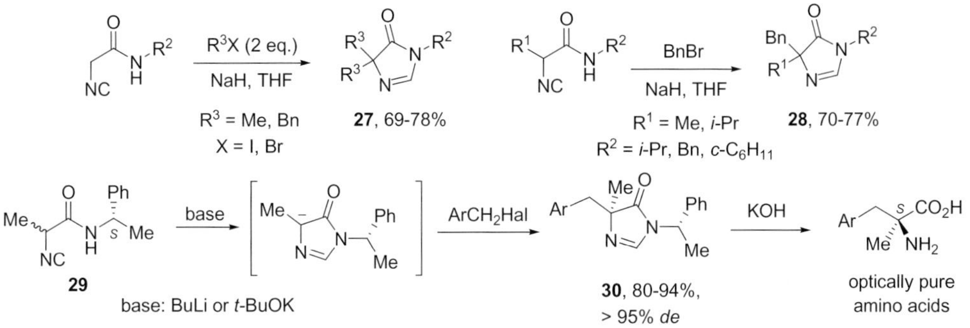

Scheme 4.10

via an intramolecular nucleophilic attack by the amide nitrogen. Hence, the alkylated imidazolinone derivatives **27** and **29** are formed as sole products (Scheme 4.10) [31]. Based on this reaction, Schöllkopf and coworkers developed the synthesis of chiral α-methyl phenylalanines, such that a diastereoselective alkylation of N-alkylisocyanoacetamides **29** bearing a chiral phenylethyl substituent at the nitrogen, accompanied by cyclization, would afford imidazolinones **30** in diastereoselective fashion, whereas a high d.e.-value was observed only for benzyl halogenides. The subsequent hydrolysis of imidazolinones **30**, under basic conditions, provided an array of chiral amino acids in an efficient manner (Scheme 4.10) [74].

In conclusion, alkylation and/or allylation represent efficient routes for the modification of isocyanoacetates. Notably, isocyanoacetates are generally more suitable for the synthesis of dialkylated products, whereas tertiary isocyanoacetamides may be monoalkylated in an effective manner.

4.4
α-Isocyanoacetates as Michael Donors

The Michael addition of ethyl isocyanoacetate **19** to acrylates and related compounds was first described by Schöllkopf *et al.* in 1970. The reaction, which was conducted in ethanol that contained catalytic amounts of sodium ethoxide, led to

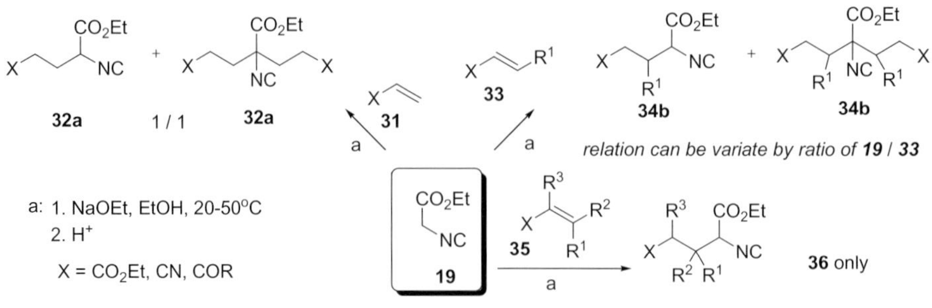

Scheme 4.11

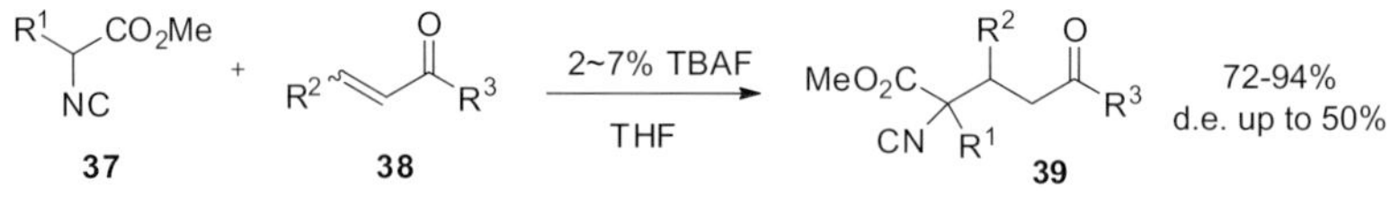

a: 1. NaOEt, EtOH, 20–50 °C
2. H⁺

X = CO_2Et, CN, COR

Scheme 4.12

the formation of Michael adducts **32a** in moderate yields. Unfortunately, the formation of bis-adducts **32b** represents a serious issue, as this becomes the main process when the isocyanoacetate **19** is treated with 3 equiv. of acrylic ester **31** in the presence of a base [75].

Other activated olefins such as acrylonitriles [76] and α,β-unsaturated ketones [77] may also be used as a Michael acceptors. Particularly reactive Michael components, such as acrylonitrile, ethyl acrylate and α,β-unsaturated carbonyls **31**, produced both mono- and bis-adducts **32** in an approximate 1:1 ratio. The β-monosubstituted Michael acceptors **33** can be selectively converted to either mono- or bis-adducts **34a** and **34b**, respectively, by varying the molar ratio of the reactants, whereas α,β-disubstituted and β,β-disubstituted olefins **35**, and those stabilized by a phenyl group, produce only the monoadducts **36** (Scheme 4.11).

In the presence of weak bases, the substituted isocyanoacetates may react with all types of activated alkenes to form mono-adducts in good yields [76]. In 1989, Ito et al. showed that the Michael reaction of substituted α-isocyanoesters **37** with α,β-unsaturated carbonyl compounds **38** could be efficiently promoted by a catalytic quantity of tetrabutylammonium fluoride, so as to produce α-isocyano-δ-ketoesters **39** in high yields [78, 79] (Scheme 4.12).

The reaction of isocyanoacetates with activated alkenes **40**, bearing good leaving groups such as nitroolefins, α,β-unsaturated sulfones and sulfides, affords the corresponding pyrroles **43** under basic conditions (Barton–Zard pyrrole synthesis) [80]. Thus, the initially formed Michael adduct **41** will cyclize spontaneously into the cyclic intermediate **42**, producing the corresponding pyrrole **43** following the elimination of HX (Scheme 4.13) (see also Chapter 11).

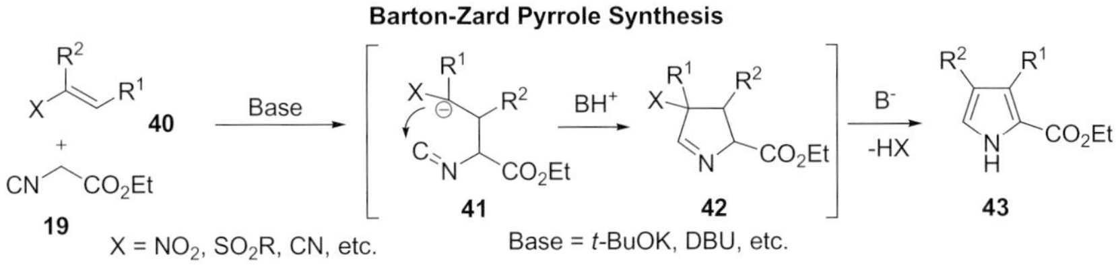

4.4 α-Isocyanoacetates as Michael Donors

Scheme 4.13

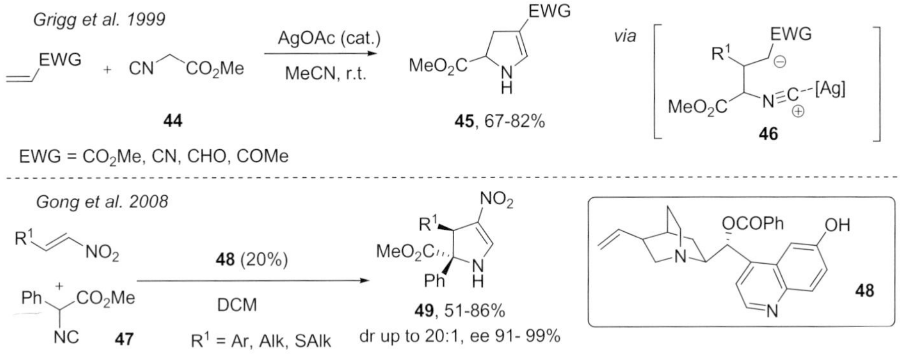

Scheme 4.14

For instance, the reaction of methyl isocyanoacetate **44** with activated alkenes catalyzed by silver acetate provides pyrrolines **45** in good yields. This Ag-catalyzed reaction proceeds smoothly with acrylates, acrylonitriles α,β-unsaturated aldehydes and ketones, via formation of the Michael adduct **46** (Scheme 4.14) [81]. Recently, Gong et al. developed an organocatalytic stereoselective variant of this transformation by using cinchona alkaloid **48** (20 mol%), such that the reaction of nitroolefins with α-aryl isocyanoacetates **47** yielded chiral 2-pyrrolines **49** in a highly stereoselective and enantioselective manner (Scheme 4.14) [82]. Subsequently, Escolano et al. applied a cooperative Ag(I)/cinchona alkaloid catalyst system to the asymmetric reaction of isocyanoacetates with α,β-unsaturated ketones, to afford the corresponding 2-pyrrolines [83].

Later, Xiu, Liu and coworkers described a tandem Michael addition/intramolecular isocyanide [3+2] cycloaddition approach towards new polycyclic heterocycles. In this case, the reaction of **50** bearing a carbonyl group and isocyanoacetate **19** led to the polycyclic scaffolds **52** diastereoselectively, via formation of the Michael adduct **51**, followed by an intramolecular cycloaddition reaction with the carbonyl group. Similarly, the reaction of alkene **53** with isocyanoacetate **19** afforded **54** in stereoselective fashion (Scheme 4.15) [84].

Xu and Liu et al. also reported the double Michael addition of unsubstituted isocyanoacetate **19** to divinyl ketone **50**, under mild conditions, to afford the cyclic

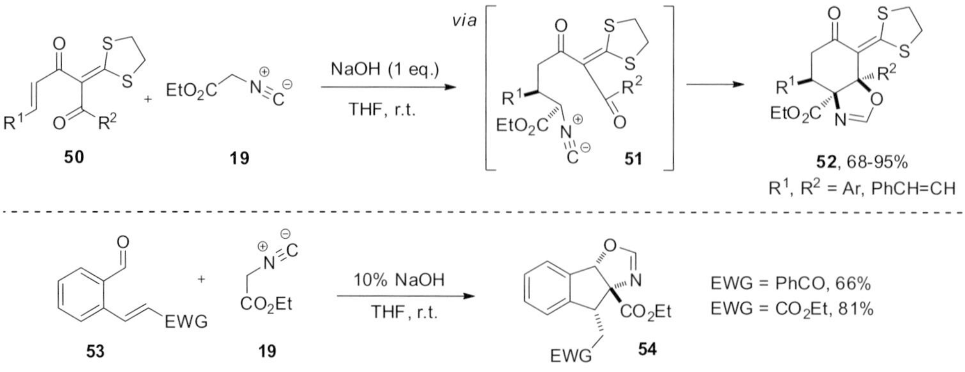

Scheme 4.15

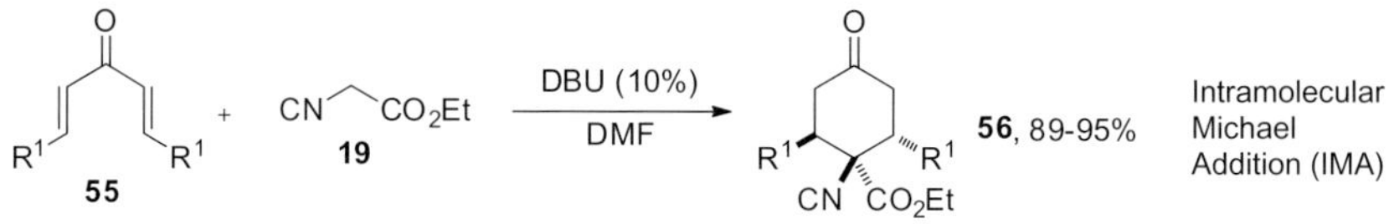

Scheme 4.16

isocyanoacetate **51** in stereoselective manner (Scheme 4.16) [85]. The insertion of additional functional groups into the molecule of divinyl ketone opens broad possibilities for tandem Michael additions/heterocyclizations processes.

Thus, the reaction of α-cyano-divinylketones **57** and ethyl isocyanoacetate **19** affords azabicyclic derivatives **61**, with the overall process involving a double Michael addition to provide the intermediate **58**, which then undergoes an intramolecular cyclization to form the intermediate **59**. A subsequent nucleophile-induced transannular attack of the imine nitrogen atom to the carbonyl atom to form **60**, followed by breaking of the C–C bond, affords the pyrrolizidine **61**. This reaction furnishes polycyclic pyrrolizidine derivatives **61** in good yields, and in both regioselective and diastereoselective manner. In similar fashion, polycyclic scaffolds **63** and **65** can be obtained via the reaction of isocyanoacetate **19** with the corresponding divinyl ketones **62** and **64**, respectively (Scheme 4.17) [86].

The same research group showed recently that the reaction of divinyl ketones **66** with ethyl isocyanoacetate **19** would furnish the 6-azabycylo [3.2.1]octane derivatives **67** under mild conditions. In this case, the reaction would proceed via a double Michael addition, followed by an intramolecular cyclization. Subsequently, derivatives **67** would undergo a rearrangement to pyrrolizidines **68** in the presence of DBU (Scheme 4.18) [87].

Very recently, the reaction was reported of unsaturated 3-ketoamide derivatives **69** with ethyl isocyanoacetate **19** to afford the derivatives **70**, containing tethered oxazole/pyrrole pairs. The reaction was shown to proceed via a double Michael

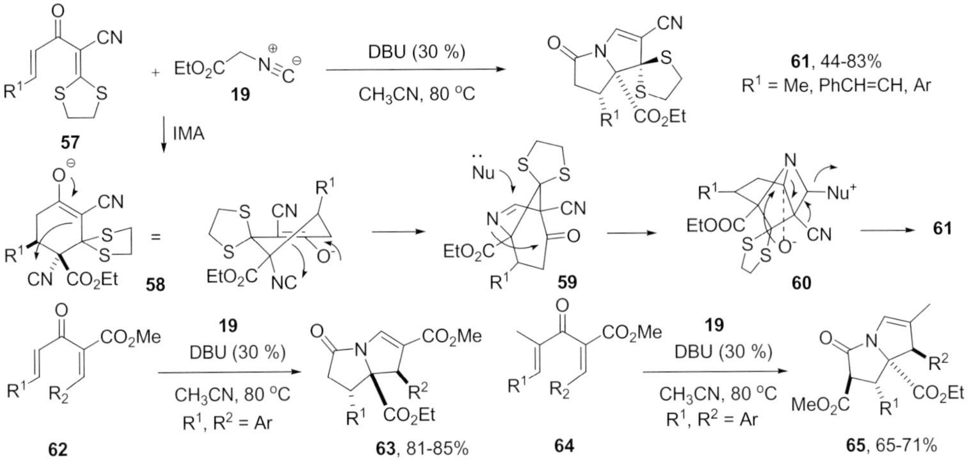

Scheme 4.17

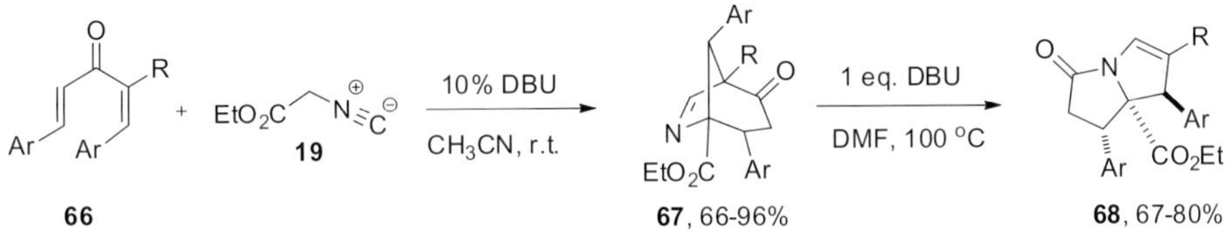

Scheme 4.18

addition to form adduct **71**, followed by an intramolecular elimination of the thiirane molecule, which led to the thioketone **72**. A subsequent preferential cleavage of the C–C bond in **73**, upon attack by the ethyl isocyanoacetate anion, produced the intermediate **74** which then underwent a double isocyanide cyclization to generate a pyrrole and oxazole ring, yielding the final product **70**. In contrast, reaction of the similar divinyl ketone **75** bearing a bulky *t*-Bu fragment with ethyl isocyanoacetate produced a bicyclic aminal **77** diastereoselectively, via cyclization of the intermediate **76** (Scheme 4.19) [88].

In conclusion, tandem reactions – including the Michael addition of isocyanoacetic acid derivatives to activated alkenes – may open broad opportunities for the synthesis of heterocyclic compounds.

4.5
Reaction of Isocyanoacetic Acids with Alkynes: Synthesis of Pyrroles

The reaction of alkynes with isocyanoacetic acid derivatives, which provides an efficient approach to pyrroles, was investigated independently by Yamamoto [89]

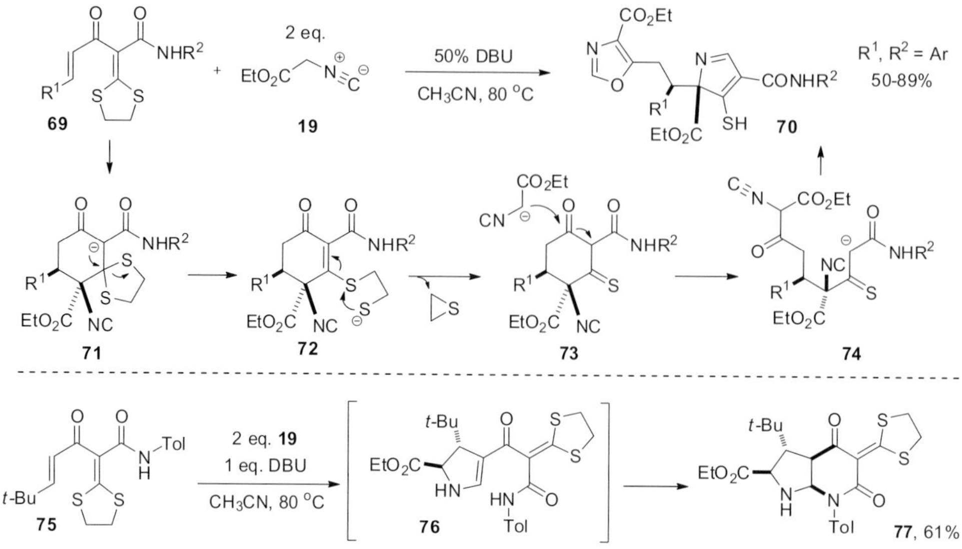

Scheme 4.19

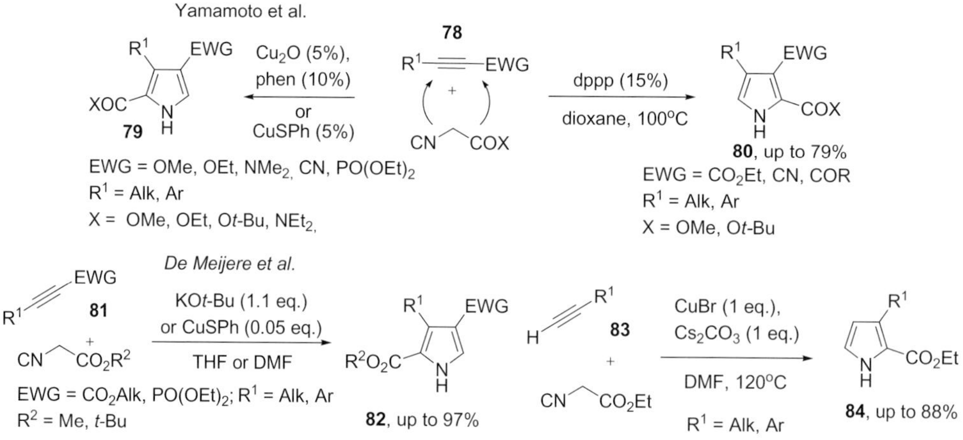

Scheme 4.20

and de Meijere [90]. In 2005, Yamamoto et al. developed the [3+2] cycloaddition of isocyanoacetate and activated alkynes, such that the reactions of activated alkynes **78** with isocyanoacetic acid derivatives would afford regioisomeric pyrroles **79** and **80** in the presence of a Cu(I)-catalyst or triphenylphosphine, respectively [89]. In contrast, De Meijere et al. employed the Cu(I) benzenethiolate or preactivated nanosized copper powder (Cu⁰–NP) catalysts, which allowed the scope of this reaction to be expanded, such that pyrroles **82** could be obtained from acetylenes **81** and isocyanoacetates (Scheme 4.20) [90]. This condensation may also be performed in the presence of KH as a base, in methyl *tert*-butyl ether (MTBE) [91].

4.6
Reaction of Isocyanoacetic Acid Derivatives with Carbonyl Compounds and Imines

In 2009, this methodology was extended by de Meijere *et al.* to include terminal alkynes as the inputs. Consequently, the reaction between substituted ethyl isocyanoacetate and nonactivated terminal alkyne **83** afforded 2,3-disubstituted pyrroles **84** in the presence of 1 equiv. of Cu(I) (Scheme 4.20) [92].

4.6
Reaction of Isocyanoacetic Acid Derivatives with Carbonyl Compounds and Imines

In general, a molecule of isocyanoacetate can primarily attack a carbonyl or imine function, via either the isocyano group or the deprotonated methylene group. Depending on the reaction conditions employed, a range of products may be obtained via the reactions of isocyanoacetic acid derivatives with carbonyl compounds and imines (Figure 4.2) [2].

The reaction of isocyanoacetates with carbonyl compounds in the presence of strong bases such as BuLi [93], *t*-BuOK, and NaH [12, 94–96], leads to α-formylaminoacrylates **87** [97, 98]. This reaction is tolerant of a wide range of aromatic or nonenolizable aldehydes and aromatic and aliphatic ketones. Likewise, tertiary α-isocyanoacetamides may react with carbonyl compounds in the presence of NaH/DMF to produce α-formylaminoacrylamides, in good yields (Scheme 4.21)

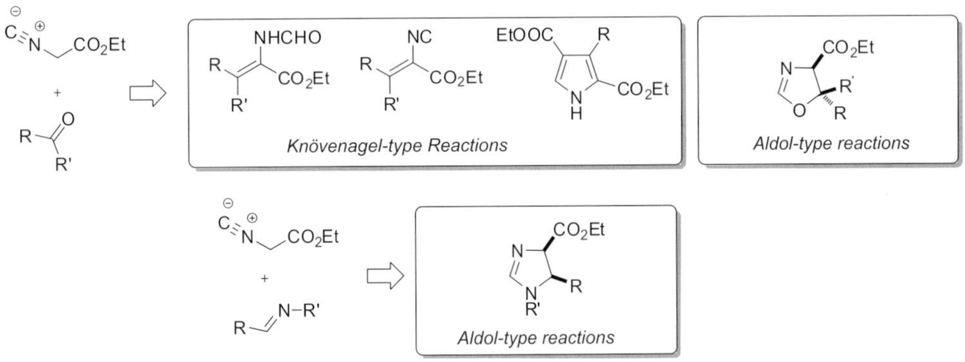

Figure 4.2 Knoevenagel-type reactions of isocyanoacetic acid derivatives.

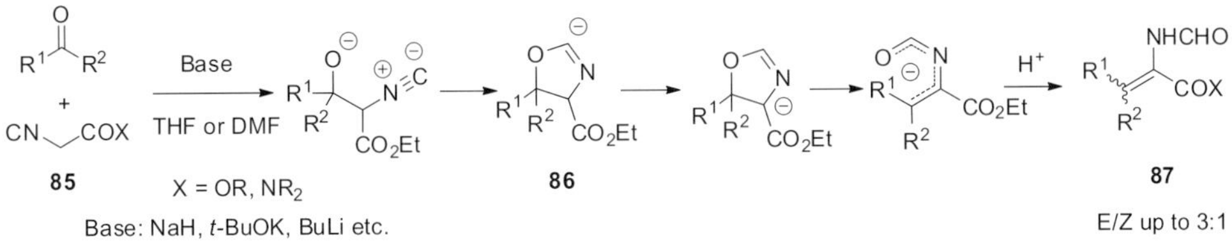

Scheme 4.21

[12, 99, 100]. When nonsymmetrically substituted carbonyl compounds are used, however, formylaminoacrylates are obtained as mixtures of E/Z-isomers, which usually may be separated using column chromatography [96, 97, 101, 102]. When the aldehydes are applied, then the Z-configured product will predominate, although the reaction of isocyanoacetates with α-trifluoromethyl ketones and 1-oxoalkylphosphonates will yield Z-isomers exclusively [103, 104]. A more flexible approach to **87** is based on the ring opening of the corresponding oxazolines **86**, which occurs in the presence of a strong base, and in a high yield. A one-pot procedure has been described for this, in which oxazolines are first generated quantitatively upon transition metal catalysis under mild conditions, after which a strong base is added to initiate the ring opening towards **87** [105].

The reaction with preservation of the isocyanide group can be performed in the presence of secondary amines, although only in selected cases. Thus, the reaction of ethyl isocyanoacetate **19** with cyclic ketones **88** in the presence of secondary amines (which are relatively weak bases) affords α-isocyanoacrylates **89** in moderate to good yield. Notably, acyclic ketones **90** under these conditions undergo amidation to afford isocyanoacetamides **91** exclusively (Scheme 4.22) [12, 106].

Generally, a regular Knoevenagel reaction is not typical for α-isocyanoacetates, due to the high reactivity of the resulting α-isocyanoacrylates towards any subsequent Michael addition. Thus, in the case of aromatic aldehydes the Knoevenagel product **92** will immediately undergo a Michael addition of the secondary amine, followed by a base-induced fragmentation into **93** [96]. At this point it could be speculated that the stability of ketone-derived Knoevenagel products **89** may be explained by their reluctance to undergo the Michael addition. Another example of the Knoevenagel-type reaction is the synthesis of pyrroles **97**, via the reaction of aldehydes with two molecules of isocyanoacetate [100, 107]. Presumably, the Knoevenagel product **95** undergoes a Michael addition of a second α-isocyanoacetate molecule, followed by cyclization of the Michael product **96** into the pyrrole **97** after the elimination of HCN (Scheme 4.22).

Special mention should be made here of the reaction that is catalyzed by the rhodium catalyst, whereby the reaction of ketones with isocyanoacetate **19** leads directly to the production of α-formylaminoacrylates **98**. Although the reaction has been suggested to proceed through Fischer carbene complexes, no mechanistic studies have yet been reported. Subsequently, cyclocondensation with various 1,3-dicarbonyl compounds **99** provides the corresponding ethyl pyrrole-2-carboxylates **102** regioselectively, on the basis of either steric or electronic effects. The reaction proceeds via the α-formylaminoacrylates **100**, which undergo decarbonylation to form the intermediate **101**, followed by subsequent cyclization into the pyrrole **102** (Scheme 4.23) [108].

4.6.1
Aldol-Type Reaction of Isocyanoacetic Acids with Aldehydes: Synthesis of Oxazolines

The reaction of α-isocyanoacetic acid derivatives **103** with carbonyl compounds, in the presence of a weak base, affords the corresponding oxazolines **104**. This

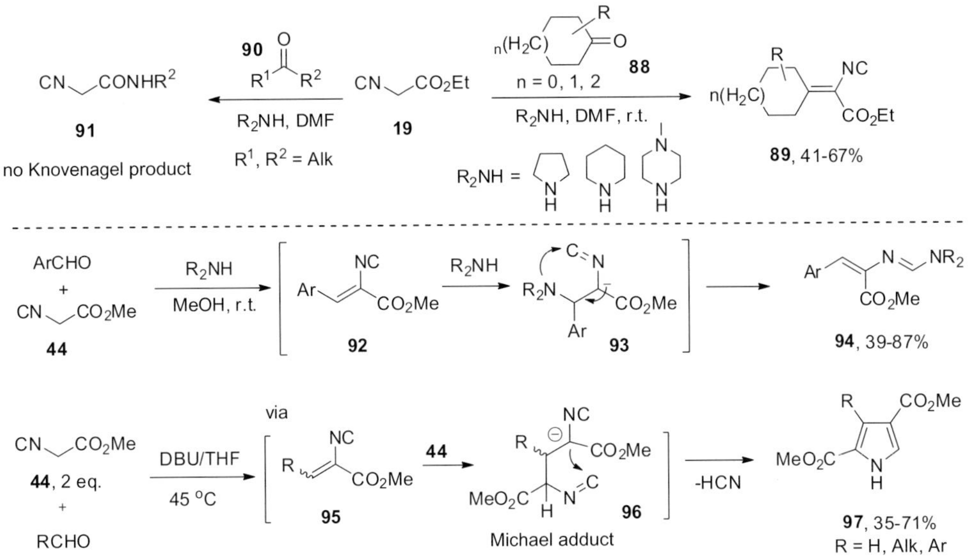

4.6 Reaction of Isocyanoacetic Acid Derivatives with Carbonyl Compounds and Imines

Scheme 4.22

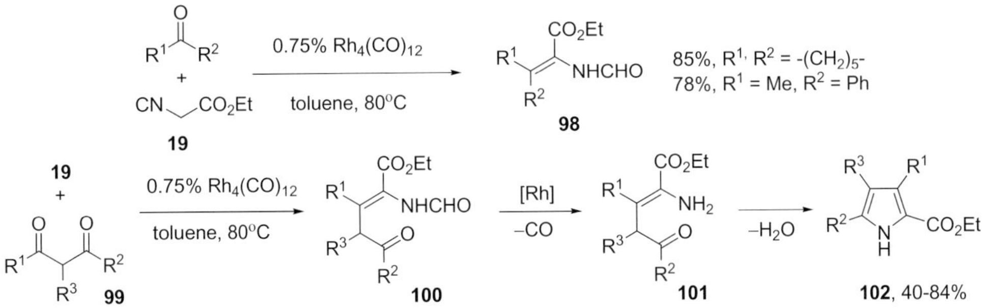

Scheme 4.23

reaction proceeds in the presence of a catalytic amount of either NaCN [109, 110], sodium ethoxide, fluoride ion [111], or Et_3N [112, 113], although at a later stage strong nonionic bases (e.g., $P(RNCH_2CH_2)_3N$; proazaphosphatranes) were also employed successfully as catalysts in this reaction [114]. In general, the *trans*-oxazoline is obtained at >90% d.e., and formed via the thermodynamically controlled process. The presence of alkaline conditions would promote an epimerization of the *cis*-isomer to the thermodynamically more stable *trans*-oxazoline during the reaction (Scheme 4.24). Whilst linear ketones require heating to achieve a good

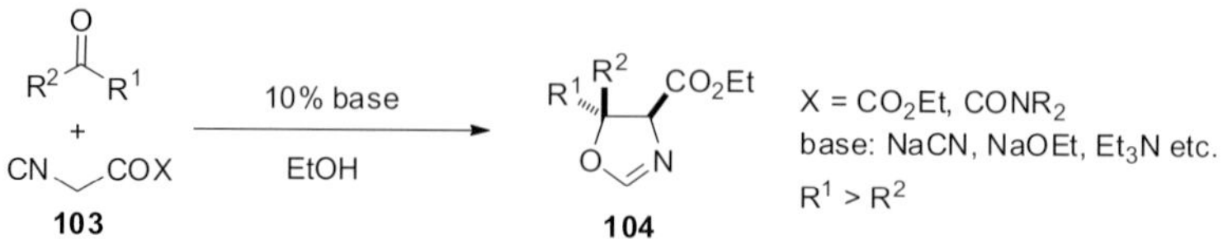

124 | *4 Synthetic Application of Isocyanoacetic Acid Derivatives*

Scheme 4.24

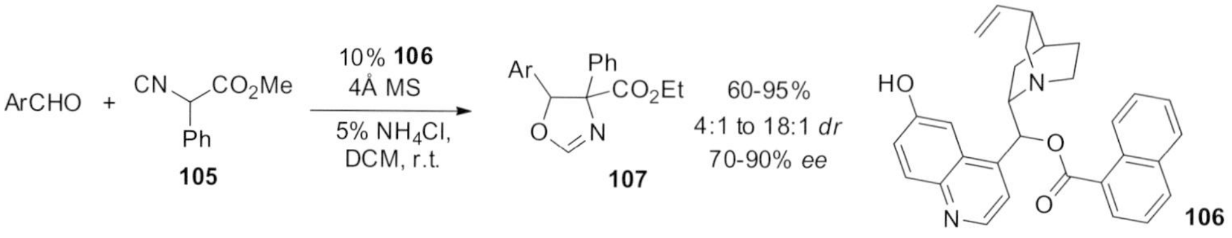

Scheme 4.25

yield of the product, the sterically hindered or aromatic ketones provide only a sluggish reaction, if at all. Primary, secondary, or tertiary α-isocyanoamides furnish the corresponding oxazolines in the presence of KOH or NaH [30, 115]. The thioketones may also react with isocyanoacetic esters to produce 2-thiazoline-4-carboxylic esters [116].

The first organocatalytic asymmetric version of the oxazoline synthesis using cinchona alkaloids was developed by Xue et al., in 2009, Thus, the reaction of α-substituted isocyanoacetates **105** with electron-poor aromatic aldehydes in the presence of **106** affords oxazolines **107** with good diaselectivity and enantioselectivity (Scheme 4.25) [117].

4.6.2
Transition Metal-Catalyzed Aldol-Type Reactions

During recent years, enormous efforts have been made to develop an efficient and stereoselective transition metal-catalyzed approach to oxazolines, via the reaction of carbonyl compounds with isocyanoacetic acid derivatives. In principle, every soft metal-based Lewis acid that is capable of forming a complex with an isocyano functionality can be used as a catalyst (e.g., Cu, Ag, Au, Pd, Zn, Ni, Ru, Rh) [2]. The catalytic effect of the metal in this transformation is based on the ability of a metal ion to form a complex with isocyanide type **108a**, with a significant enhancement of the acidity of the α-H. Moreover, it also allows a smooth deprotonation by a weak base (typically, triethylamine or i-Pr$_2$NEt are used) to produce **108b**. Following the subsequent electrophilic attack to a carbonyl compound and intramolecular cyclization of **109**, the 2-oxazoline complex **110** is formed. However, due to the low complexing ability of the oxazolines, the catalytic species will be regenerated upon liberation of the free oxazoline **107** (Scheme 4.26).

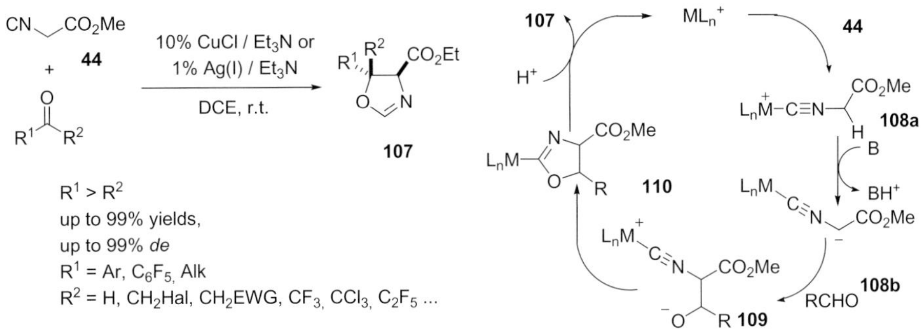

4.6 Reaction of Isocyanoacetic Acid Derivatives with Carbonyl Compounds and Imines

Scheme 4.26

Usually, the predominant *trans*-oxazoline is prepared as a main diastereomer, though on occasion the selectivity will be lower than is observed for the corresponding base-catalyzed reaction. This suggests that the transition metal-catalyzed cyclocondensation proceeds predominantly under a kinetic control.

The reaction was also applied to ketones, but their lower reactivity imposed certain restrictions on the scope of acceptable substrates [118–120]. In all such cases, the conversions using nonactivated aliphatic and aromatic ketones were less than 50% at room temperature. However, the presence of at least one electron-withdrawing substituent at the α-position of the carbonyl input drastically facilitated the reaction, such that the yield exceeded 90%. Even highly enolizable ketones that do not usually react as aldol acceptors under basic conditions can be involved in this metal-catalyzed synthesis.

Recently, highly efficient enantioselective oxazoline formation has been developed using asymmetric gold, silver, and cobalt catalysts. Although enantioselective Pd- and Pt-catalyzed reactions (using chiral pincer complexes) were each investigated, no highly efficient stereoselective catalyst systems were found. Several good examples were also noted of diastereoselective synthesis using a chiral substrate and an achiral metal catalyst [2]. For example, an enantioselective oxazoline **111** formation using chiral ferrocenylphosphine–gold(I) complexes **112** was first demonstrated by Ito and Hayashi in 1986 [121]. Subsequently, the reaction scope, and the influence of both ligand and metal on the stereoselectivity have been widely investigated, and the reaction has since become a valuable methodology for stereoselective C–C bond formation [121–126]. The most efficient catalysis was provided by gold complexes with *N*-aminoalkyl-*N*-methyl-1-[1′,2-bis(diphenylphosphino)ferrocenyl]ethylamine ligands **112a–c**, with ligands bearing a cyclic terminal amino group (e.g., morpholino or piperidino groups, **112b,c**) being generally found as superior. The high efficiency of the gold catalyst can be explained by the postulated transition state **113**, where the coordinated gold atom and the basic amine function of the pendant side chain cooperatively activate an isocyanoacetate. This allows the reaction to occur in a highly stereoselective fashion within the asymmetric complex (Scheme 4.27) [121, 127, 128].

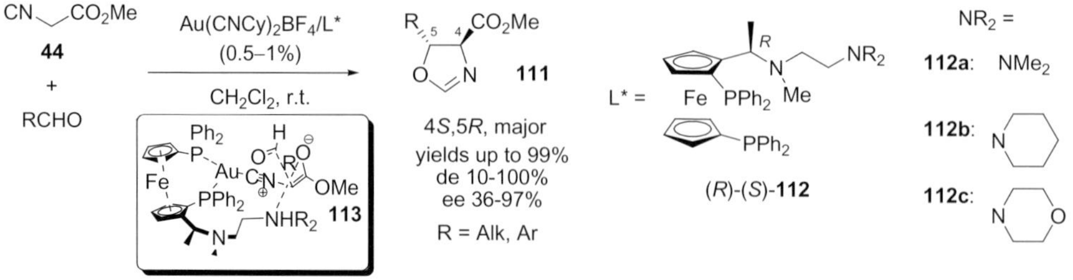

Scheme 4.27

A variety of aldehydes produce the corresponding oxazolines, although the stereoselectivity of the reaction depends on the nature of aldehyde. Thus, bulky and α,β-unsaturated aldehydes will produce the corresponding *trans*-oxazolines almost exclusively, and with high enantioselectivity. Aromatic aldehydes bearing electron-donating and neutral substituents also react in a highly stereoselective manner. In contrast, the introduction of electron-withdrawing groups into the phenyl ring will lead gradually to a decrease in diastereoselectivity and enantioselectivity [129]. This effect might be rationalized by the π–p interaction between the electron-deficient phenyl ring and the negatively charged enolate anion, which favors formation of the *cis*-isomer. In this respect, 2-heteroaromathic aldehydes are also poor substrates, due to the influence of the heteroatom [130]. Furthermore, α-isocyanoacetamides [131, 132] and α-isocyano Weinreb amides [133] represent perfect substrates for the reaction, but require longer reaction times for their complete conversion. With different stereoselectivities, however, asymmetric aldol reactions can be carried out for α-substituted isocyanoacetates with aldehydes [134, 135], and α-isocyanoacetates/amides with ketones [136].

Whilst the use of Au(I) is essential for the high stereoselectivity, the corresponding silver(I) and copper(I) catalysts are much less selective [121], as explained in stoichiometric studies of these reactions [127, 128, 137–139]. However, a good stereoselectivity of the silver-catalyzed reaction was observed at elevated temperatures [137].

Notably, in 2011 Dixon et al. developed a highly stereo- and enantioselective Ag-catalyzed aldol reaction of isocyanoacetates and aldehydes, using the aminophosphine ligand **114**. This reaction tolerated aliphatic and aromatic aldehydes, providing oxazolines in highly stereoselective fashion (Figure 4.3) [140]. Moreover, in 2011 Oh et al. reported the Co(II)-catalyzed/organocatalyzed stereo- and enantioselective reaction of isocyanoacetates and aldehydes, producing oxazolines, by using brucine amino diol **115** as a ligand, and acidic thiourea **116** as an organocatalyst (Figure 4.3) [141].

4.6.3
Reaction of Isocyanoacetic Acids with Imines: Imidazoline Formation

The reaction of ethyl isocyanoacetate with imines leading to imidazolines, which has been known since 1977 and is based on the studies of Schöllkopf et al., resem-

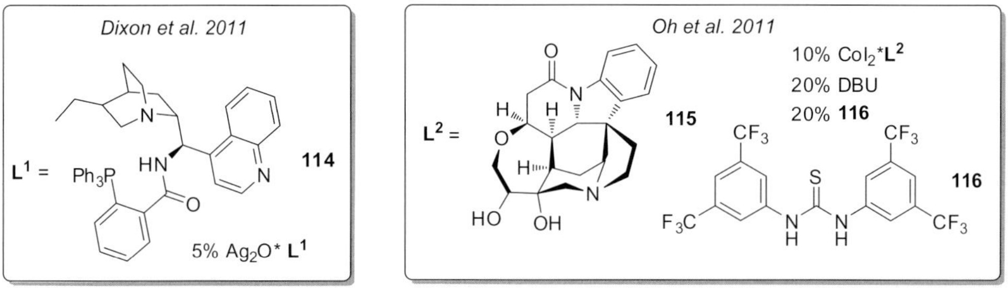

Figure 4.3 Recently developed catalytic systems for enantioselective oxazoline synthesis.

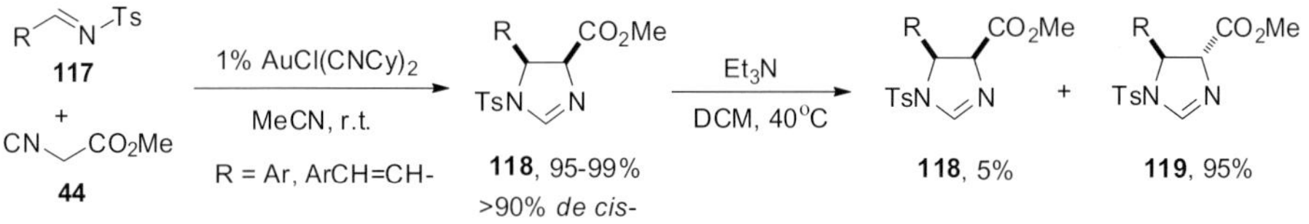

Scheme 4.28

bles the previously described oxazoline synthesis [142]. Generally, this reaction requires either a base or a metal catalyst, although in the case of aliphatic imines it may proceed even without the addition of a base, with an excellent yield; any traces of an amine present in the starting substrate may act as the base, however.

As might be expected, transition metal catalysts can greatly enhance the reactivity of the isocyanoacetate via metal coordination to the isocyano group. In 1996, Hayashi described the use of several Cu, Ag, Au, Pd, Ni, and Rh complexes in the base-free reaction of several N-sulfonylaldimines **117** with methyl isocyanoacetate **44** [143]. The best diastereoselectivity was demonstrated using the AuCl(CNCy)$_2$ catalyst, furnishing *cis*-isomers **118** with high d.e.-values, despite the fact that the *trans*-isomers were more thermodynamically stable (in contrast to the transition metal-catalyzed *trans*-selective oxazoline synthesis). Correspondingly, *trans*-isomers **119** could be obtained via the epimerization of *cis*-isomers **118** in the presence of Et$_3$N (Scheme 4.28).

One interesting peculiarity of the metal-catalyzed reaction of isocyanoacetates with imines is that the configuration of the major isomer (*trans* or *cis*) can depend dramatically on the catalyst chosen. Thus, in 1997, Lin reported a highly efficient (81–95% d.e.) *trans*-imidazoline synthesis at room temperature using a RuH$_2$(PPh$_3$)$_4$ catalyst. Likewise, IrH$_5$(*i*-Pr$_3$P)$_2$, FeH$_2$(dppe)$_2$, and some simple palladium complexes also exhibited *trans*-selectivity in catalysis, though less efficiently than their ruthenium counterpart [144]. The catalytic cycle proposed for this reaction is similar to that described above for the aldol reaction, but involves a C–H activation step (notably, a similar aldol reaction of nitriles has been reported) [145].

The reaction is also catalyzed by different (PCP), (SCS), (SeCSe), and (NCN) pincer palladium complexes, though different diastereoselectivities have been

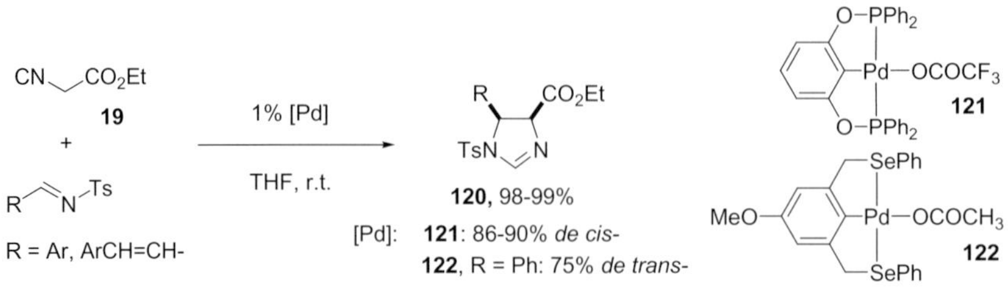

Scheme 4.29

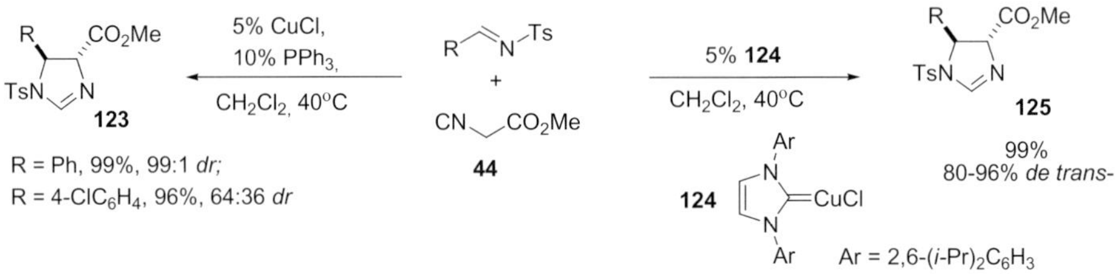

Scheme 4.30

observed, depending on the ligand [146]. Thus, an electron-deficient phosphine complex **121** displays optimal *cis*-selectivity, whereas electron-rich selenide complexes **122** showed a reversal of selectivity towards the *trans* isomer of oxazoline **120** (Scheme 4.29). These differences in stereo-directing performance by the complexes might be explained by their different behaviors in the catalytic cycle, rather than by the electronic properties of the ligand. It has been shown that the (SCS), and (NCN) complexes – but not the (PCP) pincer Pd complexes – can undergo the insertion of an isocyanide into the C–Pd bond [147, 148]. Consequently, the species that actually enter the catalytic cycle may differ in these cases, and this may explain the differences observed in diastereoselectivity.

In 2006, a highly diastereoselective imidazoline synthesis was demonstrated using copper complexes, with up to 99% d.e. of imidazolines **123** being achieved with use of the CuCl/PPh$_3$ catalyst; the diastereoselectivity was heavily dependent on the substrate utilized, however. In this case the copper carbene complex **124** proved to be more suitable, and showed an excellent *trans*-selectivity of **125** in the reaction of methyl isocyanoacetate **44** with *N*-tosylaldimines obtained from benzaldehydes and pivalaldehyde (Scheme 4.30) [119].

An enantioselective variant was accomplished using ferrocenylphosphine gold(I) complexes **112**, analogous to those used in the asymmetric aldol reaction (see Scheme 4.27). A neutral complex (AuCl)$_3$L*$_2$, formed *in situ* from an AuCl(Me$_2$S) precursor, was the catalyst of choice for this transformation (Scheme 4.31) [149, 150].

Attempts to prepare imidazolines from isocyanoacetamides have led to the observation of completely different chemistries. For example, it was found that

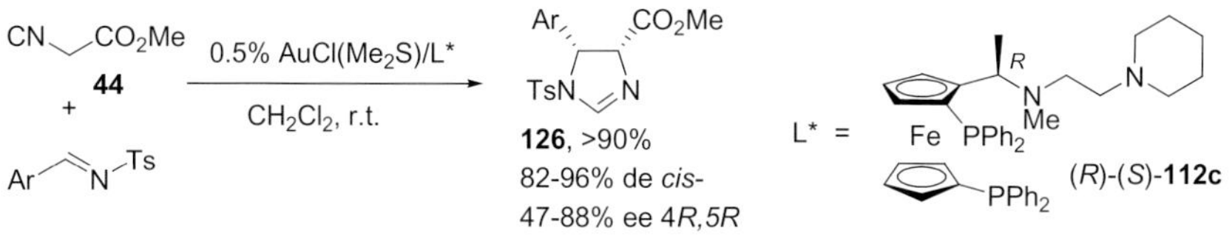

Scheme 4.31

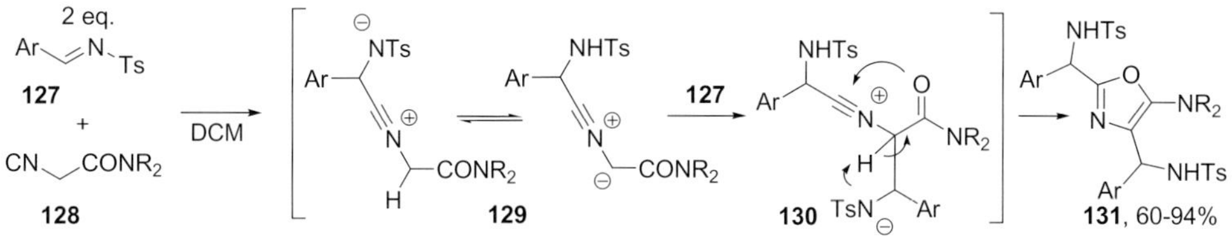

Scheme 4.32

isocyanoacetamides **128** would react with N-sulfonylimines **127** under neutral conditions without the use of an additive, a base, nor any other catalyst, to produce 2,4,5-trisubstituted oxazoles **131** in good yields [151]. The reaction was shown to proceed via an attack of isocyanide on the imine bond so as to generate intermediate **129**, which in turn attacks another molecule of imine **127**, followed by cyclization of the intermediate **130** (Scheme 4.32). Substrates bearing electron donor substituents at the aromatic ring were less reactive; typically, when methyl isocyanoacetate or tosylmethyl isocyanide were used, no reaction occurred due to the reduced nucleophilicity of the isocyanide group.

It should be noted here that, under suitable conditions, a similar nucleophilic attack of the isocyano group on activated C=O and C=N bonds will also occur, with subsequent cyclization to form an oxazole ring. The formation of imidazolines may occur competitively in the presence of a metal catalyst (see Chapter 3).

4.7
Reaction with Acylating Agents

In general, isocyanides react with acyl chlorides to afford products of α-addition (Nef reaction). However, because the nucleophilicity of the isocyano group is reduced, isocyanoacetates are able to react smoothly only with aliphatic or electron-deficient aromatic acyl chlorides to produce imidoyl chlorides **132** [152, 153]. Without isolation, the latter can be treated with a base to provide 2,5-disubstituted oxazoles **134** via the formation of nitrilium ylides **133** (Scheme 4.33) [153, 154]. The reaction of α-isocyanoacetamides with acyl chlorides, featuring ketene formation, affords a similar oxazole product [155].

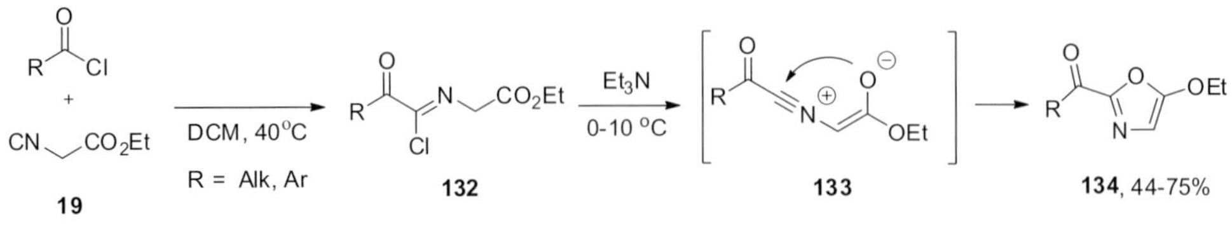

Scheme 4.33

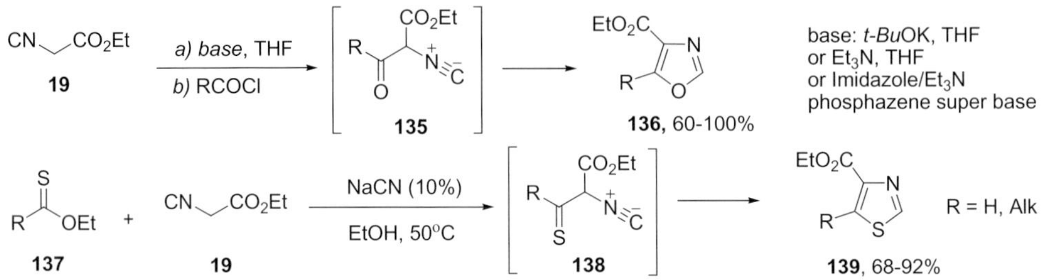

Scheme 4.34

In contrast, when isocyanoacetate is treated first with a base, followed by the addition of an acyl chloride, the formed unstable α-acyl isocyanoacetate **135** cyclizes spontaneously to 4,5-disubstituted oxazole **136** (Scheme 4.34). This reaction was discovered and developed primarily by the groups of Schöllkopf and Matsumoto during the 1970s, and constitutes a general method for the synthesis of oxazole carboxylates **136** [156–158]. In order to generate the isocyanoacetate anion, Schöllkopf used strong bases such as *t*-BuOK; however, this required the addition of 2 equiv. of both an isocyanide and a base to achieve a good yield of an oxazole. In contrast, the use of amine bases, as described by Matsumoto, offers several advantages: (i) it does not require an excess of isocyanoacetate; (ii) it occurs in milder conditions of reversible enolization; and (iii) it affords the oxazoles **136** in good yields [158–161]. The use of phosphazene super bases for this reaction was demonstrated in 1994 by Verkade *et al.*, with excellent results being obtained for aromatic acyl chlorides and anhydrides [162]. Similarly, ethyl isocyanoacetate reacts smoothly with aliphatic thioesters **137** in the presence of 10% NaCN to afford thiazole derivatives **139**, via the formation of α-thioacyl isocyanoacetate (Scheme 4.34) [163].

Very recently, Piraly *et al.* described an interesting approach to furans **144** via the formal three-component reaction of isocyanoacetamides, aliphatic acyl chlorides, and acetylenedicarboxylates. The reaction would run via a base-induced formation of ketene **141**, which would undergo reaction with isocyanoacetamide to produce the oxazole **142**. The subsequent cycloaddition of oxazole with activated alkyne **143** furnished the final furan **144**. The reaction was shown to proceed with various substrates, producing the desired oxazoles **144** in good yields (Scheme 4.35) [164].

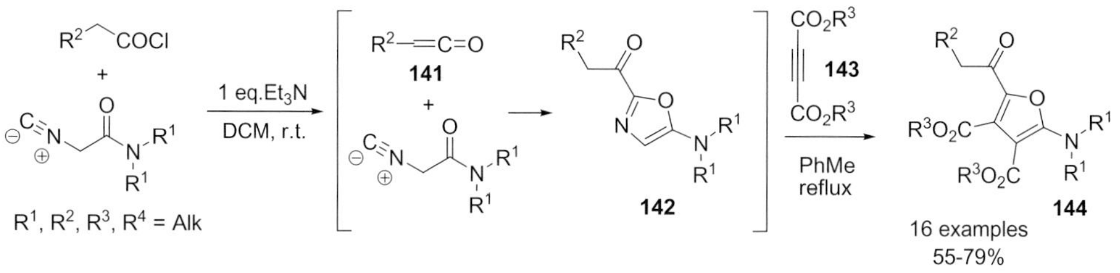

4.7 Reaction with Acylating Agents

Scheme 4.35

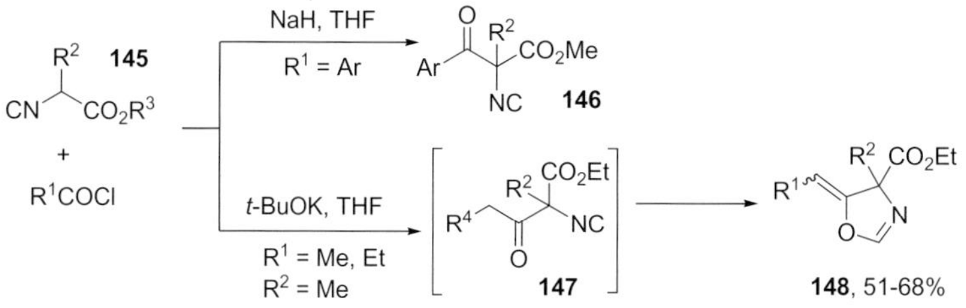

Scheme 4.36

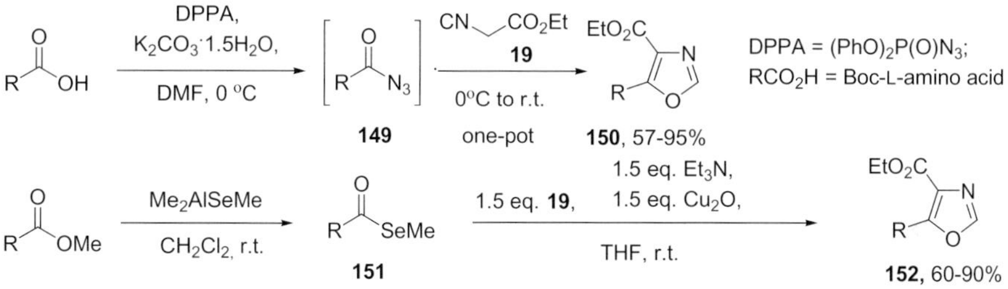

Scheme 4.37

α-Substituted isocyanoacetates **145** react with aroyl chorides **146** in the presence of base to produce stable products of C-acylation **147** that contain a quaternary carbon center [157]. In contrast, the C-acylation of isocyanoacetates **145** with aliphatic acyl chlorides produces an unstable α-acyl isocyanoacetate **147**, which then undergoes cyclization to oxazolines **148** (Scheme 4.36) [158].

Besides acyl chlorides, carboxylic acid anhydrides and N-acylimidazoles are also suitable acylating agents [165]. Notably, the *in situ* activation of free carboxylic acids or their salts with diphenyl phosphorylazide (DPPA) produces acyl azides **149** which, in turn, furnish oxazoles **150** upon reaction with ethyl isocyanoacetate (Scheme 4.37) [166–171]. This method allows the direct use of a carboxylic acid, and is easily applicable to aliphatic acids, including chiral forms. In 1980,

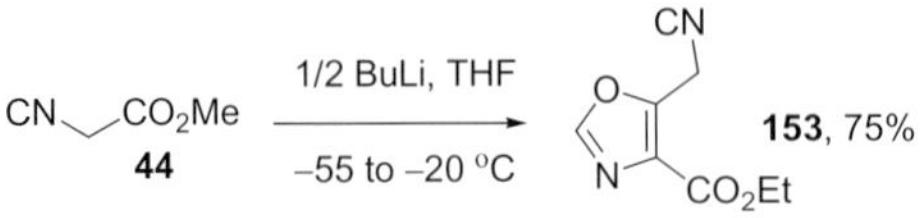

Scheme 4.38

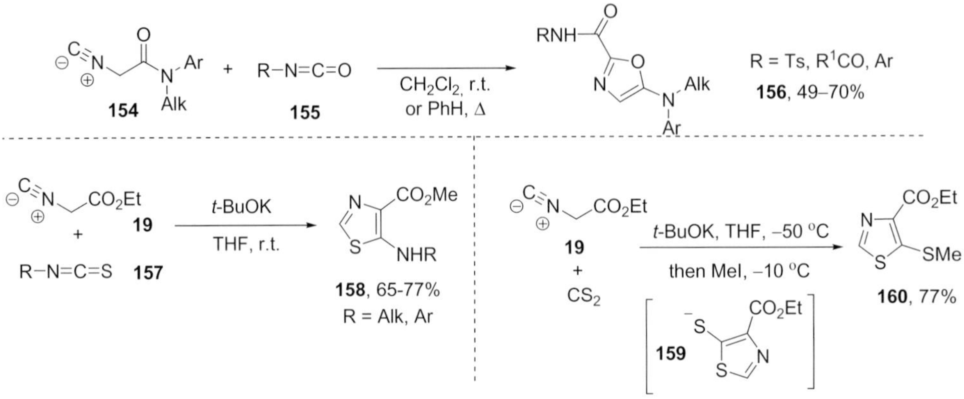

Kozikowski et al. employed selenoesters **151** as acylating agents in the reaction with isocyanoacetate (Scheme 4.37) [172, 173].

In the absence of any acylating agent, but in the presence of 0.5 equiv. of a strong base, methyl isocyanoacetate **44** will condense with itself to give oxazole **153** (Scheme 4.38).

The reactions of isocyanoacetate derivatives with heterocumulenes have been investigated only rarely. Heterocumulenes are able to react with isocyanoacetates via several routes, depending on the reaction conditions and the nature of the substrates [174]. Thus, electron-deficient isocyanates **155** react with tertiary N-alkyl-N-aryl isocyanoacetamides **154** under neutral conditions to afford 2-substituted oxazoles **156** (Scheme 4.39) [175]. Isothiocyanates **157** undergo a base-induced reaction with isocyanoacetates to afford 5-aminothiazoles **158** in good yields [176, 177]. Similarly, the reaction of isoselenocyanates with isocyanoacetic acid derivatives produces selenazoles in good yields [178]. The reaction with carbon disulfide yields thiazolethiolates **159** that may be trapped *in situ* with methyl iodide to produce the corresponding thiazole **160** (Scheme 4.39) [179].

The reaction of isocyanoacetic acid derivatives with imidoyl chlorides produces a wide variety of imidazoles. For example, trifluoromethyl imidoyl chlorides **161** (which are easily accessible from the reaction of trifluoroacetic acid, an amine, carbon tetrachloride and PPh$_3$) react with ethyl isocyanoacetate **19** to produce N-substituted imidazoles **162** in good yield (Scheme 4.40) [180]. A more general strategy consists of the activation of secondary amides, leading to an *in-situ* forma-

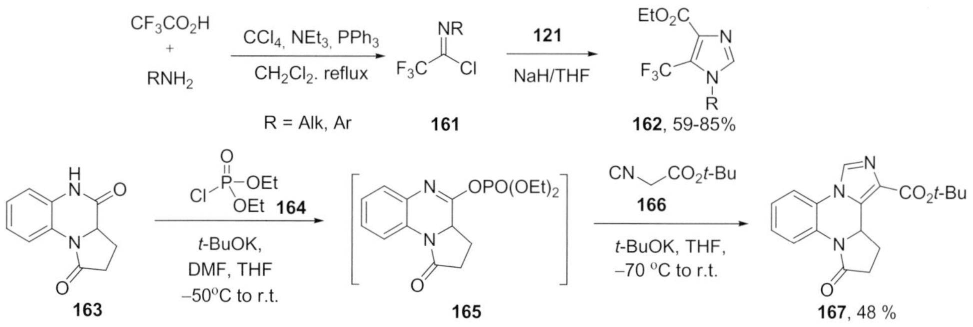

Scheme 4.40

tion of the required electrophilic agent that enters the subsequent reaction with isocyanoacetate, in one pot. As an example, cyclic amide **163** may undergo reaction with diethyl chlorophosphonate **164** to produce derivative **165**, while the subsequent base-induced reaction with isocyanoacetate **166** furnishes fused imidazole **167** (Scheme 4.40) [181]. This approach was employed by Erker et al. for the annulation of an imidazole core to different heterocyclic systems [181–189]. Other substrates which are suitable for imidazole formation include electron-deficient chlorosubstituted heterocycles [190] containing the N=C–Cl fragment, and some activated electron-deficient nitriles [191].

4.8
Multicomponent Reactions of Isocyanoacetic Acid Derivatives

Today, multicomponent reactions (MCRs) represent powerful tools in modern organic synthesis, with excellent diversity- and complexity-generating abilities [192, 193]. The isocyanide-based MCRs, such as the Passerini and Ugi reactions, are important methods for the synthesis of amino acids, peptides, and peptide-like molecules, as well as of heterocycles [194]. The α-isocyanoacetates were used in the Passerini reaction for the first time by Ugi and coworkers in 1962 [195], when they showed this reaction to offer a most convergent approach to the depsipeptides. (The latter are an important family of antitumor agents; typically, a depsipeptide bears a COO function instead of the CONH present in "normal" peptides [196]; Scheme 4.41.)

The Ugi reaction allows the construction of a broad variety of amino acid and peptide structures, with the use of isocyanoacetate derivatives (which contain an amino acid skeleton) opening up an efficient multicomponent approach to small peptides [2]. Notably, isocyanoacetates generally react in straightforward fashion in four-component Ugi reactions, with aldehydes as the carbonyl inputs. In contrast, the Ugi reactions with ketones are often accomplished by the formation of

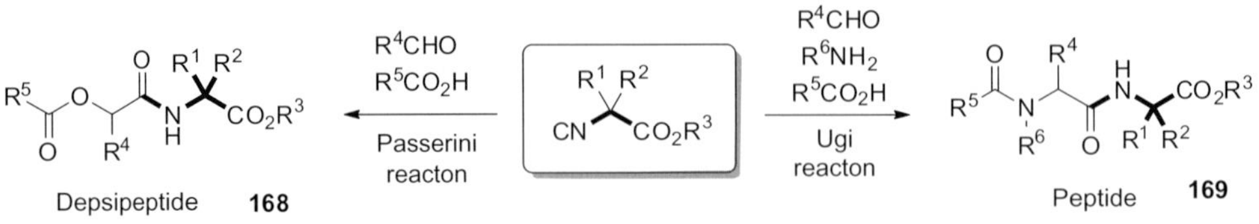

134 | *4 Synthetic Application of Isocyanoacetic Acid Derivatives*

Scheme 4.41

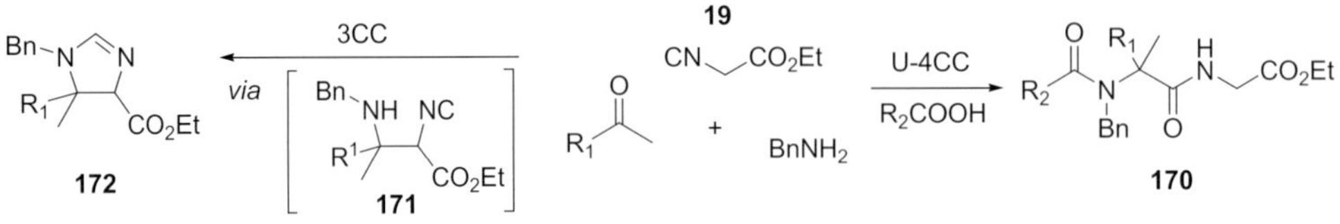

Scheme 4.42

imidazolines **172** as side products, along with the corresponding Ugi products **170** [197]. Due to the relatively high α-acidity of the isocyanoacetates, the corresponding anions can be generated even by weak bases, such as benzyl amine. The subsequent nucleophilic attack of the isocyanoacetate anion to the imine produces intermediate **171** which yields, upon cyclization, imidazoline **172** as the main side product (Scheme 4.42). Details of this three-component reaction were recently reported by Orru *et al.* as a separate synthetic method (see Chapter 3).

4.9
Chemistry of Isocyanoacetates Bearing an Additional Functional Group

β-Dimethylamino-α-isocyanoacrylates represent interesting substances that bear an isocyano functionality, a Michael acceptor, and dimethylamine as a leaving group, all in one molecule. The simplest representative of β-isocyanoacrylates **172** can be prepared by the reaction of isocyanoacetates with dimethylformamide dimethylacetal in ethanol [198]. For the preparation of other dialkylamino analogues of **175**, a versatile three-component reaction between a secondary amine, N-formylimidazole diethylacetal **174**, and isocyanoacetates is employed, whereby the products are obtained as single Z-isomers (Scheme 4.43) [199]. Resin-bound analogues of β-isocyanoacrylates have also been prepared [200].

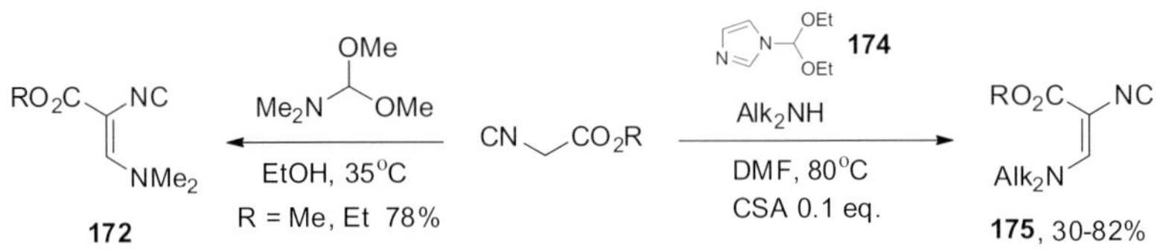

Scheme 4.43

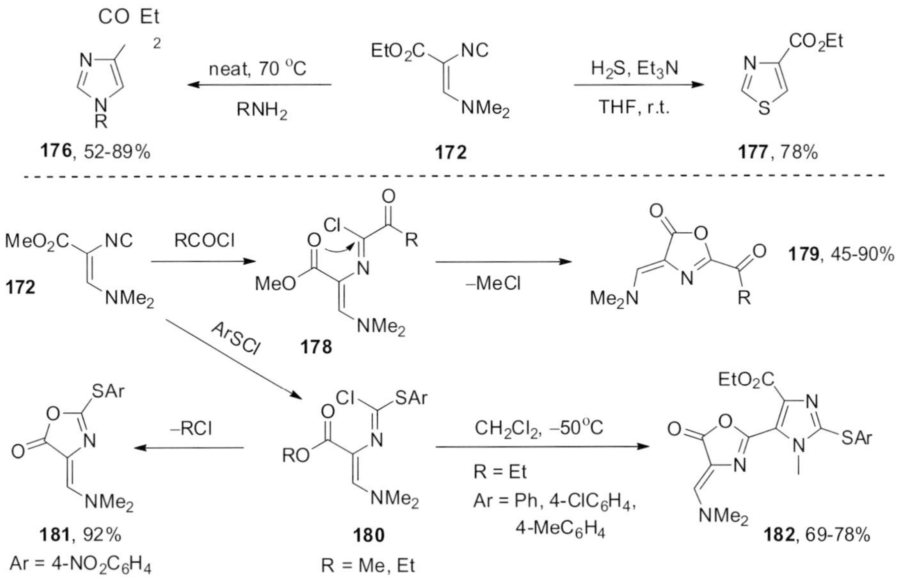

Scheme 4.44

Typically, reactions with nucleophiles begin with Michael addition–elimination with subsequent cyclization. Thus, **172** reacts with primary amines to form the corresponding imidazoles **176**, in good yields. A similar reaction with H_2S furnishes thiazole **177** as a product of the Michael addition–elimination–cyclization sequence [198, 201, 202]. Reactions of β-isocyanoacrylates **172** with acyl halides affords oxazolidinone derivatives **179**. The reaction proceeds via a Nef-type reaction to form intermediate **178**, followed by a subsequent cyclization into **179**. A reaction of β-isocyanoacrylates **172** with arenesulfenyl chlorides might produce oxazolidinone derivatives **181** or **182**, depending on the reaction conditions and the structure(s) of the starting compound(s) (Scheme 4.44) [203–205].

Dimethylamino-α-isocyanoacrylates **172** show even more interesting reactivity in the MCRs, leading to various heterocyclic scaffolds (Scheme 4.45) [206].

β-Bromo-α-isocyanoacrylates (BICAs) **190** are quite similar to β-dialkylamino-α-isocyanoacrylates, and may be prepared in two steps from α-formylaminoacrylates **188** (see Scheme 4.21 for the synthesis) via bromination and dehydration. The geometric isomers of β-bromo-α-formylaminoacrylates **189** that are obtained after the first step can be easily separated to obtain the pure isomers (the Z-isomer generally predominates). Subsequent dehydration occurs under classic conditions, usually in high yield and without any significant isomerization, which enables the preparation of BICA **190** as a single stereoisomer (Scheme 4.46) [101, 207].

The BICAs have an extra diversity point, as they contain an additional substituent (R), which can easily be varied depending on the aldehyde used to synthesize

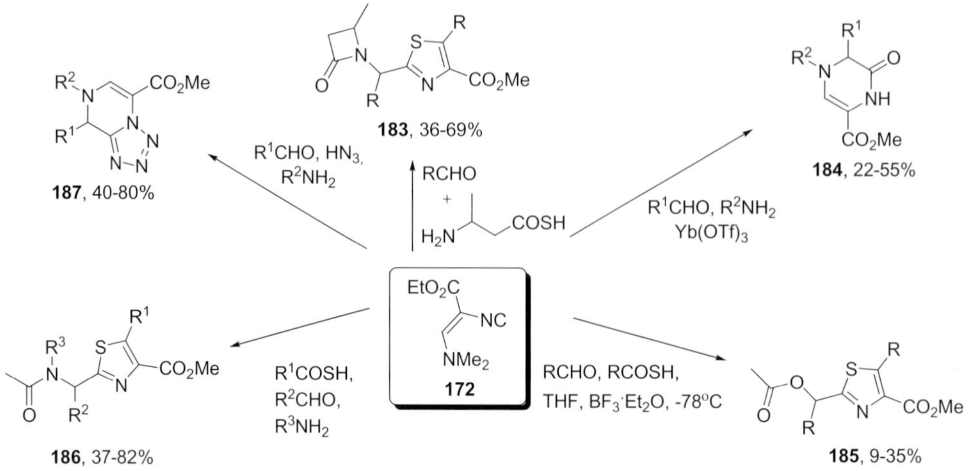

Scheme 4.45

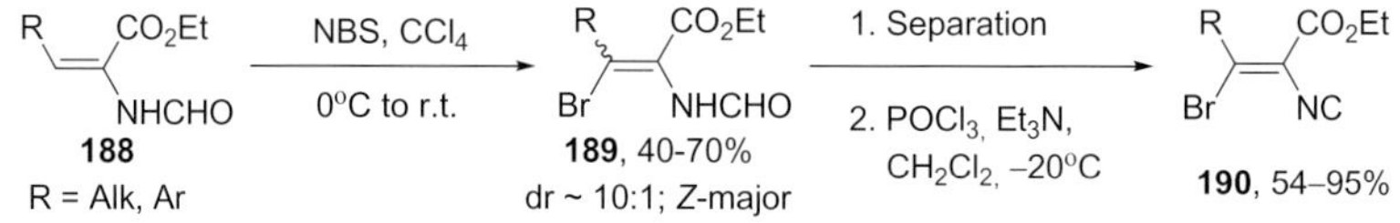

Scheme 4.46

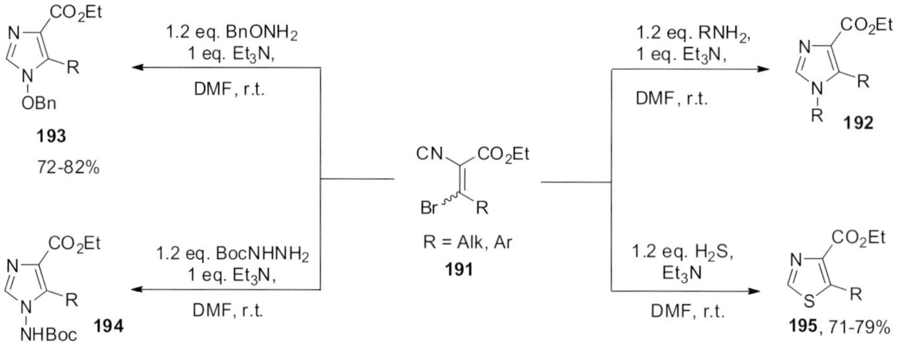

Scheme 4.47

the starting α-formylaminoacrylates. Although BICAs and their precursors can be isolated as single geometric isomers, reactions accompanied by heterocyclization (including MCRs) can be performed on the E/Z-mixtures. Previously, the BICAs have been used to synthesize a variety of heterocycles, such as imidazoles and thiazoles [207–209]. Analogously, reactions of **191** with hydroxylamine and hydrazine derivatives have led to the N-hydroxy and N-aminoimidazole derivatives **193** and **194**, respectively (Scheme 4.47) [210].

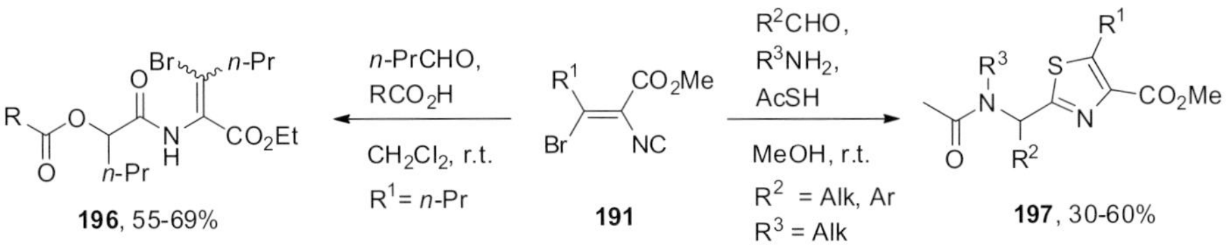

4.9 Chemistry of Isocyanoacetates Bearing an Additional Functional Group | 137

Scheme 4.48

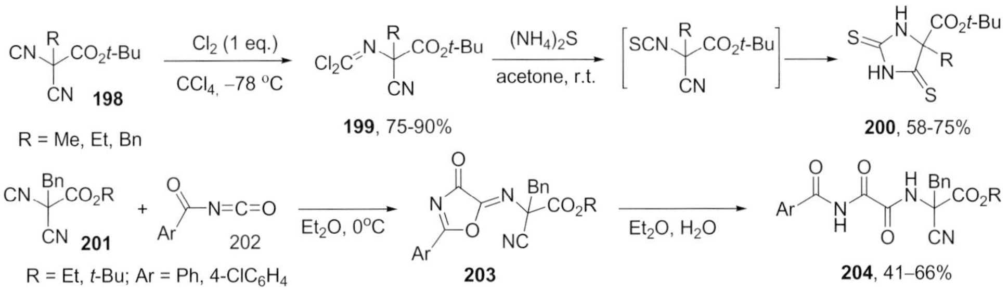

Scheme 4.49

A multicomponent synthesis of thiazoles **197** has also been described where, in contrast to the method described for β-dimethylamino-α-isocyanoacrylate, the BICA-based method permits the further introduction of diversity points in the final molecule by having an additional substituent at the 5-position of the thiazole system [211]. Several depsipeptides **196** were also obtained via a Passerini reaction in moderate yield, without affecting the bromoalkene moiety (Scheme 4.48) [212].

α-Cyano-α-isocyanoacetates **198** contain three centers that are susceptible to nucleophilic attack, namely an isocyano, a cyano, and an ester group. The selectivity of the nucleophilic attack depends on the nature of the nucleophile used. Reactions with alkoxides, thiolates, primary and secondary alkyl amines were reported to result in various products, depending on the nature of the nucleophile used [213]. α-Cyano-α-isocyanoacetates **198** react smoothly with Br_2 and Cl_2 to provide the corresponding dibromides and dichlorides [214] (Scheme 4.49).

Upon treatment with $(NH_4)_2S$, the dichlorides **199** can be transformed into 2,4-dithiohydantoins **200**. Upon treatment with highly electrophilic acyl isocyanates **202**, the α-cyano-α-isocyanoacetate **201** reacts at the isocyano group producing the corresponding oxazolin-4-one derivatives **203**, and yielding oxamides **204** upon aqueous treatment (Scheme 4.49) [213]. Multicomponent Passerini and Ugi reactions have been reported to occur only in moderate yield (<63%) [215].

4.10
Reactions of Isocyanoacetic Acids with Sulfur Electrophiles

The reaction of isocyanoacetates and amides with different sulfur electrophiles may lead to a diverse array of heterocycles. Thus, the reaction of **19** with arenesulfenyl chlorides **205** proceeds easily at low temperature to provide the corresponding α-addition products **206** that can be isolated in quantitative yield, but may also undergo cyclization upon the addition of triethylamine to yield oxazole derivatives **208** on a quantitative basis (Scheme 4.50) [216]. The tertiary isocyanoacetamide **209** then reacts with arenesulfenyl chlorides to produce trisubstituted oxazoles **213** via an addition of the second equivalent of a sulfenyl chloride to the intermediate **211**, followed by a cyclization of **212** [217]. In the case of alkyl-substituted secondary isocyanoacetamides **214**, the reaction leads to α-addition products **216** that are unstable and also suffer smooth cyclization into the oxazole **218** upon the addition of triethylamine. However, a nucleophilic attack by the nitrogen takes place to yield mesoionic 5-hydroxy imidazole derivatives **218** (Scheme 4.50) [34].

The reaction of ethyl isocyanoacetate **19** with SCl_2 and S_2Cl_2 as electrophilic reagents proceeds at low temperature, via a double α-addition to the isocyano group. Subsequent treatment with triethylamine yields bis-oxazole **221** or thiazolo[5,4-d]thiazole **226**, respectively (Scheme 4.51) [218, 219].

The use of sulfenyl thiocyanates **227** as electrophilic reagents has also been described. Reaction with secondary isocyanoacetamides **228** at low temperature leads initially to the formation of α-addition products **229** that enter a series of transformations, finally affording **230** [38, 220]. In this transformation, the

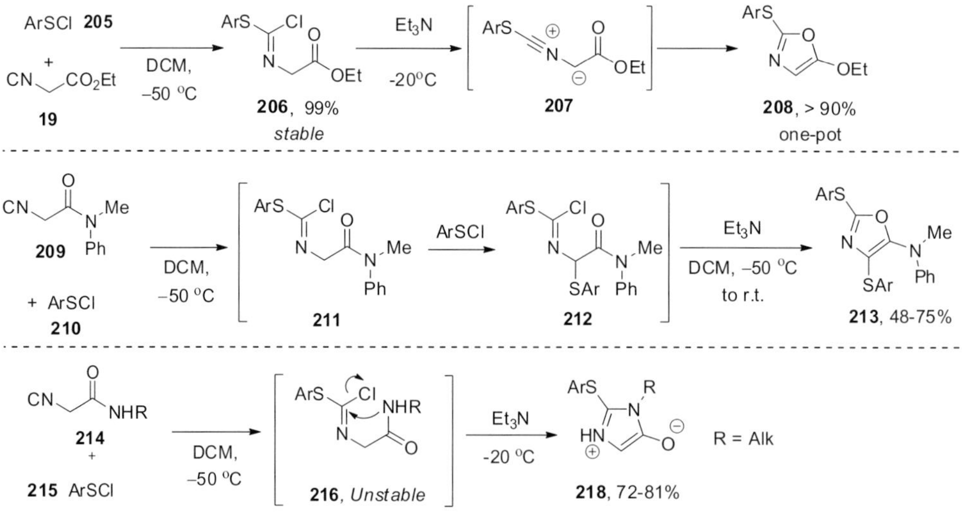

Scheme 4.50

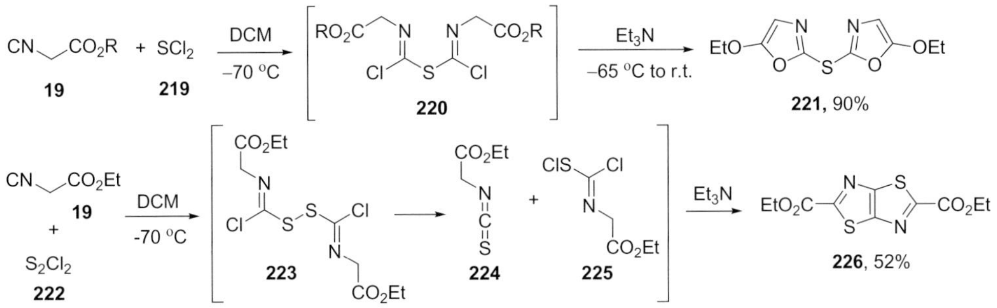

Scheme 4.51

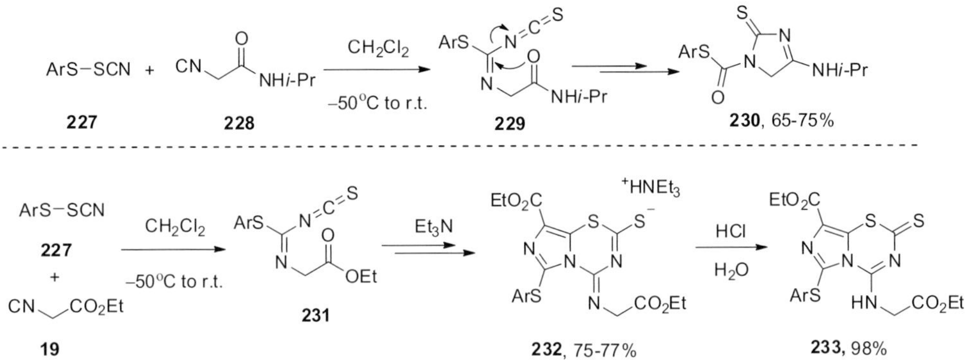

Scheme 4.52

α-methylene group is not involved and no base is required. In contrast, the α-addition products of isocyanoacetates and sulfenyl thiocyanates **227** are thought to be stable towards intramolecular cyclization, but enter a series of transformations upon the addition of triethylamine to yield imidazothiadiazine derivatives **233** (Scheme 4.52) [221].

4.11
Miscellaneous Reactions

In 1999, Grigg et al. reported the Ag-catalyzed self-condensation of isocyanoacetates into imidazoles, in excellent yields. According to the proposed mechanism, a coordination of Ag(I) to the isocyanide group increases the acidity of the methylene group; this in turn allows the formation of anion **234**, which then attacks another molecule of the coordinated isocyanoacetate to form the intermediate **235**. The latter undergoes ring closure, followed by aromatization and double protonation, to generate the corresponding imidazole **236** (Scheme 4.53) [81].

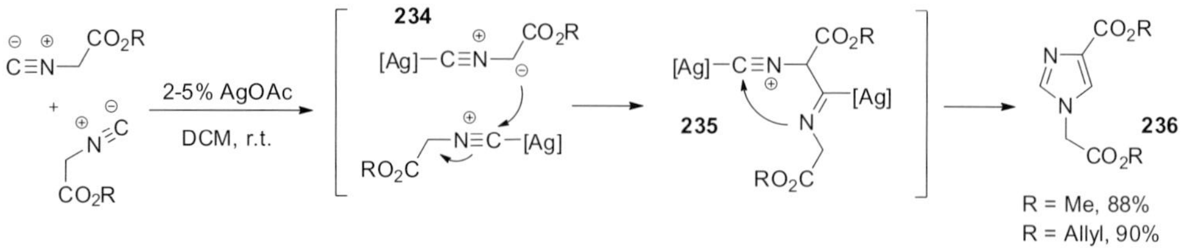

Scheme 4.53

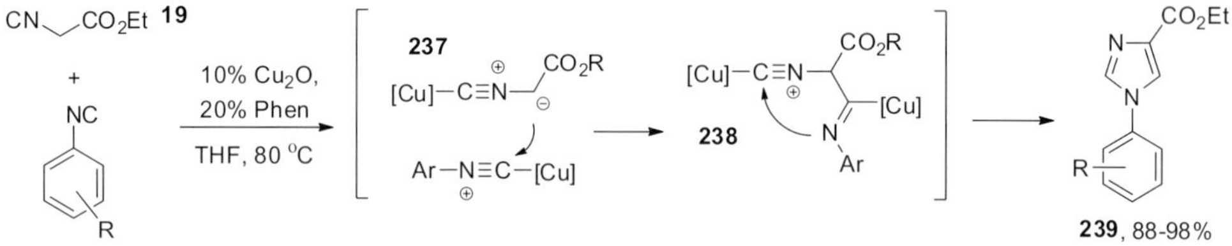

Scheme 4.54

Another very similar Cu(I)-catalyzed reaction between ethyl isocyanoacetate **19** and aryl isocyanides affords ethyl 1-arylimidazole-4-carboxylates **239**. Presumably, the reaction follows a similar mechanism, involving the reaction of two coordinated species (Scheme 4.54). Various aromatic isocyanides bearing either electron-withdrawing or electron-donating groups, and even sterically hindered 2,6-substituted substrates, were tolerated. Attempts to perform the reaction on aliphatic isocyanides proved unsuccessful, however, with the major product being the homo-cycloadduct **236**. The isocyanoacetamides produced the corresponding imidazole in moderate yields [222]. In 2007, Roy et al. described a direct semi-one-pot protocol for the synthesis of imidazoles **239**, starting from the corresponding N-formylglycine esters and N-arylformamides. At the first step, a mixture of two formamides was transformed into a crude mixture of two isocyanides; subsequent treatment of this mixture with a Cu-catalyst afforded the corresponding imidazole. However, a new catalytic system (10%Cu_2O/20% proline) allowed the reaction to proceed at room temperature, with various aromatic substrates, in high yields [223].

Depending on the nature of the substituents at the aromatic ring, the reaction of arenediazonium salts with ethyl isocyanoacetate affords two types of product. With an electron-rich aromatic ring in **240**, 1,2,4-triazole derivatives **242** are obtained via a cyclization of intermediates **241**, whereas with electron-withdrawing groups the reaction produces benzamide **244** (Scheme 4.55) [224, 225].

Under certain conditions, N-substituted isocyanoacetamides are able to cyclize either via the nitrogen or the oxygen atom, to produce imidazolinones or 5-aminooxazoles, respectively. Thus, tertiary N-alkyl-N-aryl isocyanoacetamides

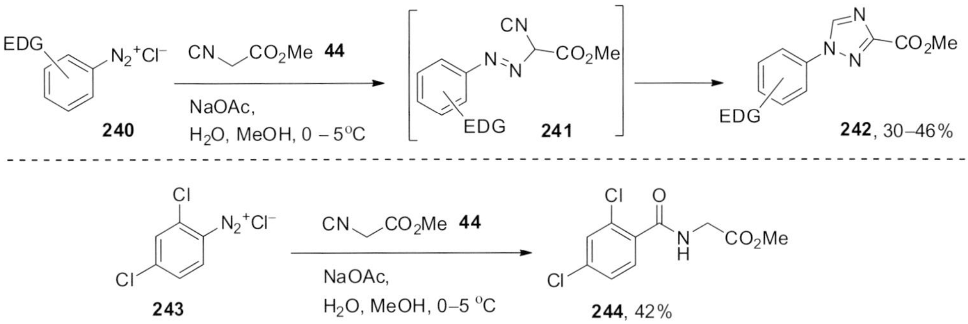

4.11 Miscellaneous Reactions

Scheme 4.55

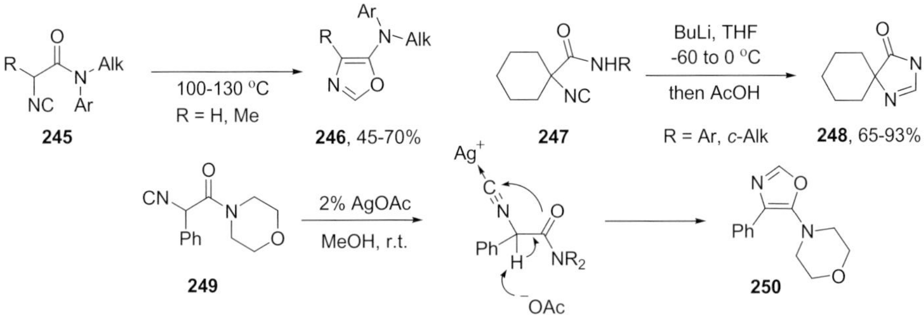

Scheme 4.56

245 will undergo thermal rearrangement to oxazoles 246 which, in certain cases, may already occur upon vacuum distillation [35]. This is a reversible process, where the equilibrium mixture consists of 97–98% of the corresponding 5-aminooxazole. Attempts to isomerize secondary α,α-unsubstituted N-alkylisocyanoacetamides were unsuccessful due to decomposition [35]. Although the behavior of the secondary N-aryl amides was not investigated, secondary α,α-disubstituted isocyanoacetamides 247 are known to undergo a smooth cyclization to 248 upon treatment with butyllithium at low temperature (details of the base-induced alkylative cyclization of secondary isocyanoacetamides into imidazolinones are shown in Scheme 4.10) [38]. More recently, the cyclization of isocyanoacetamide 249 to oxazole 250 using an Ag(I) catalyst was reported (Scheme 4.56) [226].

Zhu et al. have developed a novel method for the oxidative homologation of aldehydes 251 to amides 252. This reaction proceeds under very mild conditions, using potassium isocyanoacetate 253. In this case, p-methoxyphenyl-α-isocyanoacetic acid potassium salt 253 serves as a reducing agent and donor of the $CONH_2$ function to the aldehyde, via oxazole 255 formation. Moreover, the scope of the reaction is broad and the yields are quite good (Scheme 4.57) [17]. Thus, MCRs are effective

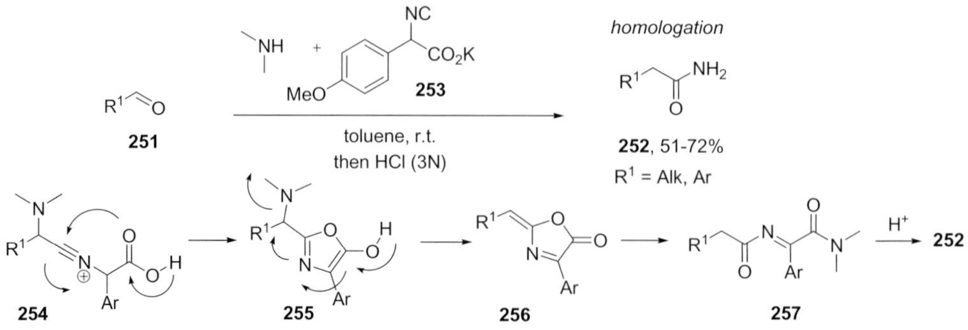

Scheme 4.57

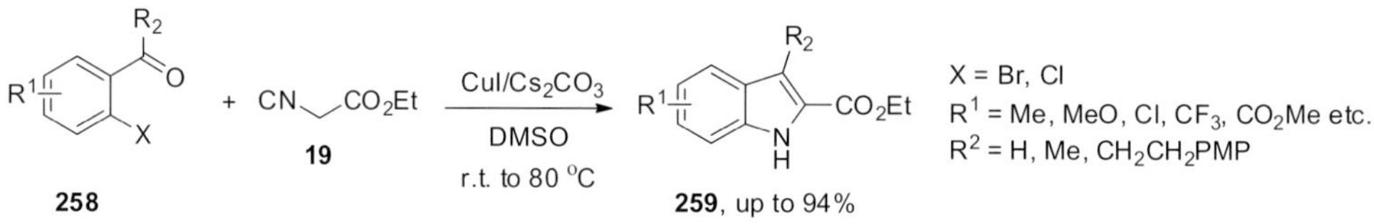

Scheme 4.58

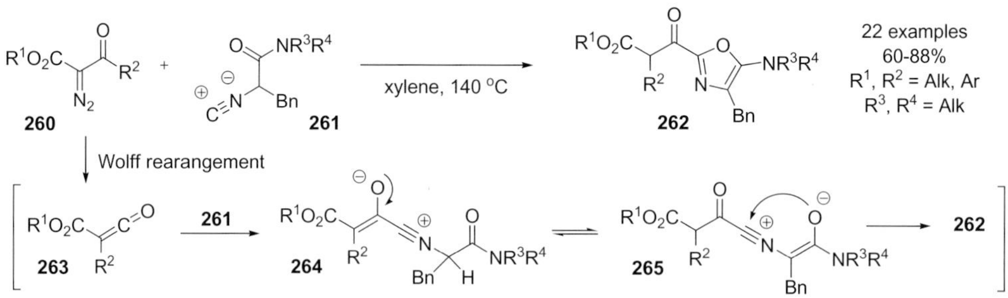

Scheme 4.59

not only for creating structural complexity and diversity but also for the discovery of new fundamental transformations.

A straightforward synthesis of indole-2-carboxylic esters **259** was developed through a ligand-free, copper-catalyzed condensation/coupling/deformylation cascade process from 2-haloaryl aldehydes or ketones **258** with ethyl isocyanoacetate **19** (Scheme 4.58) [227].

In 2011, Yu et al. described a novel synthesis of 2-keto-5-aminooxazoles **262** from α-diazocarbonyl esters **260** and isocyanoacetamides **261** upon heating in xylene. The reaction was shown to proceed via a Wolff rearrangement of α-diazocarbonyl ester **260** to form ketene **263**, followed by a reaction with the isocyanoacetamide and subsequent cyclization of **265** into the final oxazole **262** (Scheme 4.59) [228].

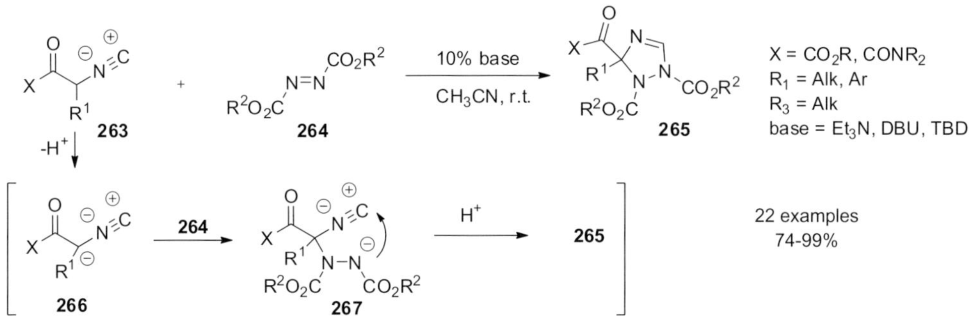

4.11 Miscellaneous Reactions

Scheme 4.60

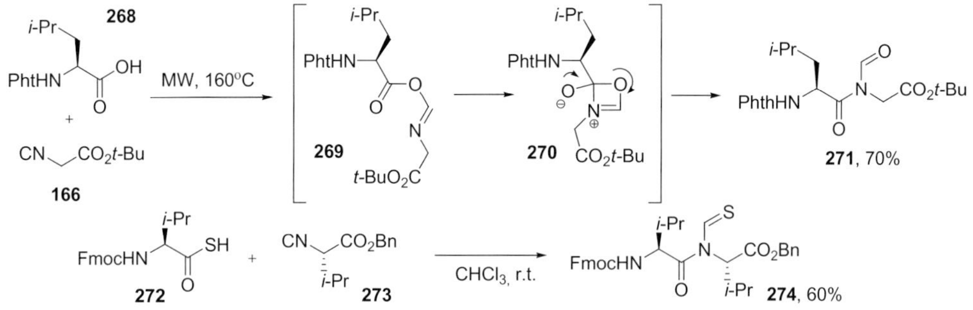

Scheme 4.61

Recently, Jørgensen et al. described the efficient synthesis of 1,2,4-triazolines **265** from α-isocyanoacetic acid derivatives **263** and azodicarboxylates **264**, in the presence of a base. The reaction proceeded via an attack of the anion **258** to azadicarboxylate **264** to form the intermediate **259**; the latter then underwent 5-*endo-dig* cyclization into the final 1,2,4-triazoline **265** (Scheme 4.60) [229].

Very recently, Danishefsky et al. developed a synthesis of N-formyl amides by a microwave-assisted, two-component coupling (2CC) of an isocyanide and a carboxylic acid [230]. Thus, as an example, the reaction of protected L-leucine **268** with isocyanoacetate **166** afforded N-formyl peptide **271** in good yield, using microwave irradiation. The reaction most likely proceeded via the formation of a mixed anhydride **269**, which underwent an intramolecular nucleophilic attack, followed by a 1,3 O→N acyl transfer in **270** (Scheme 4.61) [231]. The same reaction with thioacids was shown to proceed without irradiation at room temperature. As an example, 2CC of the protected thio L-valine **272** and chiral isocyanoacetate **273** formed **274**, in 60% yield. The reaction appeared to take place in a straightforward fashion with a variety of thioacids, whereas the yields of the desired products were increased with ascending isocyanoacetate steric hindrance (Scheme 4.61) [232]. It should be noted, that the configuration of the isocyanide center was fully retained

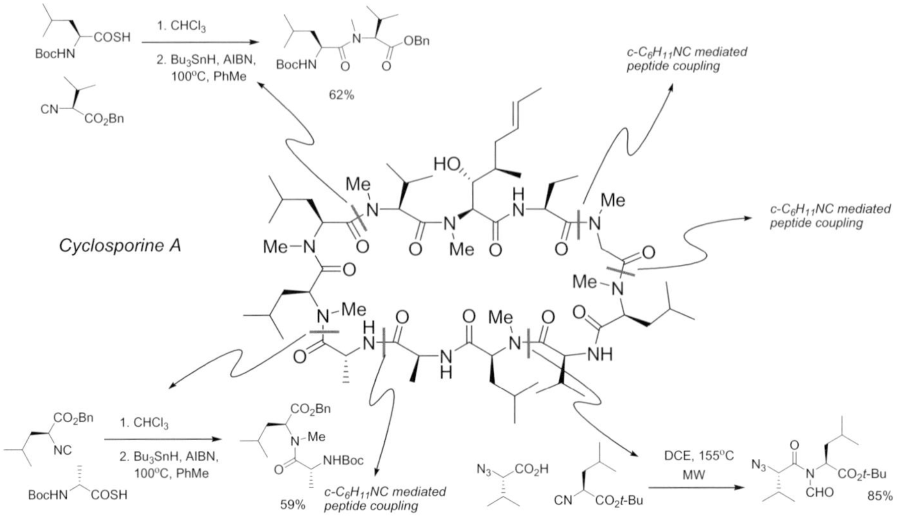

Scheme 4.62

during these transformations. Consequently, the reaction of amino acids with isocyanoacetates may represent a new route for the construction of amide bonds in peptide synthesis.

The efficiency of this approach was demonstrated in recent total synthesis of the macrocyclic peptide cyclosporine A, a reversible inhibitor of cytokines in T-helper cells that was isolated from the fungus *Tolypocladium inflatum gams* (Scheme 4.62). In this case, six out of eleven amide bonds in this macrocyclic peptide (including the last macrolactamization step) were constructed using isocyanide chemistry [233]. The synthesis of cyclosporine A by Danishefsky represents an outstanding example of the use of versatile and powerful isonitrile chemistry for the synthesis of natural products [234, 235].

4.12
Concluding Remarks

Isocyanoacetic acid derivatives occupy an important place among the isocyanides, with their unique multifunctional nature making them highly attractive objects of investigation. On this basis, a range of exciting reactions – especially tandem/cascade processes for the synthesis of a complex cyclic and macrocyclic systems – have been developed for these compounds. The multicomponent chemistry of the isocyanoacetates also represents a powerful instrument for accessing different classes of biochemically relevant compounds, such as peptides, peptide molecules, and nitrogen heterocycles. Indeed, this aspect has shown much promise from the point of view of atom-economy and protecting-group strategies. The data collected in this chapter demonstrate the extreme synthetic utility of the isocyanoacetate derivatives in modern organic, combinatorial, and medicinal chemistries and, indeed, such derivatives may in future serve as efficient building blocks in the synthesis of biologically active molecules. Clearly, further exploration of the isocyanoacetic acid derivatives should open up new frontiers, not only for organic chemistry but also for medicinal chemistry and the total synthesis of natural products.

4.13
Notes Added in Proof

During preparation of this manuscript, investigations of stability of chiral isocyanoacetates towards racemization under the conditions of the Ugi reaction was disclosed by Sello and coworkers. It was shown that racemization of the chiral center of isocyanoacetic acid in the case of the Ugi reaction with aldehydes is usually less than 5%. However, in the case of ketones, a significant racemization in the isocyanoacetate center could be observed [236].

References

1 Ugi, I., Betz, W., Fetzer, U., and Offermann, K. (1961) Notiz zur Darstellung von Isonitden aus monosubstituierten Formamiden durch Wasserabspaltung mittels Phosgen und Trialkylamine. *Chem. Ber.*, **94**, 2814.

2 Gulevich, A.V., Zhdanko, A.G., Orru, R.V.A., and Nenajdenko, V.G. (2010) Isocyanoacetate derivatives: synthesis, reactivity, and application. *Chem. Rev.*, **110**, 5235–5331.

3 Ugi, I., Fetzer, U., Eholzer, U., Knupfer, H., and Offermann, K. (1965) Isonitrile syntheses. *Angew. Chem. Int. Ed. Engl.*, **4**, 472.

4 Skorna, G. and Ugi, I. (1977) Isocyanide synthesis with diphosgene. *Angew. Chem. Int. Ed. Engl.*, **16**, 259.

5 (a) Waki, M. and Meienhofer, J. (1977) Efficient preparation of N$^\alpha$-formylamino acid *tert*-butyl esters. *J. Org. Chem.*, **42**, 2019; (b) Eckert, H. and Forster, B. (1987) Triphosgene, a crystalline phosgene substitute. *Angew. Chem. Int. Ed. Engl.*, **26**, 894.

6 Obrecht, R., Hermann, R., and Ugi, I. (1985) Isocyanide synthesis with phosphoryl chloride and diisopropylamine. *Synthesis*, 400.

7 Ugi, I. and Meyr, R. (1960) Isonitrile, I. Darstellung von Isonitrilen aus monosubstituierten Formamiden durch Wasserabspaltung. *Chem. Ber.*, **93**, 239.

8 Berłożecki, S., Szymański, W., and Ostaszewski, R. (2008) Application of isocyanides derived from α-amino acids as substrates for the Ugi reaction. *Synth. Commun.*, **38**, 2714.

9 Matsumoto, K., Suzuki, M., and Miyoshi, M. (1973) A new synthesis of α-amino acids. *J. Org. Chem.*, **38**, 2094.

10 Sai, H., Ozaki, Y., Hayashi, K., Onoda, Y., and Yamada, K. (1996) Synthesis and structure–activity relationship of trimebutine derivatives. *Chem. Pharm. Bull.*, **44**, 1168.

11 (a) Kobayashi, K., Akamatsu, H., Susumu, I., Takahashi, M., Morikawa, O., and Konishi, H. (1997) *Chem. Lett.*, 503; (b) Kobayashi, K., Irisawa, S., Akamatsu, H., Takahasi, M., Kitamura, T., Tanmatsu, M., Morikawa, O., and

Konishi, H. (1999) *Bull. Chem. Soc. Jpn*, **72**, 2307.

12 Suzuki, M., Nunami, K., Matsumoto, K., Yoneda, N., Kasuga, O., Yoshida, H., and Yamaguchi, T. (1980) *Chem. Pharm. Bull.*, **28**, 2374.

13 Ito, Y., Higuchi, N., and Murakami, M. (1988) *Tetrahedron Lett.*, **29**, 5151.

14 Henkel, B., Sax, M., and Dömling, A. (2003) *Tetrahedron Lett.*, **44**, 7015.

15 Hoppe, I. and Schöllkopf, U. (1976) *Chem. Ber.*, **109**, 482.

16 Bonne, D., Dekhane, M., and Zhu, J. (2004) *Org. Lett.*, **6**, 4771.

17 Bonne, D., Dekhane, M., and Zhu, J. (2005) *J. Am. Chem. Soc.*, **127**, 6926.

18 Vaalburg, W., Strating, J., Wikdring, M.G., and Wynberg, H. (1972) Rapid synthesis of π-phenylglycine by carboxylation of π-lithiobenzylisocyanide. *Synth. Commun.*, **2**, 423.

19 Glaser, M., Spies, H., Lügger, T., and Hahn, F.E. (1995) Formation of a rhenium(III) carbonyl complex by electrophilic attack on rhenium isocyanides. Synthesis and molecular structure of $\{Re\{NCH_2CH_2S\}_3\}$ $[CNC(CH_3)_3]\}$ and $\{Re[N(CH_2CH_2S)_3]$ $(CO)\}$. *J. Organomet. Chem.*, **503**, C32.

20 Steil, P., Nagel, U., and Beck, W. (1988) Metallorganische Lewis-Säuren XXXIV. Kationische Pentacarbonyl(isocyanid) rhenium(I)-Komplexe und deren Reaktionen mit Nucleophilen. *J. Organomet. Chem.*, **339**, 111.

21 Fehlhammer, W.P., Völkl, A., Plaia, U., and Beck, G. (1987) 1,3-Dipolare Cycloadditionen von Heteroallenen an die metallorganischen Nitrilylide $[(OC)_5M\text{-}CN\text{-}CHR]^-$ (M = Cr, W; R = CO_2Et). *Chem. Ber.*, **120**, 2031.

22 Achatz, D., Lang, M.A., Völkl, A., Fehlhammer, W.P., and Beck, W.Z. (2005) *Anorg. Allg. Chem.*, **631**, 2339.

23 Zhu, J., Wu, X., and Danishefsky, S.J. (2009) On the preparation of enantiomerically pure isonitriles from amino acid esters and peptides. *Tetrahedron Lett.*, **50**, 577.

24 (a) Kamer, P.C.J., Cleij, M.C., Nolte, R.J.M., Harada, T., Hezemans, A.M.F., and Drenth, W. (1988) Screw sense selective polymerization of achiral isocyanides catalyzed by optically active nickel(II) complexes. *J. Am. Chem. Soc.*, **110**, 6818; (b) Van Beijnen, A.J.M., Nolte, R.J.M., Naaktgeboren, A.J., Zwikker, J.W., and Drenth, W. (1983) Helical configuration of poly(iminomethy1enes). Synthesis and CD spectra of polymers derived from optically active isocyanides. *Macromolecules*, **16**, 1679.

25 Yamada, Y., Kawai, T., Abe, J., and Iyoda, T. (2002) Synthesis of polyisocyanide derived from phenylalanine and its temperature-dependent helical conformation. *J. Polym. Sci. [A]*, **40**, 399.

26 Giesemann, G., Von Hinrichs, E., and Ugi, I. (1982) *J. Chem. Res.*, 79.

27 Porcheddu, A., Giacomelli, G., and Salaris, M. (2005) Microwave-assisted synthesis of isonitriles: a general simple methodology. *J. Org. Chem.*, **70**, 2361.

28 Zhdanko, A.G. and Nenajdenko, V.G. (2009) Nonracemizable isocyanoacetates for multicomponent reactions. *J. Org. Chem.*, **74**, 884.

29 Pando, O., Stark, S., Denkert, A., Porzel, A., Preusentanz, R., and Wessjohann, L.A. (2011) The multiple multicomponent approach to natural product mimics: tubugis, N-substituted anticancer peptides with picomolar activity. *J. Am. Chem. Soc.*, **133**, 7692.

30 Ozaki, Y., Maeda, S., Miyoshi, M., and Matsumoto, K. (1979) An improved stereoselective synthesis of threo-β-hydroxyamino acids. *Synthesis*, 216.

31 Matsumoto, K., Suzuki, M., Yoneda, N., and Miyoshi, M. (1977) Alkylation of α-isocyanoacetamides; synthesis of 1,4,4-trisubstituted 5-oxo-4,5-dihydroimidazoles. *Synthesis*, 249.

32 Takiguchi, K., Yamada, K., Suzuki, M., Nunami, K., and Hayashi, K. (1989) *Agric. Biol. Chem.*, **53**, 77.

33 Housseman, C. and Zhu, J. (2006) *Synlett*, 1777.

34 Bossio, R., Marcaccini, S., Pepino, R., Polo, C., and Valle, G. (1989) A novel synthetic route to imidazole derivatives: synthesis of mesoionic 3-Alkyl-2-arylthio-1,3-diazolium-4-olates. *Synthesis*, 641.

35 Chupp, J.P. and Leschinsky, K.L. (1980) Heterocycles from substituted

amides. VII (1,2). Oxazoles from 2-isocyanoacetamides. *J. Heterocycl. Chem.*, **17**, 705.

36 Dömling, A., Beck, B., Fuchs, T., and Yazbak, A. (2006) Parallel synthesis of arrays of amino-acid-derived isocyanoamides useful as starting materials in IMCR. *J. Comb. Chem.*, **8**, 872.

37 Bossio, R., Marcaccini, S., and Pepino, R. (1990) Studies on isocyanides. synthesis of N-substituted 2-isocyanocarboxamides. *Liebigs Ann. Chem.*, 935.

38 Bossio, R., Marcaccini, S., Paoli, P., Papaleo, S., Pepino, R., and Polo, C. (1991) Studies on isocyanides and related compounds. Synthesis and cyclization of N-substituted 1-isocyano-1-cycloalkanecarboxamides. *Liebigs Ann. Chem.*, 843.

39 Bossio, R., Marcaccini, S., Papaleo, S., and Pepino, R. (1994) Studies on isocyanides and related compounds. A convenient synthesis of 2,3-disubstituted spiroimidazolones. *J. Heterocycl. Chem.*, **31**, 397.

40 Pirali, T., Tron, G.C., Masson, G., and Zhu, J. (2007) Ammonium chloride promoted three-component synthesis of 5-iminooxazoline and its subsequent transformation to macrocyclodepsipeptide. *Org. Lett.*, **9**, 5275.

41 Fayol, A. and Zhu, J. (2005) Three-component synthesis of polysubstituted 6-azaindolines and its tricyclic derivatives. *Org. Lett.*, **7**, 239.

42 Fayol, A., Housseman, C., Sun, X., Janvier, P., Bienaymé, H., and Zhu, J. (2005) Synthesis of α-isocyano-α-alkyl(aryl)acetamides and their use in the multicomponent synthesis of 5-aminooxazole, pyrrolo[3,4-b]pyridin-5-one and 4,5,6,7-tetrahydrofuro[2,3-c] pyridine. *Synthesis*, 161.

43 Pirali, T., Tron, G.C., and Zhu, J. (2006) One-pot synthesis of macrocycles by a tandem three-component reaction and intramolecular [3+2] cycloaddition. *Org. Lett.*, **8**, 4145.

44 (a) Sawamura, M., Nakayama, Y., Kato, T., and Ito, Y. (1995) Gold(1)-catalyzed asymmetric aldol reaction of N-methoxy-N-methyl-α-isocyanoacetami(dae- Isocyano Weinreb Amide). An efficient synthesis of optically active β-hydroxy α-amino aldehydes and ketones. *J. Org. Chem.*, **60**, 1727; (b) Kim, S.W., Bauer, S.M., and Armstrong, R.W. (1998) Construction of combinatorial chemical libraries using a rapid and efficient solid phase synthesis based on a multicomponent condensation reaction. *Tetrahedron Lett.*, **39**, 6993.

45 Zang, X., Zou, X., and Xu, P. (2005) Template synthesis of peptidomimetics composed of aspartic acid moiety by Ugi four-component condensation reaction. *Synth. Commun.*, **35**, 1881.

46 Lang, M.A. and Beck, W.Z. (2005) Metallkomplexe mit biologisch wichtigen Liganden. CLVII Halbsandwich-Komplexe mit Isocyanoacetylaminosäureestern und Isocyanoacetyldi- und tripeptidestern ("Isocyano-Peptide"). *Anorg. All. Chem.*, **631**, 2333.

47 Urban, R., Marquarding, D., Seidel, P., Ugi, I., and Weinelt, A. (1977) Notiz zur Synthese optisch aktiver α-Isocyancarbonsäure-Derivate für Peptidsynthesen mittels Vier-Komponenten-Kondensation (4 CC). *Chem. Ber.*, **110**, 2012.

48 Cornelissen, J.J.L.M., Sander Graswinckel, W.S., Adams, P.J.H.M., Nachtegaal, G.H., Kentgens, A.P.M., Sommerdijk, N.A.J.M., and Nolte, R.J.M. (2001) Synthesis and characterization of polyisocyanides derived from alanine and glycine dipeptides. *J. Polym. Sci. [A]*, **39**, 4255.

49 Schwartz, E., Kitto, H.J., De Gelder, R., Nolte, R.J.M., Rowan, A.E., and Cornelissen, J.J.L.M. (2007) Synthesis, characterisation and chiroptical properties of "click" able polyisocyanopeptides. *J. Mater. Chem.*, **17**, 1876.

50 Van der Eijk, J.M., Nolte, R.J.M., Drenth, W., and Hezemans, A.M.F. (1980) Optically active polyampholytes derived from L- and D-carbylanayl-L-histidine. *Macromolecules*, **13**, 1391.

51 Metselaar, G.A., Hans, M., Adams, P.J., Nolte, R.J.M., Cornelissen, J.J.L.M., and

Rowan, A.E. (2007) Polyisocyanides derived from tripeptides of alanine. *Chem. Eur. J.*, **13**, 950.

52 Visser, H.G.J., Nolte, R.J.M., Zwikker, J.W., and Drenth, W. (1985) Synthesis of copolymers of isocyanides derived from alanylserine and alanylhistidine. *J. Org. Chem.*, **50**, 3138.

53 Zhao, G., Bughin, C., Bienaymé, H., and Zhu, J. (2003) Synthesis of isocyano peptides by dehydration of the N-formyl derivatives. *Synlett*, 1153.

54 (a) Buschmann, J., Lentz, D., Luger, P., Perpetuo, G., Scharn, D., and Willemsen, S. (1995) Synthesis, structural investigation, and ligand properties of isocyanoacetonitrile. *Angew. Chem. Int. Ed. Engl.*, **34**, 914; (b) Buschmann, J., Lentz, D., Luger, P., Röttger, M., Perpetuo, G., Scharn, D., and Willemsen, S. (2000) Synthese, Strukturuntersuchung und Koordinationschemie von Isocyanacetonitril. *Z. Anorg. Allg. Chem.*, **626**, 2107.

55 Hantke, K., Schöllkopf, U., and Hausberg, H.-H. (1975) Synthesen mit α-metallierten isocyaniden, XXX. 2- Oxazolin-4-carbonitrile aus α-isocyannitrilen und carbonylverbindungen. *Liebigs Ann. Chem.*, 1531.

56 Hoppe, D. (1974) α-Metalated isocyanides in organic synthesis. *Angew. Chem. Int. Ed. Engl.*, **13**, 789.

57 Kotha, S., Shah, V.R., Halder, S., Vinodkumar, R., and Lahiri, K. (2007) Synthesis of bis-armed amino acid derivatives via the alkylation of ethyl isocyanoacetate and the Suzuki–Miyaura cross-coupling reaction. *Amino Acids*, **32**, 387.

58 Schöllkopf, U., Hoppe, D., and Jentsch, R. (1971) Higher amino acids by alkylation of α-metalated isocyano-acetic or -propionic esters. *Angew. Chem. Int. Ed. Engl.*, **10**, 331.

59 Schöllkopf, U., Hoppe, D., and Jentsch, R. (1975) Synthesen mit α-metallierten Isocyaniden, XXIX. Hohere Amhosiiuren durch Alkylieren von a-metallierten α-Isocyan-propionsaure- und -essigsaureestern. *Chem. Ber.*, **108**, 1580.

60 Kotha, S., Halder, S., Damodharan, L., and Pattabhi, V. (2002) First and unexpected synthesis of macrocyclic cyclophane-based unusual α-amino acid derivatives by phosphazene base without high dilution conditions. *Bioorg. Med. Chem. Lett.*, **12**, 1113.

61 Kotha, S. and Halder, S. (2005) Synthesis of macrocyclic cyclophane-based unusual α-amino acid derivatives. *ARKIVOC*, **3**, 56.

62 Kotha, S. and Brahmachary, E. (1997) Synthesis of unusual α-amino acids via a 2+2+2 cycloaddition strategy. *Tetrahedron Lett.*, **38**, 3561.

63 Kotha, S. and Sreenivasachary, N. (1998) Synthesis of constrained α-amino acid derivatives via ring-closing olefin metathesis. *Bioorg. Med. Chem. Lett.*, **8**, 257.

64 Kotha, S., Behera, M., and Kumar, R.V. (2002) Synthesis of highly functionalised dibenzylglycine derivatives via the Suzuki–Miyaura coupling reaction. *Bioorg. Med. Chem. Lett.*, **12**, 105.

65 Kalvin, D., Ramalingam, K., and Woodard, R. (1985) A facile procedure for the preparation of alicyclic-amino acids. *Synth. Commun.*, **15**, 267.

66 Osipova, A., Yufit, D.S., and de Meijere, A. (2007) Synthesis of new cyclopropylisonitriles and their applications in Ugi four-component reactions. *Synthesis*, 131.

67 Lin, S.S., Liu, J.Y., and Wang, J.M. (2003) The synthesis of cyclic amino acids. *Chin. Chem. Lett.*, **14**, 883.

68 Suzuki, M., Matsumoto, K., Iwasaki, K., and Okumura, K. (1972) Sterically controlled synthesis of α-methyldopa. *Chem. Ind. (London)*, 687.

69 Längström, B., Stridsberg, B., and Bergson, G. (1978) Asymmetric synthesis of alanine by methylation of chiral isocyanoacetic esters. *Chem. Scr.*, **13**, 49.

70 Ito, Y., Sawamura, M., Matsuoka, M., Matsumoto, Y., and Hayashi, T. (1987) Palladium-catalyzed allylation of α-isocyanocarboxylates. *Tetrahedron Lett.*, **28**, 4849.

71 Kazmaier, U. and Ackermann, S. (2004) An improved protocol for the synthesis of allylated isonitriles. *Synlett*, 2576.

72 Kotha, S. (2003) The building block approach to unusual α-amino acid derivatives and peptides. *Acc. Chem. Res.*, **36**, 342.

73 Kotha, S. and Halder, S. (2010) Ethyl isocyanoacetate as a useful glycine equivalent. *Synlett*, 337.

74 Schöllkopf, U., Hausberg, H.-H., Segala, M., Reiter, U., Hoppe, I., Saenger, W., and Lindner, K. (1981) Asymmetrische Synthesen uber heterocyclische Zwischenstufen, IV. Asymmetrische Synthese von α-Methylphenylalanin und seinen Analoga durch Alkylieren von 1-chiral substituiertem 4-Methyl-2-imidazolin-5-on. *Liebigs Ann. Chem.*, 439.

75 Schöllkopf, U. and Hantke, K. (1970) Preparation of glutamic acid derivatives from α-metalated ethyl isocyanoacetate and acrylic esters. *Angew. Chem. Int. Ed. Engl.*, **9**, 896.

76 Schöllkopf, U. and Porsch, P.-H. (1972) Ethyl γ-cyano-α-isocyanoalkanoates from α-metalated ethyl isocyano-acetates or -propionates and acrylonitriles. *Angew. Chem. Int. Ed. Engl.*, **11**, 429.

77 Schöllkopf, U. and Hantke, K. (1973) Synthesen mit α-metallierten Isocyaniden, XXIV. α-Isocyanglutarsaure-diathylester, Glutaminsaure-Derivate und Pyrrolincarbonsaure-athylester aus a-metalliertem Isocyanessigsaure-oder -propionsame-athylester und Acrylsaureestern. *Liebigs Ann. Chem.*, 1571.

78 Murakami, M., Hasegawa, N., Tomita, I., Inouye, N., and Ito, Y. (1989) Fluoride-catalyzed Michael reaction of α-isocyanoesters with α,β-unsaturated carbonyl compounds. *Tetrahedron Lett.*, **30**, 1257.

79 Murakami, M., Hasegawa, N., Hayashi, M., Inouye, N., and Ito, Y. (1991) Synthesis of (±)-α-allokainic acid via the zinc acetate catalyzed cyclization of γ-isocyano silyl enolates. *J. Org. Chem.*, **56**, 7356.

80 Barton, D.H.R. and Zard, S.Z. (1985) A new synthesis of pyrroles from nitroalkenes. *J. Chem. Soc. Chem. Commun.*, 1098.

81 Grigg, R., Lansdell, M.I., and Thornton-Pett, M. (1999) Silver acetate catalyzed cycloaddition of isocyanoacetates. *Tetrahedron*, **55**, 2025.

82 Guo, C., Xue, M.-X., Zhu, M.-K., and Gong, L.-Z. (2008) Organocatalytic asymmetric formal [3+2] cycloaddition reaction of isocyanoesters to nitroolefins leading to highly optically active dihydropyrroles. *Angew. Chem. Int. Ed.*, **47**, 3414.

83 Arróniz, C., Gil-González, A., Semak, V., Escolano, C., Bosch, J., and Amat, M. (2011) Cooperative catalysis for the first asymmetric formal [3+2] cycloaddition reaction of isocyanoacetates to α,β-unsaturated ketones. *Eur. J. Org. Chem.*, **2011**, 3755.

84 Zhang, D., Xu, X., Tan, J., Xia, W., and Liu, Q. (2010) Tandem Michael addition/intramolecular isocyanide [3+2] cycloaddition: highly diastereoselective one pot synthesis of fused oxazolines. *Chem. Commun.*, **46**, 3357.

85 Zhang, D., Xu, X., Tan, J., and Liu, Q. (2010) Direct stereoselective synthesis of 1-amino-2,5-diarylcyclohexanecarboxylic acid derivatives based on a [5+1] annulation of divinyl ketone and isocyanoacetate. *Synlett*, 917.

86 Tan, J., Xu, X., Zhang, L., Li, Y., and Liu, Q. (2009) Tandem double-Michael-addition/cyclization/acyl migration of 1,4-dien-3-ones and ethyl isocyanoacetate: stereoselective synthesis of pyrrolizidines. *Angew. Chem. Int. Ed.*, **48**, 2868.

87 Xu, X., Li, Y., Zhang, Y., Zhang, L., Pan, L., and Liu, Q. (2011) Direct synthesis of 6-azabicyclo[3.2.1]oct-6-en-2-ones and pyrrolizidines from divinyl ketones and observation of remarkable substituent effects. *Adv. Synth. Catal.*, **353**, 1218.

88 Li, Y., Xu, X., Xia, C., Zhang, D., and Liu, Q. (2011) Double isocyanide cyclization: a synthetic strategy for two-carbon-tethered pyrrole/oxazole pairs. *J. Am. Chem. Soc.*, **133**, 1775.

89 (a) Kamijo, S., Kanazawa, C., and Yamamoto, Y. (2005) Phosphine-catalyzed regioselective heteroaromatization between activated alkynes and isocyanides leading to pyrroles. *Tetrahedron Lett.*, **46**, 2563; (b) Kamijo, S., Kanazawa, C., and Yamamoto, Y. (2005) Copper- or

phosphine-catalyzed reaction of alkynes with isocyanides. Regioselective synthesis of substituted pyrroles controlled by the catalyst. *J. Am. Chem. Soc.*, **127**, 9260.

90 Larionov, O.V. and de Meijere, A. (2005) Versatile direct synthesis of oligosubstituted pyrroles by cycloaddition of α-metalated isocyanides to acetylenes. *Angew. Chem. Int. Ed.*, **44**, 5664.

91 Bhattacharya, A., Cherukuri, S., Plata, R.E., Patel, N., Tamez, Jr, V., Grosso, J.A., Peddicord, M., and Palaniswamy, V.A. (1981) Remarkable solvent effect in Barton–Zard pyrrole synthesis: application in an efficient one-step synthesis of pyrrole derivatives. *Tetrahedron Lett.*, **47**, 5481.

92 Lygin, A.V., Larionov, O.V., Korotkov, V.S., and de Meijere, A. (2009) Oligosubstituted pyrroles directly from substituted methyl isocyanides and acetylenes. *Chem. Eur. J.*, **15**, 227.

93 Schöllkopf, U., Gerhart, F., and Schröder, R. (1969) α-(Formylamino) acrylic esters from α-metalated isocyanoacetic esters and carbonyl compounds. *Angew. Chem. Int. Ed. Engl.*, **8**, 672.

94 Yim, N.C.F., Bryan, H., Huffman, W.F., and Moore, M.L. (1988) Facile synthesis of protected β,β-dialkylcysteine derivatives suitable for peptide synthesis. *J. Org. Chem.*, **53**, 4605.

95 Moriya, T., Matsumoto, K., and Miyoshi, M. (1981) Acidolysis of 2-formylamino-2-alkenoic ester to 2,3-dehydro amino acid esters. *Synthesis*, 915.

96 Suzuki, M., Nunami, K., Moriya, T., Matsumoto, K., and Yoneda, N. (1978) Synthesis of α-[aminomethylene)amino] acrylic acid esters. *J. Org. Chem.*, **43**, 4933.

97 Schöllkopf, U., Gerhart, F., Schröder, R., and Hoppe, D. (1972) Synthesen mit α-metallierten Isocyaniden, XVI. β-Substituierte α-Formylamino-acrylsaureathylester aus α-metallierten Isocyanessigestern und Carbonylverbindungen (Formylaminomethylenierung von Carbonylverbindungen. *Liebigs Ann. Chem.*, **766**, 116.

98 Genin, D., Andriamialisoa, R.Z., Langlois, N., and Langlois, Y. (1987) A short stereoselective synthesis of the alkaloid vincamine. *J. Org. Chem.*, **52**, 353.

99 Nunami, K., Suzuki, M., Hayashi, K., Matsumoto, K., Yoneda, N., and Takiguchi, K. (1985) Synthesis and herbicidal activities of α-isocyanocyclo-alkylidenacetamides. *Agric. Biol. Chem.*, **49**, 3023.

100 Sakai, K., Suzuki, M., Nunami, K., Yoneda, N., Onoda, Y., and Iwasawa, Y. (1980) Synthesis of 3-substituted pyrrole derivatives with antiinflammatory activity. *Chem. Pharm. Bull.*, **28**, 2384.

101 Nunami, K., Hiramatsu, K., Hayashi, K., and Matsumoto, K. (1988) Synthesis of β-functionalized α,β-dehydroamino acid derivatives. *Tetrahedron*, **44**, 5467.

102 Reetz, M., Kayser, F., and Hams, K. (1992) Cycloaddition reactions of γ-amino α,β-didehydro amino acid esters: a test case for the principle of 1,3-allylic strain. *Tetrahedron Lett.*, **33**, 3453.

103 Enders, D., Chen, Z.-X., and Raabe, G. (2005) Stereoselective synthesis of 3-substituted ethyl (Z)-4,4,4-trifluoro-2-formylamino-2-butenoates. *Synthesis*, 306.

104 Huang, W., Zhang, Y., and Yuan, C. (1997) Studies on organophosphorus compounds; 99: a novel stereoselective synthesis of dialkyl (Z)-1-alkyl-2-ethoxycarbonyl-2-formylaminoeth-1-enylphosphonates. *Synthesis*, 162.

105 Panella, L., Aleixandre, A.M., Kruidhof, G.J., Robertus, J., Feringa, B.L., de Vries, J.G., and Minnaard, A.J. (2006) Enantioselective Rh-catalyzed hydrogenation of N-formyl dehydroamino esters with monodentate phosphoramidite ligands. *J. Org. Chem.*, **71**, 2026.

106 Nunami, K., Suzuki, M., and Yoneda, N. (1982) Synthesis of α-alkylated β,γ-unsaturated α-amino acids. *Chem. Pharm. Bull.*, **30**, 4015.

107 Suzuki, M., Miyoshi, M., and Matsumoto, K. (1974) A convenient synthesis of 3-substituted pyrrole-2,4-dicarboxylic acid esters. *J. Org. Chem.*, **39**, 1980.

108 Takaya, H., Kojima, S., and Murahashi, S.-I. (2001) Rhodium complex-catalyzed reaction of isonitriles with carbonyl compounds: catalytic synthesis of pyrroles. *Org. Lett.*, **3**, 421.

109 Hoppe, D. and Schöllkopf, U. (1970) Ethyl 2-oxazoline-5-carboxylate from ethyl isocyanoacetate and carbonyl compounds. *Angew. Chem. Int. Ed. Engl.*, **9**, 300.

110 Hoppe, D. and Schöllkopf, U. (1972) Synthesen mit a-metallierten Isocyaniden, XV. 4-Äthoxycarbonyl-2-oxazoline und ihre Hydrolyse zu N-Formyl-β-hydroxy-α-aminosaureathyleste. *Liebigs Ann. Chem.*, **763**, 1.

111 Ito, Y., Higuchi, N., and Murakami, M. (2000) Fluoride-catalyzed aldol-type reactions of α-isocyano esters producing 2-oxazoline derivatives. *Heterocycles*, **52**, 91.

112 Matsumoto, K., Ozaki, Y., Suzuki, M., and Miyoshi, M. (1976) *Agric. Biol. Chem.*, **40**, 2045.

113 Matsumoto, K., Urabe, Y., Ozaki, Y., Iwasaki, T., and Miyoshi, M. (1975) A stereoselective synthesis of threo-threonine reaction of isocyanoacetate with acetaldehyde. *Agric. Biol. Chem.*, **39**, 1869.

114 Kisanga, P., Ilankumaran, P., and Verkade, J.G. (2001) P(RNCH$_2$CH$_2$)$_3$N-catalyzed diastereoselective synthesis of oxazolines. *Tetrahedron Lett.*, **42**, 6263.

115 Ozaki, Y., Matsumoto, K., and Miyoshi, M. (1978) A useful synthetic method of DL-threonine using α-isocyanoacetamides. *Agric. Biol. Chem.*, **42**, 1565.

116 Meyer, R., Schöllkopf, U., Madawinata, K., and Stafforst, D. (1978) Synthesen mit a-metallierten Isocyaniden, LII. Synthese von 2-Alkyl- und 2-Acyl-1-methylimidazol-4-carbonsauremethylestern aus (Z)-β-Dimethylamino-α-isocyanacrylsauremethylester und Alkyl- oder Acylhalogeniden. *Liebigs Ann. Chem.*, 1982.

117 Xue, M.-X., Guo, C., and Gong, L.-Z. (2009) Asymmetric synthesis of chiral oxazolines by organocatalytic cyclization of α-aryl isocyanoesters with aldehydes. *Synlett*, 2191.

118 Ito, Y., Matsuura, T., and Saegusa, T. (1985) ZnCl$_2$ and CuCl-promoted aldol reactions of isocyanoacetate with α,β-unsaturated carbonyl compounds. *Tetrahedron Lett.*, **26**, 5781.

119 Benito-Garagorri, D., Bocokić, V., and Kirchner, K. (2006) Copper(I)-catalyzed diastereoselective formation of oxazolines and N-sulfonyl-2-imidazolines. *Tetrahedron Lett.*, **47**, 8641.

120 Soloshonok, V.A., Hayashi, T., Ishikawa, K., and Nagashima, N. (1994) Highly diastereoselective aldol reaction of fluoroalkyl aryl ketones with methyl isocyanoacetate catalyzed by silver(I)/triethylamine. *Tetrahedron Lett.*, **35**, 1055.

121 Ito, Y., Sawamura, M., and Hayashi, T. (1986) Catalytic asymmetric aldol reaction: reaction of aldehydes with isocyanoacetate catalyzed by a chiral ferrocenylphosphine-gold(I) complex. *J. Am. Chem. Soc.*, **108**, 6405.

122 Pastor, S.D. and Togni, A. (1989) Asymmetric synthesis with chiral ferrocenylamine ligands: the importance of central chirality. *J. Am. Chem. Soc.*, **111**, 2333.

123 Pastor, S.D., Kesselring, R., and Togni, A. (1992) The gold(I)-catalyzed aldol reaction utilizing chiral ferrocenylamine ligands: synthesis of N-benzyl-substituted ligands. *J. Organomet. Chem.*, **429**, 415.

124 Togni, A. and Hausel, R. (1990) A new entry into sulfur-containing ferrocenylphosphine ligands for asymmetric catalysis. *Synlett*, 633.

125 Ito, Y., Sawamura, M., and Hayashi, T. (1987) Asymmetric aldol reaction of an isocyanoacetate with aldehydes by chiral ferrocenylphosphine-gold(I) complexes: design and preparation of new efficient ferrocenylphosphine ligands. *Tetrahedron Lett.*, **28**, 6215.

126 Hayashi, T., Sawamura, M., and Ito, Y. (1992) Asymmetric synthesis catalyzed by chiral ferrocenylphosphine-transition metal complexes. 10 gold(I)-catalyzed asymmetric aldol reaction of isocyanoacetate. *Tetrahedron*, **48**, 1999.

127 Togni, A. and Pastor, S.D. (1990) Chiral cooperativity: the nature of the diastereoselective and enantioselective step in the gold(I)-catalyzed aldol reaction utilizing chiral ferrocenylamine ligands. *J. Org. Chem.*, **55**, 1649.

128 Sawamura, M., Ito, Y., and Hayashi, T. (1990) NMR studies of the gold(I)-catalyzed asymmetric aldol reaction of isocyanoacetate. *Tetrahedron Lett.*, **31**, 2723.

129 Soloshonok, V.A. and Hayashi, T. (1994) Gold(I)-catalyzed asymmetric aldol reaction of methyl isocyanoacetate with fluorinated benzaldehydes. *Tetrahedron Lett.*, **35**, 2713.

130 Togni, A. and Pastor, S.D. (1989) Enantioselective synthesis: steric and electronic effects of the substrates upon stereoselectivity in the gold(I)-catalyzed aldol reaction with chiral ferrocenylamine ligands. *Helv. Chim. Acta*, **72**, 1038.

131 Ito, Y., Sawamura, M., Kobayashi, M., and Hayashi, T. (1988) Asymmetric aldol reaction of α-isocyanoacetamides with aldehydes catalyzed by a chiral ferrocenylphosphine-gold(I) complex. *Tetrahedron Lett.*, **29**, 6321.

132 Soloshonok, V.A., Kacharov, A.D., and Hayashi, T. (1996) Gold(I)-catalyzed asymmetric aldol reactions of isocyanoacetic acid derivatives with fluoroaryl aldehydes. *Tetrahedron*, **52**, 245.

133 Sawamura, M., Nakayama, Y., Kato, T., and Ito, Y. (1995) Gold(I)-catalyzed asymmetric aldol reaction of N-methoxy-N-methyl-α-isocyanoacetami(α-Isocyano Weinreb Amide). An efficient synthesis of optically active β-hydroxy α-amino aldehydes and ketones. *J. Org. Chem.*, **60**, 1727.

134 Ito, Y., Sawamura, M., Shirakawa, E., Hayashizaki, K., and Hayashi, T. (1988) Asymmetric synthesis of β-hydroxy-α-alkylamino acids by asymmetric aldol reaction of α-isocyanocarboxylates catalyzed by chiral ferrocenylphosphine-gold(I) complexes. *Tetrahedron*, **44**, 5253.

135 Ito, Y., Sawamura, M., Shirakawa, E., Hayashizaki, K., and Hayashi, T. (1988) Asymmetric aldol reaction of α-isocyanocarboxylates with paraformaldehyde catalyzed by chiral ferrocenylphosphine-gold(I) complexes: catalytic asymmetric synthesis of α-alkylserines. *Tetrahedron Lett.*, **29**, 235.

136 Ito, Y., Sawamura, M., Hamashima, H., Emura, T., and Hayashi, T. (1989) Asymmetric aldol reaction of α-ketoesters with isocyanoacetate and isocyanoacetamide catalyzed by a chiral ferrocenylphosphine-gold(I) complex. *Tetrahedron Lett.*, **30**, 4681.

137 Hayashi, T., Uozumi, Y., Yamazaki, A., Sawamura, M., Hamashima, H., and Ito, Y. (1991) Silver(I)-catalyzed asymmetric aldol reaction of isocyanoacetate. *Tetrahedron Lett.*, **32**, 2799.

138 Lianza, F., Macchioni, A., Pregosin, P., and Ruegger, H. (1994) Multidimensional ^{109}Ag, ^{31}P, and ^{1}H HMQC and HSQC NMR studies on a model homogeneous catalyst. Reactions of a chiral ferrocenylphosphine with Ag(CF$_3$SO$_3$). *Inorg. Chem.*, **33**, 4999.

139 Togni, A., Pastor, S.D., and Rihs, G. (1990) Enantioselective synthesis: catalysis of the aldol reaction by neutral gold(I)-chiral ferrocenylphosphine complexes. Crystal structure of the complex [{(η^5-C$_5$H$_4$PPh$_2$)(η5-C$_5$H$_3$(PPh$_2$)CH(Me)N(Me)CH$_2$CH$_2$NMe$_2$)Fe}$_2$(AuCl)$_3$]·Et$_2$O. *J. Organomet. Chem.*, **381**, C21.

140 Sladojevich, F., Trabocchi, A., Guarna, A., and Dixon, D.J. (2011) A new family of cinchona-derived amino phosphine precatalysts: application to the highly enantio- and diastereoselective silver-catalyzed isocyanoacetate aldol reaction. *J. Am. Chem. Soc.*, **133**, 1710.

141 Kim, H.Y. and Oh, K. (2011) Highly diastereo- and enantioselective aldol reaction of methyl α-isocyanoacetate: a cooperative catalysis approach. *Org. Lett.*, **13**, 1306.

142 Meyer, R., Schöllkopf, U., and Böhme, P. (1977) Synthesen mit α-metallierten Isocyaniden, XXXIX. 2-Imidazoline aus α-metallierten Isocyaniden und Schiff-Basen; 1,2-Diamine und 2,3-Diaminoalkansäuren. *Liebigs Ann. Chem.*, 1183.

143 Hayashi, T., Kishi, E., Soloshonok, V.A., and Uozumi, Y. (1996) Erythro-selective

aldol-type reaction of N-sulfonylaldimines with methyl isocyanoacetate catalyzed by gold(I). *Tetrahedron Lett.*, **37**, 4969.

144 Lin, Y., Zhou, X., Dai, L., and Sun, J. (1997) Ruthenium complex-catalyzed reaction of isocyanoacetate and N-sulfonylimines: stereoselective synthesis of N-sulfonyl-2-imidazolines. *J. Org. Chem.*, **62**, 1799.

145 Murahashi, S.-I., Naota, T., Taki, H., Mizuno, M., Takaya, H., Komiya, S., Mizuho, Y., Oyasato, N., Hiraoka, M., Hirano, M., and Fukuoka, A. (1995) Ruthenium-catalyzed aldol and Michael reactions of nitriles. Carbon-carbon bond formation by α-C-H activation of nitriles. *J. Am. Chem. Soc.*, **117**, 12436.

146 Aydin, J., Kumar, K.S., Eriksson, L., and Szabó, K.J. (2007) Palladium pincer complex-catalyzed condensation of sulfonimines and isocyanoacetate to imidazoline derivatives. Dependence of the stereoselectivity on the ligand effects. *Adv. Synth. Catal.*, **349**, 2585.

147 Gagliardo, M., Selander, N., Mehendale, N.C., van Koten, G., Klein Gebbink, R.J.M., and Szabó, K.J. (2008) Catalytic performance of symmetrical and unsymmetrical sulfur-containing pincer complexes: synthesis and tandem catalytic activity of the first PCSPincer palladium complex. *Chem. Eur. J.*, **14**, 4800.

148 Mehendale, N.C., Sietsma, J.R.A., de Jong, K.P., van Walree, C.A., Klein Gebbink, R.J.M., and van Koten, G. (2007) PCP- and SCS-pincer palladium complexes immobilized on mesoporous silica: application in C–C bond formation reactions. *Adv. Synth. Catal.*, **349**, 2619.

149 Zhou, X., Lin, Y., Dai, L., Sun, J., Xia, L., and Tang, M. (1999) A catalytic enantioselective access to optically active 2-imidazoline from N-sulfonylimines and isocyanoacetates. *J. Org. Chem.*, **64**, 1331.

150 Zhou, X.-T., Lin, Y.-R., and Dai, L.-X. (1999) A simple and practical access to enantiopure 2,3-diamino acid derivatives. *Tetrahedron: Asymmetry*, **10**, 855.

151 Zhou, X.-T., Lin, Y.-R., and Dai, L.-X. (1998) A novel double addition of isocyanoacetamide to N-sulfonylimines for the synthesis of trisubstituted oxazoles. *Tetrahedron*, **54**, 12445.

152 El Kaïm, L., Gaultier, L., Grimaud, L., and Vieu, E. (2004) From isocyanides to trichloropyruvamides: application to a new preparation of oxamide derivatives. *Tetrahedron Lett.*, **45**, 8047.

153 Huang, W.-S., Zhang, Y.-X., and Yuan, C.-Y. (1996) Base induced intramolecular cyclization of α-ketoimidoyl chloride–an efficient preparation of 2-acyl 5-ethoxy oxazoles. *Synth. Commun.*, **26**, 1149.

154 Santos, A., El Kaïm, L., Grimaud, L., and Ronsseray, C. (2009) Unconventional oxazole formation from isocyanides. *Chem. Commun.*, 3907.

155 Mossetti, R., Pirali, T., Tron, G.C., and Zhu, J. (2010) Efficient synthesis of r-ketoamides via 2-acyl-5-aminooxazoles by reacting acyl chlorides and α-isocyanoacetamides. *Org. Lett.*, **12**, 820.

156 Schöllkopf, U. and Schröder, R. (1971) 2-unsubstituted oxazoles from α-metalated isocyanides and acylating agents. *Angew. Chem. Int. Ed. Engl.*, **10**, 333.

157 Schröder, R., Schöllkopf, U., Blume, E., and Hoppe, I. (1975) Synthesen mit α-metallierten Isocyaniden, XXVIII1. In 2-Stellung unsubstituierte Oxazole aus α-metallierten Isocyaniden und Acylierungsreagenzien. *Liebigs Ann. Chem.*, **3**, 533.

158 Suzuki, M., Iwasaki, T., Matsumoto, K., and Okumura, K. (1972) Convenient syntheses of aroylamino acids and α-amino ketones. *Synth. Commun.*, **2**, 237.

159 Suzuki, M., Iwasaki, T., Miyoshi, M., Okumura, K., and Matsumoto, K. (1973) New convenient syntheses of α-C-acylamino acids and α-amino ketones. *J. Org. Chem.*, **38**, 3571.

160 Ozaki, Y., Maeda, S., Iwasaki, T., Matsumoto, K., Odawara, A., Sasaki, Y., and Morita, T. (1983) Reaction of phthalic anhydrides with methyl isocyanoacetate: a useful synthesis of 1,2-dihydro-1-oxoisoquinolines. *Chem. Pharm. Bull.*, **31**, 4417.

161 Maeda, S., Suzuki, M., Iwasaki, T., Matsumoto, K., and Iwasawa, Y. (1984) Synthesis of 2-mercapto-4-substituted imidazole derivatives with antiinflammatory properties. *Chem. Pharm. Bull.*, **32**, 2536.

162 Tang, J. and Verkade, J.G. (1994) Nonionic superbase-promoted synthesis of oxazoles and pyrroles: facile synthesis of porphyrins and α-C-acyl amino acid esters. *J. Org. Chem.*, **59**, 7793.

163 Hartman, G. and Weinstock, L. (1976) A novel 1,3-thiazole synthesis via α-metallated isocyanides and thiono esters. *Synthesis*, 681.

164 Mossetti, R., Caprioglio, D., Colombano, G., Tron, G.S., and Pirali, T. (2011) A novel α-isocyanoacetamide-based three-component reaction for the synthesis of dialkyl 2-acyl-5-aminofuran-3,4-dicarboxylates. *Org. Biomol. Chem.*, **9**, 1627.

165 Henneke, K., Schöllkopf, U., and Neudecker, T. (1979) Synthesen mit α-metallierten Isocyaniden, XLII. Bi-, Ter- und Quateroxazole aus α-anionisierten Isocyaniden und Acylierungsmitteln; α-Aminoketone und α,α-Diaminoketone. *Liebigs Ann. Chem.*, **9**, 1370.

166 Hamada, Y. and Shiori, T. (1982) New methods and reagents in organic synthesis, 24. A new synthesis of prumycin as an application of the direct c-acylation using diphenyl phosphorazidate (DPPA). *Tetrahedron Lett.*, **23**, 1193.

167 Hamada, Y., Kawai, A., and Shioiri, T. (1984) New methods and reagents in organic synthesis. 49. A highly efficient stereoselective synthesis of L-daunosamine through direct C-acylation using diphenyl phosphorazidate (DPPA). *Tetrahedron Lett.*, **25**, 5409.

168 Brescia, M., Rokosz, L.L., Cole, A.G., Stauffer, T.M., Lehrach, J.M., Auld, D.S., Henderson, I., and Webb, M.L. (2007) Discovery and preliminary evaluation of 5-(4-phenylbenzyl)oxazole-4-carboxamides as prostacyclin receptor antagonists. *Bioorg. Med. Chem. Lett.*, **17**, 1211.

169 Hamada, Y. and Shioiri, T. (1982) New methods and reagents in organic synthesis. 22. 1-Diphenyl phosphorazidate as a reagent for C-acylation of methyl isocyanoacetate with carboxylic acids. *Tetrahedron Lett.*, **23**, 235.

170 Kline, T., Bowman, J., Iglewski, B.H., de Kievit, T., Kakai, Y., and Passador, L. (1999) Novel synthetic analogs of the *Pseudomonas* autoinducer. *Bioorg. Med. Chem. Lett.*, **9**, 3447.

171 Hamada, Y., Kawai, A., Matsui, T., Hara, O., and Shioiri, T. (1990) 4-Alkoxycarbonyloxazoles as β-hydroxy-α-amino acid synthons: efficient stereoselective syntheses of 3-amino-2,3,6-trideoxyhexoses and a hydroxy amino acid moiety of Al-77-B. *Tetrahedron*, **46**, 4823.

172 Kozikowski, A.P. (1980) Copper(I)-promoted acylation reactions. A transition metal-mediated version of the Friedel-Crafts reaction. *J. Am. Chem. Soc.*, **102**, 860.

173 Kozikowski, A.P. and Ames, A. (1981) Dimethylaluminum methaneselenolate–a useful reagent for the preparation of selenoesters. A new Friedel-Crafts acylation procedure promoted by Cu(I). *Tetrahedron*, **41**, 4821.

174 Solomon, D.M., Rizvi, R.K., and Kaminski, J.J. (1987) Observations on the reactions of isocyanoacetane esters with isothiocyanates and isocyanates. *Heterocycles*, **26**, 651.

175 Chupp, J.P. and Leschinsky, K.L. (1980) Heterocycles from substituted amides. VIII (1,2). Oxazole derivatives from reaction of isocyanates with 2-isocyanoacetamides. *J. Heterocycl. Chem.*, **17**, 711.

176 Suzuki, M., Moriya, T., Matsumoto, K., and Miyoshi, M. (1982) A new convenient synthesis of 5-amino-1,3-thiazole-4-carboxylic acids. *Synthesis*, 874.

177 Baxendale, I.R., Ley, S.V., Smith, C.D., Tamborini, L., and Voica, A.-F. (2008) A bifurcated pathway to thiazoles and imidazoles using a modular flow microreactor. *J. Comb. Chem.*, **10**, 851.

178 Maeda, H., Kambe, N., Sonoda, N., Fujiwara, S., and Shin-ike, T. (1997) Synthesis of 1,3-selenazoles and 2-imidazolin-5-selones from isoselenocyanates and isocyanides. *Tetrahedron*, **40**, 13667.

179 Schöllkopf, U., Porsch, P., and Blume, E. (1976) Synthesen mit α-Metallierten Isocyaniden, XXXV. 5-(Methylthio) thiazole aus α-Metallierten Isocyaniden und Schwefelkohlenstoff. *Liebigs Ann. Chem.*, **11**, 2122.

180 Huang, W.S., Yuan, C.Y., and Wang, Z.Q. (1995) Facile synthesis of 1-substituted 5-trifluoromethylimidazole-4-carboxilates. *J. Fluorine Chem.*, **74**, 279.

181 TenBrink, R.E., Im, W.B., Sethy, V.H., Tang, A.H., and Carter, D.B. (1994) Antagonist, partial agonist, and full agonist imidazo[1,5-a]quinoxaline amides and carbamates acting through the $GABA_A$/benzodiazepine receptor. *J. Med. Chem.*, **37**, 758.

182 Watjen, F., Baker, R., Engelstoff, M., Herbert, R., MacLeod, A., Knight, A., Merchant, K., Moseley, J., and Saunders, J. (1989) Novel benzodiazepine receptor partial agonists: oxadiazolylimidazobenzodiazepines. *J. Med. Chem.*, **32**, 2282.

183 Fryer, R.I., Gu, Z.-Q., and Wang, C.-G. (1991) Synthesis of novel substituted 4H-imidazo[1,5-a][1,4]benzodiazepines. *J. Heterocycl. Chem.*, **28**, 1661.

184 Li, X., Yu, J., Atack, J., and Cook, J. (2004) Development of selective ligands for benzodiazepine receptor subtypes by manipulating the substituents at positions 3- and 7- of optically active BzR ligands. *Med. Chem. Res.*, 259.

185 Jacobsen, E.J., Stelzer, L.S., TenBrink, R.E., Belonga, K.L., Carter, D.B., Im, H.K., Im, W.B., Sethy, V.H., Tang, A.H., VonVoigtlander, P.F., Petke, J.D., Zhong, W., and Mickelson, J.W. (1999) Piperazine imidazo[1,5-a]quinoxaline ureas as high-affinity $GABA_A$ ligands of dual functionality. *J. Med. Chem.*, **42**, 1123.

186 Bonnett, R. (2002) Progress with heterocyclic photosensitizers for the photodynamic therapy (PDT) of tumours. *J. Heterocycl. Chem.*, **39**, 465.

187 Erker, T. and Trinkl, K. (2003) Studies on the chemistry of thienoannelated O,N- and S,N-containing heterocycles, 28 [1]. Synthesis of imidazo[1,5-d]thieno[2,3-b][1,4]thiazine derivatives as GABA-receptor ligands. *Monatsch. Chem.*, **134**, 439.

188 Erker, T. and Trinkl, K. (2001) Studies on the chemistry of thienoanellated O,N- and S,N-containing heterocycles. Part 22. Synthesis of 1,2,4-oxadiazolylimidazo[1,5-a]thieno[2,3-e]pyrazines as ligands for the γ-aminobutyric acid A/benzodiazepine receptor complex. *Heterocycles*, **55**, 1963.

189 Weber, M. and Erker, T. (2002) Synthesis of a new series of imidazo[1,5-d]pyrido[2,3-b][1,4]thiazines as potential ligands for the GABA receptor complex. *Monatsch. Chem.*, **133**, 1205.

190 Sundaram, G.S.M., Singh, B., Venkatesh, C., Ila, H., and Junjappa, H. (2007) Dipolar cycloaddition of ethyl isocyanoacetate to 3-chloro-2-(methylthio)/2-(methylsulfonyl) quinoxalines: highly regio and chemoselective synthesis of substituted imidazo[1,5-a]quinoxaline-3-carboxylates. *J. Org. Chem.*, **72**, 5020.

191 Murakami, T., Otsuka, M., and Ohno, M. (1982) An efficient synthesis of 5-diethoxymethylimidazole-4-carboxylate, a potential precursor for various imidazole derivatives. *Tetrahedron Lett.*, **23**, 4729.

192 (a) Zhu, J. and Bienaymé, H. (eds) (2005) *Multicomponent Reactions*, Wiley-VCH Verlag GmbH, Weinheim; (b) O'Neil, I.A. (2003) Isocyanides and their heteroanalogues (RZC), in *Comprehensive Organic Functional Group Transformations*, vol. 3, Part IV, 3.21 (ed. G. Pattenden), Pergamon, Oxford, pp. 693; (c) Ugi, I. (1971) *Isonitrile Chemistry*, Academic Press, New York, London.

193 See for example, recent reviews: (a) Sadjadi, S. and Heravi, M.M. (2011) Recent application of isocyanides in

synthesis of heterocycles. *Tetrahedron*, **67**, 2707; (b) Heravi, M.M. and Moghimi, S. (2010) *J. Iran. Chem. Soc.*, **8**, 306; (c) Lygin, A.V. and de Meijere, A. (2010) Isocyanides in the synthesis of nitrogen heterocycles. *Angew. Chem. Int. Ed.*, **49**, 9094; (d) Ivachtchenko, A.V., Ivanenkov, Y.A., Kysil, V.M., Krasavin, M.Y., and Ilyin, A.P. (2010) Multicomponent reactions of isocyanides in the synthesis of heterocycles. *Russ. Chem. Rev.*, **79**, 787; (e) Wu, J. and Cao, S. (2009) Multicomponent reactions based on fluoro-containing building blocks. *Curr. Org. Chem.*, **13**, 1791; (f) Hulme, C. and Dietrich, J. (2009) Emerging molecular diversity from the intra-molecular Ugi reaction: iterative efficiency in medicinal chemistry. *Mol. Divers.*, **13**, 195; (g) El Kaïm, L. and Grimaud, L. (2009) *Tetrahedron*, **65**, 2153; (h) Akritopoulou-Zanze, I. (2008) Isocyanide-based multicomponent reactions in drug discovery. *Curr. Opin. Chem. Biol.*, **12**, 324; (i) Marcaccini S. and Torroba T. (2007) The use of the Ugi four-component condensation. *Nat. Protoc.*, **2**, 632; (j) Eckert, H. (2007) From multi-component-reactions (MCRs) towards multi-function-component-reactions (MFCRs). *Heterocycles*, **73**, 149; (k) Dömling, A. (2006) *Chem. Rev.*, **106**, 17; (l) Mironov, M.A. (2006) Design of multi-component reactions: from libraries of compounds to libraries of reactions. *QSAR Comb. Sci.*, **25**, 423; (m) Zhu, J. (2003) Recent developments in the isonitrile-based multicomponent synthesis of heterocycles. *Eur. J. Org. Chem.*, 1133; (n) Ugi, I., Werner, B., and Dömling, A. (2003) The chemistry of isocyanides, their multicomponent reactions and their libraries. *Molecules*, **8**, 53; (o) Dömling, A. and Ugi, I. (2000) Multicomponent reactions with isocyanides. *Angew. Chem. Int. Ed.*, **39**, 3168.

194 (a) Passerini, M. and Simone, L. (1921) *Gazz. Chim. Ital.*, **51**, 126; (b) Passerini, M. and Ragni, G. (1921) *Gazz. Chim. Ital.*, **51**, 762.

195 Fetzer, U. and Ugi, I. (1962) Isonitrile, XII. Synthese von Depsipeptid-Derivaten Mittels der Passerini-Reaktion. *Liebigs Ann. Chem.*, **659**, 184.

196 (a) Burger, K., Schierlinger, C., and Mütze, K. (1993) 3,3,3-Trifluoro-2-isocyanopropionates, new versatile building blocks for the introduction of trifluoromethyl groups into organic molecules. *J. Fluorine Chem.*, **65**, 149; (b) Burger, K., Mütze, K., Hollweck, W., and Kocksch, B. (1998) Incorporation of α-trifluoromethyl-substituted α-amino acids into C- and N-terminal position of peptides and peptide mimetics using multicomponent reactions. *Tetrahedron*, **54**, 5915; (c) Gulevich, A.V., Shpilevaya, I.V., and Nenajdenko, V.G. (2009) The Passerini reaction with CF3-carbonyl compounds – multicomponent approach to trifluoromethyl depsipeptides. *Eur. J. Org. Chem.*, 3801.

197 Zhdanko, A.G., Gulevich, A.V., and Nenajdenko, V.G. (2009) One-step synthesis of N-acetylcysteine and glutathione derivatives using the Ugi reaction. *Tetrahedron*, **65**, 4692.

198 Schöllkopf, U., Porsch, P.-H., and Lau, H.-H. (1979) Synthesen mit α-metallierten Isocyaniden, XLIV. Notiz über β-Dimethylamino-α-isocyanacrylsäureester und ihre Verwendung in der Heterocyclenchemie. *Liebigs Ann. Chem.*, 1444.

199 Bienaymé, H. (1998) "Reagent explosion": an efficient method to increase library size and diversity. *Tetrahedron Lett.*, **39**, 4255.

200 De Luca, L., Giacomelli, G., and Porcheddu, A. (2005) Synthesis of 1-alkyl-4-imidazolecarboxylates: a catch and release strategy. *J. Comb. Chem.*, **7**, 905.

201 Helal, C.J. and Lucas, J.C. (2002) A concise and regioselective synthesis of 1-alkyl-4-imidazolecarboxylates. *Org. Lett.*, **4**, 4133.

202 Henkel, B. (2004) Synthesis of imidazole-4-carboxylic acids via solid-phase bound 3-N,N-(dimethylamino)-2-isocyanoacrylate. *Tetrahedron Lett.*, **45**, 2219.

203 Lau, H.-H. and Schöllkopf, U. (1982) Synthese von 2-Alkyl- und 2-Acyl-1-methylimidazol-4-

carbonsauremethylestern aus (2)-P-Dimethylamino-a-isocyanacrylsauremethylester und Alkyl- oder Acylhalogeniden. *Liebigs Ann. Chem.*, 2093.

204 Bossio, R., Marcaccini, S., Pepino, R., Paoli, P., and Polo, C. (1993) Studies on isocyanides and related compounds. Investigation on the reaction between (Z)-alkyl 3-dimethylamino-2-isocyanoacrylates and acyl chlorides. *J. Heterocycl. Chem.*, **30**, 575.

205 Bossio, R., Marcaccini, S., Pepino, R., and Paoli, P. (1994) Studies on isocyanides and related compounds. An unusual synthesis of imidazolyloxazolones. *J. Heterocycl. Chem.*, **31**, 729.

206 Dömling, A. and Illgen, K. (2005) 1-Isocyano-2-dimethylamino-alkenes: versatile reagents in organic synthesis. *Synthesis*, 662.

207 Nunami, K., Yamada, M., Fukui, T., and Matsumoto, K. (1994) A novel synthesis of methyl 1,5-disubstituted imidazole-4-carboxylates using 3-bromo-2-isocyanoacrylates. *J. Org. Chem.*, **59**, 7635.

208 Hiramatsu, K., Nunami, K., Hayashi, K., and Matsumoto, K. (1990) A facile synthesis of methyl 1,5-disubstituted imidazole-4-carboxylates. *Synthesis*, 781.

209 Yamada, M., Fukui, T., and Nunami, K. (1995) A novel synthesis of methyl 5-substituted thiazole-4-carboxylates using 3-bromo-2-isocyanoacrylates (BICA). *Tetrahedron Lett.*, **36**, 257.

210 Yamada, M., Fukui, T., and Nunami, K. (1995) Synthesis of 5-substituted methyl 1-hydroxyimidazole-4-carboxylates and 5-substituted methyl 1-aminoimidazole-4-carboxylates using 3-bromo-2-isocyanoacrylates (BICA). *Synthesis*, 1365.

211 Umkehrer, M., Kolb, J., Burdack, C., and Hiller, W. (2005) 2,4,5-Trisubstituted thiazole building blocks by a novel multi-component reaction. *Synlett*, 79.

212 Kim, S.W., Bauer, S.M., and Armstrong, R.W. (1998) Multicomponent solution phase synthesis of dehydroamino acid derivatives base on the Passerini reaction. *Tetrahedron Lett.*, **39**, 7031.

213 Bergemann, M., and Neidlein, R. (1999) Studies on the reactivity of α-cyano-α-isocyanoalkanoates ± versatile synthons for the assembly of imidazoles. *Helv. Chim. Acta*, **82**, 909.

214 Bergemann, M. and Neidlein, R. (1998) An efficient approach to novel 5,5-disubstituted dithiohydantoins via α-cyano-α-(dihalogenomethyleneamino) alkanoic acid esters. *Synthesis*, 1437.

215 Müller, S. and Neidlein, R. (2002) 2-Cyano-2-isocyanoalkanoates in multicomponent reactions. *Helv. Chim. Acta*, **86**, 2222.

216 (a) Bossio, R., Marcaccini, S., and Pepino, R. (1986) A novel synthetic route to oxazoles: one pot synthesis of 2-arylthio-5-alkoxyoxazoles. *Heterocycles*, **24**, 2003; (b) Berrée, F., Marchand, E., and Morel, G. (1992) Base-mediated reactions in solid-liquid media. A 1,3-dipolar cycloaddition route to pyrrolines and pyrroles from imino chlorosulfides. *Tetrahedron Lett.*, **33**, 6155.

217 Bossio, R., Marcaccini, S., Pepino, R., Polo, C., and Torroba, T. (1989) Synthesis of 2,4-diarylthio-5-N-alkyl-N-phenylaminooxazoles. A novel class of oxazole derivatives. *Heterocycles*, **29**, 1829.

218 Bossio, R., Marcaccini, S., and Pepino, R. (1986) Synthesis of 5,5'-dialkoxy-2,2'-dioxazolylsulfides. *Heterocycles*, **24**, 2411.

219 Bossio, R., Marcaccini, S., Pepino, R., Torroba, T., and Valle, G. (1987) An unusual and simple one-pot synthesis of thiazolo[5,4-d]-thiazoles. *Synthesis*, 1138.

220 Bossio, R., Marcaccini, S., Pepino, R., Polo, C., Torroba, T., and Valle, G. (1989) Synthesis of 1-arylthiocarbonyl-4-isopropylamino-2,5-dihydro-*lH*-imidazole-2-thines, a novel class of imidazole derivatives. *Heterocycles*, **29**, 1843.

221 Bossio, R., Marcaccini, S., Muratori, M., Pepino, R., and Valle, G. (1990) Studies on alkyl isocyanoacetates and related compounds. Synthesis of 6-arylthio-8-ethoxycarbonyl-4-ethoxycarbonyl methylaminoimidazo[5,1-b][1,3,5] thiadiazine-2-thiones. *Heterocycles*, **31**, 611.

222 Kanazawa, C., Kamijo, S., and Yamamoto, Y. (2006) Synthesis of imidazoles through the copper-catalyzed cross-cycloaddition between two different isocyanides. *J. Am. Chem. Soc.*, **128**, 10662.

223 Bonin, M.-A., Giguère, D., and Roy, R. (2007) N-arylimidazole synthesis by cross-cycloaddition of isocyanides using a novel catalytic system. *Tetrahedron*, **63**, 4912.

224 Matsumoto, K., Suzuki, M., Tomie, M., Yoneda, N., and Miyoshi, M. (1975) A new synthesis of 1-aryl-1H-1,2,4-triazole-3-carboxylic acid esters. *Synthesis*, 609.

225 Genin, M.J., Allwine, D.A., Anderson, D.J., Barbachyn, M.R., Emmert, D.E., Garmon, S.A., Graber, D.R., Grega, K.C., Hester, J.B., Hutchinson, D.K., Morris, J., Reischer, R.J., Ford, C.W., Zurenko, G.E., Hamel, J.C., Schaadt, R.D., Stapert, D., and Yagi, B.H. (2000) Substituent effects on the antibacterial activity of nitrogen–carbon-linked (azolylphenyl)oxazolidinones with expanded activity against the fastidious Gram-negative organisms *Haemophilus influenzae* and *Moraxella catarrhalis*. *J. Med. Chem.*, **43**, 953.

226 Elders, N., Ruijter, E., de Kanter, F.J.J., Groen, M.B., and Orru, R.V.A. (2008) Selective formation of 2-imidazolines and 2-substituted oxazoles by using a three-component reaction. *Chem. Eur. J.*, **14**, 4961.

227 Cai, Q., Li, Z., Wei, J., Ha, C., Pei, D., and Ding, K. (2009) Assembly of indole-2-carboxylic acid esters through a ligand-free copper-catalysed cascade process. *Chem. Commun.*, 7581.

228 Wu, J., Chen, W., Hu, M., Zou, H., and Yu, Y. (2010) Synthesis of polysubstituted 5-aminooxazoles from α- diazocarbonyl esters and α-isocyanoacetamides. *Org. Lett.*, **12**, 616.

229 Monge, D., Jensen, K.L., Marín, I., and Jørgensen, K.A. (2011) Synthesis of 1,2,4-triazolines: base-catalyzed hydrazination/cyclization cascade of α-isocyano esters and amides. *Org. Lett.*, **13**, 328.

230 (a) Li, X. and Danishefsky, S. J. (2008) New chemistry with old functional groups: on the reaction of isonitriles with carboxylic acids – a route to various amide types. *J. Am. Chem. Soc.*, **130**, 5446; (b) Li, X., Yuan, Y., Berkowitz, W., Todaro, L. J., and Danishefsky, S. J. (2008) On the two-component microwave-mediated reaction of isonitriles with carboxylic acids: regarding alleged formimidate carboxylate mixed anhydrides. *J. Am. Chem. Soc.*, **130**, 13222.

231 Li, X., Yuan, Y., Kan, C., and Danishefsky, S.J. (2008) Addressing mechanistic issues in the coupling of isonitriles and carboxylic acids: potential routes to peptidic constructs. *J. Am. Chem. Soc.*, **130**, 13225.

232 Yuan, Y., Zhu, J., Li, X., Wu, X., and Danishefsky, S.J. (2009) Preparation and reactions of N-thioformyl peptides from amino thioacids and isonitriles. *Tetrahedron Lett.*, **50**, 2329.

233 Wu, X., Stockdill, J.L., Wang, P., and Danishefsky, S.J. (2010) Total synthesis of cyclosporine: access to N-methylated peptides via isonitrile coupling reactions. *J. Am. Chem. Soc.*, **132**, 4098.

234 Wu, X., Park, P.K., and Danishefsky, S.J. (2011) On the synthesis of conformationally modified peptides through isonitrile chemistry: implications for dealing with polypeptide aggregation. *J. Am. Chem. Soc.*, **133**, 7700.

235 Wilson, R.M., Stockdill, J.L., Wu, X., Li, X., Vadola, P.A., Park, P.K., Wang, P., and Danishefsky, S.J. (2012) A fascinating journey into history: exploration of the world of isonitriles en route to complex amides. *Angew. Chem. Int. Ed.*, **51**, 2834.

236 Carney, D.W., Truong, J.V., and Sello, J.K. (2011) Investigation of the configurational stabilities of chiral isocyanoacetates in multicomponent reactions. *J. Org. Chem.*, **76**, 10279.

5
Ugi and Passerini Reactions with Carboxylic Acid Surrogates
Laurent El Kaïm and Laurence Grimaud

5.1
Introduction

Although the Passerini reaction was discovered in 1921, and the Ugi reaction in 1959, it was not until the late 1980s that these couplings began to attract attention from a wide range of industrial and academic chemists [1]. In association with the development of combinatorial chemistry, interest in the isocyanides and in their role in multicomponent reactions (MCRs) has advanced considerably and, as a consequence, the scope and potential of these reactions have undergone extensive reinvestigation. The initial reports on the Passerini [2] and Ugi [3] reactions involved an isocyanide, a carboxylic acid, and a carbonyl or an imine. In addition, they demonstrated a high functional tolerance, with much effort being focused on the use of correct difunctional starting reagents to generate heterocycles via post-condensation transformations.

Other extensions involved more important changes in the nature of the starting components, and this led eventually to the different reaction mechanisms involved. Whilst variations on the amino component of the Ugi reaction, as well as modifications of the isocyanides, are discussed in other chapters of this book, in this chapter attention is focused on variants concerning the acidic partner. The intention is not to cover the topic comprehensively, but rather to provide clues towards a better understanding of the synthetic possibilities offered by the substitution pattern of carboxylic acids. At this point, an apology is provided in advance for any references that might have been omitted but are, nonetheless, significant to the subject.

The acidic component plays a prominent role in both Ugi and Passerini couplings. Indeed, it ensures a formation of the most electrophilic species through protonation of the imine or carbonyl moieties; the carboxylate counteranion then traps the nitrilium that results from an attack of the isocyanide on the iminium, while the nature of the acid allows the imidate to evolve via an acyl transfer – a process known as the Mumm rearrangement [4]. This last step is important, as the formation of the C–O double bond is commonly accepted as being the "driving force" of the entire process [5]. The details of most acid surrogates were described by Ugi shortly his first report. Among these, some – such as mineral acids – were

Isocyanide Chemistry: Applications in Synthesis and Material Science, First Edition. Edited by Valentine Nenajdenko.
© 2012 Wiley-VCH Verlag GmbH & Co. KGaA. Published 2012 by Wiley-VCH Verlag GmbH & Co. KGaA.

mentioned in 1962 in *Angewandte Chemie* [6], but were scarcely used. In this chapter, the details will be presented of the different acid surrogates that have been developed since that time, for both Ugi and Passerini couplings, while various examples will be provided of the significant post-condensation transformations performed on the resulting adducts. Attention will be focused on those reactions that involve the intermolecular trapping of nitrilium intermediates. Indeed, reactions such as Groebke–Bienaymé couplings [7] and/or some preparations of oxazoles from isocyanoacetamides [8] might be considered as extensions of the Ugi and Passerini reactions, involving specific amino and isocyanide derivatives.

5.2
Carboxylic Acid Surrogates

Nitrilium salts are highly electrophilic intermediates species that can be attacked by the nucleophiles that usually are associated with the Ugi or Passerini activating acids, and this leads to the corresponding imidates (**B** in Scheme 5.1). In general, the latter are of moderate stability due to the nature of the nucleophiles, which are also good leaving groups. Consequently, efficient isocyanide-based multicomponent reactions (IMCRs) mostly involve nucleophiles that allow a rapid rearrangement of the corresponding imidate, such as those observed with the Mum rearrangement in traditional Ugi couplings. Here, the various carboxylic surrogates in the IMCRs have been classified according to the nature of the last step: (i) Mumm rearrangement for thiocarboxylic and carbonic acids; (ii) cyclizations for hydrazoic acid and cyanic derivatives; (iii) imidate–amide tautomerization for hydroselenide; and (iv) the Smiles rearrangement for phenolic derivatives.

5.2.1
Thiocarboxylic Acids

Thiocarboxylic acids are able to promote an analogous Ugi-type coupling, thus affording the corresponding α-aminoacylthioamides (Scheme 5.2) [9]. This cou-

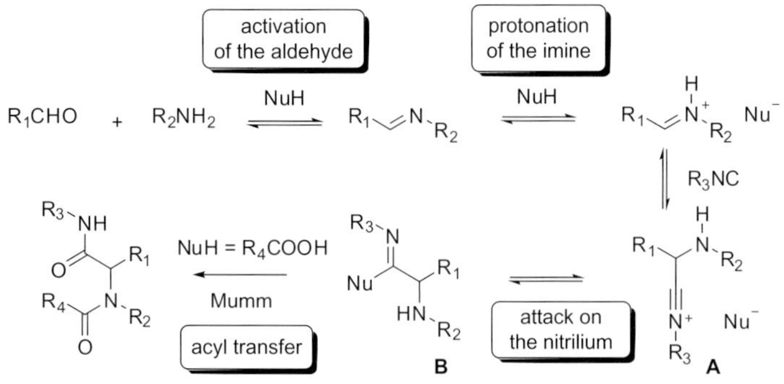

Scheme 5.1

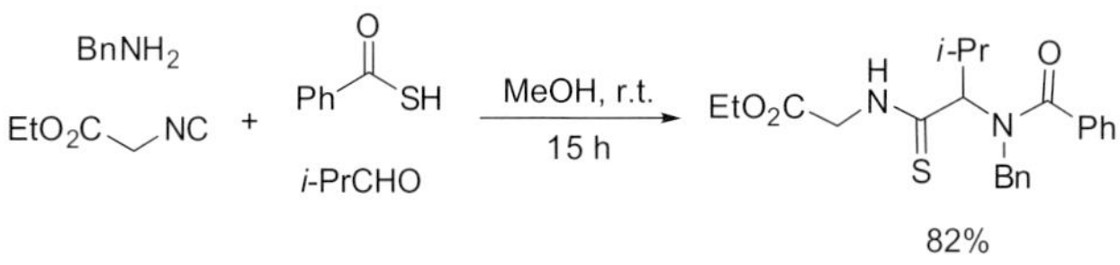

Scheme 5.2

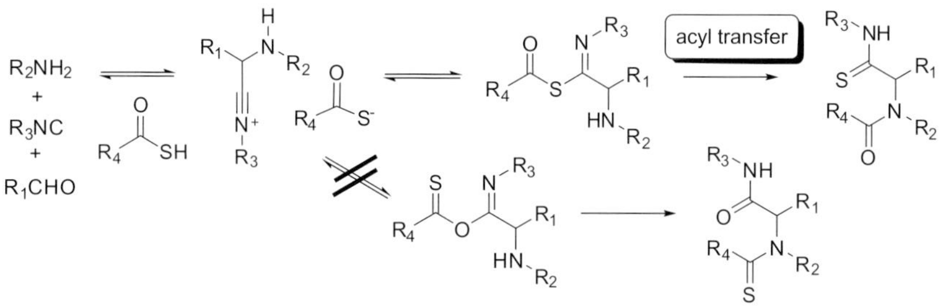

Scheme 5.3

pling proceeds much like a classical Ugi reaction, where the nitrilium is trapped by the thiocarboxylate and the resultant thioimidate further evolves via the acyl transfer to form the thioamide bearing the former isocyanide moiety (Scheme 5.3). The other regiomeric α-aminothioacylamide, which would result from a rearrangement of the imidate formed by the attack on the nitrilium by the oxygenated anion, was not detected in these couplings, most likely due to the higher nucleophilicity of the sulfur fragment of the thiocarboxylate ion.

This reaction is of great interest as it allows the synthesis of highly functionalized thioamides encompassing an amide moiety, which have been coined as "endothiopeptides" [10]. Notably, the reaction is efficient with a variety of carbonyl derivatives, although with regard to the amine partner the adducts obtained when using allylamine are rather unstable and decompose rapidly. Typically, ammonia provides moderate yields in this reaction (Scheme 5.3).

Various extensions employing bifunctional partners have been investigated. For instance, Schöllkopf isocyanide (β-dimethylamino-α-isocyanoacrylate) provides access to functionalized thiazoles following cyclization of the four-component adducts (Scheme 5.4) [9, 11, 12]. More substituted thiazoles can result from the condensation of three-substituted 3-bromo-2-isocyanoacrylate derivatives (Scheme 5.4) [13]. In this reaction, β-aminothiocarboxylic acid leads to thiazolo β-lactams in a one-step process, in an efficient manner (Scheme 5.5) [11].

The use of partners bearing a masked carbonyl moiety allows the formation of various heterocycles; for example, acetal-substituted isocyanides undergo the four-component reaction (4CR), while the resulting thioamides further cyclize under acidic conditions to form thiazolines or thiazoles [10], or imidazoles in the presence of ammonia (Scheme 5.6) [14]. A sequence Ugi–Ullmann coupling was developed to access benzothiazoles quite efficiently (Scheme 5.7) [15].

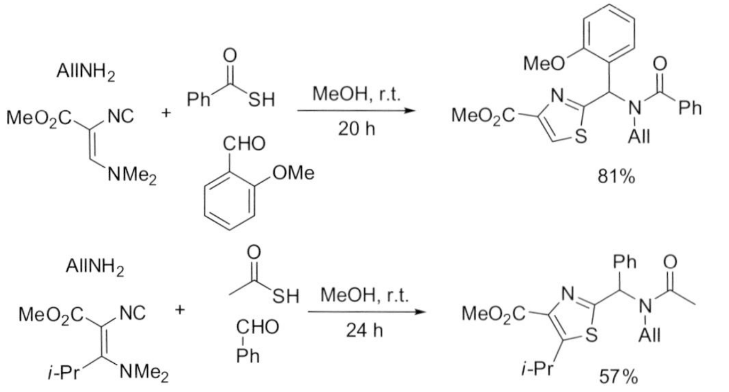

Scheme 5.4

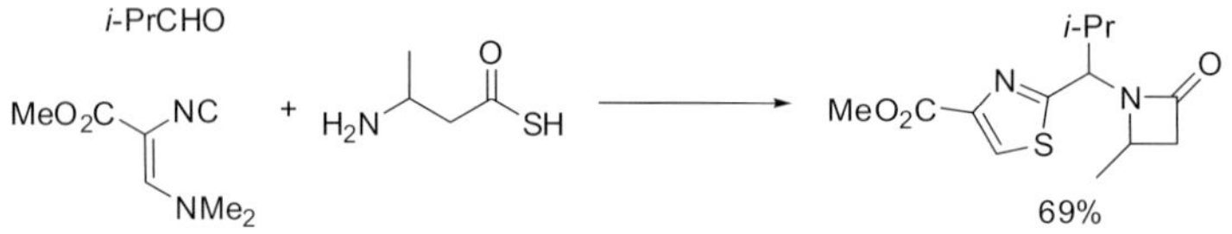

Scheme 5.5

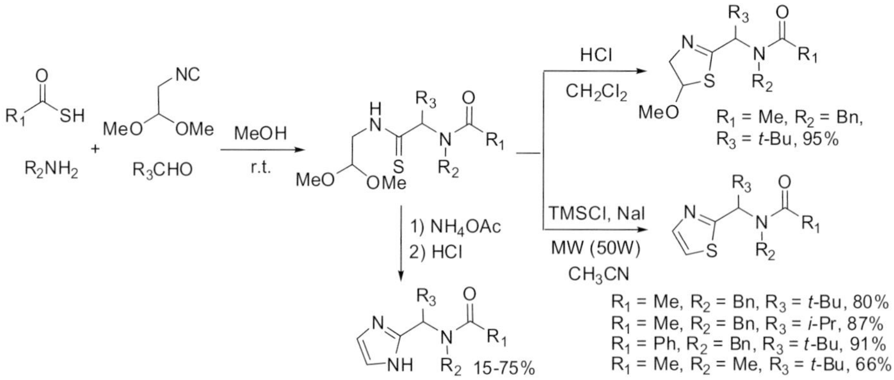

Scheme 5.6

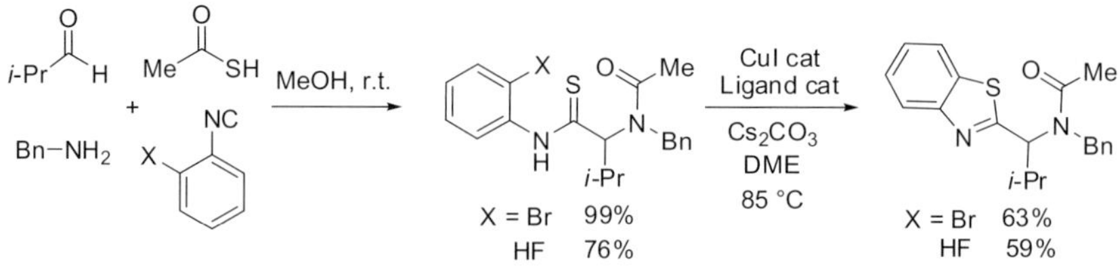

Scheme 5.7

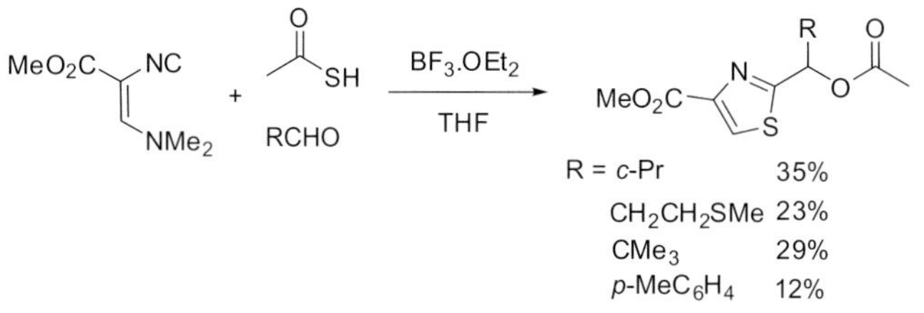

Scheme 5.8

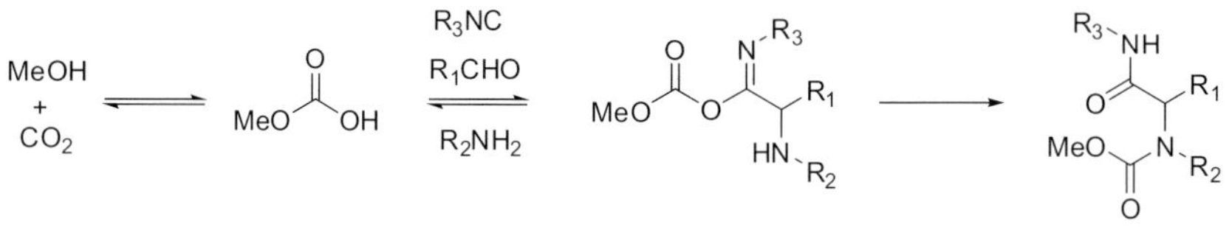

R	Yield
c-Pr	35%
CH$_2$CH$_2$SMe	23%
CMe$_3$	29%
p-MeC$_6$H$_4$	12%

Scheme 5.9

The corresponding Passerini coupling was shown to be less efficient; indeed, the reaction with the Schöllkopf isocyanide failed to provide the acyloxythiazole unless 1 equiv. of a Lewis acid, as BF$_3$.OEt$_2$, Sc(OTf)$_3$, or ZrCl$_4$, was used (Scheme 5.8) [16]. Even under these optimized conditions, however, the yields failed to exceed 40%.

5.2.2
Carbonic Acid and Derivatives

In-situ-generated methyl carbonic acids may be used as acid surrogates in Ugi-type couplings, as shown first by Ugi [17]. In this five-component coupling (5-CC), the methanol first attacks CO$_2$ to form the carbonic acid species, which then traps the nitrilium forming the imidocarbonate. The latter further evolves via a Mumm-type rearrangement to afford the corresponding carbamates (Scheme 5.9).

In 1998, this reaction was extended to a wide range of alcohols; typically, low-molecular-weight alcohols could be used as the solvent, while nonvolatile partners should be added in large excess when using chloroform as solvent (Scheme 5.10) [18]. The reaction was performed in a CO$_2$-presaturated solvent at 0 °C, and with a slight excess of amine. Methanol appeared to be the best partner in this 5-CC, but even in this case the yields varied dramatically, depending on the other inputs. These results have most likely contributed to the lack of further development of these couplings.

Some post-condensations of the adducts were described for the syntheses of various hydantoin derivatives [19], according to a UDC (Ugi/de-Boc/cyclize) strategy (Scheme 5.11).

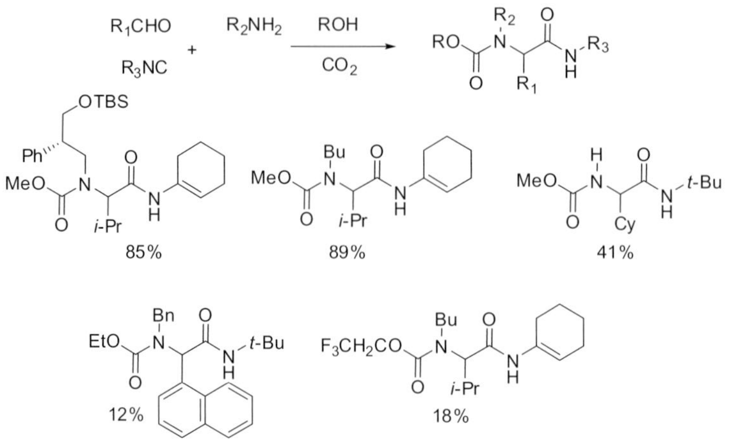

164 | *5 Ugi and Passerini Reactions with Carboxylic Acid Surrogates*

Scheme 5.10

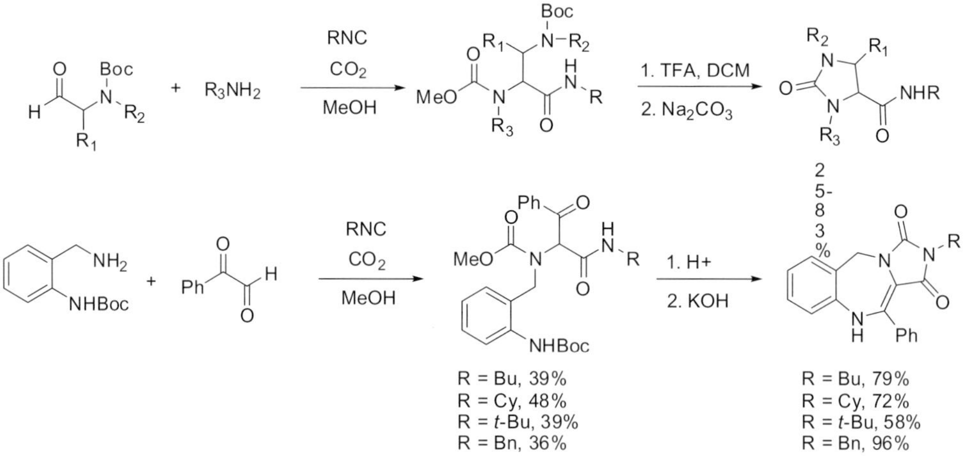

Scheme 5.11

Thiocarbonic acid derivatives are even less efficient, as they always generate a complex mixture of products, most likely due to the formation of hydrogen sulfide that results from the addition of an amine to COS or CS_2 (Scheme 5.12) [18]. When conducted in dichloromethane with 4 equiv. of the amine partner, the reactions with CS_2 proceed smoothly to provide the corresponding α-amino thioamides in good yields. Notably, these products might also be obtained from a Ugi coupling using either H_2S or $H_2S_2O_3$ as acid, thereby avoiding the thiourea byproducts.

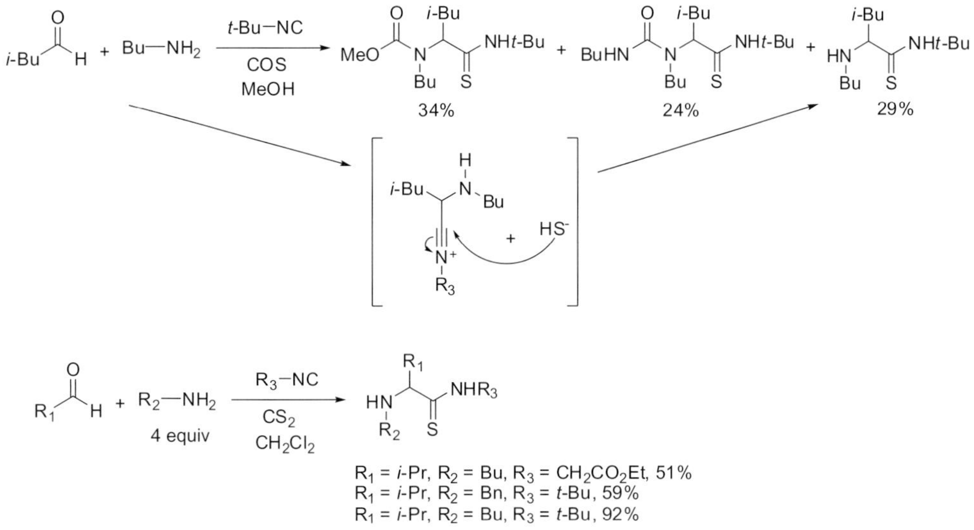

5.2 Carboxylic Acid Surrogates

[Scheme 5.12 reaction diagram]

R$_1$ = i-Pr, R$_2$ = Bu, R$_3$ = CH$_2$CO$_2$Et, 51%
R$_1$ = i-Pr, R$_2$ = Bn, R$_3$ = t-Bu, 59%
R$_1$ = i-Pr, R$_2$ = Bu, R$_3$ = t-Bu, 92%

Scheme 5.12

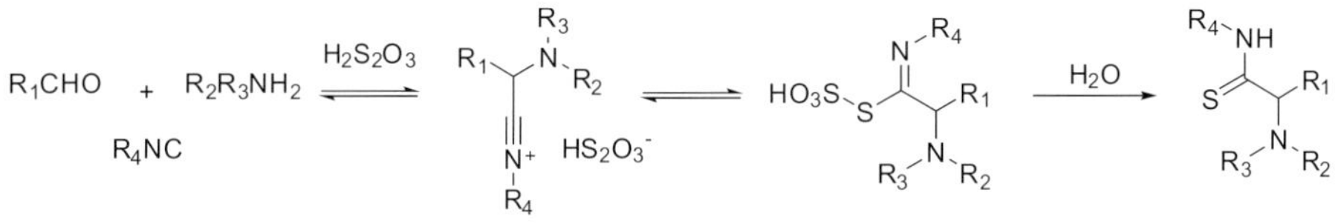

[Scheme 5.13 reaction diagram]

Scheme 5.13

5.2.3
Selenide and Sulfide

Since water is not sufficiently acidic to promote an Ugi-type reaction, additives are required for an efficient coupling (as will be seen later in this chapter). Hydrogen sulfide behaved in a similar fashion, with no coupling observed, but thiosulfuric acid was able to promote a 4CR, forming the corresponding thioamides (Scheme 5.13) [6, 20].

A similar reaction could be performed with hydroselenide, providing rapid access to functionalized selenoamides (Scheme 5.14) [6, 20].

5.2.4
Silanol

Recently, silanols were reported to promote a three-component reaction (3CR) by coupling an aldehyde with an isocyanide and triphenylsilanol to form

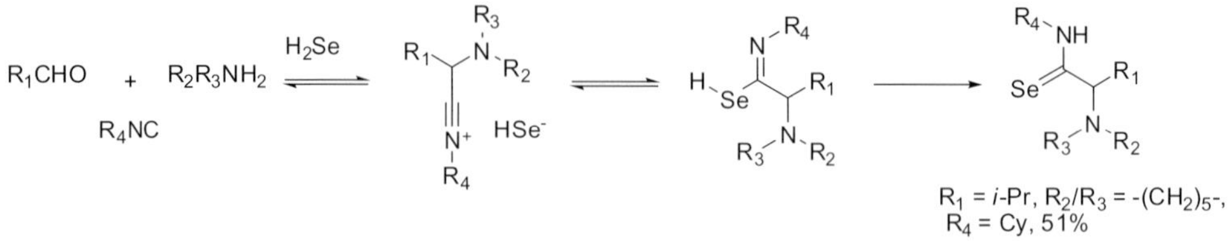

Scheme 5.14

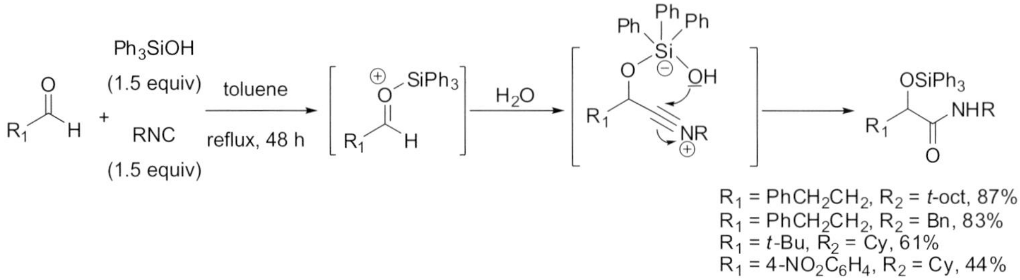

Scheme 5.15

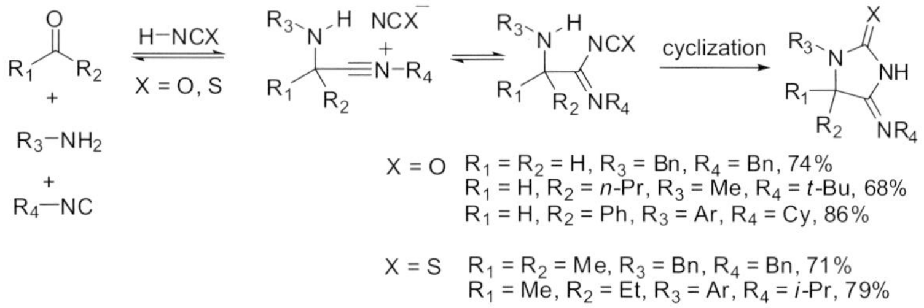

Scheme 5.16

α-siloxyamides [21]. The optimized conditions required a nonpolar solvent, a slight excess of isocyanide, and silanol (the presence of the three phenyl groups ensures a relative Lewis acidity on the silicon atom). Activation of the aldehyde by the silylated Lewis acid promotes the nucleophilic attack of the isocyanide, while the resulting nitrilium is trapped by an intramolecular migration of the hydroxy group of the silicon atom to form the corresponding siloxyamide (Scheme 5.15).

5.2.5
Isocyanic Acid and Derivatives

Isocyanic acid were described in 4CR by Ugi himself, shortly after the first report [6]. In this case, the resulting nitrilium is trapped by an isocyanate anion, while the final step consists of a cyclization of the amine by addition to the C=O to produce hydantoin derivatives (Scheme 5.16) [6, 22].

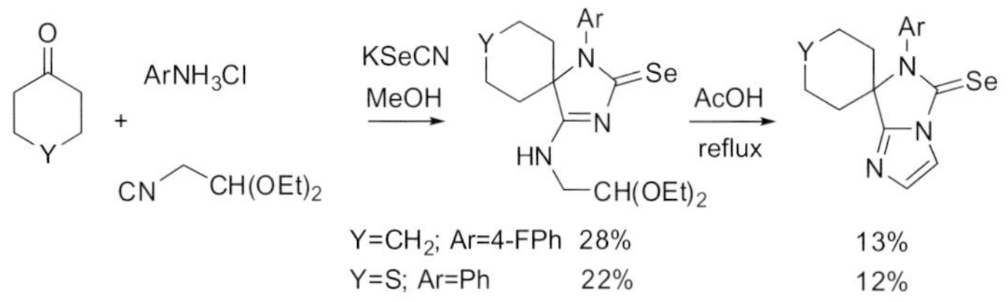

Scheme 5.17

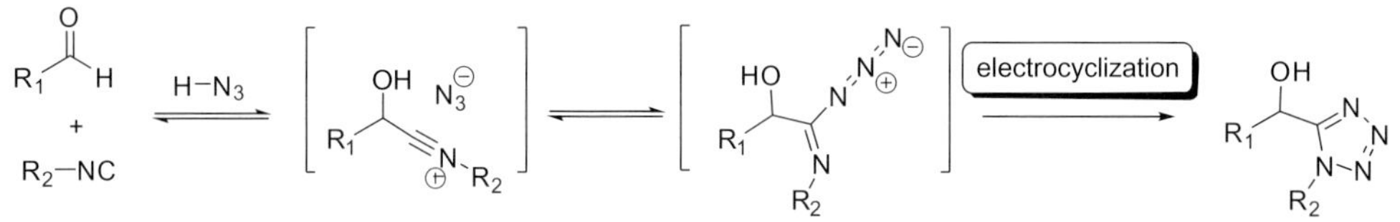

Scheme 5.18

As the starting acids are rather unstable, the reaction is performed using the potassium isocyanate salt and the amine hydrochloride, or in the presence of a stoichiometric quantity of pyridinium hydrochloride. Under these conditions, various aldehydes undergo this new 4CR, but ketones do not. The behavior of isothiocyanic acids was examined [23] in an effort to understand why they react only with ketones to form the corresponding hydantoins [22a]. Subsequently, a variety of thiohydantoins were synthesized using isocyanoesters supported on the Wang resin [24].

Selenocyanate undergoes an analogous reaction to form imidazolino selenones, which further cyclize under acidic conditions (Scheme 5.17) [25].

5.2.6
Hydrazoic Acid

Passerini conducted pioneer studies of tetrazole synthesis via IMCR, in which the first coupling of an aldehyde, an isocyanide, and hydrazoic acid was described. In this reaction, the nitrilium is trapped by the azide, and a final electrocyclization of the resulting 1,3-dipole affords the product (Scheme 5.18).

This coupling could be performed either as shown above [26] or with Lewis or Brønsted acid catalysis [27]. When using weakly reactive carbonyl compounds, the reaction requires the use of $Al(N_3)_3$ as the azide source (Scheme 5.19) [28].

However, the Passerini–azide coupling fails when using an aromatic isocyanide and benzaldehyde derivatives, even under Lewis acid activation; $AlCl_3$ provided the best results but the yields did not exceed 30% (Scheme 5.20). The desired triazoles

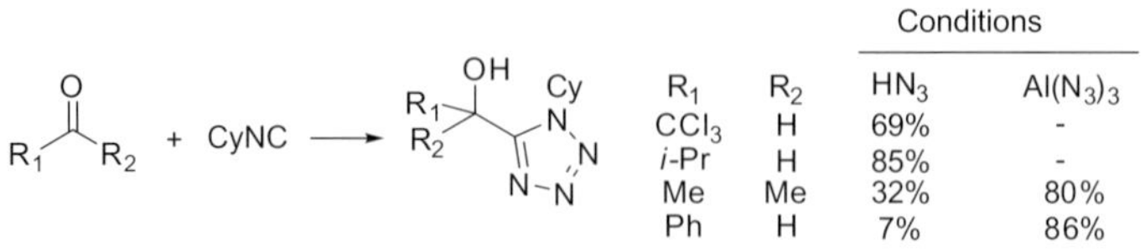

Scheme 5.19

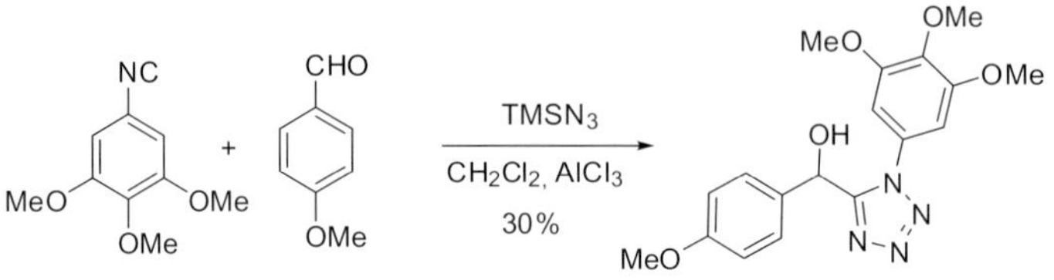

Scheme 5.20

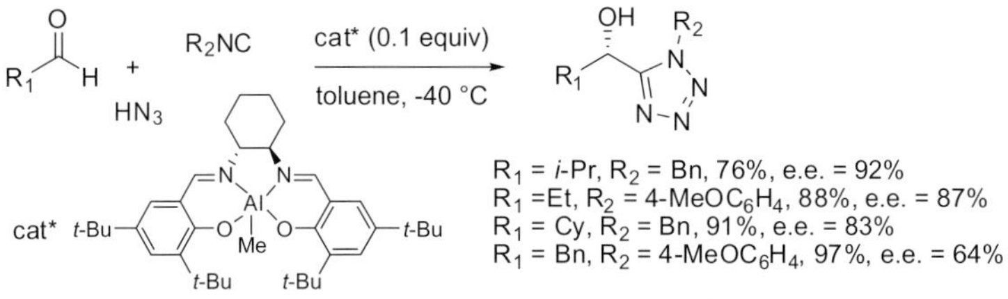

Scheme 5.21

can be obtained, however, through an Ugi–azide coupling by using benzylamine; the latter is cleaved under hydrogenolysis, and the resulting NH-compounds are oxidized before hydrolysis [29].

In 2008, an asymmetric version of this three-component coupling was developed using a 1:1.2:2.5 mixture of isocyanide/aldehyde/HN_3 and an Al–salen chiral complex. The desired α-hydroxy tetrazoles were obtained in good yields with a good enantiomeric excess (e.e.) of 64–97% when performed at −40 °C in toluene (Scheme 5.21) [30].

Shortly after reporting the first 4-CC, Ugi developed the azide version for the synthesis of α-amino-1,5-disubstituted tetrazole derivatives (Scheme 5.22) [22, 26, 31]. The reaction is rather efficient with a wide range of carbonyl compounds, even functionalized compounds, such as N-protected α-amino aldehydes [32]. Secondary amines behave similarly to their primary counterparts, as the nitrogen atom of this partner is not involved in the final step. Various imine surrogates were also described; for example, enamine reacted with the isocyanide when heated in the presence of hydrazoic acid to produce the three-component adduct (Scheme 5.23) [31h].

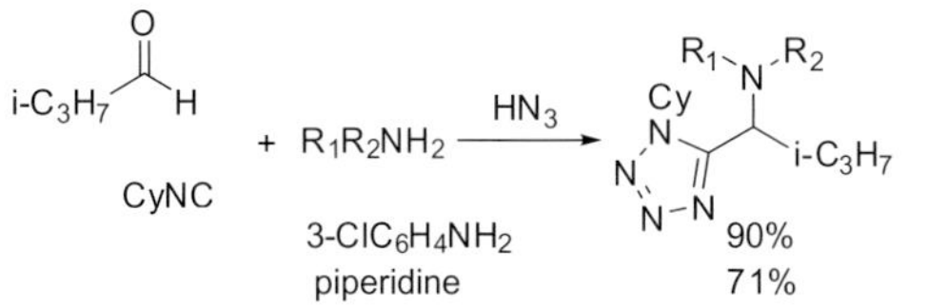

Scheme 5.22

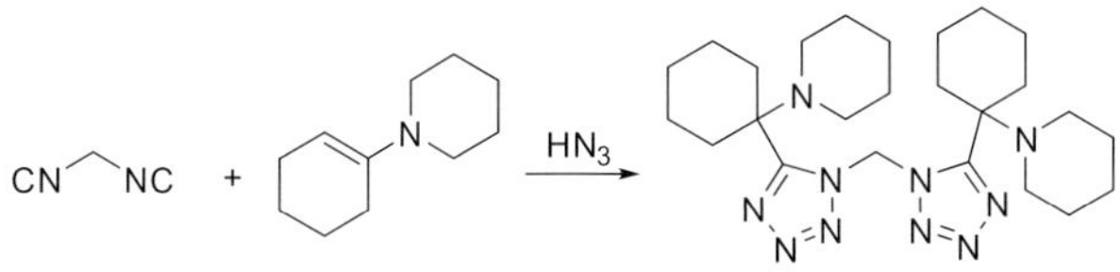

Scheme 5.23

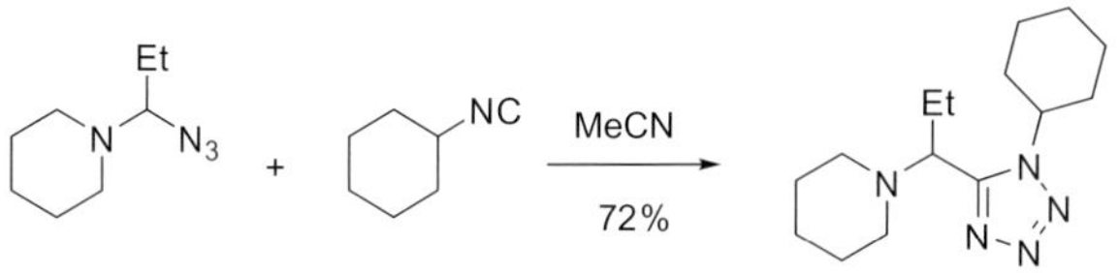

Scheme 5.24

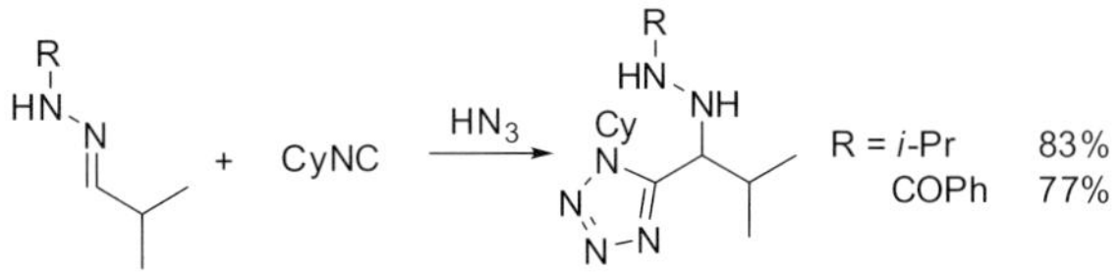

Scheme 5.25

Preformed Mannich adducts derived from the condensation of an amine, an aldehyde and hydrazoic acid can also be used (Scheme 5.24); indeed, this iminium equivalent reacts with an isocyanide in acetonitrile to form the corresponding amino tetrazole [31d]. Cyclic iminium derivatives such as N-fluoropyridinium fluoride can also be coupled to produce pyridine tetrazoles [33].

Hydrazines were also used as imine equivalents at a very early stage in the development process (Scheme 5.25) [34]. However, the extreme toxicity of hydrazoic acid, in addition to its propensity to explode, make its manipulation quite hazardous. Consequently, most of these studies involved in-situ-generated hydrazoic acid; this involved the use of equimolar quantities of sodium azide and hydrochloric acid in a water/acetone mixture or, more recently, of $TMSN_3$ in methanol, which is more easily and safely handled (Scheme 5.26) [35].

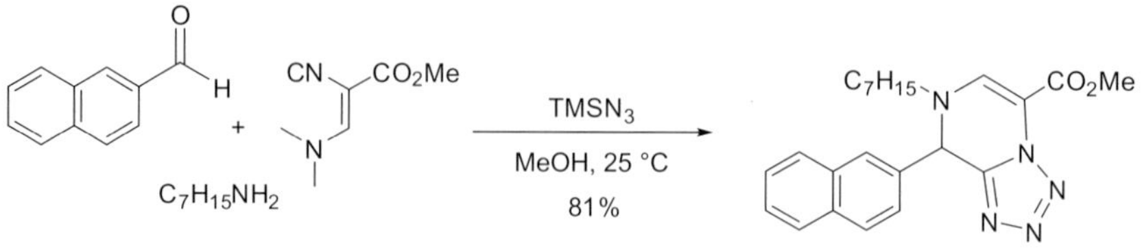

Scheme 5.26

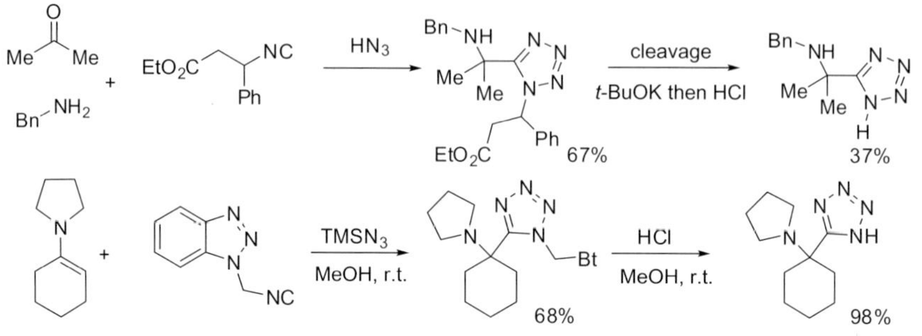

Scheme 5.27

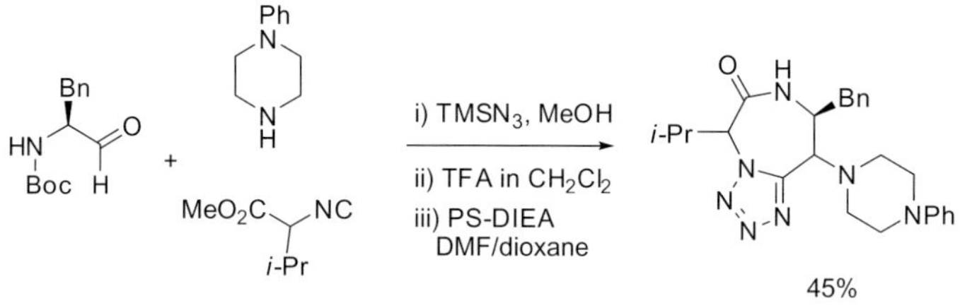

Scheme 5.28

The 4CR is rather efficient with a wide range of isocyanides [32]. Schöllköpf isocyanides provide an efficient access to the fused tetrazoyl system, with the possibility of performing the complete sequence (i.e., synthesis of the Schöllkopf isocyanide/Passerini) in one pot in dimethylformamide (isocyanoacrylates) [35, 36]. Various cleavable isocyanides have been described that lead to NH-free tetrazoles (Scheme 5.27), including 3-isocyanopropionic acid esters [37], 1-isocyano methylbenzotriazole, or 2,2,4,4-tetramethylbutyl isocyanide [38].

Based on the biological interest in tetrazoles, their synthesis underwent extensive examination. Tetrazoles are well recognized as being carboxylic acid isosteres that mimic the *cis* amide conformation which is particularly important in biological recognition. Consequently, this 4CR has attracted the interest of many medicinal chemists, such that a variety of post-condensation transformations were developed to create even more complex heterocyclic systems bearing a tetrazole core. For example, a Ugi–DBoc strategy afforded azepine tetrazoles (Scheme 5.28) [39].

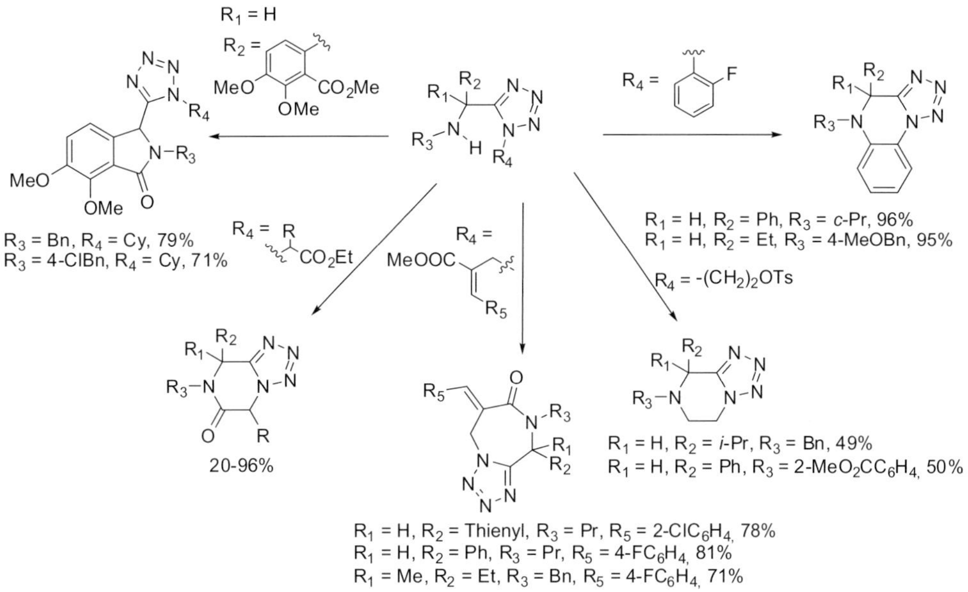

Scheme 5.29

The use of primary amines in this coupling provides secondary amino tetrazoles that can be exploited to provide more complex derivatives (Scheme 5.29). Indeed, fused tetrazoloheterocycles are obtained via cascades involving the nitrogen atom of the former amine: 4CR-SN starting from tosylate-substituted isocyanides [40]; 4CR-SNAr with fluoroaryl isocyanides [41]; 4CR-lactamization when using an ester-substituted starting material [36, 42]; or 4CR-Michael addition when using Schöllkopf isocyanides (Scheme 5.26) [35].

5.2.7
Phenols and Derivatives

Unlike alcohols, which are not sufficiently acidic to trigger any coupling of isocyanides with imines, phenols have such an ability as their acidity can be tuned by correct substitution of the aromatic ring. When considering a possible coupling of the Ugi reaction with a phenol, however, consideration should be given to a plausible final rearrangement of the resulting aryl imidate, to ensure the efficiency of the process (Scheme 5.30). The use of electron-deficient phenols in Ugi couplings was first reported in 2005 [43]. In this case, the introduction of a nitro substituent on the aromatic core ensures a correct phenol acidity to promote iminium formation, and allows a Smiles rearrangement as the final step. This reaction – which was coined as the Ugi–Smiles coupling – affords N-aryl carboxamides in one step.

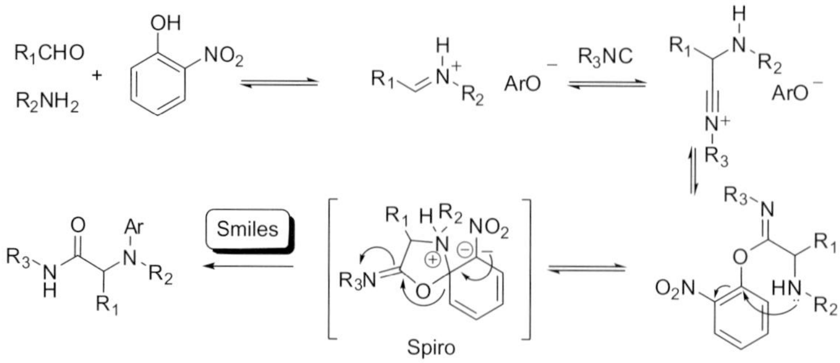

Scheme 5.30

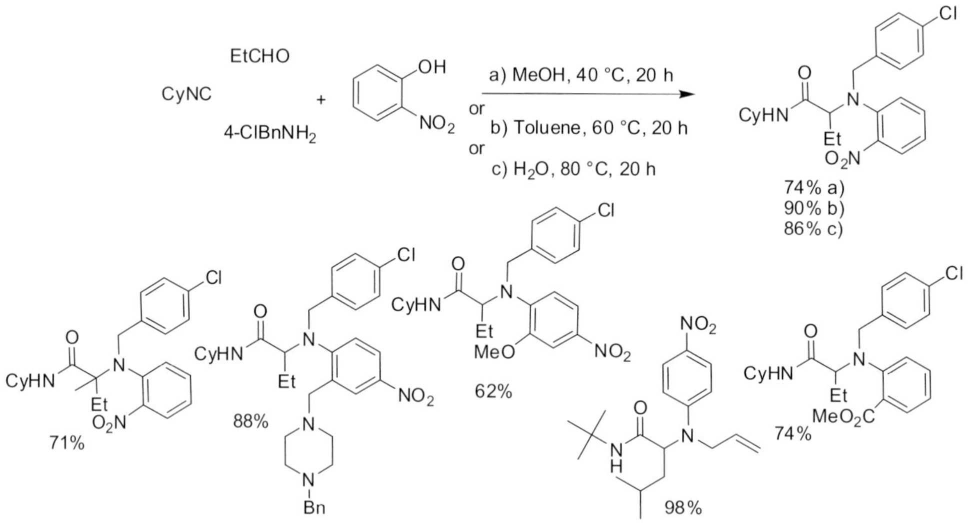

Scheme 5.31

These four-component couplings require temperatures up to 40 °C, and can be performed either in a polar solvent such as methanol, water, or in an apolar solvent such as toluene (Scheme 5.31). The coupling is successful when using a wide range of aldehydes and ketones, though the latter require a longer reaction time than usual for Ugi-type couplings. Isocyanides generally undergo a smooth coupling. However, the choice of the amine partner is still limited to primary amines or ammonia, as both anilines and hindered amines fail to react. Indeed, the Smiles aryl transfer appears to be sensitive to steric hindrance and, in the case of a secondary amine, the aryl transfer on the isocyanide nitrogen atom (Chapman rearrangement) requires harsher conditions.

The reaction is quite efficient using *ortho*- and *para*-nitrophenols as well as salicylate derivatives, although surprising results were obtained when using more

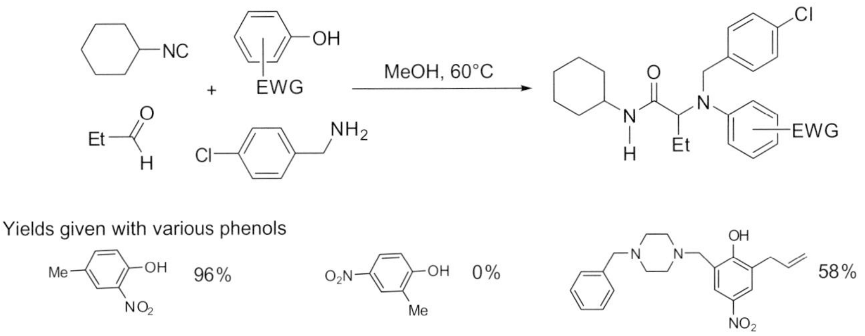

Scheme 5.32

complex nitrophenols [44]. When substituted with a methyl group, a methoxy or a chlorine atom, the *para*-nitrophenol provides the desired adducts in excellent yields. However, no coupling is observed using the 2-methyl or 2-allyl 4-nitrophenol. As these results could not be explained on the basis of steric hindrance and/or simple electronic effects, it was surmised that a possible hydrogen bond would activate the nitrogen atom for a nucleophilic attack and stabilize the spiro intermediate. However, if no hydrogen bond could develop during the spiro formation then steric repulsions would predominate and the phenols would fail to promote the reaction. More interestingly, if the heteroatom was not directly linked to the aromatic ring, a similar interaction could develop and the Ugi-Smiles coupling could proceed smoothly. An example of this was that the hindered Mannich-modified phenol proved to be rather efficient in this coupling (Scheme 5.32).

The potential of Ugi–Smiles couplings was further extended by using hydroxy heteroaromatic compounds, as the resultant adducts may serve as interesting scaffolds in medicinal chemistry. Although 2-hydroxy pyridines react if substituted by an electron-withdrawing group (e.g., a nitro), the reaction is still efficient for less-activating substituents, such as a chlorine atom or a trifluoromethyl group [45]. Furthermore, the introduction of a second nitrogen atom into the heterocycle lowers the electron density on the aromatic core, such that the Smiles rearrangement proceeds smoothly. Indeed, hydroxy-pyrimidines [45], -pyrazines [46], and -triazines [47] each trigger the 4CR in methanol with a wide range of substrates (Scheme 5.33).

Albeit less efficient, the corresponding Passerini–Smiles coupling was also developed. In this case, the *ortho*-nitrophenol reacts with an aldehyde and an isocyanide in methanol at 45 °C for three days with polar protic solvents producing better yields than toluene (in contrast to Passerini couplings) (Scheme 5.34) [48].

With regard to the carbonyl input in Passerini–Smiles couplings, whilst aliphatic and aromatic aldehydes each react smoothly, the ketones serve as poor substrates unless they are activated with electron-withdrawing substituents, such as α-chloro- or α-trifluoromethyl ketones. Acetonitrile is the solvent of choice to

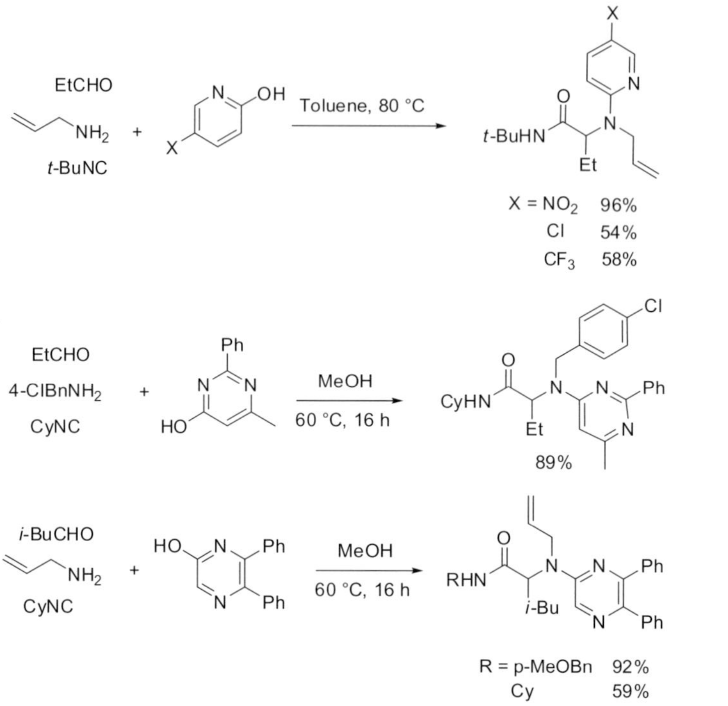

Scheme 5.33

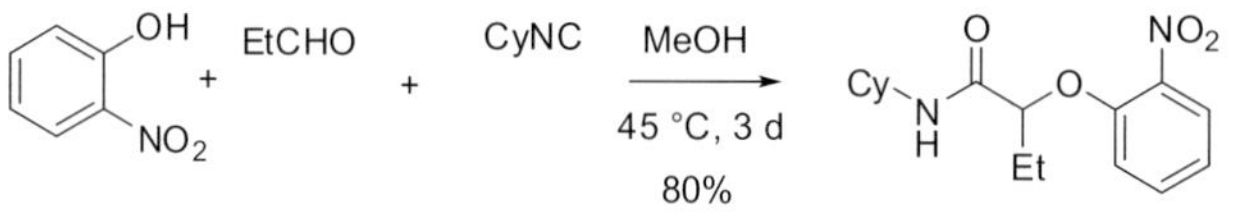

Scheme 5.34

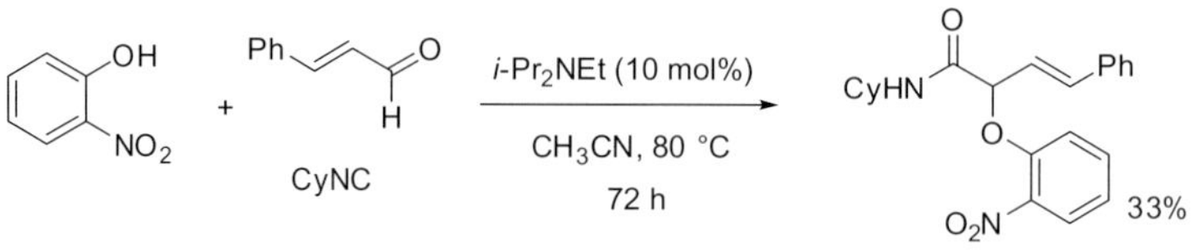

Scheme 5.35

perform the reaction with α,β-unsaturated carbonyl compounds, as the temperature must be raised to 80 °C (Scheme 5.35) [49].

However, this 3CR is quite limited in scope and other electron-deficient phenols (*para*-nitrophenol, alkyl salicylate) fail to react under these conditions. The *ortho*-effect appears to have a greater influence than in the Ugi–Smiles reactions, as the

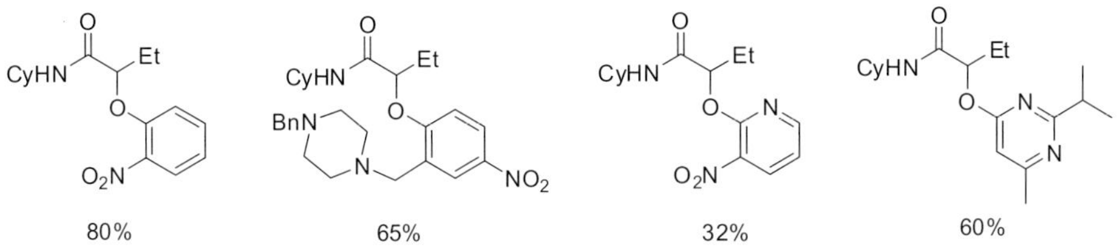

Figure 5.1 Passerini–Smiles couplings of nitrophenols and hydroxy heterocycles.

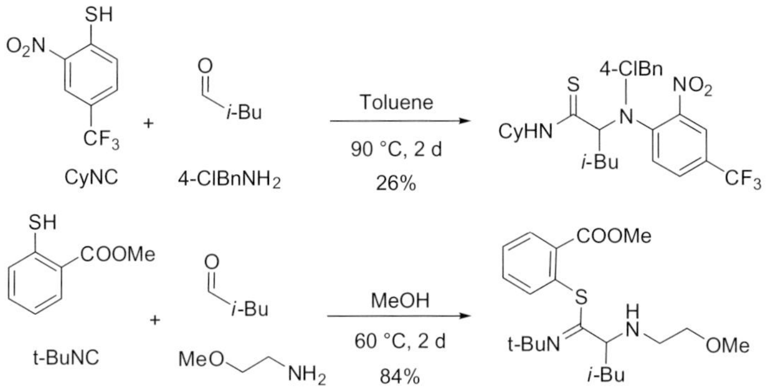

Scheme 5.36

Mannich adducts derived from *para*-nitrophenols undergo coupling. The extension to heterocyclic derivatives is rather unsuccessful, as 5-nitro-2-hydroxy pyridine and 4-hydroxypyrimidines both provide only moderate amounts of product (Figure 5.1) [44].

When considering the higher acidity of thiophenols compared to phenols, and also the better nucleophilicity of thiolates compared to alkoxides, the Ugi–Smiles should be much more efficient when using the corresponding thiols. However, yields with nitrothiophenols do not exceed 26%–whichever solvent is used–and thiosalicylate derivatives afford the corresponding thioimidates without further Smiles rearrangement (Scheme 5.36) [50].

If simple aromatics represent poor partners in these couplings, then the mercapto heterocycles are efficient and afford the corresponding *N*-aryl thiocarboxamides in good yields (Figure 5.2). Indeed, six-membered ring heterocycles such as mercapto-pyridines, -pyrimidines, -pyrazines and -triazines [45, 47] promote the reaction in moderate to good yields. Most noteworthy is the possibility of observing the 4CR by using five-membered ring heterocycles. Indeed, mercapto benzofused azoles (benzoxazoles and benzothiazoles) react to provide access to the corresponding amino benzofused heterocycles [51].

The resulting *N*-heteroaryl thiocarboxamides could be further transformed into more complex heterocycles. Indeed, when treated with 1 equiv. of copper(II) triflate, the nitrogen atom of the heterocycle (pyridine, pyrimidine, or triazine)

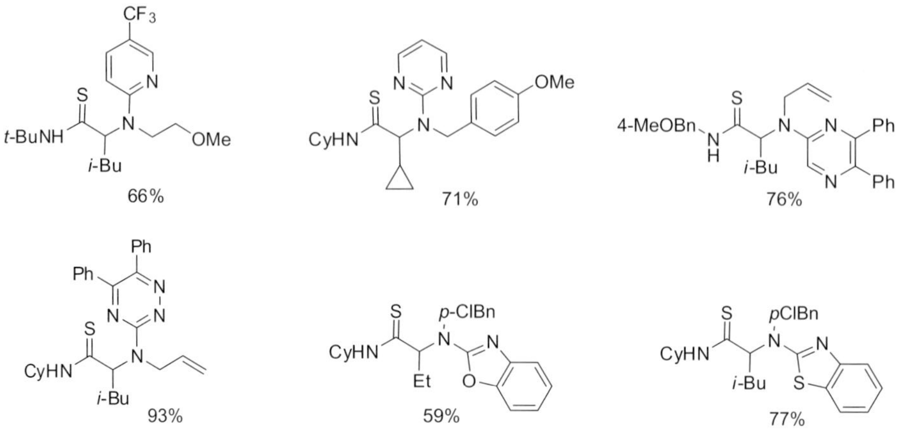

Figure 5.2 Ugi–Smiles access to thioamides.

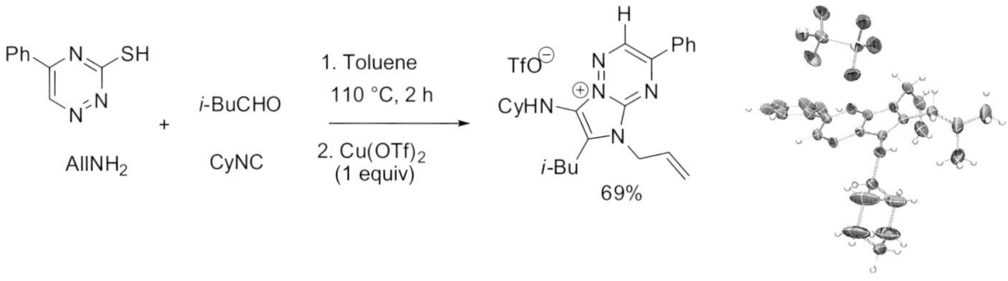

Scheme 5.37

attacks the thioamide moiety to produce fused imidazolium salts, the structure of which was confirmed with X-ray analysis (Scheme 5.37) [47].

More generally, the synthetic potential of Ugi–Smiles couplings was extended by developing various post-condensations of the adducts, and focusing on transformations involving the aromatic ring of the starting phenols. The nitro group required for this coupling can be reduced through hydrogenolysis to form *ortho*-phenylenediamine derivatives, which constitute versatile synthons in heterocyclic chemistry (Scheme 5.38). Indeed, benzopiperazinones are formed under the subsequent acidic conditions that cause the loss of the former isocyanide moiety [43, 52]; nitrosation of the diamine affords benzotriazoles [53], and various benzimidazoles can be synthesized upon oxidative treatment with an aldehyde, or with CS_2 [53].

Considering the rich chemistry of phenols, various post-condensations can be envisioned using functionalized inputs. For instance, the introduction of an allyl moiety into the starting acidic component provides access to adducts that are suited to metathesis couplings. Although the strategy failed with plain phenols,

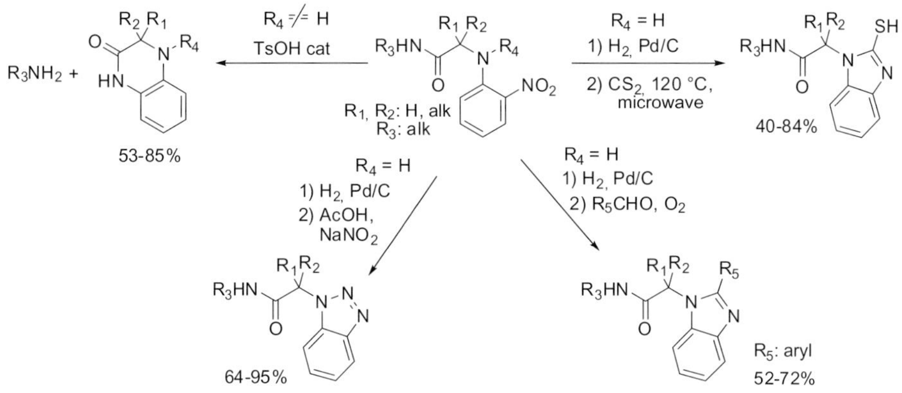

Scheme 5.38

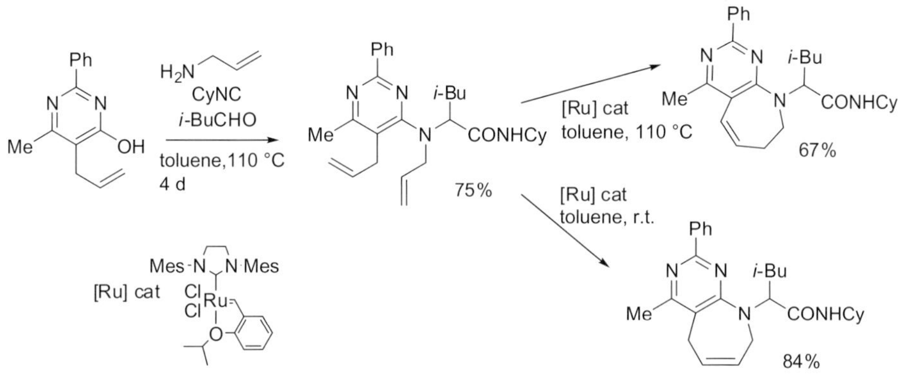

Scheme 5.39

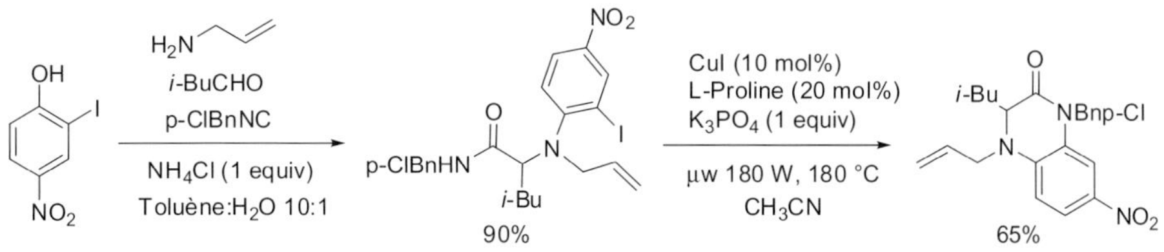

Scheme 5.40

benzofused pyrimidines could be synthesized quite efficiently via a Ugi–Smiles/ring-closing metathesis (RCM) cascade (Scheme 5.39) [54].

The introduction of a halogen atom on the starting phenol opens the way to a variety of metal-induced reactions. For instance, benzopiperazinones with four points of diversity could be synthetized through a sequence U-4CR/Ullmann cyclization (Scheme 5.40) [52].

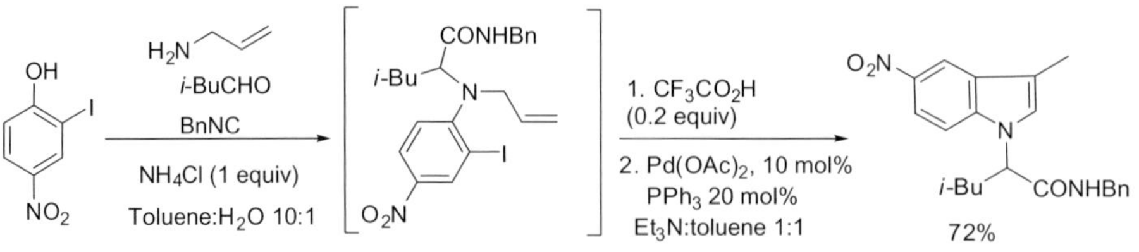

Scheme 5.41

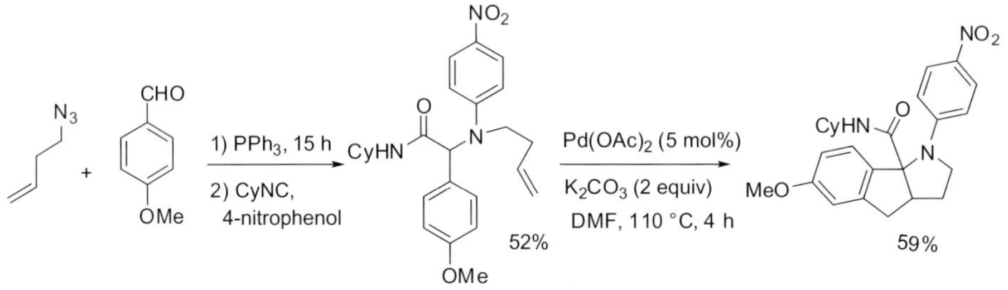

Scheme 5.42

Palladium-catalyzed transformations further enlarge the nature of the synthesized heterocycles. A new one-pot indole synthesis was discovered in which a Heck-type cyclization–isomerization cascade of the adduct could be performed, resulting from the coupling of 2-iodo-4-nitro-phenols and allyl amine (Scheme 5.41) [55].

The attempts to perform a related biaryl coupling using a benzaldehyde derivative failed, and fragmentation of the adduct is observed instead. This behavior was attributed to the acidity of the proton at the peptidic position, and the possibility of forming the corresponding enolate, which can be trapped with an olefin to produce fused heterocyclic systems. Indeed, in the case of N-homoallylamine a tricyclic compound resulting from a palladium-induced cyclization cascade may be isolated (Scheme 5.42) [56].

A related trapping is observed with alkyne located on the starting phenol. In this case, the cyclization does not require metal salts and pyrrole-fused systems are formed in good yields. This sequence is not limited to adducts derived from aromatic aldehydes, but may be extended to aliphatic partners by varying the nature of the base employed in the carbanionic cyclization (Scheme 5.43) [57].

Furthermore, the enolate could be involved in a new SNAr step to transfer the aromatic ring of the former phenol at the peptidic position. Spiroisoquinolines and spiroisoindolines were each synthesized efficiently according to a sequence lactamization–Truce–Smiles rearrangement (Scheme 5.44) [58].

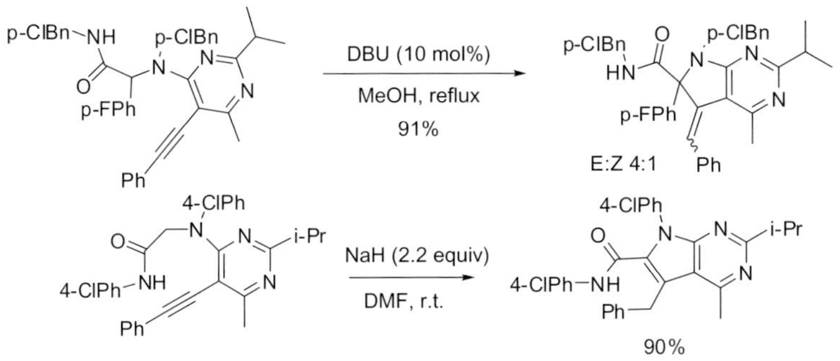

Scheme 5.43

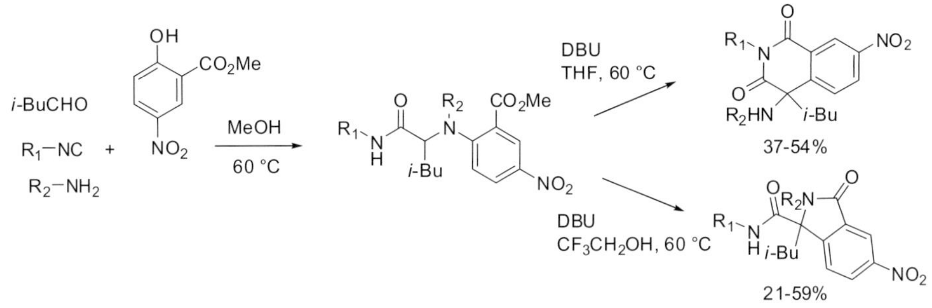

Scheme 5.44

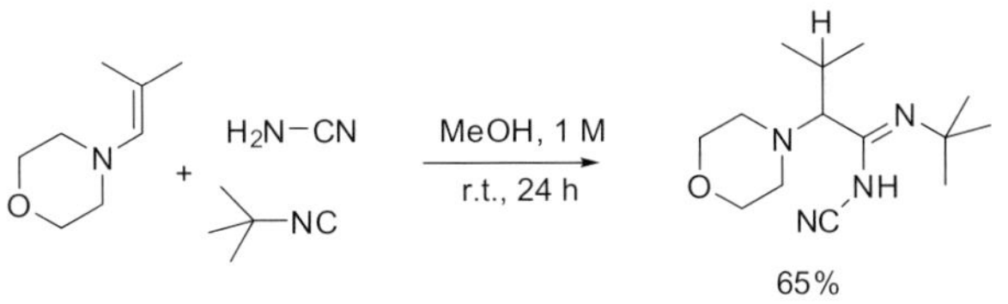

Scheme 5.45

5.2.8
Cyanamide

Cyanamide, with a relatively high pK_a value (17 in DMSO), is able to promote a Ugi-type reaction. Indeed, a methanolic solution of morpholino enamine, *tert*-butyl isocyanide and cyanamide, when stirred at room temperature for one day, affords the amidine in 65% isolated yield (Scheme 5.45) [59].

5.3
Use of Mineral and Lewis Acids

When considering the role of the acid–nucleophile pair, it is difficult to identify surrogates that might address the requirements of efficient Ugi couplings. This is the main reason why useful 4CRs have been mostly limited to carboxylic acids, hydrazoic acids and, more recently, to electron-poor phenols. These constraints are even greater for the less-efficient Passerini reaction. Hence, in order to extend the scope of these couplings, it might be interesting to devote the various tasks performed by the carboxylic acids not to a single compound, but rather to a combination of reagents. A good illustration of this can be found in the reactivity of thiocarboxylic acids in the Ugi and Passerini reactions. Whereas, the Ugi variant with these acids affords thioamides with moderate efficiency, the analogous Passerini coupling does not lead to any reaction unless a Lewis acid is added to the medium (see Scheme 5.8) [16]. Compared to carboxylic acids, the lower Brønsted acidity of thiocarboxylic acids certainly accounts for a lack of activation of the carbonyl group towards the isocyanide addition. To overcome these problems, the use of Lewis acids or strong Brønsted acids, in association with a weakly nucleophilic counteranion (HCl, RSO$_3$H ...) might allow the addition of less-reactive nucleophiles or induce the reactions of too-weakly acidic partners in the IMCR. Lewis acids are particularly interesting when the intermediate nitrilium may be trapped intramolecularly, as observed in the Groebke–Bienaymé reaction (Scheme 5.46).

In this case, the nucleophile is brought by the amino component of the Ugi reaction, though similar behavior may be observed when working on the aldehyde component, as in the interaction of salicylaldehyde with isocyanide [60]. Whilst these intramolecular versions may be considered as extensions of the Ugi and Passerini reactions with specific amino and aldehydic partners, they are beyond the scope of this chapter, which is focused on intermolecular trapping with acidic surrogates.

Despite the use of strong Brønsted acids and Lewis acids having been reported with both 4CR and 3CR, it has been described more frequently with the latter.

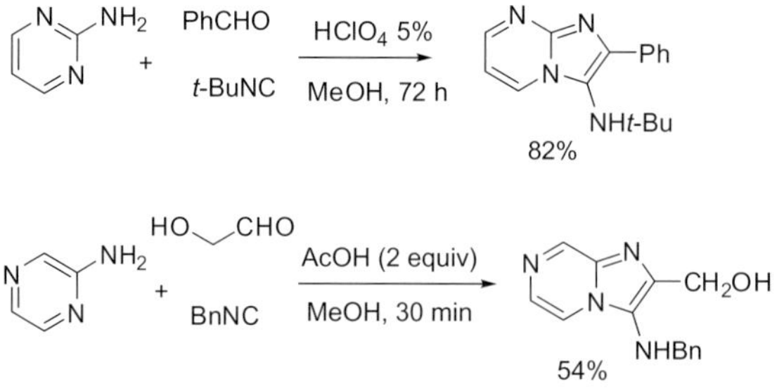

Scheme 5.46

This was because, in the Ugi reaction, the formation of a secondary amine may be associated with selectivity issues related to the potential intermolecular trapping of reactive nitrilium intermediates by both the Ugi adduct and the starting primary amine. In order to restrict these problems, the Ugi reactions triggered by Lewis and Brønsted acids are often limited to secondary amines. In the Passerini reaction, the formation of an alcohol is less problematic, and reactive intermediates may even be isolated in certain cases. Hence, the use of Brønsted acids will be examined first.

5.3.1
Ugi and Passerini Reactions Triggered by Mineral Acids

Although the Passerini reaction was discovered more than 30 years before the Ugi reaction, the first 3CR under mineral acid activation was reported with aqueous hydrochloric acid shortly after disclosure of the Ugi reaction [61]. The reaction involves the use of aqueous HCl to form α-hydroxyamide derivatives through trapping of the intermediate nitrilium with water (Scheme 5.47). These conditions allow the addition of water, which is not sufficiently acidic to induce the 3CR. Besides HCl, other acids such as nitric [28], phosphoric [28] or sulfuric [28, 62] produce the same compounds. Whilst some good yields have been reported, the scope of these reactions is often limited by the rapid hydrolysis of the isocyanides in a strong acidic medium [27d]. Hence, to limit the isocyanide solvolysis the carbonyl derivative must be used in large excess, or even as the solvent [28].

A more efficient access to hydroxyamides relies on a combination of pyridine and trifluoroacetic acid [63]. Under these buffered conditions, the isocyanide is less prone to hydrolysis, while the intermediate nitrilium is most likely trapped by the nucleophilic pyridine before hydrolysis of the pyridinyl–imidoyl adduct (Scheme 5.48).

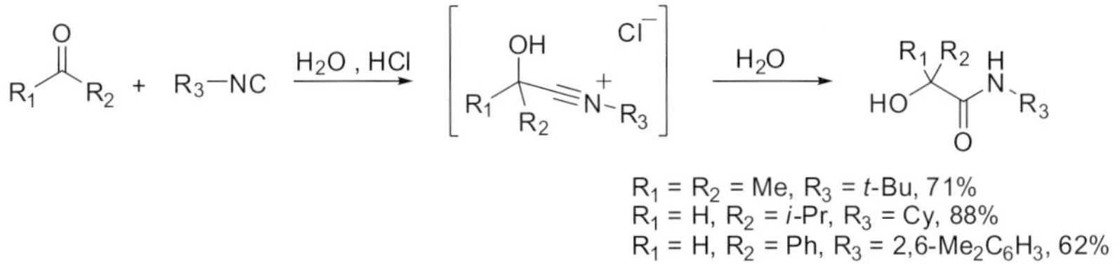

$R_1 = R_2 = Me, R_3 = t\text{-}Bu$, 71%
$R_1 = H, R_2 = i\text{-}Pr, R_3 = Cy$, 88%
$R_1 = H, R_2 = Ph, R_3 = 2,6\text{-}Me_2C_6H_3$, 62%

Scheme 5.47

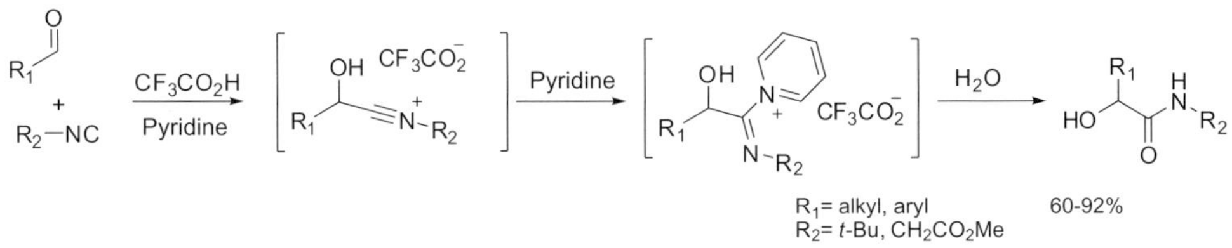

$R_1 = $ alkyl, aryl
$R_2 = t\text{-}Bu, CH_2CO_2Me$
60-92%

Scheme 5.48

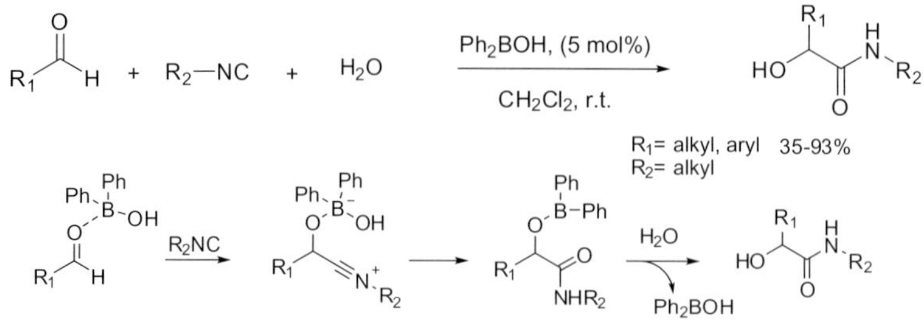

Scheme 5.49

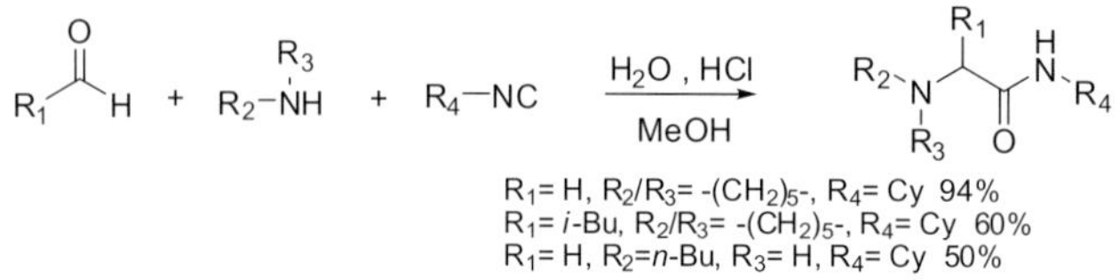

Scheme 5.50

An efficient access to hydroxyamide derivatives has been reported using boric acid in equimolar amounts with aldehydes and isocyanides [64]. Although this reaction does not suffer from the limitations observed with strong mineral acids, the effect of the boronic acid may be associated with its Lewis behavior at the boron atom, activating the aldehyde towards the isocyanide addition with an internal transfer of the hydroxy group. This effect was supported by a further report on the use of phenylboronic acids as catalysts [65]. Under these weakly acidic conditions, high yields of hydroxyamide are obtained when reacting equimolar amounts of aldehydes, isocyanides and water in dichloromethane at room temperature (Scheme 5.49).

The use of mineral acids in the Ugi reaction was proposed by Ugi himself shortly after disclosure of the 4CR with carboxylic acids. The reaction is performed with secondary amines and HCl in water and, when compared to the 3CR with HCl, the hydrolysis of the isocyanide proved to be less problematic with an amine buffer as the medium. The acidic conditions promoted formation of the iminium, with subsequent isocyanide addition; the nitrilium was then trapped by water to produce the amino carboxamide derivatives (Scheme 5.50) [31a].

In a more detailed study conducted by McFarland in anhydrous methanol [27d], it was shown that the efficiency of the process could be reduced by competing Passerini coupling and double addition of the amino derivatives. With equimolar amounts of amine hydrochloride and aldehyde, the main product is the amide I, along with the Passerini adduct II (Scheme 5.51). The formation of II may be explained by a reduced efficiency of the amine hydrochloride in forming the intermediate imine, thus allowing more time for the Passerini reaction to occur. If a

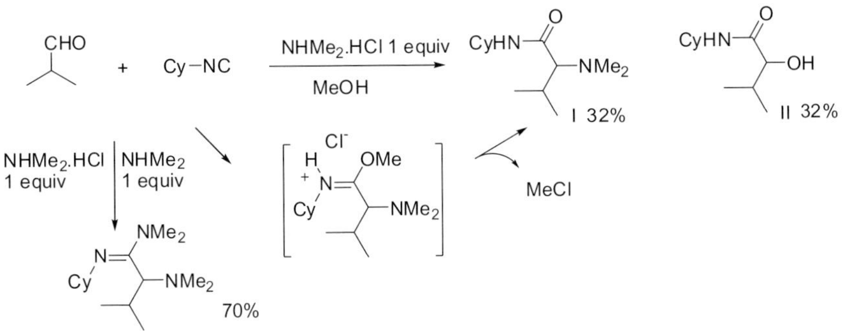

5.3 Use of Mineral and Lewis Acids | 183

Scheme 5.51

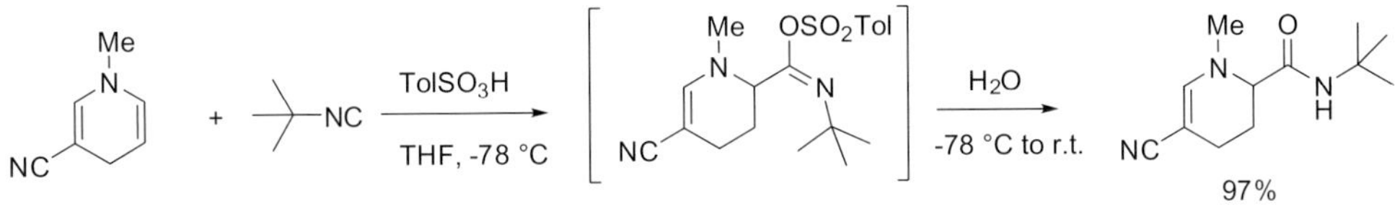

Scheme 5.52

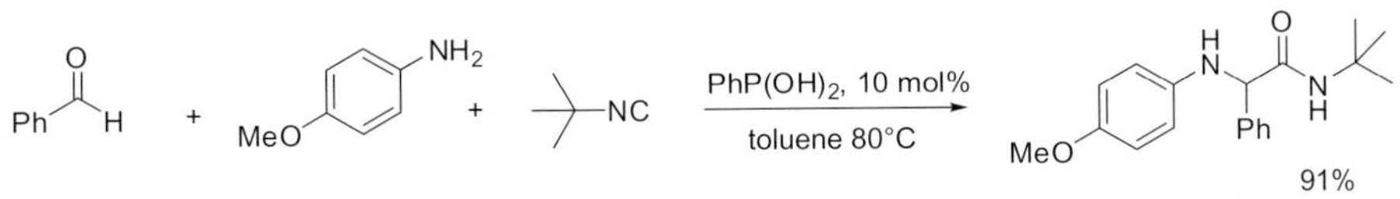

Scheme 5.53

larger amount of amine is used, however, the amidine derivatives may be obtained as the major adduct.

Better yields might be expected if the iminium formation step were to be avoided, as shown by the reaction of dihydropyridines with isocyanides and sulfonic acid (Scheme 5.52) [66]. In this case, the reaction must be performed in stepwise fashion to avoid the hydrolysis of both isocyanide and enamine, in the presence of water and tosic acid.

Efficient Ugi couplings between anilines, aldehydes, and various isocyanides have been catalyzed by phosphinic acid. This reaction is carried out in toluene with 10 mol% of acid, and without any added water. Phenylphosphonic provides a much lower conversion, while diphenylphosphinic acid is inactive under the same conditions (Scheme 5.53) [67]. Unfortunately, the chiral phosphinic derivatives tested in these studies failed to provide any substantial enantioselectivity.

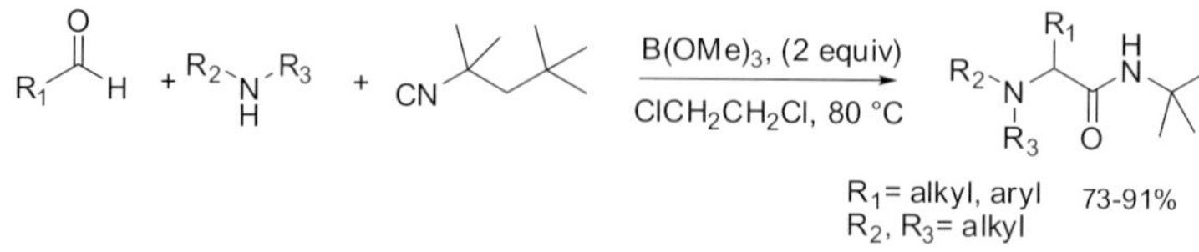

Scheme 5.54

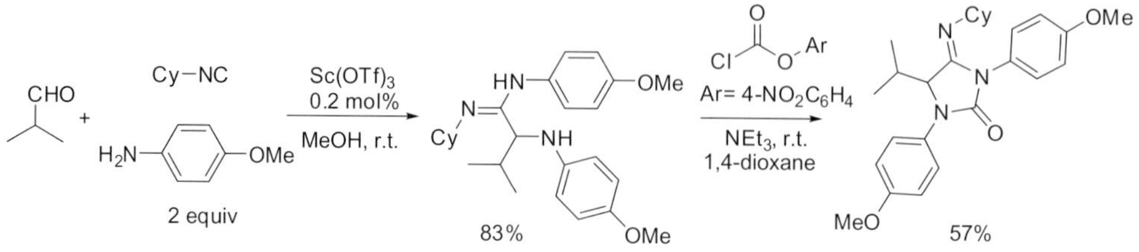

Scheme 5.55

5.3.2
Ugi and Passerini Reactions Triggered by Lewis Acids

Beside mineral acids, Lewis acids allow additions of various O- and N-centered nucleophiles. These conditions are often superior in term of yields and functional tolerance, as shown for the phenylboronic-catalyzed Passerini reaction (Scheme 5.49).

The Ugi reaction of secondary amines under activation with aminoboranes and boric esters derivatives has been reported by Suginome et al. [68]. In this case, the best activating agent in terms of scope and availability appeared to be the borate trimethyl ester (Scheme 5.54).

This McFarland amidine synthesis, using amine hydrochlorides, was reinvestigated with Lewis acids as activating agents [69]. A catalytic amount of lanthanum, yttrium, ytterbium, and scandium triflate (used in a 0.25 molar ratio) provided good yields with secondary amines, and some primary amines also behaved remarkably well under these conditions (Scheme 5.55). The adducts formed under scandium catalysis with anilines may be further transformed into hydantoin imides when treated with chloroformates.

Compared to the Ugi reaction, the effect of Lewis acid in Passerini couplings has been the object of much more attention. The case of titanium tetrachloride has been studied in detail, and X-ray analyses of the intermediates has provided a good indication of the mechanism involved in this reaction (Scheme 5.56) [70]. Following the addition of $TiCl_4$ to the aldehyde and the isocyanide, hydrolysis of the titanium–imidoyl chloride complex usually leads to the formation of hydroxy amides in good yields, although trapping with other nucleophiles in the last step of this reaction should extend its scope.

One of the most interesting uses of Lewis acids in the Passerini reaction is associated with the disclosure of enantioselective IMCRs. As observed with $TiCl_4$,

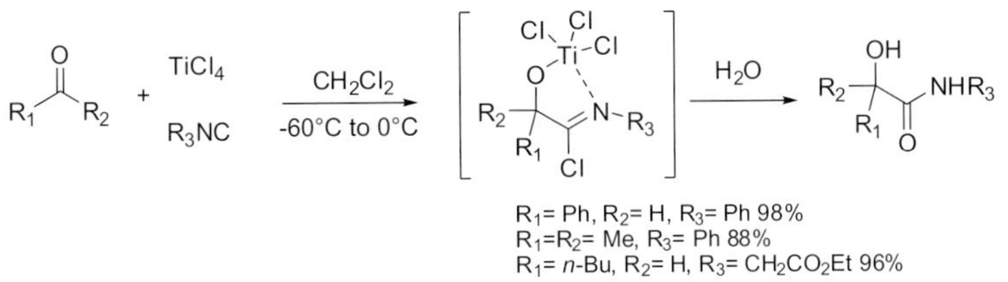
TiCl4
R3NC
CH2Cl2
-60°C to 0°C
H2O
R1= Ph, R2= H, R3= Ph 98%
R1=R2= Me, R3= Ph 88%
R1= n-Bu, R2= H, R3= CH2CO2Et 96%

Scheme 5.56

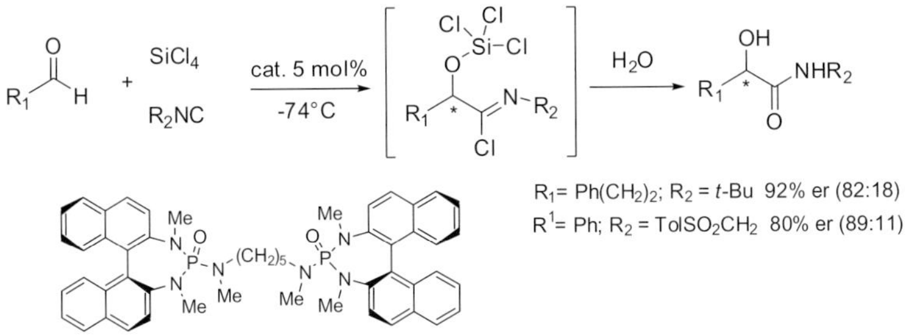
SiCl4
R2NC
cat. 5 mol%
-74°C
H2O
R1= Ph(CH2)2; R2 = t-Bu 92% er (82:18)
R1= Ph; R2 = TolSO2CH2 80% er (89:11)

Scheme 5.57

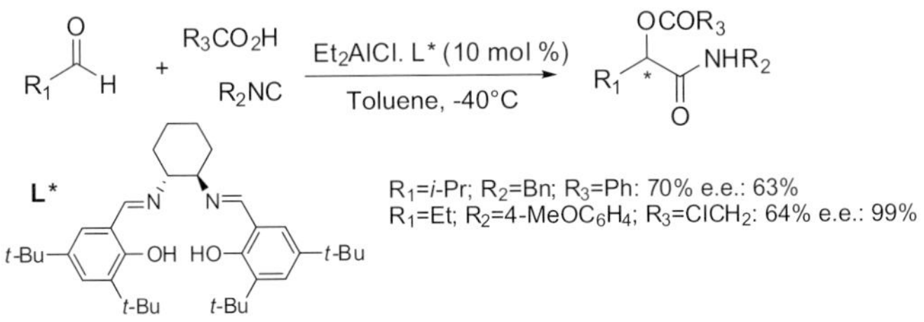
R3CO2H
R2NC
Et2AlCl. L* (10 mol %)
Toluene, -40°C
L*
R1=i-Pr; R2=Bn; R3=Ph: 70% e.e.: 63%
R1=Et; R2=4-MeOC6H4; R3=ClCH2: 64% e.e.: 99%

Scheme 5.58

$SiCl_4$ efficiently promotes the Passerini reaction forming imidoyl chlorides at low temperatures. However, if the reaction is conducted in the presence of a catalytic amount of chiral bisphosphoramides, then α-hydroxyamides are obtained after hydrolysis, in excellent yields and with good enantiomeric excesses (Scheme 5.57) [71].

Following this seminal report, the use of different chiral Lewis acids as catalysts has been disclosed for the three-component Passerini coupling with carboxylic acids. With Cu(II)-(pybox) derivatives as catalyst, the reaction is limited to α-alkoxyaldehydes as the starting carbonyl derivatives [72], whereas Al(III) salen complexes afford an enantioselective access to a wider range of α-acyloxyamides (Scheme 5.58) [30].

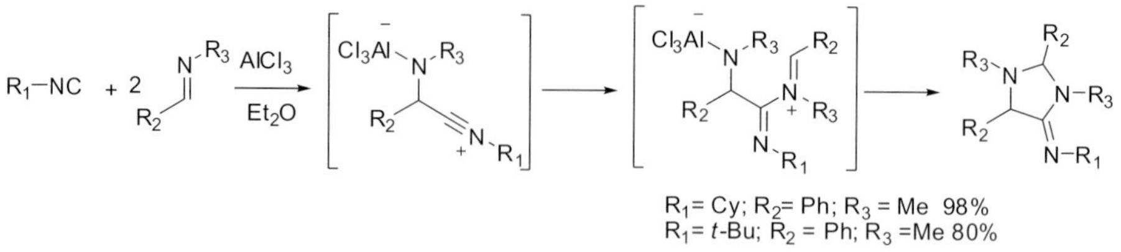

Scheme 5.59

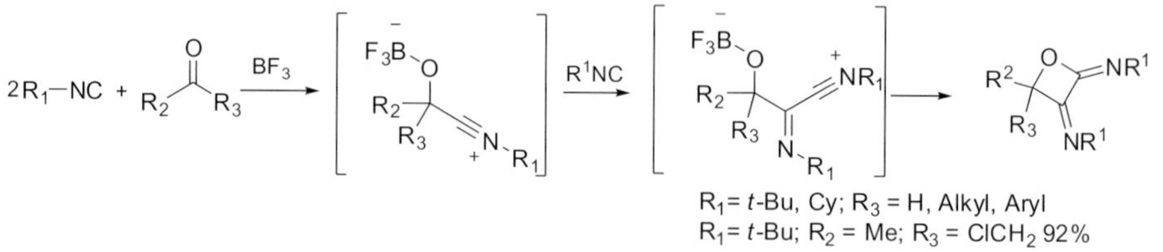

R_1= Cy; R_2= Ph; R_3 = Me 98%
R_1= t-Bu; R_2 = Ph; R_3 =Me 80%

Scheme 5.60

R_1= t-Bu, Cy; R_3 = H, Alkyl, Aryl
R_1= t-Bu; R_2 = Me; R_3 = ClCH$_2$ 92%

Most Ugi and Passerini reactions are performed with stoichiometric amounts of reagents; however, when Lewis acid conditions are used with the same components, but introduced with different stoichiometries, then less-conventional adducts may be observed. Thus, imidazolines may be formed when the isocyanide is treated with 2 equiv. of imine under aluminum trichloride conditions. In this situation, the first equivalent is involved in the initial Ugi steps, while the second equivalent traps the intermediate nitrilium to form imidazolidine derivatives (Scheme 5.59) [73].

Subsequently, if the stoichiometry is altered to perform the reaction with an excess of isocyanide, then azetidinones and oxetanes may be obtained by the sequential addition of isocyanide to the nitrilium [74]. This behavior has been documented much more often for oxetanes resulting from Passerini reactions triggered by boron trifluoride, while good yields are generally obtained for bulky isocyanides, such as t-butyl isocyanide [75] (Scheme 5.60).

According to their substitution pattern, these oxetanes may be converted into a variety of ring-opened products under acidic treatment. Aqueous HCl provides α-ketoamide derivatives [76], the formation of which has been also observed by the simple hydrolysis of a Passerini reaction performed with 2 equiv. of isocyanide and a stoichiometric amount of BF$_3$ [77] (Scheme 5.61, route a). When the oxetane-ring opening is induced by a carboxylic acid, however, a Mumm rearrangement constitutes the final step of the process, transferring the acyl moiety to a nitrogen atom (Scheme 5.61, route b) [78].

A similar sequential addition of 2 equiv. of the isocyanide component in a Passerini reaction also occurs with other Lewis acids. When t-butyl isocyanide is added to ketones or aldehydes in the presence of trimethylsilyl chloride and a catalytic

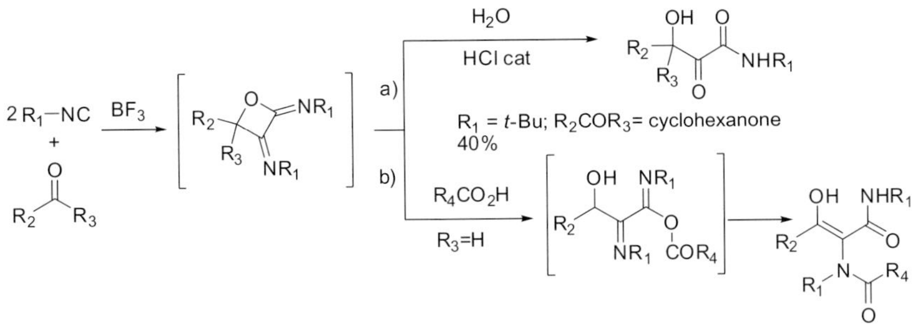

Scheme 5.61

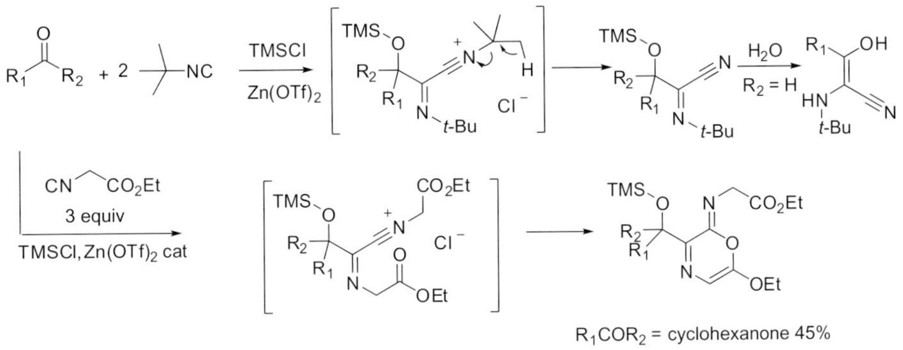

Scheme 5.62

amount of zinc(II) triflate, the formation of cyanoimines may be explained by an intermediate dealkylation, coupled with silylation of the hydroxy residue that prevents oxetane formation [79]. In the case of aldehydes, the cyano imine tautomerizes into the more stable enamine. Unfortunately, the scope of these couplings is rather limited by the nature of the isocyanide, with no coupling being observed for simple compounds, such as cyclohexyl isocyanide. However, when a neighboring donating group is present, as in isocyanoacetate, these conditions may lead to oxazine formation with a larger excess of the isocyanide (3 equiv.) (Scheme 5.62) [80].

Beside triggering the direct addition of carbonyl derivatives in the Ugi and Passerini reactions, Lewis and Brønsted acids also have the advantage of activating carbonyl surrogates such as acetals or aminals towards the direct addition of isocyanide. In the case of acetals, such strategies constitute a formal use of alcohols as carboxylic acid surrogates in Passerini reactions. The reaction of dimethyl acetals with isocyanides has been reported under trifluoroacetic activation, forming α-methoxy amides in high yields (Scheme 5.63, route A) [81] A similar reaction may be observed when acetals are treated with a stoichiometric amount of TiCl$_4$, at low temperature (Scheme 5.63, route B) [82].

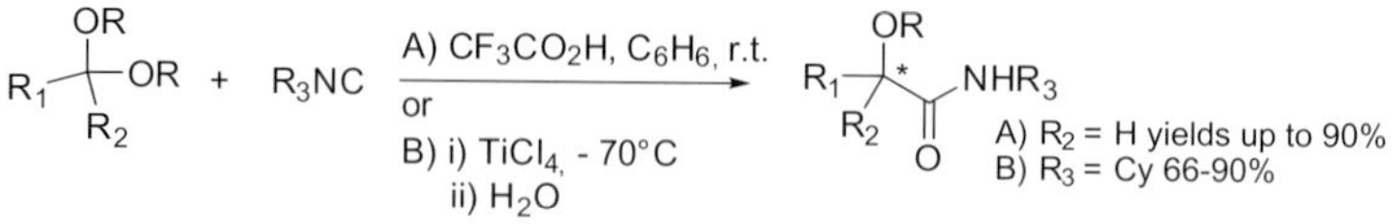

188 | *5 Ugi and Passerini Reactions with Carboxylic Acid Surrogates*

Scheme 5.63

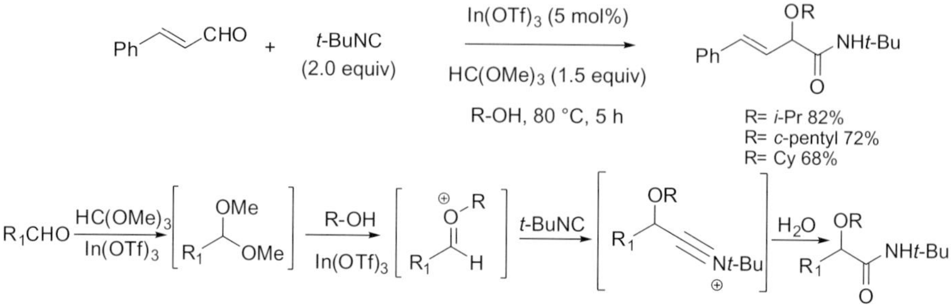

Scheme 5.64

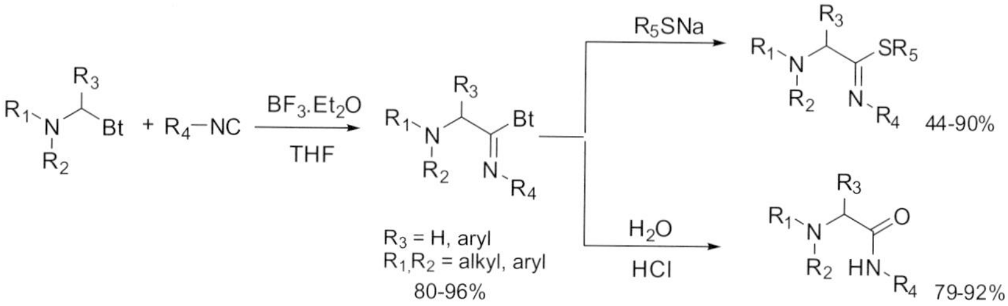

Scheme 5.65

A related alkylative Passerini coupling between an aldehyde, an isocyanide, and an alcohol (the latter served as the solvent) was reported in the presence of an excess of trimethyl orthoformate and a catalytic amount of In(OTf)$_3$ (Scheme 5.64) [83]. The mechanism most likely involves a prior formation of the dimethyl acetal of the aldehyde, promoted by the Lewis acid.

The addition of isocyanides to aminals has also been reported under Lewis acid activation, and this has opened the way to less-classical nucleophilic additions onto intermediate nitrilium salts in Ugi reactions. When the preformed Mannich base between benzotriazole and the imine is treated with the isocyanide in the presence of BF$_3$, the isocyanide inserts into the C–Bt bond, leading to adducts that may be further hydrolyzed or treated with other nucleophiles, such as mercaptans (Scheme 5.65) [84].

Following the use of acetals and aminals as substrates for ring-opening reactions, oxazolidines may be involved in isocyanide insertion reaction under tosic acid activation (Scheme 5.66) [85].

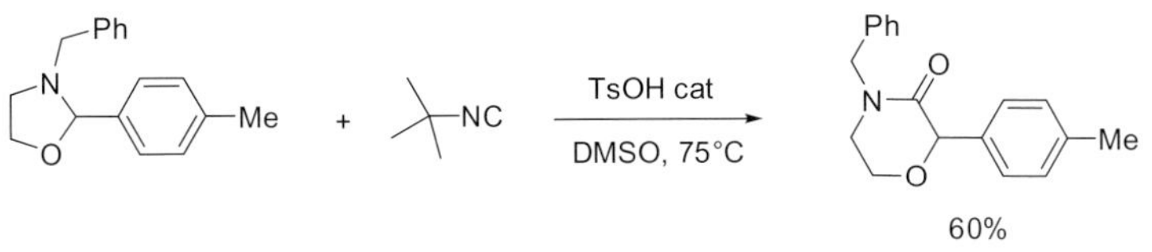

Scheme 5.66

5.4
Conclusions

Today, the Ugi reaction is widely recognized as a four-component IMCR that incorporates the use of carboxylic acids. Yet, despite this coupling having been used in an impressive number of synthetic strategies towards various heterocyclic targets, this definition might be considered somewhat restrictive, as several other Brønsted and Lewis acidic derivatives have demonstrated equal efficiencies in the creation of four-component adducts, under similar conditions. As indicated by recent surveys, these extensions share many mechanistic features, but differ mostly with regard to the fate of the intermediate imidates, including Mumm and Smiles rearrangements, and electrocylization. Undoubtedly, the current increasing demand for low-cost synthetic processes will secure a prosperous future for IMCRs, with further applications involving Ugi and Passerini reactions with carboxylic acid surrogates to be expected.

References

1. (a) Banfi, L. and Riva, R. (2005) The Passerini reaction. *Org. React.*, **65**, 1–140; (b) Dömling, A. (2006) Recent developments in isocyanide based multicomponent reactions in applied chemistry. *Chem. Rev.*, **106** (1), 17–89 and references cited therein.
2. (a) Passerini, M. (1921) Sopra gli isonitrili (I). Composto del *p*-isonitrili-azobenzole con acetone ed acido acetica. *Gazz. Chim. Ital.*, **51**, 126–129; (b) Passerini, M. (1921) Sopra gli isonitrili (II). Composti con aldeidi o con chetoni ed acidi organici monobasici. *Gazz. Chim. Ital.*, **51**, 181–188.
3. Ugi, I., Meyr, R., Fetzer, U., and Steinbrückner, C. (1959) Versuche mit Isonitrilen. *Angew. Chem.*, **71** (11), 386–388.
4. (a) Mumm, O. (1910) Umsetzung von Säureimidchloriden mit Salzen organischer Säuren und mit Cyankalium. *Ber. Dtsch Chem. Ges.*, **43** (1), 886–893; (b) Mumm, O., Hesse, H., Volquartz, H. (1915) Zur Kenntnis der Diacylamide. *Ber. Dtsch Chem. Ges.*, **48** (1), 379–391.
5. Chéron, N., Ramozzi, R., El Kaïm, L., Grimaud, L., and Fleurat-Lessard, P. (2012) Challenging 50 years of established views on Ugi reaction: a theoretical approach. *J. Org. Chem.*, **12**, 1361–1366.
6. (a) Ugi, I. (1962) Mit Sekundär-Reaktionen gekoppelte α-Additionen von Immonium-Ionen und Anionen an Isonitrile. *Angew. Chem.*, **74** (1), 9–22; (b) Ugi, I. (1962) *Angew. Chem. Int. Ed. Engl.*, **1** (1), 8–21.
7. (a) Groebke, K., Weber, L., and Mehlin, F. (1998) Synthesis of imidazo[1,2-a] annulated pyridines, pyrazines and pyrimidines by a novel three-component

condensation. *Synlett*, (6), 661–663; (b) Bienaymé, H. and Bouzid, K. (1998) A new heterocyclic multicomponent reaction for the combinatorial synthesis of fused 3-aminoimidazoles. *Angew. Chem. Int. Ed.*, **37** (16), 2234–2237; (c) Blackburn, C., Guan, B., Fleming, P., Shiosaki, K., and Tsai, S. (1998) Parallel synthesis of 3-aminoimidazo[1,2-a] pyridines and pyrazines by a new three-component condensation. *Tetrahedron Lett.*, **39** (22), 3635–3638.

8 (a) Sun, X., Janvier, P., Zhao, G., Bienaymé, H., and Zhu, J. (2001) A novel multicomponent synthesis of polysubstituted 5-aminooxazole and its new scaffold-generating reaction to pyrrolo[3,4-b]pyridine. *Org. Lett.*, **3** (6), 877–880; (b) Gonzalez-Zamora, E., Fayol, A., Bois-Choussy, M., Chiaroni, A., and Zhu, J. (2001) *Chem. Commun.*, (17), 1684–1685.

9 Heck, S. and Dömling, A. (2000) A versatile multi-component one-pot thiazole synthesis. *Synlett*, (3), 424–426.

10 Kazmaier, U. and Ackermann, S. (2005) A straightforward approach towards thiazoles and endothiopeptides via Ugi reaction. *Org. Biomol. Chem.*, **3** (17), 3184–3187.

11 Kolb, J., Beck, B., and Dömling, A. (2002) Simultaneous assembly of the β-lactam and thiazole moiety by a new multicomponent reaction. *Tetrahedron Lett.*, **43** (39), 6897–6901.

12 (a) Henkel, B., Westner, B., and Dömling, A. (2003) Polymer-bound 3-N,N-(dimethylamino)-2-isocyanoacrylate for the synthesis of thiazoles via a multicomponent reaction. *Synlett*, (15), 2410–2412; (b) Kolb, J., Beck, B., Almstetter, M., Heck, S., Herdtweck, E., and Dömling, A. (2003) New MCRs: the first 4-component reaction leading to 2,4-disubstituted thiazoles. *Mol. Div.*, **6** (3–4), 297–313.

13 Umkehrer, M., Kolb, J., Burdack, C., and Hiller, W. (2005) 2,4,5-trisubstituted thiazole building blocks by a novel multi-component reaction. *Synlett*, (1), 79–82.

14 Gulevich, A.V., Balenkova, E.S., and Nenajdenko, V.G. (2007) The first example of a diastereoselective thio-Ugi reaction: a new synthetic approach to chiral imidazole derivatives. *J. Org. Chem.*, **72** (21), 7878–7885.

15 Spatz, J.H., Bach, T., Umkehrer, M., Bardin, J., Ross, G., Burdack, C., and Kolb, J. (2007) Diversity oriented synthesis of benzoxazoles and benzothiazoles. *Tetrahedron Lett.*, **48** (51), 9030–9034.

16 Henkel, B., Beck, B., Westner, B., Mejat, B., and Dömling, A. (2003) Convergent multicomponent assembly of 2-acyloxymethyl thiazoles. *Tetrahedron Lett.*, **44** (50), 8947–8950.

17 Ugi, I. and Steinbrückner, C. (1961) Isonitrile, IX. a-Addition von Immonium-Ionen und Carbonsäure-Anionen an Isonitrile. *Chem. Ber.*, **94** (8), 2802–2814.

18 Keating, T.A. and Armstrong, R.W. (1998) The Ugi five-component condensation using CO_2, CS_2, and COS as oxidized carbon sources. *J. Org. Chem.*, **63** (3), 867–871.

19 (a) Hulme, C., Ma, L., Romano, J.J., Morton, G., Tang, S.-Y., Cherrier, M.-P., Choi, S., Salvino, J., and Labaudiniere, R. (2000) Novel applications of convertible isonitriles for the synthesis of mono and bicyclic γ-lactams via a UDC strategy. *Tetrahedron Lett.*, **41** (9), 1889–1893; (b) Gunawan, S., Nichol, G.S., Chappeta, S., Dietrich, J., and Hulme, C. (2010) Concise preparation of novel tricyclic chemotypes: fused hydantoin-benzodiazepines. *Tetrahedron Lett.*, **51** (36), 4689–4692.

20 (a) Ugi, I. and Steinbrückner, C. (1960) Über ein neues Kondensations-Prinzip. *Angew. Chem.*, **72** (7–8), 267–268; (b) Ugi, I., Lohberger, S., and Karl, R. (1991) The Passerini and Ugi reactions, in *Comprehensive Organic Synthesis*, vol. 2 (eds B.M. Trost and I. Fleming), Pergamon, New York, pp. 1083–1109.

21 Soeta, T., Kojima, Y., Ukaji, Y., and Inomata, K. (2010) O-silylative Passerini reaction: a new one-pot synthesis of α-siloxyamides. *Org. Lett.*, **12** (19), 4341–4343.

22 (a) Ugi, I., Rosendhal, F.K., and Bodesheim, F. (1963) Isonitrile, XIII. Kondensation von primären Aminen und Ketonen mit Isonitrilen und Rhodanwasserstoffsäure. *Justus Liebigs*

Ann. Chem., **666** (1), 54–61; (b) Ugi, I. and Offerman, K. (1964) Isonitrile, XVIII. Hydantoin-imide-(4). *Chem. Ber.*, **97** (8), 2276–2281.

23 Polyakov, A.I., Medvedeva, L.A., Dyachenko, O.A., Zolotoi, A.B., and Atovmyan, L.O. (1986) Preparation of 1-methyl-4-imino-5,5-dialkylthiohydantoins. *Khim. Geterotsikl. Soedin.*, **1**, 53–61.

24 Short, K.M., Ching, B.W., and Mjalli, A.M.M. (1996) The synthesis of hydantoin 4-imides on solid support. *Tetrahedron Lett.*, **37** (42), 7489–7492.

25 Bossio, R., Marcaccini, S., and Pepino, R. (1993) Studies on isocyanides and related compounds. A facile synthesis of imidazo[1,5-a]imidazoles. *Liebigs Ann. Chem.*, **11**, 1229–1231.

26 (a) Passerini, M. (1922) Sopra gli isonitrili. (III) Reazione con gli idrati di aldeidi alogenate. *Gazz. Chim. It.*, **52**, 432–435; (b) Passerini, M. (1926) Sopra gli isonitrili. (XV) Reazioni con gli isonitrili alifatici. *Gazz. Chim. It.*, **56**, 826–829; (c) Kreutzkamp, N. and Lämmerhirt, K. (1968) Phosphinylmethyl-isocyanide. *Angew. Chem.*, **80** (10), 394–395; (d) Kreutzkamp, N. and Lämmerhirt, K. (1968) *Angew. Chem. Int. Ed. Engl.*, **7** (5), 372–373.

27 (a) Passerini, M. (1924) Sopra gli isonitrili. (VIII) Ancora della reazione con aldeidi o chetoni in presenza di acidi organici. *Gazz. Chim. It.*, **54**, 529–540; (b) Müller, E. and Zeeh, B. (1966) Lewissäure-katalysierte Umsetzung von Carbonylverbindungen mit tert.-Butylisonitril. *Justus Liebigs Ann. Chem.*, **696** (1), 72–80; (c) Zeeh, B. and Müller, E. (1968) Lewissäure-katalysierte Umsetzung von aliphatischen Ketonen mit tert.-Butyl-isonitril. *Justus Liebigs Ann. Chem.*, **715** (1), 47–51; (d) Mcfarland, J.W. (1963) Reactions of cyclohexylisonitrile and isobutyraldehyde with various nucleophiles and catalysts. *J. Org. Chem.*, **28** (9), 2179–2181.

28 (a) Ugi, I. and Meyr, R. (1961) Erweiterter Anwendungsbereich der Passerini-Reaktion. *Chem. Ber.*, **94** (8), 2229–2233; (b) Hagedorn, I. and Eholzer, U. (1965) Einstufige Synthese von α-Hydroxysäure-amiden durch Abwandlung der Passerini-Reaktion. *Chem. Ber.*, **98** (3), 936–940.

29 Giustiniano, M., Pirali, T., Massarotti, A., Biletta, B., Novellino, E., Campiglia, P., Sorba, G., and Tron, G.C. (2010) A practical synthesis of 5-aroyl-1-aryltetrazoles using an Ugi-like 4-component reaction followed by a biomimetic transamination. *Synthesis*, **23**, 4107–4118.

30 Yue, T., Wang, M.-X., Wang, D.-X., and Zhu, J. (2008) Asymmetric synthesis of 5-(1-hydroxyalkyl)tetrazoles by catalytic enantioselective Passerini-type reactions. *Angew. Chem. Int. Ed.*, **47** (49), 9454–9457.

31 (a) Ugi, I. and Steinbruckner, C. (1961) Reaktion von Isonitrilen mit Carbonylverbondungen, Aminen und Stickstoffwasserstoffsäure. *Chem. Ber.*, **94** (3), 734–742; (b) Ugi, I. and Bodesheim, F. (1963) Isonitrile, XIV. Umsetzung von Isonitrilen mit Hydrazonen und Carbonsäuren. *Justus Liebigs Ann. Chem.*, **666** (1), 61–64; (c) Opitz, G., Griesinger, A., and Schubert, H.W. (1963) Enamine, XII. Umsetzung von tertiären Enamin-Salzen mit nucleophilen Verbindungen. *Justus Liebigs Ann. Chem.*, **665** (1), 91–101; (d) Opitz, G. and Merz, W. (1962) Enamine, IX. Addition von Stickstoffwasserstoffsäure an Dien-amine. *Justus Liebigs Ann. Chem.*, **652** (1), 158–162; (e) Neidlein, R. (1965) Reaktionen mit Poly-isonitrilen. *Arch. Pharm.*, **298**, 491–497; (f) Neidlein, R. (1964) Reaktionen mit Poly-isonitrilen. *Arch. Pharm.*, **297**, 589–592; (g) Neidlein, R. (1964) Synthese von 1.1-Diisonitrilomethan. *Angew. Chem.*, **76** (10), 440; (h) Neidlein, R. (1964) *Angew. Chem. Int. Ed. Engl.*, **3** (5), 384.

32 Nixey, T., Kelly, M., and Hulme, C. (2002) Rapid generation of cis-constrained norstatine analogs using a $TMSN_3$-modified Passerini MCC/N-capping strategy. *Tetrahedron Lett.*, **43** (38), 6833–6835.

33 Kiselyov, A.S. (2005) Reaction of N-fluoropyridinium fluoride with isonitriles and $TMSN_3$: a convenient one-pot synthesis of tetrazol-5-yl pyridines. *Tetrahedron Lett.*, **46** (29), 4851–4854.

34 (a) Ugi, I. and Bodesheim, F. (1961) Umsetzung von Isonitrilen mit Hydrazonen und Stickstoffwasserstoffsäure. *Chem. Ber.*, **94** (10), 2797–2801; (b) Zinner, G. and Bock, W. (1971) Zur Kenntnis des Ugi-Reaktion mit Hydrazinen, II. *Arch. Pharmaz.*, **304**, 933–943; (c) Zinner, G. and Bock, W. (1973) Notiz über die UGI-Reaktion mit Diaziridinen. *Arch. Pharmaz.*, **306**, 94–96.

35 Bienaymé, H. and Bouzid, K. (1998) Synthesis of rigid hydrophobic tetrazoles using an Ugi multi-component heterocyclic condensation. *Tetrahedron Lett.*, **39** (18), 2735–2738.

36 Nayak, M. and Batra, S. (2010) Isonitriles from the Baylis–Hillman adducts of acrylates: viable precursor to tetrazolo-fused diazepinones via post-Ugi cyclization. *Tetrahedron Lett.*, **51** (3), 510–516.

37 Mayer, J., Umkehrer, M., Kalinski, C., Ross, G., Kolb, J., Burdack, C., and Hiller, W. (2005) New cleavable isocyanides for the combinatorial synthesis of α-amino acid analogue tetrazoles. *Tetrahedron Lett.*, **46** (43), 7393–7396.

38 Dömling, A., Beck, B., and Magnin-Lachaux, M. (2006) 1-Iso cyanomethylbenzotriazole and 2,2,4,4-tetr amethylbutylisocyanide – cleavable isocyanides useful for the preparation of α-aminomethyl tetrazoles. *Tetrahedron Lett.*, **47** (25), 4289–4291.

39 Nixey, T., Kelly, M., Semin, D., and Hulme, C. (2002) Short solution phase preparation of fused azepine-tetrazoles via a UDC (Ugi/de-Boc/cyclize) strategy. *Tetrahedron Lett.*, **43** (20), 3681–3684.

40 Umkehrer, M., Kolb, J., Burdack, C., Rossa, G., and Hiller, W. (2004) Synthesis of tetrazolopiperazine building blocks by a novel multi-component reaction. *Tetrahedron Lett.*, **45** (34), 6421–6424.

41 Kalinski, C., Umkehrer, M., Gonnard, S., Jäger, N., Rossa, G., and Hiller, W. (2006) A new and versatile Ugi/SNAr synthesis of fused 4,5-dihydrotetrazolo[1,5-a] quinoxalines. *Tetrahedron Lett.*, **47** (12), 2041–2044.

42 (a) Nixey, T., Kelly, M., and Hulme, C. (2000) The one-pot solution phase preparation of fused tetrazole-ketopiperazines. *Tetrahedron Lett.*, **41** (45), 8729–8733; (b) Marcos, C.F., Marcaccini, S., Menchi, G., Pepino, R., and Torroba, T. (2008) Studies on isocyanides: synthesis of tetrazolyl-isoindolinones via tandem Ugi four-component condensation/intramolecular amidation. *Tetrahedron Lett.*, **49** (1), 149–152.

43 El Kaïm, L., Grimaud, L., and Oble, J. (2005) Phenol Ugi–Smiles systems: strategies for the multicomponent N-arylation of primary amines with isocyanides, aldehydes, and phenols. *Angew. Chem. Int. Ed.*, **44** (44), 7165–7169.

44 El Kaïm, L., Gizolme, M., Grimaud, L., and Oble, J. (2007) Smiles rearrangements in Ugi- and Passerini-type couplings: new multicomponent access to O- and N-arylamides. *J. Org. Chem.*, **72** (6), 4169–4180.

45 El Kaïm, L., Gizolme, M., Grimaud, L., and Oble, J. (2006) Direct access to heterocyclic scaffolds by new multicomponent Ugi-Smiles couplings. *Org. Lett.*, **8** (18), 4019–4021.

46 Barthelon, A., Dos Santos, A., El Kaïm, L., and Grimaud, L. (2008) Ugi/Smiles access to pyrazine scaffolds. *Tetrahedron Lett*, **49** (20), 3208–3211.

47 Barthelon, A., Legoff, X.-F., El Kaïm, L., and Grimaud, L. (2010) Four-component synthesis of imidazolinium-fused heterocycles from Ugi–Smiles couplings. *Synlett*, (1), 153–157.

48 El Kaïm, L., Gizolme, M., and Grimaud, L. (2006) O-arylative Passerini reactions. *Org. Lett.*, **8** (22), 5021–5023.

49 Dai, W.-M. and Li, H. (2007) Lewis acid-catalyzed formation of Ugi four-component reaction product from Passerini three-component reaction system without an added amine. *Tetrahedron*, **63** (52), 12866–12876.

50 Barthelon, A., El Kaïm, L., Gizolme, M., and Grimaud, L. (2008) Thiols in Ugi- and Passerini–Smiles-type couplings. *Eur. J. Org. Chem.*, **35**, 5974–5987.

51 El Kaïm, L., Gizolme, M., Grimaud, L., and Oble, J. (2007) New benzothiazole and benzoxazole scaffolds from the Ugi–Smiles couplings of heterocyclic thiols. *Synlett*, **3**, 465–469.

52 Oble, J., El Kaïm, L., Gizzi, M., and Grimaud, L. (2007) Ugi–Smiles access to quinoxaline derivatives. *Heterocycles*, **73** (1), 503–517.

53 Coffinier, D., El Kaïm, L.M., and Grimaud, L. (2009) New benzotriazole and benzimidazole scaffolds from Ugi–Smiles couplings of isocyanides. *Org. Lett.*, **11** (4), 995–997.

54 El Kaïm, L., Gizolme, M., Grimaud, L., and Oble, J. (2007) New Ugi–Smiles–metathesis strategy toward the synthesis of pyrimido azepines. *J. Org. Chem.*, **72** (15), 5835–5838.

55 El Kaïm, L., Gizzi, M., and Grimaud, L. (2008) New MCR–Heck-isomerization cascade towards indoles. *Org. Lett.*, **10** (16), 3417–3419.

56 El Kaïm, L., Grimaud, L., Ibarra, T., and Montano-Gamez, R. (2008) New palladium-catalyzed aerobic oxidative cleavage and cyclization of N-aryl peptide derivatives. *Chem. Commun.*, (3), 350–352.

57 El Kaïm, L., Grimaud, L., and Wagschal, S. (2010) Toward pyrrolo[2,3-d]pyrimidine scaffolds. *J. Org. Chem.*, **75** (15), 5343–5346.

58 El Kaïm, L., Legoff, X.-F., Grimaud, L., and Schiltz, A. (2011) Smiles cascades toward heterocyclic scaffolds. *Org. Lett.*, **13** (3), 534–536.

59 Dömling, A., Herdtweck, E., and Heck, S. (2006) Cyanamide in isocyanide-based MCRs. *Tetrahedron Lett.*, **47** (11), 1745–1747.

60 Bossio, R., Marcaccini, S., Paoli, P., Pepino, R., and Polo, C. (1991) Studies on isocyanides and related compounds. Synthesis of benzofuran derivatives. *Synthesis*, (11), 999–1000.

61 Hagedorn, I., Eholzer, U., and Winkelmann, H.D. (1964) Beitrag zur Isonitril-Chemie. *Angew. Chem.*, **76** (13), 583–584.

62 (a) Konig, S., Lohberger, S., and Ugi, I (1993) Synthese von N-tert-Alkylglyoxylsäureamiden. *Synthesis*, (12), 1233–1234; (b) Zeeh, B. (1968) Heterocyclen Aus Isocyaniden-IV. 3H-Benzo[g]indole Aus 1-Naphthylisocyanid Und Aliphatischen Ketonen. *Tetrahedron*, **24** (23), 6663–6669.

63 (a) Lumma, W.C.J. (1981) Modification of the Passerini reaction: facile synthesis of analogues of isoproterenol and (aryloxy) propanolamine P-adrenergic blocking agents. *J. Org. Chem.*, **46** (18), 3668–3671; (b) Semple, J.E., Owens, T.D., Nguyen, K., and Levy, O.E. (2000) New synthetic technology for efficient construction of α-hydroxy-β-amino amides via the Passerini reaction. *Org. Lett.*, **2** (18), 2769–2772.

64 Kumar, J.S., Jonnalagadda, S.C., and Mereddy, V.R. (2010) An efficient boric acid-mediated preparation of α-hydroxyamides. *Tetrahedron Lett.*, **51** (5), 779–782.

65 Soeta, T., Kojima, Y., Ukaji, Y., and Inomata, K. (2011) Borinic acid catalyzed α-addition to isocyanide with aldehyde and water. *Tetrahedron Lett.*, **52** (20), 2557–2559.

66 Masdeu, C., Diaz, J.L., Miguel, M., Jimenez, O., and Lavilla, R. (2004) Straightforward α-carbamoylation of NADH-like dihydropyridines and enol ethers. *Tetrahedron Lett.*, **45** (42), 7907–7909.

67 Chandra Pan, S. and List, B. (2008) Catalytic three-component Ugi reaction. *Angew. Chem. Int. Ed.*, **47** (19), 3622–3625.

68 (a) Tanaka, Y., Hidaka, K., Hasui, T., and Suginome, M. (2009) B(OMe)$_3$ as a nonacidic iminium ion generator in Mannich- and Ugi-type reactions. *Eur. J. Org. Chem.*, (8), 1148–1151; (b) Tanaka, Y., Hidaka, K., Hasui, T., and Suginome, M. (2007) Acid-free, aminoborane-mediated Ugi-type reaction leading to general utilization of secondary amines. *Org. Lett.*, **9** (22), 4407–4410.

69 Keung, W., Bakir, F., Patron, A.P., Rogers, D., Priest, C.D., and Darmohusodo, V. (2004) Novel α-amino amidine synthesis via scandium(III) triflate mediated 3CC Ugi condensation reaction. *Tetrahedron Lett.*, **45** (4), 733–737.

70 (a) Schiess, M. and Seebach, D. (1983) N-Methyl-C-(trichlortitanio) formimidoylchlorid. Ein effizientes Reagenz zur Homologisierung von Aldehyden und Ketonen zu a-Hydroxy-carbonsäureamiden. *Helv. Chem. Acta*, **66**

(5), 1618–1623; (b) Carofiglio, T., Cozzi, P.G., Floriani, C., Chiesi-Villa, A., and Rizzoli, C. (1993) Nonorganometallic pathway of the Passerini reaction assisted by titanium tetrachloride. *Organometallics*, **12** (7), 2726–2736; (c) Seebach, D., Adam, G., Gees, T., Schiess, M., and Weigand, W. (1988) Scope and limitations of the TiCl$_4$-mediated additions of isocyanides to aldehydes and ketones with formation of α-hydroxycarboxylic acid amides. *Chem. Ber.*, **121** (3), 507–517.

71 (a) Denmark, S.E. and Fan, Y. (2003) The first catalytic, asymmetric α-additions of isocyanides. Lewis-base-catalyzed, enantioselective Passerini-type reactions. *J. Am. Chem. Soc.*, **125** (26), 7825–7827; (b) Denmark, S.E. and Fan, Y. (2005) Catalytic, enantioselective α-additions of isocyanides: Lewis base-catalyzed Passerini-type reactions. *J. Org. Chem.*, **70** (24), 9667–9676.

72 Andreana, P.R., Liu, C.C., and Schreiber, S.L. (2004) Stereochemical control of the Passerini reaction. *Org. Lett.*, **6** (23), 4231–4233.

73 Saegusa, T., Takishi, N., Tamura, I., and Fuji, H. (1969) Acid-catalyzed reaction of isocyanide with a Schiff base. New and facile syntheses of imidazolidines. *J. Org. Chem.*, **34** (4), 1145–1147.

74 For a review, see Moderhack, D. (1985) Four-membered rings from isocyanides – recent advances. *Synthesis*, (12), 1083–1096.

75 (a) Kabbe, H.J. (1969) Isonitrile, II. 2.3-Bis-alkylimino-oxetane aus Carbonylverbindungen und Isonitrilen. *Chem. Ber.*, **102** (4), 1404–1409; (b) Zeeh, B. (1969) Additions Reaktionen zwischen lsocyaniden und Doppelbin-dungssystemen. *Synthesis*, (2), 65–73.

76 Saegusa, T., Taka-Ishi, N., and Fujii, H. (1968) Reaction of carbonyl compound with isocyanide. *Tetrahedron*, **24** (10), 3795–3798.

77 Muller, E. and Zeeh, B. (1966) Lewissäure-katalysierte Umsetzung von Carbonylverbindungen mit *tert.*-Butylisonitril. *Justus Liebigs Ann. Chem.*, **696** (1), 72–80.

78 Kabbe, H.J. (1969) Isonitrile, III. Umsetzungen von 2.3-Bis-alkylimino-oxetanen. *Chem. Ber.*, **102** (4), 1410–1417.

79 Xia, Q. and Ganem, B. (2002) Metal-mediated variants of the Passerini reaction: a new synthesis of 4-cyanooxazoles. *Synthesis*, (14), 1969–1972.

80 Xia, Q. and Ganem, B. (2002) Metal-promoted variants of the Passerini reaction leading to functionalized heterocycles. *Org. Lett.*, **4** (9), 1631–1634.

81 Barrett, A.G.M., Barton, D.H.R., Falck, J.R., Papaioannou, D., and Widdowson, D.A. (1979) Phenol oxidation and biosynthesis. Part 26. Isonitriles in the synthesis of benzylisoquinoline derivatives. *J. Chem. Soc., Perkin Trans. 1*, 652–661.

82 Mukaiyama, T., Watanabe, K., and Shiono, M. (1974) A convenient method for the synthesis of α-alkoxycarboxamide derivatives. *Chem. Lett.*, **3** (12), 1457–1458.

83 Yanai, H., Oguchi, T., and Taguchi, T. (2009) Direct alkylative Passerini reaction of aldehydes, isocyanides, and free aliphatic alcohols catalyzed by indium(III) triflate. *J. Org. Chem.*, **74** (10), 3927–3929.

84 (a) Katritzky, A.R., Button, M.A.C., and Busont, S. (2001) *J. Org. Chem.*, **66**, 2865–2868; (b) Katritzky, A.R., Mohapatra, P.P., Singh, S., Clemens, N., and Kirichenko, K. (2005) Synthesis of α-amino amides via α-amino imidoylbenzotriazoles. *J. Serb. Chem. Soc.*, **70** (3), 319–327.

85 Waller, R.W., Diorazio, L.J., Taylor, B.A., Motherwell, W.B., and Sheppard, T.D. (2010) Isocyanide based multicomponent reactions of oxazolidines and related systems. *Tetrahedron*, **66** (33), 6496–6507.

6
Amine (Imine) Component Surrogates in the Ugi Reaction and Related Isocyanide-Based Multicomponent Reactions

Mikhail Krasavin

6.1
Introduction

The Ugi reaction [1] has manifested itself as a remarkably powerful synthetic tool that allows the unification – both efficiently and predictably – of the four reaction components, namely an amine, a carbonyl compound, an isocyanide, and a carboxylic acid, within the structure of a single product **1**. The reaction provides an exemplary showcase of atom economy [2], offering access to amino acid derivatives.

From the mechanistic perspective [3], the reaction proceeds due to the ability of the isocyanide to react at the same (carbenoid) atom with both nucleophiles and electrophiles. Specifically, the imine resulting from the condensation of the amine and carbonyl components, in its protonated form, reacts as an electrophile with the isocyanide that is, in turn, intercepted by the nucleophilic carboxylate anion. This leads to the so-called α-adduct **2** which undergoes O→N acyl migration (termed the Mumm rearrangement) to deliver the final compound **1** (Scheme 6.1).

The Ugi reaction opened a technically simple and straightforward access to medicinally relevant peptide-like compounds **1**, in which the diversity of peripheral groups is controlled unequivocally by the matrix of the four reagents. This aspect also made this reaction one of the favorite tools for combinatorial library development [4].

During the past 10–15 years, significant efforts in both academic and industrial research laboratories have been directed towards applying the extremely efficient Ugi chemistry to the preparation of new organic compounds that are structurally different from the initially lookalike dipeptoid products [5]. Numerous successes in this area can be attributed to the use of bifunctional reagents [6, 7] that contain any two of the four reactive functionalities participating in the Ugi event. This strategy, which typically has employed keto or aldehydo carboxylic acids as well as amino acids, has delivered a range of important – and often unique – heterocyclic frameworks with an unmatched convergent efficiency [8].

Isocyanide Chemistry: Applications in Synthesis and Material Science, First Edition. Edited by Valentine Nenajdenko.
© 2012 Wiley-VCH Verlag GmbH & Co. KGaA. Published 2012 by Wiley-VCH Verlag GmbH & Co. KGaA.

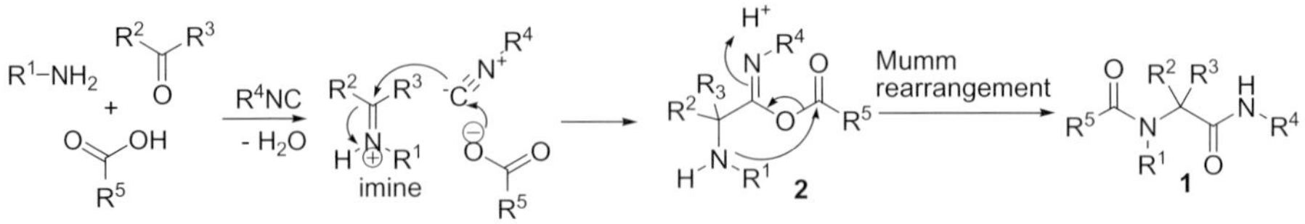

6 Amine (Imine) Component Surrogates

Scheme 6.1

Another rich area of research that continues to yield skeletally complex and otherwise inaccessible compounds is the use of so-called post-Ugi modifications [9]. Considering the remarkably broad scope of the Ugi reaction, it is possible to include in the reagent structure additional functional groups (or even latently reactive moieties) that are compatible with the Ugi event itself. Once the two or more mutually reactive substituents have been incorporated into the structure on the Ugi reaction product, a post-condensational transformation can either occur spontaneously [10] or be triggered by an added reagent [11] or by the application of forcing reaction conditions [12].

Finally, the product diversity of the Ugi and related reactions (broadly termed isocyanide-based multicomponent reactions; IMCRs) can be extended quite remarkably by using various surrogates for its components (excluding, of course, the critical isocyanide). The most commonly replaced component of the Ugi reaction is the carboxylate anion, which intercepts the isocyanide during the course of its interaction with the protonated imine component. Numerous examples of C-, N-, O-, and S-nucleophiles enabling isocyanide multicomponent chemistry in lieu of a carboxylic acid have been recently summarized in an excellent review by El Kaïm and Grimaud [13]. The use of surrogate replacements for the amine (or, more broadly, the imine) input for the Ugi reaction is far less common, however.

In this chapter, a survey is presented of the many reports that have described the use of this strategy, in addition to details of some recent contributions made by the author's research group.

6.2
Hydroxylamine Components in the Ugi Reaction

The first investigations of the possibility to use hydroxylamine as the amine replacement in the Ugi reaction were made by Zinner et al. in 1969 [14]. Initially, it was shown that N,O-dimethylhydroxylamine (3) is capable of reacting in this reaction in similar fashion to a regular amine, thereby providing fair to excellent yields of the Ugi adducts 4a–d with cyclohexylisocyanide and a range of aldehydes, both aliphatic and aromatic (Scheme 6.2).

Notably, in this case the fourth component of the Ugi reaction – the carboxylic acid – was omitted. Instead, in the presence of an equivalent amount of hydrochloric acid, the water served essentially as the isocyanide-intercepting nucleophile.

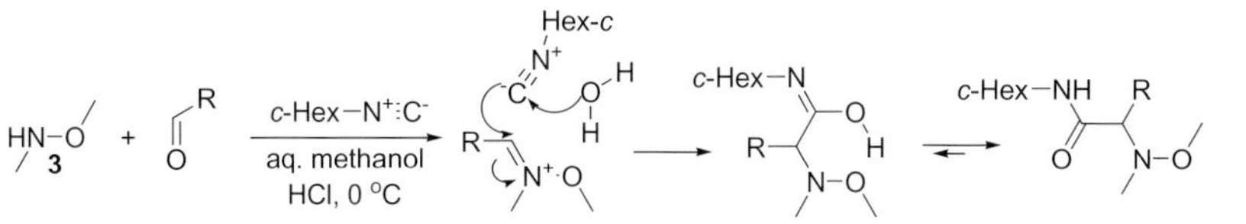

Scheme 6.2

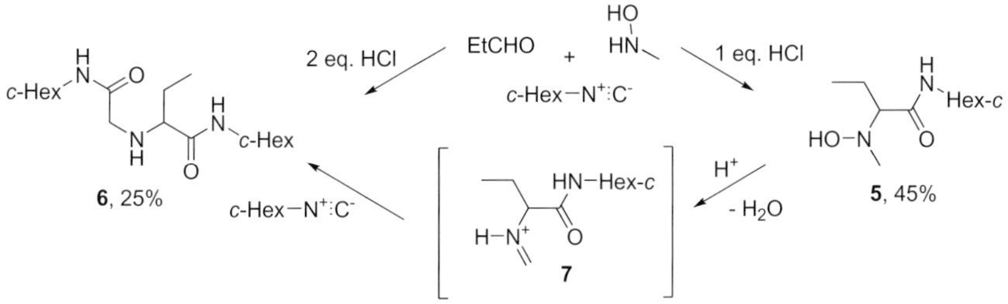

Scheme 6.3

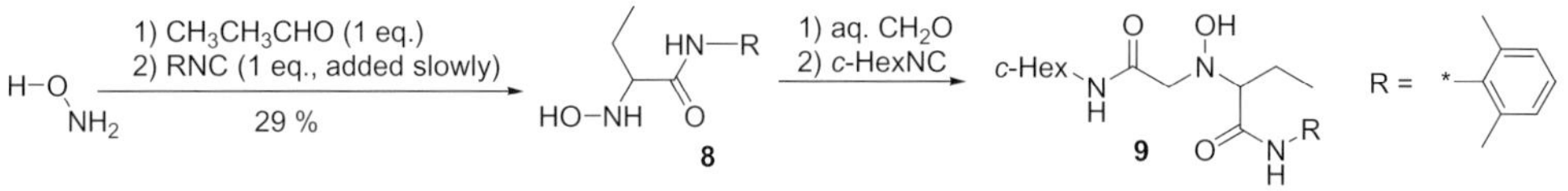

Scheme 6.4

Under the same conditions, *N*-alkylhydroxylamines provided lower yields of the Ugi adducts **5** containing a free hydroxyl group. In subsequent studies, Moderhack described [15] an unexpected outcome of this reaction, when it was conducted in the presence of 2 equiv. of a mineral acid, when the predominant product was the 2,2′-iminodicarboxamide **6** (Scheme 6.3)! It is quite likely that, in the presence of an excess amount of hydrochloric acid, dehydration of the hydroxylamine moiety in the initial product **5** takes place, followed by an interaction of the intermediate imine **7** with the residual amount of isocyanide, leading to **6**.

Regardless of the O-substitution, a simple mixing of N-unsubstituted hydroxylamines with an aldehyde and an isocyanide provided only moderate yields of the expected products **8**, and also led to the products of a double condensation **9** [14]. Premixing equimolar amounts of a hydroxylamine and an aldehyde, followed by the slow, dropwise addition of an isocyanide, increased the chances of single adduct formation. A single adduct **8** can be isolated and introduced in the Ugi reaction again, thus leading to an unsymmetrical bis-amide **9**. This early realization of the "sequential IMCR" concept is illustrated by the example in Scheme 6.4.

In the same studies, the Ugi reaction with hydroxylamine was conducted in a four-component format, using benzoic acid as the fourth component. In this case, it was established that the rearrangement of the α-adduct **10** proceeded exclusively via N→O acyl migration, leading to the "internal" hydroxamic acids **11a–b** in low yield (Scheme 6.5).

A remarkable extension of hydroxylamine usage in the Ugi reaction is provided by the combination with sodium azide (as a carboxylate anion replacement) [16]. The reaction with N-alkylhydroxylamines gave good to excellent yields of tetrazoles **12a–c**, with the reaction presumably proceeding via an intramolecular dipolar cycloaddition of intermediate **13** (Scheme 6.6). Unsubstituted hydroxylamine gives rise to dimeric adducts, albeit in very low yields.

The reaction may also involve cyclic ketones as the carbonyl component. In this case, the best (though still quite modest) yield of the Ugi adduct **14** was observed for cyclohexanone (Scheme 6.7).

The hydroxylaminomethyl tetrazoles (e.g., **12c**) can be converted into the carbamate derivative **15** which, on pyrolysis in methanol, provided an excellent yield of the valuable tetrazole carboxaldehyde building block **16** (isolated as a hydrate; Scheme 6.8).

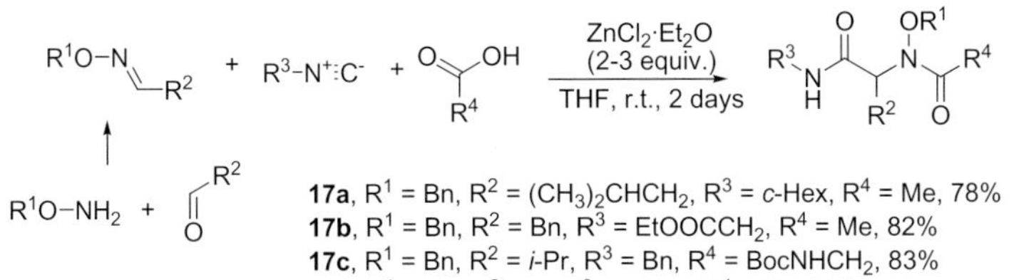

6.2 Hydroxylamine Components in the Ugi Reaction | 199

Scheme 6.9

17a, R¹ = Bn, R² = (CH₃)₂CHCH₂, R³ = c-Hex, R⁴ = Me, 78%
17b, R¹ = Bn, R² = Bn, R³ = EtOOCCH₂, R⁴ = Me, 82%
17c, R¹ = Bn, R² = i-Pr, R³ = Bn, R⁴ = BocNHCH₂, 83%
17d, R¹ = Bn, R² = H, R³ = c-Hex, R⁴ = Ph, 60%
17e, R¹ = TBDMS, R² = i-Pr, R³ = c-Hex, R⁴ = Me, 33% (desilylated product)

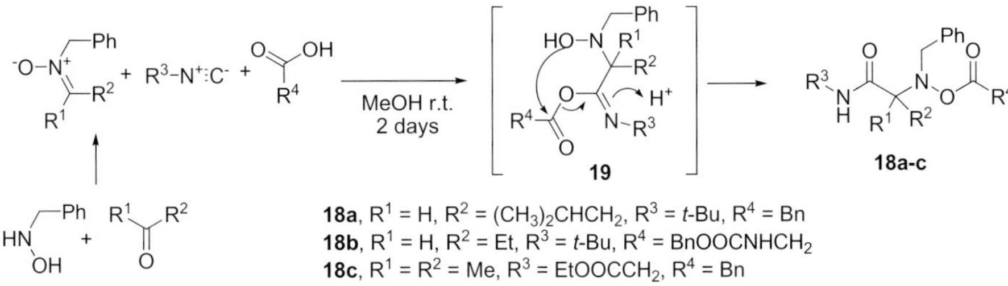

Scheme 6.10

18a, R¹ = H, R² = (CH₃)₂CHCH₂, R³ = t-Bu, R⁴ = Bn
18b, R¹ = H, R² = Et, R³ = t-Bu, R⁴ = BnOOCNHCH₂
18c, R¹ = R² = Me, R³ = EtOOCCH₂, R⁴ = Bn

The use of hydroxylamine inputs in the Ugi reaction was revisited some 30 years later by Basso and coworkers, who reported [17] on the poor reactivity of the O-protected oximes of aliphatic aldehydes towards isocyanides. However, when the reaction was performed in the presence of $ZnCl_2$, quite respectable yields of the target O-protected hydroxamic acid derivatives **17** were obtained. This Lewis acid also provided superior results compared to other promoters (e.g., $MgBr_2$, BF_3, $TiCl_4$), and the optimized procedure (2–3 equiv. of the Lewis acids and the use of a preformed oxime in the reaction) was found to be applicable to a range of aliphatic aldehydes and carboxylic acids (Scheme 6.9).

These authors were unable to induce the same reaction with oximes of *aromatic* aldehydes, however. Notably, the O-benzyl products **17** could be cleanly and selectively debenzylated via hydrogenolysis on Pd/$BaSO_4$, providing good to excellent yields of the respective "internal" hydroxamic acids (R^1 = H).

The same group extended this reaction to N-benzylhydroxylamine [18], a reagent capable of forming nitrones with carbonyl compounds that would be expected to be more reactive towards isocyanides than towards the respective oximes. Indeed, the simple mixing of N-benzylhydroxylamine with an aliphatic aldehyde or a ketone, an isocyanide, and a carboxylic acid in methanol provided, after 2 days at room temperature, very respectable yields of α-acyloxyamino acetamides **18**. Clearly, in this case the rearrangement of the α-adduct **19** can proceed only via the O→O acyl migration (Scheme 6.10).

The adducts **18** can be converted into medicinally important α-hydroxylamino acetamides via removal of the R_4CO group, with mild alkaline hydrolysis. Likewise,

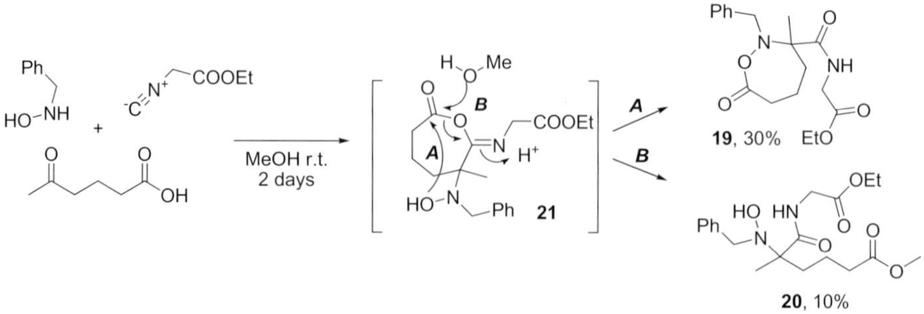

Scheme 6.11

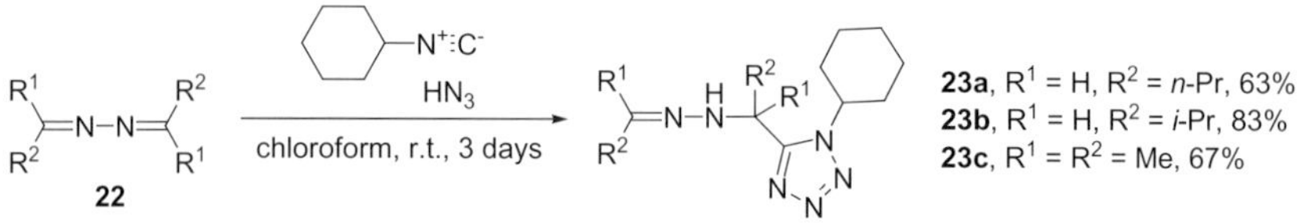

Scheme 6.12

when 4-acetylbutyric acid (a bifunctional reagent with respect to the Ugi reaction) was used in combination with *N*-benzylhydroxylamine and ethyl isocyanoacetate, a mixture of oxazepinone **19** and methyl ester **20** was formed. Mechanistically, this was not an unexpected result, as the putative α-adduct **21** may undergo either an intramolecular O→O acyl migration (Scheme 6.11, path A) or interact with the nucleophilic solvent molecule (path B). As will be discussed below, it might be possible to guard against side reactions leading to esters such as **20**, by employing less-nucleophilic solvents (e.g., isopropanol).

6.3
Hydrazine Components in the Ugi Reaction

The first applications of hydrazine and its derivatives in the Ugi reaction were reported shortly after the discovery of the reaction was disclosed. As reported by the Ugi group, azines **22** reacted with 1 equiv. of cyclohexyl isocyanide in the presence of an equimolar amount of hydrazoic acid, to provide good yields of tetrazoles **23** (Scheme 6.12) [19]. Similarly, *N*-acylhydrazones **24** reacted with isocyanides and hydrazoic acid (Scheme 6.13) to give excellent isolated yields of the tetrazoles **25**.

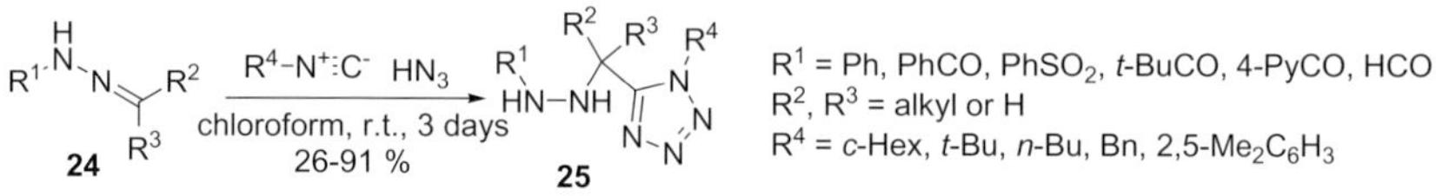

Scheme 6.13

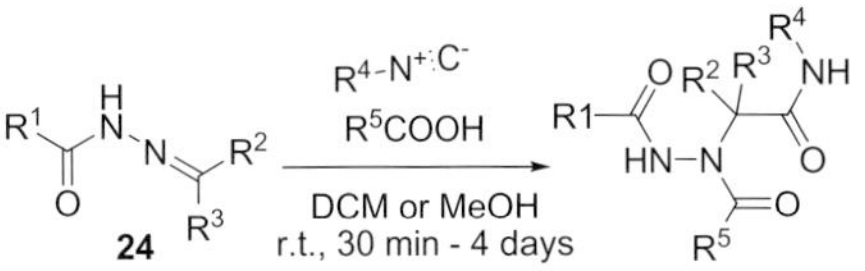

6.3 Hydrazine Components in the Ugi Reaction | 201

Scheme 6.14

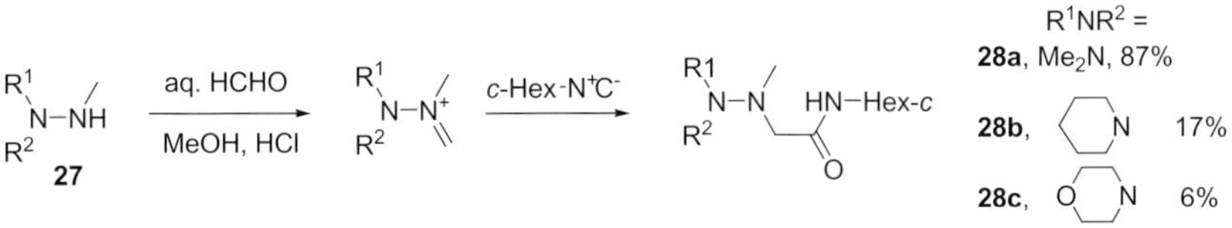

26a, R¹ = EtO, R² = H, R³ = i-Pr, R⁴ = t-Bu, R⁵ = CHCl₂, 80%
26b, R¹ = Ph, R² = H, R³ = i-Pr, R⁴ = t-Bu, R⁵ = CH₂Cl, 79%
26c, R¹ = Ph, R² = H, R³ = c-Hex, R⁵ = H, 80%
26d, R¹ = Ph, R² = R³ = Me, R⁴ = 2,6-Me₂C₆H₃, R⁵ = CH₂Cl, 25%
26e, R¹ = Ph, R² = R³ = Me, R⁴ = t-Bu, R⁵ = NCCH₂, 45%

Scheme 6.15

R¹NR² =
28a, Me₂N, 87%
28b, (pyrrolidine) 17%
28c, (morpholine) 6%

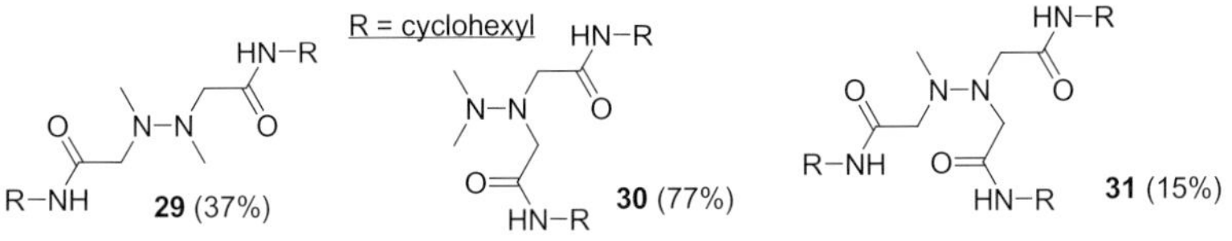

Figure 6.1 Products from the reaction of 1,2-dimethylhydrazine and 1,1-dimethylhydrazine with formaldehyde and cyclohexylisocyanide.

In the traditional, four-component format, including the carboxylic acid component, the reaction furnished a range of N, N'-bis(acyl)-N'-alkyl hydrazines **26** in good yields (Scheme 6.14) [20]. It should be noted that the scope of the reaction appeared to be limited with respect to the carboxylic acid component (only those more acidic inputs appeared to function well).

At about the same time, Zinner investigated the outcome of an Ugi reaction that included an excess of formaldehyde and cyclohexyl isocyanide with hydrazines of various degrees of N-alkylation [21]. The products **28** resulted from the addition of isocyanide onto the iminium intermediate (Scheme 6.15). Under the same conditions, 1,2-dimethylhydrazine and 1,1-dimethylhydrazine (with 2 equiv. of the isocyanide) provided the double-condensation products **29** and **30**, respectively. Not unexpectedly, the reaction of methylhydrazine with excess formaldehyde and 3 equiv. of the isocyanide produced the triple-condensation product **31** (Figure 6.1).

The outcome of reacting an unsubstituted hydrazine with an excess of formaldehyde and cyclohexylisocyanide in acidic methanol was, however, quite surprising. Notably, the symmetrical tetrahydro[1,2,4]triazino[2,1-a][1,2,4]triazine-3,8(4H,9H)-dione **32** was formed in 40% yield, instead of the expected (at least in this case) quadruple-condensation product. Such a result may be mechanistically justified as shown in Scheme 6.16.

A similar reactivity was demonstrated by the same group for various alkylated hydrazine inputs in the reaction of carbonyl compounds with cyclohexyl isocyanide and hydrazoic acid, leading to products **33a–c** (Scheme 6.17) [22]. Notably,

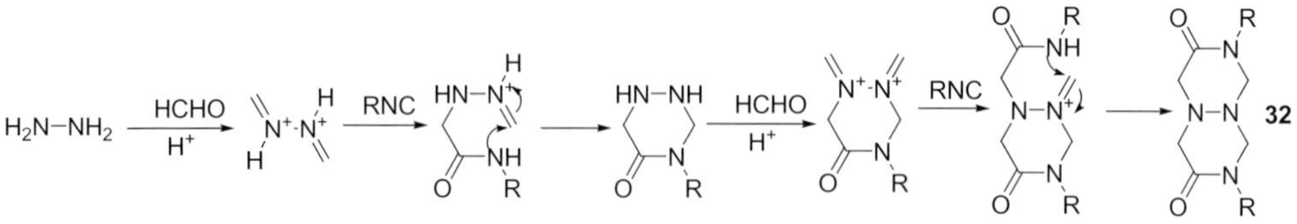
HCHO
H+
RNC
HCHO
H+
RNC
32

Scheme 6.16

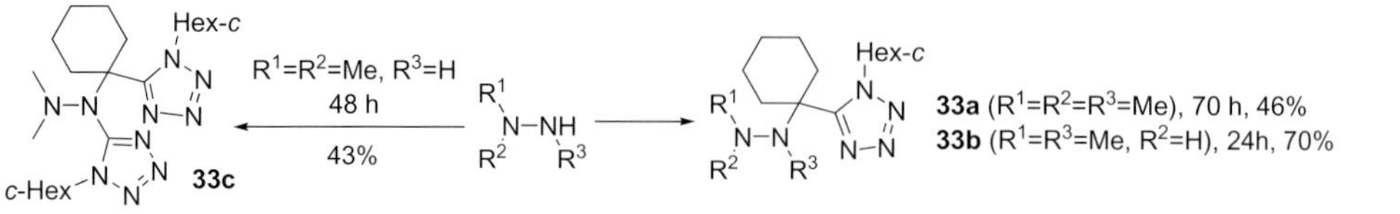
R1=R2=Me, R3=H
48 h
43%
33c
33a (R1=R2=R3=Me), 70 h, 46%
33b (R1=R3=Me, R2=H), 24h, 70%

Conditions: cyclohexanone (1 eq.), cyclohexyl isocyanide (1 eq.), HN_3 (1 eq., solution in benzoic acid), aq. methanol, r.t.

Scheme 6.17

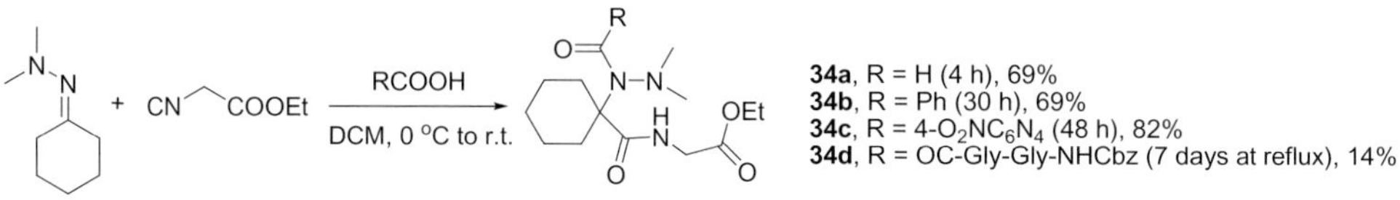
RCOOH
DCM, 0 °C to r.t.
34a, R = H (4 h), 69%
34b, R = Ph (30 h), 69%
34c, R = 4-O2NC6N4 (48 h), 82%
34d, R = OC-Gly-Gly-NHCbz (7 days at reflux), 14%

Scheme 6.18

despite the availability of the second reactive nitrogen atom in product **33b**, only traces of the dimeric product were detected in the reaction of 1,2-dimethylhydrazine (most likely, due to the steric hindrance introduced by the first Ugi reaction). At the same time, the similar steric hindrance did not preclude product **33c** from forming in essentially quantitative yield (considering the use of equimolar amounts of the four reagents in this reaction).

The potential of using various hydrazine inputs in the Ugi reaction to access, in rapid fashion, a variety of unnatural peptoid fragments, was recognized very soon after the discovery of the "hydrazo-Ugi" reaction. Generally, interest in such structures is driven by the quest for more hydrolytically stable analogues of the bioactive peptides that would preserve a similar or (acceptably) only a slightly decreased level of the biological activity of their natural congeners. The first report to appear on the hydrazo-Ugi reaction in the context of this application related to the synthesis of *N*-aminopeptides, by Götz and coworkers. In these model studies [23], the research group showed that the dimethylhydrazone of cyclohexanone yielded (somewhat reluctantly, and more so with the increasing bulk of the carboxylate component) *N*-dimethylamino peptoids **34** (Scheme 6.18).

The successful use of ethyl isocyanoacetate as a glycine synthon inspired the same group to use an optically active phenylalanine methyl ester **35**. However, despite the resultant compound (**36**) having a high isolated yield, it was obtained

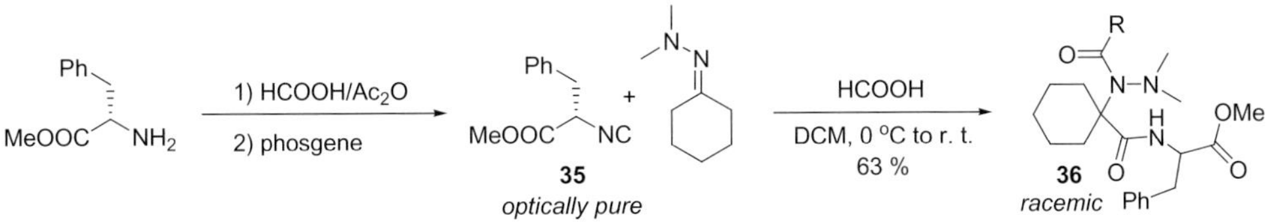

Scheme 6.19

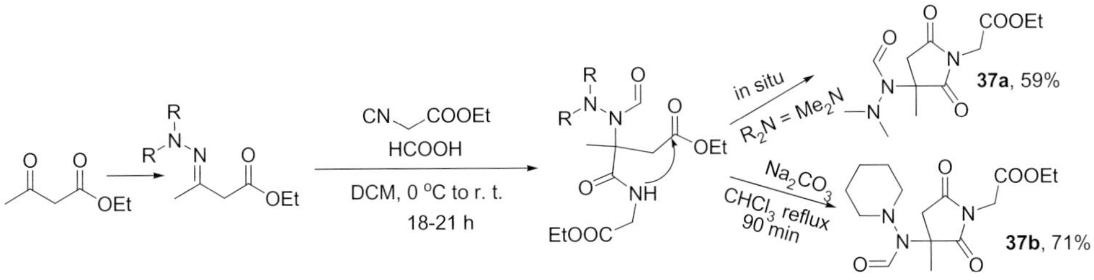

Scheme 6.20

in racemic form (Scheme 6.19). At this point, it should be noted that racemization of the C–H acidic α-amino acid-derived isocyanides (e.g., ethyl isocyanoacetate or **35**) during the Ugi reaction is a common obstacle [24, 25] that has been the subject of much synthetic methodological research [26].

When ethyl acetoacetate was used as the carbonyl component in the hydrazo-Ugi reaction (Scheme 6.20) [23], an unexpected outcome was the formation of cyclized imide products **37**. Such cyclization occurred either spontaneously (**37a**) or on mildly basic treatment of the crude product (**37b**). This, in fact, can be regarded as the first example of post-condensational hydrazo-Ugi modification, an approach that is used widely to increase the complexity of the traditional Ugi reaction products, but is rather scarce in hydrazo-Ugi chemistry.

The results obtained in these model studies were subsequently applied to the preparation of a number of N-amino peptide analogues of hexapeptide H-Ala-Phe-Val-Gly-Leu-Met-NH$_2$, which is a partial pharmacophoric sequence of eledoisin, a naturally occurring hypotensive peptide [27]. The introduction of these N-amino modifications to the peptide backbone led subsequently to an approximately 10-fold improvement in hydrolytic stability compared to chemotrypsin. However, the effective dose of the resulting analogues was, unfortunately, reduced by more than 100-fold compared to their natural congeners.

Somewhat surprisingly, since this seminal report on the application of the hydrazo-Ugi reaction to peptidomimetic design, no research data relating to the use of hydrazine inputs in the Ugi reaction have emerged for over two decades. Burger and colleagues [28] showcased the power of the reaction in preparing peptidomimetic structures of general interest by reacting the N-benzoylhydrazone of isobutyraldehyde with a special type of trifluoromethyl-containing isocyanide

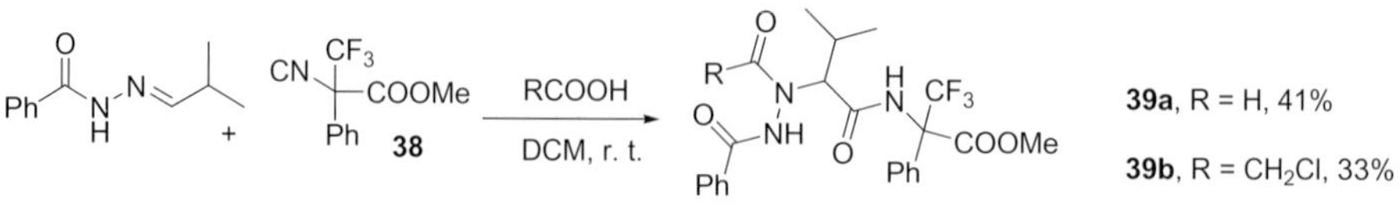

Scheme 6.21

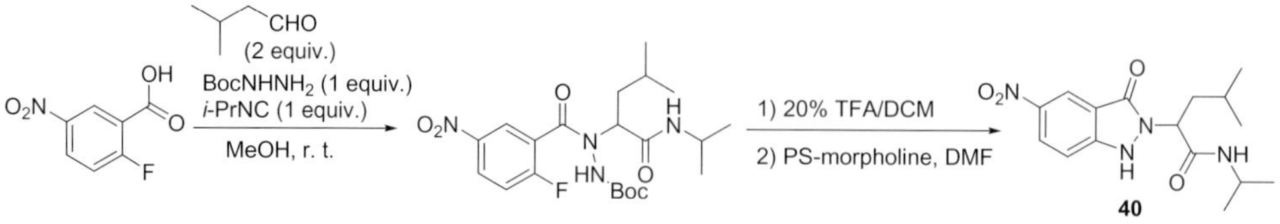

Scheme 6.22

(38) and either formic or chloroacetic acid (Scheme 6.21). In this way, moderate yields of hydrazinopeptoids 39 were provided as 1 : 1 mixtures of diastereomers (for 39b the individual diastereomer was separable via crystallization, however).

An elegant variant of the hydrazo-Ugi reaction has been applied by Hulme and Tempest, in the context of generating combinatorial libraries of compounds with diverse heterocyclic cores [29]. By using Boc-hydrazine in combination with 2-fluoro-5-nitrobenzoic acid (and a diversity of aldehydes and isocyanides), the so-called Ugi/de-Boc/cyclize (UDC) strategy was realized and a good yield of indazolones obtained; this was similar to the example (40) shown in Scheme 6.22, and required only two chemical stages and one product purification. On completion of the Ugi step, the polymer-supported tosylhydrazine and diisopropylethylamine were applied sequentially to scavenge any excess aldehyde (added to ensure complete formation of the hydrazone intermediate) and any unreacted acid, respectively; this prevented the latter materials from interfering with the subsequent cyclization step. Following treatment with trifluoroacetic acid (TFA) to remove the Boc group, the S_NAr-type cyclization occurred efficiently, aided by the addition of resin-bound morpholine to scavenge any TFA and hydrofluoric acid that was present.

Marcaccini and coworkers applied aldazines and ketazines 41 as the imine surrogates in the Ugi reaction with cyclohexyl isocyanide and 2-benzoyl benzoic acid [30]. Since only monoadducts 42 were observed to result from this hydrazino-Ugi reaction, the unreacted hydrazone moiety in 42 might be viewed as a protecting group for the otherwise reactive hydrazine nitrogen atom. This protection was subsequently removed on treatment with aqueous ethanolic HCl, and the 4-phenyl-1-phthalazinones 43 (which were "alkylated" regiospecifically at only one nitrogen atom with substituted acetamide side chains) were isolated in good yields (Scheme 6.23). This regiospecific "decoration" of ambiphilic phthalazinones with these side chains is immensely important when designing biologically active compounds, as various side chains can be introduced via direct alkylation strategies. When taking into consideration the specialized steric situation around the side-chain α-carbon

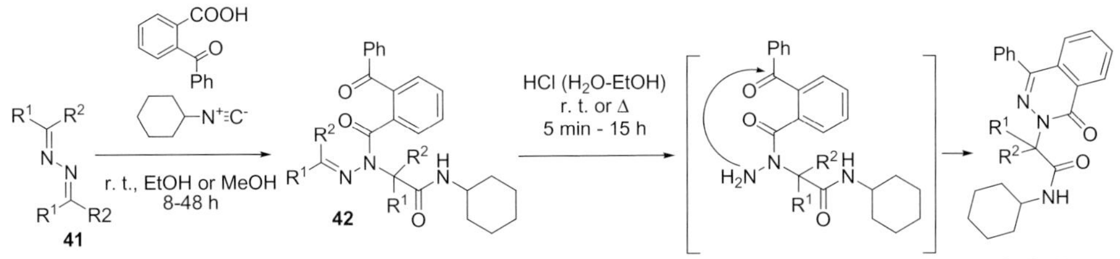

43a, R^1 = Me, R^2 = H; 43b, R^1 = Bn, R^2 = H; 43c, R^1 = R^2 = $(CH_2)_4$; 43d, R^1 = R^2 = $(CH_2)_5$; 43e, R^1 = R^2 = $CH_2CH_2OCH_2CH2$; 43f, R^1 = R^2 = $CH_2CH_2N(Ac)CH_2CH_2$

Scheme 6.23

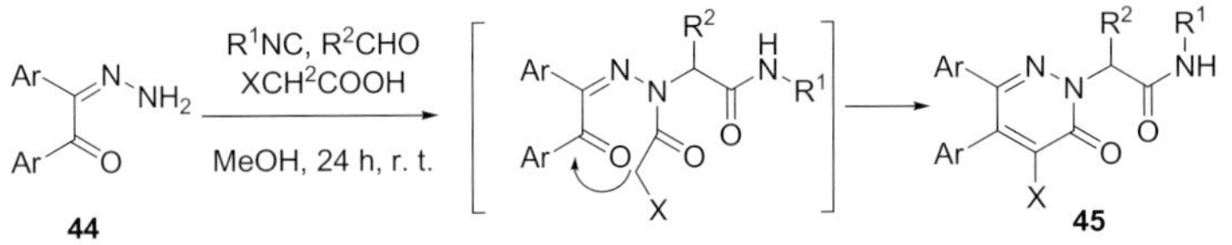

Ar = Ph, 4-MeC_6H_4, 4-$MeOC_6H_4$
R^1 = cyclohexyl, 4-MeC_6H_4
R^2 = Me. Et, Bn
X = COOEt, CN, SO_2Tol

Scheme 6.24

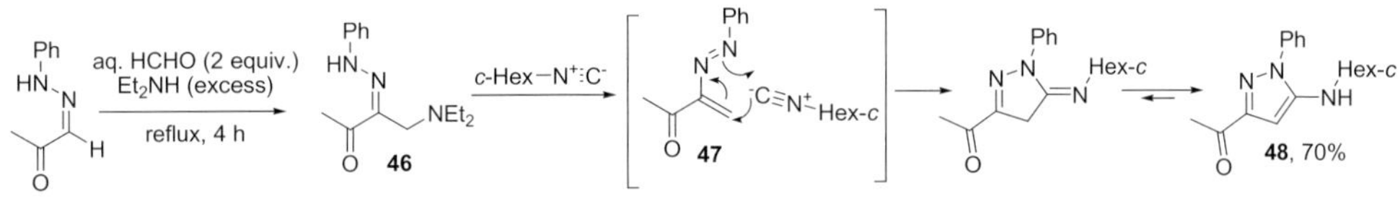

Scheme 6.25

atom of the products **43**, the present methodology might well represent the only means of accessing these compounds.

The monohydrazones of diarylglyoxal **44** can be viewed as both a protected and a functionalized form of a hydrazine. In further studies conducted by Marcaccini and coworkers [31], **44** was used as an amine replacement for the Ugi reaction of an aldehyde, isocyanide, and C–H-acidic carboxylic acids (e.g., cyanoacetic, tosylacetic, monoethyl malonate). When the four components had been united in the hydrazo-Ugi event, the product underwent a spontaneous Knoevenagel-type cyclization of the CH-acidic side chain onto the keto group of the diarylglyoxal moiety, thus providing good yields of pyridazinones **45**, "monoalkylated" with bulky acetamide side chains (Scheme 6.24). This proved to be an outstanding example of the judicious design of Ugi precursors, and led to a significant build-up of the product's molecular complexity in a single chemical operation!

El Kaim reported an interesting case of a hydrazone partner (**46**) for the reaction with isocyanides that could be conveniently prepared via a Mannich protocol [32]. Under thermal conditions, **46** essentially acted as a synthetic equivalent of (or was, in fact, a source of [33]) the azoalkene **47** and gave, on reaction with cyclohexyl isocyanide, a good yield of pyrazole **48** (Scheme 6.25).

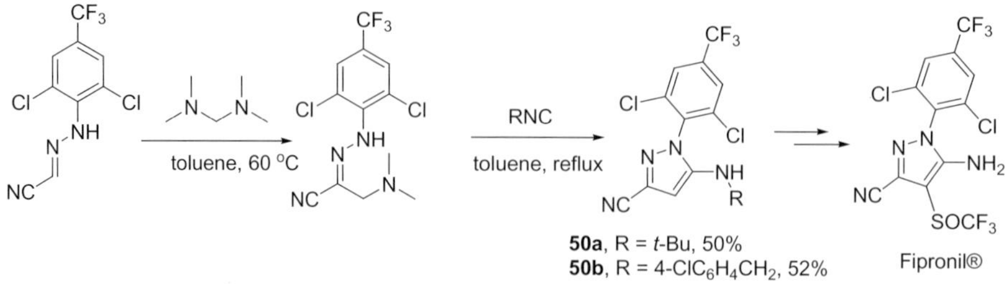

Scheme 6.26

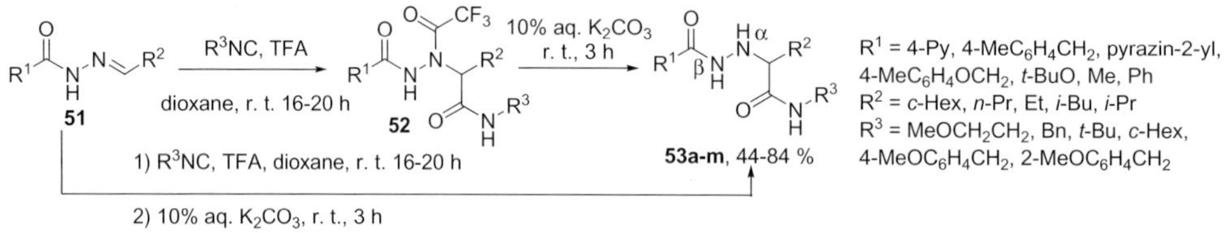

Scheme 6.27

In this process, the azoalkene **47** performed as a bifunctional reagent towards the isocyanide, providing both the electrophilic moiety (which can be viewed as an imine replacement in the Ugi reaction) and the isocyanide-intercepting N-nucleophile (which can be thought of as the caboxylate anion surrogate in the Ugi reaction). An analogous chemistry cascade can be triggered from the cyanohydrazone **49** (also Mannich reaction-derived), as demonstrated by El Kaim and colleagues [34] in their synthesis of the key pyrazole intermediates **50a–b**, *en route* to the potent insecticide Fipronil®, and its analogues (Scheme 6.26).

The synthesis of N,N'-bis(acyl)-N'-alkyl hydrazines **26**, as developed during the 1960s [20] (*vide supra*), suffered from a relatively narrow range of the workable reaction components; moreover, the resultant compounds did not offer much scope for their further modification. Recently, however, this methodology was modified and TFA employed as the carboxylic acid partner in the reaction of hydrazones **51** with isocyanides [35]. In the initial hydrazo-Ugi products **52**, the trifluoroacetyl group was found to be quite labile: in fact, it could be removed by mild alkaline hydrolysis (10% aqueous sodium carbonate solution) of either the isolated trifluoroacetyl adducts **52** (that are stable to chromatography) or *in situ* (Scheme 6.27). This seemingly small modification to the original protocol by Ugi created an avenue to the far more promising N^α-alkyl-N^β-acylhydrazine synthons **53**. Indeed, compounds **53** not only represented hydrazinopeptide-like structures but also contained a potentially reactive α-nitrogen atom that could be used to introduce further modifications of the newly formed hydrazinopeptide backbone.

The hydrazinopeptides [36] represent a less-extensively studied class of peptidomimetics that are useful for preparing more proteolytically stable [37] analogues

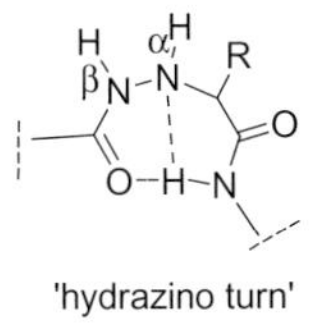

Figure 6.2 The structure of the "hydrazino turn."

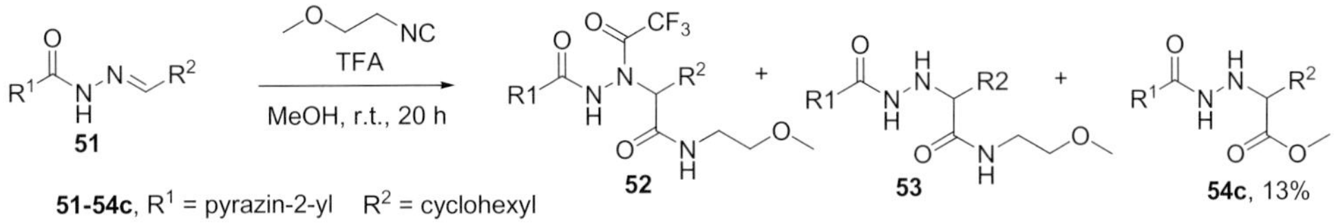

Scheme 6.28

51-54c, R¹ = pyrazin-2-yl R² = cyclohexyl
51-54d, R¹ = pyrid-4-yl   R² = n-propyl

54c, 13%
54d, 24%

of natural peptides. Occasionally, the insertion of a hydrazino linkage into the peptide backbone is not accompanied by any loss of biological activity [38].

From the points of view of peptidomimetic design and chemical biology, however – and in particular when targeting various physiologically important protein–protein interactions – the hydrazinopeptides are rather intriguing as they are known to contain a unique secondary structure bias that is referred to as a "hydrazino turn" [39]. This effect is caused by the bifurcated hydrogen bond involving the sp3-hybridized α-nitrogen atom (Figure 6.2), a situation that is quite similar to the natural peptide β-turn.

A number of diversely substituted hydrazinopeptide motifs **53** can be synthesized according to the modified protocol for the hydrazo-Ugi reaction, in good to excellent yields. Moreover, the mild basic work-up of the reaction mixture would lead to the clean and complete removal of the trifluoroacetyl group.

The choice of dioxane as a medium for the hydrazo-Ugi reactions shown in Scheme 6.27 is worthy of mention, as the same reactions when conducted in methanol produced more complex product mixtures (Scheme 6.28). Even before the basic work-up, the reactions were found to contain both the expected (**52**) and the de-trifluoroacetylated (**53**) products. In addition, substantial amounts of the methyl esters **54** were present and isolated in moderate yields, as indicated. The formation of products **53** and **54** in methanol can be rationalized by a possible interference of the nucleophilic solvent molecule during the course of the reactions, as indicated in Scheme 6.29.

One important limitation of the hydrazo-Ugi reaction with TFA has been established in that, similar to the nitrones of aromatic aldehydes (see Section 6.2), the hydrazones of aromatic aldehydes fail to undergo any reaction with isocyanides, under a variety of conditions that include the use of aqueous medium and Lewis acid catalysis [40].

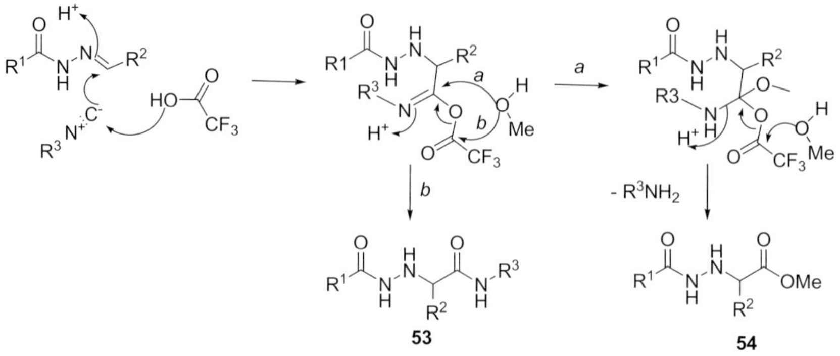

208 | *6 Amine (Imine) Component Surrogates*

Scheme 6.29

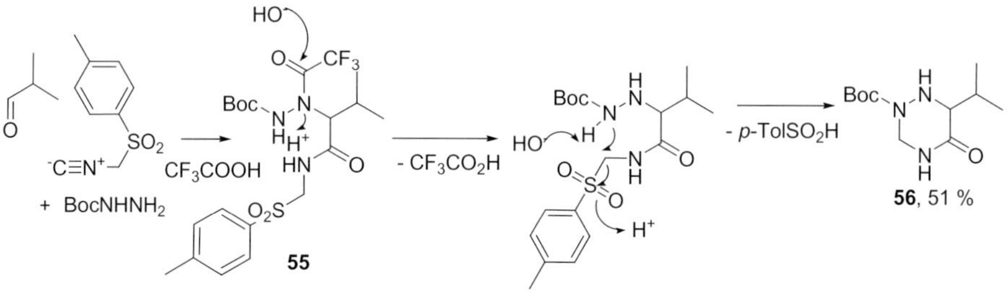

Scheme 6.30

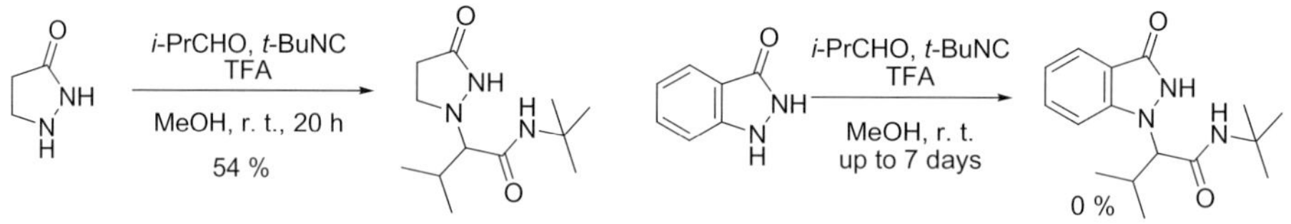

Scheme 6.31

An interesting result was obtained in the hydrazo-Ugi reaction of Boc-hydrazine with isobutyraldehyde, TFA, and tosylmethyl isocyanide (TosMIC). While the desired adduct **55** was indeed detected in the reaction mixture (using liquid chromatography–mass spectrometry), the alkaline treatment of the reaction mixture led not only to the expected removal of the trifluoroacetyl group but also to an intramolecular substitution of the tosyl group to provide the substituted 1,2,4-triazinane **56** (Scheme 6.30) [40].

A similar hydrazo-Ugi reaction protocol was applied to the two cyclic *N*-acylhydrazines, pyrazolidin-3-one and 1,2-dihydro-3*H*-indazol-3-one (Scheme 6.31). Whereas the former compound was found to be quite a workable partner

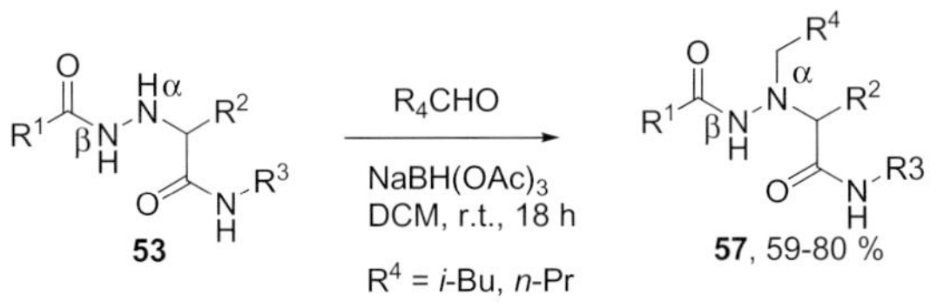

Scheme 6.32

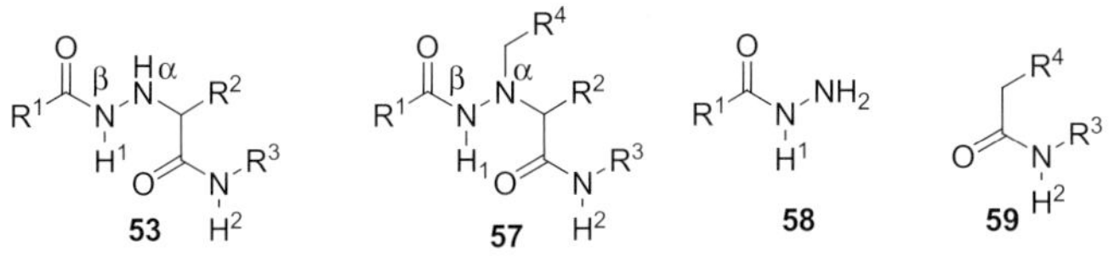

Figure 6.3 Hydrazinopeptide and reference fragments for the ^{1}H NMR study.

in this reaction (and also in the method itself, providing a facile access to monoalkylated pyrazolydin-3-ones), the latter compound was shown to be relatively inert, most likely due to the lower reactivity of the "anilinic" nitrogen atom of the hydrazine [40].

As mentioned above, the α-nitrogen atom of the compounds **53** may serve as a potential reactive center, and used for the introduction of further diversity elements of the hydrazine moiety. Although reductive alkylation was found not to be feasible with aromatic or heteroaromatic aldehydes (thus substantially limiting the scope of this important hydrazinopeptide modification), $N^α$, $N^α$-dialkyl, and $N^β$-acyl hydrazines **57** were successfully prepared in good to excellent yields, using isovaleraldehyde and butyraldehyde (Scheme 6.32).

Having prepared two sets of hydrazinopeptide fragments (**53** and **57**), curiosity took over to determine if the notorious structural bias – the "hydrazino turn" – might be present and detectable in these short, hydrazinodipeptide motifs. To this end the following, rather straightforward, biophysical model was applied. First, the changes in the ^{1}H NMR spectroscopic chemical shift of the protons in **53** and **57** that were suspected of participating in the intramolecular hydrogen bonding, were compared with the same changes in the reference fragments **58** and **59** (Figure 6.3), in which no such bonding was possible, in deuterated dimethylsulfoxide (DMSO-d_6) and chloroform-d (CDCl$_3$).

Indeed, as can be seen from the diagram in Figure 6.4, the use of DMSO-d_6 as a solvent capable of accepting a hydrogen bond from the solute caused a downfield shift of the acidic, amide-type protons H^1 and H^2 in the reference fragments **58** and **59**, respectively [35]. Roughly the same downfield shift of the ^{1}H NMR spectroscopic signals was observed for the $N^β$–H^1 protons of compounds **53** compared to those of **58**, but this was not the case for the terminal amide protons H^2, which were significantly less sensitive to the solvent change than the corresponding H^2

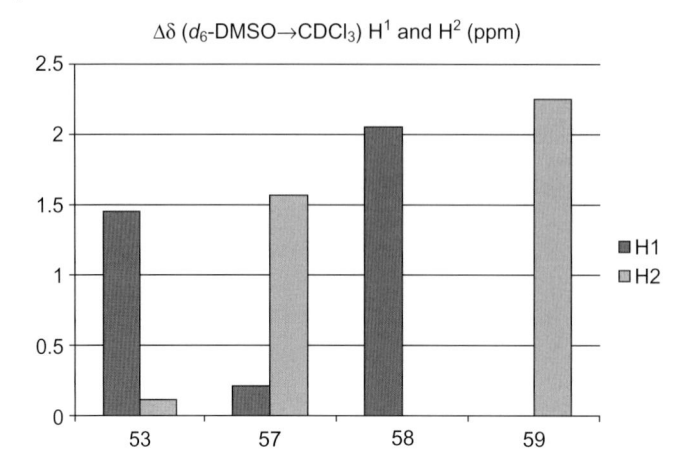

Figure 6.4 Differences in chemical shift observed for **53, 57–59** upon solvent change.

protons in **59**. This could be explained only by an involvement of the H^2 in the terminal amide of the compounds **53** in an intramolecular hydrogen bond. The situation was reversed for compounds **57**, which prompted the conclusion that the introduction of a second alkyl group at the α-nitrogen created an alternative hydrogen-bonding pattern that involved H^1 (but not H^2) and which, on this occasion, was more favorable.

The above conclusions were fully confirmed by single-crystal X-ray analyses of compounds **53i** and **57h** [35], as representatives of each set of the hydrazinopeptide units studied. Indeed, while compound **53i** adopted the typical hydrazino turn secondary structure, this structural bias was not observed for the N^α,N^α-dialkyl derivative **57h** (Figure 6.5). These observations are of great importance when selecting appropriate hydrazinopeptide inserts for natural peptide analogue design.

The reactive α-nitrogen atom of the hydrazinodipeptides **53** can, in turn, be used as a surrogate amine input for the second hydrazo-Ugi reaction. This led to a practical realization of the sequential IMCR concept, which provides a powerful means of building up a product's molecular diversity in an operationally simple, reliable, predictable, and streamlined manner. For example, compounds **53h** and **53c** may be reacted with a new combination of an isocyanide and an aldehyde to produce compounds **60** (Scheme 6.33).

These compounds contain two stereocenters, and were formed with no appreciable stereocontrol [41]. It would appear, however, that the formation of difficult-to-separate diastereomeric mixtures can be avoided if nonprochiral carbonyl inputs are used for the second hydrazo-Ugi reaction. As shown in Scheme 6.34, these products (e.g., **61**) can be obtained in good yields if excess amounts of either paraformaldehyde or acetone (combined with an equimolar quantity of an isocyanide) is used. In particular, the second hydrazo-Ugi reaction proceeds quite well with methanol as solvent.

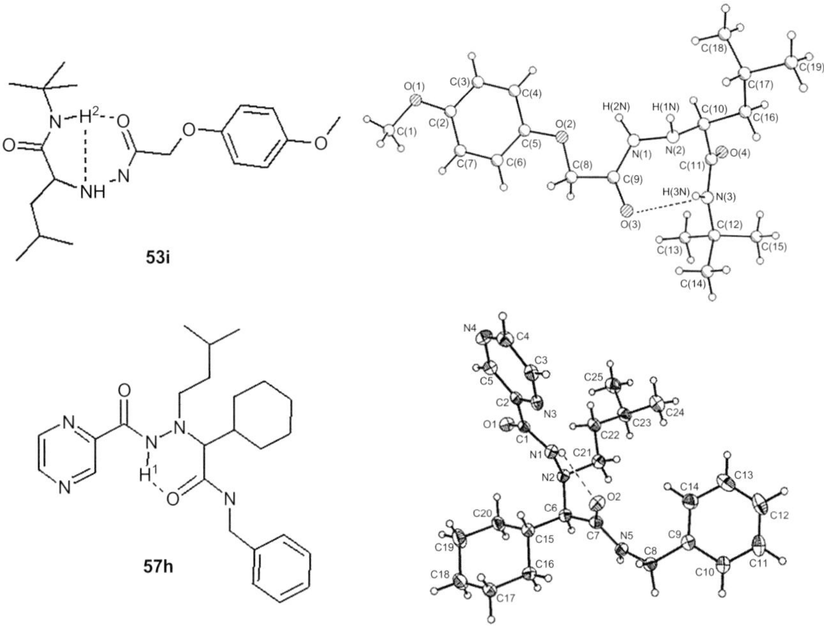

6.3 Hydrazine Components in the Ugi Reaction | 211

Figure 6.5 Predicted secondary structures of **53i** and **57h**, as confirmed by X-ray analysis.

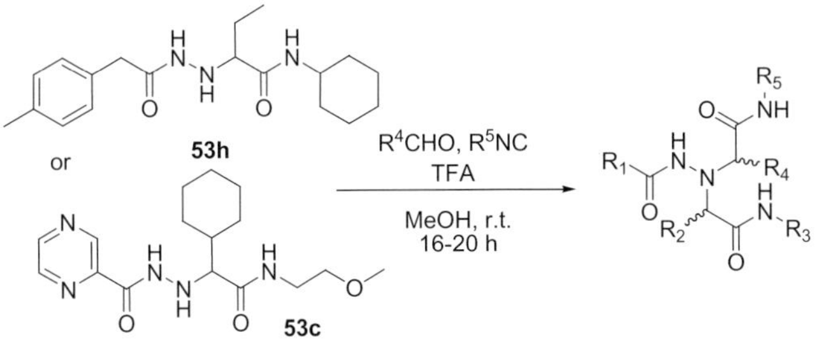

Scheme 6.33

60a, R^1 = 4-MePhCH2, R^2 = Et, R^3 = cyclohexyl, R^4 = *i*-Pr, R^5 = *t*-Bu, 62%

60b, R^1 = pyrazinyl, R^2 = cyclohexyl, R^3 = MeOCH$_2$CH$_2$, R^4 = *n*-Pr, R^5 = 4-FPhCH$_2$, 56%

60c, R^1 = pyrazinyl, R^2 = cyclohexyl, R^3 = MeOCH$_2$CH$_2$, R^4 = Et, R^5 = 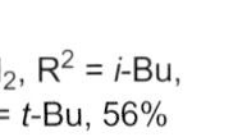 , 48%

Scheme 6.34

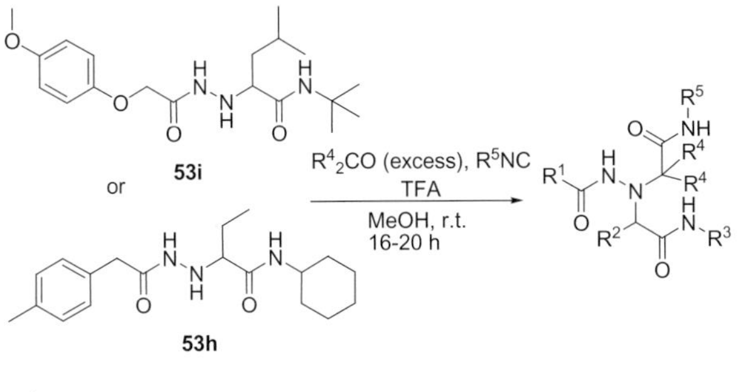

60d, R^1 = 4-MeOPhCH$_2$, R^2 = *i*-Bu, R^3 = *t*-Bu, R^4 = H, R^5 = *t*-Bu, 56%

60e, R^1 = 4-MeOPhCH$_2$, R^2 = *i*-Bu, R^3 = *t*-Bu, R^4 = Me, R^5 = MeOCH$_2$CH$_2$CH$_2$, 68%

60f, R^1 = 4-MePhCH$_2$, R^2 = Et, R^3 = cyclohexyl, R^4 = Me, R^5 = *t*-Bu, 60%

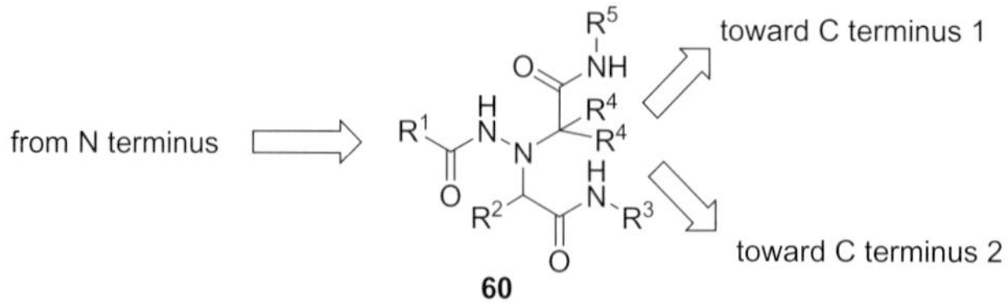

Figure 6.6 New types of "forked" peptidomimetics.

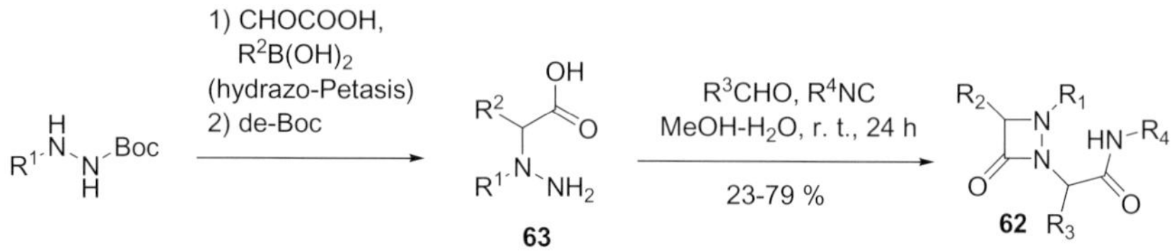

Scheme 6.35

Although synthesized in only two chemical operations from simple precursors, compounds **60** contain *five* elements of diversity that originate from an acylhydrazine, two different carbonyl components, and two different isocyanide components, all of which are introduced sequentially and in a regiochemically controlled manner. The hydrazinopeptide units **60** represent a conceptually novel class of peptidomimetic structures that are capable of introducing a "fork" into the polypeptide chain, after which the latter may take two alternative courses towards the C terminus (Figure 6.6).

If two reaction partners of the Ugi process are combined within the structure of a single reactant, then a ring-forming process may possibly occur, and this strategy has been exploited very successfully in traditional Ugi reactions [7, 42–45]. The use of bifunctional reagents in the hydrazo-Ugi reaction is far less common, however.

Recently, the Chembiotek group in India, in collaboration with Procter & Gamble Pharmaceuticals, disclosed an elegant synthesis of aza-β-lactams **62** via an "intramolecular" hydrazo-Ugi reaction, using α-hydrazino acids **63** as the Ugi amine surrogates [46]. The latter compounds had been prepared previously by the same research group [47], using a "hydrazo-Petasis" approach. As shown in Scheme 6.35, those α-hydrazino acids which already contained two elements of diversity yielded a library of aza-β-lactams **62** on exposure to a matrix of aldehydes and isocyanides. However, no control of the relative stereochemistry could be achieved in these reactions.

Subsequently, compound **64** was designed which contained both the hydrazone and carboxylic moieties, separated by a flexible and sufficiently long linker to allow for the intramolecular hydrazo-Ugi process. In analogy to the synthesis of aza-β-lactams from α-hydrazino acids (as described above), compound **64** produced a series of N-monoalkylated tetrahydropyridazine-3,6-diones **65a–c** in good yields

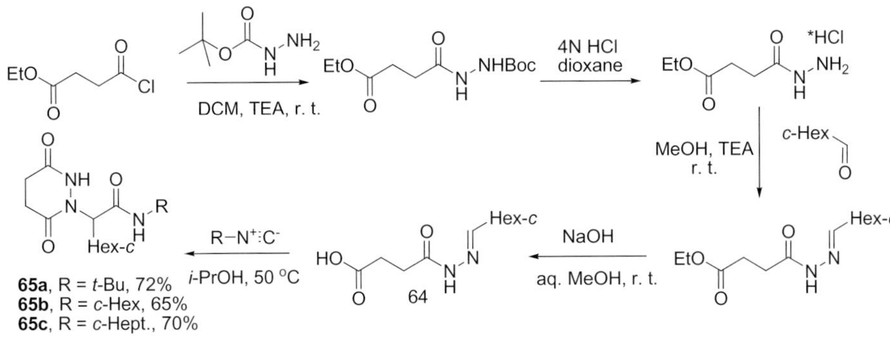

Scheme 6.36

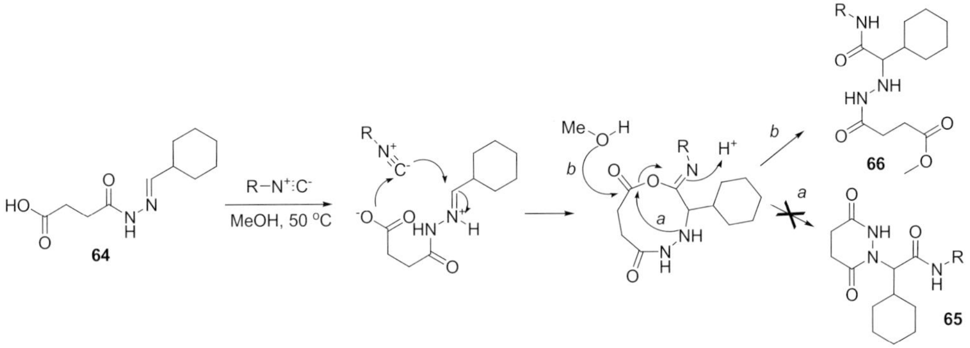

Scheme 6.37

upon an "intramolecular" hydrazo-Ugi reaction with isocyanides, in isopropyl alcohol as solvent (Scheme 6.36).

Notably, the use of isopropyl alcohol as solvent led to very low yields of the target cyclic products **65**, though when the same reaction was performed in methanol the methyl ester **66** was identified as the main product. Formation of the ester **66** can be rationalized by the interference of methanol during the course of the intramolecular hydrazo-Ugi reaction (Scheme 6.37). Thus, any unwanted reaction pathways in the intramolecular MCR processes can be suppressed by adjusting the nucleophilicity of the reaction medium.

Another example of applying bifunctional reagents in the hydrazo-Ugi reaction has been recently reported [48]. In this case, a series of ketocarboxylic acids **66** underwent a facile "intramolecular" hydrazo-Ugi reaction with monoprotected hydrazines and isocyanides, under novel reaction conditions (NH$_4$Cl in aqueous methanol) that had been identified previously in an optimization study (Scheme 6.38). Mechanistically, the hydrazo-Ugi reaction towards **67** is thought to follow the expected course.

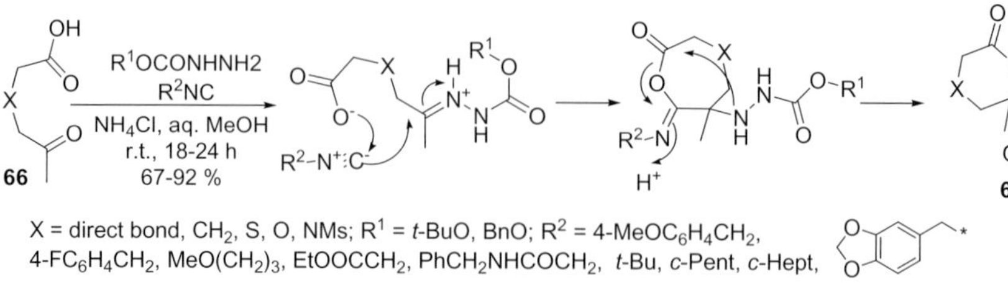

Scheme 6.38

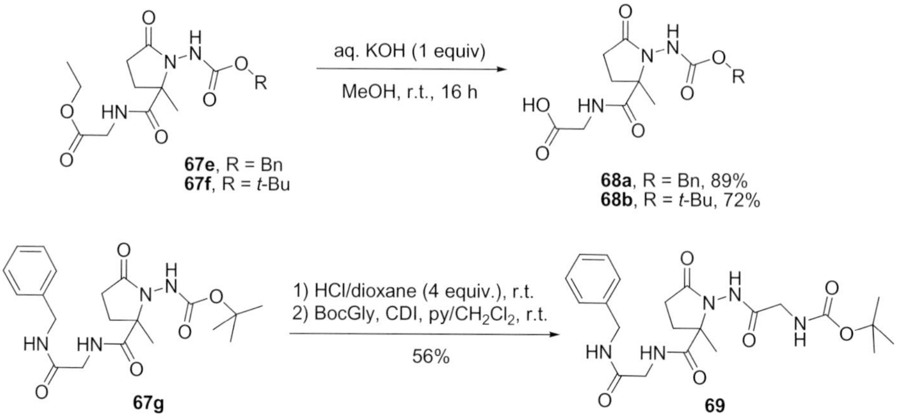

X = direct bond, CH_2, S, O, NMs; R^1 = t-BuO, BnO; R^2 = 4-$MeOC_6H_4CH_2$, 4-$FC_6H_4CH_2$, $MeO(CH_2)_3$, $EtOOCCH_2$, $PhCH_2NHCOCH_2$, t-Bu, c-Pent, c-Hept,

Scheme 6.39

The presence of a pronounced hydrogen bond between the carbonyl oxygen atom of the alkoxycarbonyl group and the exocyclic secondary amide side chain renders compounds **67** similar to proline, not only from the viewpoints of their shape and chemical structure, but also in terms of their potential ability to mimic β-turns when introduced into a polypeptide chain – a secondary bias that typically is associated with proline in natural peptides.

N-aminolactams **67** may also be elaborated into useful proline-like building blocks, as well as being incorporated into short peptide structures. As shown in Scheme 6.39, compounds **67e** and **67f** were hydrolyzed by aqueous KOH in methanol into the respective "glycine-ΨPro" carboxylic acids **68a–b**, without disruption of the lactam ring or the exocyclic amide bond. Likewise, the *N*-Boc group was easily removed from the *N*-terminus of the *N*-aminolactam, which was then acylated with Boc-protected glycine, as illustrated by the conversion of **67g** into the "BnNHGly–ΨPro–GlyBoc" tripeptoid **69**. Moreover, it was shown that the intramolecular hydrogen bonding pattern, which led to a β-turn-like secondary structure, also continued to persist in **69**.

In addition to being used in peptidomimetic design, the hydrazinopeptoid fragments synthesized via the hydrazo-Ugi reaction (as dipeptoids **53**) may also be

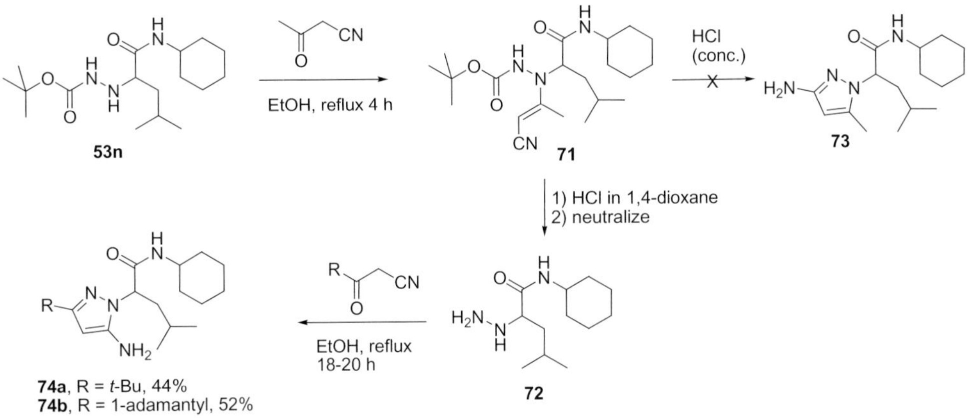

Scheme 6.40

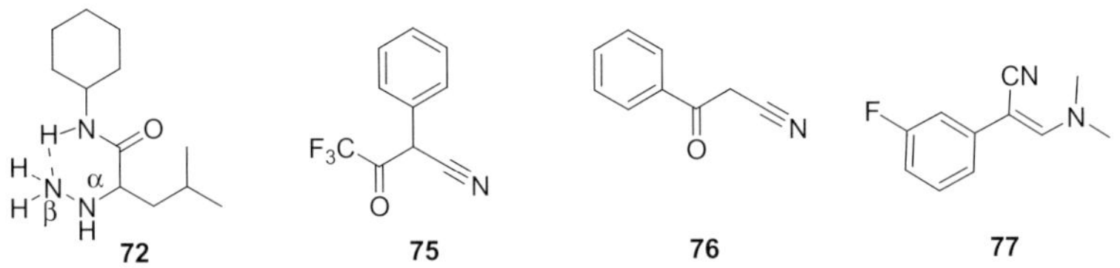

Figure 6.7 Inert reaction partners in attempted pyrazole synthesis.

viewed as highly elaborate but easily synthesized hydrazine synthons. Such fragments can, in principle, be used as prefunctionalized reagents in heterocycle syntheses that involve hydrazine units.

Initially, significant limitations were encountered to this approach [49]. For example, when attempts were made to condense a hydrazine fragment **53n** with cyanoacetone, the addition of an equimolar amount of concentrated HCl led to the formation of the enamine adduct **71**, and also to (DL)-N'-cyclohexyl-N-aminoleucinamide **72**, rather than to the expected aminopyrazole **73**. Nonetheless, it proved possible to condense **72** directly with aliphatic α-cyanoketones to produce moderate yields of the single regioisomers of pyrazoles **74a,b** (Scheme 6.40).

The range of suitable condensation partners, even for compound **72**, appeared rather limited, however. For instance, no reaction was observed with the bis-electrophiles **75–77** that previously had been condensed quite efficiently with hydrazine. The reasons for the inert character of **72** in these condensations were unclear; it may have been due to the steric bulk on one of the hydrazine nitrogen atoms, but this should not have precluded the compound from reacting at N^β. It is also possible that the secondary amide functionality also deactivated N^β, via an intramolecular hydrogen bonding (Figure 6.7).

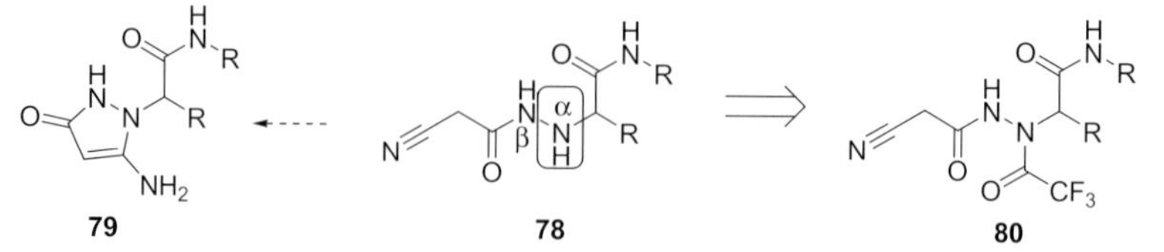

Figure 6.8 Investigated synthetic approach to **79** via hydrazino-Ugi reaction products **80**.

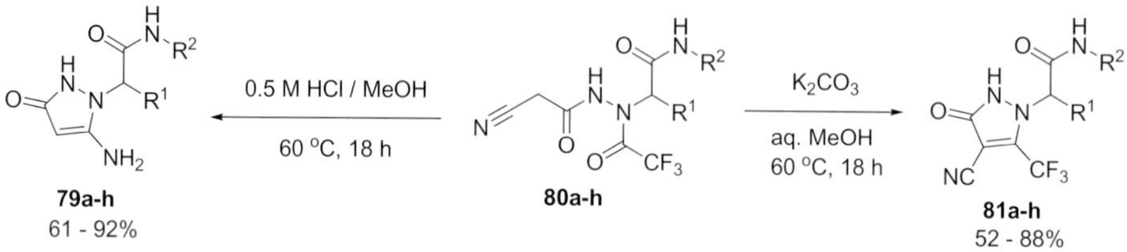

Scheme 6.41

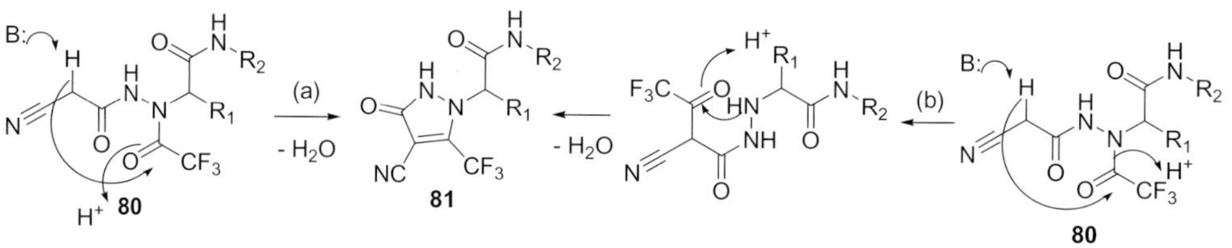

Scheme 6.42

Prompted by these observations, the pyrazole precursors **78** were designed that contained both an electrophilic nitrile functionality and a reactive N^α atom [49]. These precursors for intramolecular cyclization (presumably, into 1-alkyl-5-amino-1,2-dihydro-3*H*-pyrazol-3-ones; **79**) could be derived from the hydrazino-Ugi reaction products **80** by a simple removal of the trifluoroacetyl group (Figure 6.8). This methodology was expected to explore the potential of post-condensation modifications of the hydrazino-Ugi core, an approach that previously had yielded numerous novel heterocyclic scaffolds when applied to classical Ugi products. However, when a removal of the trifluoroacetyl group in **80** was attempted, the isolated products were the unexpected (and hitherto undescribed) 1-alkyl-3-oxo-5-(trifluoromethyl)-2,3-dihydro-1*H*-pyrazole-4-carbonitriles **81**. In contrast, an acidic treatment led to the expected de-trifluoroacetylation and formation of the pyrazol-3-ones **79** (Scheme 6.41) [49].

Mechanistically, the unexpected formation of **81** under basic conditions can be rationalized by a deprotonation at the activated methylene group of **80**, with subsequent cyclization onto the CF$_3$CO group to give, after dehydration, the observed pyrazol-3-one **81**. Given the labile character of the trifluoroacetamide, the cyclization may be preceded by the N → C migration of the CF$_3$CO group (Scheme 6.42).

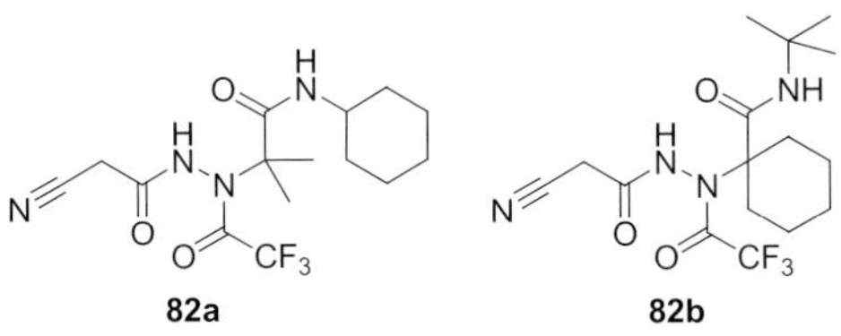

Figure 6.9 Sterically hindered hydrazines.

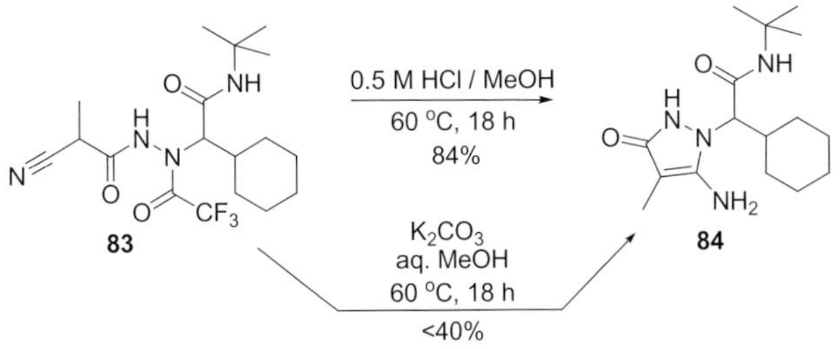

Scheme 6.43

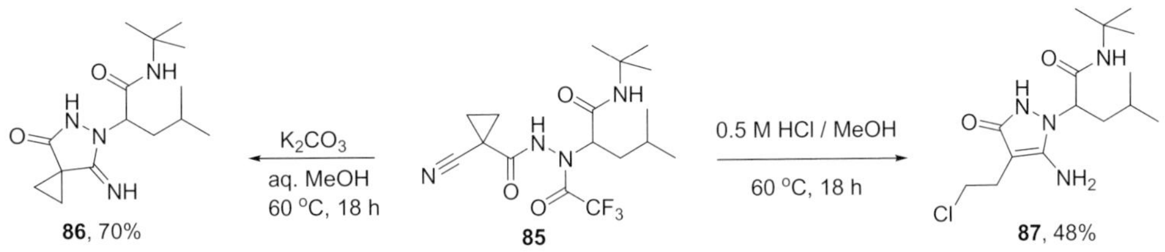

Scheme 6.44

The ability of **80** (or **78**) to undergo cyclization into the pyrazol-3-one should depend on the steric bulk, both at the activated methyl group of the cyanoacetyl side chain and at N^α. The products **82a,b** (Figure 6.9), when synthesized via the hydrazo-Ugi reaction from acetone and cyclohexanone as the carbonyl component, respectively, were completely inert towards such cyclization under either acidic or basic treatment, and provided only the respective de-trifluoroacetylated products.

Moreover, even a small substituent such as a methyl group, when located at the methylene group of **79** (which, presumably, was involved in a cyclization path leading to **81**) can affect the outcome of further cyclizations. The compound **83** (also prepared via a hydrazino-Ugi reaction) still provided an excellent yield of pyrazol-3-one **84** when treated with dilute acid (Scheme 6.43).

An interesting result was obtained with the cyclopropane derivative **85** which, under basic conditions, cyclized to provide a good yield of 6-alkyl-7-imino-5,6-diazaspiro[2.4]heptan-4-one **86**. An acidic treatment of **85** resulted in cyclopropane ring-opening and the isolation of 2-chloroethyl derivative **87**, in fair yield (Scheme 6.44) [49].

Thus, it became apparent that those hydrazino dipeptoids which are accessible in one-step fashion via the hydrazo-Ugi reaction may find a remarkably broad usage in Knorr-type pyrazolone syntheses, delivering skeletally diverse compounds with high efficiency.

6.4
Miscellaneous Amine Surrogates for the Ugi Reaction

During the early studies conducted by Zinner and Bock, diaziridine **88** was shown to react with formaldehyde, hydrazoic acid, and cyclohexyl isocyanide to produce a low yield of the diaziridine tetrazole derivative **89** [50]. The latter compound, not unexpectedly, is prone to diaziridine ring rupture on acidic treatment, providing a quantitative yield of the hydrazone **90** (Scheme 6.45).

An intriguing – and, to date, a unique – example of urea being used as an amine component in the Ugi reaction was described [51], whereby an unspecified set of an isocyanide and a carboxylic acid, urea and glyoxylic acid hemiacetal **91** yielded a respectable yield of N-aminocarbonyl amide **92** (Scheme 6.46). The straightforward formation of this product is, in the present author's opinion, difficult to justify in view of the generally accepted mechanism of the reaction (see Scheme 6.2).

Similarly, when considering the mechanism of the Ugi reaction, it would be surprising if primary sulfonamides, as poor N-nucleophiles, were to react as valid alternatives for an amine component in the Ugi reaction. Indeed, *p*-toluenesulfonamide was shown to be almost completely inert towards a combination of an aldehyde, an isocyanide, and acetic acid in methanol–THF (tetrahydrofuran) solution [52]. However, when a Rink resin-bound benzenesulfonamide **93** was reacted with an excess quantity of aldehydes, isocyanides, and acetic acid in a THF–methanol solvent system, a mixture of monoacetylated resin-

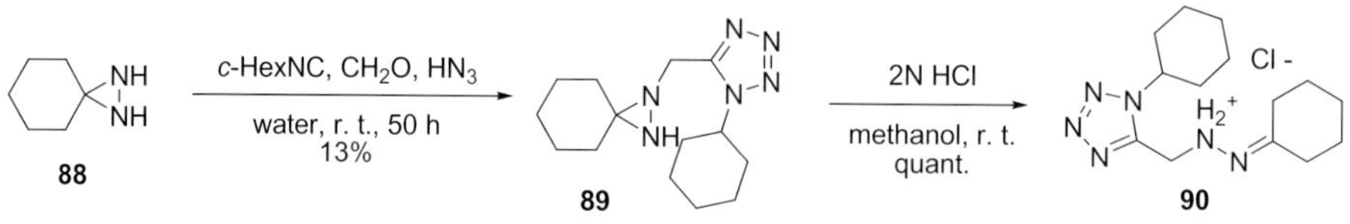

Scheme 6.45

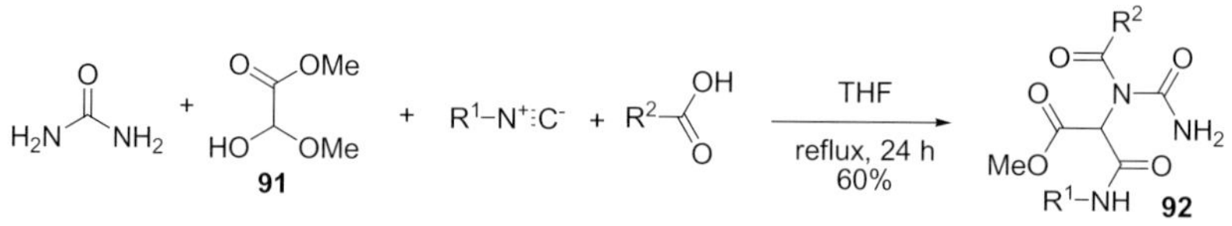

Scheme 6.46

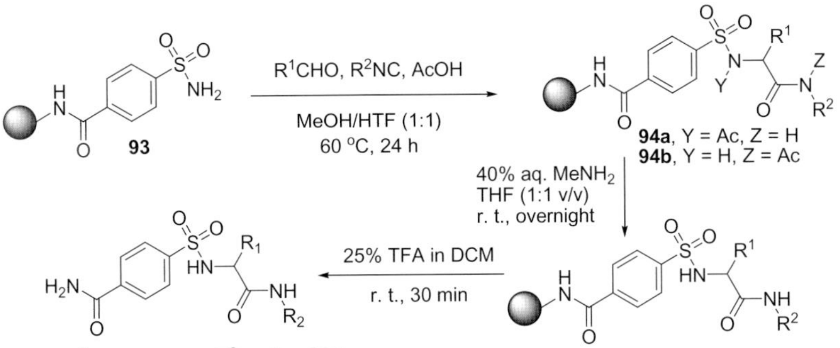

Scheme 6.47

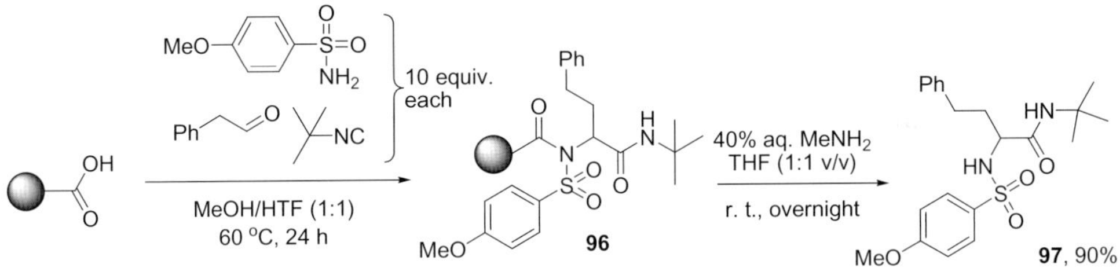

Scheme 6.48

bound products **94a** and **94b** was formed. The latter material was deacetylated with aqueous methylamine solution (in THF) and subjected to TFA cleavage, thus providing excellent yields of the Ugi reaction products (a representative example **95** is shown in Scheme 6.47) [52].

By using carboxypolystyrene as the carboxylic acid input for the Ugi reaction, polymer-bound products **96** were obtained and converted, via traceless cleavage with methylamine in aqueous THF, into pure products **97**. These were also isolated in excellent yields, as shown by the example in Scheme 6.48. The reason for using a polymer support to recover any otherwise unreactive sulfonamides as valid inputs for the Ugi reaction, as investigated by the Advanced SynTech research group, remains unclear however.

The semicarbazone of cyclohexanone (as well as other ketones and aldehydes) was used successfully by Marcaccini and Torroba as a surrogate imine for an Ugi-type reaction (Scheme 6.49) [53]. In this case, and according to single-crystal X-ray analyses, the resulting compounds (e.g., **98**) contained a pronounced intramolecular hydrogen bond, quite similar to certain hydrazinodipeptide fragments as described above (see Section 6.3). The choice of benzoylformic acid as a carboxylic acid input towards **98** (as well as eight other examples) was strategic. On treatment with an ethanolic sodium ethoxide solution, the terminal semicarbazide moiety

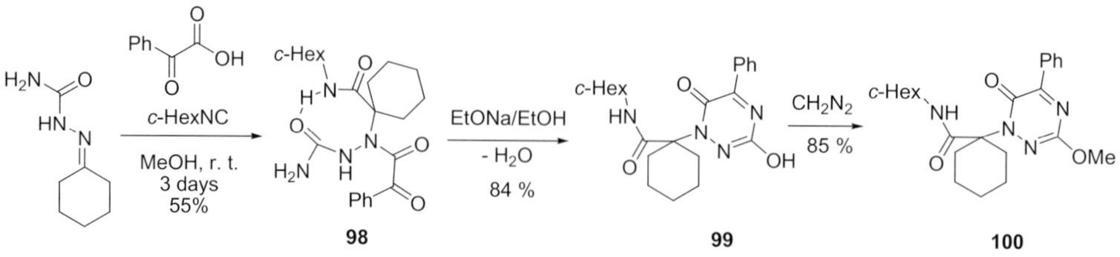

Scheme 6.49

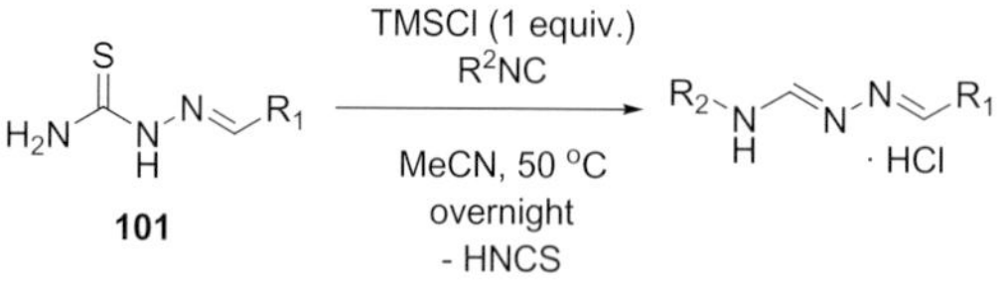

102a, R^1 = 2-MeC_6H_4, R^2 = $C(CH_3)_2CH_2C(CH_3)_3$, 53%
102b, R^1 = 3-MeC_6H_4, R^2 = $C(CH_3)_2CH_2C(CH_3)_3$, 61%
102c, R^1 = 4-MeC_6H_4, R^2 = $C(CH_3)_2CH_2C(CH_3)_3$, 73%
102d, R^1 = 4-$MeOC_6H_4$, R^2 = *t*-Bu, 85%

Scheme 6.50

became cyclized onto the keto group of the benzoylformyl moiety, so as to provide the 1,2,4-triazine derivative **99** (the latter was O-methylated, for easier characterization, to yield 3-methoxy-1,2,4-triazine, **100**). Compounds **99** and **100** represent a novel class of pseudopeptidic 1,2,4-triazines that are most likely to be used in the design of peptidomimetic compounds with biological activity.

A strikingly different result was observed, however, when the thiosemicarbazones of aromatic aldehydes **101** were reacted with isocyanides in the presence of an equimolar amount of trimethylsilyl chloride. (This silicon Lewis acid has been used successfully as an effective promoter of IMCRs [54, 55].) The resulting products were the poorly described [56] azadienes **102**, isolated in good to excellent yields as the hydrochloride salts (Scheme 6.50).

6.5
Activated Azines in Reactions with Isocyanides

When the aromaticity of azines is perturbed by interactions of the ring nitrogen atom with various electrophilic agents, these heterocycles become activated towards nucleophiles in a fashion similar to the activation of imine species for nucleophilic addition by Lewis or Brønsted acids. Thus, it is to be expected that activated azines would be reactive towards isocyanides (in the presence of suitable isocyanide-intercepting nucleophiles) and thus essentially to serve as viable surrogates for an imine component for the Ugi reaction.

The first example [57] of such a reaction design was reported by Ugi, whereby *N*-alkylated quinolines **103** were reacted with isocyanides and carboxylic acids to

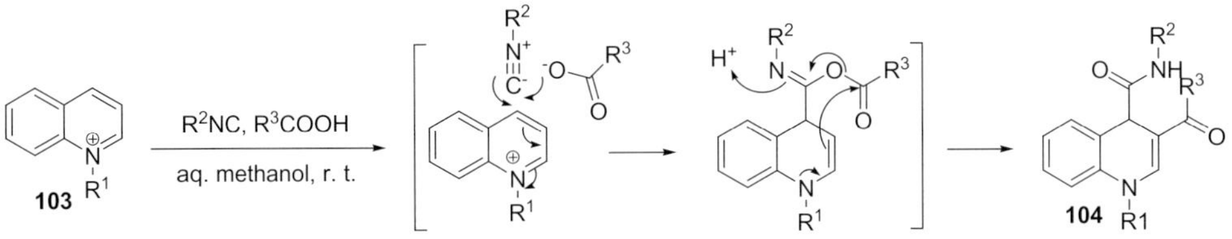

6.5 Activated Azines in Reactions with Isocyanides | 221

Scheme 6.51

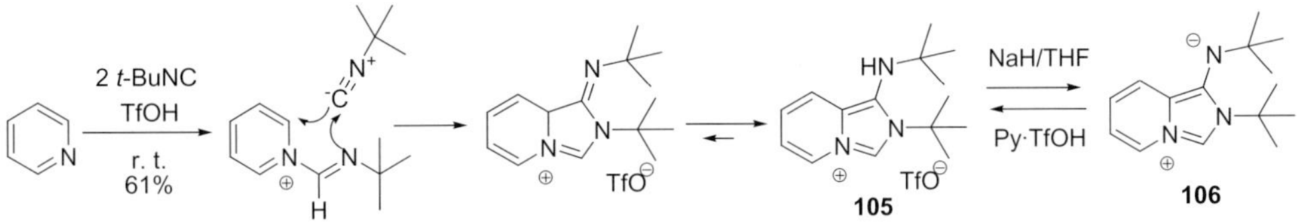

Scheme 6.52

provide **104** – that is, the products of isocyanide addition at the position 4 of the quinoline nucleus (Scheme 6.51).

Berthet et al. [58] established that the addition of triflic acid to a mixture of pyridine and an excess amount of *tert*-butyl isocyanide would lead to the formation (in good isolated yield) of imidazo[1,5-a]pyridin-4-ium triflate **105**. Mechanistically, the α-addition of pyridinium species onto the first molecule of isocyanide (i.e., azine activation) followed by a formal [4+1] cycloaddition with the second molecule of isocyanide, is believed to be key (Scheme 6.52). The latter event is, essentially, an α-addition that is typical for isocyanides, while the product can be converted to the mesoionic compound **106** on treatment with sodium hydride in THF at room temperature. In turn, **106** can easily be converted back to **105** on the addition of pyridinium triflate.

An interesting example of azine activation was discovered by Mironov et al., who described an excellent approach to the combinatorial search for new MCRs [59]. These authors showed that the mixing of isoquinoline with benzylidenemalonodinitrile and adamant-1-yl isocyanide in diethyl ether provided, after two days of stirring, an 82% yield of dihydropyrrolo[2,1-a]isoquinoline-2,2(3H)-dicarbonitrile **107**. This reaction, which subsequently was extended to a range of isocyanides and electron-poor olefins, is believed to involve an isoquinoline activation by benzylidenemalonodinitrile towards interaction with the isocyanide that is, in turn, intercepted by the neighboring dinitrile anion (Scheme 6.53).

A more systematic exploration of isoquinoline activation towards the reaction with isocyanides by acylating agents was undertaken by Lavilla and coworkers [60]. In an initial optimization study, a stoichiometric amount of *tert*-butyl isocyanide was added to a solution of isoquinoline and methyl chloroformate in dichloromethane, at room temperature. Compound **10** – which was essentially the product of the

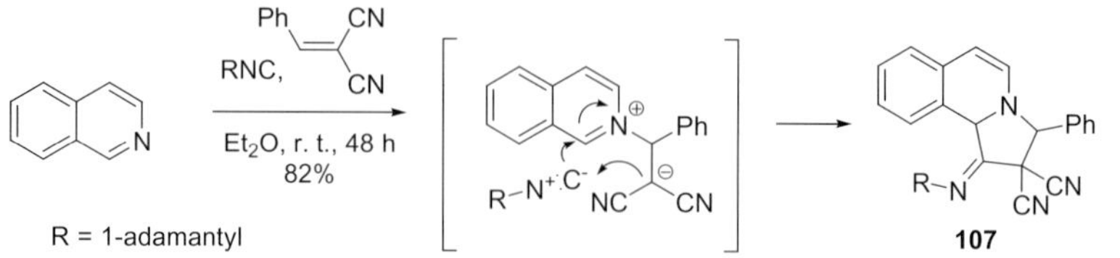

Scheme 6.53

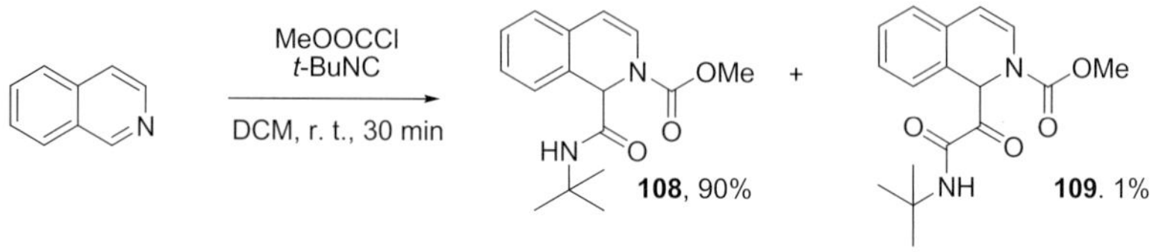

Scheme 6.54

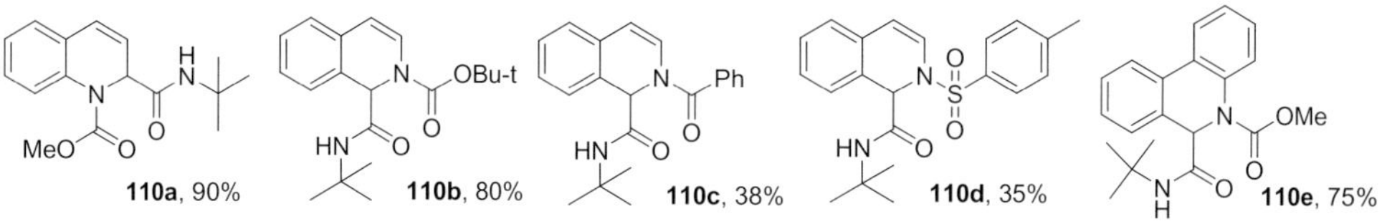

Figure 6.10 Products of the reaction of *t*-BuNC with various activated azines.

intended 2-carbamoylation – was isolated in 90% yield, along with a small amount (1%) of oxoamide **109** that presumably arose from the double addition of isocyanide and imine hydrolysis (Scheme 6.54).

This protocol was found to be applicable to a variety of quinolines, isoquinolines, and phenanthridine as azine substrates, and to a range of activating agents (e.g., benzyl and allyl chloroformates, Boc$_2$O, benzoyl chloride, tosyl chloride), thus providing a wide diversity of carbamoylated dihydroazines **110a–e** (Figure 6.10). Moreover, this elegant azine carbamoylation strategy could be amended for solid-phase-supported formats.

Pyridine can also be activated towards the reaction with isocyanides via N-fluorination, as demonstrated by Kiselyov [61]. In this case, N-fluoropyridinium fluoride reacted with a range of isocyanides and trimethylsilyl azide (as a source of isocyanide-intercepting azide nucleophiles) to provide good yields of tetrazol-5-yl pyridines **111**, along with the simple 2-carbamoylation products **112** and a minor tetrazolo[1,5-a]pyridine byproducts **113** (Scheme 6.55).

This protocol was also shown to be applicable to quinoline and isoquinoline activation, providing a straightforward (but only moderately yielding) access to 2-(tetrazol-5-yl)quinolines and isoquinolines, respectively.

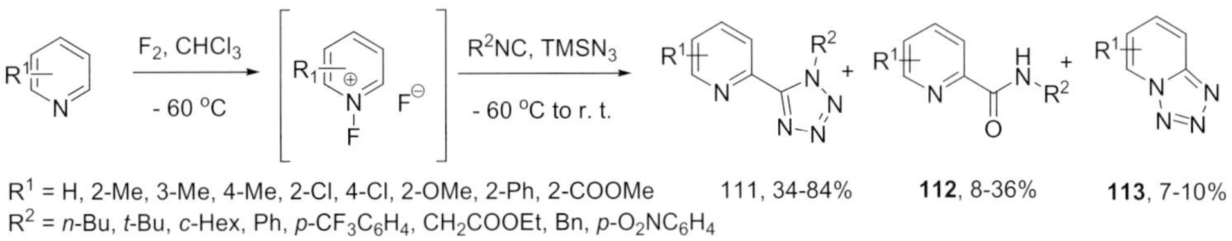

6.6 Enamines, Masked Imines, and Cyclic Imines in the Ugi Reaction

R^1 = H, 2-Me, 3-Me, 4-Me, 2-Cl, 4-Cl, 2-OMe, 2-Ph, 2-COOMe   **111**, 34–84%   **112**, 8–36%   **113**, 7–10%
R^2 = n-Bu, t-Bu, c-Hex, Ph, p-CF$_3$C$_6$H$_4$, CH$_2$COOEt, Bn, p-O$_2$NC$_6$H$_4$

Scheme 6.55

6.6
Enamines, Masked Imines, and Cyclic Imines in the Ugi Reaction

Lavilla and coworkers used readily available N-alkyldihydropyridines **114** as enamines (or, under acidic conditions, masked versions of cyclic iminium ions, **115**) that were capable of reacting with isocyanides in the presence of a stoichiometric quantity of sulfonic acid (e.g., methanesulfonic, (±)-camphorsulfonic, p-toluenesulfonic) to provide good to excellent yields of carbamoylated products **116** (Scheme 6.56). Carbamoylation, as expected, occurred at the more electron-rich enamine moiety. A similar reactivity was observed for 2H-dihydropyrans serving as carbonyl compound surrogates for the Passerini reaction [62].

Similar to the protonation of **114**, cyclic iminium species (e.g., **115**) may also be generated on the interaction of **114** with other electrophilic species (e.g., Br$_2$, ICl, PhSeCl). Some examples of the successful halo- and seleno-carbamoylation of a representative dihydropyridine **114**, as achieved by Lavilla and colleagues, are shown in Scheme 6.57 [63].

Oxazolidines, imidazolidines, and 1,3-thiazolidines (**117–119**) that are easily obtained via the condensation of a carbonyl compound with β-aminoalcohols,

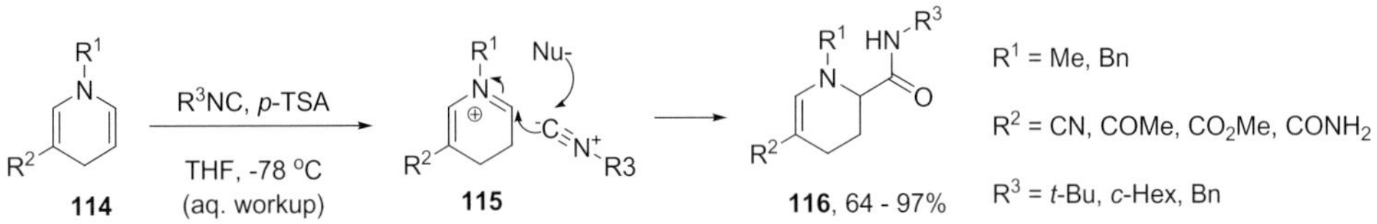

Scheme 6.56

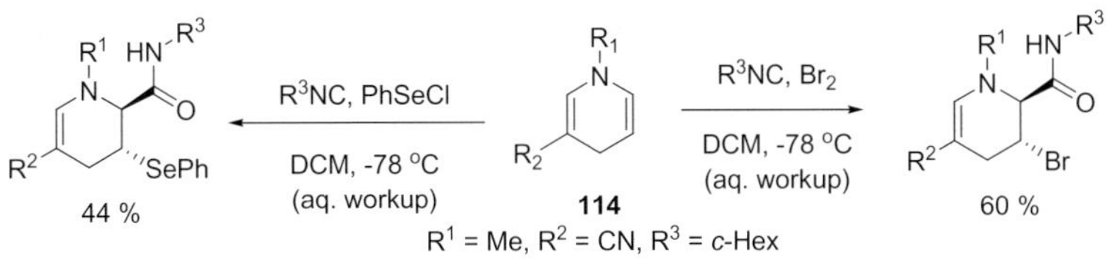

Scheme 6.57

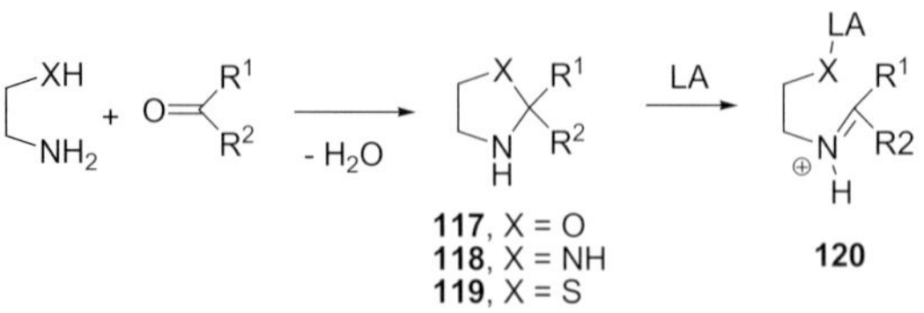

Scheme 6.58

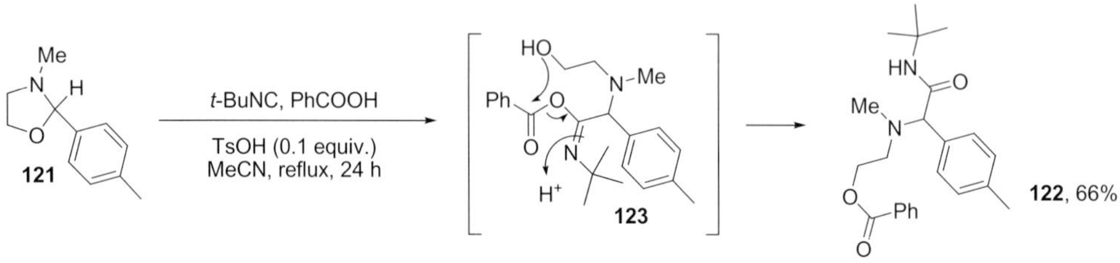

Scheme 6.59

1,2-diamines and β-aminothiols, respectively, can be considered as masked versions of iminium ions **120** when an acidic catalysis is applied (Scheme 6.58). Currently, the details are available of several examples of the successful use of these masked imines in Ugi-type reactions with isocyanides.

Motherwell and coworkers have reported [64] on the successful use of oxazolidines (e.g., **121**) in reactions with isocyanides and carboxylic acids. Based on a traditional mechanistic understanding of this Ugi-type reaction, the predominant formation of **122** in a representative reaction of this type is not surprising. In fact, the initial α-adduct **123** would most likely be transformed into **122** via an intramolecular O→O acyl migration (Scheme 6.59).

Imidazolidines **124** are the implied intermediates in the elegant approach to tetrahydropyrazines **125–126** via an IMCR of ethylenediamine (Scheme 6.60), as recently disclosed by Kysil and coworkers [55, 65]. The preformation of **124** increases the yield of tetrahydropyrazines, as opposed to a simple mixing of the carbonyl and isocyanide reactants with ethylenediamine. Indeed, the intermediacy of **124** in these or similar reactions, prior to addition of the Lewis acid (TMSCl) and an isocyanide, has been detected using ^{1}H NMR spectroscopy [66]. Notably, in these four-center, three-component reactions, one of the amino groups of the bifunctional ethylenediamine serves as an amine component for the Ugi-type IMCR, while the other serves as the isocyanide-intercepting (internal) nucleophile. Overall, this leads to the ring-forming process.

Whilst, to date, no reports have been made on the successful use of 1,3-thiazolidines in IMCRs, 2,3-dihydro-1,3-benzothiazoles **127** and their like have been found [67] to be ambiguous partners in reactions with isocyanides. This was in striking contrast with an earlier optimistic report [68] in which, out of 10 reac-

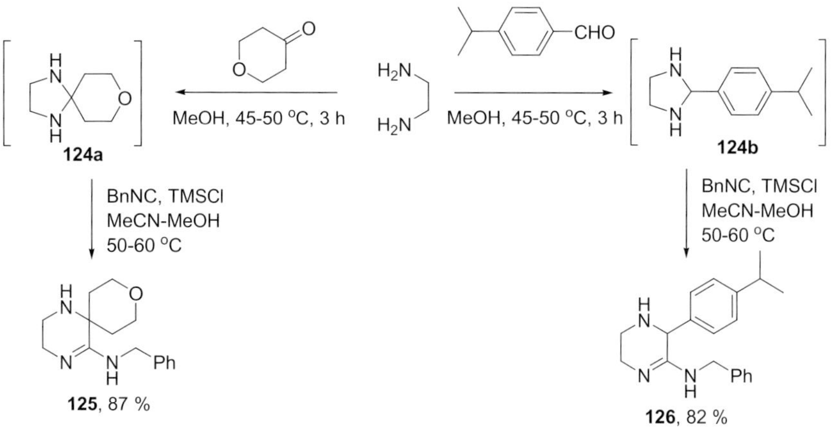

6.6 Enamines, Masked Imines, and Cyclic Imines in the Ugi Reaction | 225

Scheme 6.60

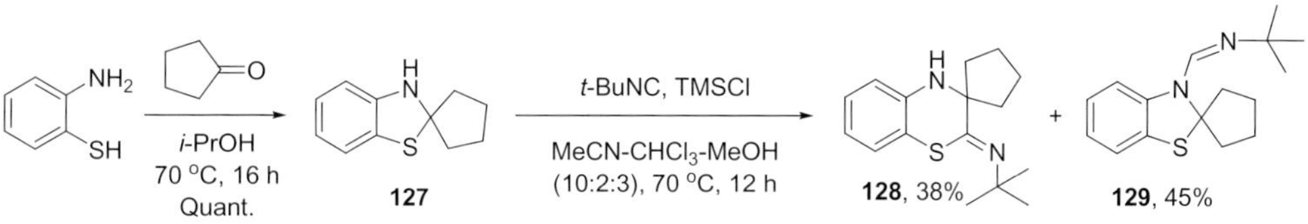

Scheme 6.61

tions of **127** with *tert*-butyl isocyanide that were studied, only in one case was the expected IMCR product **128** (similar to **125**) formed in appreciable yield in the presence of TMSCl as the IMCR promoter, along with a comparable amount of amidine **129** that had resulted from a direct reaction of **127** with the isocyanide (Scheme 6.61). In all other cases, amidines similar to **129** were the predominant products.

The use of cyclic imines in IMCRs (termed the Joullie–Ugi reaction [69]) leads to significantly greater skeletal product diversity compared to formation of the imine intermediate from a carbonyl component and an amine. However, this is most likely due to the plethora of opportunities that exist for generating cyclic imine species.

Davis and coworkers [70] have reported the successful generation of azasugar imine **130** via the treatment of *N*-chloramine precursor with DBU (1,8-Diazabicyclo[5.4.0]undec-7-ene). Subsequently, **130** was reacted with acetylglycine (as well as other 10 carboxylic acids) and a series of isocyanides (e. g., *t*-BuNC) to produce, after acetonide deprotection, a single diastereomer of the C-carbomoylated/N-acylated products **131** in 43–77% yield (Scheme 6.62). After further synthetic manipulation, these glycomimetic and peptidomimetic compounds gave rise to effective inhibitors of the enzyme glucosylceramide synthase, which is a promising target for the treatment of Gaucher's disease, a genetically

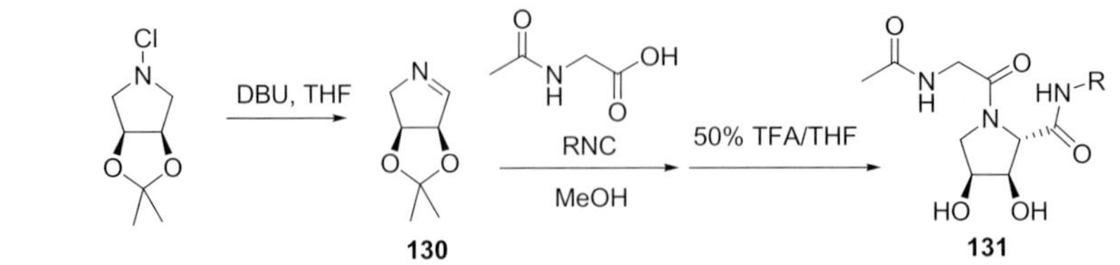

Scheme 6.62

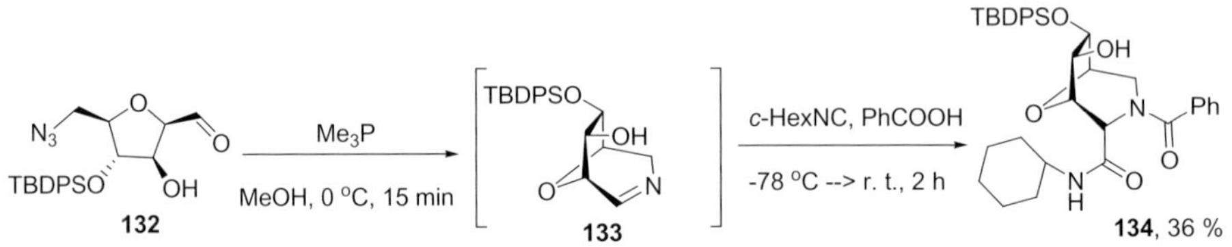

Scheme 6.63

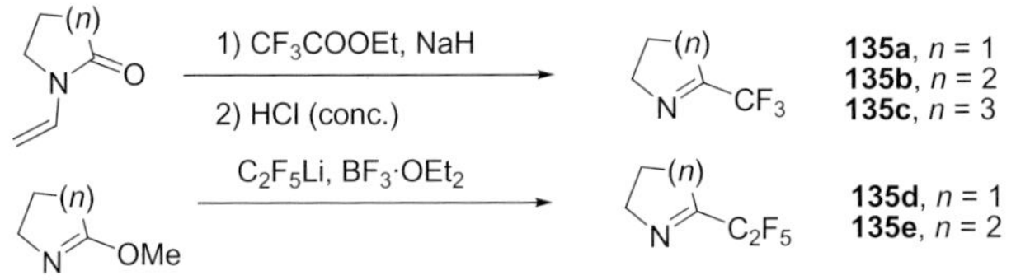

Scheme 6.64

linked condition in which lipids accumulate abnormally in the tissue cells and in certain organs.

In addition to the secondary amine chlorination–elimination strategy, as exemplified above, the Staudinger/Aza–Wittig reaction can be also used to generate cyclic amine precursors for the Ugi-type IMCRs [71]. A remarkable example from the research group of van Boom has demonstrated the power of this approach in generating product's skeletal complexity [72]. In this case, a carbohydrate-derived azide **132** generated the bicyclic imine **133** on treatment with trimethylphosphine. Subsequently, without isolating **133**, cyclohexyl isocyanide and benzoic were added, after which an unusually rapid and facile Ugi reaction, that reached completion in only 2 h at room temperature, occurred to provide (after purification) product **134** as a single diastereomer (Scheme 6.63).

Cyclic perfluoroalkyl imines **135a–e** have also been prepared by the group of Nenajdenko [73, 74], as shown in Scheme 6.64.

Imines **135** can be effectively used as IMCR partners to prepare biomedically important dipeptide compounds that contain the perfluoroalkyl prolinomimetic

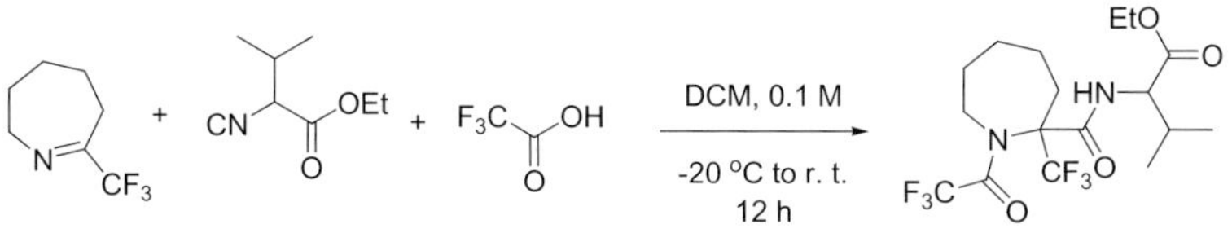

Scheme 6.65

moiety within their structure [75]. This approach is illustrated by the synthesis of the azepan carboxylic acid dipeptide **136**, as shown in Scheme 6.65.

6.7
Concluding Remarks

The identification of various workable replacements to the amine (or imine) components for the Ugi reaction and related IMCRs has proven also to serve as a remarkably efficient strategy for extending the skeletal diversity of the products that result from these efficient and atom-economical processes. It is strongly believed that this area will continue to provide conceptually novel opportunities for accessing a variety of medicinally promising scaffolds, and also enrich both corporate and academic compound collections with novel, drug-like compounds.

Acknowledgments

The author is indebted to Drs Ekaterina Lakontseva, Sergey Tsirulnikov, Mikhail Nikulnikov, and Vladislav Parchinsky of the Chemical Diversity Research Institute, Khimki, Russia, for their invaluable experimental contributions that are reflected in this chapter.

References

1 Ugi, I., Meyr, R., Fetzer, U., and Steinbrückner, C. (1959) Versuche mit Isonitrilen. *Angew. Chem.*, **71** (11), 386.
2 Trost, B.M. (1995) Atom economy. A challenge for organic synthesis. *Angew. Chem. Int. Ed. Engl.*, **34** (3), 259–281.
3 Dömling, A. and Ugi, I. (2000) Multicomponent reactions with isocyanides. *Angew. Chem. Int. Ed. Engl.*, **39** (18), 3168–3210.
4 Ugi, I., Werner, B., and Dömling, A. (2003) The chemistry of isocyanides, their multicomponent reactions and their libraries. *Molecules*, **8**, 53–66.
5 Dömling, A. (2006) Recent developments in isocyanide-based multicomponent reactions in applied chemistry. *Chem. Rev.*, **106**, 17–89.
6 Tempest, P.A. (2005) Recent advances in heterocycle generation using the efficient Ugi multiple-component condensation reaction. *Curr. Opin. Drug Discov. Dev.*, **8** (6), 776–788.

7 Hulme, C. and Dietrich, J. (2009) Emerging molecular diversity from the intra-molecular Ugi reaction: iterative efficiency in medicinal chemistry. *Mol. Divers.*, **13**, 195–207.

8 Ivachtchenko, A.V., Ivanenkov, Y.A., Kysil, V.M., Krasavin, M.Y., and Ilyin, A.P. (2010) Multicomponent reactions of isocyanides in the synthesis of heterocycles. *Russ. Chem. Rev.*, **79**, 787–817.

9 Akritopoulou-Zanze, I. and Djuric, S.W. (2007) Recent advances in the development and applications of post-Ugi transformations. *Heterocycles*, **73**, 125–147.

10 Ilyin, A., Kysil, V., Krasavin, M., Kurashvili, I., and Ivachtchenko, A.V. (2006) Complexity-enhancing acid-promoted rearrangement of tricyclic products of tandem Ugi 4CC/intramolecular Diels–Alder reaction. *J. Org. Chem.*, **71**, 9544–9547.

11 Krasavin, M. and Parchinsky, V. (2010) Thiophene-containing products of the Ugi reaction in oxidation-triggered IMDA/aromatization cascade: a facile access to 3-oxoisoindolines. *Tetrahedron Lett.*, **51**, 5657–5661.

12 Tsirulnikov, S., Nikulnikov, M., Kysil, V., Ivachtchenko, A., and Krasavin, M. (2009) Streamlined access to 2,3-dihydropyrazino[1,2-a]indole-1,4-diones via Ugi reaction followed by microwave-assisted cyclization. *Tetrahedron Lett.*, **50**, 5529–5531.

13 El Kaim, L. and Grimaud, L. (2009) Beyond the Ugi reaction: less conventional interactions between isocyanides and iminium species. *Tetrahedron*, **65** (11), 2153–2171.

14 Zinner, G., Moderhack, D., and Kliegel, W. (1969) Hydroxylamine in der Vierkomponenten-Kondensation nach Ugi. *Chem. Ber.*, **102**, 2536–2546.

15 Moderhack, D. (1973) Unerwarterer Verlauf der Ugi-Reaktion mit N-Alkylhydroxylaminen. *Liebigs Ann. Chem.*, 359–364.

16 Zinner, G., Moderhack, D., Hantelmann, O., and Bock, W. (1974) Hydroxylamine in der Vierkomponenten-Kondensation nach Ugi, II. *Chem. Ber.*, **107**, 2947–2955.

17 Basso, A., Banfi, L., Guanti, G., Riva, R., and Riu, A. (2004) Ugi multicomponent reaction with hydroxylamines: an efficient route to hydroxamic acid derivatives. *Tetrahedron Lett.*, **45**, 6109–6111.

18 Basso, A., Banfi, L., Guanti, G., and Riva, R. (2005) One-pot synthesis of α-acyloxyaminoamides via nitrones as imine surrogates in the Ugi MCR. *Tetrahedron Lett.*, **46**, 8003–8006.

19 Ugi, I. and Bodesheim, F. (1961) Umsetzung von Isonitrilen mit Hydrazonen und Stickstoffwasserstoffsäure. *Chem. Ber.*, **94**, 2797–2801.

20 Ugi, I. and Bodesheim, F. (1963) Umsetzung von Isonitrilen mit Hydrazonen und Carbonsäuren. *Liebigs Ann. Chem.*, **166**, 61–64.

21 Zinner, G. and Kliegel, W. (1966) Zur Erkenntnis der Ugi-Reaktion mit Hydrazinen, I. *Arch. Pharm.*, **299**, 746–756.

22 Zinner, G. and Bock, W. (1971) Zur Erkenntnis der Ugi-Reaktion mit Hydrazinen, II. *Arch. Pharm.*, **304**, 933–943.

23 Failli, A., Nelson, V., Immer, H., and Götz, M. (1973) Model experiments directed towards the synthesis of N-aminopeptides. *Can. J. Chem.*, **51**, 2769–2775.

24 Elders, N., Ruijter, E., Nenajdenko, V.G., and Orru, R.V.A. (2010) α-Acidic isocyanides in multicomponent chemistry, in *Topics in Heterocyclic Chemistry 23; Synthesis of Heterocycles via Multicomponent Reactions I* (eds R. Orru and E. Ruijter), Springer-Verlag, Berlin, Heidelberg, pp. 129–159.

25 Gulevich, A.V., Zhdanko, A.G., Orru, R.V.A., and Nenajdenko, V.G. (2010) Isocyanoacetate derivatives: synthesis, reactivity, and application. *Chem. Rev.*, **110**, 5235–5331.

26 Zhdanko, A.G. and Nenajdenko, V.G. (2009) Nonracemizable isocyanoacetates for multicomponent reactions. *J. Org Chem.*, **74**, 884–887 and references cited therein.

27 Immer, H., Nelson, V., Robinson, W., and Götz, M. (1973) Anwendung der Ugi-Reaktion zur Synthese modifizierter

Eledoisin-Teilsequenzen. *Liebigs Ann. Chem.*, 1789–1796.

28 Burger, K., Mütze, K., Hollweck, W., and Koksch, B. (1998) Incorporation of α-trifluoromethyl substituted α-amino acids into C- and N-terminal position of peptides and peptide mimetics using multicomponent reactions. *Tetrahedron*, **54**, 5915–5928.

29 Tempest, P., Ma, V., Kelly, M.G., Jones, W., and Hulme, C. (2001) MCC/S_NAR methodology. Part I: novel access to a range of heterocyclic cores. *Tetrahedron Lett.*, **42**, 4963–4968.

30 Marcaccini, S., Pepino, R., Polo, C., and Pozo, M.C. (2001) Studies on isocyanides and related compounds: a facile synthesis of 4-phenyl-1-(2H)phthalazinone-2-alkanoic acid amides. *Synthesis*, **1**, 85–88.

31 Marcos, C.F., Marcaccini, S., Pepino, R., Polo, C., and Torroba, T. (2003) Studies on isocyanides and related compounds: a facile synthesis of 3(2H)-pyridazinones via Ugi four-component condensation. *Synthesis*, **5**, 691–694.

32 El Kaïm, L., Gautier, L., Grimaud, L., and Michault, V. (2003) New insight into the azaenamine behaviour of N-arylhydrazones: first aldol and improved Mannich reactions with unactivated aldehydes. *Synlett*, **12**, 1844–1846.

33 Attanasi, O.A. and Filippone, P. (1997) Working twenty years on conjugated azo-alkenes (and environs) to find new entries in organic synthesis. *Synlett*, **10**, 1128–1140.

34 Ancel, J.E., El Kaïm, L., Gadras, A., Grimaud, L., and Jana, N.K. (2002) Studies towards the synthesis of Fipronil® analogues: improved decarboxylation of α-hydrazonoacid derivatives. *Tetrahedron Lett.*, **43**, 8319–8321.

35 Krasavin, M., Bushkova, E., Parchinsky, V., and Shumsky, A. (2010) Hydrazinopeptide motifs synthesized via the Ugi reaction: an insight into the secondary structure. *Synthesis*, **6**, 933–942.

36 Gruppe, R., Baeck, B., and Niedrich, H. (1972) Hydrazinverbindungen als Heterobestandteile in Peptiden. XIV. $N^α$-Acylierungreaktionen und $N^β$-acylierten Derivaten und Peptiden von α-Hydrazino-β-phenylpropionsäure. *J. Prakt. Chem.*, **314**, 751–758.

37 Lelais, G. and Seebach, D. (2003) Synthesis, CD spectra, and enzymatic stability of b2-oligoazapeptides prepared from (S)-2-hydrazino carboxylic acids carrying the side chains of Val, Ala, and Leu. *Helv. Chim. Acta*, **86**, 4152–4168.

38 Guy, L., Vidal, J., and Collet, A. (1998) Design and synthesis of hydrazinopeptides and their evaluation as human leukocyte elastase inhibitors. *J. Med. Chem.*, **41**, 4833–4843.

39 Aubury, A., Mangeot, J.-P., Vidal, J., Collet, A., Zerkout, S., and Marraud, M. (1994) Crystal structure analysis of a β-turn mimic in hydrazino peptides. *Int. J. Pept. Protein Res.*, **43**, 305–311.

40 Lakontseva, E.E. (2010) Ugi reaction of acylhydrazines: practical applications in peptidomimetic design and synthesis of heterocycles. Dissertation, People's Friendship University of Russia, Moscow.

41 Bushkova, E.E., Parchinsky, V.Z., and Krasavin, M. (2010) Efficient entry into hydrazinopeptide-like structures via sequential Ugi reactions. *Mol. Diversity*, **14**, 493–399.

42 Marcaccini, S., Miguel, D., Torroba, T., and Garcia-Valverde, M. (2003) 1,4-Thiazepines, 1,4-benzothiazepin-5-ones,and 1,4-benzothioxepin orthoamides via multicomponent reactions of isocyanides. *J. Org. Chem.*, **68**, 3315–3318.

43 Marcaccini, S., Pepino, R., Torroba, T., Miguel, D., and Garcia-Verde, M. (2002) Synthesis of thiomorpholines by an intramolecular Ugi reaction. *Tetrahedron Lett.*, **43**, 8591–8593.

44 Ilyin, A.P., Trifilenkov, A.S., Kurashvili, I.D., Krasavin, M., and Ivachtchenko, A.V. (2005) One-step construction of peptidomimetic 5-carbamoyl-4-sulfonyl-2-piperazinones. *J. Comb. Chem.*, **7**, 360–363.

45 Ilyin, A.P., Loseva, M.V., Vvedensky, V.Y., Putsykina, E.B., Tkachenko, S.E., Kravchenko, D.V., Khvat, A.V., Krasavin, M.Y., and Ivachtchenko, A.V. (2006) One-step assembly of carbamoyl-substituted heteroannelated [1,4] thiazepines. *J. Org. Chem.*, **71**, 2811–2819.

46 Naskar, D., Roy, A., Seibel, W.L., West, L., and Portlock, D.E. (2003) The synthesis of aza-β-lactams via tandem Petasis–Ugi multi-component condensation and 1,3-diisopropylcarbodiimide (DIC) condensation reaction. *Tetrahedron Lett.*, **44**, 6297–6300.

47 Portlock, D.E., Naskar, D., West, L., and Li, M. (2002) Petasis boronic acid–Mannich reactions of substituted hydrazines: synthesis of α-hydrazinocarboxylic acids. *Tetrahedron Lett.*, **43**, 6845–6847.

48 Krasavin, M., Parchinsky, V., Shumsky, A., Konstantinov, I., and Vantskul, A. (2010) Proline-like β-turn mimics accessed via Ugi reaction involving monoprotected hydrazines. *Tetrahedron Lett.*, **51**, 1367–1370.

49 Lakontseva, E. and Krasavin, M. (2010) Diversity-oriented pyrazol-3-one synthesis based on hydrazinopeptide-like units via the Ugi reaction. *Tetrahedron Lett.*, **51**, 4095–4099.

50 Zinner, G. and Bock, W. (1972) Notiz über die UGI-Reaktion mit Diaziridinen. *Arch. Pharm.*, **306**, 94–96.

51 von Zychlinski, A. and Ugi, I. (1998) MCR IX: a new and easy way for the preparation of piperazine-2-keto-3-carboxamides. *Heterocycles*, **49**, 29–32.

52 Campian, E., Lou, B., and Saneii, H. (2002) Solid-phase synthesis of α-sulfonylamino amide derivatives based on Ugi-type condensation reaction using sulfonamides as amine input. *Tetrahedron Lett.*, **43**, 8467–8470.

53 Sañudo, M., Marcaccini, S., Basurto, S., and Torroba, T. (2006) Synthesis of 3-hydroxy-6-oxo[1,2,3]triazin-1-yl alaninamides, a new class of cyclic dipeptidyl ureas. *J. Org. Chem.*, **71**, 4578–4584.

54 Krasavin, M., Tsirulnikov, S., Nikulnikov, M., Kysil, V., and Ivachtchenko, A. (2008) Poorly reactive 5-piperazin-1-yl-1,3,4-thiadiazol-2-amines rendered as valid substrates for Groebke–Blackburn-type multicomponent reaction with aldehydes and isocyanides using TMSCl as a promoter. *Tetrahedron Lett.*, **49**, 5241–5243.

55 Kysil, V., Tkachenko, S., Khvat, A., Williams, C., Tsirulnikov, S., Churakova, M., and Ivachtchenko, A. (2007) TMSCl-promoted isocyanide-based MCR of ethylenediamines: an efficient assembling of 2-aminopyrazine core. *Tetrahedron Lett.*, **48**, 6239–6244.

56 Neunhoeffer, H. and Hennig, H. (1968) Formamidrazone. *Chem. Ber.*, **101**, 3947–3951.

57 Ugi, I. and Bottner, E. (1963) Isonitrile, XVI. Mit O→C-Acyl-Wanderung gekoppelte -Additionen von N-Alkylchinolinium- und Carboxylat-Ionen der Isonitrile. *Liebigs Ann. Chem.*, **670**, 74–80.

58 Berthet, J.-C., Nierlich, M., and Ephritikhine, M. (2002) Reactions of isocyanides and pyridinium triflates – a simple and efficient route to imidazopyridinium derivatives. *Eur. J. Org. Chem.*, 375–378.

59 Mironov, M.A., Mokrushin, M.S., and Maltsev, S.S. (2003) New method for the combinatorial search of multi component reactions. *Synlett*, 943–945.

60 Diaz, J.L., Miguel, M., and Lavilla, R. (2004) N-acylazinium salts: a new source of iminium ions for Ugi-type processes. *J. Org. Chem.*, **69**, 3550–3553.

61 Kiselyov, A.S. (2005) Reaction of N-fluoropyridinium fluoride with isonitriles and $TMSN_3$: a convenient one-pot synthesis of tetrazol-5-yl pyridines. *Tetrahedron Lett.*, **46**, 4851–4854.

62 Masdeu, C., Diaz, J.L., Miguel, M., Jimenez, O., and Lavilla, R. (2004) Straightforward α-carbamoylation of NADH-like dihydropyridines and enol ethers. *Tetrahedron Lett.*, **45**, 7907–7909.

63 Masdeu, C., Gomez, E., Williams, N.A., and Lavilla, R. (2006) Hydro-, halo- and seleno-carbamoylation of cyclic enol ethers and dihydropyeridines: new mechanistic pathways for Passerini- and Ugi-type multicomponent reactions. *QSAR Comb. Sci.*, **25**, 465–473.

64 Diorazio, L.J., Motherwell, W.B., Sheppard, T.D., and Waller, R.W. (2006) Observations on the reaction of N-alkyloxazolidines, isocyanides, and carboxylic acids: a novel three-component

reaction leading to N-acyloxyethylamino acid amides. *Synlett*, 2281–2283.

65 Kysil, V., Khvat, A., Tsirulnikov, S., Tkachenko, S., Williams, C., Churakova, M., and Ivachtchenko, A. (2010) General multicomponent strategy for the synthesis of 2-amino-1,4-diazaheterocycles: scope, limitations, and utility. *Eur. J. Org. Chem.*, 1525–1543.

66 Tsirulnikov, S.A. (2010) Reactions of N,N-dinucleophiles with carbonyl compounds and isocyanides. Dissertation, The Zelinsky Institute of Organic Chemistry, Moscow.

67 Tsirulnikov, S., Dmitriev, D., and Krasavin, M. (2010) O-aminothiophenol in reactions with carbonyl compounds and isocyanides: a word of caution. *Synlett*, 1935–1938.

68 Heravi, M.M., Baghernejad, B., and Oskooie, H.A. (2009) A novel and facile one-pot synthesis of 3-aryl-4H-benzo[1,4]thiazin-2-amine. *Synlett*, 1123–1125.

69 Bowers, M.M., Carroll, P., and Joullie, M.M. (1989) Model studies directed toward the total synthesis of 14-membered cyclopeptide alkaloids: synthesis of prolyl peptides via a four-component condensation. *J. Chem. Soc., Perkin Trans. 1*, 857–865.

70 Chapman, T.M., Davies, I.G., Gu, B., Block, T.M., Scopes, D.I.C., Hay, P.I., Courtney, S.M., McNeill, L.A., Schofield, C.J., and Davies, B.G. (2006) Glyco- and peptidomimetics from three-component Joullie–Ugi coupling show selective antiviral activity. *J. Am. Chem. Soc.*, **127**, 506–507.

71 Banfi, L., Basso, A., Guanti, G., and Riva, R. (2004) Enantio- and diastereoselective synthesis of 2,5-disubstituted pyrrolidines through a multicomponent Ugi reaction and their transformation into bicyclic scaffolds. *Tetrahedron Lett.*, **45**, 6637–6640.

72 Timmer, M.S.M., Risseeuw, M.D.P., Verdoes, M., Filippov, D.V., Plaisier, J.R., van der Marel, G.A., Overkleeft, H.S., and van Boom, J.H. (2005) Synthesis of functionalized heterocycles via a tandem Staudinger/aza-Wittig/Ugi multicomponent reaction. *Tetrahedron: Asymmetry*, **16**, 177–185.

73 Shevchenko, N.E., Roeschenthaler, G.-V., Mitiaev, A.S., Lork, E., and Nenajdenko, V.G. (2008) Diastereoselective synthesis of cyclic 1,3-amino alcohols bearing CF_3 (CCl_3)-groups. *J. Fluorine Chem.*, **129**, 637–644.

74 Gulevich, A.V., Shevchenko, N.E., Balenkova, E.S., Roeschenthaler, G.-V., and Nenajdenko, V.G. (2008) The Ugi reaction with CF_3-carbonyl compounds: effective synthesis of α-trifluoromethyl amino acid derivatives. *Tetrahedron*, **64**, 11706–11712.

75 Gulevich, A.V., Shevchenko, N.E., Balenkova, E.S., Roschenthaler, G.-V., and Nenajdenko, V.G. (2009) Efficient multicomponent synthesis of α-trifluoromethyl proline, homoproline, and azepan carboxylic acid dipeptides. *Synlett*, 403–406.

7
Multiple Multicomponent Reactions with Isocyanides

Ludger A. Wessjohann, Ricardo A.W. Neves Filho, and Daniel G. Rivera

7.1
Introduction

Multicomponent reactions (MCRs) are one-pot processes that combine three or more building blocks to produce chemical moieties containing atoms from all of the reactants [1, 2]. Notably, MCRs are almost unrivaled processes in terms of atom economy and the rapid generation of molecular complexity; consequently, they have been widely used in the total synthesis of complex natural products [3] and also in drug discovery and development [4]. Those MCRs which include the use of isocyanides (isocyanide-based multicomponent reactions; IMCRs) as one of the components usually belong to type II of a commonly accepted classification of MCRs; that is, they are concluded by a final, irreversible step [5]. Among reactions of this type are included the Passerini three-component reaction (Passerini-3CR; Scheme 7.1a) [5, 6] and the Ugi four-component reaction (Ugi-4CR; Scheme 7.1b) [5, 7]. Although isocyanides are also known for chemistry based on their induction of α-acidity, for radical-induced reactions, or for their ability to be metalated [5, 8], they are mostly recognized by the synthetic potential derived from the addition of the isocyanide carbon atom to electrophiles and nucleophiles [5, 9]. The remarkable discovery made by Passerini, who described the condensation of a carbonyl compound, a carboxylic acid, and an isocyanide to furnish an α-acyloxycarboxamide, paved the way towards the further development of isocyanide-driven MCRs [10]. Moreover, Ugi's report of his famous four-component reaction in 1959 [11] – as well as all variations introduced by him during the following years [9, 12] – was not only a confirmation of the synthetic potential of isocyanides, but also the initiation of a new era in the field of MCRs. This received a further boost with the advent of combinatorial chemistry during the early 1990s. Since these initial discoveries, the IMCRs have become one of the dominant approaches to heterocycle syntheses, especially through the early developments made by the groups of Schöllkopf [13] and van Leusen [14] and, more recently, by Marcaccini [15], Dömling [16], Zhu [17], and many others.

Because of its high chemical efficiency, easy implementation and the high level of molecular diversity that is accessible in a one-pot process, the Ugi-4CR is seen

Isocyanide Chemistry: Applications in Synthesis and Material Science, First Edition. Edited by Valentine Nenajdenko.
© 2012 Wiley-VCH Verlag GmbH & Co. KGaA. Published 2012 by Wiley-VCH Verlag GmbH & Co. KGaA.

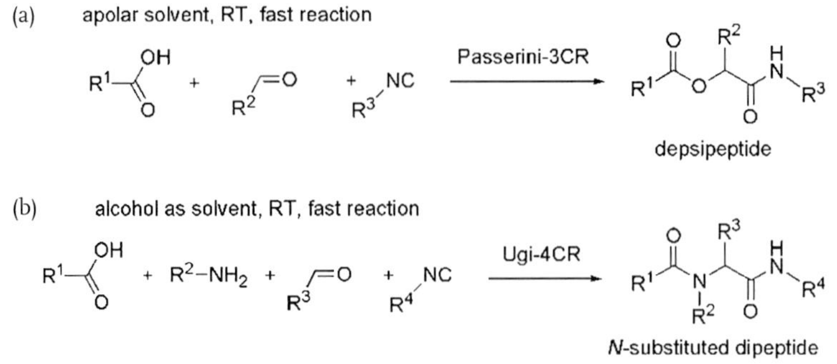

(a) apolar solvent, RT, fast reaction

Passerini-3CR

depsipeptide

(b) alcohol as solvent, RT, fast reaction

Ugi-4CR

N-substituted dipeptide

Scheme 7.1 (a) The Passerini three-component reaction (3CR); (b) The Ugi four-component reaction (4CR).

as a landmark for the utilization of MCRs in combinatorial and medicinal chemistry. Ugi's ideas of using combinations of each one of the four components to generate large sets of compounds (now referred to as "compound libraries") were revolutionary for that time [18]. In addition, the remarkable scope derived from the variation of the nature of the components, as well as the development of intramolecular versions, makes Ugi-4CRs and the many related and devised multicomponent processes the most important ones among isocyanide-based chemical transformations [5].

Many years ago, Ugi himself envisioned the potential of using polyfunctional building blocks for the production of complex, eventually high-molecular-weight and polymer(-like) molecules through the performance of sequential Ugi-4CRs [19]. However, only during the past decades have powerful methodologies emerged that make use of multifold IMCRs towards applications.

In this chapter, the latest results on the development and applications of the "Multiple Multicomponent Approach" are described. These are based on the chemistry of the isocyanide functional group, in which multiple MCRs based on polyfunctional building blocks allow for the assembly or modification of particularly complex, or at least multimeric, molecules. The approaches included here are those which rely exclusively on the implementation of multiple IMCRs, either in one-step or as sequential (multistep) procedures, though in some instances even the multistep procedures were performed in one pot.

7.2
One-Pot Multiple IMCRs

The IMCRs are among the most powerful procedures in terms of complexity generation with low synthetic cost. However, this aspect can be improved by implementing multiple MCRs that require at least one polyfunctional building

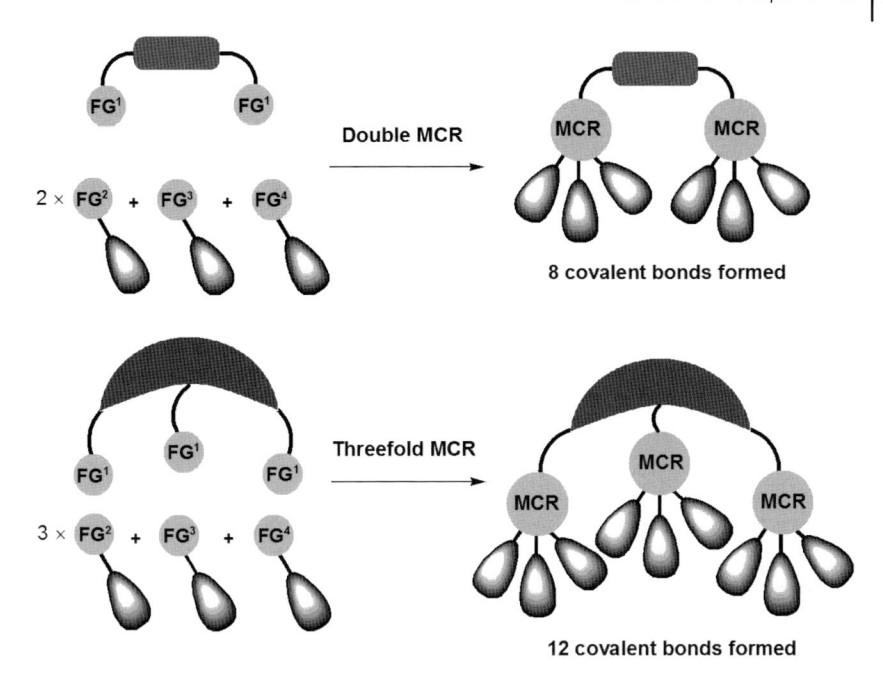

Scheme 7.2 The multiple multicomponent approach including bifunctional and trifunctional core building blocks, exemplified with a four-component reaction. The lobes, though shown identical, may be differential in size and chemical nature for each type of functional group (FG).

block. To avoid uncontrolled intramolecular reactions, and also to provide a better synthesis and storage, it is strategically advisable to utilize building blocks that are homo-polyfunctionalized; that is, blocks which possess the same type of functional group (FG^1-R-FG^1) such as di-amines, instead of a hetero polyfunctionalization (FG^1-R-FG^2) such as amino-aldehydes. A schematic representation of the multiple multicomponent approaches with a bi- and trifunctional building block is shown in Scheme 7.2. This leads to the formation of very complex molecular assemblies in one pot, despite the homo-polyfunctionalization of the core unit. When considering the possibility of employing MCR-reactive moieties with catalytic or molecular/ion-pairing recognition capabilities, their multiple incorporation into a single skeleton foresees important applications. The IMCRs most widely exploited for this goal are the Passerini-3CR and the Ugi-4CR (*vide infra*), because of their simple implementation compared to other isocyanide-based procedures, and the large number of chemically or biologically relevant building blocks bearing one or more carbonyl, carboxylic, and amino functionalities. Moreover, polyisocyano compounds are easily accessible from the corresponding polyamines, thus allowing access to multivalent architectures not accessible from the other components.

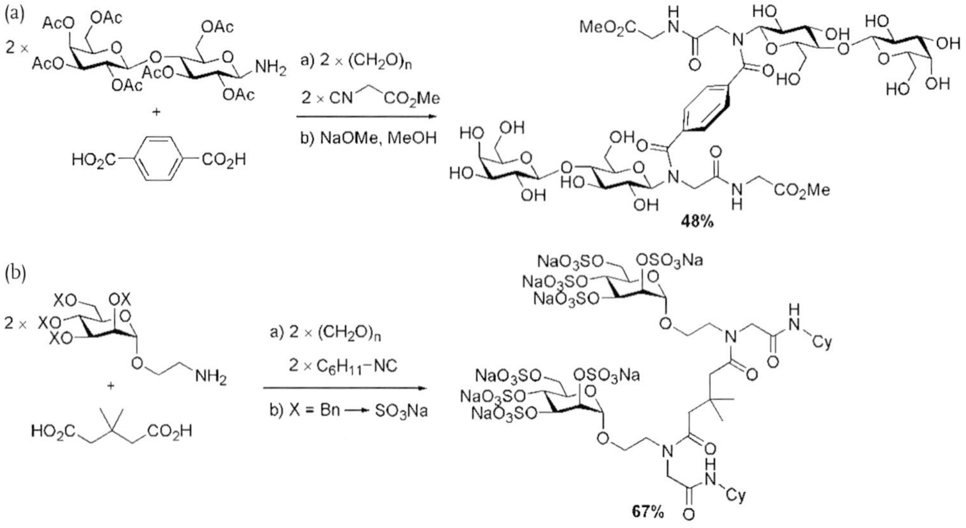

Scheme 7.3 Synthesis of bivalent glycoconjugates by double Ugi-4CR with dicarboxylic acids and amino-sugars.

7.2.1
Synthesis of Multivalent Glycoconjugates

Murphy and coworkers employed a double Ugi-4CR-based approach for the assembly of bivalent lactosides derived from terephthalic acid [20]. In their effort to assess the inhibitory activity of this type of neoglycoconjugate on the recognition of glycans by lectins, the group prepared a variety of rigid and flexible oligosaccharide scaffolds for biological evaluation. As depicted in Scheme 7.3a, the Ugi-4CR was utilized to build up the tertiary terephthalamide moieties joining the two lactoside residues. Previously, in conjunction with the Kunz research group [21], the synthetic scope of the multiple multicomponent approach, as well as the conformational analysis of the resulting glycoconjugates, had been investigated using galactosyl amine rather than lactosyl amine as the amino component. The moderate to good yields achieved in the one-pot construction of the bivalent galactosides, and the nuclear magnetic resonance (NMR)-based evidence that the sugars are restricted in conformation, were the keys to expanding this strategy for the discovery of new inhibitors of medicinally relevant lectins.

Ferro and coworkers also utilized this approach for the synthesis of sulfated glycoconjugates designed as heparin sulfate mimics [22]. As shown in Scheme 7.3b, two per-benzylated aminomannosides were assembled into dimeric neoglycoconjugates via a double Ugi-4CR, and subsequently deprotected and sulfated to the final polyanionic target molecules. In the same study, uronic acids were dimerized via a double Ugi-4CR using diamine building blocks, and these dimers were transformed into the sulfated glycoconjugates.

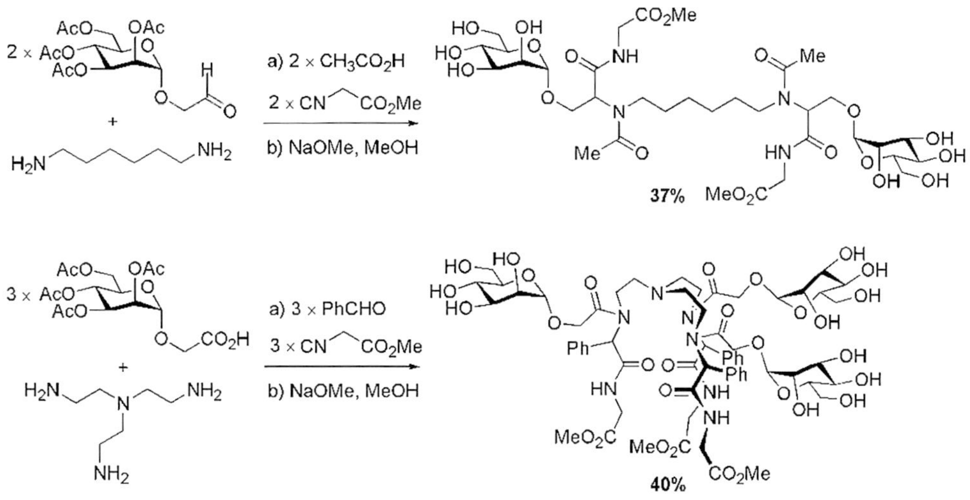

Scheme 7.4 Synthesis of bivalent and trivalent cluster mannosides by double and threefold Ugi-4CRs on polyamino building blocks.

Li et al. have performed multiple Ugi-4CRs with polyamino compounds to obtain cluster mannosides with inhibitory activity on the yeast mannan/concanavalin A binding process [23]. As depicted in Scheme 7.4, carbonyl- and carboxymethyl-mannosides were employed as the oxo and acid components, respectively, of the multiple multicomponent approach, and this led to bivalent and trivalent glycoconjugates in moderate yields. Interestingly, the yields were not lowered when passing from the double to the threefold Ugi-4CR-based process, whereas the inhibitory activity was higher for the divalent than for the trivalent glycoconjugate. The influence of the flexibility and length of the polyamine linker on the reaction efficiency and the bioactivity of the clusters was not evaluated, however.

7.2.2
Synthesis of Hybrid Peptide–Peptoid Podands

Wessjohann and coworkers have recently implemented the multiple multicomponent approach for the one-pot construction of podand architectures, by incorporating several binding and catalytic elements into a single platform (L.A. Wessjohann, D.G. Rivera, F. León, O. Concepción, and F.E. Morales, unpublished results). The focus of these investigations was on the use of tripodal, bowl, and concave-shaped scaffolds for positioning the multiple peptidic chains in an organized form. In addition, the use of α-amino acids allowed for the assembly of podands bearing hybrid peptide–peptoid chains as appended arms. Peptoids are non-natural surrogates of peptides that contain the side-chain derivatization at the nitrogen of the amide instead of the α-carbon. Thus, peptoids can mimic many of the well-known recognition and coordination properties of peptides, including the chelation of

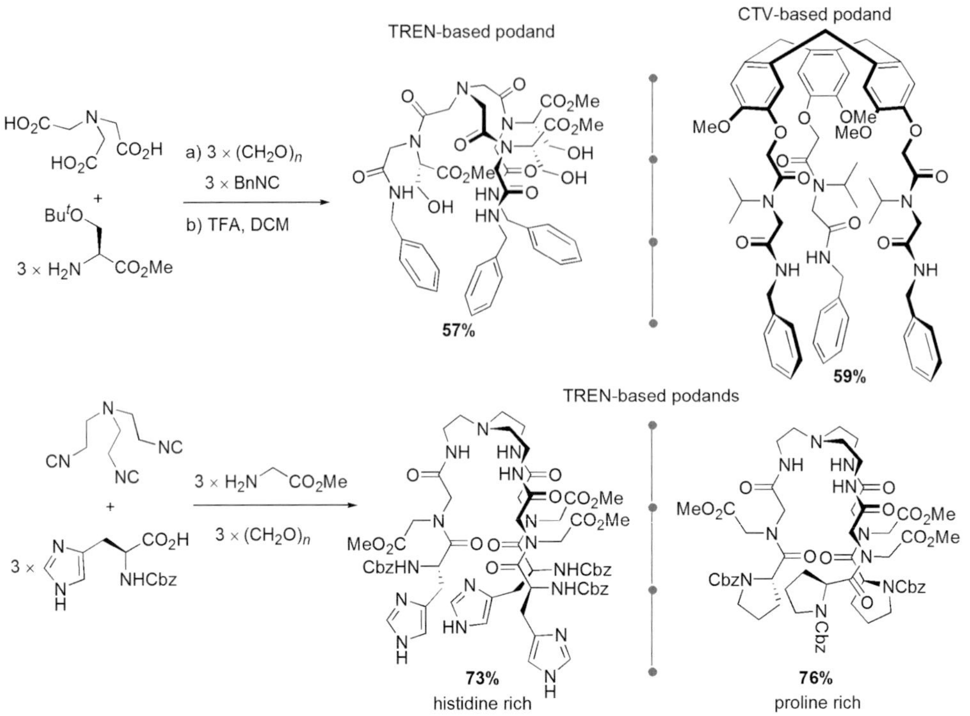

Scheme 7.5 Synthesis of hybrid peptide–peptoid podands by threefold Ugi-4CRs on TREN- and CTV-based triacid and triisocyanide building blocks.

cations by the amide's free electron pairs, and binding to neutral molecules by means of hydrogen bonding [24]. The advantages of assembling chimeric peptide–peptoid moieties over the all-peptide or all-peptoid backbones have been discussed recently [25].

The multiple multicomponent approach to hybrid peptide–peptoid podands, as developed by Wessjohann and coworkers, is shown in Scheme 7.5. Here, two different types of organizing platform with previously proven utility in supramolecular and coordination chemistry are highlighted: (i) the tripodal scaffold derived from tris(2-aminoethyl)amine (TREN); and (ii) the bowl-shaped scaffold derived from cyclotriveratrylene (CTV). Many other polyfunctional building blocks have also been utilized in multiple MCRs by this research group, including citric acid and Kemp's triacid, as well as various polyamines derived from calixarene and trialkylbenzene. The experience of this research group was that the yields of multiple Ugi-4CR-based approaches were usually higher when polycarboxylic acids and polyisocyanides were used, than with polyamino building blocks.

The use of the tris(2-isocyanoethyl)amine and related polyisocyanides offers an additional possibility, namely, the incorporation of almost all proteinogenic amino acids – either as the amino or as the carboxylic component – into hybrid peptide–peptoid architectures. The TREN-based podands are of special interest as they show potential for the chelation of main group and transition metal ions, due to their favorable arrangement of multiple amides and participation of the central tertiary amine in the coordination process.

Previously, Wessjohann and coworkers have pioneered the use of polyisocyanides for the construction of pseudopeptides with complex architectures, by performing multiple MCRs in one pot [26]. Such preparations included, for example, the first polyisocyano-steroids (L.A. Wessjohann, D.G. Rivera, F. León, O. Concepción, and F.E. Morales, unpublished results) derived from the corresponding spirostanic [27] and cholanic [28] polyamines. As shown in Scheme 7.6, steroidal building blocks with axially oriented isocyano groups were employed for the incorporation of several peptidic chains onto the rigid platform, thus providing

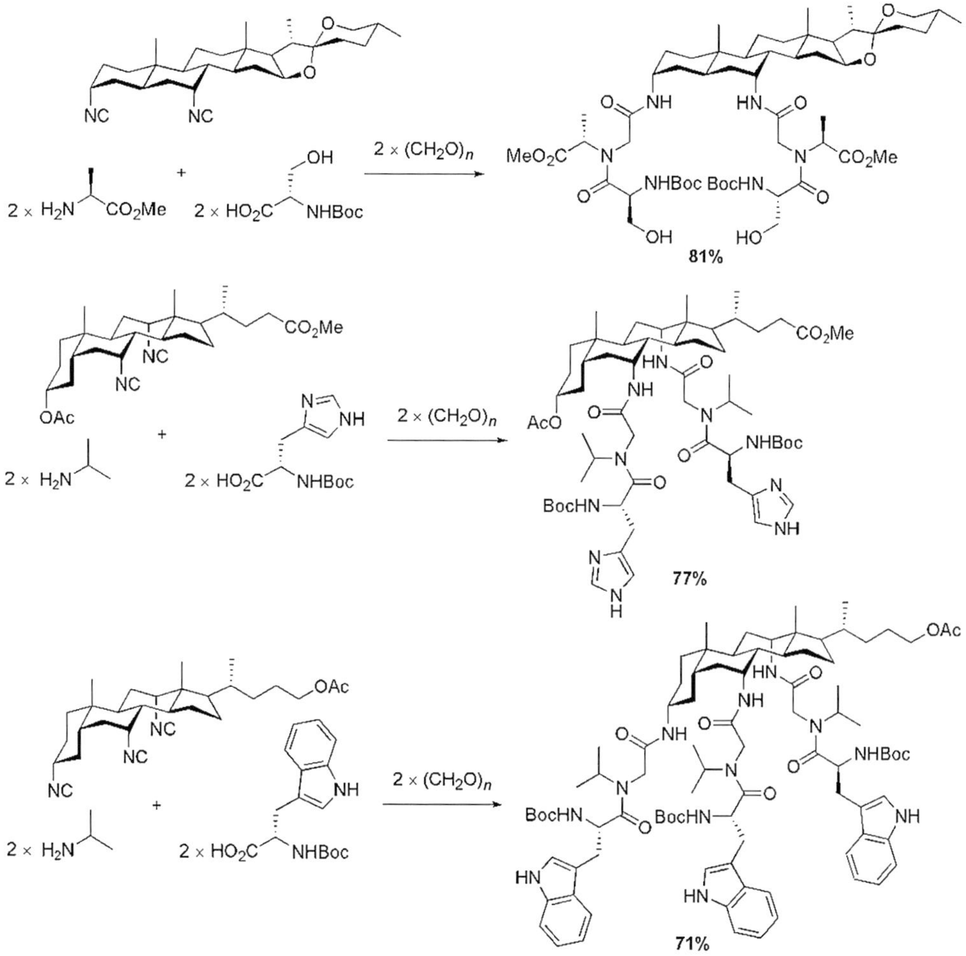

Scheme 7.6 Synthesis of spirostanic and cholanic peptido-steroidal hybrids by double and threefold Ugi-4CRs, including monoprotected α-amino acids.

a very straightforward approach to peptide–steroidal hybrids. When compared to the traditional peptide coupling protocol used to produce peptido-steroidal receptors, the multiple multicomponent approach shows a greater efficiency in terms of the rapid generation of multiple binding motifs. It must be noted that the latter approach not only enables the incorporation of a two- or threefold number of amino acids in each condensation step, but also offers access to branched peptidic moieties that are not easily accessible by the classic peptide synthesis. The shown availability of the TREN and steroid-based polyisocyanides opens up an avenue of possibilities for the combinatorial production of multi-armed, peptidic scaffolds that might act either as ligands or receptors.

Previously, combinatorial procedures based on multiple IMCRs have been implemented with the use of a dicarboxylic acid as the branching scaffold [29]. The use of pyridine-2,6-dicarboxyclic acid, together with four different amines, aldehydes, and isocyanides (13 components in total) in a combinatorial protocol, led to the formation of 8256 *pseudo*-symmetric peptidomimetics. Such a compound library was successfully evaluated for the discovery of protease inhibitors, thus proving the potential of the multiple multicomponent approach in combinatorial chemistry applications.

7.2.3
Covalent Modification and Immobilization of Proteins

Proteins are naturally polyfunctionalized, thereby exhibiting two of the four components taking part in the Ugi-4CR – that is, the amino and carboxylic groups. Accordingly, IMCRs can be considered as suitable procedures for the conjugation of small molecules to medicinally relevant proteins and enzymes. Bioconjugates such as antibody–enzyme, oligosaccharide antigen–protein, and label/tag–protein – to mention but a few examples – are of paramount importance in such fields as immunology, vaccination, and enzymology.

To the present authors' knowledge, the first reports on the use of proteins in isocyanide-based chemistry were concerned with enzyme immobilization to solid supports, functionalized with isocyanide groups [30]. Ugi and coworkers also described the immobilization of enzymes in an alginate network by multiple Ugi-4CRs [31]. Nevertheless, forthcoming applications of these enzyme immobilization protocols have, until very recently, remained elusive.

In 2000, Ziegler *et al.* undertook an extensive study on the conjugation of small molecules to proteins, by means of a multiple multicomponent approach [32] (a schematic representation of the procedures implemented in these studies is shown in Scheme 7.7). In this case, bovine serum albumin and horseradish peroxidase (HRP) were conjugated to a variety of small molecules, including carbohydrates, fluorescent dyes, and biotin. An alternative bioconjugation process was also examined, whereby oxidized HRP served as the carbonyl component of the multiple IMCRs. The best parameters for the multicomponent bioconjugation were found to range from 0.01 M to 0.1 M in pH 7.5 phosphate buffer, and produced an average epitope density of between 1 and 3, in very good yields. Lower pH values resulted in less-efficient bioconjugation processes, most likely due to

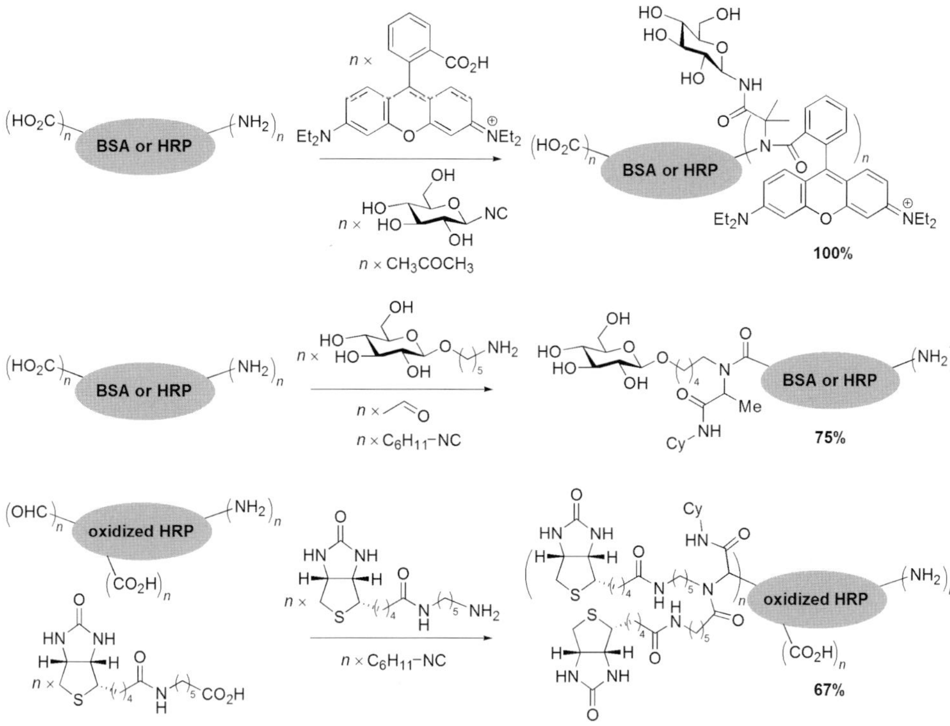

Scheme 7.7 Conjugation of small molecules to bovine serum albumin (BSA) and horseradish peroxidase (HRP) by multiple Ugi-4CRs.

decomposition of the isocyanide. The retained enzymatic activity of the modified HRPs ranged between 62% and 95% of the initial activity, though in some cases it was significantly increased.

Inspired by their success in utilizing the multiple multicomponent approach for the modification of enzymes, Rivera and Villalonga developed an important application of this procedure by creating an amperometric enzyme biosensor for hydrogen peroxide [33]. For this, HRP was immobilized on sodium alginate-coated gold electrodes, via multiple Ugi-4CRs (see Scheme 7.8). Again, the conjugation process proved to be very efficient, such that the HRP-modified electrode was applied successfully to the construction of a biosensor for hydrogen peroxide. Moreover, the electrode-immobilized enzyme displayed a full enzymatic activity in the electrochemical measurements that was retained after a 30-day storage period in a pH 7 phosphate buffer solution.

7.2.4
Assembly of Polysaccharide Networks as Synthetic Hydrogels

Crescenzi and coworkers developed an important application of the multiple multicomponent approach, by utilizing multifold Passerini and Ugi reactions for the

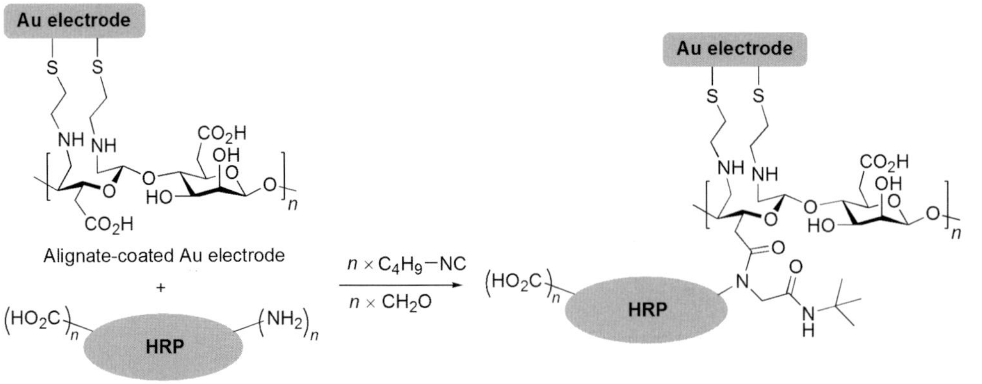

Scheme 7.8 Multicomponent immobilization of horseradish peroxidase (HRP) on an alginate-coated gold electrode for the construction of a hydrogen peroxide biosensor.

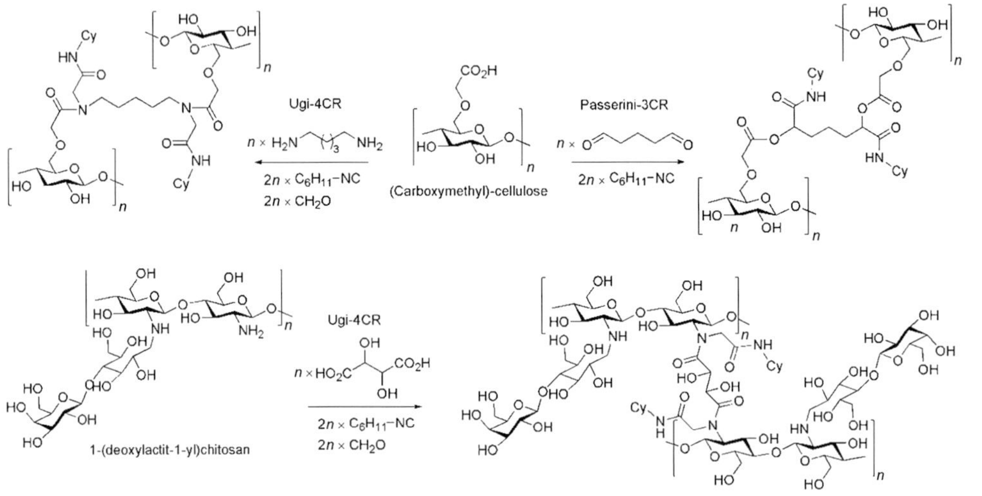

Scheme 7.9 Multicomponent assembly of polysaccharide networks as synthetic hydrogels.

production of biocompatible, synthetic hydrogels derived from polysaccharide networks [34]. Such interest in hydrophilic polymeric networks derives from their remarkable applications as drug-delivery systems, for cell growth scaffolding, and tissue engineering. Originally, (carboxymethyl)-cellulose and 1-(deoxylactityl)-chitosan were employed as polycarboxylic and polyamino building blocks, respectively, for the assembly of hydrophilic networks. As an example, the use of (carboxymethyl)-cellulose in the Passerini-3CR- and Ugi-4CR-based approaches, to incorporate glutaraldehyde and 1,5-diaminopentane as crosslinking elements for the polysaccharide, is shown in Scheme 7.9. Alternatively, tartaric acid was employed as the crosslinking unit for the preparation of polysaccharide networks

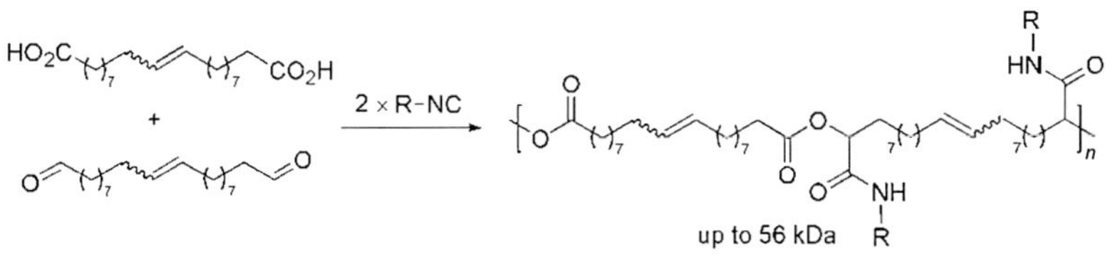

Scheme 7.10 Synthesis of polymers by multiple Passerini-3CRs, including bifunctional building blocks.

derived from 1-(deoxylactidyl)chitosan. All procedures were carried out in an aqueous medium, using in all cases cyclohexyl-isocyanide as the isonitrile component, to produce satisfactory levels of crosslinking (up to 80%) that enabled the formation of stable hydrogels.

Further examples include the use of various carboxylated polysaccharides, such as sodium hyaluronate and sodium alginate, as well as carboxymethylation-derived scleroglucan and dextran [35]. Hydrogels prepared via the Passerini-3CR were transparent, alkali-labile materials, whereas the transparency of the Ugi-4CR-based hydrogels depended on the polysaccharide, the crosslinker, and the degree of crosslinking.

7.2.5
Synthesis of Macromolecules by Multicomponent Polymerization

While many of the examples described above have involved the conjugation, immobilization, or crosslinking of bio(macro)molecules, the multiple multicomponent approach itself was not at the root of the construction of the polymeric scaffolds, but rather applied to their modification. Very recently, Meier and coworkers reported the first example of multiple Passerini-3CRs for the polymerization of bifunctional building blocks [36]. As shown in Scheme 7.10, long-chain dialdehydes and dicarboxylic acids were reacted with a variety of isocyanides to assemble high-molecular-weight compounds having the repeating α-acyloxycarboxamide units. As expected, an increment of the concentration of the bifunctional building blocks led to the formation of higher-molecular-weight polymers (up to 56 kDa).

7.3
Isocyanide-Based Multiple Multicomponent Macrocyclizations

While high concentrations are usually employed in reactions aimed at the generation of linear homopolymers, the opposite is necessary when a cyclization is desired. Accordingly, multiple IMCRs are suitable not only for the one-pot synthesis of high-molecular-weight polymers, but also of macrocycles – if the appropriate reaction conditions are implemented. Over the past ten years, Wessjohann and

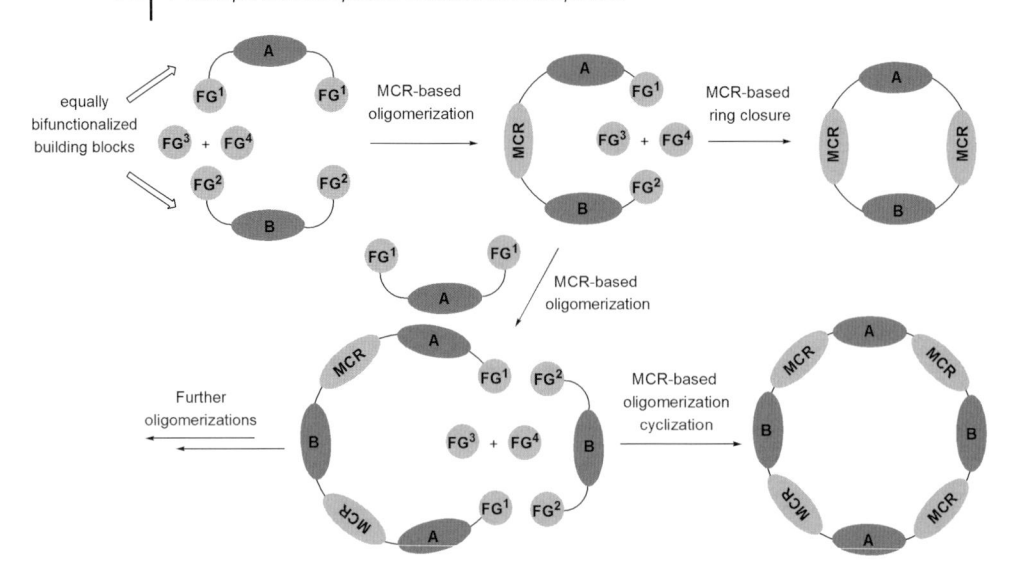

Scheme 7.11 Conceptual representation of the multiple multicomponent macrocyclization (MiB) methodology, as exemplified by an isocyanide-based four-component reaction (e.g., Ugi-4CR).

colleagues have developed different applications of a methodology termed "multiple multicomponent macrocyclization" (MiB) [26], the conceptual description of which is depicted in Scheme 7.11. The fundamental principle of this approach encompasses the use of two different bifunctional building blocks, which under high (pseudo) dilution conditions perform a tandem oligomerization/cyclization process to afford either a single macrocycle or a mixture of homo-oligomeric macrocycles. Accordingly, the key feature of this approach is that the IMCRs are responsible for not only the oligomerization(s) but also the ring-closure step, all of which takes place as a one-pot process. Another important characteristic of this method is the capability to generate hybrid macrocycles, owing to the different nature of the building blocks A and B. Thus, this approach differs conceptually from the traditional cyclo-oligomerization processes that usually lead to highly repetitive, symmetric macrocyclic scaffolds. Details of this concept, and its stereochemical and combinatorial consequences, have been reviewed recently [26].

7.3.1
Synthesis of Hybrid Macrocycles by Double Ugi-4CR-Based Macrocyclizations

Examples of the multiple multicomponent macrocyclization methodology have been described for a variety of IMCRs, including the Passerini-3CR [37], the Zhu three-component reaction (Zhu-3CR) for oxazoles [38], and (especially) the Ugi-4CR [39, 40]. Some examples of macrocycle syntheses endowed with hybrid structures, including aromatic, aliphatic, polyether, heterocyclic, steroidal, and

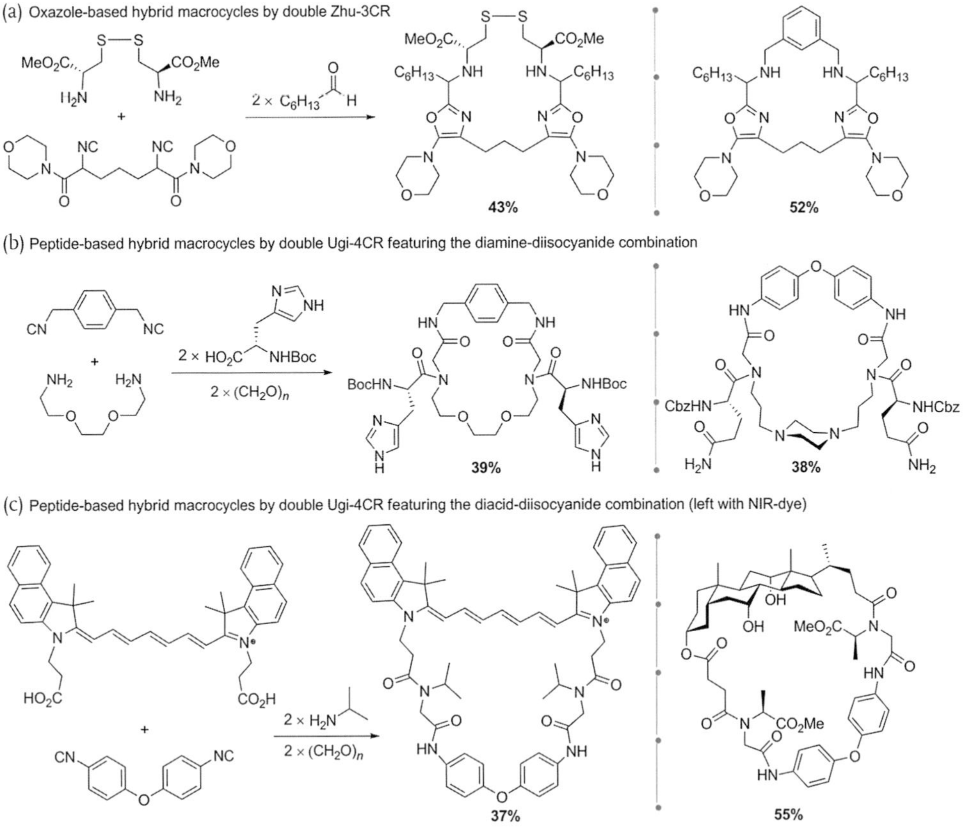

7.3 Isocyanide-Based Multiple Multicomponent Macrocyclizations | 245

(a) Oxazole-based hybrid macrocycles by double Zhu-3CR

(b) Peptide-based hybrid macrocycles by double Ugi-4CR featuring the diamine-diisocyanide combination

(c) Peptide-based hybrid macrocycles by double Ugi-4CR featuring the diacid-diisocyanide combination (left with NIR-dye)

Scheme 7.12 Synthesis of hybrid macrocycles by double Zhu-3-R and double Ugi-4CR-based macrocyclizations.

near-infrared dyes, are shown in Scheme 7.12. Zhu et al. employed their previously developed three-component reaction for the production of oxazole-based macrocycles [39]. Alternatively, Wessjohann and coworkers have developed several protocols – including template-driven procedures – for the synthesis of peptide-based hybrid macrocycles with potential in medicinal and supramolecular chemistry [40].

Variation of the macrocycle size based on the different multiplicities of the macrocyclization can be achieved using the building blocks A and B with mismatching size and length. For example, if a very long building block A and a too-short B are subjected to multiple multicomponent macrocyclization, this disparity favors the macrocycle based on fourfold rather than double MCRs [40b, 40c]. Another possibility of shifting the reaction outcome towards the fourfold-MCR-based macrocycle would be to use building blocks with a low conformational flexibility. In this case, the acyclic intermediate derived from the first MCR may be too strained to perform the ring closure itself; thus, the system may escape by

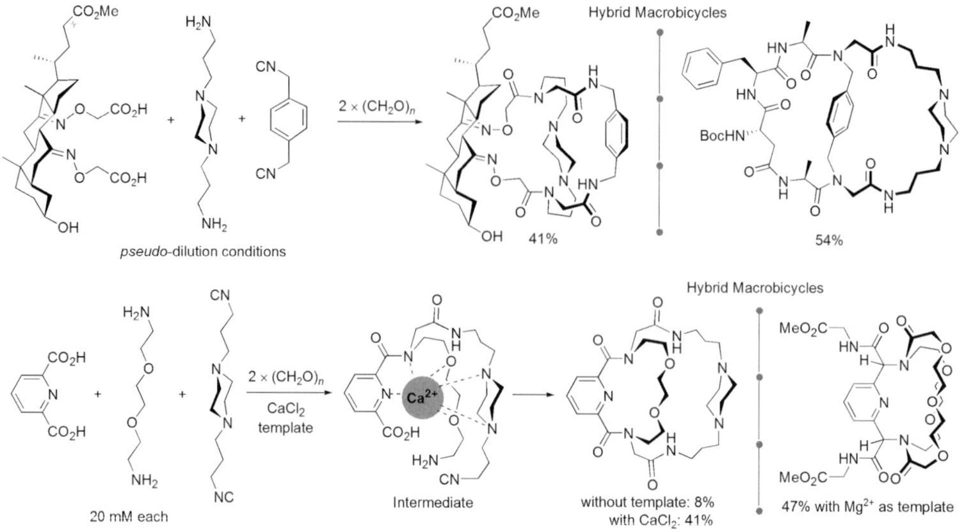

Scheme 7.13 Synthesis of hybrid macrobicycles by double Ugi-4CR-based macrocyclizations, incorporating three bifunctional building blocks.

a competitive, tandem-oligomerization/cyclization pathway, leading to the fourfold (and eventually, the sixfold) MCR-based macrocycle. This is accessible with very rigid building blocks, such as steroidal and *para*-substituted arylic building blocks [40]. Templating may also be used to form, preferentially, a certain oligomeric macrocycle, either from a static or a combinatorially dynamic preorganization [39a, 40b].

One important achievement in the field of multiple multicomponent macrocyclizations has been the recent development of one-pot procedures towards hybrid cryptands and cages. This unique procedure relies on the utilization of three different bifunctional building blocks which, either under *pseudo*-dilution or template-driven conditions, perform a double Ugi-4CR-based macrocyclization to furnish the corresponding macrobicycles (L.A. Wessjohann, O. Kreye, and D.G. Rivera, unpublished results).

As shown in Scheme 7.13, peptide-based cryptands and cages were assembled in one pot by the incorporation of three dissimilar building blocks into the hybrid macrobicyclic scaffolds. Attempts have also been made to assemble macrotricycles by the incorporation of four different bifunctional building blocks [41].

7.3.2
Synthesis of Macrobicycles by Threefold Ugi-4CR-Based Macrocyclization

Macrobicycles are usually endowed with improved encapsulating properties compared to analogous simple macrocycles, although the affinity towards a guest depends not only on the cavity size and shape but also on the host–guest complementarity. In the cryptands shown above, the hybrid macrobicyclic skeletons

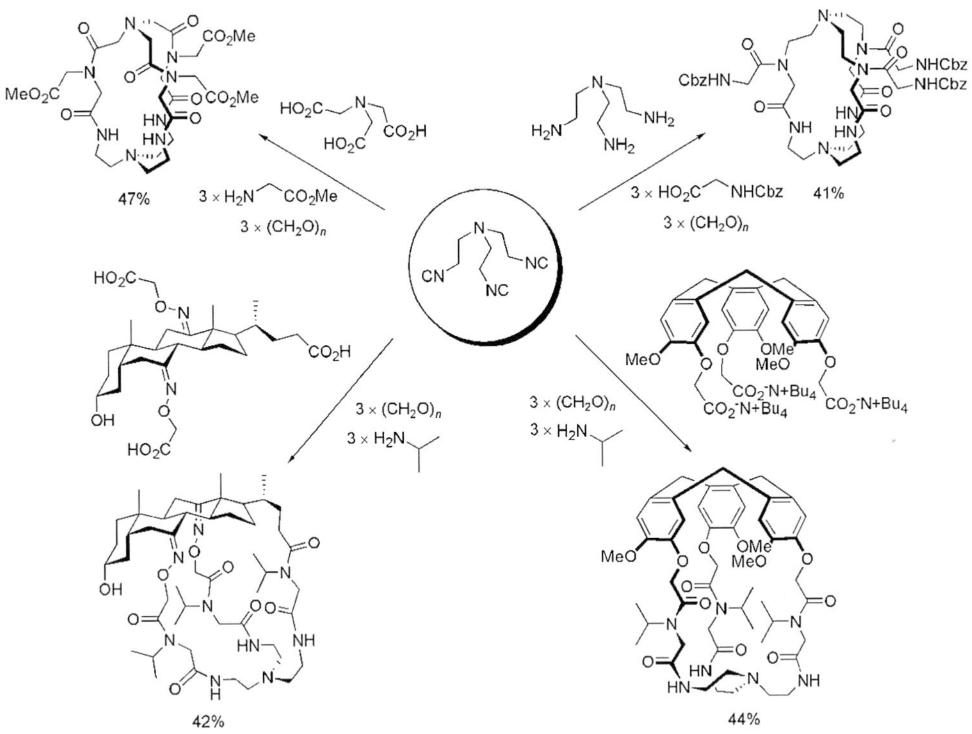

Scheme 7.14 Synthesis of macrobicycles by threefold Ugi-4CR-based macrocyclizations, including trifunctional building blocks.

include three dissimilar tether chains, whereas the peptidic moieties formed during the Ugi-4CRs appear as bridgeheads of the bicyclic structures. A few years ago, Wessjohann and Rivera devised an alternative strategy for the one-pot assembly of macrobicycles, wherein the Ugi-derived peptidic moieties emerge as tethers and not as bridge heads [41]. For this, the initial concept of utilizing bifunctional building blocks towards macrocycles was extended to the third dimension, thus giving rise to macrobicycles by the incorporation of two trifunctional building blocks.

As depicted in Scheme 7.14, a wide variety of macrobicyclic cavities – for example, cryptands, hemicryptophanes, and steroidal cages – were created in a one-pot process featuring the direct threefold Ugi-4CR-based macrocyclization. As shown previously for macrocycles, skeletal diversity could be generated either by varying the combination of Ugi-reactive functional groups (e.g., triacid/triisocyanide versus triamine/triisocyanide) or by variation of the nature of the building blocks. Further protocols were developed whereupon the nature of both the endocyclic and exocyclic elements was varied in a combinatorial manner [40b, 41]. Interestingly, the reaction yields were not decreased when passing from the double to the threefold Ugi-4CR-based macrocyclizations, though much care must be taken to select building blocks that match in size as well as being complementary in terms of their conformational flexibility.

7.4
Sequential Isocyanide-Based MCRs

The multiple multicomponent approaches described above allow an unprecedented diversity and complexity generation, despite being mostly homo-oligomeric in nature. Nevertheless, in areas such as target-oriented synthesis and drug development, the main challenge generally relates to the assembly of elaborated architectures, such that the functional groups and structural moieties must be located at precise positions, and with the correct stereochemistry. Consequently, the implementation of multiple IMCRs in one pot faces many difficulties, and most multiple multicomponent approaches related to medicinally relevant compounds involve IMCRs that are performed in a consecutive manner rather than in one pot. Because solid-phase synthesis is well suited for the implementation of sequential procedures, it becomes clear that many examples of sequential MCRs are carried out on solid phases. Solution-phase synthesis has also provided interesting inputs to this field, as it is more suitable for larger amounts or for the construction of branched scaffolds not available via linear, solid-phase protocols. Finally, the sequential implementation of IMCRs has provided one of the most remarkable examples of atom economy and creative synthetic design within the area of natural product-like syntheses. To date, several natural product syntheses have included one MCR step in their synthetic pathways, though until very recently none has been capable of assembling either a natural product scaffold or a very potent analog utilizing exclusively IMCRs. Rather, this was achieved for the first time in the synthesis of extremely active analogs of natural anticancer peptides, with only IMCRs being employed.

7.4.1
Sequential Approaches to Linear and Branched Scaffolds

As in the traditional peptide-coupling protocols, the sequential MCR-based approach comprises the use of a bifunctional component, in which one of the MCR-reactive groups remains protected during the first condensation cycle. After each MCR, a simple deprotection step is utilized to activate the remaining MCR-functional group, which reacts in the next MCR with a further mono-protected bifunctional component. The concept of a sequential IMCR was introduced in 1998 by Dömling [42], who proposed a solution-phase strategy based on consecutive Ugi-4CRs towards peptide nucleic acid (PNA) oligomers. Initially, the protocol was described for short sequences, but not for the long PNA oligomers as described previously by Nielsen [43]. Following publication of the sequential protocol towards PNA oligomers employing multiple Ugi-4CRs [43], several research groups applied the same concept to the solution-phase and [44, 45] solid-phase [46] syntheses of PNA monomers, amino-nucleobase chimeras [47], and chromium tricarbonyl labeled thymine PNA monomers [48, 49]. The most detailed investigation of using the Ugi-4CR to synthesize PNA oligomers was reported by Xu and coworkers in 2003 [50]. In this case, as depicted in Scheme 7.15, [N^4-(benzyloxy-carbonyl)

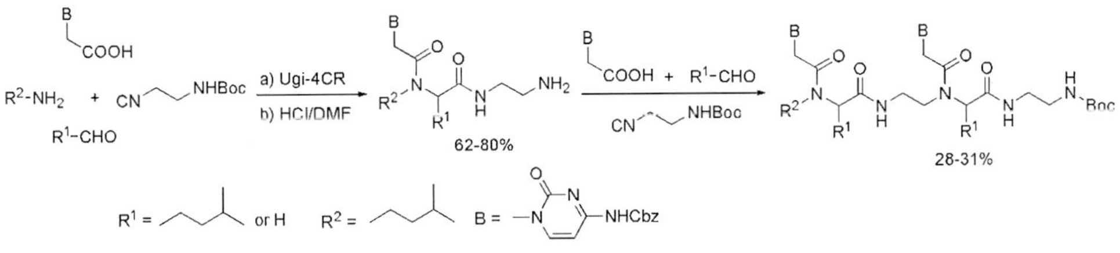

Scheme 7.15 Solution-phase sequential Ugi-4CRs to PNA dimers.

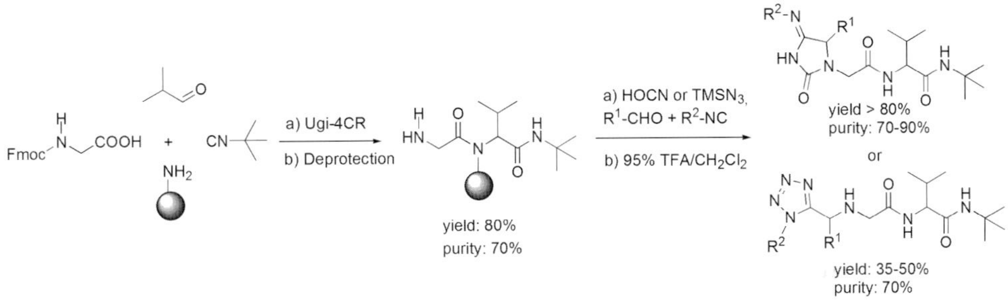

Scheme 7.16 Solid-phase sequential Ugi-4CRs to peptidic tetrazoles and hydantonimides.

cytosine-1-yl] acetic acid, 3-methylbutylamine, formaldehyde, or 3-methylbutanal and Boc-protected 2-isocyanoethylamine were employed to furnish the first-generation PNA monomers. The acidic cleavage of the Boc group and subsequent Ugi-4CR afforded the corresponding PNA dimers in 28–31% yields. Subsequently, the use of TFA in the deprotection step was shown to lead to an undesired side reaction during the next Ugi-4CRs.

In 2001, Constabel and Ugi described a detailed experimental protocol for the solid-phase implementation of repetitive Ugi-4CRs [51]. The first Ugi-4CR was performed on modified polystyrene or TentaGel which had free amino groups, while utilizing Fmoc-glycine, isobutyraldehyde, and *tert*-butylisocyanide to afford the corresponding resin-bound first-generation Ugi product (see Scheme 7.16). Subsequent deprotection of the Fmoc group enabled the second Ugi-4CR, which included diversification of the acid component leading to different reaction products. When seeking biologically active compounds, trimethylsilylazide and cyanic acid were chosen as the acid components, so as to afford peptidic tetrazoles and hydantonimides, respectively, after release from the solid support.

An alternative sequential approach to the same type of molecule was also developed by Ugi [52]. In this case, glycine was reacted with propanal and methyl isocyanide to afford the product of the Ugi-five-center-four-component reaction (Ugi-5C-4CR) (see Scheme 7.17), which bears a secondary amino group that is reactive in a further Ugi-3CR with parallel tetrazole formation. However, the intermediate did not react in the absence of an acidic catalyst, such as Dowex®.

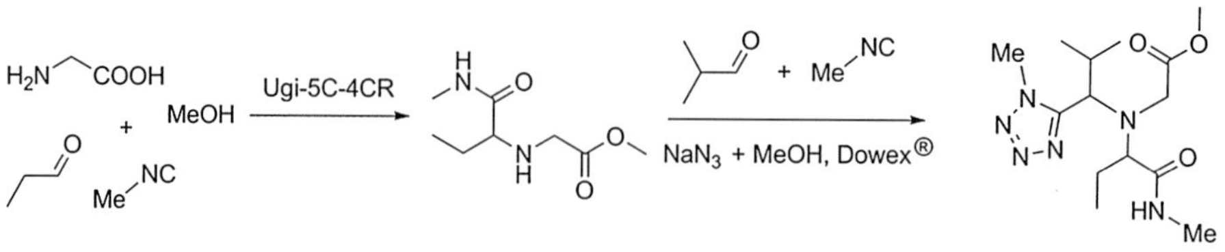

Scheme 7.17 Solution-phase sequential Ugi-4CRs to peptidic tetrazoles.

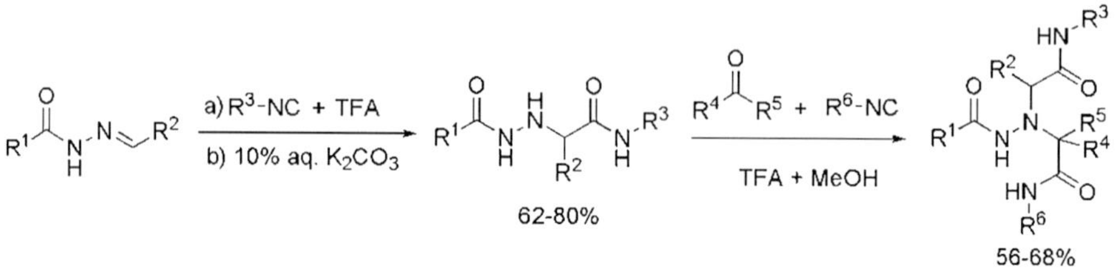

Scheme 7.18 Solution-phase sequential Ugi-4CRs to hydrazinopeptides.

As a consequence, this property was used by the authors to selectively produce a high skeletal diversity with the second Ugi-3CR.

Sequential Ugi-4CRs have also been explored in the synthesis of hydrazinopeptide-like scaffolds [53]. To achieve reactivity of the otherwise sparsely reactive hydrazones, trifluoroacetic acid (TFA) was employed as a temporary acid component, in order to form the trifluoroacetamide that is easily hydrolyzed *in situ* and under mild conditions, without affecting the peptidic skeleton. Here, trifluoroacetamide cleavage led to an Ugi product bearing a N^β-monoacylated hydrazine group which was suitable for an additional Ugi-4CR (see Scheme 7.18). The first IMCR was carried out in 1,4-dioxane in the presence of equimolar amounts of TFA, isocyanide, and pre-formed N-acyl hydrazones, after which the deprotected products were used in the second Ugi-4CR, together with a variety of ketones and aldehydes, TFA, and isocyanides. Although TFA acts in this later step as the acid component, it is cleaved *in situ* by the nucleophilic attack of methanol, producing an Ugi-3CR formally.

Westermann and coworkers reported the synthesis of a small library of peptoids by employing sequential, chemoselective, Ugi–Mumm/Ugi–Smiles reactions [54]. Generally, when 3-nitro-4-hydroxy-benzoic acid was used as the acid component in an Ugi-4CR, a product distribution derived from the competition between the Mumm and the Smiles rearrangements was found. However, by using controlled reaction conditions it was possible to determine the system's behavior and to achieve a reasonable level of selectivity. Thus, the addition of 1 equiv. of isopropylamine, isobutyraldehyde, and *tert*-butylisocyanide to 3-nitro-4-hydroxy-benzoic acid caused the product distribution to be 91% of the Ugi–Mumm product, and 7% of the Ugi–Smiles product (see Scheme 7.19). Subsequently, another set of components (i.e., amine, aldehyde, and isocyanide) was added to this mixture in order to proceed with the subsequent Ugi–Smiles reaction. This, in turn, led to

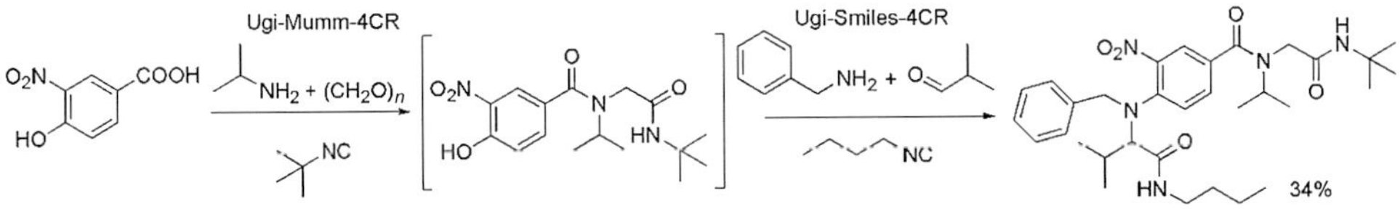

Scheme 7.19 Sequential Ugi–Mumm/Ugi–Smiles-4CRs to functionalized peptoids.

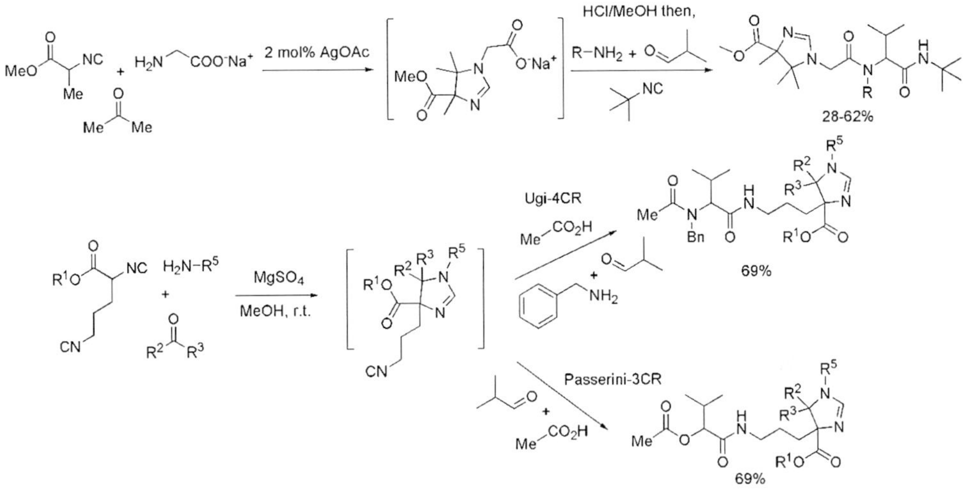

Scheme 7.20 Sequential isocyanide-based MCRs to *pseudo*-peptidic imidazolines.

the corresponding sequential Ugi–Mumm/Ugi–Smiles compounds as the major reaction products. This versatile method was also shown to tolerate the use of unprotected 1-β-amino-GlcNAc as the amino component, thus providing a new entry to the field of peptide-glycoconjugates. Although the overall reaction was conducted in one pot, this type of procedure has been included here because the two different MCRs occur in a sequential, time-resolved manner.

The synthesis of highly functionalized 2*H*-2-imidazolines using sequential MCRs was developed by Orru and coworkers in 2009 [55]. Among the various reactions investigated, the silver acetate-catalyzed reaction of an isocyanoacetate, acetone, and sodium glycinate was implemented to afford the corresponding imidazoline bearing a carboxylate function. Following acidification of the intermediate, the resulting carboxylic acid reacted with a variety of amines in the Ugi-4CR to furnish the final peptidic imidazolines, in moderate to good reaction yields (Scheme 7.20).

An alternative procedure that is shown in Scheme 7.20 utilizes the different reactivity between an isocyanoacetate and an alkyl isocyanide. Thus, the ornithine-derived diisocyanide could be selectively subjected to an initial 2*H*-2-imidazoline synthesis by taking advantage of the α-acidity of the isocyanoacetate moiety. Subsequently, the remaining isocyanide group was submitted to an Ugi-4CR or a Passerini-3CR to afford the peptidic or depsipeptidic-imidazolines, respectively.

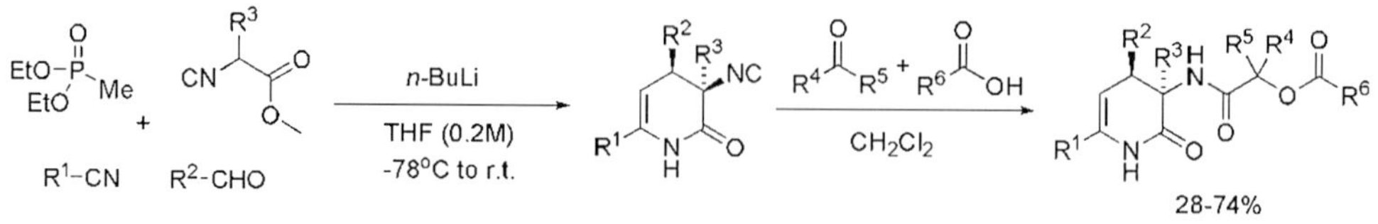

Scheme 7.21 Sequential Horner–Wadsworth–Emmons cyclocondensation/Passerini reactions to conformationally constrained peptidomimetics.

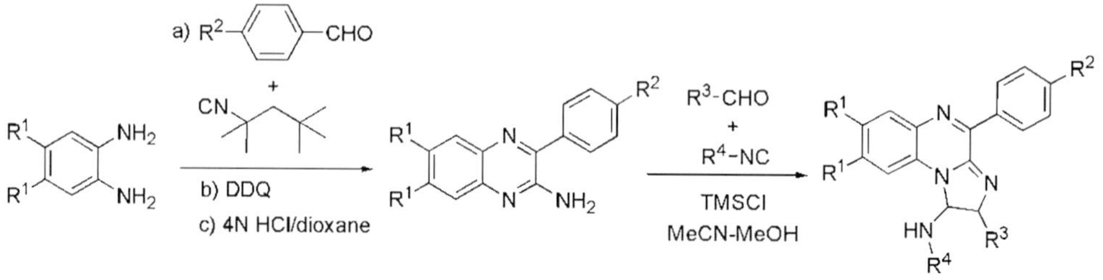

Scheme 7.22 Sequential isocyanide-based MCRs to imidazolo[1,2-a]quinoxalines.

Additional experiments to investigate variations of the first- and second-generation MCRs were also performed in this study [56].

Owing to their great feasibility and synthetic scope, most multiple MCRs have been based on IMCRs exclusively. Nevertheless, multiple MCRs that in addition to an IMCR also include non-isocyanide MCRs, have shown increasing successes during recent years. For example, constrained depsipeptides based on the dihydropyridone core were synthesized by a sequential Horner–Wadsworth–Emmons cyclocondensation/Passerini reaction sequence [56]. Initially, the sequential protocol was implemented as a two-step procedure, aiming at the substrate scope within the reaction sequence. Although the first MCR was found to tolerate both aliphatic and aromatic aldehydes and isocyanides, it transpired that the aromatic isocyanides provided a higher *syn* diastereoselectivity. As shown in Scheme 7.21, although both processes include isocyanide building blocks, the isocyanide functional group does not participate directly in the first condensation step. The isocyano-dihydropyridones arising from the first step were then subjected to Passerini-3CRs with a variety of aldehydes and ketones, to furnish conformationally constrained ester-peptides in moderate to good yields. Typically, the one-pot procedure achieved yields comparable to those of the two-step protocol.

Another interesting multiple multicomponent approach that makes use of a convertible isocyanide was recently reported [57]. In this case, new imidazolo[1,2-a]quinoxalines were accessible by a sequential protocol based on two different IMCRs (see Scheme 7.22). The reaction of 1,2-diamino benzenes and aldehydes with the Walborsky reagent, followed by tandem 2,3-dichloro-5,6-dicyanobenzoquinone (DDQ) oxidation/acidic cleavage, afforded quinoxalines in very good yields. The resulting intermediates were next employed as the amino component in a Groebke–

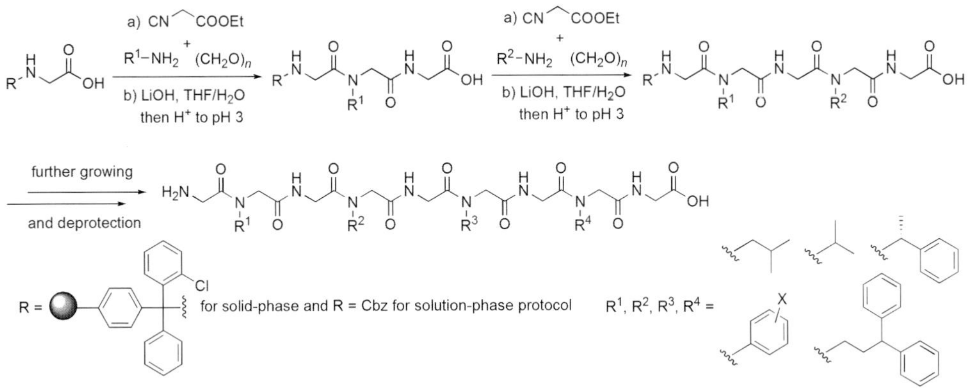

Scheme 7.23 Solid- and solution-phase syntheses of alternating peptoid–peptide chimeras by sequential Ugi-4CRs.

Blackburn reaction to furnish the desired polyheterocyclic scaffold with complete regioselectivity, as confirmed with nuclear Overhauser effect spectroscopy (NOESY) analysis and X-ray diffractometry.

Although most of the sequential procedures reported to date have relied on the implementation of two IMCRs, some examples exist of multiple multicomponent approaches that are based on three or more consecutive MCRs [51]. Notably, Wessjohann and coworkers have developed sequential, solid- and solution-phase procedures towards chimeric peptoid–peptide oligomers [58], for which the protocols include the consecutive implementation of several Ugi-4CRs such that elongation of the chain occurs at the C terminus – that is, the acid component. For this, either Cbz-glycine or resin-bound glycine (2-chlorotrityl chloride polystyrene resin) was subjected to the first Ugi-4CR by treatment with an amine, paraformaldehyde, and ethyl isocyanoacetate, to furnish the first-generation Ugi product (Scheme 7.23). Further saponification of the terminal ester moiety allowed for the next Ugi-4CR to be implemented, and this led to the second-generation Ugi product. The iterative cycles could be repeated up to five times in the solution-phase protocol, thus producing alternating peptoid–peptide chimeras in 52–90% yield. The solid-phase procedure also proved suitable for implementation of the repetitive Ugi-4CR/deprotection iterations, thus producing chimeric oligomers in 93–100% purity after cleavage from the solid support. The solution-phase variant was applied in the synthesis of cyclic peptides mimicking RGD-peptides (*vide infra*) [59–61].

One very important application of this multiple multicomponent approach has been recently described for the construction of functional dendrimers [60]. As shown in Scheme 7.24, the so-called MCR-dendrimers can be assembled by repetitive MCR/deprotection protocols to furnish three-dimensional (3-D), branched systems with high levels of external functionalization. This strategy unifies the concept of one-pot multiple MCRs with that of sequential MCRs, as the IMCRs are both multiple and sequential at each stage (generation) of the growing

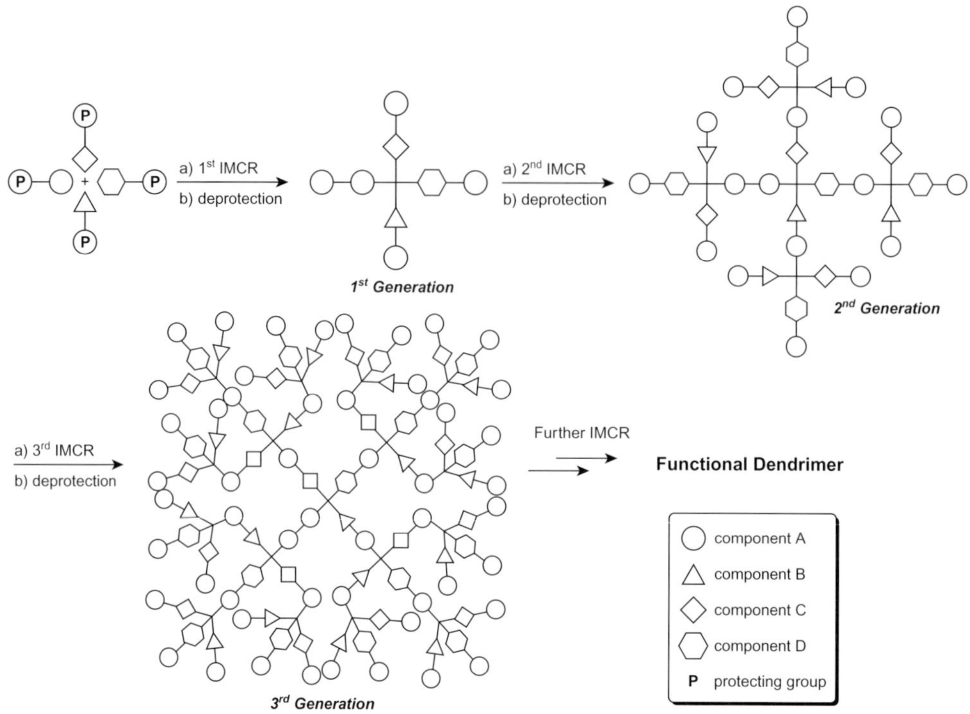

Scheme 7.24 Sequential/multiple MCR-based approach to functional dendrimers.

hyperbranched polymer. When utilizing the Ugi-4CR, the method includes the use of up to four mono-protected bifunctional building blocks, which undertake the first Ugi-4CR to produce a peptidic tetrafunctional core. Deprotection, followed by a fourfold Ugi-4CR-based procedure, affords a polypeptidic scaffold bearing up to 12 Ugi-reactive functional groups. A subsequent deprotection/Ugi-4CR-based protocol multiplies the number of internal peptidic moieties, and thus the outer functional groups by three, which makes this process one of the most efficient, bottom-up strategies towards biomimetic, functional dendrimers. The system also enables a total control of the size and molecular weight of the target dendrimer, while being well suited to the incorporation of almost every type of chemical functionalization. Notably, these functional dendrimers may be endowed with specific properties towards diverse applications in the fields of material chemistry, nanotechnology, molecular probing/recognition, and medical diagnostics.

7.4.2
Sequential Approaches to Macrocycles

The same strategy, of using sequential Ugi-4CRs for the assembly of linear peptidic scaffolds, may be combined with the Ugi-4C-macrocyclization as the last step, and has been applied to the solution-phase synthesis of RGD cyclopeptoids featur-

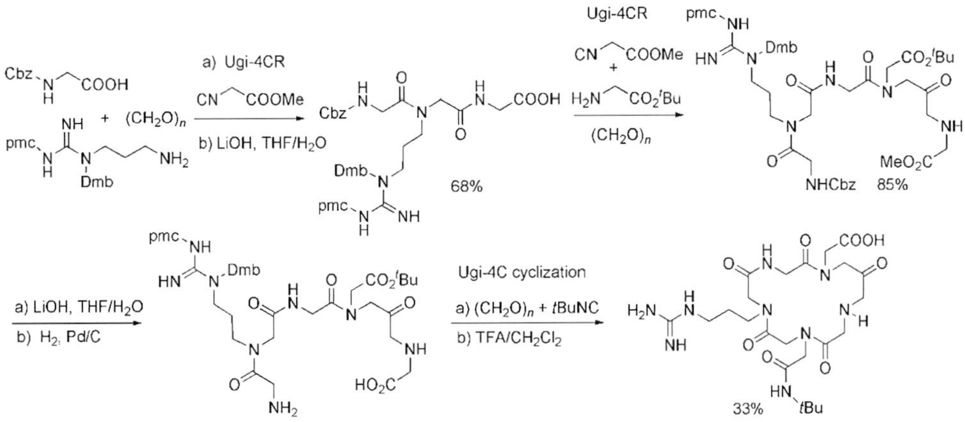

Scheme 7.25 Sequential elongation/cyclization in an Ugi-4CR-based approach to cyclopeptoids.

ing three IMCRs [61]. These unique cyclopeptoids were designed to behave as mimics of the arginine–glycine–aspartic acid loop sequence, which constitutes a major system for cell-adhesion processes. In this case, the sequential approach encompasses the use of Cbz-glycine and an arginine mimetic as the amino component, paraformaldehyde, and methyl isocyanoacetate to furnish the first-generation peptoid (Scheme 7.25). Carboxylate deprotection, followed by the second Ugi-4CR, led to an incorporation of the aspartate peptoid-mimic, the last acyclic intermediate, in very good yield. Finally, Ugi-4C-3CR macrocyclization allowed production of the RGD cyclopeptoids in a good yield for a macrocyclization, thus confirming the potential of consecutive Ugi-4CRs, not only for acyclic scaffolds but also for medicinally relevant cyclopeptide mimetics. This strategy towards cyclopeptide mimetics was subsequently extended to other examples and conditions, one example being the use of microwave-heated Ugi-4CRs to evaluate the substrate scope for peptoid elongation and macrocyclization [59].

Another sequential methodology to produce topologically diverse macromulticycles was recently described by Rivera and Wessjohann [62]. In this case, hybrid macrobicycles were prepared by consecutive double Ugi-4CR-based macrocyclizations that included diisocyanides at each step of the reaction sequence. The initial reaction between the dicarboxylic and diisocyanide building blocks enabled a synthesis of the first-generation macrocycles, which bear protected Ugi-reactive functionalities arising from the bifunctional amino component (Scheme 7.26). Deprotection of the intermediate macrocycles was then followed by the second double Ugi-4CR-based macrocyclization(s), using different diisocyanides to furnish highly complex macrobicycles, and with the overall formation of 16 new covalent bonds. By using this methodology a wide variety of hybrid macromulticycles was produced, including nonsymmetric cryptands, steroid-based clams, and igloo-shaped molecular cages. Conceptually, this approach is related to the direct assembly of hybrid macrobicycles by double Ugi-4CR-based macrocyclization (see Scheme 7.13). However, whilst the former method is more straightforward, the

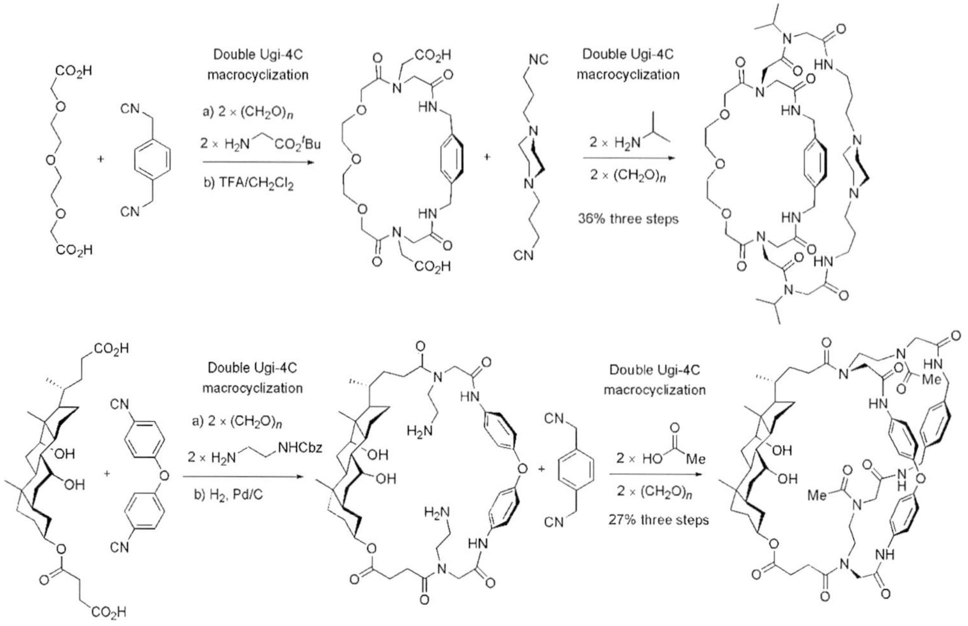

Scheme 7.26 Sequential double Ugi-4CR-based macrocyclizations to hybrid macrobicycles.

latter gives rise to higher levels of topological diversity, as the six different combinations of bifunctional building blocks can be implemented in a sequential, combinatorial manner [26, 39c, 62]. Indeed, the ability to readily synthesize such intriguing molecules, without involving previously required exhaustive protocols, represents a remarkable improvement towards synthetic tailor-made macrocycles and cages of as-yet unrivaled diversity and complexity.

7.4.3
Convergent Approach to Natural Product Mimics

Although the impact of IMCRs in target-oriented syntheses has improved over the past few years, most of the examples reported to date have incorporated one (or at most two) MCR step(s) towards the target compound [2c, 3]. However, the total synthesis of complex natural products (or their analogues), based exclusively on multiple MCRs, has remained elusive. Very recently, Wessjohann and colleagues described an important application of the multiple multicomponent approach for the synthesis of mimetics of the anticancer peptide tubulysin [63]. This approach has proved highly successful, with the designed mimetics—termed tubugis—displaying picomolar anticancer activities comparable to that of the natural product. Moreover, the analogues also incorporated an isosteric tertiary amide substitution that provided a greater accessibility and an improved hydrolytic stability compared to the "mother" compounds.

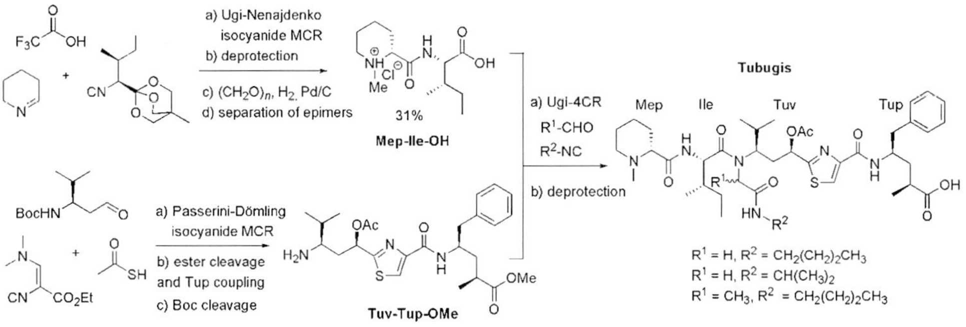

Scheme 7.27 Convergent approach to anticancer peptide mimetics, using three different sequential isocyanide-based MCRs.

The convergent approach employs three different sequential IMCRs to construct the natural product mimetics (Scheme 7.27). The strategy here is to divide the molecule into two different building blocks that can be prepared by separate MCRs and eventually unite them in a fragment condensation step based on a conventional Ugi-4CR ligation, ultimately to assemble the desired peptoid scaffold. For this, the Mep-Ile-OH dipeptide was prepared in 31% overall yield via an initial Ugi-4CR according to the amino acid-*ortho* ester isonitrile protocol of Nenajdenko [64]. The use of TFA as the acid component allowed for a mild cleavage of the trifluoroacetamide, and this was followed by a reductive amination to furnish the desired methylpipecolic moiety as the N-terminus. In a separate procedure, the Passerini–Dömling MCR was applied to furnish the tubuvaline amino acid, which was then used to prepare the Tuv-Tup-OMe dipeptide fragment, as described previously [65]. The key feature of this strategy was the use of the Ugi-4CR for the final condensation of Mep-Ile-OH and Tuv-Tup-OMe; in this way, a peptoid moiety could be produced that behaved as a surrogate of the labile *N,O*-acetal-ester moiety found in the natural peptide. Moreover, variation of the isocyanide component provided an easy access to an entire family of analogues for biological evaluation. The tubugis represent the first examples of highly bioactive natural product analogues produced via sequential multiple IMCRs. Indeed, their remarkable anticancer activities against human prostate and colon cancer cell lines, coupled to their availability, underline their great promise for further development as anticancer drugs.

7.5 Conclusions

Today, MCRs are considered to be among the most powerful processes for the production of complex and diverse chemical architectures. Taking into account the high levels of chemical efficiency and atom economy of most IMCR-based approaches, the tremendous potential derived from not only one, but rather two, three, or multiple MCRs, can easily be foreseen. As the IMCRs give rise to a wide variety of heterocycles and *pseudo*-peptides, the assembly of multiple peptidic and heterocyclic moieties at low synthetic cost provides one of the fastest and most efficient approaches to multivalent scaffolds. Accordingly, the application of multiple multicomponent approaches and products in the materials and life sciences is expected to increase in the near future, especially in areas of molecular architecture and supramolecular chemistry, or molecular sensing. As noted above, acyclic or macrocyclic multivalent platforms that are suitable for molecular interaction and recognition studies can be designed and produced in an efficient manner with (poly)isocyanides applied in multiple Ugi-4CRs and other reactions. Protein modification, enzyme immobilization and metal-free bioconjugation strategies represent other topics wherein the IMCRs continue to show great promise, owing to their high efficiency and the biocompatible nature of the resulting peptide-like skeletons. In addition, with the recent development of methodologies towards new, fully defined monodisperse polymers and dendrimers, those chemists involved in the (bio) materials field now possess a powerful, alternative strategy for the construction of highly functionalized macromolecules via solid- and solution-phase protocols based on sequential MCRs.

Finally, further applications of the multiple multicomponent approach for the total synthesis of natural products, and/or of their analogues, are to be expected, although at present a lack of stereoselective control poses major problems for most IMCRs.

With the advent of new IMCRs, however, polyfunctional isocyanides will become increasingly relevant and common among day-to-day investigations. Yet, very few of the above-described applications would have been possible without the appropriate isocyanides. Clearly, the synthesis, chemical behavior, tunable reactivity and, eventually, the commercial availability of these materials represent issues of utmost importance for further developments in this area.

References

1 Zhu, J. and Bienyamé, H. (eds) (2005) *Multicomponent Reactions*, Wiley-VCH Verlag GmbH, Weinheim.

2 (a) Ruijter, E., Scheffelaar, R., and Orru, R.V.A. (2011) Multicomponent reaction design in the quest for molecular complexity and diversity. *Angew. Chem. Int. Ed.*, **50**, 6234–6246; (b) El Kaïm, L. and Grimaud, L. (2009) Beyond the Ugi reaction: less conventional interactions between isocyanides and iminium species. *Tetrahedron*, **65**, 2153–2171; (c) Dömling, A. (2006) Recent developments in isocyanide based multicomponent reactions in applied chemistry. *Chem. Rev.*, **106**, 17–89; (d) Kappe, C.O. (2003) The generation of dihydropyrimidine libraries utilizing Biginelli multicomponent chemistry. *QSAR Comb. Sci.*, **22**, 630–645; (e) Dömling, A. (2002) Recent advances in isocyanide-based multicomponent chemistry. *Curr. Opin. Chem. Biol.*, **6**, 306–313; (f) Ugi, I. (2001) Recent progress in the chemistry of multicomponent reactions. *Pure Appl. Chem.*, **73**, 187–191; (g) Weber, L. (2000) High-diversity combinatorial libraries. *Curr. Opin. Chem. Biol.*, **4**, 295–302.

3 Toure, B.B. and Hall, D.G. (2009) Natural product synthesis using multicomponent reaction strategies. *Chem. Rev.*, **109**, 4439–4486.

4 (a) Hulme, C. (2005) *Multicomponent Reactions* (eds J. Zhu and H. Bienyamé),

Wiley-VCH Verlag GmbH, Weinheim, pp. 311–341; (b) Hulme, C. and Gore, V. (2003) "Multi-component reactions: emerging chemistry in drug discovery" "From xylocain to crixivan". *Curr. Med. Chem.*, **10**, 51–80.

5 Dömling, A. and Ugi, I. (2000) Multicomponent reactions with isocyanides. *Angew. Chem. Int. Ed.*, **39**, 3169–3210.

6 Banfi, L. and Riva, R. (2005) The Passerini reaction. *Org. React.*, **65**, 1–140.

7 Marcaccini, S. and Torroba, T. (2007) The use of the Ugi four-component condensation. *Nat. Protoc.*, **2**, 632–639.

8 Gulevich, A.V., Zhdanko, A.G., Orru, R.V.A., and Nenajdenko, V.G. (2010) Isocyanoacetate derivatives: synthesis, reactivity, and application. *Chem. Rev.*, **110**, 5235–5331.

9 Ugi, I. (1962) The α-addition of immonium ions and anions to isonitriles accompanied by secondary reactions. *Angew. Chem. Int. Ed. Engl.*, **1**, 8–21.

10 (a) Passerini, M. (1921) Sopra gli isonitril (I). Composto del p-isonitril-azobenzole con acetone ed acido acetica. *Gazz. Chim. Ital.*, **51**, 126–129; (b) Passerini, M. (1931) Sopra gli isonitrile. *Gazz. Chim. Ital.*, **61**, 964–969.

11 (a) Ugi, I., Meyr, R., Fetzer, U., and Steinbrückner, C. (1959) Versuche mit Isonitrilen. *Angew. Chem. Int. Ed. Engl.*, **71**, 386; (b) Ugi, I. and Steinbrückner, C. (1960) Über ein neues Kondensations-Prinzip. *Angew. Chem. Int. Ed. Engl.*, **72**, 267–268.

12 Ugi, I., Rosendahl, F.K., and Bodesheim, F. (1963) Isonitrile XIII. Kondensation von primären Aminen und Ketonen mit Isonitrilen und Rhodanwasserstoffsäure. *Justus Liebigs Ann. Chem.*, **666**, 54–61.

13 Hoppe, D. and Schöllkopf, U. (1972) Synthesen mit α-metallierten Isocyaniden, XV – 4-Äthoxycarbonyl-2-oxazoline und ihre Hydrolyse zu N-Formyl-β-hydroxy-α-aminosäureäthylestern. *Justus Liebigs Ann. Chem.*, **763**, 1–16.

14 (a) van Leusen, D. and van Leusen, A.M. (1991) Chemistry of sulfonylmethyl isocyanides 34. A new synthesis of progesterone from 17-[isocyano(tosyl)methylene]-3-methoxyandrosta-3,5-diene. *Synthesis*, 531–532; (b) van Leusen, A.M., van Leusen, D., Siderius, H., and Hoogenbo, B.E. (1972) New and simple synthesis of pyrrole ring-system from Michael acceptors and tosylmethylisocyanides. *Tetrahedron Lett.*, **13**, 5337–5340; (c) van Leusen, A.M., Wildeman, J., and Oldenziel, O.H. (1977) Base-induced cycloaddition of sulfonylmethyl isocyanides to C,N double-bonds – synthesis of 1,5-disubstituted and 1,4,5-trisubstituted imidazoles from aldimines and imidoyl chlorides. *J. Org. Chem.*, **42**, 1153–1159.

15 (a) Marcaccini, S. and Torroba, T. (2005) *Multicomponent Reactions* (eds J. Zhu and H. Bienyamé), Wiley-VCH Verlag GmbH, Weinheim, pp. 33–75; (b) Marcaccini, S. and Torroba, T. (1993) The use of isocyanides in heterocyclic synthesis – a review. *Org. Prep. Proced. Int.*, **25**, 141–208.

16 Dömling, A. (2005) *Multicomponent Reactions* (eds J. Zhu and H. Bienyamé), Wiley-VCH Verlag GmbH, Weinheim, pp. 76–94.

17 (a) Zhu, J.P. (2003) Recent developments in the isonitrile-based multicomponent synthesis of heterocycles. *Eur. J. Org. Chem.*, 1133–1144; (b) Janvier, P., Bienayme, H., and Zhu, J.P. (2002) A five-component synthesis of hexasubstituted benzene. *Angew. Chem. Int. Ed.*, **41**, 4291–4294; (c) Sun, X.W., Janvier, P., Zhao, G., Bienayme, H., and Zhu, J.P. (2001) A novel multicomponent synthesis of polysubstituted 5-aminooxazole and its new scaffold-generating reaction to pyrrolo[3,4-b]pyridine. *Org. Lett.*, **3**, 877–880; (d) Fayol, A. and Zhu, J.P. (2002) Synthesis of furoquinolines by a multicomponent domino process. *Angew. Chem. Int. Ed.*, **41**, 3633–3635; (e) Janvier, P., Sun, X.W., Bienayme, H., and Zhu, J.P. (2002) Ammonium chloride-promoted four-component synthesis of pyrrolo[3,4-b]pyridin-5-one. *J. Am. Chem. Soc.*, **124**, 2560–2567.

18 Ugi, I. and Steinbrückner, C. (1961) Isonitrile II. Reaktion von Isonitrilen mit Carbonylverbindungen, Aminen und Stickstoffsäure. *Chem. Ber.*, **64**, 734.

19 Ugi, I., Marquarding, D., and Urban, R. (1982) Chemistry and biochemistry of amino acids, in *Peptides and Proteins* (ed. B. Weinstein), Marcel Dekker, New York, pp. 246–289.

20 Leyden, R., Velasco-Torrijos, T., Andre, S., Gouin, S., Gabius, H.J., and Murphy, P.V. (2009) Synthesis of bivalent lactosides based on terephthalamide, N,N′-diglucosylterephthalamide, and glycophane scaffolds and assessment of their inhibitory capacity on medically relevant lectins. *J. Org. Chem.*, **74**, 9010–9026.

21 Bradley, H., Fitzpatrick, G., Glass, W.K., Kunz, H., and Murphy, P.V. (2001) Application of Ugi reactions in the synthesis of divalent neoglycoconjugates: evidence that the sugars are presented in restricted conformation. *Org. Lett.*, **3**, 2629–2632.

22 Liu, L.G., Li, C.P., Cochran, S., and Ferro, V. (2004) Application of the four-component Ugi condensation for the preparation of sulfated glycoconjugate libraries. *Bioorg. Med. Chem. Lett.*, **14**, 2221–2226.

23 Li, Y.X., Zhang, X.R., Chu, S.D., Yu, K.Y., and Guan, H.S. (2004) Synthesis of cluster mannosides via a Ugi four-component reaction and their inhibition against the binding of yeast mannan to concanavalin A. *Carbohydr. Res.*, **339**, 873–879.

24 Patch, J.A., Kirshenbaum, K., Seurynck, S.L., Zuckermann, R.N., and Barron, A.E. (2004) Versatile oligo(N-substituted) glycines: the many roles of peptoids in drug discovery, in *Pseudo-Peptides in Drug Discovery* (ed. P.E. Nielsen), Wiley-VCH Verlag GmbH, Weinheim, pp. 1–30.

25 Olsen, C.A. (2010) Peptoid-peptide hybrid backbone architectures. *Chembiochem*, **11**, 152–160.

26 Wessjohann, L.A., Rivera, D.G., and Vercillo, O.E. (2009) Multiple multicomponent macrocyclizations (MiBs): a strategic development toward macrocycle diversity. *Chem. Rev.*, **109**, 796–814.

27 Rivera, D.G., Concepcion, O., Perez-Labrada, K., and Coll, F. (2008) Synthesis of diamino-furostan sapogenins and their use as scaffolds for positioning peptides in a preorganized form. *Tetrahedron*, **64**, 5298–5305.

28 (a) Li, C.H., Atiq, U.R., Dalley, N.K., and Savage, P.B. (1999) Short syntheses of triamine derivatives of cholic acid. *Tetrahedron Lett.*, **40**, 1861–1864; (b) Davis, A.P. and Perez-Payan, M.N. (1999) The "triamino-analogue" of methyl cholate: a practical, large-scale synthesis. *Synlett*, 991–993.

29 Ugi, I., Goebel, M., Gruber, B., Heilingbrunner, M., Heiss, C., Horl, W., Kern, O., Starnecker, M., and Dömling, A. (1996) Molecular libraries in liquid phase via Ugi-MCR. *Res. Chem. Intermed.*, **22**, 625–644.

30 (a) Marek, M., Jary, J., Valentova, O., and Vodrazka, Z. (1983) Immobilization of glycoenzymes by means of their glycosidic components. *Biotechnol. Lett.*, **5**, 653–658; (b) Marek, M., Valentova, O., and Kas, J. (1984) Invertase immobilization via its carbohydrate moiety. *Biotechnol. Bioeng.*, **26**, 1223–1226; (c) Axen, R., Vretblad, P., and Porath, J. (1971) The use of isocyanides for the attachment of biologically active substances to polymers. *Acta. Chem. Scand.*, **25**, 1129–1132.

31 König, S. and Ugi, I. (1991) Crosslinking of aqueous alginic acid by four component condensation with inclusion immobilization of enzymes. *Z. Naturforsch. B*, **46b**, 1261–1265.

32 Ziegler, T., Gerling, S., and Lang, M. (2000) Preparation of bioconjugates through an Ugi reaction. *Angew. Chem. Int. Ed.*, **39**, 2109–2112.

33 Camacho, C., Matias, J.C., García-Rivera, D., Simpson, B.K., and Villalonga, R. (2007) Amperometric enzyme biosensor for hydrogen peroxide via Ugi multicomponent reaction. *Electrochem. Commun.*, **9**, 1655–1660.

34 de Nooy, A.E.J., Masci, G., and Crescenzi, V. (1999) Versatile synthesis of polysaccharide hydrogels using the Passerini and Ugi multicomponent condensations. *Macromolecules*, **32**, 1318–1320.

35 (a) De Nooy, A.E.J., Capitani, D., Masci, G., and Crescenzi, V. (2000) Ionic polysaccharide hydrogels via the Passerini

and Ugi multicomponent condensations: synthesis, behavior and solid-state NMR characterization. *Biomacromolecules*, **1**, 259–267; (b) Crescenzi, V., Francescangeli, A., Capitani, D., Mannina, L., Renier, D., and Bellini, D. (2003) Hyaluronan networking via Ugi's condensation using lysine as cross-linker diamine. *Carbohydr. Polym.*, **53**, 311–316.

36 Kreye, O., Toth, T., and Meier, M.A.R. (2011) Introducing multicomponent reactions to polymer science: Passerini reactions of renewable monomers. *J. Am. Chem. Soc.*, **133**, 1790–1792.

37 Leon, F., Rivera, D.G., and Wessjohann, L.A. (2008) Multiple multicomponent macrocyclizations including bifunctional building blocks (MiBs) based on Staudinger and Passerini three-component reactions. *J. Org. Chem.*, **73**, 1762–1767.

38 Janvier, P., Bois-Choussy, M., Bienayme, H., and Zhu, J.P. (2003) A one-pot four-component (ABC(2)) synthesis of macrocycles. *Angew. Chem. Int. Ed.*, **42**, 811–814.

39 (a) Wessjohann, L.A., Rivera, D.G., and Leon, F. (2007) Freezing imine exchange in dynamic combinatorial libraries with Ugi reactions: Versatile access to templated macrocycles. *Org. Lett.*, **9**, 4733–4736; (b) Wessjohann, L.A., Rivera, D.G., and Coll, F. (2006) Synthesis of steroid-biaryl ether hybrid macrocycles with high skeletal and side chain variability by multiple multicomponent macrocyclization including bifunctional building blocks. *J. Org. Chem.*, **71**, 7521–7526; (c) Wessjohann, L.A., Voigt, B., and Rivera, D.G. (2005) Diversity oriented one-pot synthesis of complex macrocycles: very large steroid-peptoid hybrids from multiple multicomponent reactions including bifunctional building blocks. *Angew. Chem. Int. Ed.*, **44**, 4785–4790.

40 (a) Rivera, D.G., Vercillo, O.E., and Wessjohann, L.A. (2008) Rapid generation of macrocycles with natural product-like side chains by multiple multicomponent macrocyclizations (MiBs). *Org. Biomol. Chem.*, **6**, 1787–1795; (b) Rivera, D.G., Vercillo, O.E., and Wessjohann, L.A. (2007) Combinatorial synthesis of macrocycles by multiple multicomponent macrocyclization including bifunctional building blocks (MiB). *Synlett*, 308–312; (c) Rivera, D.G. and Wessjohann, L.A. (2007) Synthesis of novel steroid-peptoid hybrid macrocycles by multiple multicomponent macrocyclizations including bifunctional building blocks (MiBs). *Molecules*, **12**, 1890–1899; (d) Michalik, D., Schaks, A., and Wessjohann, L.A. (2007) One-step synthesis of natural product-inspired biaryl ether-cyclopeptoid macrocycles by double Ugi multiple-component reactions of bifunctional building blocks. *Eur. J. Org. Chem.*, 149–157; (e) Kreye, O., Westermann, B., Rivera, D.G., Johnson, D.V., Orru, R.V.A., and Wessjohann, L.A. (2006) Dye-modified and photoswitchable macrocycles by multiple multicomponent macrocyclizations including bifunctional building blocks (MiBs). *QSAR Comb. Sci.*, **25**, 461–464.

41 Rivera, D.G. and Wessjohann, L.A. (2006) Supramolecular compounds from multiple Ugi multicomponent macrocyclizations: peptoid-based cryptands, cages, and cryptophanes. *J. Am. Chem. Soc.*, **128**, 7122–7123.

42 Dömling, A. (1998) A novel concept for the combinatorial synthesis of peptide nucleic acids. *Nucleosides, Nucleotides, Nucleic Acids*, **17**, 1667–1670.

43 Nielsen, P.E. (ed.) (2004) *Pseudo-Peptides in Drug Discovery*, Wiley-VCH Verlag GmbH, Weinheim.

44 Maison, W., Schlemminger, I., Westerhoff, O., and Martens, J. (1999) Modified PNAs: a simple method for the synthesis of monomeric building blocks. *Bioorg. Med. Chem. Lett.*, **9**, 581–584.

45 Dömling, A., Chi, K.Z., and Barrere, M. (1999) A novel method to highly versatile monomeric PNA building blocks by multi component reactions. *Bioorg. Med. Chem. Lett.*, **9**, 2871–2874.

46 Wang, W.H., Zou, X.M., Zhang, X., Fu, Y.Q., and Xu, P. (2005) Solid-phase synthesis of PNA monomer by Ugi four-component condensation. *Chin. Chem. Lett.*, **16**, 585–588.

47 Maison, W., Schlemminger, I., Westerhoff, O., and Martens, J. (2000) Multicomponent synthesis of novel

amino acid-nucleobase chimeras: a versatile approach to PNA-monomers. *Bioorg. Med. Chem.*, **8**, 1343–1360.
48. Baldoli, C., Maiorana, S., Licandro, E., Zinzalla, G., and Perdicchia, D. (2002) Synthesis of chiral chromium tricarbonyl labeled thymine PNA monomers via the Ugi reaction. *Org. Lett.*, **4**, 4341–4344.
49. Baldoli, C., Giannini, C., Licandro, E., Maiorana, S., and Zinzalla, G. (2004) A thymine-PNA monomer as new isocyanide component in the Ugi reaction: a direct entry to PNA dimers. *Synlett*, 1044–1048.
50. Xu, P., Zhang, T., Wang, W.H., Zou, X.M., Zhang, X., and Fu, Y.Q. (2003) Synthesis of PNA monomers and dimers by Ugi four-component reaction. *Synthesis*, 1171–1176.
51. Constabel, F. and Ugi, I. (2001) Repetitive Ugi reactions. *Tetrahedron*, **57**, 5785–5789.
52. Ugi, I.K., Ebert, B., and Horl, W. (2001) Formation of 1,1′-iminodicarboxylic acid derivatives, 2,6-diketo-piperazine and dibenzodiazocine-2,6-dione by variations of multicomponent reactions. *Chemosphere*, **43**, 75–81.
53. Bushkova, E., Parchinsky, V., and Krasavin, M. (2010) Efficient entry into hydrazinopeptide-like structures via sequential Ugi reactions. *Mol. Divers.*, **14**, 493–499.
54. Brauch, S., Gabriel, L., and Westermann, B. (2010) Seven-component reactions by sequential chemoselective Ugi-Mumm/Ugi-Smiles reactions. *Chem. Commun.*, **46**, 3387–3389.
55. Elders, N., van der Born, D., Hendrickx, L.J.D., Timmer, B.J.J., Krause, A., Janssen, E., de Kanter, F.J.J., Ruijter, E., and Orru, R.V.A. (2009) The efficient one-pot reaction of up to eight components by the union of multicomponent reactions. *Angew. Chem. Int. Ed.*, **48**, 5856–5859.
56. Paravidino, M., Scheffelaar, R., Schmitz, R.F., de Kanter, F.J.J., Groen, M.B., Ruijter, E., and Orru, R.V.A. (2007) A flexible six-component reaction to access constrained depsipeptides based on a dihydropyridinone core. *J. Org. Chem.*, **72**, 10239–10242.
57. Krasavin, M., Shkavrov, S., Parchinsky, V., and Bukhryakov, K. (2009) Imidazo[1,2-a]quinoxalines accessed via two sequential isocyanide-based multicomponent reactions. *J. Org. Chem.*, **74**, 2627–2629.
58. (a) Wessjohann, L.A., Phuong, T.T., and Westermann, B. (2008) PTC Int. Appl. 2008022800; (b) Wessjohann, L.A., Phuong, T.T., and Westermann, B. (2008) *Chem. Abstr.*, **148**, 285479.
59. Barreto, A.D.S., Vercillo, O.E., Birkett, M.A., Caulfield, J.C., Wessjohann, L.A., and Andrade, C.K.Z. (2011) Fast and efficient microwave-assisted synthesis of functionalized peptoids via Ugi reactions. *Org. Biomol. Chem.*, **9**, 5024–5027.
60. Wessjohann, L.A., Henze, M., Kreye, O., and Rivera, D.G. (2010) Patent Application: MCR-Dendrimere. EP 10 018882.4.
61. Vercillo, O.E., Andrade, C.K.Z., and Wessjohann, L.A. (2008) Design and synthesis of cyclic RGD pentapeptoids by consecutive Ugi reactions. *Org. Lett.*, **10**, 205–208.
62. Rivera, D.G. and Wessjohann, L.A. (2009) Architectural chemistry: synthesis of topologically diverse macromulticycles by sequential multiple multicomponent macrocyclizations. *J. Am. Chem. Soc.*, **131**, 3721–3732.
63. Pando, O., Stark, S., Denkert, A., Porzel, A., Preusentanz, R., and Wessjohann, L.A. (2011) The multiple multicomponent approach to natural product mimics: tubugis, N-substituted anticancer peptides with picomolar activity. *J. Am. Chem. Soc.*, **133**, 7692–7695.
64. Zhdanko, A.G. and Nenajdenko, V.G. (2009) Nonracemizable isocyanoacetates for multicomponent reactions. *J. Org. Chem.*, **74**, 884–887.
65. Dömling, A., Beck, B., Eichelberger, U., Sakamuri, S., Menon, S., Chen, Q.Z., Lu, Y.C., and Wessjohann, L.A. (2006) Total synthesis of tubulysin U and V. *Angew. Chem. Int. Ed.*, **45**, 7235–7239.

8
Zwitterions and Zwitterion-Trapping Agents in Isocyanide Chemistry

Ahmad Shaabani, Afshin Sarvary, and Ali Maleki

8.1
Introduction

The combinatorial chemistry, sequential transformations, and one-pot multicomponent reactions (MCRs), in which three or more starting materials react to form a product, are always resource-effective and environmentally acceptable, and thus "greener" as compared to multistep reactions. Moreover, MCRs offer significant advantages over conventional linear-step syntheses, by reducing time and also saving money, energy, and raw materials. Consequently, both economical and environmental benefits are achieved. Yet, at the same time, diversity can be achieved for building up libraries, simply by varying each component. Because of the unique reactivity of the isocyanide functional group, which undergoes facile additions with nucleophiles and electrophiles, isocyanide-based MCRs (IMCRs) are among the most versatile, in terms of the number and variety of compounds with potential biological and medicinal activities which can be generated. As a result of these benefits, the chemistry of the isocyanides and the use of IMCRs have been the subject of many reviews [1–10].

The term "zwitterion" is derived from the German word "zwitter" (hybrid); hence, a zwitterion is a neutral molecule with a positive and a negative electrical charge at different locations within that molecule; occasionally, zwitterions may be referred to as "inner salts."

Usually, dipolar compounds are not classified as zwitterions. For example, amine oxides are not zwitterions in true terms of the definition, which specifies that there must be unit electrical charges on the atoms. The distinction lies in the fact that the plus and minus signs on the amine oxide signify formal charges, but not electrical charges. Other compounds which sometimes are referred to as zwitterions (mistakenly according to the definition above) include nitrones and 1,2- or 1,3-dipolar compounds [11].

It was in 1969 that Winterfeld described, for the first time, the reaction between isocyanides **1** and acetylenic compounds **2** [12]. This chemistry is based on the

Isocyanide Chemistry: Applications in Synthesis and Material Science, First Edition. Edited by Valentine Nenajdenko.
© 2012 Wiley-VCH Verlag GmbH & Co. KGaA. Published 2012 by Wiley-VCH Verlag GmbH & Co. KGaA.

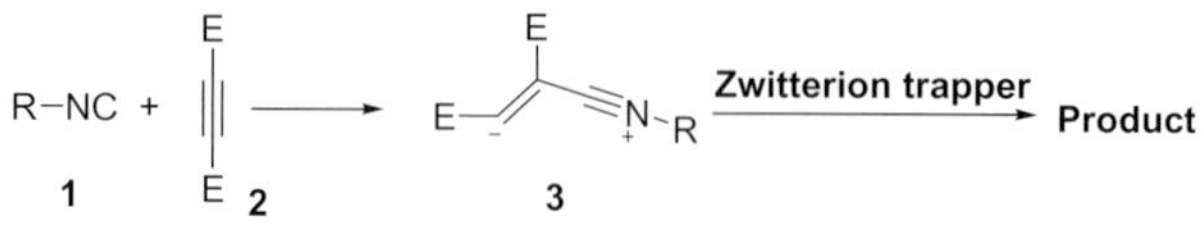

Scheme 8.1

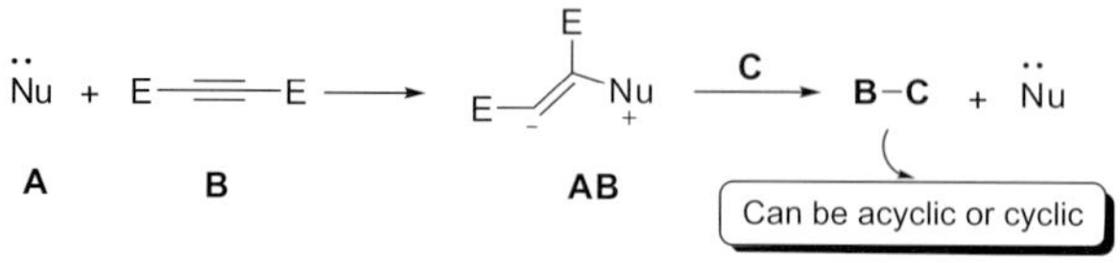

Scheme 8.2

initial formation of a zwitterionic adduct **3**, which further undergoes cycloaddition-type reactions (Scheme 8.1).

A variety of nucleophiles such as triphenyl phosphine (PPh$_3$) [13], pyridine [14], tertiary amine [15], and dimethyl sulfoxide [16] has been known to generate zwitterionic species via this pathway. These reactive intermediates can be captured by suitable substrates and, after a series of transformations, the nucleophile will be eliminated from the system. Thus, the effective role played by the nucleophile is to "conduct" a reaction involving two components (see Scheme 8.1). Such two-component reactions may be catalyzed by a variety of different nucleophilic species, among which phosphines, pyridines, and tertiary amines have been the most extensively studied. In contrast, nucleophiles such as isocyanides and nucleophilic carbenes generally lead to multicomponent reactions that involve an incorporation of the nucleophile. A generalized representation of the rationale behind these reactions is shown in Scheme 8.2.

In this class of reactions, a dipolar/zwitterionic intermediate **III** (or **AB**) is generated by the addition of an aprotic nucleophile **I** to activated π systems, such as alkynes, alkenes, and their diaza analogues (collectively represented as **II**) (Scheme 8.3). The zwitterion may have a 1,3 or 1,4 disposition, depending on the nature of the nucleophile. A subsequent interception of the zwitterion **III** with **IV**, which represents a large variety of compounds, can take different routes. Overall, three-component dipolar cycloadditions will result when the nucleophile is retained in the product (as in paths **A** and **C**), whereas the nucleophile plays a catalytic role en route to **VI**, **VII**, or other products (Scheme 8.3, path **B**). The nature of the product formed in path **B** depends largely on the nucleophile **I** and the substituents R^1 and R^2 on receptor **II**. It is evident from this scheme that path **B** surpasses both the other options possible for **III** in terms of variety and, in principle, falls under the realm of organocatalysis.

Similar zwitterionic species have been known to result from the addition of an isocyanide to dimethyl acetylenedicarboxylate (DMAD). Isocyanides, however, differ in their reactivity profile due to the presence of a formal divalent carbon. In transformations involving these entities, this divalent carbon becomes converted

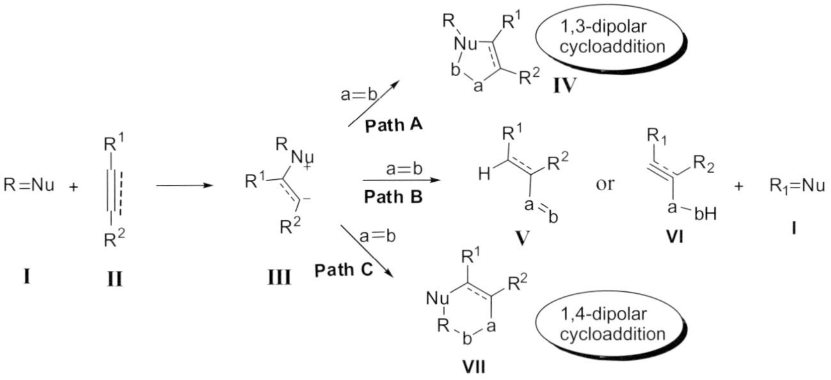

Scheme 8.3

to a tetravalent state, which causes the addition to become irreversible. It might be recalled that this property of isocyanides has been crucial in the successful development of classical MCRs such as the Passerini three-component reaction (P-3CR) [17] and Ugi four-component reaction (U-4CR) [9].

8.2
Generation of Zwitterionic Species by the Addition of Isocyanides to Alkynes

The preliminary studies on the reaction of cyclohexyl isocyanide with DMAD were conducted by Huisgen and coworkers, in Munich [18]; indeed, the advent of the Huisgen reactions [19, 20] heralded a new era in the synthesis of heterocyclic compounds. Consequent to the development of this reaction, the concept of generating and trapping zwitterionic species, leading to heterocycles, appeared highly attractive, and one of the most important means of generating zwitterionic species was shown to involve the addition of an isocyanide to an electron-deficient alkyne, such as DMAD. Subsequently, the zwitterionic intermediate could be captured efficiently by employing various CH-, NH-, or OH-acids, electron-deficient olefins and other systems, leading in turn to five- and six-membered heterocycles such as carbonyl compounds.

A correct choice of the third component is crucial; notably, it should be inert towards both DMAD and isocyanide yet, at the same time, it should be more reactive towards the zwitterion than DMAD itself [21]. Several compounds, including CH-acids, NH-acids, OH-acids, carbonyl and imine compounds, and electron-deficient olefins were shown to satisfy both of these criteria. In the absence of a third component, however, the reaction of isocyanides with alkynes provided a mixture of products, all of which essentially involved isocyanide and DMAD, albeit in different proportions [22]. Thus, it has been reported that, in aprotic solvents, the reaction of several isocyanides with hexafluoro-2-butyne **4** can provide interesting cyclopropene derivatives **5** (1 : 2 adducts) (Scheme 8.4) [23].

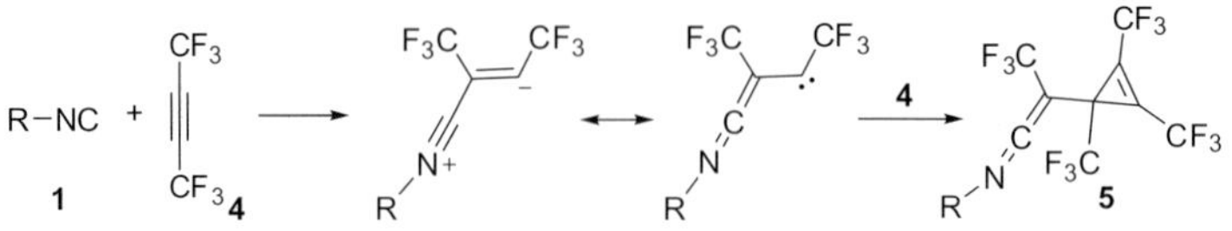

Scheme 8.4

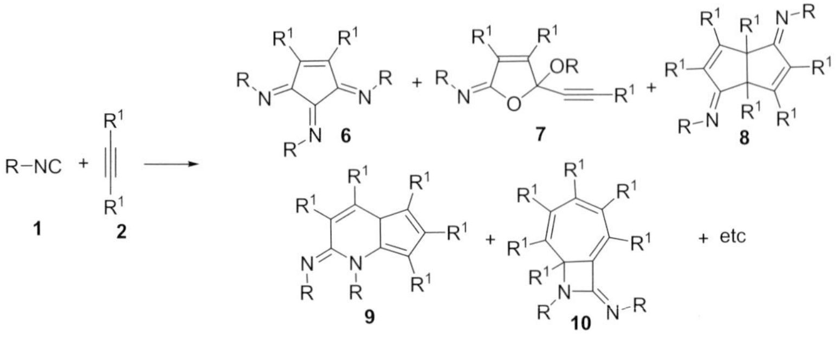

Scheme 8.5

The reaction of isocyanides with DMAD, in the absence of a trapper, has afforded several adducts that include the cyclopent-4-ene-1,2,3-triylidene **6**, the iminolactone **7**, the pentalene **8**, the cyclopenta[b]pyridine derivatives **9**, and the azabicyclononatriene **10** [24] (Scheme 8.5). More recently, George et al. established the structure of these adducts by using single-crystal X-ray analysis [18].

8.2.1
CH-Acids as Zwitterion-Trapping Agents

The great versatility of CH-acids as reagents in organic synthesis is based on the simplicity of their structures which, when combined with their unique properties and complex reactivities, has led to them becoming molecules of great synthetic and theoretical interest. Today, wide arrays of synthetic transformations have emerged as a result of the intrinsic electrophilic and nucleophilic properties of CH-acids. Moreover, because of these features the CH-acids have been used extensively to trap zwitterions created by the addition of isocyanides to alkynes and systems, leading to five- and six-membered heterocycles.

The trapping of zwitterions **3** with various annular CH-acids via three-component, one-pot reactions has been reported for the synthesis of oxocyclohepta[b]pyrans **12** [25], pyranophenalenones **14** [26], bis(4H-chromene-) **16** and 4H-benzo[g]chromenes **18** [27], annulated 4H-pyrans **20** and **22** (the synthesis of **20** and **22** in water, under phase-transfer conditions, has also been developed) [28], 5-oxo-4,5-dihydroindeno[1,2-b]pyrans **24** [29], 4H-pyrans **26** and **28** [30], tetrahydrocyclopenta[b]pyrans **30** [31], pyrano[3,2-d]isoxazoles **32** [32], pyrano[2,3-c]pyrazoles **34** [33], and

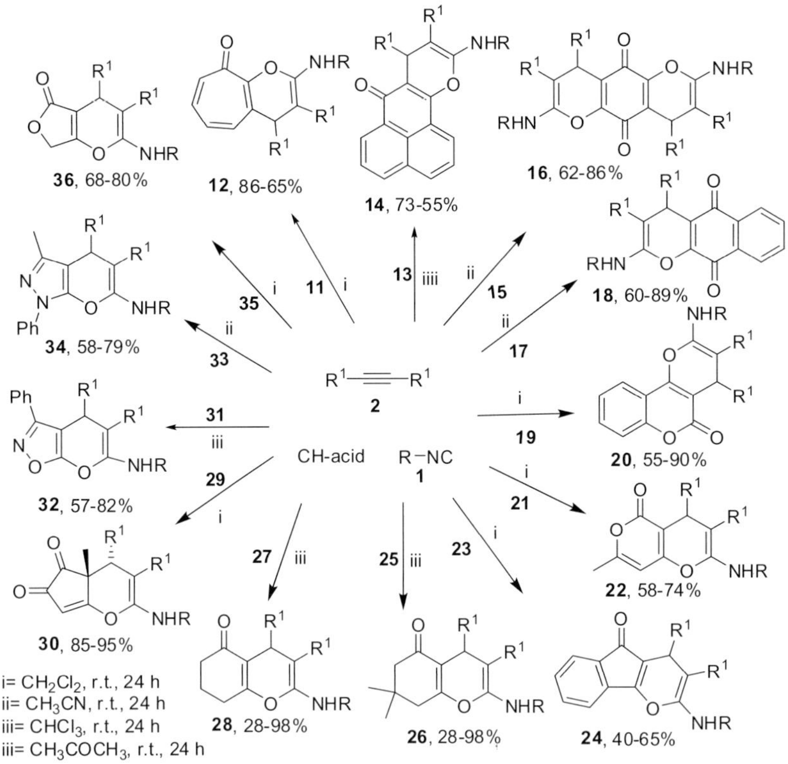

8.2 Generation of Zwitterionic Species by the Addition of Isocyanides to Alkynes | 267

Figure 8.1 Products obtained from the trapping of zwitterions with various CH-acids: **11**, α-Tropolone; **13**, 3-Hydroxy-1H-phenalene-1-one; **15**, 2,5-Dihydroxycyclohexa-2,5-diene-1,4-dione; **17**, 2-Hydroxynaphthalene-1,4-dione; **19**, 4-Hydroxy-2H-chromen-2-one; **21**, 4-Hydroxy-6-methyl-2H-pyran-2-one; **23**, 1H-indene-1,3(2H)-dione; **25**, 5,5-Dimethylcyclohexane-1,3-dione; **27**, Cyclohexane-1,3-dione; **29**, 3-Methylcyclopentane-1,2,4-trione; **31**, Isoxazol-4-(5H)-one; **33**, 1H-pyrazol-5(4H)-one; **35**, Furan-2,4(3H,5H)-dione.

i= CH_2Cl_2, r.t., 24 h
ii= CH_3CN, r.t., 24 h
iii= $CHCl_3$, r.t., 24 h
iiii= CH_3COCH_3, r.t., 24 h

4H-furo[3,4-b]pyrans **36** [34]. Some of the reaction details for these products are shown in Figure 8.1.

One possible mechanism for the formation of these heterocycles can be rationalized by the nucleophilic addition of isocyanides to the acetylenic system, followed by protonation of the 1:1 adduct by the CH-acids. The positively charged ions **A** are then attacked by the anion of the CH-acids to form ketenimines **B**. Under the reaction conditions employed, such an addition product may tautomerize and then cyclize to produce **12–36** (Scheme 8.6).

The synthesis of highly functionalized 2H-pyran-2-one derivatives from the reaction of *tert*-butyl isocyanide **37** and dialkyl acetylenedicarboxylates **2** in the presence of strong CH-acids **38** afforded the corresponding 2H-pyran-2-one derivatives **39** (Scheme 8.7) [35].

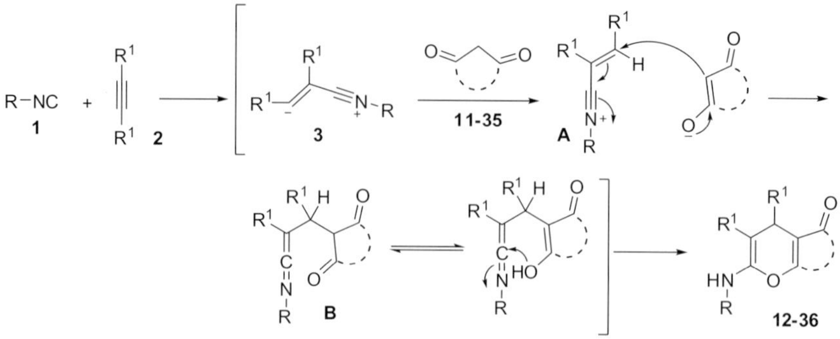

Scheme 8.6 The possible mechanism for the formation of **12–36**.

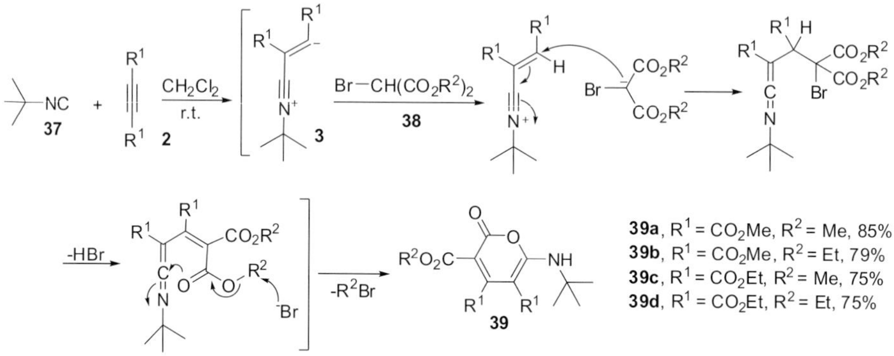

Scheme 8.7

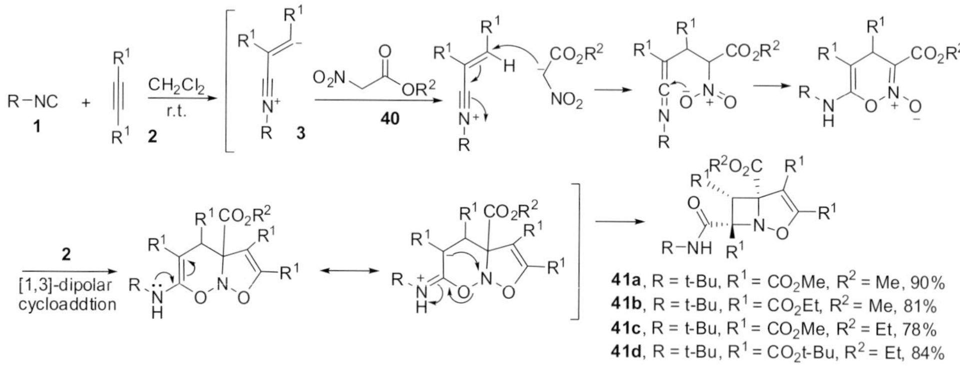

Scheme 8.8

Another interesting reaction of the acetylenic esters with alkyl isocyanides was reported in the presence of alkyl 2-nitroethanoates **40** as CH-acids in CH_2Cl_2 to afford the target compounds **41** in 78–90% yield (Scheme 8.8) [36].

The synthesis of trialkyl N-alkyl-6-methyl-2-pyridone-3,4,5-tricarboxylates **43** from a three-component reaction between the zwitterions **3** with alkyl 2-

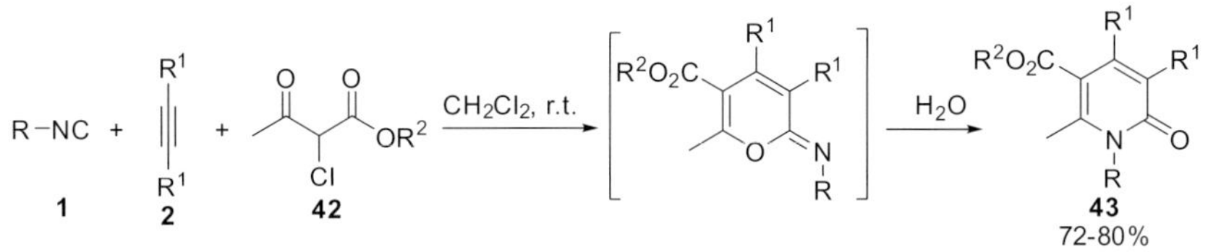

Scheme 8.9

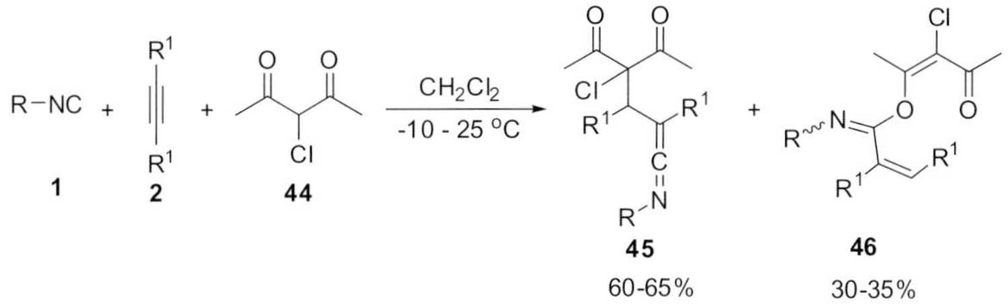

Scheme 8.10

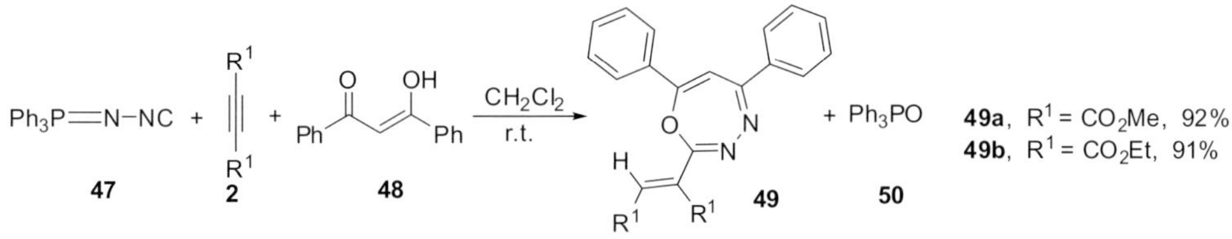

Scheme 8.11

chloroacetoacetates **42** has been described by Yavari and Zare. The resulting intermediate reacted with water to afford **43** (Scheme 8.9) [37].

Polyfunctional ketenimines **45** and 1-azadienes **46** are generated by the addition of isocyanides to dialkyl acetylenedicarboxylates and trapping the 3-chloropentane-2,4-dione **44**; C-alkylation of the latter leads to ketenimines **45**, and O-alkylation to 1-azadienes **46** (Scheme 8.10) [38].

Ramazani and coworkers have also described the stereoselective reactions of **2** with (N-isocyanimino) triphenylphosphorane **47** in the presence of 1,3-diphenyl-1,3-propanedione **48** at room temperature, to afford oxadiazepin **49** in high yields (>90%). Triphenylphosphine oxide **50** was the byproduct of this reaction (Scheme 8.11) [39].

The reactive 1:1 adduct, produced from the reaction between **2** and 2,6-dimethylphenyl isocyanide **51**, was trapped by methyl 2,4-dioxopentanoate **52** yielding **53** in moderate yields (Scheme 8.12) [40].

The use of dimethyl 3-oxopentanedioate as a trapper of zwitterionic species produced from the reaction of alkyl isocyanides with dimethyl 1,3-acetonedicarboxylate **2** in the presence of **54** in CH_2Cl_2 at room temperature

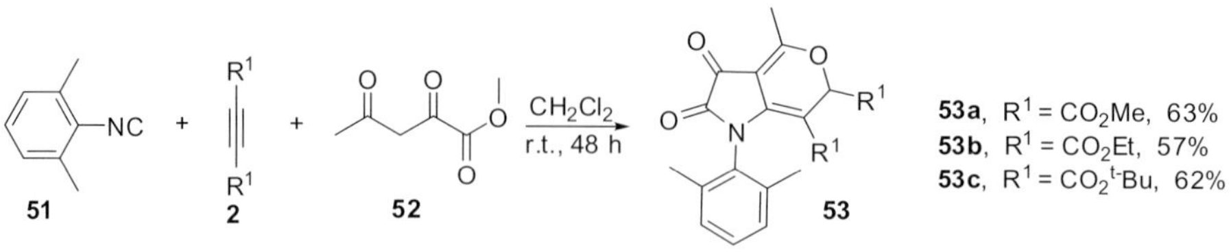
51
52
53
CH2Cl2
r.t., 48 h
53a, R1 = CO2Me, 63%
53b, R1 = CO2Et, 57%
53c, R1 = CO2tBu, 62%

Scheme 8.12

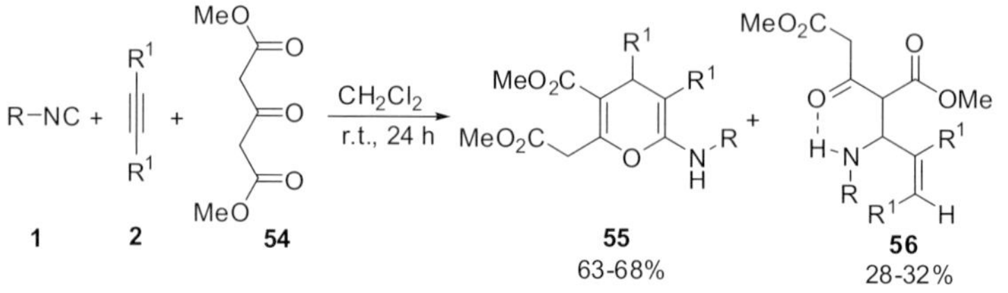
R–NC
CH2Cl2
r.t., 24 h
55
63-68%
56
28-32%

Scheme 8.13

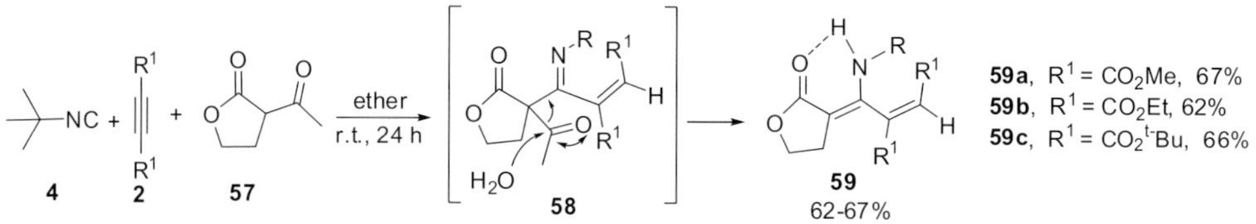
ether
r.t., 24 h
58
59
62-67%
59a, R1 = CO2Me, 67%
59b, R1 = CO2Et, 62%
59c, R1 = CO2tBu, 66%

Scheme 8.14

led to highly functionalized 2-amino-4*H*-pyrans **55** and 1,2-dialkyl 4,6-dimethyl-(1*E*, 3*E*)-3 (alkylamino)-5-oxo-1,3-hexadiene- 1,2,4,6-tetracarboxylates **56** (Scheme 8.13) [41].

The reaction of *tert*-butyl isocyanide with acetylenic diesters **2** in the presence of 2-acetylbutyrolactone **57** produced compounds **58** [42] which, under the same reaction conditions, were able to react with H_2O, thereby losing the acetyl group as acetic acid and resulting in compounds **59** (Scheme 8.14).

In the reaction between isocyanides **1** and **2**, a reactive intermediate **3** was produced that could be trapped with *N,N*-dimethylbarbituric acid **60** to yield the isomeric products dimethyl 7-(2,6-dimethylphenylamino)-1,3-dimethyl-2,4-dioxo-4*H*-pyrano[3,2-*d*]pyrimidine-5,6-dicarboxylate (**61**) and dimethyl (*E*)-2-((2,6-dimethylphenylamino)-(1,3-dimethyl-2,4,6-trioxo-pyrimidine-5-ylidene)-methyl)-butyl-2-enedioate (**62**) in an almost 1 : 1 ratio and with an overall yield of 85% (Scheme 8.15) [43]. In contrast, the use of ethynyl phenyl ketone instead of **2** produced only compounds **63** [44], while ethyl propiolate produced only compounds **61** [45].

A one-pot, three-component condensation reaction between **1** with **2** and (ethoxycarbonylmethyl)triphenylphosphonium bromide **64** as a CH-acid, led to the efficient provision of the fully substituted glutarimides **65** (Scheme 8.16) [46].

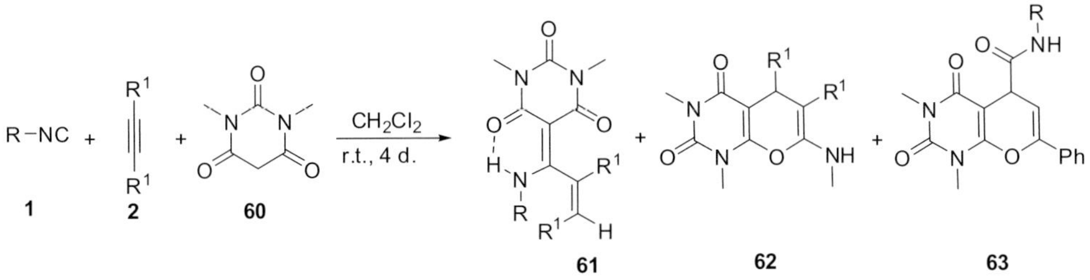

8.2 Generation of Zwitterionic Species by the Addition of Isocyanides to Alkynes

Scheme 8.15

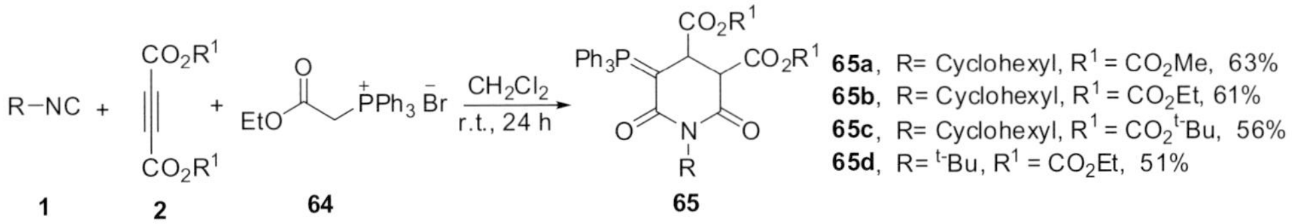

Scheme 8.16

65a, R= Cyclohexyl, R^1 = CO_2Me, 63%
65b, R= Cyclohexyl, R^1 = CO_2Et, 61%
65c, R= Cyclohexyl, R^1 = $CO_2{}^tBu$, 56%
65d, R= ^{t}-Bu, R^1 = CO_2Et, 51%

8.2.2
NH-Acids as Zwitterion-Trapping Agents

The amide group is a suitable agent for the trapping of zwitterion intermediates generated by the addition of isocyanides to dialkyl acetylenedicarboxylates. As shown in Figure 8.2, many NH-acids have been used as zwitterion-trapping agents in order to create functionalized heterocyclic compounds connecting nitrogen; these include functionalized pyrroles 67 [47], pyrazoles 69 [48], dialkyl 5-(alkylamino)-1-aryl-1H-pyrazole-3,4-dicarboxylates 71 [49], pyrazolo-[1,2-a][1,2,4]triazoles 73 [50], pyrazolo[1,2-a]pyrazoles 75 [51], imidazo[2,1-b][1,3]oxazines 77 [52], thiazolo[3,2-a]pyrimidines 79 [53], pyrido[1,2-a]pyrimidines 81 [54], pyrazino[1,2-a]pyrimidines 83 [55], and 2,6-dioxohexahydropyrimidines 85 [56].

When the zwitterion intermediates generated by the reaction of 1 and 2 were trapped by NH-acids, this led to formation of stable ketenimines 86. Various NH-acids have also been used as zwitterion-trapping agents, including sulfonamide derivatives 87 [57], methyl 2-oxo-2-(phenylamino) acetate 88 [58], trifluoro-N-arylacetamides 89 [59], benzoyl hydrazones 90 [60], 5,5-dimethylimidazolidine-2,4-dione 91 [61], succinimide 92 [62], pyrrole 93 [63], imidazole derivatives 94 [62], indole 95 [63], carbazole 96 [63], isatin 97 [62], phthalimide 98 [62], and pyridin-2(1H)-one or isoquinolin-1(2H)-one 99 [64] (Figure 8.3).

Protonation of the highly reactive 1:1 intermediates produced in the reaction between 1 and dibenzoylacetylene 100 by phthalimide 98, led to the creation of vinylnitrilium cations; the latter then undergo a carbon-centered Michael-type addition with the conjugate base of the NH-acid to produce highly functionalized aminofuran derivatives 101. Compound 102 was also the byproduct of this reaction (Scheme 8.17) [65].

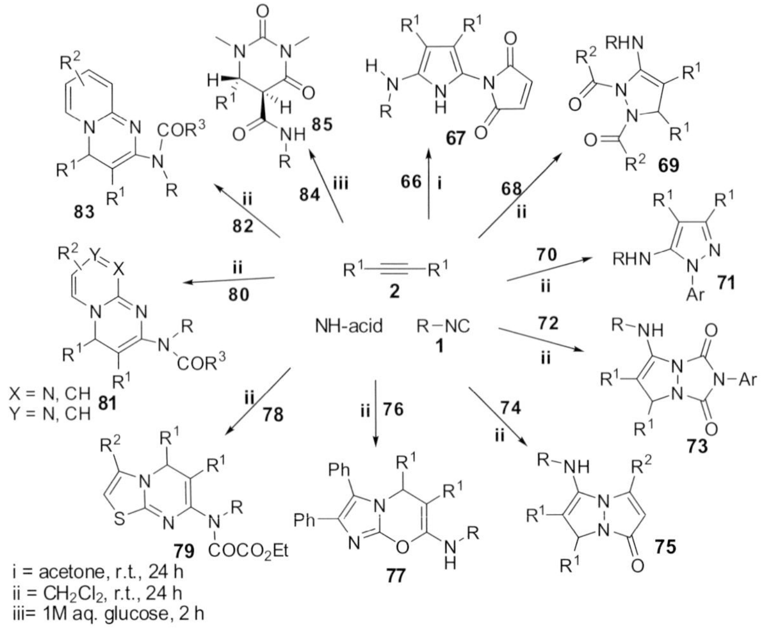

Figure 8.2 Products obtained from the trapping of zwitterions with various NH-acids: **66**, 1-(Aminomethyl)-1H-pyrrole-2,5-dione; **68**, N′-Acetylacetohydrazide; **70**, Phenylhydrazine; **72**, 4-Argio-1,2,4-triazolidine-3,5-dione; **74**, 3-Methyl-1H-pyrazol-5(4H)-one; **76**, 4,5-Diphenyl-1H-imidazol-2(3H)-one; **78**, Ethyl 2-(4-methylthiazol-2-ylamino)-2-oxoacetate; **80**, N-(Pyrimidin-2-yl)acetamide and N-(pyrazin-2-yl)acetamide; **82**, N-(Pyridin-2-yl)acetamide; **84**, N,N′-Dimethylurea.

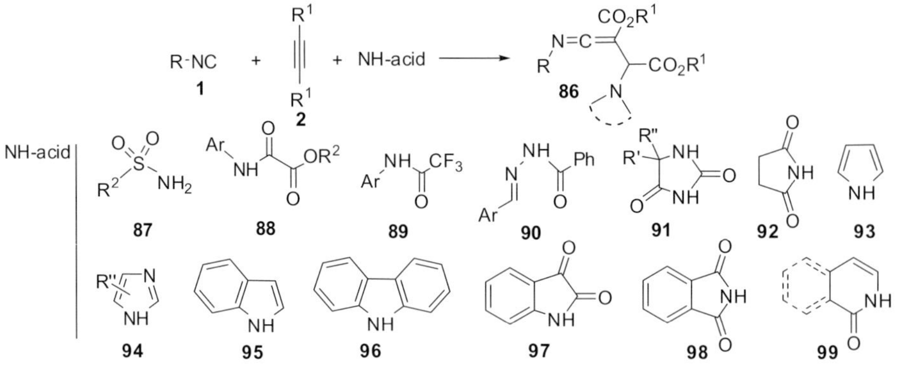

Figure 8.3 Synthesis of stable ketenimines obtained from the trapping of zwitterions with various NH-acids.

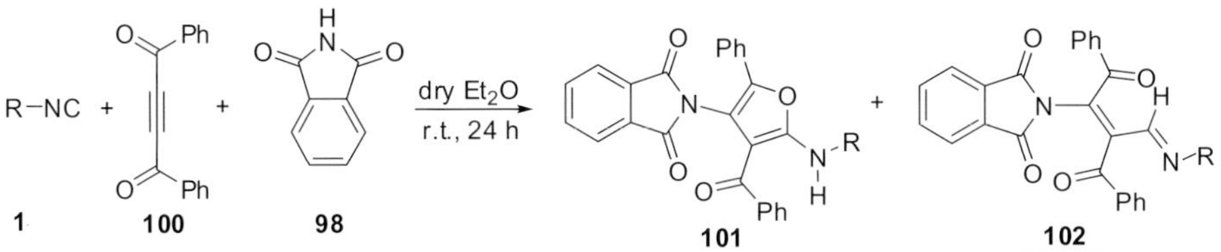

1 100 98 101 60% 102 35-38%

Scheme 8.17

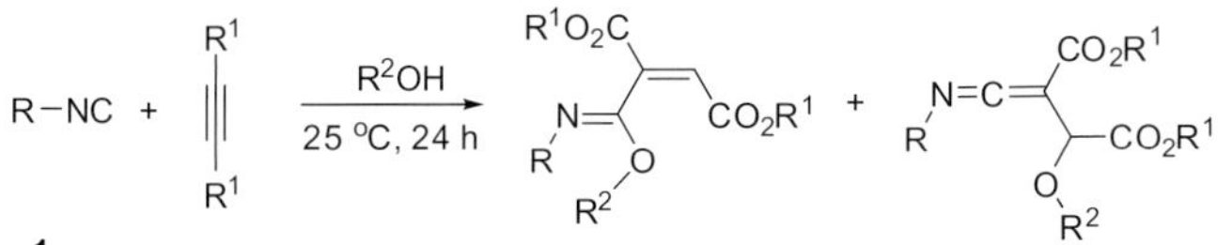

1 2 103 104

Scheme 8.18

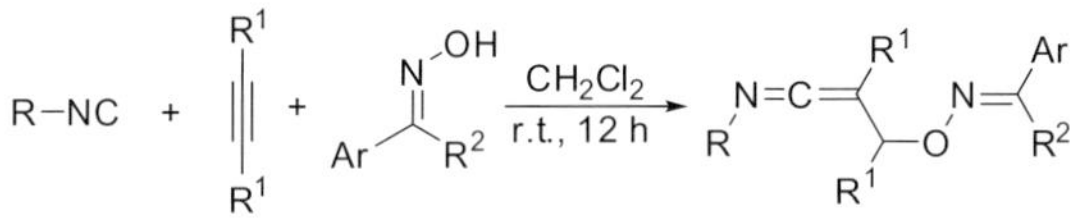

1 2 105 106

106a, $R^1 = CO_2Me$, R^2 = Me, Ar = Ph, 97%
106b, $R^1 = CO_2Et$, R^2 = Me, Ar = Ph, 95%
106c, $R^1 = CO_2{}^tBu$, R^2 = Me, Ar = Ph, 90%
106d, $R^1 = CO_2Et$, R^2 = H, Ar = 4-ClC_6H_4, 97%
106e, $R^1 = CO_2Me$, R^2 = H, Ar = 4-ClC_6H_4, 91%
106f, $R^1 = CO_2Me$, R^2 = H, Ar = 4-ClC_6H_4, 93%

Scheme 8.19

8.2.3
OH-Acids as Zwitterion-Trapping Agents

The reaction of isocyanides with activated acetylenes in alcoholic solvents has been shown to produce a mixture of two different 1:1:1 adducts (isocyanide:acetylene:alcohol), an unsaturated imino ester, and a ketenimine [66]. The alcohols behave as zwitterion-trapping agents in these reactions (Scheme 8.18). In the case when R was an aryl group (e.g., phenyl, *o*-tolyl, or *p*-nitrophenyl), the only product obtained is the imino ester **103**, whereas when R was alkyl (e.g., cyclohexyl or *tert*-butyl), both the imino ester **103** and the ketenimine **104** are obtained. In the case of cyclohexyl, the ratio of **103**:**104** was approximately 1:1, whereas in the case of *tert*-butyl isocyanide it was about 1:9. Thus, the attack by alkoxide may be governed by the steric effect of R in a zwitterion intermediate.

It has been shown that *N*-alkyl-substituted ketenimine derivatives **106** can be efficiently prepared by the three-component coupling of **1**, **2**, and oximes **105** [67]. The reaction has been reported under mild conditions, with a high selectivity, and a tolerance to various functional groups in near-quantitative yields (Scheme 8.19).

The reactive zwitterionic intermediate **3** was trapped by the OH group of the naphthols **107** [68], phenols **109** [69] or 3,6-dihydroxypyridazine **111** [70], to produce

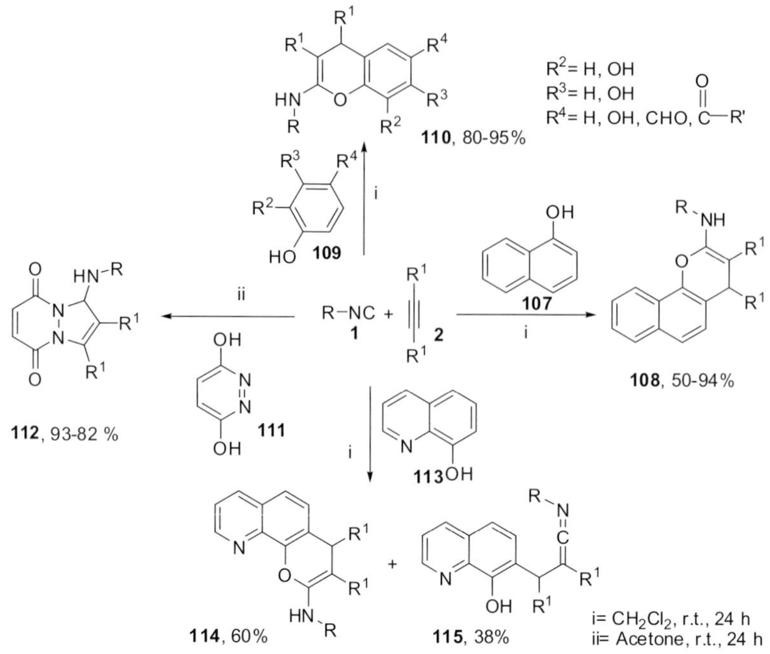

Scheme 8.20

benzochromenes **108**, chromenes **110**, or pyrazolo[1,2-a]pyridazines **112**, respectively. When the OH group was on the naphthol, however, the only product obtained was the chromene **108**; however, when the OH group was on the 8-hydroxyquinoline **113**, both 4H-pyrano-[3,2,h]-quinoline **114** and the ketenimine **115** were obtained [69b] (Scheme 8.20).

Several important syntheses of furans via the trapping of electron-deficient acetylenic ester–isocyanide zwitterions with various carboxylic acids are shown in Scheme 8.21. A facile IMCR of highly functionalized 2,5-diaminofuran derivatives **117** is reported by using 2-phenylacetic acid or 2,2-diphenylacetic acid **116** [71]. Furthermore, nicotinamides **119** [72] and bis(aminofuryl)bicinchoninic amides **121** [73] are produced from the reaction of **3** with nicotinic acid **118** or bicinchoninic acid **120**, respectively.

The reaction of isocyanide **1** with dialkyl acetylenedicarboxylates **2** in the presence of aromatic carboxylic acids **122** undergoes a smooth 1:1:1 addition reaction in dichloromethane at ambient temperature, to produce dialkyl (E)-2-{[benzoyl(tert-butyl)amino]carbonyl}-2-butenedioate derivatives **123** [74]. When this reaction was extended to sulfonic acids **124**, instead of aromatic carboxylic acids, it led to the production of intermediate **125** [75], which was subsequently trapped with H_2O under these reaction conditions to produce the corresponding dialkyl 2-({alkyl[(4-methylphenyl)sulfonyl]amino}carbonyl)-3-hydroxysuccinate derivatives **126** (Scheme 8.22).

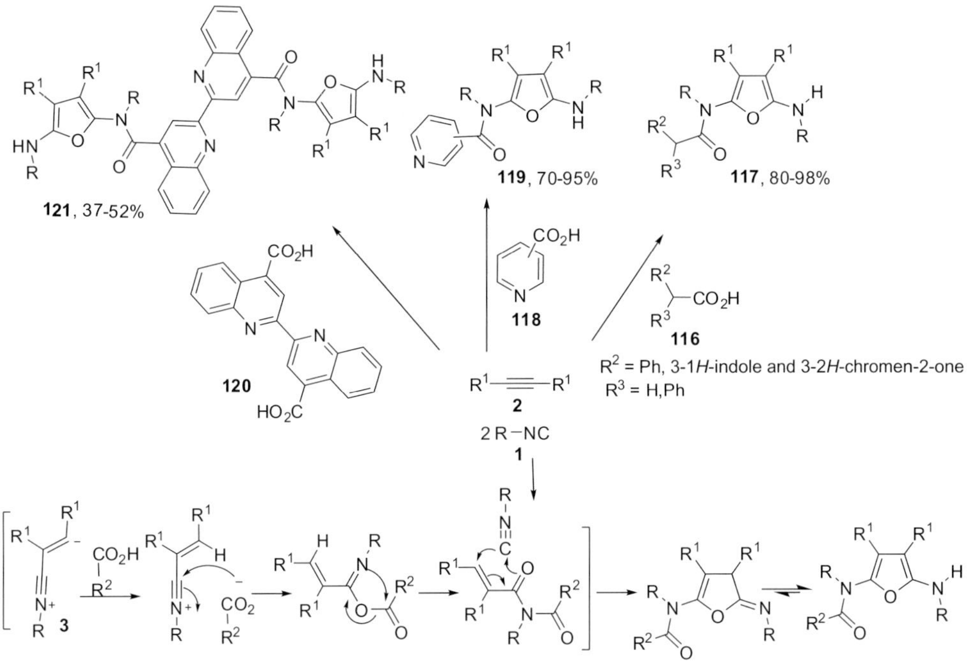

Scheme 8.21

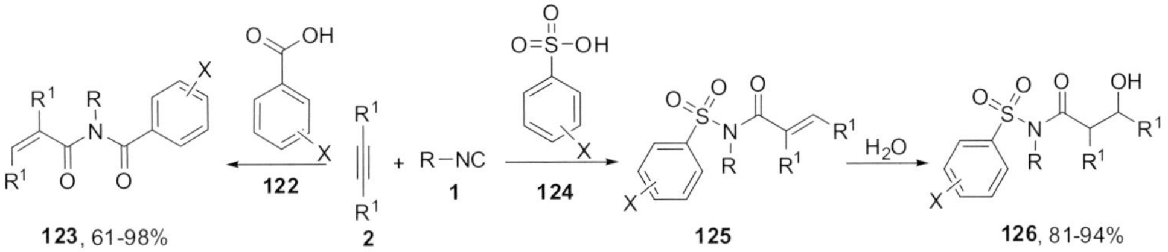

Scheme 8.22

8.2.4
Carbonyl Compounds as Zwitterion-Trapping Agents

Furan-containing products may be obtained by trapping the zwitterions of **3** with various carbonyl compounds [76]. In this case, the *in situ*-generated reactive 1:1 zwitterionic intermediates **3** were trapped by aromatic aldehydes **127** to produce highly functionalized furans **128** (Scheme 8.23). A "green" methodology for the preparation of polysubstituted furans **128** has been introduced by the reaction of aromatic aldehydes **127**, acetylenic esters, and alkyl isocyanides in water [77] and [bmim]BF$_4$ ionic medium at room temperature [78].

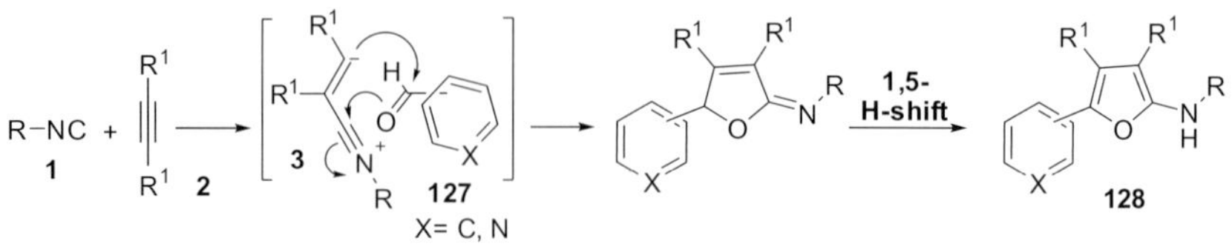

Scheme 8.23

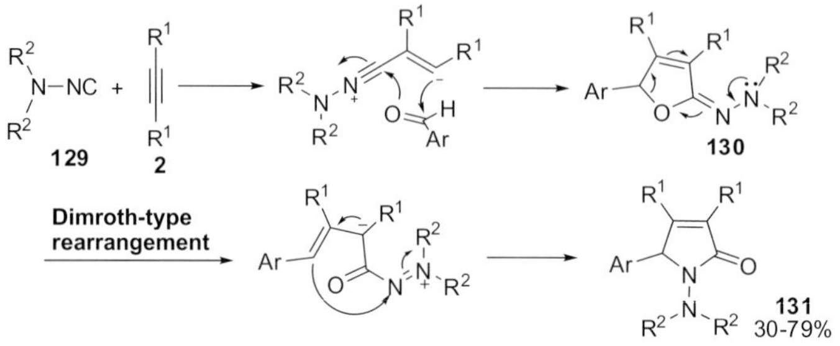

Scheme 8.24

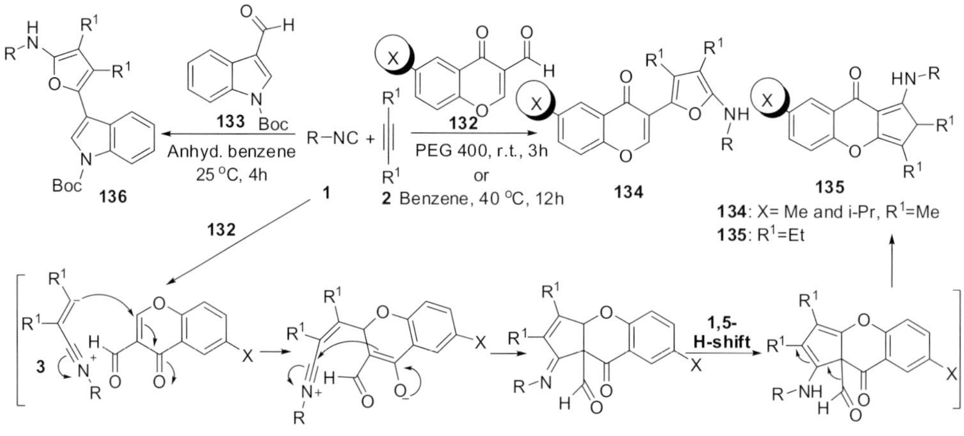

Scheme 8.25

The reaction of diisopropylaminoisocyanides **129** with DMAD and carbonyl compounds in benzene under an atmosphere of argon, when heated for 3 h under reflux conditions, afforded the 1-aminopyrrolin-2-one derivatives **131** [79]. These pyrrolin-2-one derivatives presumably arise from a Dimroth-type rearrangement of the primary adducts **130**, through a furanone hydrazone pathway (Scheme 8.24).

The zwitterions **3** were trapped by α,β-unsaturated aldehydes [80] such as 3-formylchromones **132** and 3-formylindole **133** to afford the corresponding styrylfuran derivatives **134** and **136** (Scheme 8.25). It was confirmed that electron-

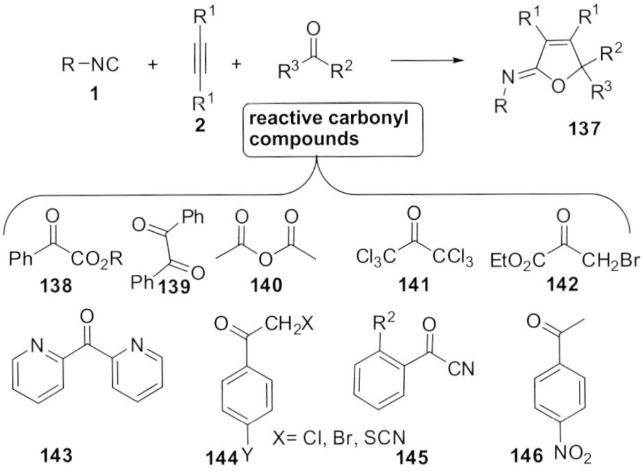

8.2 Generation of Zwitterionic Species by the Addition of Isocyanides to Alkynes

Scheme 8.26

donating substituents (such as Me and *i*-Pr) in the 6-position of the chromone moiety favor reaction with the aldehyde carbonyl, leading to chromenylfurandicarboxylates **134**. Unexpectedly, by changing the ester from DMAD to diethyl acetylenedicarboxylate, in all cases, the cyclopentachromenedicarboxylates **135** were formed preferentially [81].

The reaction between alkyl (aryl) isocyanides and acetylenedicarboxylates in the presence of various reactive carbonyl compounds such as alkyl phenylglyoxylates **138** [82], benzyl **139** [76a], acetic anhydride **140** [83], hexachloroacetone **141** [84], ethyl bromopyruvate **142** [85], *di*-(2-pyridyl) ketone **143** [86], phenacyl halides or phenacyl thiocyanate **144** [87], benzoyl cyanide **145** [88], and 1-(4-nitrophenyl) ethanone **146** [87a] led to fully substituted iminolactone derivatives. The results showed that reactive zwitterionic intermediate had been trapped by carbonyl compounds involving an electron-withdrawing substituent in the *para* position of the phenyl group, or an electron-withdrawing group in the α position of the carbonyl group (Scheme 8.26).

Some spirocyclic iminolactones containing products **147** have been obtained by trapping zwitterions **3** with various cyclic carbonyl compounds. In this regard, maleic anhydride or citraconic anhydride **148** [89], N-arylmaleimide **149** [90], isobenzofuran-1,3-dione **150** [83], N-arylphthalimide **151** [90], 1-methylindoline-2,3-dione **152** [91], 9H-fluoren-9-one **153** [92], 1,8-diazafloren-9-one **154** [93], indolo[1,2-b]isoquinoline-6,12-dione **155** [94], quinines **156** and **157** [95], and acenaphthene quinines **158** [95] are used as zwitterion-trapping agents (Scheme 8.27).

The reactive zwitterionic intermediates **3** were trapped with cyclobutenediones **159** [96]. In this reaction, a few novel spirocyclic compounds **160** were afforded following the addition of a second molecule of the isocyanide (Scheme 8.28).

The vicinal tricarbonyl system **161** participates in a reaction with the zwitterions **3** to form fully substituted furan **162** and iminopyrone **163** derivatives [97]. The

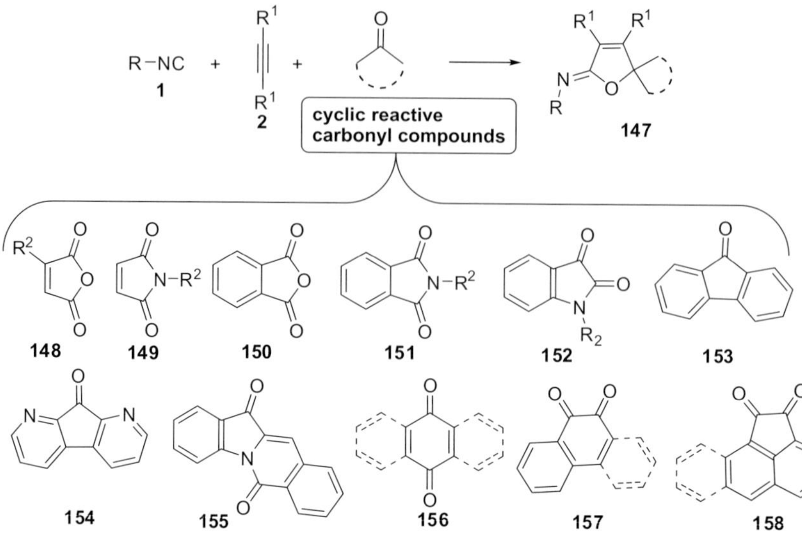

Scheme 8.27

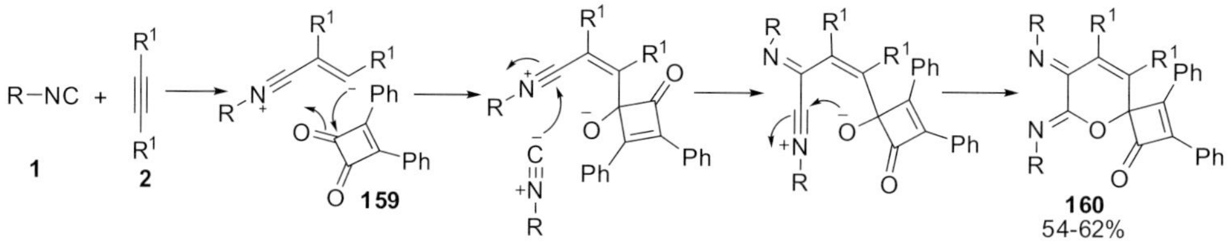

Scheme 8.28

formation of **163** in this reaction may be rationalized as occurring via path B in Scheme 8.29. Presumably, due to the steric demands imposed by the two benzoyl groups, path A – which leads to the furan derivative – was not operative. Thus, the participation of another molecule of the isocyanide was allowed [97b].

The three-component reaction of the zwitterions **3** with isocyanates **164** has been described [98]. The reaction afforded the corresponding special type of ethylene-tetracarboxylic acid derivatives **165**, in good yields (Scheme 8.30).

8.2.5
Imine Compounds as Zwitterion-Trapping Agents

The possibility to trap the zwitterionic intermediate **3** with imines appeared attractive when devising a novel IMCR [99]. Reaction of the zwitterions with N-tosylimines **166** afforded aminopyrrole derivatives **167**, while γ-spiroiminolactams **169** were formed via the three-component reaction of zwitterions with various

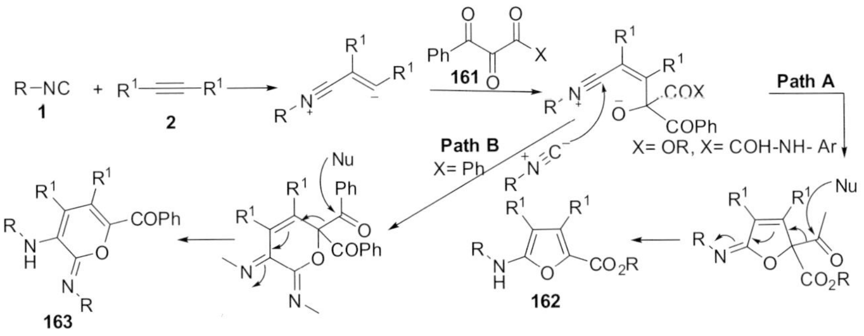

8.2 Generation of Zwitterionic Species by the Addition of Isocyanides to Alkynes | 279

Scheme 8.29

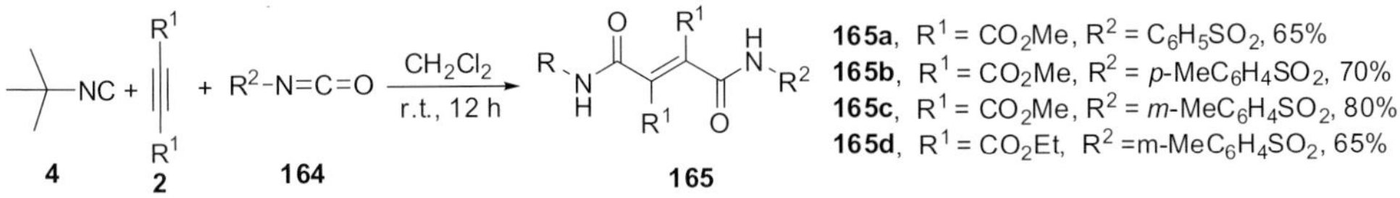

Scheme 8.30

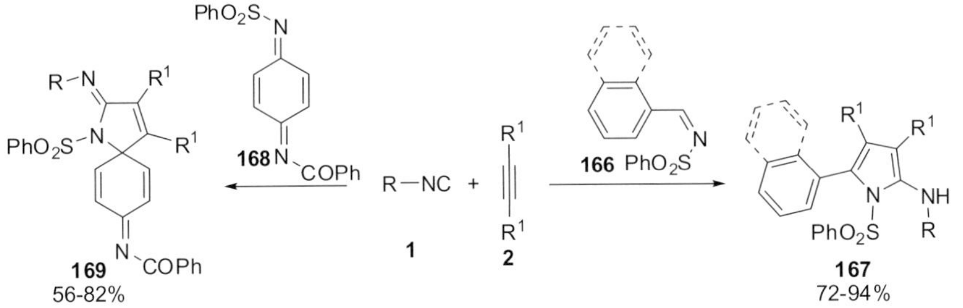

165a, $R^1 = CO_2Me$, $R^2 = C_6H_5SO_2$, 65%
165b, $R^1 = CO_2Me$, $R^2 = p\text{-}MeC_6H_4SO_2$, 70%
165c, $R^1 = CO_2Me$, $R^2 = m\text{-}MeC_6H_4SO_2$, 80%
165d, $R^1 = CO_2Et$, $R^2 = m\text{-}MeC_6H_4SO_2$, 65%

Scheme 8.31

quinoneimides **168** [100]. This reaction was chemoselective in nature, as the C=N of the sulfonamide acted as a dipolarophile (Scheme 8.31).

The reaction between the zwitterionic intermediates **3** with iminium ion intermediates **171** formed from aromatic aldehydes **127** and diethylamine **170** in the presence of silica gel led to dialkyl 2-[(alkylamino)-carbonyl]-3-[(Z)-1-(diethylamino)-1-arylmethylidene]succinates **172** via a one-pot, four-component regioselective and stereoselective process (Scheme 8.32) [101].

8.2.6
Electron-Deficient Olefins as Zwitterion-Trapping Agents

Zwitterionic species **3** were trapped with dipolarophiles to generate cyclic compounds. Activated alkenes were used as dipolarophiles, and underwent a

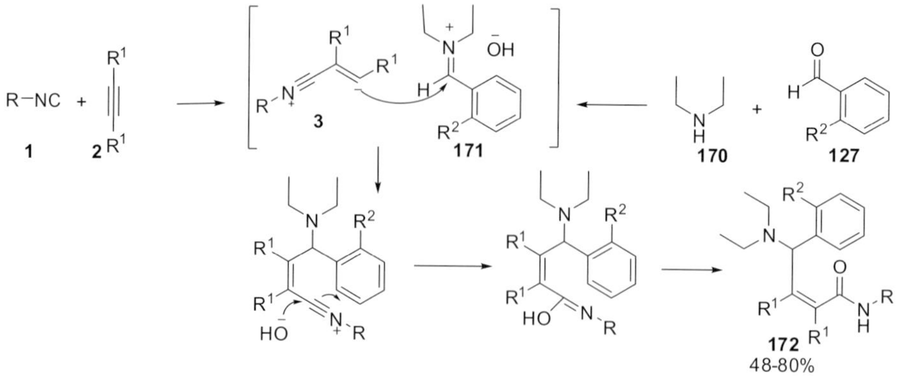

Scheme 8.32

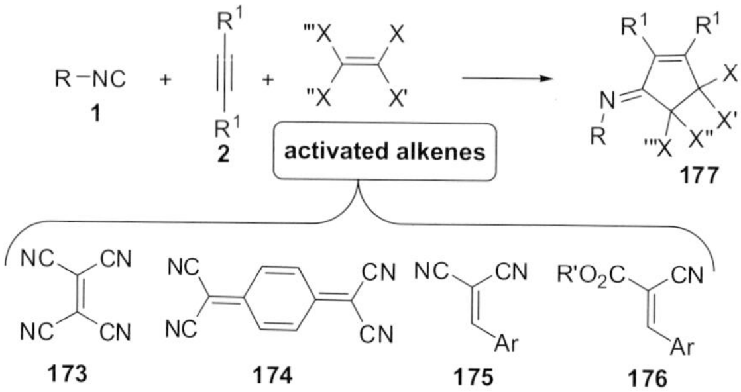

Scheme 8.33

cycloaddition reaction with **3** to produce highly substituted cyclopentadienoid systems **177**. In this regard, tetracyanoethylene **173** [102], 7,7,8,8-tetracyanoquinodimethane **174** [102] and various activated styrenes **175** and **176** [103] were used as dipolarophiles (Scheme 8.33).

Functionalization of the fullerenes (C_{60}) **178** via a three-component reaction has been described [104]. In this reaction, trapping of the zwitterions **3** with C_{60} afforded the fully substituted bicyclo[2.1.0]-pentanes **179** and the cyclopentenimine fullerene compounds **180** (Scheme 8.34).

8.2.7
Miscellaneous Compounds as Zwitterion-Trapping Agents

Alkyl (aryl) isocyanides react with acetylenic compound in the presence of benzoyl isothiocyanate **181** to afford highly substituted 4,7-bis[alkyl(aryl)imino]-2-phenyl-3-oxa-6-thia-1-azaspiro[4.4]nona-1,8-dienes **183**, with double insertion of the isocyanide [105]. The reaction involves two cyclizations: first, a 1,3-dipolar cycloaddition between the zwitterion intermediate **3** with **181**, followed by a [4+1] cycloaddition reaction between intermediate **182** and isocyanide (Scheme 8.35).

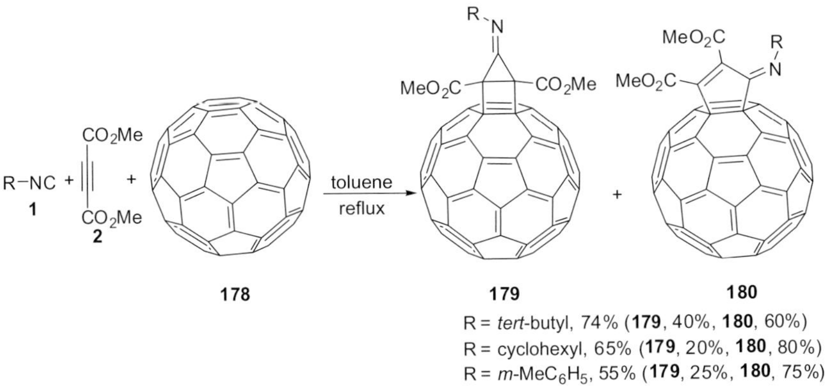
R−NC + 1
CO2Me
CO2Me
2
+
toluene
reflux
MeO2C
CO2Me
MeO2C
R
N
+
178
179
180
R = tert-butyl, 74% (179, 40%, 180, 60%)
R = cyclohexyl, 65% (179, 20%, 180, 80%)
R = m-MeC6H5, 55% (179, 25%, 180, 75%)

Scheme 8.34

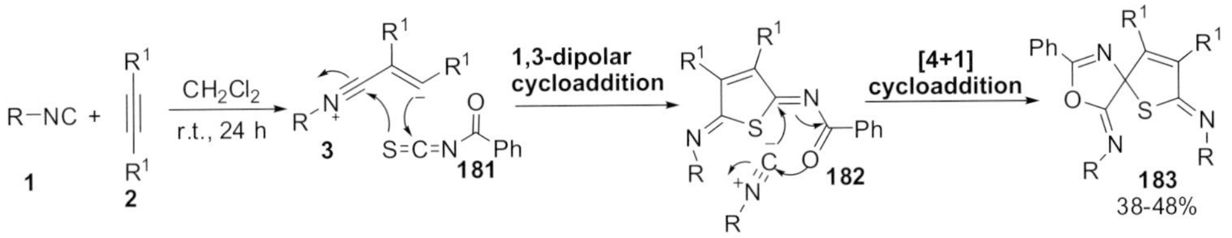
R−NC +
1
R1
R1
2
CH2Cl2
r.t., 24 h
3
S=C=N
Ph
181
1,3-dipolar
cycloaddition
182
[4+1]
cycloaddition
Ph
183
38-48%

Scheme 8.35

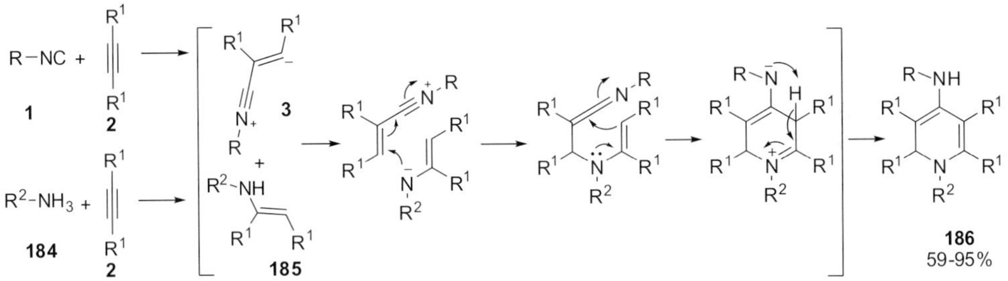
R−NC +
1
2
R2-NH3 +
184
2
3
185
R2-NH
186
59-95%

Scheme 8.36

A cascade, four-component reaction between primary alkylamines **184**, acetylenic esters **2**, and alkyl isocyanides **1** led to tetraalkyl-1-alkyl(aryl)-4-alkylamino-1,2-dihydropyridine-2,3,5,6-tetracarbxylates **186** [106]. In this reaction, the zwitterionic intermediates **3** are trapped by enaminoesters **185** that had been generated *in situ* from primary amines and dialkyl acetylenedicarboxylates (Scheme 8.36).

The reaction of **1** with **2** in the presence of elemental sulfur, proceeded spontaneously without activation of elemental sulfur **187** at room temperature in an

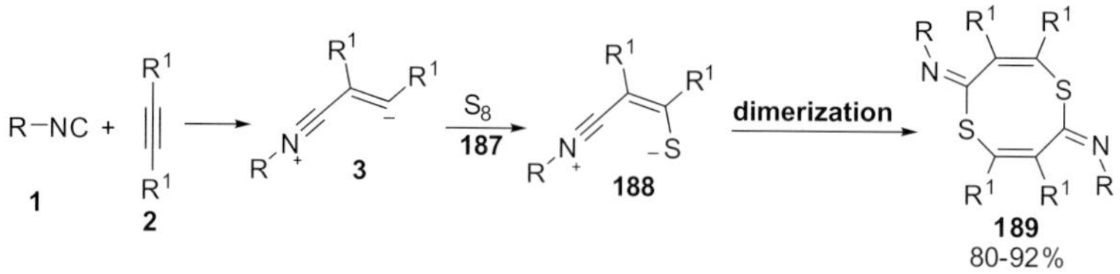

Scheme 8.37

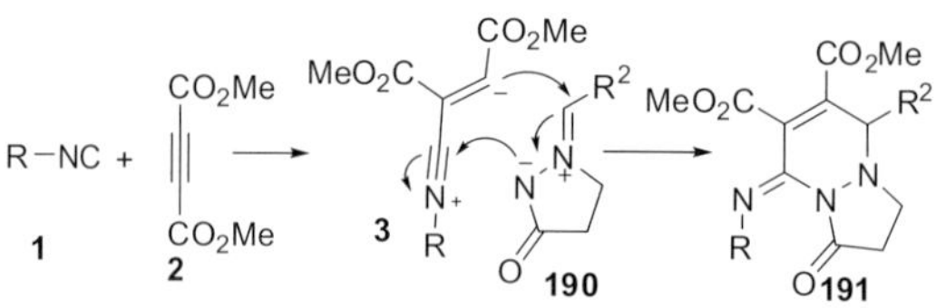

Scheme 8.38

191a, R = t-Bu, R² = C₆H₅, 80%
191b, R = t-Bu, R² = p-ClC₆H₄, 77%
191c, R = t-Bu, R² = m-BrC₆H₄, 79%
191d, R = t-Bu, R² = m-MeC₆H₄, 70%
191g, R = t-Bu, R² = p-O₂NC₆H₄, 67%
191e, R = t-Bu, R² = p-MeC₆H₄, 35%
191f, R = t-Bu, R² = p-MeOC₆H, 33%
191g, R = Cyclohexyl, R² = p-O₂NC₆H₄, 70%
191h, R = Bu, R² = C₆H₅, 63%

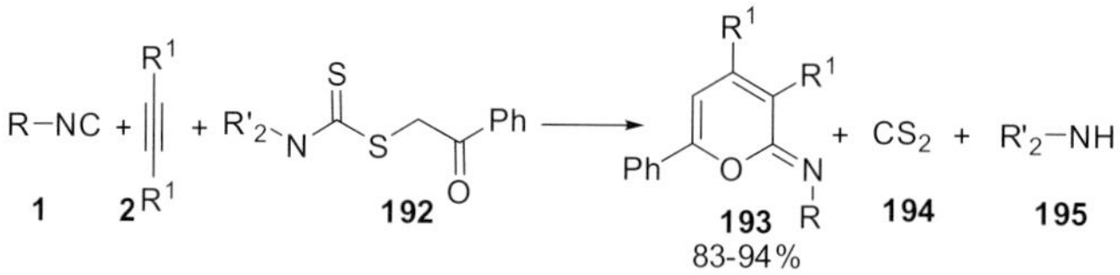

Scheme 8.39

anhydrous dichloromethane–carbon disulfide mixture (1 : 10) to produce cyclic thioimidic esters **189** [107]. A possible mechanism was proposed via the concomitant addition of **3** to **187**, which would lead to a charge transfer from the negative carbon atom of zwitterions **3** onto the sulfur atom; this would provide new zwitterions **188**, with the subsequent dimerization producing cyclic thioimidic esters **189** (Scheme 8.37).

The zwitterionic intermediate **3** reacted with 1,3-dipolar compounds **190** under mild conditions to produce 5-imino-2,3,5,8-tetrahydropyrazolo[1,2-a]pyridazin-1-one derivatives **191** (Scheme 8.38) [108].

The synthesis of 2H-pyran-3,4-dicarboxylates **193** using the three-component reaction of dithiocarbamates **192**, dialkyl acetylenedicarboxylates, and isocyanides in solvent-free conditions, was described [109]. In these reactions, the synthesis of 2H-pyran-3,4-dicarboxylates may have been based on the trapping of zwitterionic intermediate **3** with dithiocarbamate and the loss of carbon disulfide **194** and a secondary amine **195** (Scheme 8.39).

8.3
Generation of Zwitterionic Species by the Addition of Isocyanides to Arynes

Here, the term "arynes" is used when referring both to derivatives of 1,2-dehydrobenzene (benzyne) and their heterocyclic analogues (heteroarynes). The arynes **196** are powerful and useful reactive intermediates in synthetic organic chemistry that can be readily converted into polysubstituted arenes and benzoannulated structures yet are barely accessible by conventional methods [110]. Because of the highly electrophilic character of arynes, which arises from their lowest unoccupied molecular orbital (LUMO) [111], even neutral nucleophiles of diminished nucleophilicity can be added quite simply to arynes to produce zwitterions, which serve as key intermediates in three-component coupling reactions (Scheme 8.40) [112]. The addition of isocyanide to arynes produced a zwitterion reactive intermediate **197**, which could be trapped with a third component.

The reactive zwitterionic intermediates generated *in situ* from the reaction between a series of isocyanides and arynes were trapped by carbonyl compounds that included aromatic aldehydes **127** [113], phenyl esters **138** [114], in addition to several ketones such as acetophenone derivatives **198**, trifluoroacetophenone **199**, and benzyl **139** [115], to produce benzoannulated iminofurans **200** (Scheme 8.41).

A three-component coupling reaction of arynes, isocyanides and N-tosylaldimines **166** as zwitterion-trapping agents has been developed to offer modest to high yields of diversity 2-iminoisoindolines **201**, in one step (Scheme 8.42) [116].

Phenyl acetylene, methyl propiolate, and dimethyl acetylenedicarboxylate are used to trap zwitterion intermediates **197** that have been generated by the addition

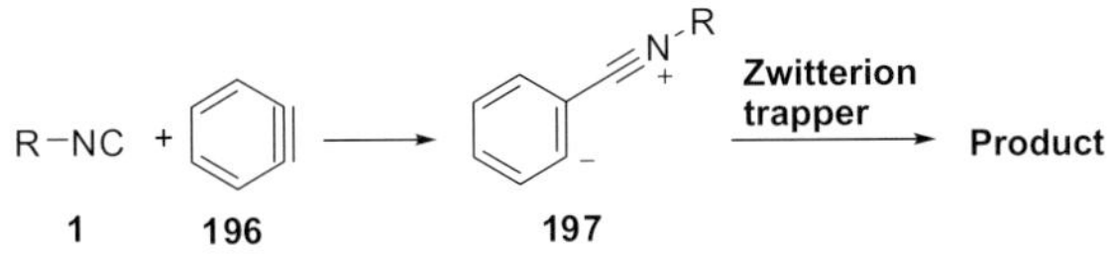

Scheme 8.40

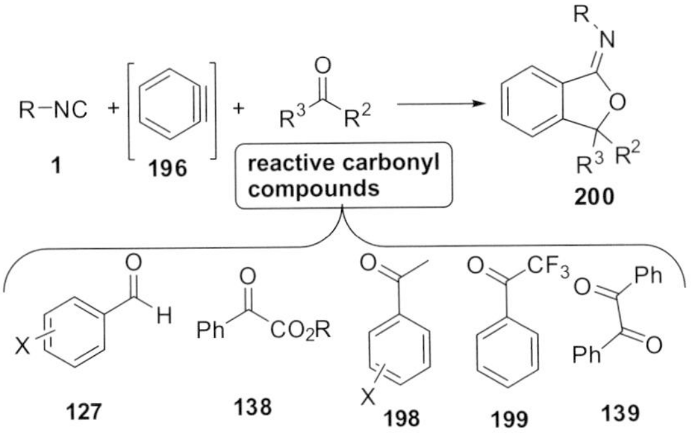

Scheme 8.41

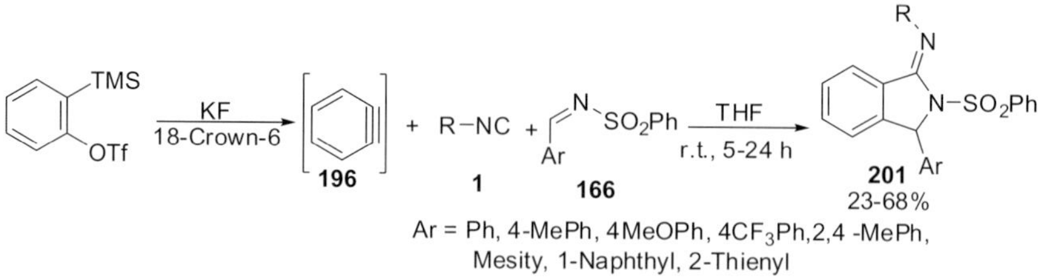

Scheme 8.42

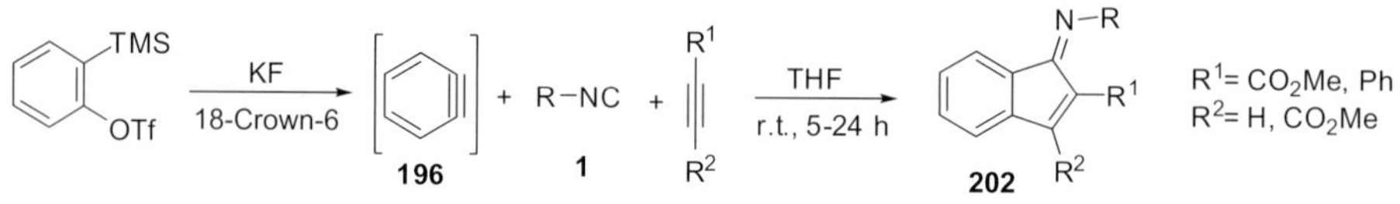

Scheme 8.43

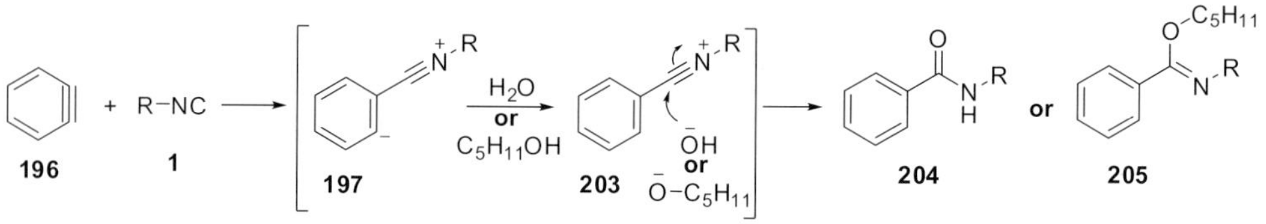

Scheme 8.44

of isocyanides to arynes, to afford the substituted iminoindenones **202** [115]. Efforts to employ additional alkynes lacking conjugate acceptors, such as cyclohexylacetylene, ethoxyacetylene, and diphenylacetylene, failed to generate the corresponding three-component adducts (Scheme 8.43).

Protonation of the highly reactive intermediates **197** produced in the reaction between **1** and arynes **196** (obtained from anthranilic acid) by H_2O, leading to N-alkylnitrilium ion **203** which then undergo a carbon-centered, Michael-type addition with the OH^- to produce benzamide derivatives **204** (Scheme 8.44). The nitrilium ion can be trapped with isoamyl alcohol to afford the corresponding N-substituted imidate ester **205** [117].

8.4
Generation of Zwitterionic Species by the Addition of Isocyanides to Electron-Deficient Olefins

Isocyanides form zwitterions with electron-deficient olefins to develop novel protocols for the synthesis of cyclic compounds. The addition of isocyanide to electron-

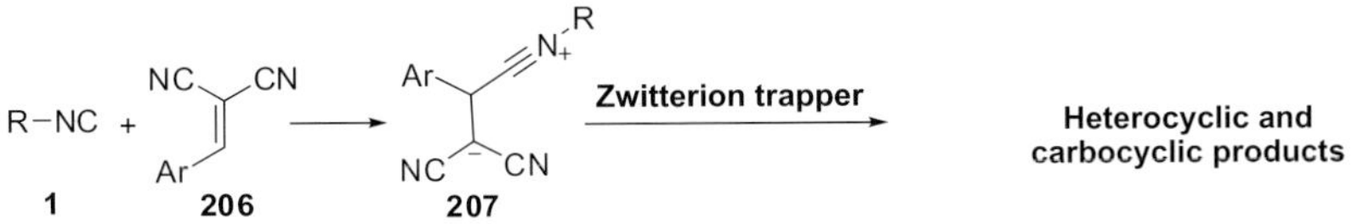

8.4 Generation of Zwitterionic Species by the Addition of Isocyanides to Electron-Deficient Olefins

Scheme 8.45

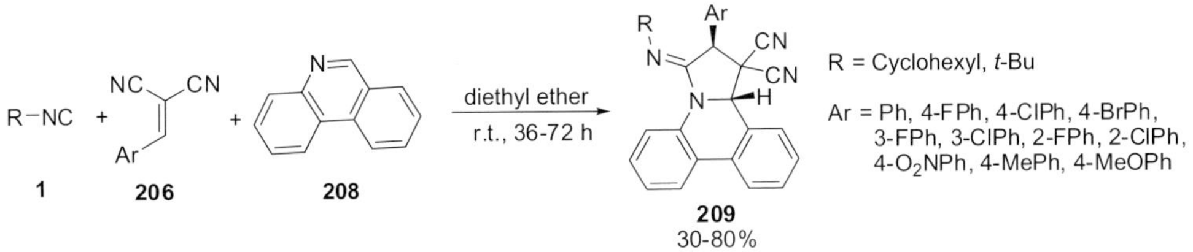

Scheme 8.46

deficient olefins, such as arylidenemalononitriles **206**, produced zwitterion intermediate **207**; these reactive intermediates could be trapped with a third component to provide the final product (Scheme 8.45).

A three-component reaction of zwitterions generated *in situ* from the reaction of isocyanides and 2-arylidenemalononitriles with phenanthridine **208** via 1,3-dipolar cycloaddition has been described [118]. The reactions afforded the corresponding dihydropyrrolo[1,2-*f*]phenanthridine derivatives **209** in good yields, without employing any catalyst and/or activation. The arylidenemalononitriles, which included electron-withdrawing groups such as fluoro, chloro, bromo, and nitro, reacted faster and produced higher yields than which included electron-rich groups, such as methyl and methoxyl (Scheme 8.46).

The direct construction of imino-pyrrolidine-thione **211** scaffolds via the coupling of isocyanides, heterocyclic thiols **210**, and *gem*-dicyano olefins **206** has been developed [119]. A plausible mechanism was proposed for the formation of the thioxopyrrolidine derivatives by analogy with the Ugi multicomponent condensation. The first step most likely involves reaction of the *gem*-dicyano olefin with isocyanide, followed by an attack of the thiol on the resulting intermediate. A subsequent Ugi–Smiles-type rearrangement, followed by nucleophilic addition of the amino group onto the cyano group, afforded the final product **211** (Scheme 8.47).

The intermolecular [2+2+1] multicomponent cycloaddition reactions of allenoates **212**, activated olefins **206** and isocyanides **1**, catalyzed by silver hexafluoroantimonate (AgSbF$_6$), has been reported [120]. This protocol allowed the syntheses of highly functionalized five-membered carbocycles **215** with exclusive regioselectivity and stereoselectivity, in an efficient and atom-economic manner. In this case, the AgSbF$_6$ behaved as a δ-Lewis acid and activated the allenoate. The nucleophilic attack between isocyanide **1** and activated allenoate **213** essentially led to the

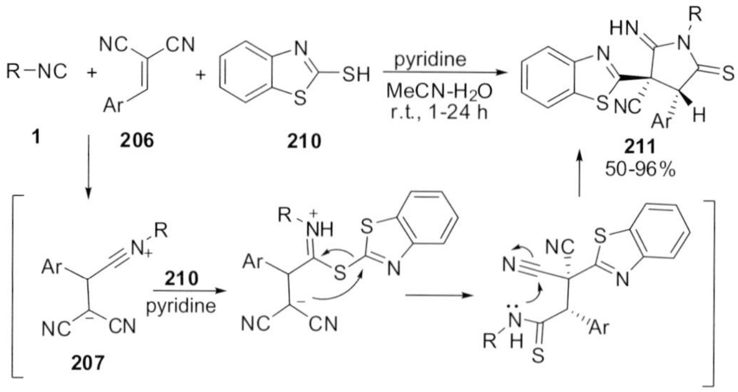

286 | *8 Zwitterions and Zwitterion-Trapping Agents in Isocyanide Chemistry*

Scheme 8.47

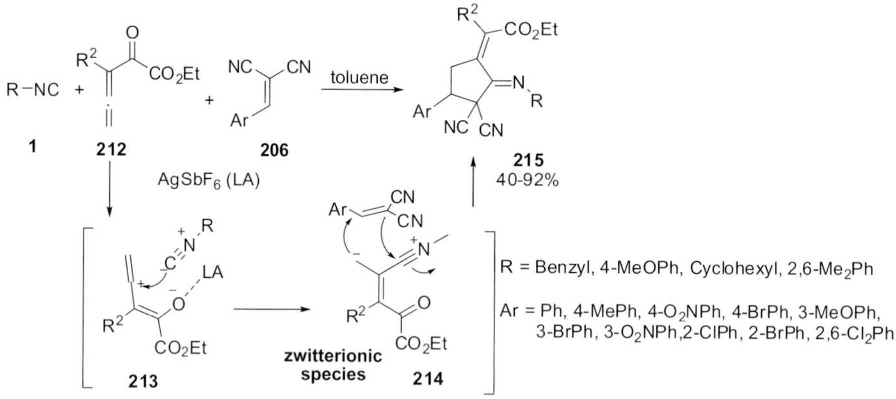

Scheme 8.48

formation of zwitterionic species intermediate **214**, which experienced a further Michael addition followed by intramolecular annulation to yield the product **215** (Scheme 8.48).

8.5
Miscellaneous Reports for the Generation of Zwitterionic Species

The nucleophilic addition of isocyanides to the *ortho* hydroxy- or fluoro-substituted benzaldehydes produces the zwitterionic species **218**. These reactive intermediates are then trapped with H_2O and phenanthroline **217** to produce 2-oxoacetamides **218** [121] and oxazolo[3,2-a][1,10]phenanthrolines **219** [122], respectively (Scheme 8.49).

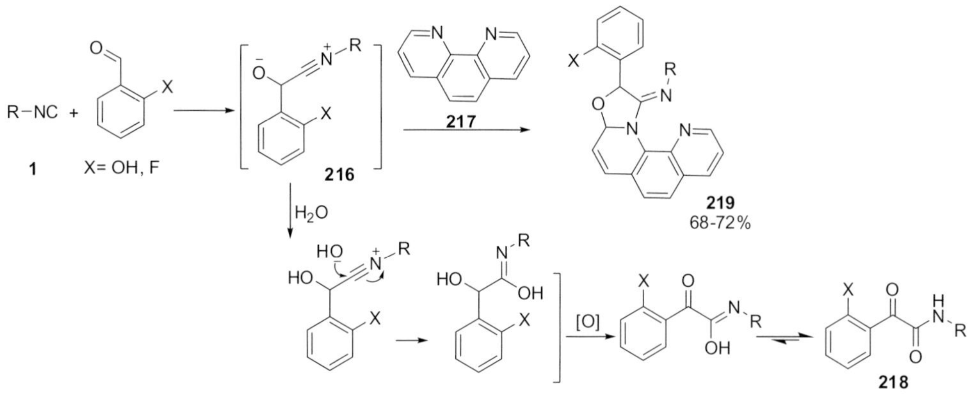

Scheme 8.49

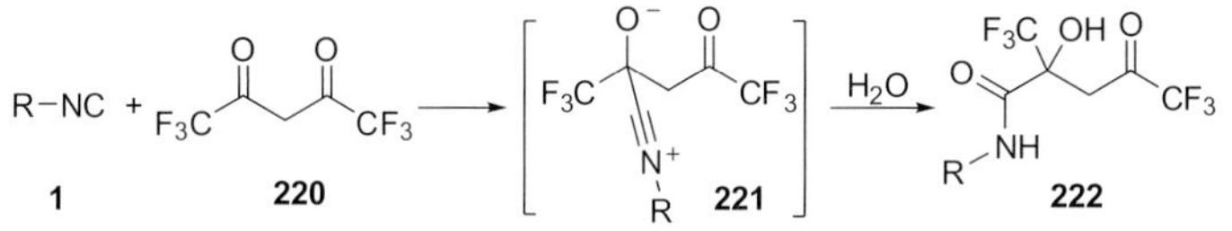

Scheme 8.50

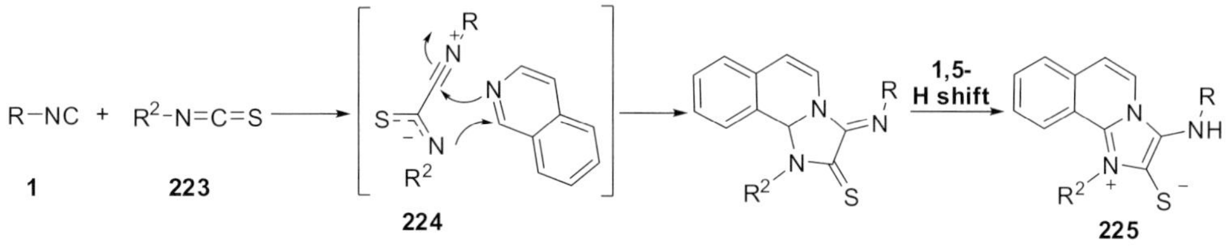

Scheme 8.51

The reaction of alkyl isocyanides **1** with 1,1,1,5,5,5-hexafluoropentane-2,4-dione **220**, afforded zwitterionic species **221**. The latter intermediate was then trapped with H$_2$O to produce γ-keto-α-hydroxy amide compounds **222** (Scheme 8.50) [123].

The three-component reaction of isoquinoline, isothiocyanates **223**, and isocyanides produces the zwitterionic imidazoisoquinolines **225** [124]. This MCR involves the formation of a zwitterionic intermediate **224** from isocyanide and isothiocyanate that has been trapped by isoquinoline (Scheme 8.51).

8.6
Isocyanides as Zwitterion-Trapping Agents

Few reports exist for the trapping of zwitterionic intermediates with isocyanides. These intermediates may be generated *in situ* from the nucleophilic additions of

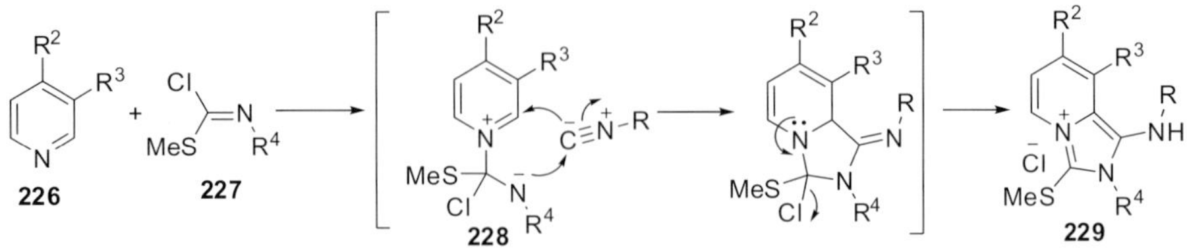

Scheme 8.52

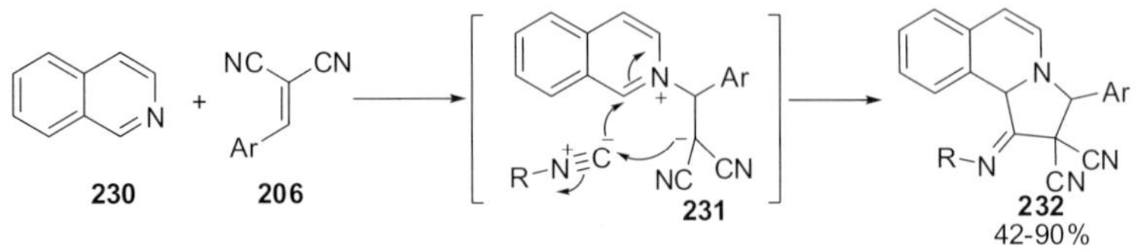

Scheme 8.53

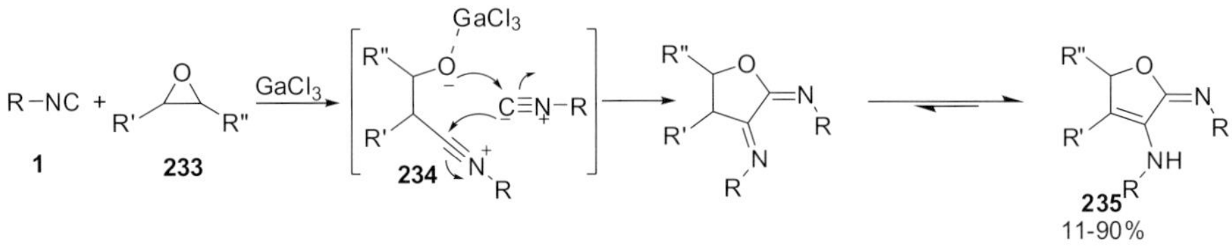

Scheme 8.54

compounds such as isoquinoline, pyridine, and/or unsaturated compounds. In this way, zwitterionic intermediates **228** generated from the reaction of pyridine **226** and chlorothioimidates **227**, were trapped by isocyanides to afford 1-amino-3-(methylthio) imidazolium chlorides **229** (Scheme 8.52) [125].

A three-component reaction between isoquinoline **230**, *gem*-diactivated olefins **206** and isocyanides **1** has afforded substituted 2,3-dihydro-10*H*-pyrrolo[2,1-*a*]isoquinoline-1-ones **232** [126]. In this reaction, the zwitterionic intermediates **231** were generated *in situ* from the reaction of isoquinoline and *gem*-diactivated olefins. Subsequently, **231** was trapped with isocyanides to obtain the product **232** (Scheme 8.53).

The formation of the α,β-unsaturated α-amino iminolactone products **235** can be rationalized by a $GaCl_3$-catalyzed double insertion of isocyanide into the epoxides **233** [127]. Ring-opening of the epoxide is mediated by the $GaCl_3$ catalyst to produce the corresponding carbocation, which reacts with a molecule of the isocyanide to produce zwitterionic intermediates **234**. This latter was then trapped with another molecule of the isocyanide to provide the final products **235** (Scheme 8.54).

The ring-opening of cyclopropane **236** is achieved with a suitable Lewis acid (such as the triflate of a rare earth metal [Yb(OTf)$_3$] and lanthanides Ln(OTf)$_3$, [Ln = Pr and Gd]) catalyst to provide the corresponding zwitterionic intermediate

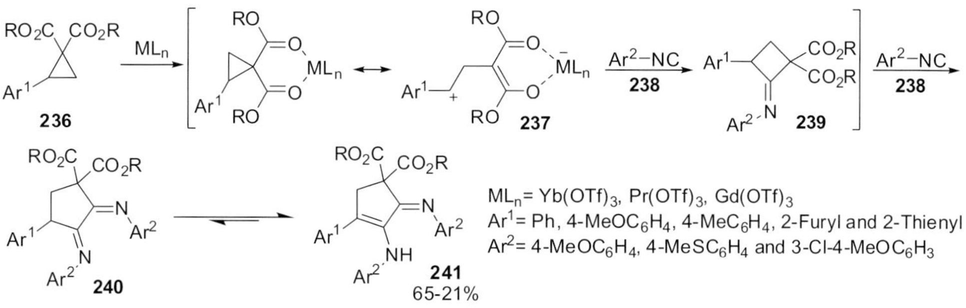

Scheme 8.55

237 [128]. The latter would then be capable of inserting an isocyanide 238 to produce a 2-iminocyclobutane-1,1-dicarboxylate 239 that might be prone to undergo the insertion of a second isocyanide molecule to furnish a more stable 2,3-diiminocyclopentane-1,1-dicarboxylate 240, or its tautomer 241 (Scheme 8.55).

8.7
Conclusions

The details of essentially all reported studies featuring zwitterionic species based on the isocyanide functional group have been included in this chapter. Attention has been focused mainly on the trapping of zwitterionic species generated by the addition of isocyanides to alkynes, arynes, and electron-deficient olefins, using a variety of trapping reagents such as CH-, NH-, and OH-acids, carbonyl and imine compounds, and electron-deficient olefins, via various MCRs. In addition, the details of some reactions in which isocyanides acted as zwitterion-trapping agents, and of miscellaneous reports involving zwitterionic species and isocyanides, have been included.

Acknowledgments

The authors gratefully acknowledge support from the Shahid Beheshti University, and are also indebted to all of their colleagues whose names are listed in the references.

References

1 Shaabani, A., Maleki, A., Rezayan, A.H., and Sarvary, A. (2011) Recent progress of isocyanide-based multicomponent reactions in Iran. *Mol. Divers.*, **15**, 41–68.

2 Kaim, L.E. and Grimaud, L. (2009) Beyond the Ugi reaction: less conventional interactions between isocyanides and iminium species. *Tetrahedron*, **65**, 2153–2171.

3 Dömling, A. (2006) Recent developments in isocyanide based multicomponent reactions in applied chemistry. *Chem. Rev.*, **106**, 17–89.

4 Zhu, J. and Bienayme, H. (2005) *Multicomponent Reactions*, Wiley-VCH Verlag GmbH, Weinheim.

5 Dömling, A. (2002) Recent advances in isocyanide-based multicomponent chemistry. *Curr. Opin. Chem. Biol.*, **6**, 306–313.

6 Ugi, I. and Heck, S. (2001) The multicomponent reactions and their libraries for natural and preparative chemistry. *Comb. Chem. High-Throughput Screen.*, **4**, 1–34.

7 Dömling, A. and Ugi, I. (2000) Multicomponent reactions with isocyanides. *Angew. Chem. Int. Ed.*, **39**, 3168–3210.

8 Ugi, I. (1971) *Isonitrile Chemistry*, Academic Press, New York.

9 Ugi, I., Meyr, R., Fetzer, U., and Steinbrückner, C. (1959) Versuche mit Isonitrilen. *Angew. Chem.*, **71**, 386–388.

10 Gulevich, A.V., Zhdanko, A.G., Orru, R.V.A., and Nenajdenko, V.G. (2010) Isocyanoacetate derivatives: synthesis, reactivity and application. *Chem. Rev.*, **110**, 5425–5446.

11 IUPAC (1997) *Gold Book: Zwitterionic Compounds/Zwitterions, IUPAC Compendium of Chemical Terminology*, 2nd edn, Blackwell Scientific Publications, Oxford, UK.

12 (a) Winterfeldt, E. (1967) Additions to the activated CC triple bond. *Angew. Chem. Int. Ed. Engl.*, **6**, 423–434; (b) Winterfeldt, E., Schumann, D., and Dillinger, H.J. (1969) Additionen an Die Dreifachbindung XI Struktur und Reaktionen des 2:1-Adduktes Aus Acetylenedicarbonester und Isonitrilen. *Chem. Ber.*, **102**, 1656–1664.

13 (a) Johnson, A.W. and Tebby, J.C. (1961) The adducts from triphenylphosphine and dimethyl acetylenedicarboxylate. *J. Chem. Soc.*, 2126–2130; (b) Tebby, J.C., Wilson, I.F., and Grifiths, D.V. (1979) Reactions of phosphines with acetylenes. Part 18. The mechanism of formation of 1,2-alkylidene phosphoranes. *J. Chem. Soc., Perkin Trans. I*, 2133–2135.

14 (a) Diels, O. and Alder, K. (1932) Synthesen in der Hydroaromatischen Reihe. *Liebigs Ann. Chem.*, **498**, 16–49; (b) Acheson, R.M. (1963) Reactions of acetylene carboxylic acids and their esters with nitrogen-containing heterocyclic compounds. *Adv. Heterocycl. Chem.*, **1**, 125–165.

15 Winterfeldt, E. (1964) Reaktionen des Propiolsaureesters Mit Tertiaren Aminen. *Chem. Ber.*, **97**, 1952–1958.

16 Winterfeldt, E. (1965) Additionen and die dreifachbindung III die reaktion von acetylendicarbonsaure-dimethyl ester mit dimethylsulfoxyd. *Chem. Ber.*, **98**, 1581–1587.

17 (a) Passerini, M. (1921) Composto Del *p*-Isonitrilazobenzolocon Acetone ed Acido Acetico. *Gazz. Chim. Ital.*, **51**, 126–129; (b) Passerini, M. (1921) Sopra gli isonitrili(I). Sopra Gli Isonitrili(II): Composti Con Aldeidi o Con Chetoni ed Acidi Organic Monobasici. *Gazz. Chim. Ital.*, **51**, 181–188.

18 Junjappa, H., Saxena, M.K., Ramaiah, D., Loharay, B.B., Rath, N.P., and George, M.V. (1998) Structure and thermal isomerization of the adducts formed in the reaction of cyclohexyl isocyanide with dimethyl acetylenedicarboxylate. *J. Org. Chem.*, **63**, 9801–9805.

19 (a) Huisgen, R. (1963) 1,3-Dipolare Cycloadditionen Rückschau und Ausblick. *Angew. Chem.*, **75**, 604–637; (b) Huisgen, R. (1961) Centenary Lecture: 1,3-Dipolar cycloadditions. *Proc. Chem. Soc.*, 357–369.

20 Huisgen, R., Grashey, R., and Sauer, J. (1964) *Chemistry of the Alkenes*, Interscience, New York.

21 Nair, V., Rajesh, C., Vinod, A.U., Bindu, S., Sreekanth, A.R., Mathen, J.S., and Balagopal, L. (2003) Strategies for heterocyclic construction via novel multicomponent reactions based on isocyanides and nucleophilic carbenes. *Acc. Chem. Res.*, **36**, 899–907.

22 (a) Winterfeldt, E., Schumann, D., and Dillinger, H.J. (1969) Additionen an die Dreifachbindung XI Struktur und Reaktionen des 2:1-Adducktes aus Acetylenedicarbonester und Isonitrilen. *Chem. Ber.*, **102**, 1656–1664;

(b) Dillinger, H.J., Fengler, G., Schumann, D., and Winterfeldt, E. (1974) Additionen an die Dreifachbindung-XXI: Das Kinetisch Kontrollierte Addukt aus tert-Butylisonitril und Acetylendicarbonester. *Tetrahedron*, **30**, 2553–2559; (c) Dillinger, H.J., Fengler, G., Schumann, D., and Winterfeldt, E. (1974) Additionen an die Dreifachbindung-XXII: Das Thermodynamisch Kontrollierte Adduct aus tert-butylisonitril und Acetylendicarbonester. *Tetrahedron*, **30**, 2561–2564.

23 (a) Oakes, T.R., David, H.G., and Nagel, F.J. (1969) Small ring systems from isocyanides I: reaction of isocyanides with hexafluorobutyne. *J. Am. Chem. Soc.*, **91**, 4761–4765; (b) Oakes, T.R. and Donovan, D.J. (1973) Synthesis of aryl isocyanates from nitro compounds and carbon monoxide. *J. Org. Chem.*, **38**, 1319–1325.

24 (a) Takizawa, T., Obata, N., Suzuki, T., and Yanagida, T. (1969) Conformations de Furannes-2-carbonyles. *Tetrahedron Lett.*, **11**, 3407–3410; (b) Suzuki, Y., Obata, N., and Takizawa, T. (1970) Novel synthesis of azaindolizines by reaction of isonitrosoflavanone esters with pyridine bases. *Tetrahedron Lett.*, **12**, 2667–2670.

25 Azizian, J., Ramazani, A., and Haji, M. (2011) Synthesis of dialkyl 2-(alkylamino)-4,9-dihydro-9-oxocyclohepta[b]pyran-3,4-dicarboxylates. *Helv. Chim. Acta*, **94**, 371–375.

26 Teimouri, M.B. and Bazhrang, R. (2009) The synthesis of functionalized pyranophenalenones. *Monatsh. Chem.*, **140**, 513–517.

27 Shaabani, A., Ghadari, R., Sarvary, A., and Rezayan, A.H. (2009) A simple and efficient method for the synthesis of highly functionalized bis(4H-chromene-) and 4H-benzo[g]chromene derivatives via an isocyanide-based multicomponent reaction. *J. Org. Chem.*, **74**, 4372–4374.

28 (a) Teimouri, M.B., Bazhrang, R., Eslamimanesh, V., Nouri, A. (2006) Reaction between isocyanides and dialkyl acetylenedicarboxylates in the presence of strong CH-acids: one-pot synthesis of highly functionalized annulated 4H-pyrans. *Tetrahedron*, **62**, 3016–3020; (b) Sarma, R., Sarmah, M.M., Lekhok, K.C., and Prajapati, D. (2010) Organic reactions in water: an efficient synthesis of pyranocoumarin derivatives. *Synlett*, 2847–2852.

29 Yavari, I., Sirouspour, M., and Souri, S. (2006) Three-component synthesis of functionalized 5-oxo-4,5-dihydroindeno[1,2-b]pyrans. *Mol. Divers.*, **10**, 265–270.

30 Maghsoodlou, M.T., Yavari, I., Nassiri, F., Djahaniani, H., and Razmjoo, Z. (2003) Reaction between alkyl isocyanides and cyclic 1,3-diketones: a convenient synthesis of functionalized 4H-pyrans. *Monatsh. Chem.*, **134**, 1585–1591.

31 Yavari, I., Adib, M., and Sayahi, M.H. (2002) Reaction between isocyanides and dialkyl acetylenedicarboxylates in the presence of 3-methylcyclopentane-1,2,4-trione. One-pot diastereoselective synthesis of tetrahydrocyclopenta[b] pyran derivatives. *J. Chem. Soc., Perkin Trans. I*, 2343–2346.

32 Esmaeili, A., Hosseinabadi, R., and Habibi, A. (2010) An efficient synthesis of highly functionalized 4H-pyrano[3,2-d]isoxazoles via isocyanide-based three-component reaction. *Synlett*, 477–1480.

33 Shaabani, A., Sarvary, A., Rezayan, A.H., and Keshipour, S. (2009) Synthesis of pyrano[2,3-c]pyrazole derivatives via a multicomponent reaction of isocyanides. *Tetrahedron*, **65**, 3492–3495.

34 Shaabani, A., Soleimani, E., Sarvary, A., and Rezayan, A.H. (2008) A simple and efficient approach to the synthesis of 4H-furo[3,4-b]pyrans via a three-component reaction of isocyanides. *Bioorg. Med. Chem. Lett.*, **18**, 3968–3970.

35 Yavari, I., Esmaili, A.A., Asghari, S., and Bijanzadeh, H.R. (1999) A new and efficient one-pot synthesis of trialkyl 6-tert-butylamino-2H-pyran-2-one-3,4,5-tricarboxylates. *J. Chem. Res. (S)*, 368–369.

36 Yavari, I. and Moradi, L. (2006) One-pot synthesis of pentaalkyl-7-[(alkylamino) carbonyl]-2-oxa-1-azabicyclo[3.2.0] hept-3-ene-3,4,5,6,7-pentacarboxylate. *Helv. Chim. Acta.*, **89**, 1942–1946.

37 Yavari, I. and Zare, H. (2007) An efficient synthesis of trialkyl N-alkyl-6-methyl-2-pyridone-3,4,5-tricarboxylates. *Monatsh. Chem.*, **138**, 787–790.

38 Asghari, S., Qandalee, M., and Bijnzadeh, H.R. (2006) Synthesis of polyfunctional ketenimines and 1-azadienes use of *tert*-butyl isocyanide and acetylenic esters in the presence of 3-chloropentane-2,4-dione. *J. Chem. Res. (S)*, 233–235.

39 Souldozi, A., Ramazani, A., Bouslimani, N., and Welter, R. (2007) The reaction of (N-isocyanimino)triphenylphosphorane with dialkyl acetylenedicarboxylates in the presence of 1,3-diphenyl-1,3-propanedione: a novel three-component reaction for the stereoselective synthesis of dialkyl (Z)-2-(5,7-diphenyl-1,3,4-oxadiazepin-2-yl)-2-butenedioates. *Tetrahedron Lett.*, **48**, 2617–2620.

40 Yavari, I. and Esnaasharia, M. (2005) Three-component one-pot synthesis of functionalized 1,2,3,6-tetrahydropyrano[4,3-b]pyrroles. *Synthesis*, 1049–1051.

41 Nasiri, F., Nazem, F., and Pourdavaie, K. (2007) Chemo- and stereoselective reaction between alkyl isocyanides and dimethyl 1,3-acetonedicarbocxylate in the presence of acetylenic esters. *Mol. Diversity*, **11**, 101–105.

42 Asghari, S. and Mohammadi, L. (2006) Reaction of *tert*-butyl isocyanide and dialkyl acetylenedicarboxylates in the presence of 2-acetylbutyrolactone: synthesis of functionalized α-methylene-γ-butyrolactones. *Tetrahedron Lett.*, **47**, 4297–4299.

43 Yavari, I., Hazeri, N., Maghsoodlou, M.T., and Zabarjad-Shiraz, N. (2001) Dynamic ^{1}H NMR study of aryl-nitrogen single bond and carbon-carbon double bond rotational energy barriers in two highly functionalized pyranopyrimidines. *Monatsh. Chem.*, **132**, 683–687.

44 Yavari, I., Anary-Abbasinejad, M., and Alizadeh, A. (2002) On the reaction between alkyl isocyanides and ethynyl phenyl ketone in the presence of N,N-dimethylbarbituric acid. *Monatsh. Chem.*, **133**, 1221–1224.

45 Maghsoodlou, M.T., Khorassani, S.M.H., Hazeri, N., Marandi, G., and Bijanzadeh, H.R. (2006) Synthesis and dynamic ^{13}C NMR study of new system containing polarized carbon–carbon double bonds from reaction between cyclohexyl isocyanide and ethyl propiolate in the presence of N,N-dimethylbarbituric acid. *J. Chem. Res. (S)*, 73–74.

46 Shaabani, A., Soleimani, E., Khavasi, H.R., Hoffmann, R.D., Rodewald, U., and Pottgen, R. (2006) An isocyanide-based three-component reaction: synthesis of fully substituted N-alkyl-2-triphenylphosphoranylidene glutarimides. *Tetrahedron Lett.*, **47**, 5493–5496.

47 Shaabani, A., Teimouri, M.B., and Arab-Ameri, S. (2004) A novel pseudo four-component reaction: unexpected formation of densely functionalized pyrroles. *Tetrahedron Lett.*, **45**, 8409–8413.

48 Adib, M., Sayahi, M.H., and Rahbari, S. (2005) Reactions between isocyanides and dialkyl acetylenedicarboxylates in the presence of 1,2-diacylhydrazines: one-pot synthesis of highly functionalized. *Tetrahedron Lett.*, **46**, 6545–6547.

49 Adib, M., Mohammadi, B., and Bijanzadeh, H.R. (2008) A novel, one-pot, three-component synthesis of dialkyl 5-(alkylamino)-1-aryl-1H-pyrazole-3,4-dicarboxylates. *Synlett*, 3180–3182.

50 Adib, M., Sayahi, M.H., Mohammadi, B., and Bijanzadeh, H.R. (2006) One-pot three-component synthesis of highly functionalized 2,3-dihydro-1,3-dioxo-1H,5H-pyrazolo[1,2-a][1,2,4]triazoles. *Helv. Chim. Acta*, **89**, 1176–1180.

51 Adib, M., Sayahi, M.H., Aghaaliakbarib, B., and Bijanzadeh, H.R. (2005) Reaction between isocyanides and dialkyl acetylenedicarboxylates in the presence of 2,4-dihydro-3H-pyrazol-3-ones: one-pot synthesis of highly functionalized 7-oxo-1H,7H-pyrazolo[1,2-a]pyrazoles. *Tetrahedron*, **61**, 3963–3966.

52 Adib, M., Ghanbary, K., Mostofia, M., and Bijanzadeh, H.R. (2005) Reaction between isocyanides and dialkyl

acetylenedicarboxylates in the presence of 4,5-diphenyl-1,3-dihydro-2H-imidazol-2-one. One-pot synthesis of 5H-imidazo[2,1-b][1,3]oxazine derivatives. *Tetrahedron*, **61**, 2645–2648.

53 Adib, M., Nosrati, M., Mahdavi, M., Zhu, L.G., and Mirzaei, P. (2007) A novel, one-pot, three-component synthesis of 5H-[1,3]thiazolo[3,2-a] pyrimidine derivatives. *Synlett*, 2703–2706.

54 Adib, M., Sayahi, M.H., Nosrati, M., and Zhu, L.G. (2007) A novel, one-pot, three-component synthesis of 4H-pyrido[1,2-a]pyrimidines. *Tetrahedron Lett.*, **48**, 4195–4198.

55 Adib, M., Sayahi, M.H., Ziyadi, H., Bijanzadeh, H.R., and Zhu, L.G. (2007) A new, one-pot, three-component synthesis of 4H-pyrido[1,2-a]-pyrimidines, 4H-pyrimido[1,2-a] pyrimidines, and 4H-pyrazino[1,2-a] pyrimidines. *Tetrahedron*, **63**, 1135–11140.

56 Yavari, I., Karimi, E., and Djahaniani, H. (2007) Synthesis of some novel γ-spiroiminolactones from reaction of cyclohexyl isocyanide and dialkyl acetylene dicarboxylates with 1-benzylisatin and tryptanthrine. *Synth. Commun.*, **33**, 387–391.

57 Shaabani, A., Sarvary, A., Ghasemi, S., Rezayan, A.H., Ghadari, R., and Ng, S.W. (2011) An environmentally benign approach for the synthesis of bifunctional sulfonamide-amide compounds via isocyanide-based multicomponent reactions. *Green Chem.*, **13**, 582–585.

58 Yavari, I., Nasiri, F., and Djahaniani, H. (2004) Synthesis and dynamic NMR study of ketenimines derived from *tert*-butyl isocyanide, alkyl 2-arylamino-2-oxo-acetates, and dialkyl acetylenedicarboxylates. *Mol. Divers.*, **8**, 431–435.

59 Anary-Abbasinejad, M., Moslemine, M.H., and Anaraki-Ardakani, H. (2009) One-pot synthesis of highly functionalized stable ketenimines of 2,2,2-trifluoro-N-aryl-acetamides. *J. Fluor. Chem.*, **125**, 1497–1500.

60 (a) Anary-Abbasinejad, M., Ghanea, F., and Anaraki-Ardakani, H. (2009) One-pot synthesis of highly functionalized stable ketenimines by three-component reaction of cyclohexyl isocyanide, dialkyl acetylenedicarboxylates, and benzoyl hydrazones. *Synth. Commun.*, **39**, 544–551; (b) Anary-Abbasinejad, M., Anaraki-Ardakani, H., and Ghanea, F. (2009) Facile synthesis of highly functionalized stable ketenimines by a three-component reaction of alkyl isocyanides, dialkyl acetylenedicarboxylates and ethyl carbazones. *Monatsh. Chem.*, **140**, 397–400.

61 Adib, M., Sayahi, M.H., Behnam, B., and Sheibani, E. (2006) Reaction between isocyanides and dialkyl acetylenedicarboxylates in the presence of hydantoins: a one-pot synthesis of stable ketenimines. *Monatsh. Chem.*, **137**, 191–196.

62 Bayat, M., Imanieh, H., and Hossieninejad, E. (2008) Simple synthesis of highly functionalized ketenimines. *Synth. Commun.*, **38**, 2567–2574.

63 Yavari, I., Djahaniani, H., and Nasiri, F. (2004) Synthesis of highly functionalized 1-azadienes and ketenimines. *Monatsh. Chem.*, **135**, 543–548.

64 Mohtat, B., Djahaniani, H., Khorrami, R., Mashayekhi, S., and Yavari, I. (2011) Three-component synthesis of functionalized 1-azabuta-1,3-dienes from alkyl isocyanides, activated acetylenes, and pyridin-2(1H)-one or isoquinolin-1(2H)-one. *Synth. Commun.*, **41**, 784–791.

65 Yavari, I., Alizadeh, A., Anary-Abbasinejad, M., and Bijanzadeh, H.R. (2003) Reaction between alkyl isocyanides and dibenzoylacetylene in the presence of strong NH-acids: synthesis of highly functionalized aminofurans. *Tetrahedron*, **59**, 6083–6086.

66 Oakes, T.S. and Donovan, A.D. (1973) Reactions of isocyanides with activated acetylenes in protic solvents. *J. Org. Chem.*, **38**, 1319–1325.

67 Alizadeh, A. and Rostamnia, S. (2008) Facile synthesis of highly functionalized

68 Yavari, I., Anary-Abbasinejad, M., Alizadeh, A., and Hossaini, Z. (2003) A simple and efficient approach to the synthesis of highly functionalized fused benzochromenes. *Tetrahedron*, **59**, 1289–1292.

69 (a) Baharfar, R., Vahdat, S.M., Ahmadian, M., and Taghizadeh, M.J. (2010) An efficient multicomponent transformation of alkyl isocyanides, dialkyl acetylenedicarboxylates, and 2,4-dihydroxybenzophenones or 2,4-dihydroxyacetophenones into 2-amino-4*H*-chromene derivatives. *Monatsh. Chem.*, **141**, 213–218; (b) Yavari, I., Djahaniani, H., and Nasiri, F. (2003) Reaction between alkyl isocyanides and dimethyl acetylenedicarboxylate in the presence of polyhydroxybenzenes: synthesis of 4,*H*-chromene derivatives. *Tetrahedron*, **59**, 9409–9412.

70 Teimouri, M.B., Mansouri, F., and Bazhrang, R. (2010) Facile synthesis of 1*H*-pyrazolo[1,2-a]pyridazine-5,8-dione derivatives by a one-pot, three-component reactions. *Tetrahedron*, **66**, 259–264.

71 (a) Alizadeh, A., Rostamnia, S., and Zhu, L.G. (2008) Competition of the R₃P/DAAD and RNC/DAAD zwitterions in their production and reaction with aromatic carboxylic acids: a novel binucleophilic system for three-component synthesis of 2-aminofurans. *Synthesis*, 1788–1792; (b) Alizadeh, A., Rostamnia, S., and Hu, M.L. (2006) A novel four-component reaction for the synthesis of 2,5-diaminofuran derivatives. *Synlett*, 1592–159421; (c) Adib, M., Sheikhi, E., Kavoosi, A., and Bijanzadeh, H.R. (2010) Synthesis of 2-(alkylamino)-5-{alkyl[(2-oxo-2*H*-chromen-3-yl)carbonyl]amino}-3,4-furandicarboxylates using a multi-component reaction in water. *Tetrahedron*, **66**, 9263–9269.

72 Alizadeh, A., Oskueyan, Q., and Rostamnia, S. (2007) Synthesis of nicotinamide and isonicotinamide derivatives via multicomponent reaction of alkyl asocyanides and acetylenic compounds in the presence of nicotinic or isonicotinic acid. *Synthesis*, 2637–2640.

73 Alizadeh, A., Oskueyan, Q., Rostamnia, S., Ghanbari-Niaki, A., and Mohebbi, A.R. (2008) Synthesis of bis(aminofuryl) bicinchoninic amides by a one-pot three-component reaction of isocyanides, acetylenic esters, and bicinchoninic acid. *Synthesis*, 2929–2932.

74 (a) Alizadeh, A., Rostamnia, S., and Zhu, L.G. (2006) Reaction between *tert*-butyl isocyanide, dialkyl acetylenedicarboxylates, and aromatic carboxylic acids: an efficient method for the synthesis of dialkyl (*E*)-2-{[benzoyl(*tert*-butyl)amino]carbonyl}-2-butenedioate derivatives. *Tetrahedron*, **62**, 5641–5644; (b) Sha, F., Lin, Y., and Huang, X. (2009) High regio- and stereoselective synthesis of (*Z*)- or (*E*)-*N*-acryl butenedioic monoimide derivatives by a multicomponent reaction. *Synthesis*, 424–430.

75 Alizadeh, A., Rostamnia, S., and Esmaili, A.A. (2007) Synthesis of functionalized sulfonamides via multicomponent reaction of alkyl isocyanide and dialkyl acetylenedicarboxylate with 4-methylbenzenesulfonic acid monohydrate. *Synthesis*, 709–712.

76 (a) Nair, V., Vinod, A.U., Abhilash, N., Menon, R.S., Santhi, V., Varma, R.L., Viji, S., Mathew, S., and Srinivas, R. (2003) Multicomponent reactions involving zwitterionic intermediates for the construction of heterocyclic systems: one-pot synthesis of aminofurans and iminolactones. *Tetrahedron*, **59**, 10279–10286; (b) Adib, M., Sayahi, M.H., Koloogani, S.A., and Mirzaei, P. (2006) Facile one-pot three-component synthesis of functionalized pyridylfuran-2-amines. *Helv. Chim. Acta*, **89**, 299–303; (c) Nair, V. and Vinod, A.U. (2000) The reaction of cyclohexyl isocyanide and dimethyl acetylenedicarboxylate with aldehydes: a novel synthesis of 2-aminofuran derivatives. *Chem. Commun.*, 1019–1020.

77 (a) Azizian, J., Mohammadizadeh, M.R., Mohammadi, A.A., and Karimi, A.R. (2005) A modified and green

methodology for preparation of polysubstituted furans. *Heteroat. Chem.*, 259–262; (b) Yadav, J.S., Reddy, B.V.S., Shubashree, S., Sadashiv, K., and Rao, D.K. (2007) Organic synthesis in water: green protocol for the synthesis of 2-aminofuran derivatives. *J. Mol. Catal., A Chem.*, **272**, 128–131.

78 Yadav, J.S., Reddy, B.V.S., Shubashree, S., Sadashiv, K., and Naidu, J.J. (2004) Ionic liquids-promoted multi-component reaction: green approach for highly substituted 2-aminofuran derivatives. *Synthesis*, 2376–2380.

79 Nair, V., Mathen, J.S., Viji, S., Srinivas, R., Nandakumar, M.V., and Varma, L. (2002) Diisopropylaminoisocyanide and DMAD in Multiple Component Reactions (MCRs): novel synthesis of substituted 1-amino-3-pyrrolin-2-ones by reaction with aldehydes and dicarbonyl compounds. *Tetrahedron*, **58**, 8113–8118.

80 (a) Reddy, B.V.S., Somashekar, D., Reddy, A.M., Yadav, J.S., and Sridhar, B. (2010) PEG 400 as a reusable solvent for 1,4-dipolar cycloadditions via a three-component reaction. *Synthesis*, 2069–2074; (b) Reddy, B.V.S., Somashekar, D., Reddy, A.M., Yadav, J.S., and Sridhar, B. (2010) Three-component one-pot synthesis of 3-(2-furanyl)indoles from acetylenedicarboxylate, isocyanide, and 3-formylindole. *Synthesis*, 2571–2576.

81 Terzidis, M.A., Stephanidou-Stephanatou, J., and Tsoleridis, C.A. (2010) One-pot synthesis of chromenylfurandicarboxylates and cyclopenta[b]chromenedicarboxylates involving zwitterionic intermediates: a DFT investigation on the regioselectivity of the reaction. *J. Org. Chem.*, **75**, 1948–1955.

82 Esmaeili, A.A. and Zendegani, H. (2005) Three-component reactions involving zwitterionic intermediates for the construction of heterocyclic systems: one-pot synthesis of highly functionalized γ-iminolactones. *Tetrahedron*, **61**, 4031–4034.

83 Shaabani, A., Rezayan, A.H., Ghasemi, S., and Sarvary, A. (2009) A mild and efficient method for the synthesis of 2,5-dihydro-2-methylfuran-3,4-dicarboxylates via an isocyanide-based multicomponent reaction. *Tetrahedron Lett.*, **50**, 1456–1458.

84 Yavari, I., Sabbaghan, M., and Hossaini, Z. (2008) Efficient synthesis of functionalized 2,5-dihydrofurans and 1,5-dihydro-2H-pyrrol-ones by reaction of isocyanides with activated acetylenes in the presence of hexachloroacetone. *Monatsh. Chem.*, **139**, 625–628.

85 Yavari, I., Hossaini, Z., and Sabbaghan, M. (2006) Synthesis of functionalized 5-imino-2,5-dihydro-furans through the reaction of isocyanides with activated acetylenes in the presence of ethyl bromopyruvate. *Mol. Divers.*, **10**, 479–482.

86 Hazeri, N., Maghsoodlou, M.T., Habibi-Khorassani, S.M., Marandi, G., Khandan-Barani, K., Ziyaadini, M., and Aminkhani, A. (2007) Synthesis of novel 2-pyridyl-substituted 2,5-dihydro-2-imino and 2-amino-furan derivatives via a three component condensation of alkyl isocyanides and acetylenic esters with di-(2-pyridyl)ketone or 2-pyridinecarboxaldehyde. *ARKIVOC I*, 173–179.

87 (a) Sarvary, A., Shaabani, S., Shaabani, A., and Ng, S.W. (2011) Synthesis of functionalized iminolactones via an isocyanide-based three-component reaction. *Tetrahedron*, **67**, 3624–3630; (b) Ramazani, A., Rezaei, A., Mahyari, A.T., Rouhani, M., and Khoobi, M. (2010) Three-component reaction of an isocyanide and a dialkyl acetylenedicarboxylate with a phenacyl halide in the presence of water: an efficient method for the one-pot synthesis of γ-iminolactone derivatives. *Helv. Chim. Acta*, **93**, 2033–2036.

88 Teimouri, M.B., Shaabani, A., and Bazhrang, R. (2006) Reaction between alkyl isocyanides and dialkyl acetylenedicarboxylates in the presence of benzoyl cyanides: one-pot synthesis of highly functionalized iminolactones. *Tetrahedron*, **62**, 1845–1848.

89 Bayat, M., Imanieh, H., and Abbasi, H. (2010) Chemoselective and regiospecific synthesis of iminospiro-γ-lactones from maleic anhydride or citraconic anhydride and alkyl isocyanides with dialkyl

acetylenedicarboxylates. *Helv. Chim. Acta*, **93**, 757–762.

90 Adib, M., Moghimi, S., Sayahi, M.H., and Bijanzadeh, H.R. (2009) One-pot three-component synthesis of 2-(alkylimino)-7-oxo-1-oxa-6-azaspiro[4.4]nona-3,8-diene-3,4-dicarboxylates. *Helv. Chim. Acta*, **92**, 944–949.

91 Esmaeili, A.A. and Darbanian, M. (2003) Reaction between alkyl isocyanides and dialkyl acetylenedicarboxylates in the presence of N-alkyl isatins: convenient synthesis of γ-spiro-iminolactones. *Tetrahedron*, **59**, 5545–5548.

92 Maghsoodlou, M.T., Hazeri, N., Habibi-Khorasani, S.M., Heydari, R., Marandi, G., and Nassiri, M. (2005) Reaction of alkyl and aryl isocyanide with fluren-9-ones in the presence of acetylenic esters: preparation of γ-spiroiminolactones. *Synth. Commun.*, **35**, 2569–2574.

93 Maghsoodlou, M.T., Hazeri, N., Habibi-Khorasani, S.M., Marandi, G., and Nassiri, M. (2006) 1,8-Diazafloren-9-one with alkyl and aryl isocyanides in the presence of acetylenic esters: a facile synthesis of γ-spiroiminolactones. *J. Heterocycl. Chem.*, **43**, 481–484.

94 Azizian, J., Karimi, A.R., and Mohammadi, A.A. (2003) Synthesis of some novel γ-spiroiminolactones from reaction of cyclohexyl isocyanide and dialkyl acetylene dicarboxylates with 1-benzylisatin and tryptanthrine. *Synth. Commun.*, **33**, 387–391.

95 (a) Maghsoodlou, M.T., Hazeri, N., Habibi-Khorasani, S.M., Heydari, R., Marandi, G., and Nassiri, M. (2005) γ-Spiroiminolactones synthesis by reaction of acetylenic esters and dicarbonyl compounds in the presence of aryl isocyanide. *Synth. Commun.*, **35**, 2771–2777; (b) Nair, V., Vinod, A.U., Nair, J.S., Sreekanth, A.R., and Rath, N.P. (2000) The reaction of cyclohexyl isocyanide and dimethyl acetylenedicarboxylate with o- and p-quinones: a novel synthesis of iminolactones. *Tetrahedron Lett.*, **41**, 6675–6679.

96 Nair, V., Menon, R.S., Deepthi, A., Devi, B.R., and Biju, A.T. (2005) One-pot, four-component reaction of isocyanides, dimethyl acetylenedicarboxylate, and cyclobutene-1,2-diones: a synthesis of novel spiroheterocycles. *Tetrahedron Lett.*, **46**, 1337–1339.

97 (a) Nair, V. and Deepthi, A. (2006) A novel reaction of vicinal tricarbonyl compounds with the isocyanide–DMAD zwitterion: formation of highly substituted furan derivatives. *Tetrahedron Lett.*, **47**, 2037–2039; (b) Mosslemin, M.H., Anary-Abbasinejad, M., and Anaraki-Ardakani, H. (2009) Reaction between isocyanides, dialkyl acetylenedicarboxylates and 2-hydroxy-1-aryl-2-(arylamino)ethanones: one-pot synthesis of highly functionalized 2-aminofurans. *Synlett*, 2676–2678.

98 Alizadeh, A., Rostamnia, S., Zohreh, N., and Bijanzadeh, H.R. (2008) Synthesis of ethylenetetracarboxylic acid derivatives. *Monatsh. Chem.*, **139**, 49–52.

99 Nair, V., Vinod, A.U., and Rajesh, C. (2001) A novel synthesis of 2-aminopyrroles using a three-component reaction. *J. Org. Chem.*, **66**, 4427–4429.

100 Nair, V., Dhanya, R., and Viji, S. (2005) The three-component reaction involving isocyanides, dimethyl acetylenedicarboxylate and quinoneimides: a facile synthesis of spirofused γ-iminolactams. *Tetrahedron Lett.*, **51**, 5143–5148.

101 Ahmadi, E., Ramazani, A., and Haghighi, M.N. (2007) A novel four-component reaction of diethylamine, an aromatic aldehyde and an alkyl isocyanide with dialkyl acetylenedicarboxylates in the presence of silica gel: an efficient route for the regio- and stereoselective synthesis of sterically congested alkenes. *Tetrahedron Lett.*, **48**, 6954–6957.

102 Shaabani, A., Rezayan, A.H., Rahmati, A., and Sarvary, A. (2007) A novel isocyanide-based three-component condensation reaction: synthesis of fully substituted imino- and spiroiminocyclopentenes. *Synlett*, 1458–1460.

103 Nair, V., Menon, R.S., Beneesh, P.B., Sreekumar, V., and Bindu, S. (2004) A novel multicomponent reaction involving isocyanide, dimethyl

104 Zhou, Z. and Magriotis, P.A. (2005) A new method for the functionalization of [60]fullerene: an unusual 1,3-dipolar cycloaddition pathway leading to a C_{60} housane derivative. *Org. Lett.*, **7**, 5849–5851.

105 Yavari, I. and Djahaniani, H. (2005) One-step synthesis of substituted 4,7-bis[alkyl(aryl)imino]-3-oxa-6-thia-1-azaspiro[4.4]nona-1,8-dienes. *Tetrahedron Lett.*, **46**, 7491–7493.

106 Yavari, I., Bayat, M.J., Sirouspour, M., and Souri, S. (2010) One-pot synthesis of highly functionalized 1,2-dihydropyridines from primary alkylamines, alkyl isocyanides, and acetylenic esters. *Tetrahedron*, **66**, 7995–7999.

107 Alizadeh, A. and Hosseinpour, R. (2009) An unprecedented synthesis of eight-membered-ring cyclic thioimidic esters by a three-component reaction. *Synthesis*, 2733–2736.

108 Qian, B., Fan, M.J., Xie, Y.X., Wu, L.Y., Shi, Y., and Liang, Y.M. (2009) A novel one-pot, three-component synthesis of 5-imino-2,3,5,8-tetrahydropyrazolo[1,2-a]pyridazin-1-one derivatives. *Synthesis*, 1689–1693.

109 Khalilzadeh, M.A., Hossaini, Z., Baradarani, M.M., and Hasanni, A. (2010) A novel isocyanide-based three-component reaction: a facile synthesis of substituted 2H-pyran-3,4-dicarboxylates. *Tetrahedron*, **66**, 8464–8467.

110 Pellissier, H. and Santelli, M. (2003) The use of arynes in organic synthesis. *Tetrahedron*, **59**, 701–730.

111 Rondan, N.G., Domelsmith, L.N., Houk, K.N., Bowne, A.T., and Levin, R.H. (1979) The relative rates of electron-rich and electron-deficient alkene cycloadditions to benzyne: enhanced electrophilicity as a consequence of alkyne bending distortions. *Tetrahedron Lett.*, **20**, 3237–3240.

112 Gilchrist, T.L. (1983) *In the Chemistry of Functional Groups Supplement C*, John Wiley & Sons, Ltd, Chichester.

113 Yoshida, H., Fukushima, H., Ohshita, J., and Kunai, A. (2004) Arynes in a three-component coupling reaction: straightforward synthesis of benzoannulated iminofurans. *Angew. Chem. Int. Ed.*, **43**, 3935–3938.

114 Allan, K.M., Gilmore, C.D., and Stoltz, B.M. (2011) Benzannulated bicycles by three-component aryne reactions. *Angew. Chem. Int. Ed.*, **50**, 4488–4491.

115 Yoshida, H., Fukushima, H., Morishita, T., Ohshita, J., and Kunai, A. (2007) Three-component coupling using arynes and isocyanides: straightforward access to benzo-annulated nitrogen or oxygen heterocycles. *Tetrahedron*, **63**, 4793–4805.

116 Yoshida, H., Fukushima, H., Ohshita, J., and Kunai, A. (2004) Straightforward access to 2-iminoisoindolines via three-component coupling of arynes, isocyanides and imines. *Tetrahedron Lett.*, **45**, 8659–8662.

117 Rigby, J.H. and Laurent, S. (1998) Addition of alkyl and aryl isocyanides to benzyne. *J. Org. Chem.*, **63**, 6742–6744.

118 Li, M., Qiu, Z.X., Wen, L.R., and Zhou, Z.M. (2011) Novel regio- and stereo-selectivity: synthesis of dihydropyrrolo[1,2-f]phenanthridines via isocyanide-based multicomponent reaction. *Tetrahedron*, **67**, 3638–3648.

119 Zhu, X., Xu, X.P., Sun, C., Wang, H.Y., Zhao, K., and Ji, S.J. (2010) Direct construction of imino-pyrrolidine-thione scaffold via isocyanide-based multicomponent reaction. *J. Comb. Chem.*, **12**, 822–828.

120 Li, J., Liu, Y., Li, C., and Jia, X. (2011) Silver hexafluoroantimonate-catalyzed three-component [2+2+1] cycloadditions of allenoates, dual activated olefins, and isocyanides. *Adv. Synth. Catal.*, **353**, 913–917.

121 Yavari, I. and Djahaniani, H. (2006) One-step synthesis of N-alkyl-2-aryl-2-oxoacetamides and N^2,N^4-dialkyl-2-aryl-4H-1,3-benzodioxine-2,4-dicarboxamides. *Tetrahedron Lett.*, **47**, 1477–1481.

122 Maghsoodlou, M.T., Marandi, G., Hazeri, N., Aminkhani, A., and Kabiri, R. (2007) A facile synthesis of oxazolo[3,2-a][1,10]phenanthrolines via a new multicomponent reaction. *Tetrahedron Lett.*, **48**, 3197–3199.

123 Shaabani, A., Bazgir, A., Soleimani, K., and Bijanzahdeh, H.R. (2002) Reaction between alkyl isocyanides and 1,1,1,5,5,5-hexafluoropentane-2,4-dione in the presence of water: one-pot synthesis of highly fluorinated γ-dihydroxy-α-hydroxy amides and γ-keto-α-hydroxy amides. *J. Fluorine Chem.*, **116**, 93–95.

124 Mironov, M.A., Maltsev, S.S., Mokrushin, V.S., and Bakulev, V.A. (2005) A novel three-component reaction designed by the combinatorial method: heteroarenes, isothiocyanates and isocyanides. *Mol. Divers.*, **9**, 221–227.

125 Marchand, E. and Morel, G. (1993) Three-component cyclocondensations: a convenient access to fused imidazolium and dihydropyrimidinium salts via the reaction of methyl chlorothioimidates with azines and isocyanides. *Tetrahedron Lett.*, **34**, 2319–2322.

126 Mironov, C.A., Mokrushin, V.S., and Maltsev, S.S. (2003) New method for the combinatorial search of multi component reactions. *Synlett*, 943–946.

127 Bez, G. and Zhao, C.G. (2003) Gallium(III) chloride-catalyzed double insertion of isocyanides into epoxides. *Org. Lett.*, **5**, 4991–4993.

128 Korotkov, V.S., Larionov, O.V., and de Meijere, A. (2006) Ln(OTf)$_3$-catalyzed insertion of aryl isocyanides into the cyclopropane ring. *Synthesis*, 3542–3546.

9
Recent Progress in Nonclassical Isocyanide-Based MCRs
Rosario Ramón, Nicola Kielland, and Rodolfo Lavilla

9.1
Introduction

Today, the use of isocyanides in organic chemistry is continuing to exert a huge impact on the general approach to the synthesis of bioactive compounds [1, 2]. The unique "carbene-like" structure of this functional group allows the incorporation of two other species (normally an electrophile and a nucleophile) in the same process, which makes it a privileged reactant in multicomponent reactions (MCRs) [3]. This strategy not only allows a large number of chemically diverse drug-like scaffolds to be prepared quickly and under mild conditions, but also enables diversity-oriented synthesis to compete with and complement standard multistep target-oriented synthesis. The most classical use of isocyanides involves their interaction with iminium or oxocarbenium ions as the electrophilic species, combinations that lead to the well-known Ugi-, van Leusen-, and Passerini-like processes, as well as mechanistically related variations [1, 2, 4]. Apart from these, the recent exploration of many other electrophilic partners has generated a large number of MCRs, often involving complex reaction cascades and leading to a variety of new adducts of great pharmaceutical interest. A schematic outline of the reactivity trends of isocyanides, and the types of process described in this chapter, are depicted in Scheme 9.1. The mechanistic guidelines used in the chapter are based on well-stated concepts, and serve mainly to classify the reactions:

- Type I processes are those in which the isocyanide attacks the complex formed between the electrophilic and nucleophilic partner.
- Type II processes include activation of the isocyanide by electrophilic agents to generate the reactive species towards the nucleophiles.
- Type III reactions deal with insertion processes not described above, mainly metalation-driven.

As the intention here was not to provide a comprehensive coverage, reference is made to reviews and major studies for more detailed explanations.

Isocyanide Chemistry: Applications in Synthesis and Material Science, First Edition. Edited by Valentine Nenajdenko.
© 2012 Wiley-VCH Verlag GmbH & Co. KGaA. Published 2012 by Wiley-VCH Verlag GmbH & Co. KGaA.

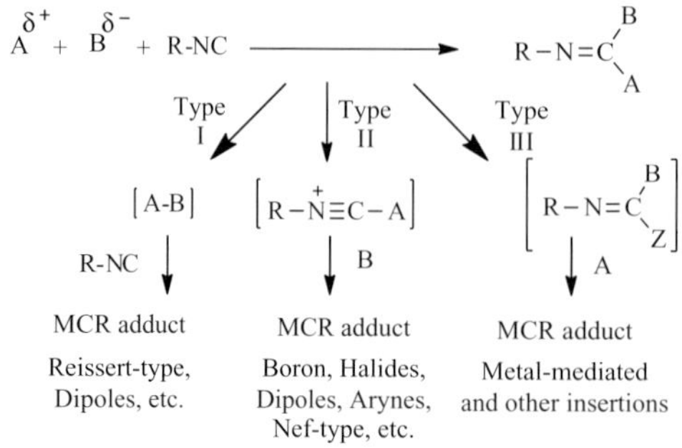

Scheme 9.1 Isocyanide reactivity: classical and alternative activating agents.

9.2
Type I MCRs: Isocyanide Attack on Activated Species

At this point, a broad family of reactions is analyzed that feature as their key step isocyanide addition on activated species generated by the interaction of the two complementary reactants. One relevant class of reactions that belong to this type involves the use of azine-derived cations as surrogates of the iminium intermediates in Ugi MCRs. These Reissert-type processes, where the azine formally replaces the imine, have been recently reviewed in the context of MCRs [5]. In this way, azines can be activated by a variety of electrophilic reactants, generating an azinium salt, the target of the isocyanide attack in a second step, to be followed by the final trapping of the nitrilium ion. Some mechanistic detours, however, may lead to alternative connectivity patterns (Scheme 9.2).

Several activating agents have been studied in this context. Due to the coexistence of two nucleophilic species in the reaction mixture (azine and isocyanide), more than one reaction pathway is theoretically possible. The most commonly found scenario occurs where the azine is activated by the electrophile (the activation of isocyanides is discussed in Section 9.3) to generate intermediate **B**, often in a reversible process. The isocyanide subsequently attacks at the α position of

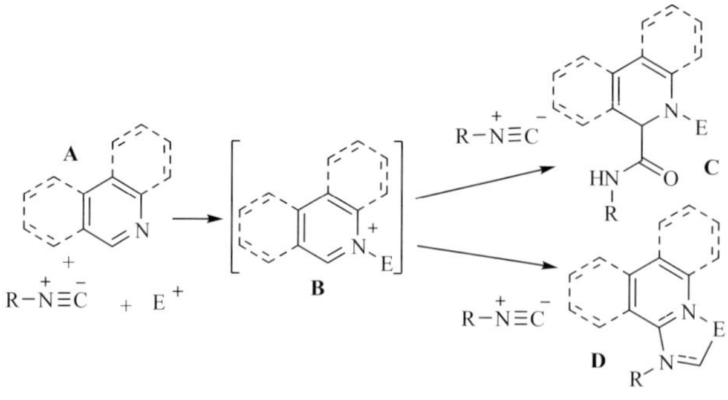

Scheme 9.2 Azine–isocyanide MCRs.

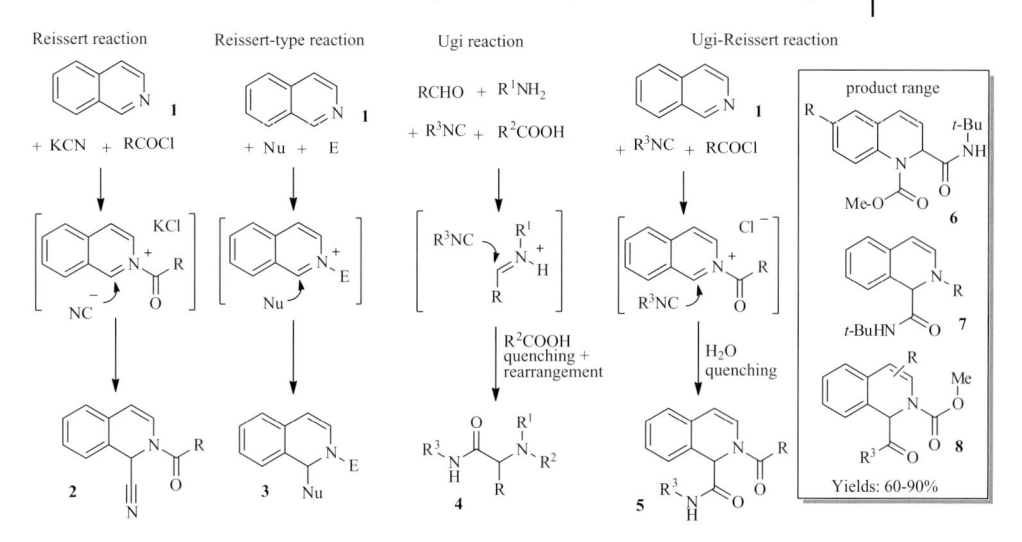

Scheme 9.3 Analogies between Reissert and Ugi MCRs.

these cationic species and generates a product of general structure **C**, after being intercepted by a quencher (normally water). An alternative pathway involves an isocyanide attack upon the electrophile-derived moiety in **B**, its nucleophilic carbon becoming attached to this residue. After a final cyclization, compounds of general structure **D** are formed (Scheme 9.2).

In this way, when isoquinoline **1** is activated by a chloroformate, isocyanides attack at the α position of the azine and, after aqueous trapping of the nitrilium intermediate, the carbamoylated dihydroisoquinolines **5** are obtained in a single step [6]. This process has been termed the Ugi–Reissert reaction, because of its mechanistic similarity to the parent transformations (Scheme 9.3).

The Ugi–Reissert reaction is quite general, and affords α-carbamoylated dihydroazines **6-8** in good to high yields. The range of isocyanides is diverse, and the activating electrophilic reagents includes chloroformates, acid chlorides, tosyl chloride, and even (Boc)$_2$O. In contrast, the choice of azines is restricted to diversely substituted isoquinolines and quinolines. Interestingly, the resulting products can be used as starting materials for Povarov reactions, leading to highly complex products in a two-step tandem reaction to afford the polyheterocyclic adduct **10** by interaction with aldehydes, anilines, and Sc(OTf)$_3$ (Scheme 9.4) [7]. A solid-phase Ugi–Reissert reaction on a chloroformate resin has also been described. In this case, the α-carbamoylated isoquinoline **11** is cleaved from the resin by oxidation with 2,3-dichloro-5,6-dicyano-1,4-benzoquinone (DDQ) [6]. Zhu recently reported an alternative method to obtain products with a similar structure by a Ugi reaction. The novelty in this case is the use of 2-iodoxybenzoic acid (IBX) to generate imine **14** *in situ* by the oxidation of tetrahydroisoquinoline **13**. Remarkably, IBX has proven to be chemically compatible with all the other components, and the process takes place as a true MCR (Scheme 9.4) [8].

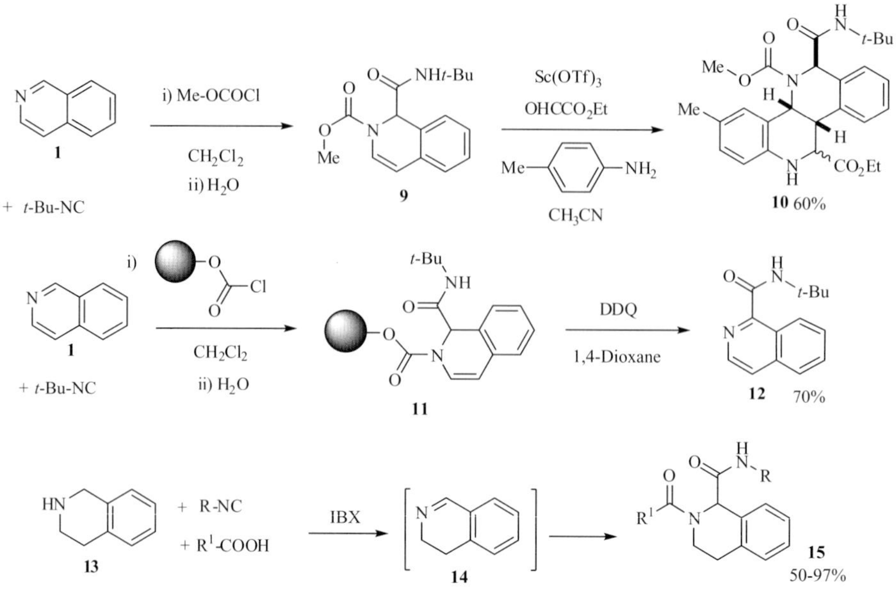

Scheme 9.4 Ugi–Reissert MCRs: a solid-phase version, post-synthetic transformations, and analogous processes.

With respect to the trapping of the intermediate nitrilium ions, Zhu designed a new class of MCRs in which these species are intramolecularly trapped by a carboxamido group to generate substituted oxazoles [9–11]. On applying this strategy, an interesting transformation was reported using the functionalized isocyanide **16** which, on interaction with isoquinoline and methylchloroformate, afforded oxazole **17** in good yields [12]. The latter compound can finally react in several cycloadditions to yield furan **19** after the rearrangement of intermediate **18** (Scheme 9.5). The use of the same approach allowed the direct addition of isocyanides to nicotinamide salt **20** (incidentally, the direct isocyanide addition to other pyridinium salts is not feasible). However, in this case the different substitution pattern of the carboxamido group led to an isomerization of the putative intermediate **21** to the cyano-carbamoyl derivative **22**. Interestingly, this process is also efficient in a *one-pot* Reissert–Ugi reaction (Scheme 9.5) [13].

The Arndtsen group described a related process involving imines or isoquinolines, acid chlorides, alkynes and isocyanides to yield highly substituted pyrroles in a single step [14]. Although, ultimately, none of the isocyanide atoms is attached to the final product, their role in the mechanism appears to be crucial. After activation of the imine (or isoquinoline), the isocyanide attack generates a nitrilium ion that is quickly trapped by the carbonyl group of the activating agent to produce the cationic intermediate **32**. The dipole **33** is then formed by proton loss, and reacts with dimethylacetylene dicarboxylate to yield pyrrole **30** through a

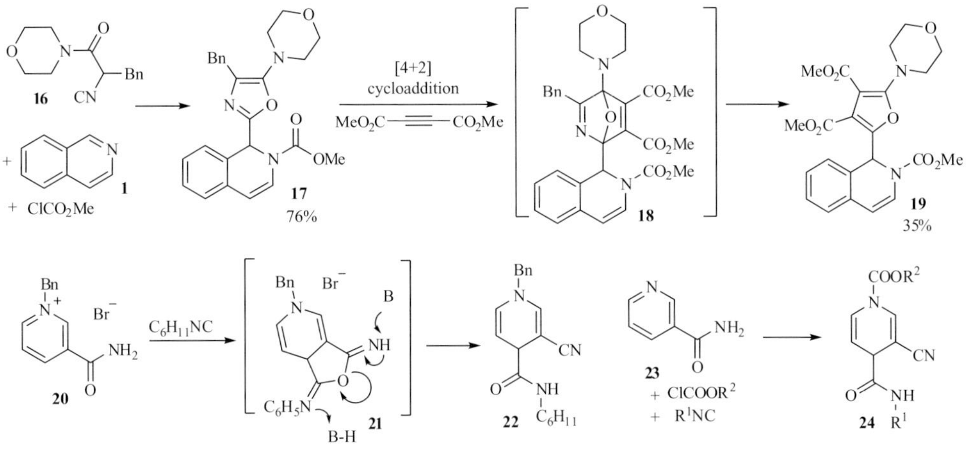

Scheme 9.5 Internal trapping of the nitrilium intermediate by carboxamido groups.

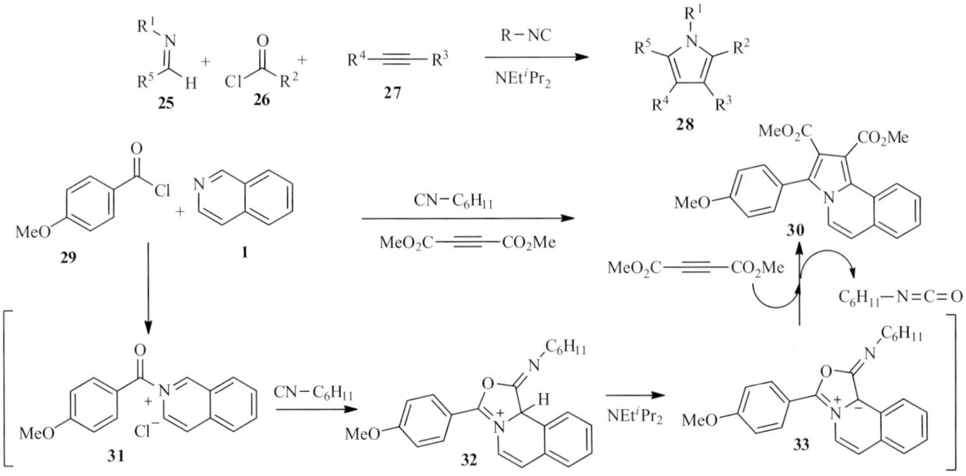

Scheme 9.6 Arndtsen pyrrole synthesis.

cycloaddition–cycloelimination process in which an equivalent of isocyanate is released (Scheme 9.6).

One of the main restrictions of the Reissert-type processes is the limited reactivity of pyridines. These azines, in sharp contrast to (iso)quinolines, are practically inert, and special conditions are required for their activation. Recently, Corey and coworkers investigated the use of triflic anhydride for this purpose [15] when, under these conditions, several nucleophiles were shown capable of reacting in Reissert-type processes with pyridine to afford the γ-substituted dihydropyridines **37** (Scheme 9.7). The same reaction in the presence of isocyanides affords a

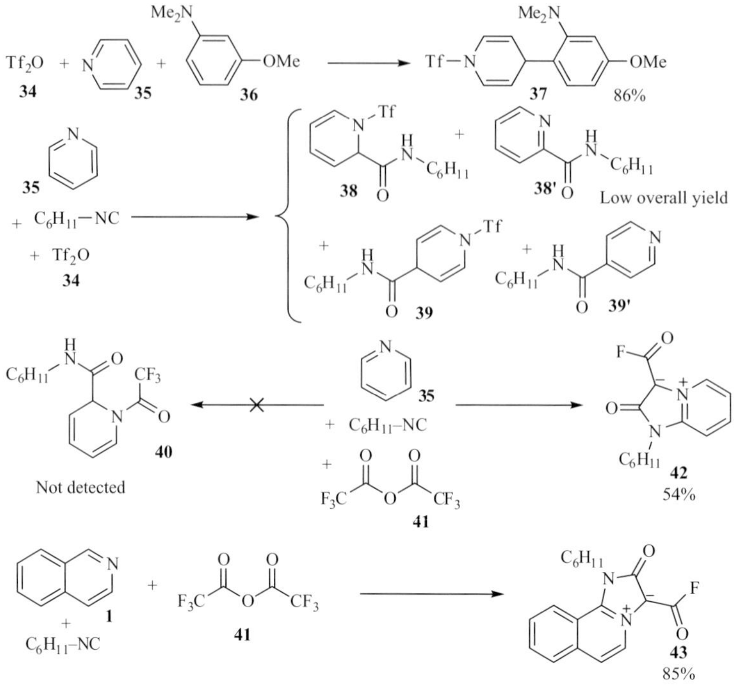

Scheme 9.7 Tf₂O- and TFAA-promoted isocyanide–azine MCRs.

mixture of γ-carbamoylated dihydropyridines **38** and **39**, together with the corresponding oxidized products **38'** and **39'** in a low overall yield. A major problem found here was the massive degradation of isocyanide, as a consequence of which milder activating agents were tested, and the use of trifluoroacetic anhydride (TFAA) was considered. Surprisingly, when pyridine was mixed with TFAA and cyclohexylisocyanide the expected α-carbamoylated dihydropyridine **40** was not detected, while the mesoionic acid fluoride **42** was isolated. Isoquinolines proved to be even better reactants under these conditions, affording the corresponding dipoles in high yields [16]. However, the unusual connectivity of the atoms in these products suggests a different type of mechanism at work (Scheme 9.8).

A reasonable mechanistic explanation involves activation of the azine by TFAA to generate the isoquinolinium salt **A** (Scheme 9.8). Here, the expected isocyanide attack at the α-position of the isoquinoline should generate a nitrilium ion, finally leading to dipoles **45**, which have been obtained as byproducts in some of these reactions. This type of cascade has been described previously in the Arndtsen pyrrole synthesis [14]. However, the main route here passes through an isocyanide addition on the carbonyl group of the trifluoroacetyl moiety, thus generating adduct **B**. This intermediate triggers a domino process that begins with an attack of the isocyanide nitrogen on the α-position of the isoquinoline, and is followed by closure of the epoxide ring with the loss of a fluorine anion, leading to **C**. Sub-

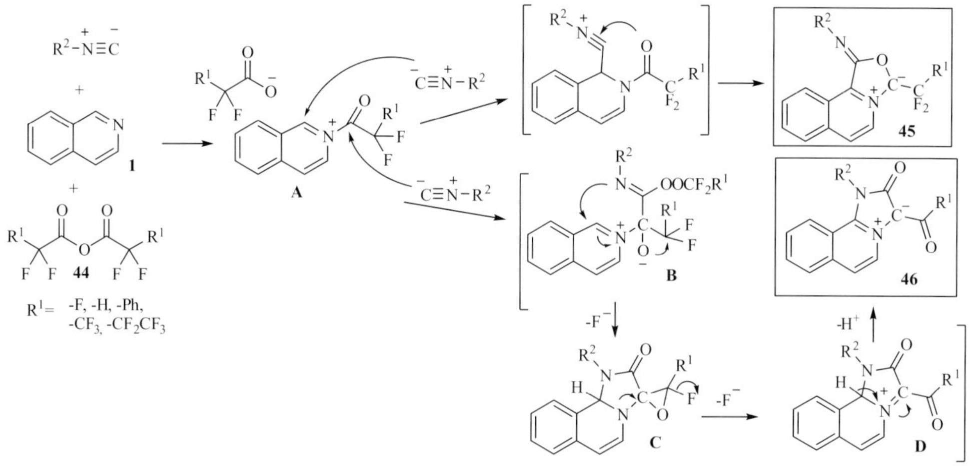

Scheme 9.8 Mechanistic proposals for MCR of azines, isocyanides, and TFAA.

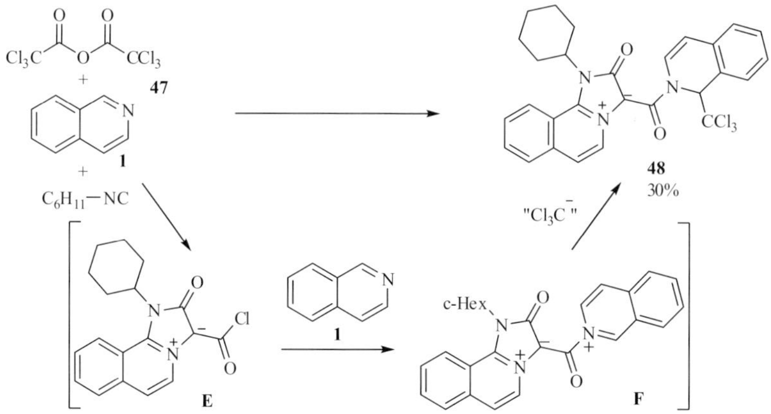

Scheme 9.9 Cascade reaction of azines and isocyanides promoted by trichloroacetic anhydride (TCAA).

sequently, a second fluorine atom is lost by opening of the epoxide intermediate and the formation of a new carbonyl group (**D**), while a final proton loss yields dipole **46** (Scheme 9.8).

The scope of the reaction is quite general. Apart from TFAA, diversely substituted difluoroacetic anhydrides are also reactive, yielding the corresponding carbonyl compounds, including ketones and aldehydes. TCAA leads to dipole **48** through an even more complex reaction cascade; in this case, the expected mesoionic acid chloride **E** is more reactive than the acid fluoride analogue, being able to activate a second equivalent of isoquinoline. The new isoquinolinium ion (**F**) is then attacked by a trichloromethyl anion, leading to dipole **48** (Scheme 9.9) [16, 17].

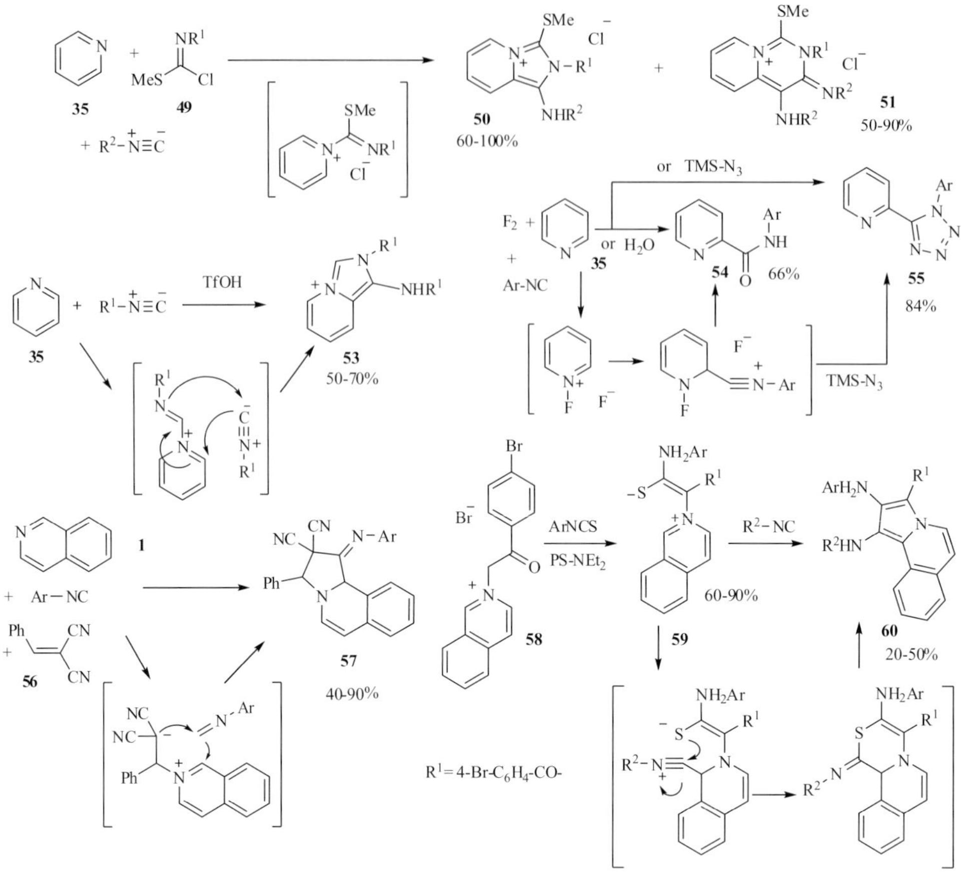

Scheme 9.10 Isocyanide–azine MCRs promoted by various activating agents.

The dipolar acid fluorides **46**, although sufficiently stable to be isolated and characterized, are reactive towards nucleophiles and conveniently afford amides, esters, and thioesters on interaction with amines, alcohols, and thiols, respectively [16]. This property, together with their natural fluorescence, has been successfully applied to the selective labeling of oligonucleotides [18].

Distinct activating agents can also promote Reissert-type reactions with isocyanides. The reaction of chlorothioimidates **49** with pyridines and isocyanides leads to the fused imidazolium salt **50** (Scheme 9.10). This reaction proceeds through the usual activation–isocyanide attack pathway to generate a nitrilium intermediate that is intramolecularly trapped to afford the imidazolium salt **50**. An analogous reaction with double isocyanide insertion yielding derivative **51** has also been described [19]. In a related study, Berthet reported the double incorporation of isocyanides into pyridine through triflic acid activation to yield the bicyclic azolium salt **53** [20]. Fluorine has also been used as an activating agent in a two-component

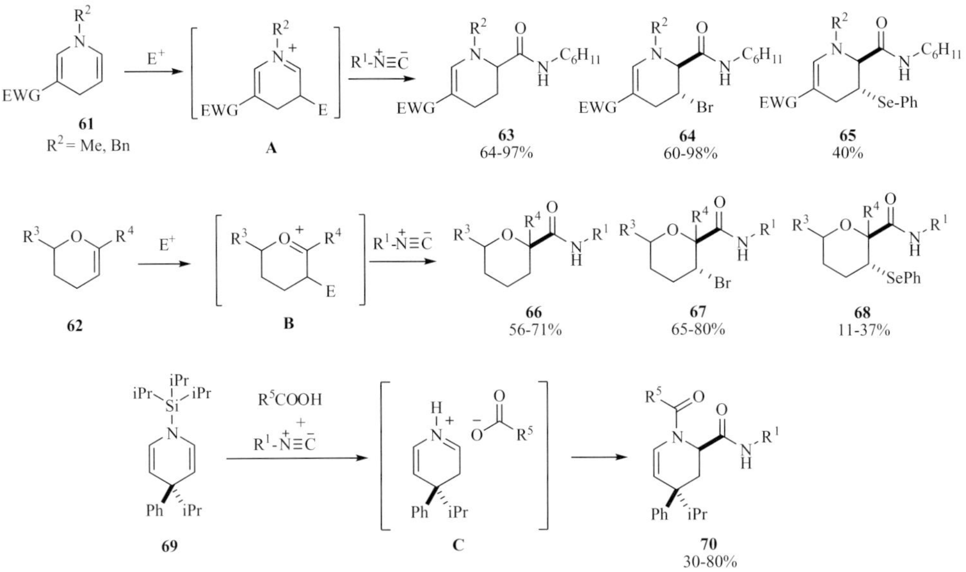

Scheme 9.11 Hydro-, halo-, and seleno-carbamoylation of DHPs and cyclic enol ethers. EWG, electron-withdrawing group.

reaction leading to α-carbamoylated pyridine **54**, or triazole **55** on trapping with trimethylsilyl (TMS)-azide [21, 22]. Electron-deficient alkenes are also suitable activating agents in Reissert-type reactions; for example, the Michael acceptor **56** activates isoquinoline, while the isocyanide attack at the α-position of the azine, followed by intramolecular trapping of the resulting nitrilium ion, leads to the polycycle **57** in average to good yields [23–25]. Analogously, the reaction of isocyanide with isoquinolinium dipole **59** (obtained by treating salt **58** with isothiocyanate) triggers a reaction cascade yielding pyrroloisoquinoline **60**, pyrroloquinolines, and indolizines by sulfur elimination (Scheme 9.10) [26].

The same reactivity principle has been applied to electron-rich double-bond functionalities. Thus, the interaction of electrophilic reagents with dihydropyridines and cyclic enol ethers generates reactive iminium species that can react with isocyanides. This enables the hydro-, halo-, and seleno-carbamoylation of activated olefin moieties present in dihydropyridines (DHPs) **61** and cyclic enol ethers **62** to be conveniently carried out with *p*-toluenesulfonic acid (as the proton source), and bromine and phenylselenyl chloride as electrophilic inputs, to afford heterocyclic systems **63–68** (Scheme 9.11) [27, 28]. Wanner *et al.* reported a straightforward method for the transformation of *N*-silyl DHPs **69** to polysubstituted tetrahydropyridines **70** using a carboxylic acid and an isocyanide (Scheme 9.11) [29]. The mechanism involves an initial protodesilylation to form the dihydropyridinum salt **C** which, once attacked by the isocyanide, yields the MCR adduct.

Interestingly, the use of iodine in an attempted iodocarbamoylation of DHPs led to benzimidazolium salts **71**; in sharp contrast, iodine monochloride (ICl)

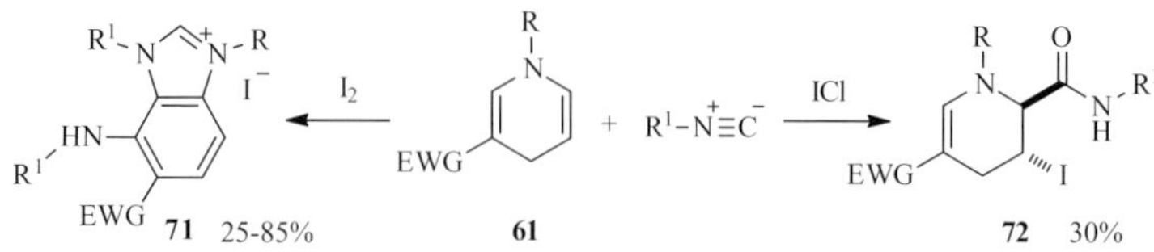

Scheme 9.12 Reactions of isocyanides, DHPs, and different iodonium sources. EWG, electron-withdrawing group.

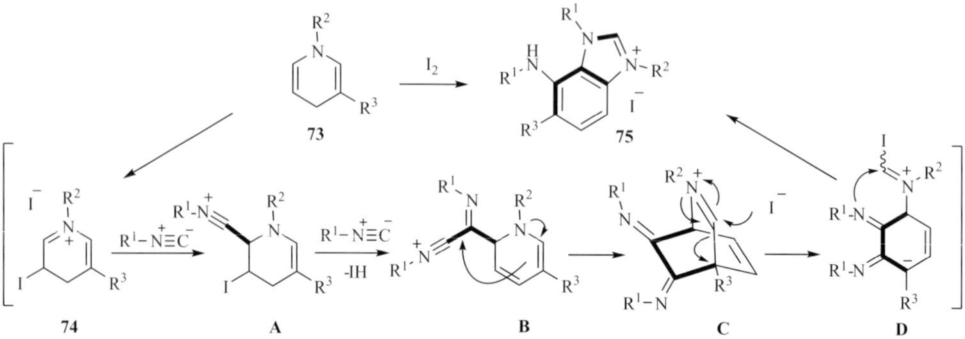

Scheme 9.13 Mechanistic hypothesis for the generation of benzimidazolium salts **75**.

afforded the expected β-iodo-α-carbamoylated tetrahydropyridine **72** (Scheme 9.12) [30]. The former transformation is unprecedented and quite general in scope, allowing several substituents at the nitrogen and β-carbon of the DHP, as well as a wide range of isocyanides.

Although the mechanism is far from being well established, a reasonable pathway starts with the formation of the iodo-dihydropyridinium ion **74** (Scheme 9.13). The consecutive attack of two isocyanide units, followed by HI elimination, gives rise to the nitrilium ion **B**, which then suffers an intramolecular trapping by the enamine unit to form the bicyclic system **C**. An iodide-mediated fragmentation may generate the iodoiminium **D** which, through ring closure and aromatization steps, yields the final benzimidazolium salts **75**. These compounds have been identified as moderate inhibitors of the human enzyme prolyl oligopeptidase, which is involved in the metabolism of neuropeptides and experiences an altered activity in several mental illnesses. Consequently, these compounds constitute a promising new scaffold for targeted drug design [31].

9.3
Type II MCRs: Isocyanide Activation

The processes described here include transformations triggered by the initial activation of isocyanides, and the subsequent reaction of these species with a third component. In this context, the interaction between boranes and isocyanides was

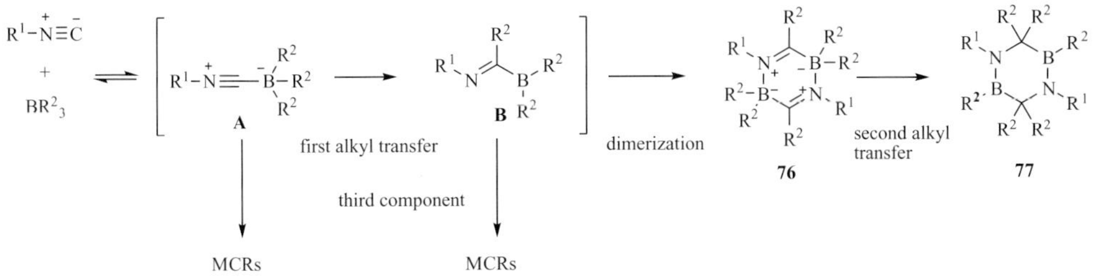

9.3 Type II MCRs: Isocyanide Activation

Scheme 9.14 Isocyanide–borane interaction.

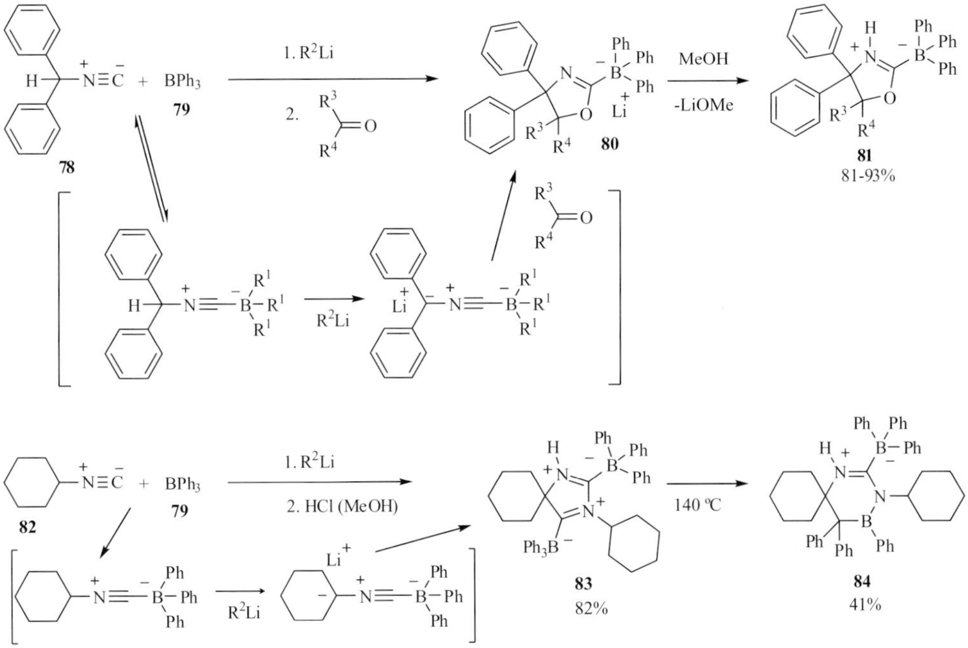

Scheme 9.15 Dipolar species from isocyanides and BPh$_3$.

first studied during the 1960 and showed promising results, but then lay dormant until recently. The process is triggered by a nucleophilic attack of the isocyanide on the borane, such that the initial complex **A** quickly evolves by an alkyl transfer from the boron atom to the isonitrile carbon to generate the iminoborane **B** (Scheme 9.14). In the absence of other components, this intermediate dimerizes to generate diazadiborinine **76** as a stable compound; a second alkyl transfer may then be induced by heating to afford the saturated analogue **77** [32–37]. The reactive nature of intermediates **A** and **B** enables the participation of appropriate components to yield a rich variety of MCRs, as discussed below.

When using isocyanide **78**, intermediates of type **A** are generated on interaction with BPh$_3$, and these can be further activated by the addition of organolithium compounds (Scheme 9.15). The resulting dipolar lithium salt can then react with

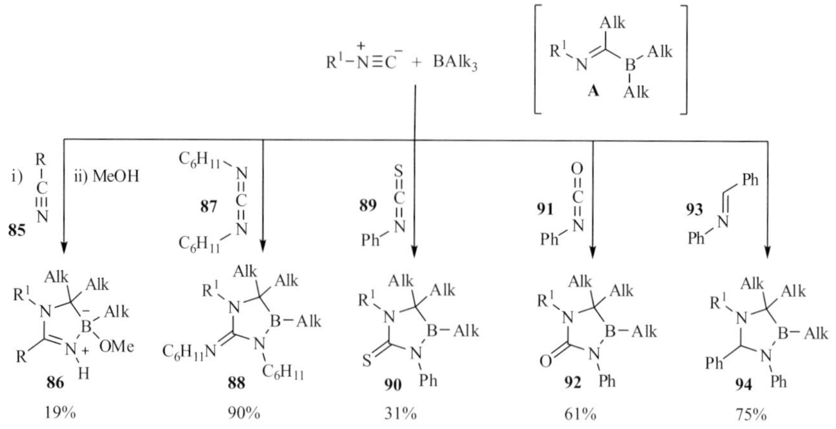

Scheme 9.16 Formal [3+2] cycloadditions with iminoboranes.

ketones or aldehydes in formal [3+2] cycloadditions, to afford the cyclic iminoborane salt **80** or its protonated analogue **81** after treatment with methanol. The same reaction, using cyclohexylisocyanide **82**, evolves through a dimerization step to yield the spiro compound **83**, which then rearranges to dipole **84** on heating [38].

In the presence of distinct substrates, the reaction pathway diversifies into different reaction mechanisms. In this way, iminoboranes **A** (Scheme 9.16), arising from the reaction between an isocyanide and trialkylboranes, are able to interact with several reactive species through formal [3+2] cycloadditions.

Therefore, when nitrile **85** is mixed with an isocyanide and trialkylborane, diazaborolidine **86** is generated in a single step through the intermediacy of iminoborane **A**. The analogous products **88**, **90**, and **92** can be synthesized using carbodiimide **87**, isothiocyanate **89**, or isocyanate **91**, respectively; the saturated analogue **94** can be obtained with imines. During these processes, a second alkyl group is transferred from boron to the terminal carbon of the isocyanide (Scheme 9.16) [39, 40]. On the other hand, alcohols, thiophenols, and anilines may also react with iminoborane **A** to afford the acyclic borane adducts **95–97** in good yields (Scheme 9.17) [41].

Interestingly, when aldehydes are used, a more complex reaction cascade takes place. Thus, by mixing aromatic isocyanides, alkylboranes, and aldehydes, the oxazolidines **98** (Scheme 9.18) are synthesized under mild conditions. This reaction was first reported by Hesse in 1964 [42], but not revisited until 2010, when the original discovery was expanded into a whole family of MCRs, leading to structurally diverse drug-like scaffolds [43]. The first steps of this process closely resemble the previous reactions involving boranes and isocyanides, and result in the formation of iminoborane **B**. Likewise, a formal cycloaddition with a molecule of aldehyde and the transfer of a second alkyl group from the boron atom leads to an oxazaborolidine **C**, which further evolves to an azomethine ylide intermediate **D** with the loss of boroxine. In the final step, a second molecule of aldehyde reacts

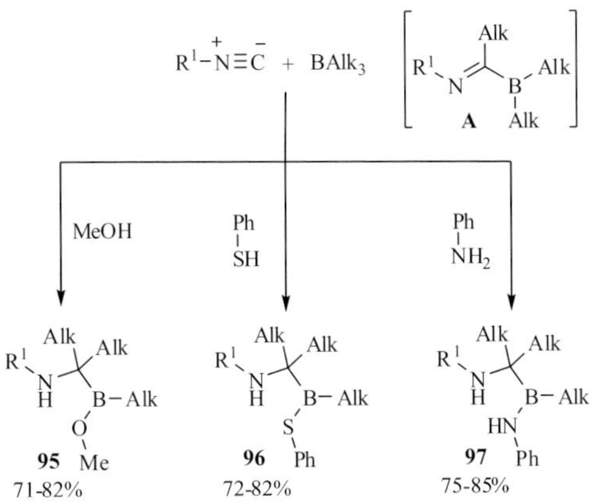

Scheme 9.17 Reaction of isocyanides and alkylboranes with methanol, phenol, or aniline.

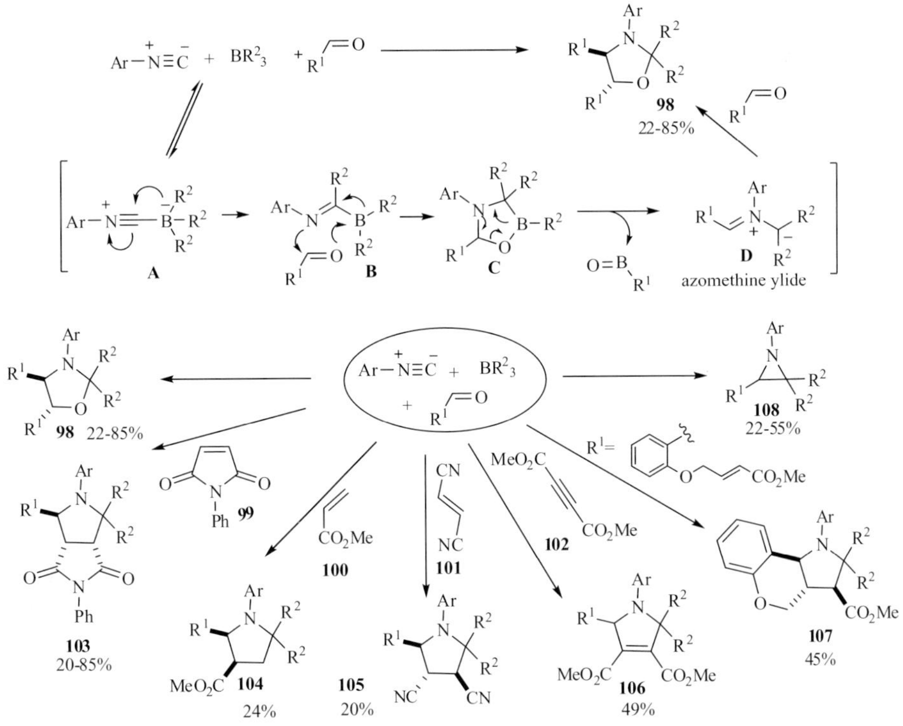

Scheme 9.18 MCRs involving isocyanides, alkylboranes, aldehydes, and dipolarophiles (the major diastereomers are shown).

in a cycloaddition to yield oxazolidine **98** with excellent *trans* diastereoselectivity. In the presence of a dipolarophile as a fourth component, the azomethine ylide reacts to afford different types of scaffolds. Under these conditions, phenylmaleimide **99**, methyl acrylate **100**, fumaronitrile **101**, and dimethyl acetylenedicarboxylate **102** yield the respective pyrrolidine derivatives **103–106**. The processes shows a moderate diastereoselectivity, with typical isomer distribution lying in the range of a 4:1 ratio. When an aromatic aldehyde bearing a dipolarophile moiety was used, an intramolecular version of the process took place, affording the fused heterocyclic system **107** in a single step. When aldehydes bearing electron-donating groups are used, the affinity of the generated azomethine ylide intermediates for dipolarophiles is reduced; under these conditions the formation of aziridine **108** is predominant with respect to the formation of oxazolidine **98** (Scheme 9.18).

As several routes can compete for the final distribution of the possible products, the overall yield of these processes is highly dependent on the electronic properties of the components, and some combinations afford complex mixtures of which the purification is difficult. To overcome this problem, a solid-phase version of these processes was developed in which the isocyanide resin **109** was reacted with phenylmaleimide **99**, triethylborane, and *p*-chlorobenzaldehyde **110** (Scheme 9.19); this led to the pure four-component reaction (4CR) adduct **111** after washing of

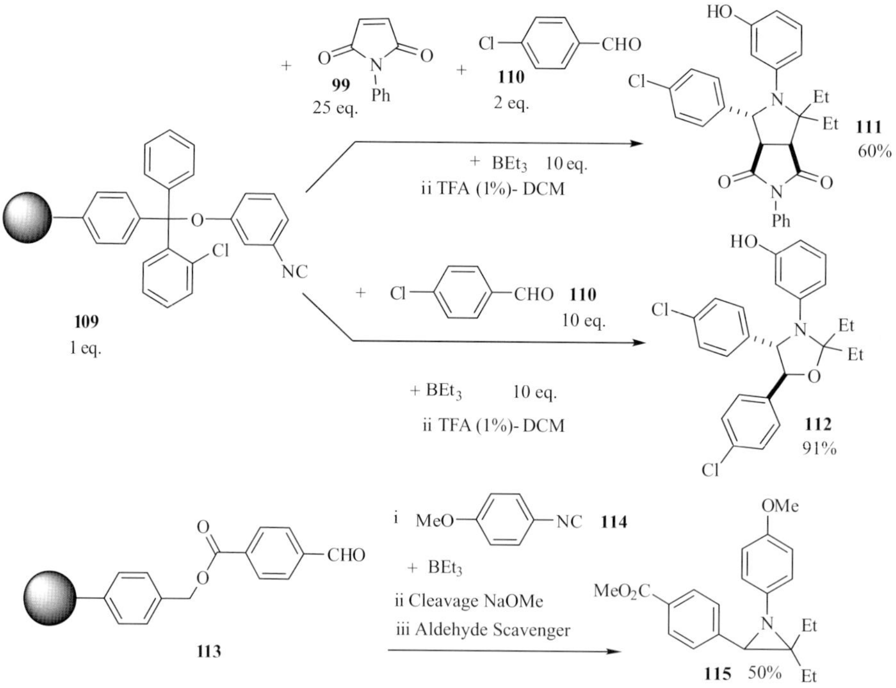

Scheme 9.19 Solid-supported boron-based MCRs.

the resin and subsequent cleavage with trifluoroacetic acid (TFA) [43]. Analogously, oxazolidine **112** was obtained in high yield without further purification. Finally, aziridine **115** was obtained by reaction of the solid-supported aldehyde **113** with *p*-methoxyisocyanide **114** and triethylborane, after cleavage with sodium methoxide and removal of the unreacted aldehyde with a scavenger. The use of solid-supported reagents not only allows the expected products to be acquired in a pure form, but also significantly improves the range of substituents compatible with the process. For instance, the use of unprotected phenols in solution would not be feasible in the presence of borane reagents, while an attempt to synthesize aziridine **115** in solution using 4-formylbenzoate would lead to formation of the corresponding oxazolidine. Yet, both problems can be resolved by using the solid-phase approach, and the expected adducts **111**, **112**, and **115** are conveniently formed.

A wide range of MCRs has been described based on the Nef isocyanide reaction [44, 45]. The α-addition of acyl chlorides to isocyanides leads to the formation of imidoyl chlorides (Nef adducts), the interaction of which with different nucleophiles form the final adducts [46]. The synthesis of a large number of heterocycles based on the intramolecular cyclization of *C*-acylnitrilium ions, derived from imidoyl chlorides, has been exhaustively reviewed by Livinghouse [47, 48]. The treatment of imidoyl chloride **116** with a base gives rise to the dipolar intermediate **A**, which can be trapped by a dipolarophile to afford the corresponding pyrroline derivatives **117** (Scheme 9.20) [49]. Finn *et al.* described the synthesis of formamidine urea salts **119** via the trapping of the imidoyl chloride **116** by substituted ureas **118**. Notably, this reaction is somewhat restricted to acid chlorides, and either does not proceed or affords lower yields with other electrophilic activating agents [50]. A wide application of these formamidine salts **119** has since been described with the development of useful protocols involving the exchange reaction of the imine fragment with nitrogen nucleophiles (aromatic and aliphatic primary amines, hydrazines, hydrazides, and hydroxylamines) [51]. In this context, the addition of thiols to generate thiocarbamates, deprotonation/alkylation sequences or amide fragment exchange have also been reported (Scheme 9.20) [52].

The scope of the Nef-MCRs has been expanded by the group of El Kaïm. Thus, the reactivity of the Nef adducts with tetrazoles **121** and the subsequent Huisgen rearrangement under Lewis acid activation lead to 1,2,4-triazol derivatives **122** (Scheme 9.20) [53]. Another interesting 3CR was described as an application of the Perkow reaction in which the addition of trialkylphosphites **123** to imidoyl chloride adducts **116** afforded the keteneimine adducts **124**. The reactivity of such compounds allowed several post-condensations [54], including the formation of phosphorylated tetrazoles **125** from the reaction of keteneimine derivatives **124** and trimethylsilyl azide, and the generation of phosphorylated triazoles **126** using trimethylsilyl diazomethane [55]. Additionally, keteneimines undergo dipolar cycloadditions with pyridinium ylides **D** generated *in situ* to afford indolizine derivatives **127** (Scheme 9.20) [56]. The same group reported the first MCR between nitro compounds **128**, acylating agents and isocyanides to yield α-oximinoamides **129** [57, 58].

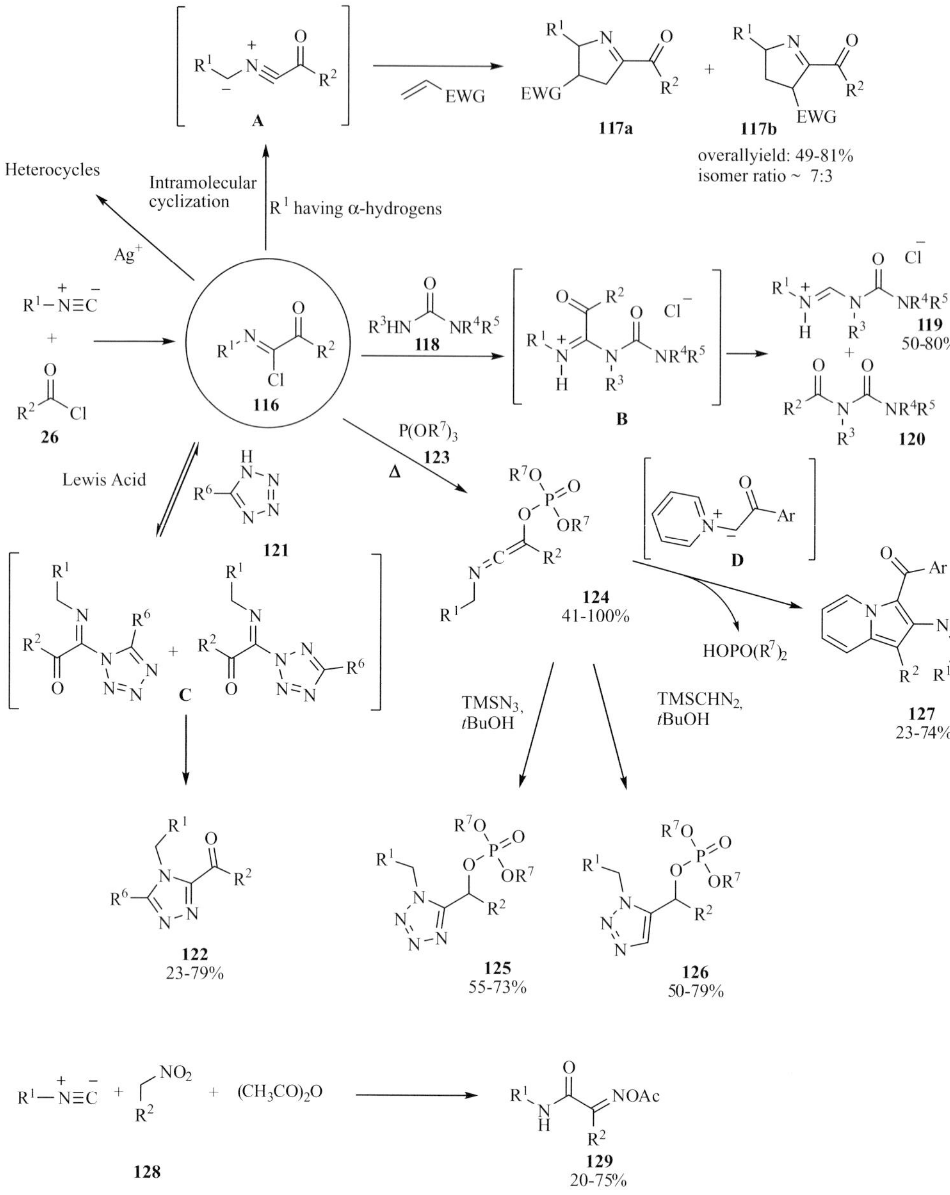

Scheme 9.20 Reactions of imidoyl chlorides arising from a Nef-isocyanide process.

Bossio *et al.* reported the α-addition of *N*-chloramines, generated from aromatic amines and chloramine T **130**, to isocyanides to produce *N*-tosylguanidines **131** (Scheme 9.21). Subsequently, imidazole derivatives were obtained after the cyclization of *N*-tosylguanidines in acidic media [59, 60]. Similarly, isocyanide dihalides can be regarded as synthetic equivalents of imidoyl chlorides in reactions with

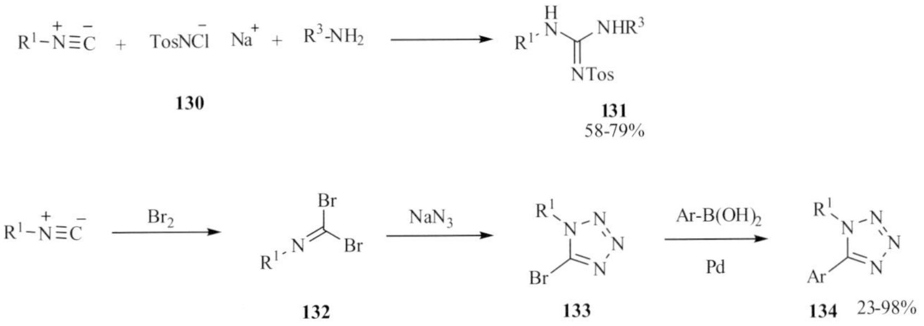

Scheme 9.21 Reactions of isocyanide dihalides and related compounds.

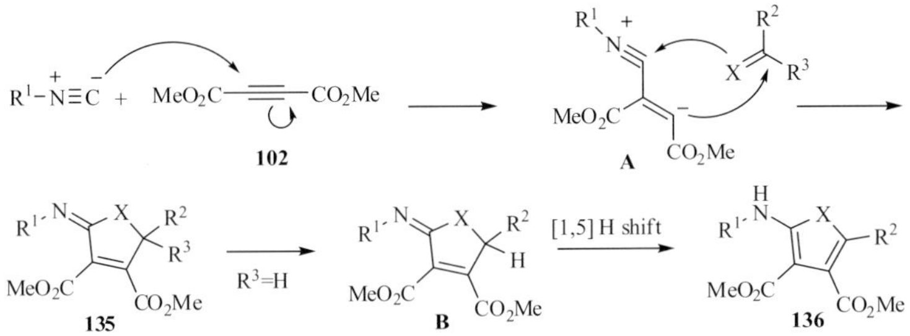

Scheme 9.22 Mechanistic proposal for the formation of compounds **135** and **136**.

nucleophiles, or in MCRs. The synthesis of these compounds, which proceeds via the halogenation of isocyanides or isothiocyanates, has been reviewed [61]. Different reactions of dihalogenated isocyanides have been extensively described, resulting in carbodiimides, chloroformamidines, thioesters, guanidines, isoureas, isothioureas, and five- and six-membered heterocycles [62–64]. More recently, some innovative applications of dihalogenated isocyanides have been reported in the synthesis of natural products [65, 66]. An illustrative example of this chemistry is the substituted tetrazole synthesis based on the addition of azide to the dihalogenated isocyanide **132**, followed by the incorporation of a second component in a palladium-catalyzed coupling [67].

A rich chemistry arises from the interaction of aromatic and aliphatic isocyanides with dimethyl acetylenedicarboxylate (DMAD) [68]. Nair et al. described a family of 3CRs based on the trapping of the resulting zwitterionic intermediate **A** with different electrophilic species to afford compounds **135** (Scheme 9.22) [69]. Interestingly, when R^3 = H, a [1,5] hydrogen shift takes place to yield the heteroaromatic amino derivative **136**; this process is extremely versatile and distinct electrophiles are engaged in these transformations. Apart from aldehydes, the addition of 1,2- and 1,4-quinones yielded dihydrofuranone derivatives **140** and **142**, respectively (Scheme 9.23), while iminolactones **144** were obtained from acyclic 1,2-diones **143** [70]. Several spirofused γ-iminolactams (**146**, **148** and **150**) were

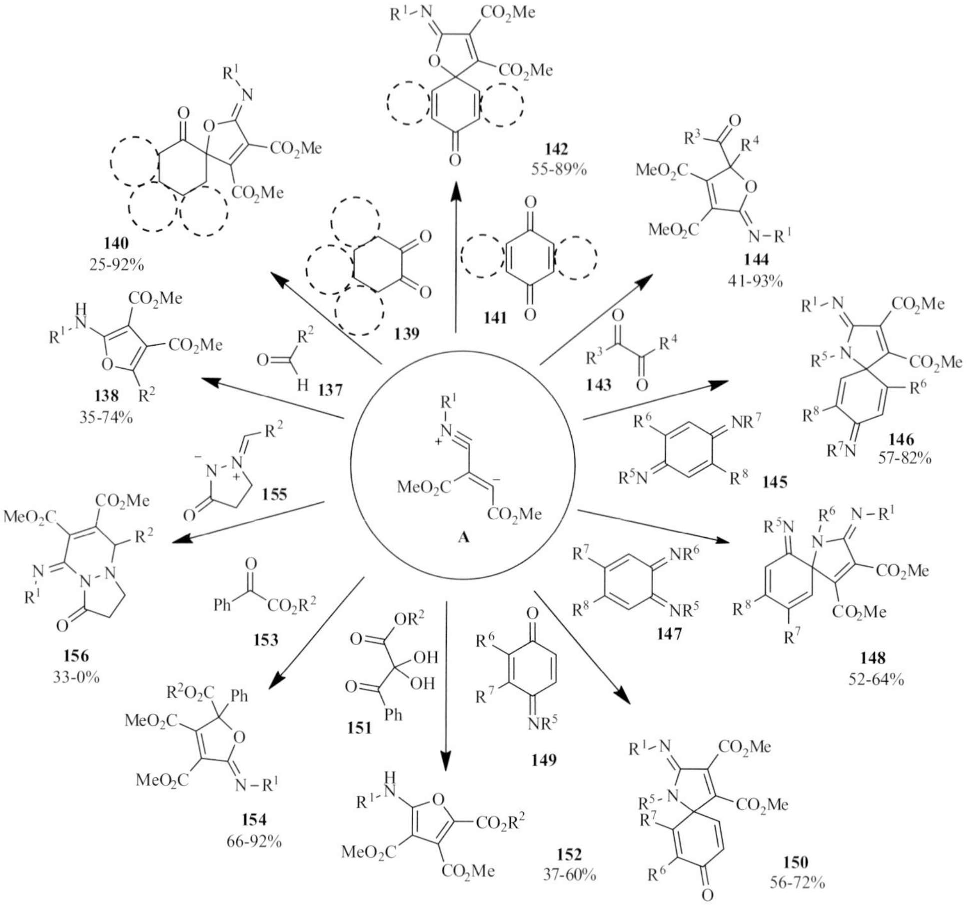

Scheme 9.23 Reactions of isocyanide–DMAD zwitterion with a range of electrophiles.

synthesized following addition of the zwitterionic intermediates to quinonediimides **145** or **147**, and to quinonemonoimides **149**, respectively [71]. The use of vicinal tricarbonyl compounds **151** as electrophiles was reported in the synthesis of furan derivatives **152** [72]. Alkyl phenylglyoxylates **153** were employed to trap zwitterionic intermediates and a subsequent cycloaddition yielded highly functionalized derivatives **154** [73]. The reaction of intermediate **A** with 1,3-dipolar compounds **155** produced pyrazolo[1,2-a]pyrazinone derivatives **156** (Scheme 9.23) [74]. A more detailed account of this chemistry is available in Chapter 8 of this book.

The highly electrophilic character of arynes, which is due to the low-lying lowest unoccupied molecular orbital (LUMO), enables isocyanide addition and the subsequent formation of a zwitterionic nitrilium intermediate **A**, which may trap a variety of electrophiles to yield a structurally diverse array of compounds (Scheme 9.24). This fruitful research line has been developed by Yoshida and coworkers,

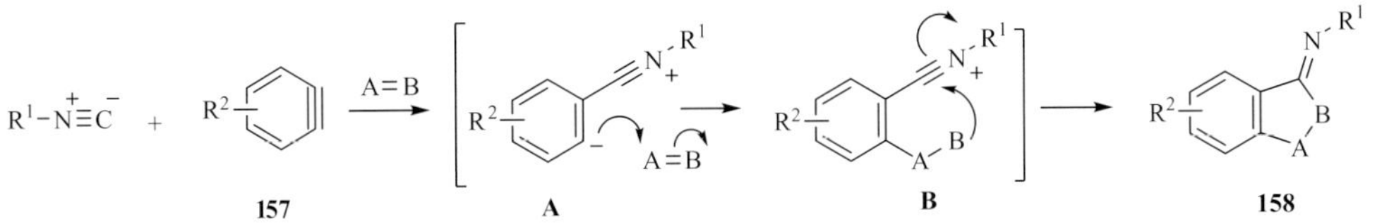

Scheme 9.24 Mechanistic rationale for the aryne–isocyanide MCRs.

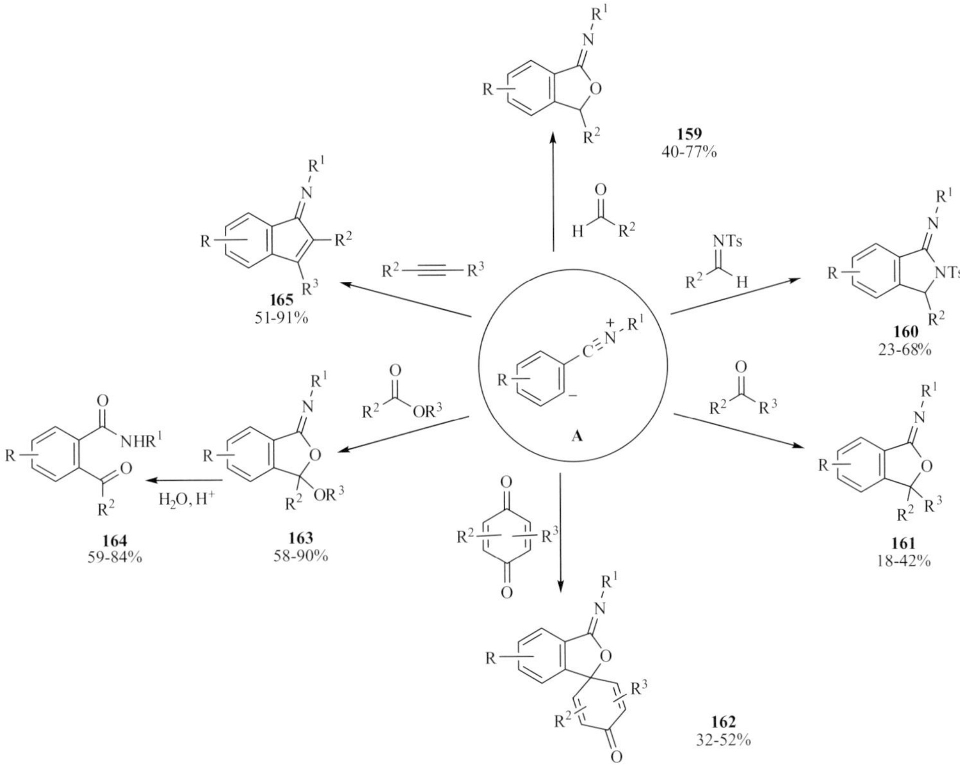

Scheme 9.25 MCRs between isocyanides, arynes, and electrophiles.

and resulted in the creation of several benzoannulated nitrogen or oxygen heterocycles **158**. Thus, arynes **157**, prepared *in situ* from 2-(trimethylsilyl)phenyl triflate derivatives, react with isocyanides and aliphatic or aromatic aldehydes to afford benzoiminofurans **159** (Scheme 9.25) [75]. The use of sulfonylimines gives rise to 2-iminoisoindolines **160** [76], whereas ketones and benzoquinones have been employed in the synthesis of iminofurans **161** and spiro-benzofuran derivatives **162** [77].

Interestingly, Allan *et al.* expanded the scope of the electrophiles by using substituted phenyl esters (lineal, branched, and cyclic), leading to phenoxyiminoisobenzofurans **163** (Scheme 9.25). A final treatment of the reaction with an aqueous solution of oxalic acid promoted the hydrolysis of these adducts and afforded the

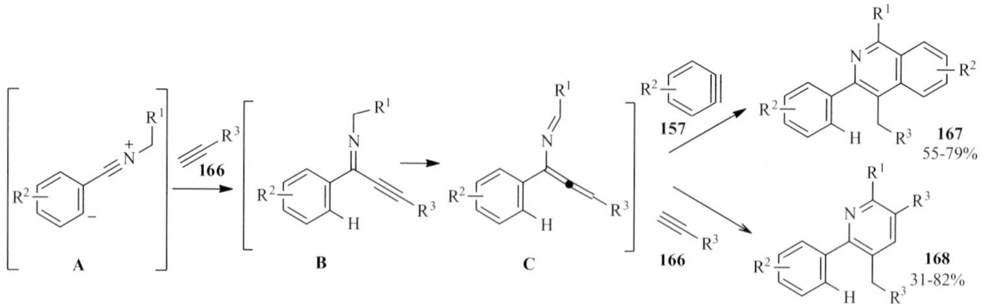

Scheme 9.26 Benzyne–isocyanide-mediated synthesis of isoquinolines and pyridines.

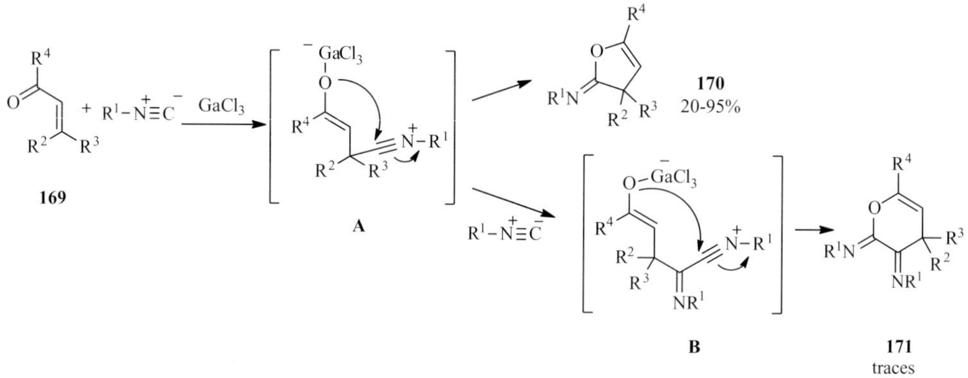

Scheme 9.27 GaCl$_3$-catalyzed single- and double-isocyanide insertion processes.

o-ketobenzamides **164**. Finally, the same authors described the use of electron-deficient alkynes as an electrophilic third component of this MCR to produce iminoindenones **165** [78]. A regioselective synthesis of polysubstituted isoquinolines **167** and pyridines **168** though this methodology was described by Sha et al. (Scheme 9.26). Trapping of the zwitterionic intermediate **A** with a terminal alkyne **166** affords imine species that undergo rearrangement to **C** via a [1,5]-hydride shift. Finally, intermediate **C** is able to react with another unit of benzyne **157** to produce isoquinolines **167**, or with a second molecule of alkyne **166** to yield pyridines **168** [79]. Additional results on aryne–isocyanide interactions are available in Chapter 8 of this book.

Chatani et al. have described a family of Lewis acid-catalyzed formal [4+1] cycloadditions of α,β-unsaturated carbonyls **169** and isocyanides [80, 81] and mechanistically related insertions of isocyanides into C–O and C–S bonds of cyclic and acyclic ketals and dithioacetals (Scheme 9.27) [82–85]. In all cases, the adducts arising from the incorporation of one unit of isocyanide (**170, 173, 176, 179, 182**) were isolated as the major products, although compounds displaying a double isocyanide insertion (**171, 174, 177, 180, 183**) were also identified (see Schemes 9.27 and 9.28) [86]. Notably, the latter compounds may predominate under certain

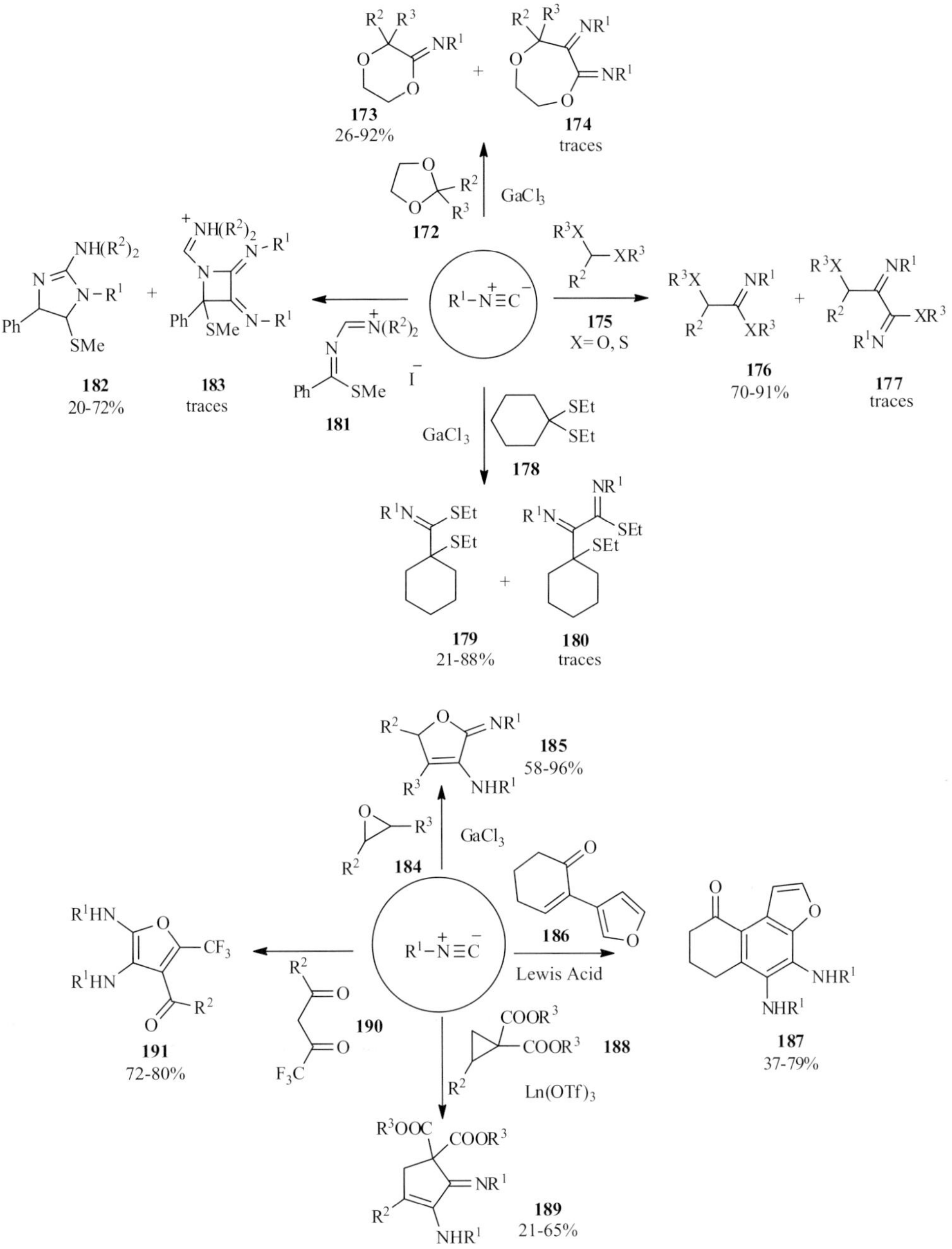

Scheme 9.28 Single- and double-isocyanide insertion processes.

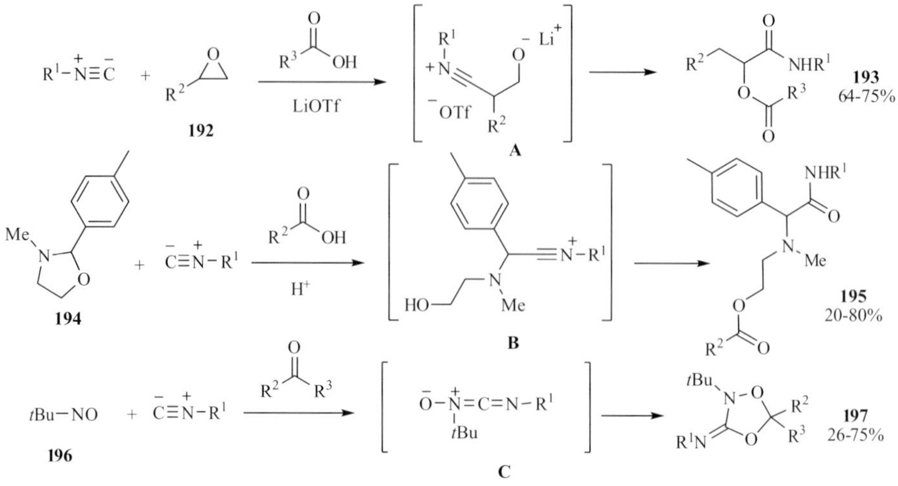

Scheme 9.29 Isocyanide-MCRs with epoxides, oxazolidines, and nitroso compounds.

circumstances; for example, 2,5-dihydrofurans **185** were obtained from disubstituted epoxides **184** [87]; 7,8-dihydronaphtho[2,1-b]furan-9(6H)-one derivatives **187** from enones **186** [88]; cyclopentenedicarboxylates **189** from cyclopropane rings **188** [89]; and diaminofuran derivatives **191** from trifluorobutane-2,4-diones **190** (Scheme 9.28) [90].

A MCR between epoxides **192**, isocyanides, and carboxylic acids under LiOTf catalysis furnishes the α-acyloxyamide adduct **193** (Scheme 9.29) [91, 92]. The chemistry of oxazolidines **194** provides an interesting 3CR to afford N-acyloxyethylamino acid amides **195** [93]. In both cases, the isocyanide attack gives rise to nitrilium ions that ultimately are trapped by the carboxylates, as in the Ugi and Passerini MCRs. More recently, a modified version of this process employed trifluoromethanesulfonic acid (TfOH) as the catalyst, and thiols and tetrazoles instead of carboxylic acids. A formal four-component version was described through the *in-situ* preparation of N-alkyloxazolidines from N-alkylethanolamines and carbonyl compounds [94]. Additionally, the interaction of nitroso compound **196**, isocyanides, and ketones has been reported to yield 3-imino-1,4,2-dioxazolidines **197** by activation of the isonitrile (Scheme 9.29) [95].

9.4
Type III MCRs: Formal Isocyanide Insertion Processes

In the past, an enormous number of studies have been devoted to the metal-mediated transformations of isocyanides. Although the nature of many of these processes is sequential, the overall synthetic outcome is typical of MCRs and, for this reason, some representative cases are presented in this section. Also included are other processes not clearly assigned to previous reaction types, especially insertions into heteroatom–heteroatom bonds. In this context, Meijere *et al.* have recently carried out a comprehensive overview of the use of isocyanides in the synthesis of a wide range of nitrogen heterocycles, based on the cyclization of metalated isocyanides [96]. The synthesis of indoles via isocyanide-mediated transformations has also been reviewed, and includes many useful MCRs involving metal-based or radical reactions [97]. For instance *o*-methylphenyl isocyanides **198** and *o*-bromophenyl isocyanides **201** are lithiated and engaged in sequential reactions with different electrophiles to trap the initially formed organometallic intermediate (Scheme 9.30). Selective metalation of isocyanide **198** with lithium diisopropylamide (LDA) in the presence of 2,2,6,6-tetramethylpiperidide affords the 3-unsubstituted indole anions **199**, which can be selectively alkylated with a variety of electrophiles, such as epoxides, alkyl halides, or acid chlorides, to yield the N-substituted derivatives **200** [98]. *o*-Lithiophenyl isocyanides **202**, obtained from the lithiation of **201** with *n*-BuLi, react with isocyanates, aldehydes, and ketones, and then with a second electrophile to afford quinazolin-4-ones **203**, benzoxazines **205**, and *o*-isocyanobenzylalcohol derivatives **206**, respectively. The use of functionalized isocyanates triggers a domino cyclization to yield N-functionalized intermediates **204** [99, 100]. The α-addition of organolithium reagents to isocyanides provides metalated aldimines **208** or **210**, which can undergo subsequent cyclization to heteroaromatic derivatives **209** or substituted indolenines **211**, after the customary trapping with metalodihalides and related species (PhPCl$_2$, SCl$_2$, MeAsCl$_2$, Me$_2$SnCl$_2$, etc.) [101, 102] or carbon monoxide/ MeI [103], respectively (Scheme 9.30).

Fujiwara *et al.* described an innovative synthesis of selenium derivatives via an isocyanide–MCR protocol promoted by several nucleophiles, such as lithiated secondary amines **212**, organolithium derivatives **213**, or alcohols **214**. These species, on interaction with elemental selenium and isocyanides, triggered a selenoimidoylation process (Scheme 9.31) [104–106]. Afterwards, in a sequential manner, the generated intermediate was trapped with butyl iodide to provide isoselenoureas **215**, selenoimidates **216**, and selenocarbonimidates **217**, respectively, in suitable yields. Recently, the synthesis of benzoselenazoles **218** from *o*-arylisocyanides **201**, selenium and several nucleophiles under copper(I) catalysis was described. It should be noted that, when the same reaction was carried out in the absence of the catalyst, selenoureas **219** were obtained instead. However, in the presence of tellurium, the use of lithiated secondary amines is required to obtain the desired benzotellurazoles **220** [107]. The synthesis of selenoureas **219** following a sequential strategy from amines, isocyanides, and selenium has also been described (Scheme 9.31) [108, 109].

Ogawa *et al.* have conducted extensive explorations into the incorporation of substituted thio, telluro, and seleno groups into isocyanides. An interesting MCR between isocyanides, electron-deficient alkynes (such as ethyl propiolate **221**) and diphenyl diselenide afforded adducts **222** (Scheme 9.32) [110]. A selective double chalcogenation of isocyanides with disulfides, diselenides, and ditellurides was similarly carried out. In this way, bisthiolated derivatives **223** were directly obtained from aromatic isocyanides, while the corresponding aliphatic derivatives

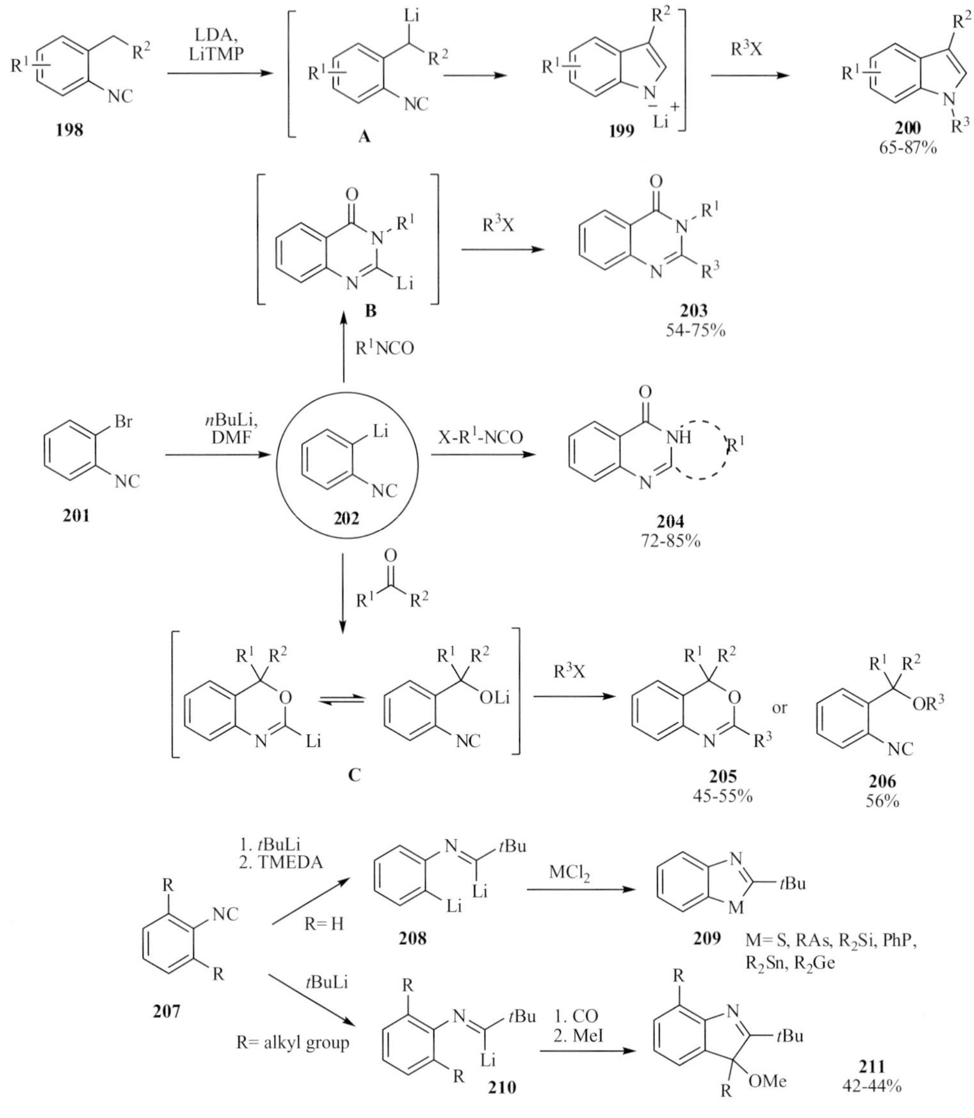

Scheme 9.30 Representative MC processes involving metalated isocyanides.

224 required a modified method using a disulfide–diselenide mixture. In contrast, the reaction of aromatic isocyanides with these reactants selectively yielded thioselenatated adducts **225** [111]. Analogously, the corresponding thiotellurated products **226** were obtained only when aryl isocyanides bearing electron-withdrawing groups were employed (Scheme 9.32) [112]. The intramolecular cyclization of *o*-vinyl- or *o*-alkynyl-arylisocyanides with organic dichalcogenides represents an ef-

Scheme 9.31 Se- and Te-promoted MCRs with isocyanides and nucleophiles.

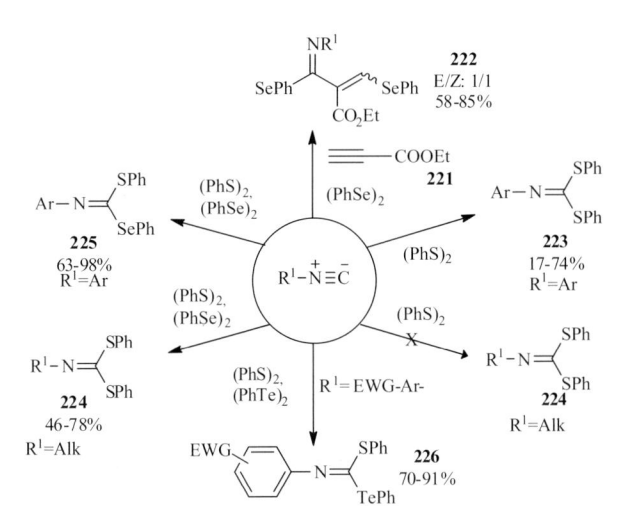

Scheme 9.32 Chalcogenation of isocyanides.

ficient method for the synthesis of interesting and useful heterocyclic compounds [113–115].

The ring-opening of silaaziridines **228**, generated from the interaction between isocyanides and silenes **227**, was carried out with another equivalent of the starting isocyanide or with an α,β-unsaturated aldehyde, to afford the corresponding heterocyclic systems **229** and **230**, respectively (Scheme 9.33) [116, 117]. The insertion of isocyanides into silicon–silicon bonds was also reported [118, 119]. A novel 3CR

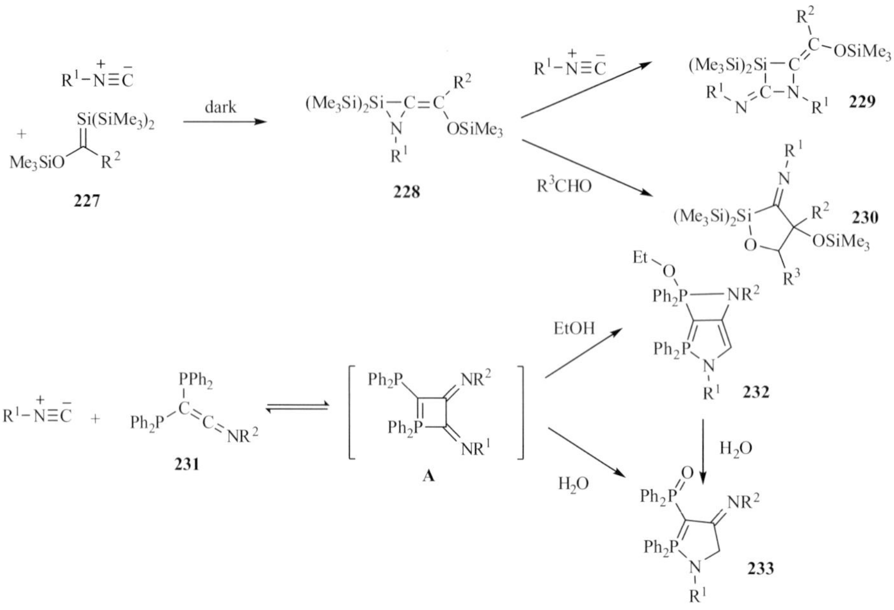

Scheme 9.33 Isocyanide-MCRs involving Si and P reactants.

between isocyanides, diphosphinoketenimines **231**, and ethanol or water affords the azaphosphaheterocycles **232** and **233**, respectively (Scheme 9.33); this mechanism most likely involves formation of the unstable four-membered ring **A** through a formal [1+3] cycloaddition. In the presence of ethanol, compound **232** is obtained, whereas water leads to the related system **233** (Scheme 9.33) [120].

Several MCRs involving the use of isocyanides in transition metal-catalyzed processes have been reported [121]. For example, Whitby et al. disclosed the synthesis of amidines, imidates, and thioimidates through the interaction between isocyanides, aryl halides, and nucleophilic species under palladium catalysis. The synthesis of amidines **234** and **235**, from secondary and primary amines respectively, was carried out in this way, the process being almost limited to tert-butylisocyanide (Scheme 9.34) [122]. The reaction with sodium ethoxide afforded imidates **236** in moderate to high yields; analogously, the use of thiols allowed the formation of thioimidates **237**. The treatment of imidates **236** with amines and acetic acid conveniently afforded amidines **234**, and overcame the limitations of the different isocyanides used in the syntheses described above [123]. Subsequently, α,β-unsaturated amidines **238-239** and imidates **240** were prepared from alkenylbromides, the resulting imidates **240** proving unstable and spontaneously evolving to amides **241** through hydrolysis. Double isocyanide insertion compounds **242** were also isolated in these transformations [124]. Due to the coordinating nature of the isocyanide, these processes (which are catalytic in the transition metal) are especially delicate, and optimization of the reaction conditions becomes a critical issue [125]. In a related study, the synthesis of 2,3-disubstituted indoles

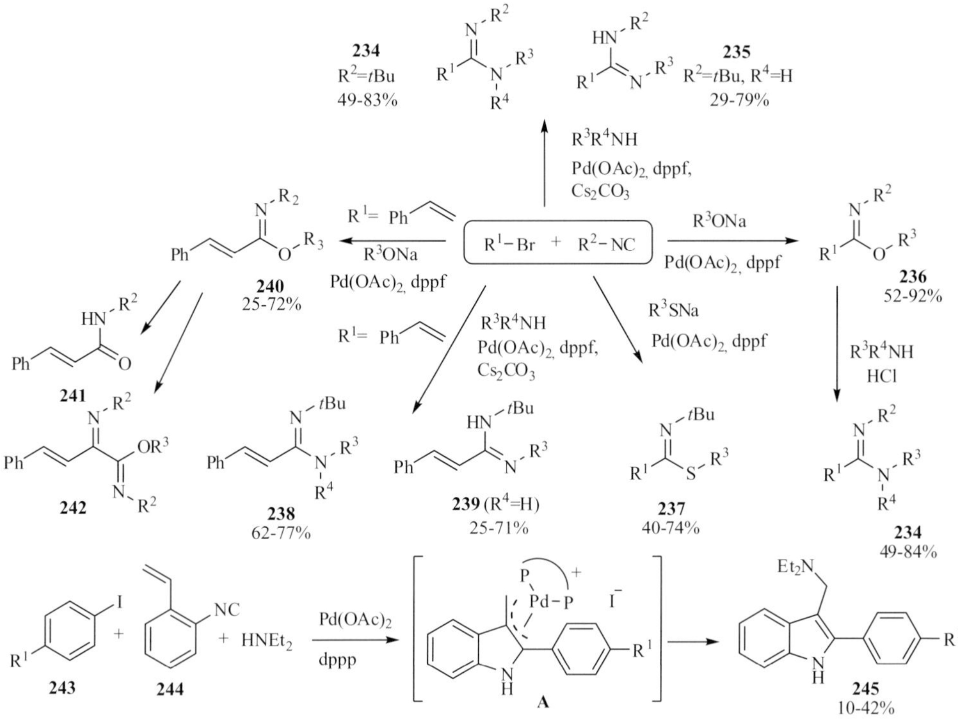

Scheme 9.34 Palladium-catalyzed isocyanide-MCRs.

245 in a 3CR was carried out via the interaction of *o*-alkenylphenyl isocyanide 244 with aryliodide 243. The coupling of the alkenyl and isocyano groups afforded the 3-(indolylmethyl)palladium complex **A** which, after the amine attack, yielded the expected adduct 245 (Scheme 9.34) [126].

Besides palladium, other transition metals have been used in isocyanide-MCRs, For instance, Odom *et al.* described 3CRs and 4CRs between isocyanides, alkynes, and alkyl- or arylamines under titanium catalysis to yield diaminopyrroles 247 and iminoenamines 246 [127] (Scheme 9.35). The process may take place through a [2+2] cycloaddition between a titanium imido complex **A** and an alkyne, giving rise to the azatitanacyclobutene intermediate **B**, in which the insertion of isocyanide leads to the corresponding complex **C**. Cleavage of this species with an amine equivalent regenerates the iminotitanium intermediate and releases the α,β-unsaturated β-aminoimines 246. A competitive pathway involves the insertion of a second unit of isocyanide into the aza-titanacycle intermediate **C**, leading to the 2,3-diaminopyrroles derivatives 247 after the corresponding demetalation and cyclization steps. The latter MCR is predominant when using an excess of isocyanide [128].

Recently, Angelici *et al.* described an interesting process involving the reactions of isocyanides with amines and oxygen under gold-catalysis [129–131]. In this case,

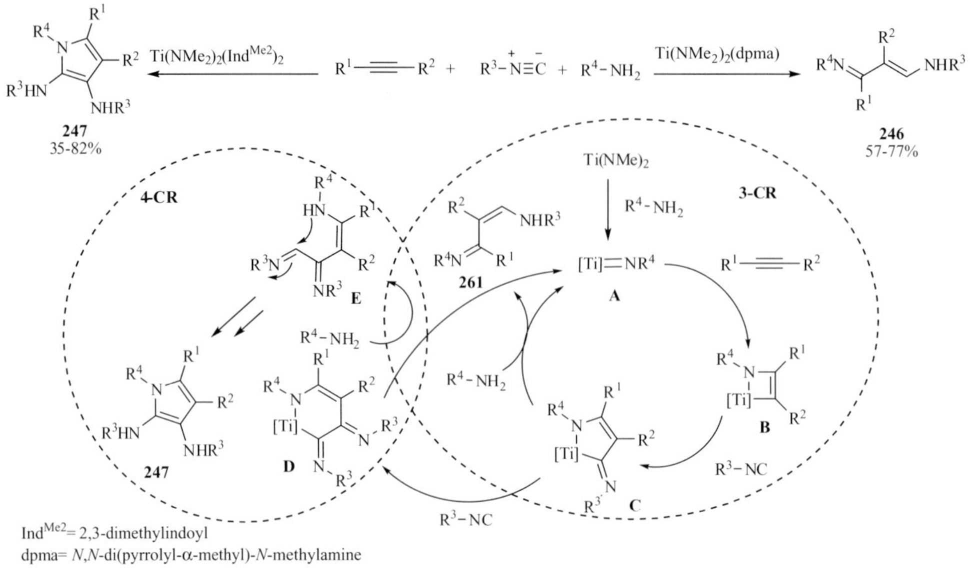

Scheme 9.35 Titanium-catalyzed isocyanide-MCRs.

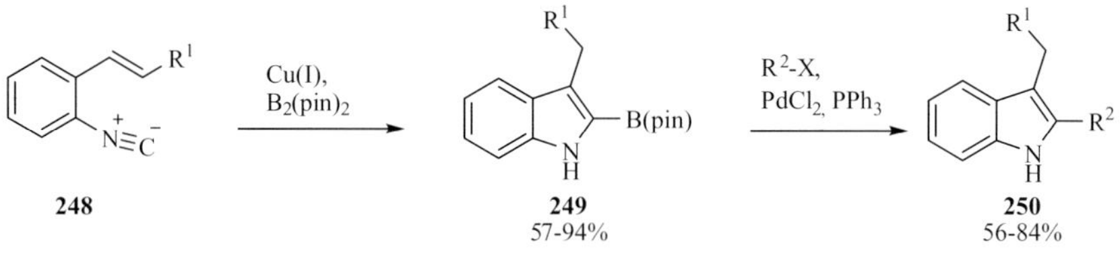

Scheme 9.36 Copper-catalyzed borylative cyclization.

a MC tandem reaction was employed by Chatani *et al.* to obtain diversely substituted indoles from arylisocyanides. The process takes place via the Cu(I)-catalyzed synthesis of 2-boroindole derivatives **249**, and this borylative cyclization is linked in tandem to a Pd-mediated Suzuki coupling with arylhalides to afford the indole derivatives **250** (Scheme 9.36) [132]. A silylative termination was also described.

A few multicomponent processes involving isocyanides and samarium have also been described. The reaction of 2,6-xylylisocyanide **251**, aldehydes or ketones and alkylbromides leads to α-hydroxy imines **252**, in good to excellent yields. The process begins with the formation of organosamarium species, arising from an interaction of the metal with the alkylbromide and isocyanide to generate intermediate **A**. The carbonyl compound is then added to the reaction mixture to yield the final adduct (Scheme 9.37) [133, 134].

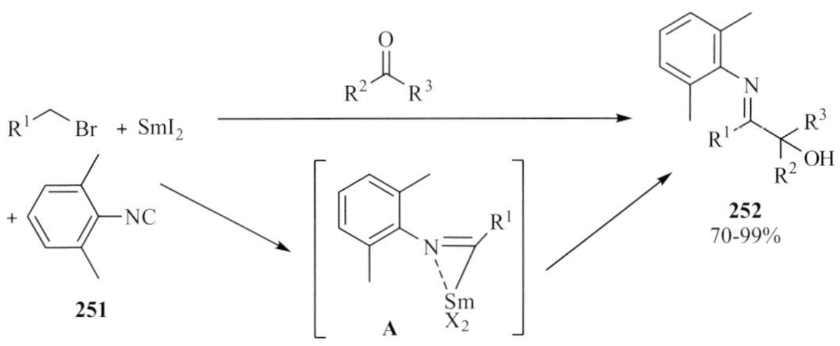

Scheme 9.37 Tandem multicomponent process promoted by SmI_2.

9.5
Conclusions

It has been shown in this chapter that, today, the rich chemistry of isocyanides goes far beyond the famous Ugi and Passerini MCRs and, in a representative overview, the useful transformations that exploit the inherent reactivity of this functional group have been outlined. Currently, the potential of isocyanide-MCRs is being explored in different directions, and at a very good pace. Although, interestingly, some recent investigations have focused on old and neglected results, the application of modern techniques and approaches to achieve a completely renewed reactivity has resulted in a substantially enhanced synthetic value. It is reasonable to expect that the already high impact of isocyanide-MCRs will be further developed in association with novel reactant combinations. This, in turn, will undoubtedly facilitate the preparation of complex organic compounds in just a few steps, the aim being to achieve the ideal synthesis [135].

Acknowledgments

The authors warmly thank all members of their research group who were involved in this project, for their enthusiasm and dedication. Financial support is acknowledged from the DGICYT (Spain, project BQUCTQ2009-07758), Generalitat de Catalunya (project 2009SGR 1024), and the Barcelona Science Park. In particular, N.K. is grateful to the Spanish Ministry of Science and Education for a grant, and all of the authors thank Grupo Ferrer (Barcelona) for their support.

References

1 Domling, A. and Ugi, I. (2000) Multicomponent reactions with isocyanides. *Angew. Chem. Int. Ed. Engl.*, **29**, 3168–3210.

2 Domling, A. (2006) Recent developments in isocyanide based multicomponent reactions in applied chemistry. *Chem. Rev.*, **106**, 17–89.

3 Zhu, J. and Bienaymé, H. (eds) (2005) *Multicomponent Reactions*, Wiley-VCH, Weinheim.
4 For a recent and alternative review, see: El Kaïm, L. and Grimaud, L. (2009) Beyond the Ugi reaction: less conventional interactions between isocyanides and iminium species. *Tetrahedron*, **65**, 2153–2171.
5 Kielland, N. and Lavilla, R. (2010) Recent developments in Reissert-type multicomponent reactions, in *Synthesis of Heterocycles via Multicomponent Reactions II, Topics in Heterocyclic Chemistry*, vol. 25 (eds R.V.A. Orru and E. Ruijter), Springer-Verlag, Berlin, Heidelberg, pp. 127–168.
6 Diaz, J.L., Miguel, M., and Lavilla, R. (2004) N-acylazinium salts: a new source of iminium ions for Ugi-type processes. *J. Org. Chem.*, **69**, 3550–3553.
7 Lavilla, R., Carranco, I., Díaz, J.L., Bernabeu, M.C., and de la Rosa, G. (2003) Dihydropyridines in MCRs. Tandem processes leading to modular tetrahydroquinoline systems with up to 6 diversity elements. *Mol. Divers.*, **6**, 171–175.
8 Ngouansavanh, T. and Zhu, J. (2007) IBX-mediated oxidative Ugi-type multicomponent reactions: application to the N and C1 functionalization of tetrahydroisoquinoline. *Angew. Chem. Int. Ed. Engl.*, **46**, 5775–5778.
9 Janvier, P., Sun, X., Bienayme, H., and Zhu, J. (2002) Ammonium chloride-promoted four-component synthesis of pyrrolo[3,4-b]pyridin-5-one. *J. Am. Chem. Soc.*, **124**, 2560–2567.
10 Gamez-Montano, R., Gonzalez-Zamora, E., Potier, P., and Zhu, J. (2002) Multicomponent domino process to oxa-bridged polyheterocycles and pyrrolopyridines, structural diversity derived from work-up procedure. *Tetrahedron*, **58**, 6351–6358.
11 Gonzalez-Zamora, E., Fayol, A., Bois-Choussy, M., Chiaroni, A., and Zhu, J. (2001) Three component synthesis of oxa-bridged tetracyclic tetrahydroquinolines. *J. Chem. Soc. Chem. Commun.*, **47**, 1684–1685.
12 Tron, G.C. and Zhu, J. (2005) A three-component synthesis of (1,3-oxazol-2-yl)-1,2-dihydro(iso)quinoline and its further structural diversifications. *Synlett*, **3**, 532–534.
13 Williams, N.A.O., Masdeu, C., Diaz, J.L., and Lavilla, R. (2006) Isocyanide addition to pyridinium salts. Efficient entry into substituted nicotinonitrile derivatives. *Org. Lett.*, **8**, 5789–5792.
14 St Cyr, D.J., Martin, N., and Arndtsen, B.A. (2007) Direct synthesis of pyrroles from imines, alkynes, and acid chlorides: an isocyanide-mediated reaction. *Org. Lett.*, **9**, 449–452.
15 Corey, E.J. and Tian, Y. (2005) Selective 4-arylation of pyridines by a nonmetalloorganic process. *Org. Lett.*, **7**, 5535–5537.
16 Arévalo, M.J., Kielland, N., Masdeu, C., Miguel, M., Isambert, N., and Lavilla, R. (2009) Multicomponent access to functionalized mesoionic structures based on TFAA activation of isocyanides: novel domino reactions. *Eur. J. Org. Chem.*, **2009**, 617–625.
17 Grignon-Dubois, M., Diaba, F., and Grellier-Marly, M.-C. (1994) Convenient synthesis of trichloromethyldihydroquinolines and -isoquinolines. *Synthesis*, 800–804.
18 Perez-Rentero, S., Kielland, N., Terrazas, M., Lavilla, R., and Eritja, R. (2010) Synthesis and properties of oligonucleotides carrying isoquinoline imidazo[1,2-a]azine fluorescent units. *Bioconjugate Chem.*, **21**, 1622–1628.
19 Marchand, E. and Morel, G. (1993) Three-component cyclocondensations. A convenient access to fused imidazolium and dihydropyrimidinium salts via the reaction of methyl chlorothioimidates with azines and isocyanides. *Tetrahedron Lett.*, **34**, 2319–2322.
20 Berthet, J.-C., Nierlich, M., and Ephritikhine, M. (2002) Reactions of isocyanides and pyridinium triflates – a simple and efficient route to imidazopyridinium derivatives. *Eur. J. Org. Chem.*, 375–378.
21 Kiselyov, A.S. (2005) Reaction of N-fluoropyridinium fluoride with isonitriles: a convenient route to picolinamides. *Tetrahedron Lett.*, **46**, 2279–2282.

22 Kiselyov, A.S. (2005) Reaction of N-fluoropyridinium fluoride with isonitriles and TMSN$_3$: a convenient one-pot synthesis of tetrazol-5-yl pyridines. *Tetrahedron Lett.*, **46**, 4851–4854.

23 Mironov, M.A. (2006) Design of multi-component reactions: from libraries of compounds to libraries of reactions. *QSAR Comb. Sci.*, **25**, 423–431.

24 Mironov, M.A., Maltsev, S.S., Mokrushin, V.S., and Bakulev, V.A. (2005) A novel three-component reaction designed by the combinatorial method: heteroarenes, isothiocyanates and isocyanides. *Mol. Divers.*, **9**, 221–227.

25 Mironov, M.A., Mokrushin, V.S., and Maltsev, S.S. (2003) New method for the combinatorial search of multi-component reactions. *Synlett*, 943–946.

26 Hopkin, M.D., Baxendale, I.R., and Ley, S.V. (2008) A new focused microwave approach to the synthesis of amino-substituted pyrroloisoquinolines and pyrroloquinolines via a sequential multi-component coupling process. *Synthesis*, 1688–1702.

27 Masdeu, C., Diaz, J.L., Miguel, M., Jimenez, O., and Lavilla, R. (2004) Straightforward α-carbamoylation of NADH-like dihydropyridines and enol ethers. *Tetrahedron Lett.*, **45** (42), 7907–7909.

28 Masdeu, C., Gomez, E., Williams, N.A., and Lavilla, R. (2006) Hydro-, halo- and seleno-carbamoylation of cyclic enol ethers and dihydropyridines: new mechanistic pathways for Passerini- and Ugi-type multicomponent reactions. *QSAR Comb. Sci.*, **25**, 465–473.

29 Sperger, C.A., Mayer, P., and Wanner, K.T. (2009) Application of an Ugi-type reaction to an N-silyl-4,4-disubstituted 1,4-dihydropyridine. *Tetrahedron*, **65** (50), 10463–10469.

30 Masdeu, C., Gomez, E., Williams, N.A., and Lavilla, R. (2007) Double insertion of isocyanides into dihydropyridines: direct access to substituted benzimidazolium salts. *Angew. Chem. Int. Ed. Engl.*, **46** (17), 3043–3046.

31 Tarrago, T., Masdeu, C., Gomez, E., Isambert, N., Lavilla, R., and Giralt, E. (2008) Benzimidazolium salts as small, nonpeptidic and BBB-permeable human prolyl oligopeptidase inhibitors. *ChemMedChem*, **3** (10), 1558–1565.

32 Hesse, G. and Witte, H. (1963) Umsetzung von Boralkylen mit Isonitrilen. *Angew. Chem.*, **75**, 791–792.

33 Bresadola, S., Carraro, G., Pecile, C., and Turco, A. (1964) The reaction of alkylisocyanides with trialkylborons. *Tetrahedron Lett.*, **43**, 3185–3188.

34 Casanova, J. Jr and Schuster, R.E. (1964) New class of compounds. Isocyanide-borane adducts. *Tetrahedron Lett.*, **8**, 405–409.

35 Casanova, J. Jr, Kiefer, H.R., Kuwada, D., and Boulton, A.H. (1965) 1,3-Diboretidines. Isocyanide–borane adducts. II. *Tetrahedron Lett.*, **12**, 703–714.

36 Casanova, J. Jr and Kiefer, H.R. (1969) 1,3-Diaza-2,4-diborolidines. Isocyanide–borane adducts. III. *J. Org. Chem.*, **34**, 2579–2583.

37 Tamm, M., Lugger, T., and Hahn, F.E. (1996) Isocyanide and ylidene complexes of boron: synthesis and crystal structures of (2-(trimethylsiloxy)phenyl isocyanide)-triphenylborane and (1,2-dihydrobenzoxazol-2-ylidene)-triphenylborane. *Organometallics*, **15**, 1251–1256.

38 Bittner, G., Witte, H., and Hesse, G. (1968) Nitril-ylide aus Isonitril-Triphenylboran-Addukten. *Liebigs Ann. Chem.*, **713**, 1–11.

39 Hesse, G., Witte, H., and Gulden, W. (1966) Uber die umsetzung von tri-n-butylboran mit phenylisonitril in gegenwart von benzanilin. *Tetrahedron Lett.*, **24**, 2707–2710.

40 Witte, H., Gulden, W., and Hesse, G. (1968) 1.3-Cycloadditionen mit α-N-Phenyl-iminoalkylboranen. *Liebigs Ann. Chem.*, **716**, 1–10.

41 Witte, H., Mischke, P., and Hesse, G. (1969) Reaktionen von α-Phenylimino-alkylboranen mit protonenaciden Verbindungen. *Liebigs Ann. Chem.*, **722**, 21–28.

42 Hesse, G., Witte, H., and Gulden, W. (1965) 1.3-Oxazolidine durch Mehrkomponentenreaktion aus

Trialkylboran, Isonitril und. *Aldehyd. Angew. Chem.*, **13**, 591.

43 Kielland, N., Catti, F., Bello, D., Isambert, N., Soteras, I., Luque, F.J., and Lavilla, R. (2010) Boron-based dipolar multicomponent reactions: simple generation of substituted aziridines, oxazolidines and pyrrolidines. *Chem. Eur. J.*, **16**, 7904–7915.

44 Nef, J.U. (1892) Ueber das zweiwerthige Kohlenstoffatom. *Justus Liebigs Ann. Chem.*, **270** (3), 267–335.

45 Ugi, I. (ed.) (1971) *Isonitrile Chemistry*, Academic Press, New York.

46 Alonso, E., Ramon, D.J., and Yus, M. (1998) Imidoyl chlorides as starting materials for the preparation of masked acyllithium intermediates: synthetic applications. *Tetrahedron*, **54** (39), 12007–12028.

47 Livinghouse, T. (1999) C-acylnitrilium ion initiated cyclizations in heterocycle synthesis. *Tetrahedron*, **55** (33), 9947–9978.

48 Luedtke, G., Westling, M., and Livinghouse, T. (1992) Acylnitrilium ions. Versatile new intermediates for the synthesis of highly functionalized heterocycles. *Tetrahedron*, **48** (11), 2209–2222.

49 Tian, W.S. and Livinghouse, T. (1989) An efficient synthesis of Δ^1-pyrrolines and related heterocycles via the base induced cyclocondensation of α-ketoimidoyl chlorides with electron deficient alkenes. *J. Chem. Soc. Chem. Commun.*, 819–821.

50 Ripka, A.S., Diaz, D.D., Sharpless, K.B., and Finn, M.G. (2003) First practical synthesis of formamidine ureas and derivatives. *Org. Lett.*, **5** (9), 1531–1533.

51 Diaz, D.D. and Finn, M.G. (2004) Formamidine ureas as tunable electrophiles. *Chem. Eur. J.*, **10** (1), 303–309.

52 Diaz, D.D. and Finn, M.G. (2004) Expanded chemistry of formamidine ureas. *Org. Lett.*, **6** (1), 43–46.

53 El Kaïm, L., Grimaud, L., and Wagschal, S. (2009) Three-component Nef-Huisgen access to 1,2,4-Triazoles. *Synlett*, 1315–1317.

54 Krow, G.R. (1971) Synthesis and reactions of keteniimines. *Angew. Chem. Int. Ed. Engl.*, **10** (7), 435–449.

55 Coffinier, D., El Kaïm, L., and Grimaud, L. (2009) Isocyanide-based two-step three-component keteneimine formation. *Org. Lett.*, **11** (8), 1825–1827.

56 Coffinier, D., El Kaïm, L., and Grimaud, L. (2010) Nef-Perkow access to indolizine derivatives. *Synlett*, 2474–2476.

57 Dumestre, P., Kaïm, L.E., and Gregoire, A. (1999) A new multicomponent reaction of nitro compounds with isocyanides. *Chem. Commun.*, 775–776.

58 Dumestre, P. and El Kaïm, L. (1999) Dramatic solvent effect in the multicomponent reaction of nitro compounds with isocyanides. *Tetrahedron Lett.*, **40** (45), 7985–7986.

59 Bossio, R., Marcaccini, S., and Pepino, R. (1995) Studies on isocyanides. Synthesis of N-tosylguanidines. *Tetrahedron Lett.*, **36** (13), 2325–2326.

60 Bossio, R., Marcaccini, S., Pepino, R., and Torroba, T. (1996) Studies on isocyanides and related compounds. Synthesis of 1-aryl-2-(tosylamino)-1H-imidazoles, a novel class of imidazole derivatives. *J. Org. Chem.*, **61** (6), 2202–2203.

61 Kuehle, E., Anders, B., and Zumach, G. (1967) Syntheses of isocyanide dihalides. *Angew. Chem. Int. Ed. Engl.*, **6** (8), 649–665.

62 Kuehle, E., Anders, B., Klauke, E., Tarnow, H., and Zumach, G. (1969) Reactions of isocyanide dihalides and their derivatives. *Angew. Chem. Int. Ed. Engl.*, **8** (1), 20–34.

63 Hashida, Y., Imai, A., and Sekiguchi, S.J. (1989) Preparation and reactions of isocyano-1,3,5-triazines. *J. Heterocycl. Chem.*, **26** (4), 901–905.

64 Bergemann, M. and Neidlein, R. (1998) An efficient approach to novel 5,5-disubstituted dithiohydantoins via α-cyano-α-(dihalomethylenamino) alkanoic acid esters. *Synthesis*, 1437–1441.

65 Baeza, A., Mendiola, J., Burgos, C., Alvarez-Builla, J., and Vaquero, J.J. (2008) Palladium-mediated C–N, C–C, and C–O functionalization of

azolopyrimidines: a new total synthesis of variolin B. *Tetrahedron Lett.*, **49** (25), 4073–4077.

66 Baeza, A., Mendiola, J., Burgos, C., Alvarez-Builla, J., and Vaquero, J.J. (2010) Application of selective palladium-mediated functionalization of the pyrido[3′,2′:4,5]pyrrolo[1,2-c]pyrimidine heterocyclic system for the total synthesis of Variolin B and Deoxyvariolin B. *Eur. J. Org. Chem.*, **29**, 5607–5618.

67 El Kaïm, L., Grimaud, L., and Patil, P. (2011) Three-component strategy toward 5-membered heterocycles from isocyanide dibromides. *Org. Lett.*, **13** (5), 1261–1263.

68 Junjappa, H., Saxena, M.K., Ramaiah, D., Loharay, B.B., Rath, N.P., and George, M.V. (1998) Structure and thermal isomerization of the adducts formed in the reaction of cyclohexyl isocyanide with dimethyl acetylenedicarboxylate. *J. Org. Chem.*, **63** (26), 9801–9805.

69 Nair, V., Rajesh, C., Vinod, A.U., Bindu, S., Sreekanth, A.R., Mathen, J.S., and Balagopal, L. (2003) Strategies for heterocyclic construction via novel multicomponent reactions based on isocyanides and nucleophilic carbenes. *Acc. Chem. Res.*, **36** (12), 899–907.

70 Nair, V., Vinod, A.U., Abhilash, N., Menon, R.S., Santhi, V., Varma, R.L., Viji, S., Mathew, S., and Srinivas, R. (2003) Multicomponent reactions involving zwitterionic intermediates for the construction of heterocyclic systems: one pot synthesis of aminofurans and iminolactones. *Tetrahedron*, **59** (51), 10279–10286.

71 Nair, V., Dhanya, R., and Viji, S. (2005) The three component reaction involving isocyanides, dimethyl acetylenedicarboxylate and quinoneimides: a facile synthesis of spirofused γ-iminolactams. *Tetrahedron*, **61** (24), 5843–5848.

72 Nair, V. and Deepthi, A. (2006) A novel reaction of vicinal tricarbonyl compounds with the isocyanide-DMAD zwitterion: formation of highly substituted furan derivatives. *Tetrahedron Lett.*, **47** (12), 2037–2039.

73 Esmaeili, A.A. and Zendegani, H. (2005) Three-component reactions involving zwitterionic intermediates for the construction of heterocyclic systems: one pot synthesis of highly functionalized γ-iminolactones. *Tetrahedron*, **61** (16), 4031–4034.

74 Qian, B., Fan, M., Xie, Y., Wu, L., Shi, Y., and Liang, Y. (2009) A novel one-pot, three-component synthesis of 5-imino-2,3,5,8-tetrahydropyrazolo[1,2-a]pyridazin-1-one derivatives. *Synthesis*, 1689–1693.

75 Yoshida, H., Fukushima, H., Ohshita, J., and Kunai, A.A. (2004) Arynes in a three-component coupling reaction: straightforward synthesis of benzoannulated iminofurans. *Angew. Chem. Int. Ed. Engl.*, **43** (30), 3935–3938.

76 Yoshida, H., Fukushima, H., Ohshita, J., and Kunai, A. (2004) Straightforward access to 2-iminoisoindolines via three-component coupling of arynes, isocyanides and imines. *Tetrahedron Lett.*, **45** (47), 8659–8662.

77 Yoshida, H., Fukushima, H., Morishita, T., Ohshita, J., and Kunai, A. (2007) Three-component coupling using arynes and isocyanides: straightforward access to benzo-annulated nitrogen or oxygen heterocycles. *Tetrahedron*, **63** (22), 4793–4805.

78 Allan, K.M., Gilmore, C.D., and Stoltz, B.M. (2011) Benzannulated bicycles by three-component aryne reactions. *Angew. Chem. Int. Ed. Engl.*, **50** (19), 4488–4491.

79 Sha, F. and Huang, X. (2009) A multicomponent reaction of arynes, isocyanides, and terminal alkynes: highly chemo- and regioselective synthesis of polysubstituted pyridines and isoquinolines. *Angew. Chem. Int. Ed. Engl.*, **48** (19), 3458–3461.

80 Chatani, N., Oshita, M., Tobisu, M., Ishii, Y., and Murai, S. (2003) A $GaCl_3$-catalyzed [4+1] cycloaddition of α,β-unsaturated carbonyl compounds and isocyanides leading to unsaturated γ-lactone derivatives. *J. Am. Chem. Soc.*, **125** (26), 7812–7813.

81 Oshita, M., Yamashita, K., Tobisu, M., and Chatani, N. (2005) Catalytic [4+1]-cycloaddition of α,β-unsaturated

carbonyl compounds with isocyanides. *J. Am. Chem. Soc.*, **127** (2), 761–766.

82 Yoshioka, S., Oshita, M., Tobisu, M., and Chatani, N. (2005) GaCl$_3$-catalyzed insertion of isocyanides into a C-O bond in cyclic ketals and acetals. *Org. Lett.*, **7** (17), 3697–3699.

83 Tobisu, M., Kitajima, A., Yoshioka, S., Hyodo, I., Oshita, M., and Chatani, N. (2007) Brønsted acid-catalyzed formal insertion of isocyanides into a C–O bond of acetals. *J. Am. Chem. Soc.*, **129** (37), 11431–11437.

84 Tobisu, M., Ito, S., Kitajima, A., and Chatani, N. (2008) GaCl$_3$- and TiCl$_4$-catalyzed insertion of isocyanides into a C–S bond of dithioacetals. *Org. Lett.*, **10** (22), 5223–5225.

85 Morel, G., Marchand, E., and Malvaut, Y. (2000) Addition of isocyanides to α-(methylthio)benzylidenamidinium iodides: a surprising access to 2-(dialkylamino)imidazoles and 3,5-diamino-2*H*-pyrrolium salts. *Heteroat. Chem.*, **11** (5), 370–376.

86 For a general review on this type of transformations, see: Tejedor, D. and Garcia-Tellado, F. (2007) Chemo-differentiating ABB' multicomponent reactions. Privileged building blocks. *Chem. Soc. Rev.*, **36**, 484–491.

87 Bez, G. and Zhao, C. (2003) Gallium(III) chloride-catalyzed double insertion of isocyanides into epoxides. *Org. Lett.*, **5** (26), 4991–4993.

88 Winkler, J.D. and Asselin, S.M. (2006) Synthesis of novel heterocyclic structures via reaction of isocyanides with *S-trans*-enones. *Org. Lett.*, **8** (18), 3975–3977.

89 Korotkov, V.S., Larionov, O.V., and Meijere, A. (2006) Ln(OTf)$_3$-catalyzed insertion of aryl isocyanides into the cyclopropane ring. *Synthesis*, 3542–3546.

90 Mosslemin, M.H., Yavari, I., Anary-Abbasinejad, M., and Nateghi, M.R. (2004) Reaction between *tert*-butyl isocyanide and 1,1,1-trifluoro-4-aryl-butane-2,4-diones: synthesis of new trifluoromethylated furan derivatives. *J. Fluorine Chem.*, **125** (10), 1497–1500.

91 Kern, O.T. and Motherwell, W.B. (2003) A novel isocyanide based three component reaction. *Chem. Commun.*, 2988–2989.

92 Kern, O.T. and Motherwell, W.B. (2005) A novel isocyanide based three component reaction. (Erratum to document cited in Ref. [91]). *Chem. Commun.*, 1787.

93 Diorazio, L.J., Motherwell, W.B., Sheppard, T.D., and Waller, R.W. (2006) Observations on the reaction of *N*-alkyloxazolidines, isocyanides and carboxylic acids: a novel three-component reaction leading to *N*-acyloxyethylamino acid amides. *Synlett*, **14**, 2281–2283.

94 Waller, R.W., Diorazio, L.J., Taylor, B.A., Motherwell, W.B., and Sheppard, T.D. (2010) Isocyanide based multicomponent reactions of oxazolidines and related systems. *Tetrahedron*, **66** (33), 6496–6507.

95 Moderhack, D. and Stolz, K. (1986) 3-Imino-1,4,2-dioxazolidines by [1+2+2] cycloaddition of an isocyanide, 2-methyl-2-nitrosopropane, and a carbonyl compound. *J. Org. Chem.*, **51** (5), 732–734.

96 Lygin, A.V. and Meijere, A. (2010) Isocyanides in the synthesis of nitrogen heterocycles. *Angew. Chem. Int. Ed. Engl.*, **49** (48), 9094–9124.

97 Campo, J., Garcia-Valverde, M., Marcaccini, S., Rojo, M.J., and Torroba, T. (2006) Synthesis of indole derivatives via isocyanides. *Org. Biomol. Chem.*, **4** (5), 757–765.

98 Ito, Y., Kobayashi, K., and Saegusa, T. (1977) An efficient synthesis of indole. *J. Am. Chem. Soc.*, **99** (10), 3532–3534.

99 Lygin, A.V. and Meijere, A. (2009) *ortho*-Lithiophenyl isocyanide: a versatile precursor for 3*H*-quinazolin-4-ones and 3*H*-quinazolin-4-thiones. *Org. Lett.*, **11** (2), 389–392.

100 Lygin, A.V. and Meijere, A. (2009) Reactions of *ortho*-lithiophenyl (-hetaryl) isocyanides with carbonyl compounds: rearrangements of 2-metalated 4*H*-3,1-benzoxazines. *J. Org. Chem.*, **74** (12), 4554–4559.

101 Walborsky, H.M. and Ronman, P.J. (1978) α-Addition and *ortho* metalation of phenyl isocyanide. *J. Org. Chem.*, **43** (4), 731–734.

102 Heinicke, J. (1989) Zur Synthese und Pyrolyse von Organoelement-benzasolderivaten des Phosphors, Arsens, Siliciums und Zinns. *J. Organomet. Chem.*, **364** (3), C17–C21.

103 Orita, A., Fukudome, M., Ohe, K., and Murai, S. (1994) Reactions via carbonyl anions. [4+1]cyclocoupling of azadienyllithium with carbon monoxide. *J. Org. Chem.*, **59** (2), 477–481.

104 Maeda, H., Matsuya, T., Kambe, N., Sonoda, N., Fujiwara, S., and Shin-Ike, T. (1997) A new synthesis of isoselenoureas by imidoylation of amines with selenium and isocyanides. *Tetrahedron*, **53** (36), 12159–12166.

105 Fujiwara, S., Maeda, H., Matsuya, T., Shin-Ike, T., Kambe, N., and Sonoda, N. (2000) Imidoylation of acidic hydrocarbons with selenium and isocyanides: a new synthetic method for preparation of selenoimidates. *J. Org. Chem.*, **65** (16), 5022–5025.

106 Asanuma, Y., Fujiwara, S., Shin-ike, T., and Kambe, N. (2004) Selenoimidoylation of alcohols with selenium and isocyanides and its application to the synthesis of selenium-containing heterocycles. *J. Org. Chem.*, **69** (14), 4845–4848.

107 Fujiwara, S., Asanuma, Y., Shin-Ike, T., and Kambe, N. (2007) Copper(I)-catalyzed highly efficient synthesis of benzoselenazoles and benzotellurazoles. *J. Org. Chem.*, **72** (21), 8087–8090.

108 Zakrzewski, J. and Krawczyk, M. (2008) Reactions of nitroxides, part 7: Synthesis of novel nitroxide selenoureas. *Heteroat. Chem.*, **19** (6), 549–556.

109 Zakrzewski, J. and Krawczyk, M. (2009) Synthesis and pesticidal properties of thio and seleno analogs of some common urea herbicides. *Phosphorus Sulfur Silicon Relat. Elem.*, **184** (7), 1880–1903.

110 Ogawa, A., Doi, M., Tsuchii, K., and Hirao, T. (2001) Selective sequential addition of diphenyl diselenide to ethyl propiolate and isocyanides upon irradiation with near-UV light. *Tetrahedron Lett.*, **42** (12), 2317–2319.

111 Tsuchii, K., Kawaguchi, S., Takahashi, J., Sonoda, N., Nomoto, A., and Ogawa, A. (2007) Highly selective double chalcogenation of isocyanides with disulfide-diselenide mixed systems. *J. Org. Chem.*, **72** (2), 415–423.

112 Mitamura, T., Tsuboi, Y., Iwata, K., Tsuchii, K., Nomoto, A., Sonoda, M., and Ogawa, A. (2007) Photoinduced thiotelluration of isocyanides by using a $(PhS)_2$-$(PhTe)_2$ mixed system, and its application to bisthiolation via radical cyclization. *Tetrahedron Lett.*, **48** (34), 5953–5957.

113 Mitamura, T., Iwata, K., and Ogawa, A. (2009) $(PhTe)_2$-mediated intramolecular radical cyclization of o-ethynylaryl isocyanides leading to bistellurated quinolines upon visible-light irradiation. *Org. Lett.*, **11** (15), 3422–3424.

114 Mitamura, T., Iwata, K., and Ogawa, A. (2011) Photoinduced intramolecular cyclization of o-ethenylaryl isocyanides with organic disulfides mediated by diphenyl ditelluride. *J. Org. Chem.*, **76** (10), 3880–3887.

115 Mitamura, T., Iwata, K., Nomoto, A., and Ogawa, A. (2011) Photochemical intramolecular cyclization of o-alkynylaryl isocyanides with organic dichalcogenides leading to 2,4-bischalcogenated quinolines. *Org. Biomol. Chem.*, **9**, 3768–3775.

116 Brook, A.G., Azarian, D., Baumegger, A., Hu, S.S., and Lough, A.J. (1993) Ring insertion reactions of silaaziridines with aldehydes and isocyanates. *Organometallics*, **12** (2), 529–534.

117 Brook, A.G., Saxena, A.K., and Sawyer, J.F. (1989) 1-Sila-3-azacyclobutanes: the insertion of isocyanides into silaaziridines. *Organometallics*, **8** (3), 850–852.

118 Ito, Y., Matsuura, T., and Murakami, M. (1988) Palladium-catalyzed regular insertion of isonitriles into silicon-silicon linkage of polysilane. *J. Am. Chem. Soc.*, **110** (11), 3692–3693.

119 Takeuchi, K., Ichinohe, M., and Sekiguchi, A. (2008) Reactivity of the disilyne RSi≡SiR (R=Si^{i}Pr[CH(SiMe$_3$)$_2$]$_2$) toward silylcyanide: two pathways to form the bis-adduct [RSiSiR(CNSiMe$_3$)$_2$] with some silaketenimine character and a 1,4-diaza-2,3-disilabenzene analogue. *J. Am. Chem. Soc.*, **130** (50), 16848–16849.

120. Ruiz, J., Gonzalo, M.P., Vivanco, M., Rosario Diaz, M., and Garcia-Granda, S. (2011) A three-component reaction involving isocyanide, phosphine and ketenimine functionalities. *Chem. Commun.*, 4270–4272.

121. Rieger, D., Lotz, S.D., Kernbach, U., Andre, C., Bertran-Nadal, J., and Fehlhammer, W.P. (1995) Synthesis of organic heterocycles via multicomponent reactions with cyano transition metal complexes. *J. Organomet. Chem.*, **491** (1–2), 135–152.

122. Saluste, C.G., Whitby, R., and Furber, M. (2000) A palladium-catalyzed synthesis of amidines from aryl halides. *Angew. Chem. Int. Ed. Engl.*, **39** (22), 4156–4158.

123. Saluste, C.G., Whitby, R.J., and Furber, M. (2001) Palladium-catalysed synthesis of imidates, thioimidates and amidines from aryl halides. *Tetrahedron Lett.*, **42** (35), 6191–6194.

124. Tetala, K.K., Whitby, R.J., Light, M.E., and Hurtshouse, M.B. (2004) Palladium-catalysed three component synthesis of α,β-unsaturated amidines and imidates. *Tetrahedron Lett.*, **45** (38), 6991–6994.

125. Whitby, R.J., Saluste, C.G., and Furber, M. (2004) Synthesis of α-iminoimidates by palladium-catalyzed double isonitrile insertion. *Org. Biomol. Chem.*, **2** (14), 1974–1976.

126. Onitsuka, K., Suzuki, S., and Takahashi, S. (2002) A novel route to 2,3-disubstituted indoles via palladium-catalyzed three-component coupling of aryl iodide, o-alkenylphenyl isocyanide and amine. *Tetrahedron Lett.*, **43** (35), 6197–6199.

127. Cao, C., Shi, Y., and Odom, A.L. (2003) A titanium-catalyzed three-component coupling to generate α,β-unsaturated β-iminoamines. *J. Am. Chem. Soc.*, **125** (10), 2880–2881.

128. Barnea, E., Majumder, S., Staples, R.J., and Odom, A.L. (2009) One-step route to 2,3-diaminopyrroles using a titanium-catalyzed four-component coupling. *Organometallics*, **28** (13), 3876–3881.

129. Lazar, M. and Angelici, R.J. (2006) Gold metal-catalyzed reactions of isocyanides with primary amines and oxygen: analogies with reactions of isocyanides in transition metal complexes. *J. Am. Chem. Soc.*, **128** (32), 10613–10620.

130. Lazar, M., Zhu, B., and Angelici, R.J. (2007) Non-nanogold catalysis of reactions of isocyanides, secondary amines, and oxygen to give ureas. *J. Phys. Chem. C*, **111** (11), 4074–4076.

131. Angelici, R.J. (2008) Organometallic chemistry and catalysis on gold metal surfaces. *J. Organomet. Chem.*, **693** (5), 847–856.

132. Tobisu, M., Fujihara, H., Koh, K., and Chatani, N. (2010) Synthesis of 2-boryl- and silylindoles by copper-catalyzed borylative and silylative cyclization of 2-alkenylaryl isocyanides. *J. Org. Chem.*, **75** (14), 4841–4847.

133. Murakami, M., Kawano, T., Ito, H., and Ito, Y. (1993) Synthesis of α-hydroxy ketones by samarium(II) iodide-mediated coupling of organic halides, an isocyanide, and carbonyl compounds. *J. Org. Chem.*, **58**, 1458–1465.

134. Curran, D.P. and Totleben, M.J. (1992) The samarium Grignard reaction. *In situ* formation and reactions of primary and secondary alkylsamarium(III) reagents. *J. Am. Chem. Soc.*, **114**, 6050–6058.

135. Gaich, T. and Baran, P.S. (2010) Aiming for the ideal synthesis. *J. Org. Chem.*, **75** (14), 4657–4673. and references cited therein.

10
Applications of Isocyanides in IMCRs for the Rapid Generation of Molecular Diversity

Muhammad Ayaz, Fabio De Moliner, Justin Dietrich, and Christopher Hulme

10.1
Introduction

Isocyanide-based multicomponent reactions (IMCRs) have been recognized for over 90 years, the first having been described in 1921 and named after its founder, Passerini [1]. Indeed, isocyanides have roots that date back further to the 1850s, the first being discovered serendipitously in 1859 by Lieke who, akin to many practicing chemists of today, was struck by their repulsive odor. However, this branch of chemistry is negative only when the applications of low-molecular-weight isocyanide congeners are evaluated, since fortunately the majority of higher-molecular-weight isocyanides are solid and odorless. The mass production of molecules derived from isocyanide chemistry can generally be performed by taking into account routine safety precautions, the odor being contained in a regular fume-hood and any residual isocyanide being destroyed with methanolic HCl.

Interestingly, in 1998, only 12 isocyanides were available from the Available Chemicals Directory but today, as the reagents' utility has blossomed dramatically, this number has increased to more than 500. Also noteworthy has been the development of several so-called universal isocyanides, which are conceptually similar to Ugi's original convertible isocyanide [2]. The isocyanide functional group can, in fact, be considered as a synthetic equivalent to vinylidene carbene; indeed, this property is a cornerstone of the compounds' successes in the Ugi reaction and also of many other isocyanide-based methodologies that currently are utilized in drug-discovery process, agrochemicals, and the materials sciences.

In being complementary to Chapter 13, the aim of the present chapter is to provide examples of isocyanide applications for the formation of heterocyclic ring systems. More specifically, the chapter mines the existing knowledge of the use of IMCRs in the creation of molecular diversity and, in particular, for use with small-molecule approaches to drug discovery. With regards to this latter subject, there is an on-going need to assemble screening collections, constructed with a

Isocyanide Chemistry: Applications in Synthesis and Material Science, First Edition. Edited by Valentine Nenajdenko.
© 2012 Wiley-VCH Verlag GmbH & Co. KGaA. Published 2012 by Wiley-VCH Verlag GmbH & Co. KGaA.

"rear-view mirror" approach of established chemical and biological space, complemented by diversity collections to probe new chemical space, both of which are vital components of file-enhancement strategies. Such knowledge-based front loading [3] systems may be heightened by compound selection criteria, that include not only "compound chemical tractability" but also details of compounds derived from methodologies that allow the simultaneous investigation of multiple sites of diversification in two or fewer synthetic steps, in an operationally friendly manner.

The Ugi IMCR [4] and several closely related IMCRs [5, 6] fall into the latter category and, as such, deliver compounds with built-in "iterative efficiency potential," thus enabling a rapid value chain progression [7]. Consequently, since the first realization of the unique efficiency of the Ugi reaction, an ever-growing number of research groups have investigated the development and application of like-methodologies to construct compound libraries for screening purposes. Along with "iterative efficiency potential," "bond-forming efficiency," and "high exploratory power" can be included the additional characteristics of the IMCRs [8]. Hence, the present chapter can be broadly classified into three areas: (i) UDC (Ugi/Deprotect/Cyclize) methodology; (ii) the miscellaneous secondary reactions of Ugi products; and (iii) bifunctional IMCR reagent chemistry.

Before moving forwards, it seems appropriate at this point briefly to recall the mechanism of the remarkably powerful Ugi reaction. The process occurs due to an ability of the isocyanide to react at the same (carbenoid) atom, with both nucleophiles and electrophiles. Specifically, condensation of the amine and carbonyl components affords the imine **1** which, in its protonated form, reacts with the isocyanide to give the so-called nitrilium ion **2** that is, in turn, intercepted by the carboxylate anion. This leads to the α-adduct **3**, which undergoes acyl migration (termed the Mumm rearrangement) to deliver the final compound **4** (Scheme 10.1). This mechanistic sequence enables both intermolecular (with more than 20 different known nucleophiles) and intramolecular trapping of the nitrilium ion, in addition to post-condensation modifications of **4**.

As such, the operationally friendly reaction enables access to peptide-like compounds **4**, with four points of potential diversity. This latter point has led to the Ugi reaction and subsequent modifications of the peptide-like product being positioned as a tremendously powerful tool in parallel synthesis, library development, and file enhancement. In fact, the Ugi reaction represents probably the most well-known and widely used application of isocyanide chemistry.

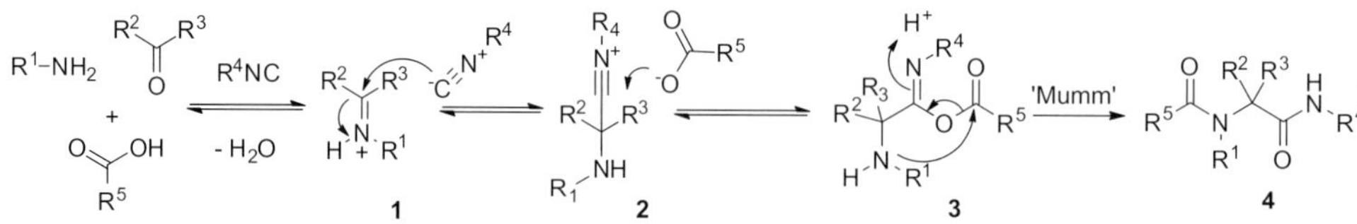

Scheme 10.1

10.2
Ugi/Deprotect/Cyclize (UDC) Methodology

10.2.1
Ugi-4CC: One Internal Nucleophile

Initially, the UDC methodology was introduced for the synthesis of arrays of benzodiazepines, which enjoy the status of "privileged structures" as they have demonstrated a multitude of biological activities [5]. 1,4-Benzodiazepine-2,5-diones (BZDs) are of particular interest as they have exhibited anticonvulsant, anxiolytic, and antitumor activities [9–16], and have also been reported as antagonists of platelet glycoprotein IIb-IIIa [17]. Despite their phenomenal biological relevance, until the mid-1990s the synthetic routes to this class of heterocycle were limited to methods that relied heavily on commercially available amino acids (mostly racemic), followed by tedious and low-yielding alkylation protocols [9]. During 1996, while addressing this problem, Armstrong and coworkers made two reports in which the convertible isocyanide **6** was introduced in a four-component Ugi reaction that was coupled to a subsequent intramolecular cyclization of an internal amine nucleophile of an anthranilic acid **5**, and which led in turn to a variety of diverse BZDs [9, 18, 19]. Following generation of the Ugi product **7**, acid treatment activated the convertible isocyanide (to possibly a munchnone **8**, an N-acyliminium ion **9**, or a methyl ester **10**), which facilitated cyclization to the final 1,4-benzodiazepine-2.5-dione **11** (Scheme 10.2). The low yields for this two-step process (reported as 15–30%) were attributed to an undesired interference of the anthranilic acid free amino group. Subsequently, Hulme and coworkers reported a series of post-Ugi transformations utilizing cyclohexenyl isocyanide, and introduced the "UDC" concept [20–22], which employs one, two, or even three N-Boc-protected internal amino-nucleophiles. As such, N-Boc-protection eliminates internal amine competition in the Ugi reaction.

Indeed, without the need for any intermediary work-up, solvent evaporation followed by acid treatment (trifluoroacetic acid (TFA)/dichloroethane (DCE)) unmasked the internal amine, thus invoking cyclization to BZDs with significantly higher overall yields. Moreover – and highly significantly – the methodology was

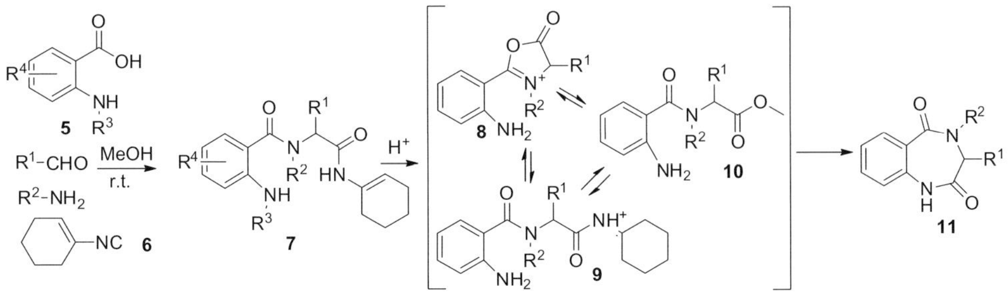

Scheme 10.2

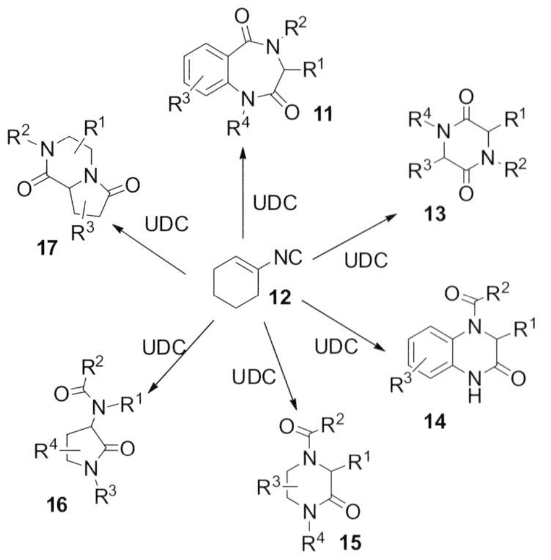

Scheme 10.3

compatible with plate-based protocols, such that a collection of 192 1,4-benzodiazepine-2,5-diones was produced. The UDC strategy was subsequently extended by repositioning the internal amino nucleophile to produce a number of well-known diverse chemotypes in succinct fashion, exemplified by benzimidazoles [23], benzodiazepines [20], quinazolines [24], and γ-lactams [25], to name but a few. Akin to the benzodiazepines, these transformations were sufficiently robust to be used for library syntheses in 96-well plates, thus enabling the rapid construction of libraries containing up to 10 000 members. A pictorial depiction of these chemotypes is shown in Scheme 10.3, where N-Boc-protected α-amino acids are used alongside the convertible isocyanide enable access to diketopiperazines **13** [26]. Likewise, the condensation of mono-N-Boc-protected phenylenediamine and N-Boc-protected ethylenediamine under UDC conditions [27, 28] produced arrays of dihydroisoquinoxalines **140** and ketopiperazines **15**. The condensation of resin-bound convertible isocyanides with N-Boc-protected β-amino aldehydes generated [25] monocyclic lactams **16**, whereas using a bifunctional input (β-keto acid; see Section 10.4.5) with cyclohexnyl isocyanide enabled the formation of the bicyclic lactam **17**.

To further enhance the utility of UDC strategy, resin-bound convertible isocyanides were introduced by Piscopio **18**, Ugi, and Kennedy **20** [29–31]. Furthermore, in 2001 a Rink isocyanide resin **19** was also reported by a group at Procter & Gamble [32] (Figure 10.1). A unique development here was the introduction of a "safety-catch linker isocyanide" **21** which, through Boc-activation (**22** → **23**) and resin cleavage with sodium methoxide, could subsequently be manipulated to generate a variety of heterocycles that previously were accessible only via cyclohexenyl isocyanide (Scheme 10.4) [22, 33, 34].

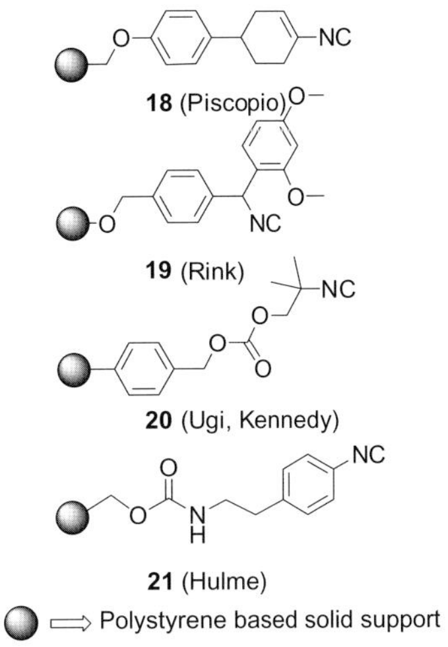

10.2 Ugi/Deprotect/Cyclize (UDC) Methodology

Figure 10.1 Polymer-supported isonitriles.

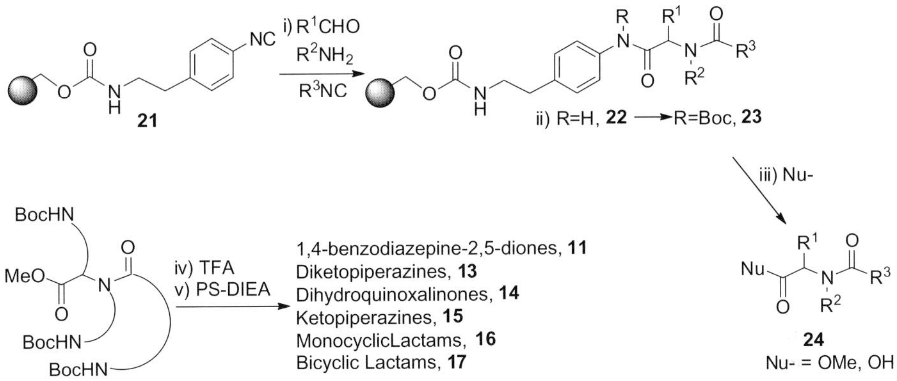

Scheme 10.4

In order to broaden the scope of the UDC strategy, Ugi inputs with tethered electrophilic sites (complementary to the internal nucleophile) were introduced (Scheme 10.5). For example, employing the UDC protocol with ethyl glyoxylate **25** in combination with either N-Boc-protected α-amino acids, N-Boc anthranilic acids, mono-Boc-protected ethylenediamines, or N-Boc-protected o-phenylenediamines, resulted in the formation of diketopiperazines **26**, benzodiazepines **27**, ketopiperazines **28**, and dihydroquinoxalinones **29**, respectively [35]. The resulting products were similar to those obtained using cyclohexenyl isocyanide, but they contained an additional amide linked to a diversity position derived from the isocyanide input.

Libraries of all four chemotypes were prepared in good to excellent yields in a mere two weeks, as exemplified by 12 480 diketopiperazines (analogues of **26**:

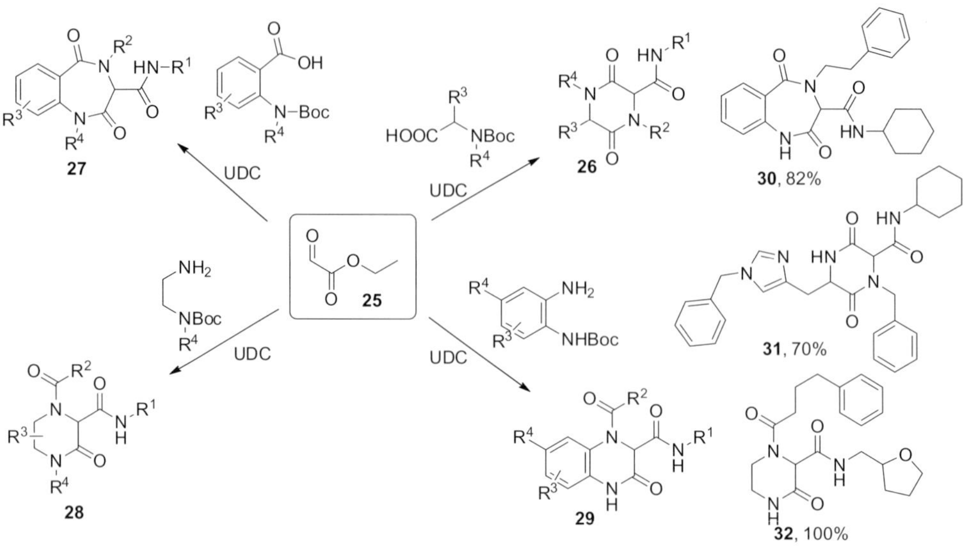

Scheme 10.5

30 amines × 16 isocyanides × 26 N-Boc-α-amino acids full matrix) [35]. The *drug-likedness* of the members of the library was ensured, as the majority of the compounds were found to be in line not only with the Lipinski "rule of five" but also oral absorption criteria [36]. Some examples from these libraries are shown in Scheme 10.5.

Subsequently, the UDC procedure was coupled with resin-bound amines **33** and the use of Rapid Plate™ liquid dispensers and Charybdis™ blocks, to produce a library of 9600 benzodiazepines **34**, in good to excellent overall purity (Scheme 10.6) [37]. Diketopiperazines **35** were also accessible by utilizing similar production protocols.

The use of these isocyanide-derived protocols has been demonstrated by Szardenings and coworkers [38] who, by using solid-supported α-amino acids **33**, in combination with a UDC strategy, discovered the highly potent and selective inhibitor **36** of the matrix metalloproteinase (MMP), collagenase-I (Figure 10.2).

Interestingly, the carbonyl derived from the carboxylic acid was also amenable to cyclodehydration of an internal amine under typical UDC conditions. Indeed, the strategy was subsequently and successfully employed to produce a collection of 10 000 imidazolines [39]. In this way, the use of N-Boc-protected α-amino aldehydes **37** in the Ugi reaction, with subsequent TFA treatment, led to the production of imidazolines of generic structure **38** (Scheme 10.7). On completion of the reaction, any unused starting materials and noncyclized free amine Ugi products were scavenged out, using polymer-supported isocyanate and tris-amine in tandem.

Likewise, the benzimidazoles **39** proved to be accessible from N-Boc-protected phenylenediamines **41** and supporting Ugi reagents. Indeed, a plate-based library

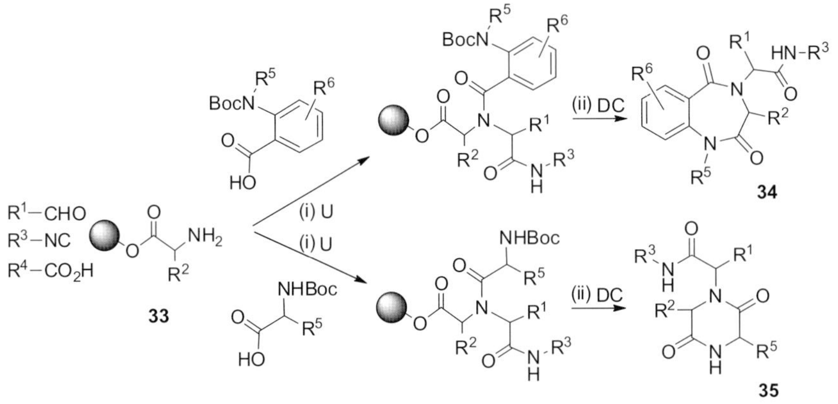

Scheme 10.6 Reagents and conditions: (i) R¹CHO (3 equiv.), **33** (3 equiv. – Wang resin), R³NC (3 equiv.), R⁴CO₂H = N-Boc anthranilic acid (3 equiv.), all 0.5 M solutions (MeOH/CH₂Cl₂, 1:1), room temperature, 24 h. Wash resin with CH₂Cl₂ (×3) and MeOH (×3); (ii) 10% TFA in CH₂Cl₂, wash resin with CH₂Cl₂ (×2).

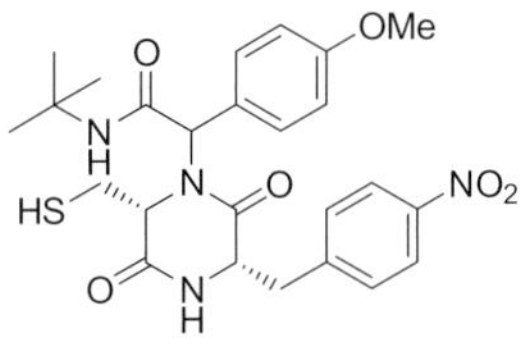

Figure 10.2 MMP collagenase-I inhibitor **36**.

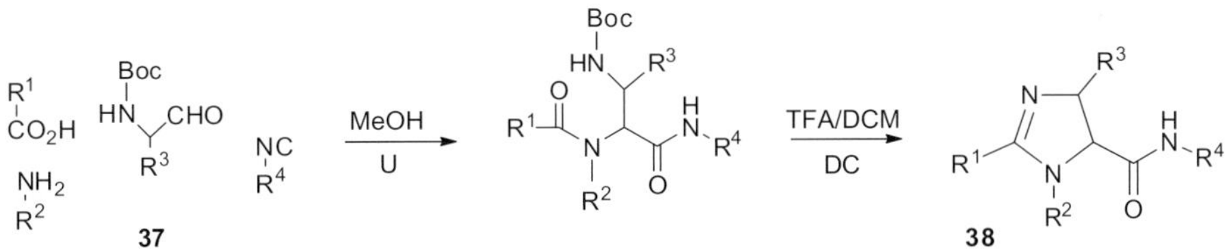

Scheme 10.7

of in excess of 10 000 products was reported, in excellent overall yields (Scheme 10.8) [23]. In analogous fashion, the use of glyoxylic acids **42** enabled access to libraries of quinoxalinones **40**, in excellent yields [24].

Subsequently, the syntheses of both **39** and **40** were further expedited [40] by combining microwave heating with fluorous technologies (**43a**, **44**, **45a**), thereby reducing the overall reaction time and simplifying the purification process by solid-phase extraction (SPE) over FluoroFlash cartridges. A representative set of examples is shown in Figure 10.3.

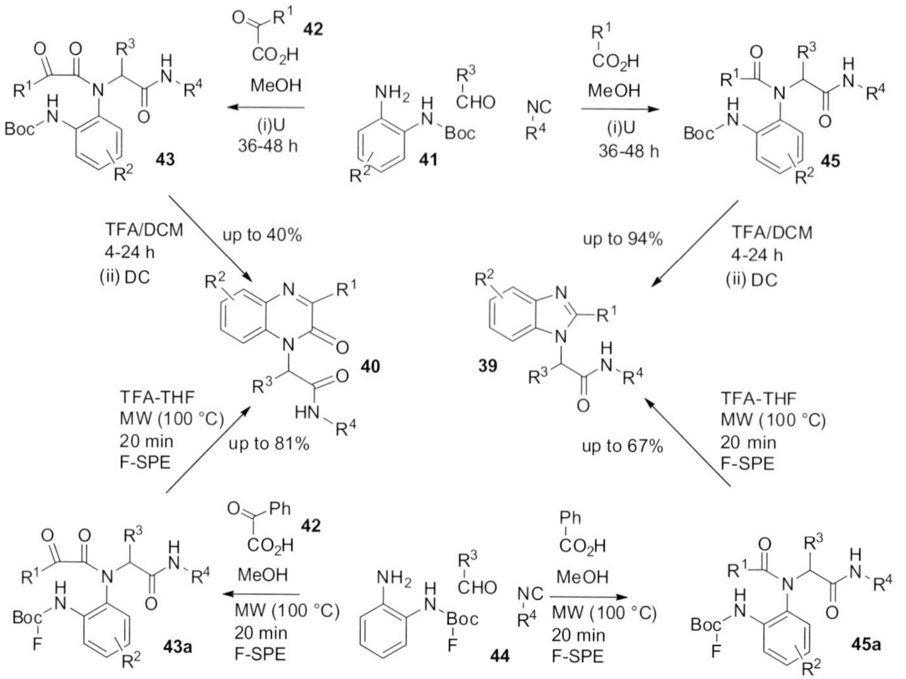

Scheme 10.8

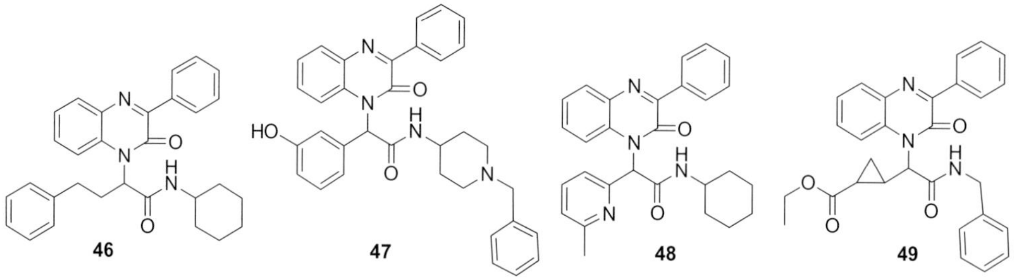

Figure 10.3 Representative examples of quinoxalinones.

By combining the UDC strategy with S_N-aromatic methodology, Tempest and coworkers reported the Ugi reaction of commercially available 2-fluoro-5-nitrobenzoic acid **50**, a primary amine tethered to a Boc-protected internal amino or hydroxyl nucleophile and supporting reagents. An acid treatment of the Ugi product and proton scavenging resulted in the cyclization of internal nucleophiles to indazolinones **51**, benzazepines **52**, and benzoxazepines **53** (Scheme 10.9). Based on this protocol, a library of 960 benzoxazepines was reported where over 75% of the compounds produced exhibited a purity in excess of 75% (as judged

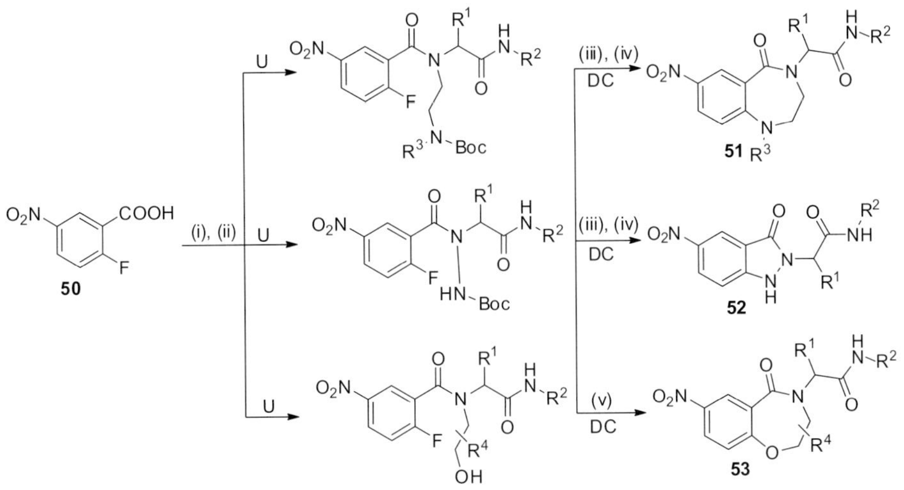

10.2 Ugi/Deprotect/Cyclize (UDC) Methodology

Scheme 10.9 Reagents and conditions: (i) R¹CHO (0.2 M, 200 µl, MeOH), R²NC (0.1 M, 200 µl, MeOH), **50** (0.1 M, 200 µl, MeOH), Boc-protected diamine or amino-alcohol (0.1 M, 200 µl, MeOH), r.t., 48 h; (ii) PS-tosylhydrazine (3 equiv.), PS-diisopropylethylamine (3 equiv.), THF : CH$_2$Cl$_2$ (600 µl, 1 : 1), 24 h; (iii) 20% TFA/CH$_2$Cl$_2$ (600 µl), 4 h; (iv) PS-morpholine (3 equiv.), DMF (600 µl), 36 h; (v) PS-TBD (7-methyl-1,5,7-triazabicyclo[4.4.0]dec-5-ene), DMF (600 µl), r.t., 36 h.

using liquid chromatography/mass spectrometry with ultraviolet detection at 215 nm) [41].

10.2.2
TMSN$_3$-Modified Ugi-4CC: One Internal Nucleophile

Possibly the most widely known alternative nucleophile to a carboxylic acid is TMSN$_3$, which produces tetrazoles through azide trapping of the intermediate Ugi nitrilium ion [42] Thus, the reaction of N-Boc-amino aldehydes **55**, secondary amines, methyl isocyanoacetate **54** and trimethylsilylazide in methanol, followed by acid treatment, proton scavenging and reflux, affords bicyclic azepine-tetrazoles **57** with three diversity points and in good to excellent yields (Scheme 10.10) (Figure 10.4) [43]. The use of standard aldehydes and primary amines with methyl isocyanoacetate and trimethylsilylazide afforded fused tetrazolo-ketopiperazines **59**, in good reported yield [44].

Marcaccini and coworkers addressed the issue of synthesizing derivatives of 4,5-dihydro-3H-1,4-benzodiazepin-5-one bearing aryl groups in the 2-position by using masked internal nucleophiles in the form of a nitro group. After condensing aminoacetophenone hydrochloride **63**, isocyanides, oxo-components, and o-nitrobenzoic acids **64** in a basic environment, the Ugi products were obtained in good yields [45]. A subsequent iron-mediated reduction of the NO$_2$ group and

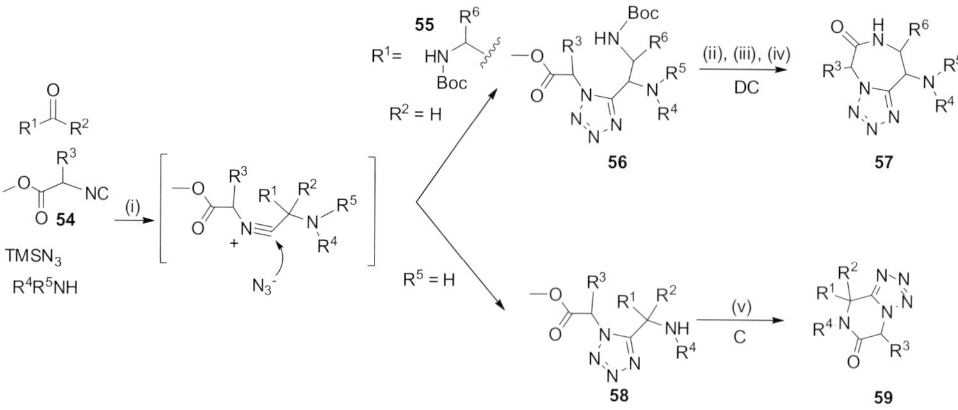

Scheme 10.10 Reagents and conditions: (i) R¹R²C=O (1.5 equiv., 0.1 M in MeOH), TMSN₃ (1 equiv., 0.1 M in MeOH), R³NH₂ (1 equiv., 0.1 M in MeOH), **54** (1 equiv., 0.1 M in MeOH), 24 h, r.t.; (ii) 10% TFA in CH₂Cl₂; (iii) PS-DIEA, DMF/dioxane, 1:1, reflux; (iv) PS-NCO, PS-TsNHNH₂ THF/DCE, 1:1; (v) Reflux, MeOH.

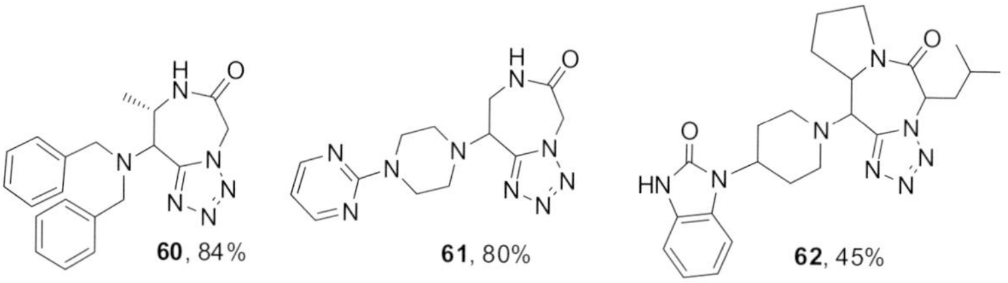

Figure 10.4 Examples of bicyclic azepine-tetrazoles.

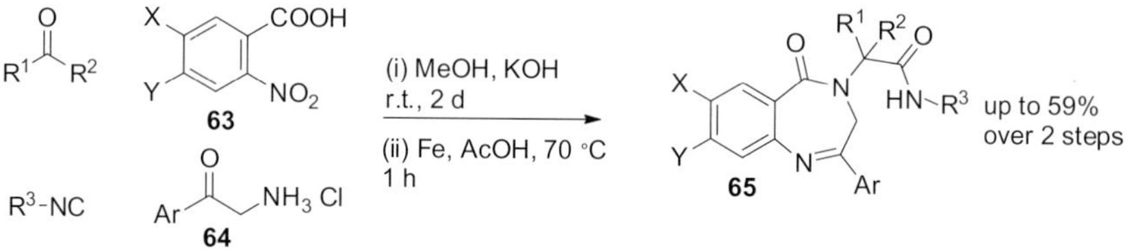

Scheme 10.11

simultaneous cyclization led to the generation of benzodiazepinones **65** in excellent overall yields (up to 50% over two steps) (Scheme 10.11).

10.2.3
Ugi-4CC: Two Internal Nucleophiles

In taking the UDC strategy to another level, Hulme and coworkers [46] employed two masked internal nucleophiles in the Ugi condensation (Scheme 10.12).

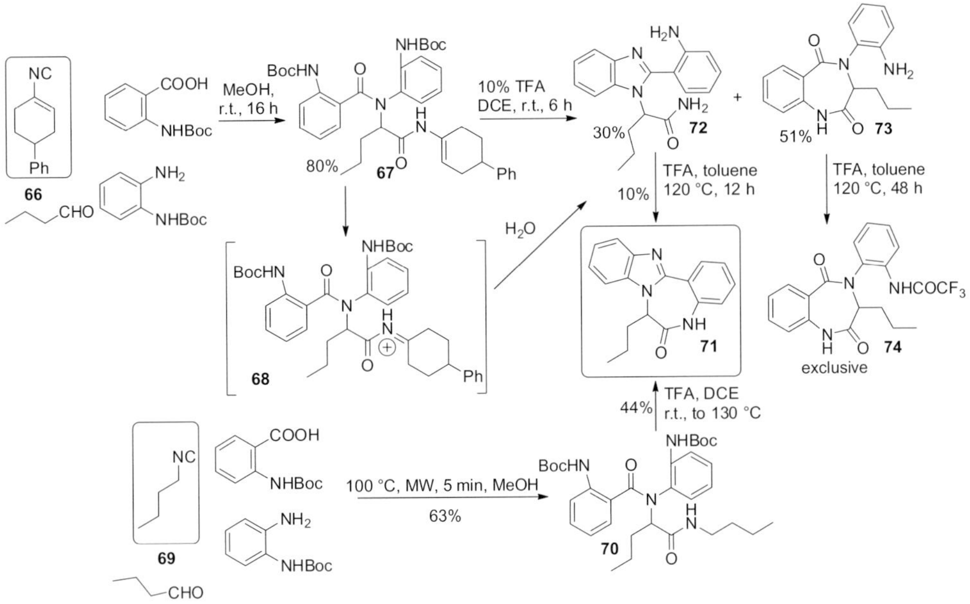

10.2 Ugi/Deprotect/Cyclize (UDC) Methodology

Scheme 10.12

Although the eventual successful goal of these investigations was the preparation of triazadibenzoazulenones **71** via a UDC protocol, the initial studies using the convertible isocyanide **66** showed that acid treatment of the Ugi adduct produced only 10% of the desired triazadibenzoazulenone **71**, while the majority of the remaining mass balance was present as a 1,4-benzodiazepine-2,5-dione **74**, due to the second internal amine nucleophile having failed to cyclize onto the benzylic tertiary amide carbonyl [47]. The formation of this inert product was circumvented, however, by altering the order of ring-forming cyclizations; this was made possible by lowering the electrophilicity of the carbonyl derived from the isocyanide, which in turn promoted initial benzimidazole formation. Such modification was achieved by using n-butyl isocyanide **69**, and with the two-step protocol being optimized to deliver a library of 12 triazadibenzoazulenones in excellent overall yields (up to 72% over two steps). Moreover, the process proved to be not only robust but also highly economical, given the fact that n-butyl isocyanide **69** was 10-fold cheaper than the convertible isocyanide **66**.

Subsequently, the use of this cheap alternative rather than a "designer convertible" isocyanide was extended [48] towards the syntheses of other biologically relevant scaffolds, such as diketopiperazine **77** and benzodiazepines **78** (Scheme 10.13). Both of these chemotypes were obtained in excellent yields in a one-pot, two-step procedure, with the process being expedited by microwave heating.

More recently [24], two internally masked amine nucleophiles were employed in the preparation of dihydroquinazoline–benzodiazepine tetracycles **83** and **84**. By using diamine **79** or **80** in combination with other supporting reagents, the

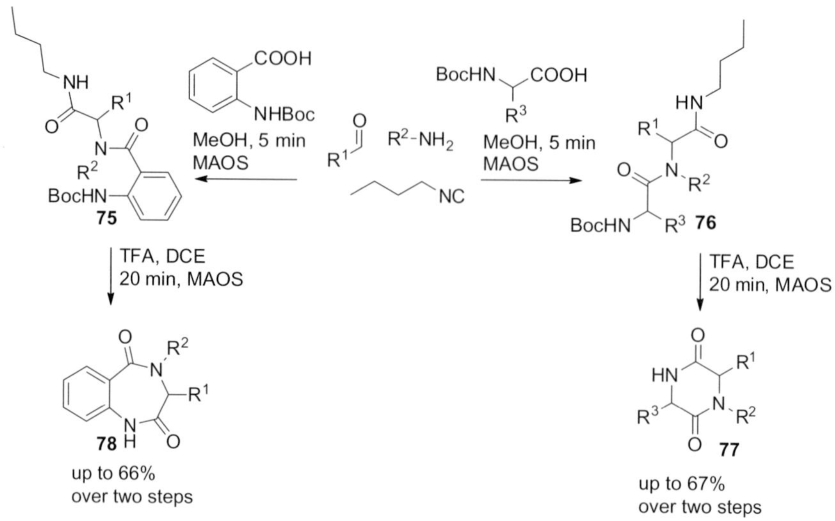

Scheme 10.13

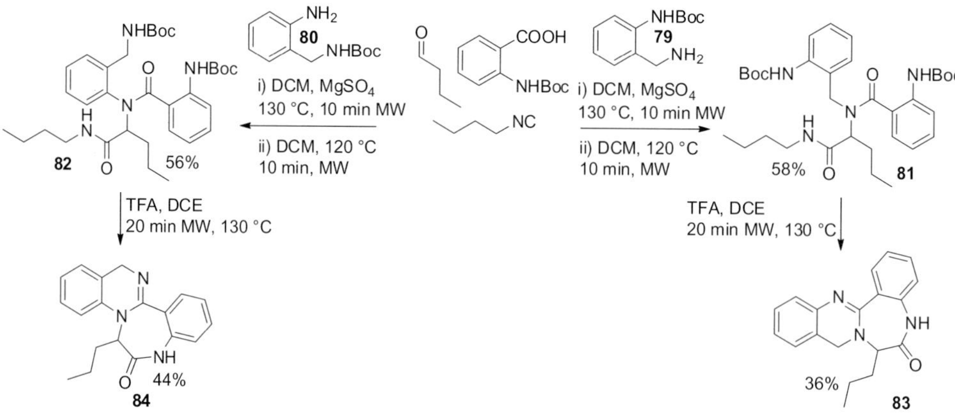

Scheme 10.14

Ugi products **81** and **82** were obtained in reasonable yields, although it was necessary to pre-form the imines by using microwave irradiation and a dehydrating agent. The Ugi products were then subjected to acid treatment at elevated temperature to produce the tetracyclic scaffolds **83** and **84**, albeit in low to moderate yields. Notably, when acting as a cleavable unit the *n*-butyl isocyanide opened up many opportunities for the further functionalization of the chemotypes **83** and **84** through, for example, amide alkylation (Scheme 10.14).

The same strategy was equally successful when applied to the synthesis of dihydroquinazolines **87** and **88**. When using one masked internal nucleophile (teth-

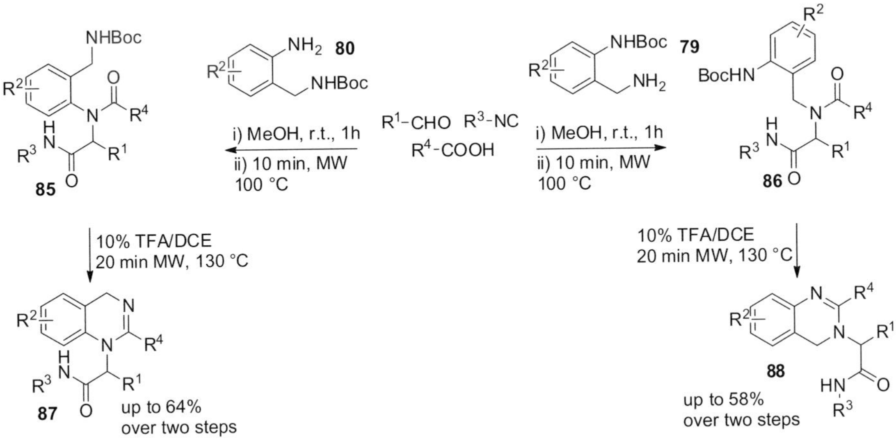

Scheme 10.15

ered to the amine input), both the Ugi reaction and the subsequent deprotection/cyclization step progressed smoothly, generating the dihydroquinazolines **87** and **88** in excellent yields (Scheme 10.15). The generality of the process was confirmed by an ability to produce small libraries of both chemotypes, in good to excellent yields.

In an effort to extend the two internal nucleophile concept, Hulme and coworkers applied the process to the synthesis of fused quinoxalinone–benzodiazepines **91** and bis-benzodiazepines **92** [49]. When using convertible cyclohexenyl isocyanide **18** alongside the preformed imines **89** and *N*-Boc-anthranilic acids, the Ugi product **90** was obtained in excellent yield (Scheme 10.16). An acidic treatment of the Ugi product, under microwave conditions, resulted in the formation of a unique quinoxalinone–benzodiazepine product **91**, again in excellent yield. Encouragingly, changing the amine input (using amine **79**) led to improved Ugi yields **93**, thus setting the scene for subsequent acid-mediated deprotection/cyclization to afford bis-benzodiazepines **92**.

10.2.4
Ugi-4CC: Three Internal Nucleophiles

Recently, Hulme and coworkers have been investigating cascade reactions of the Ugi adduct containing three internal nucleophiles and three complementary electrophilic carbonyl groups (Xu, Z. and Hulme, C., unpublished results). Gratifyingly, by employing *N*-Boc-α-amino-acids **94**, *ortho*-*N*-Boc-benzylamines **95** and *ortho*-*N*-Boc-phenyl isocyanide **96** in conjunction with ethyl glyoxalate **25**, the Ugi product folds into one scaffold in good yield (Scheme 10.17). The cascade of ring-forming events is promoted by microwave irradiation (120 °C, 10 min), producing the generic structure shown **99**.

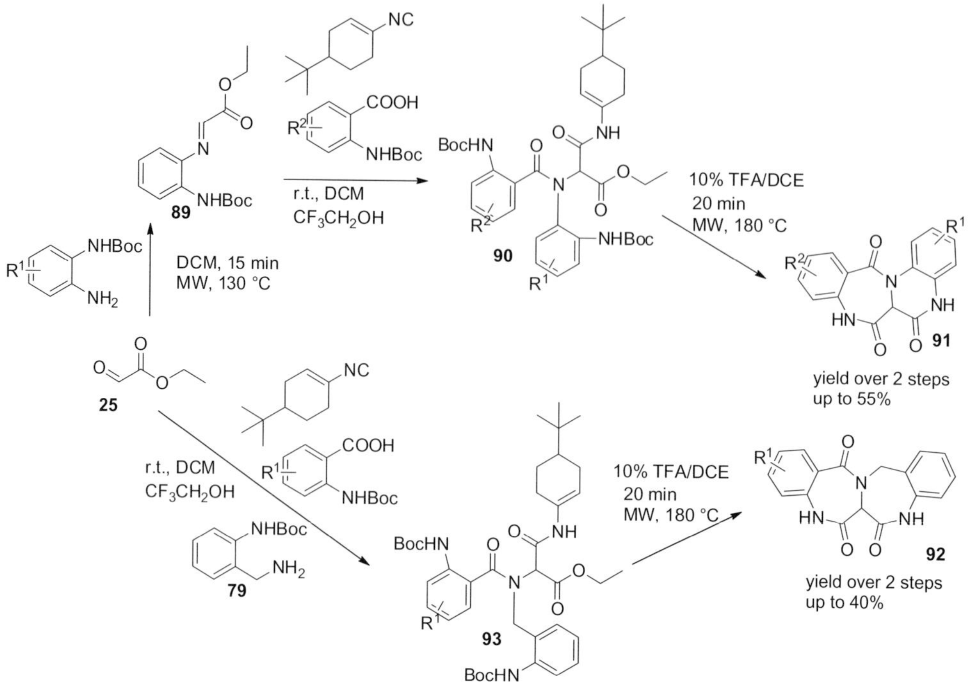

Scheme 10.16

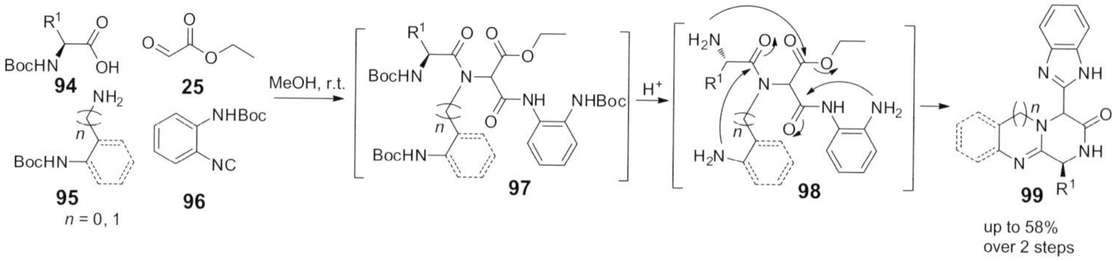

Scheme 10.17

10.2.5
Ugi-5CC: One Internal Nucleophile

In extending the strategy beyond the Ugi-4CC, the UDC strategy was applied to a relatively ignored variant of the Ugi reaction, namely the Ugi-5CC [50], where the acid input for the condensation is generated *in situ* from methanol and CO_2, and then allowed to react with either *N*-Boc-α-amino aldehydes or *N*-Boc-protected 1,2-diamines and supporting Ugi reagents. Acid treatment of the resulting Ugi products **101** and **103** led to the formation of two distinct classes of medicinally

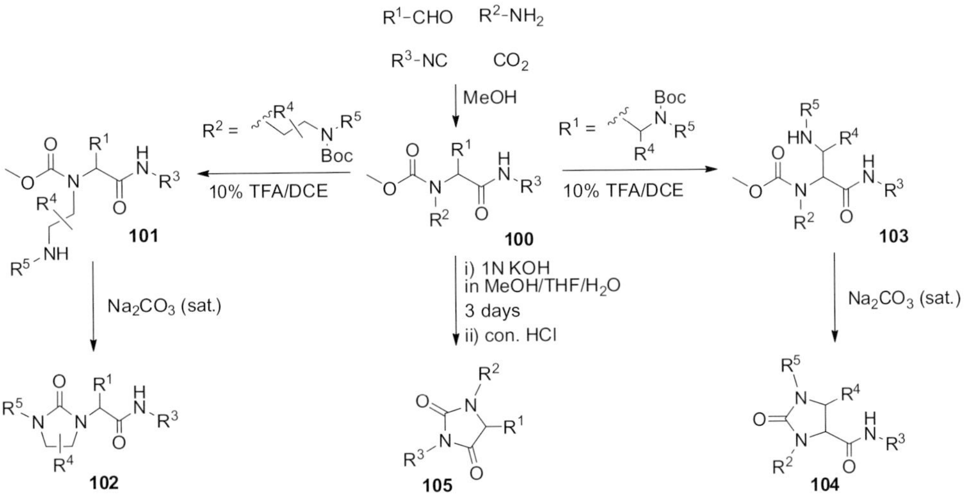

Scheme 10.18

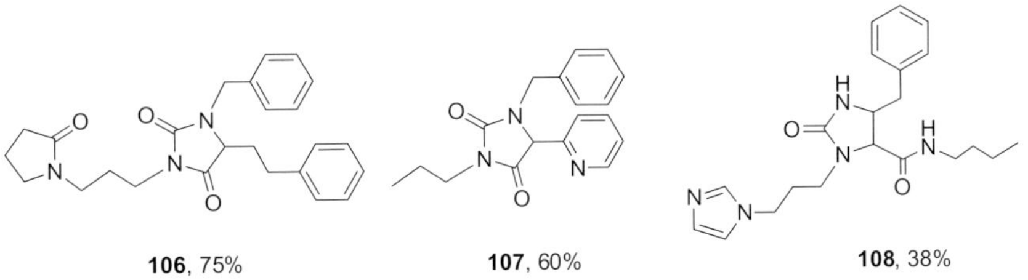

Figure 10.5 Examples of urea and urethanes.

relevant compounds, cyclic ureas **102** and **104**, though the yield for cyclic urea **102** was very low (<10%) (Scheme 10.18).

Intriguingly, in the absence of an amino internal nucleophile, exposure of the Ugi-5CC product **100** to strongly basic conditions gave rise to fully functionalized hydantoins **105**, via a cyclization of the isocyanide-derived amidic NH onto the urethane carbonyl in a mere two steps [51, 52]. Representative examples of these three chemotypes are shown in Figure 10.5.

The pool of accessible chemotypes derived from this methodology was subsequently increased by employing one internal nucleophile, and simultaneously taking advantage of the reactivity of the amidic isocyanide-derived NH. Thus, preformed imines (via reaction of glyoxaldehydes **109** and *mono*-Boc-protected diamines **79**) and carbonic acids (by bubbling CO_2 in methanol) were mixed to produce the Ugi adduct **110**, in excellent yields (Scheme 10.19) [53]. Subsequent acidic and basic treatments led to the formation of hydantoin–benzodiazepines of the generic structure **112**, in excellent yields. Very interestingly, dissolution of the

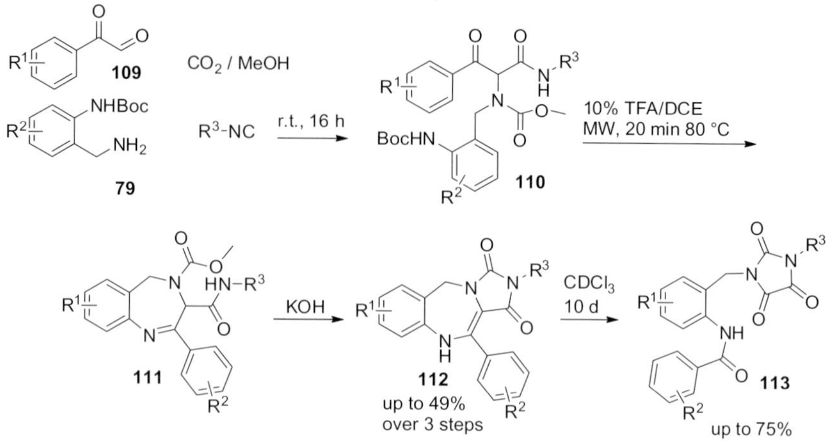

Scheme 10.19

chemotype **112** in CDCl$_3$ resulted in an oxidative cleavage which produced pharmacologically relevant imidazolidinetriones **113** [54, 55].

To summarize, the past 13 years have witnessed an evolution of the UDC strategy into a well-established discipline of synthetic chemistry, in turn to rigidify the Ugi product into pharmacologically desirable constrained scaffolds. In fact, two such protocols have resulted in the creation of Phase II small-molecule drugs which existed in the initial virtual space made accessible by the methodology, and equivalent to a "hit" in Phase II, with no scaffold hopping [56]. In particular, the simplicity, practicality, and amenability of the procedure to automation mean that it is highly preferable to accessing libraries of multicyclic, complex, and constrained drug-like compounds in a couple of synthetic steps within a short time span. Clearly, there exists a huge potential to further this field in many ways, and it is expected that the manipulation of other IMCRs will increase such access to a growing number of chemotypes [57].

10.3
Secondary Reactions of Ugi Products

Another widely exploited approach to rigidify or modify acyclic peptide-like Ugi products into a wide number of nitrogen-containing heterocycles is to couple the Ugi IMCR with a subsequent transformation or a secondary reaction [58]. Such a cyclization step is made possible by incorporating functional groups into the four Ugi reagents that do not interfere with the one-pot condensation. From a conceptual point of view, there are almost no limitations to the secondary modifications which IMCR adducts can undergo, if they are endowed with an appropriate functionality and reacted under the appropriate conditions. However, within this huge toolbox or arsenal of synthetic methodologies, a few reactions have predominated over the past 15 years, and these are the methods most commonly used [5] to build up molecular diversity while building in desirable drug-like properties during file enhancement. The main attributes of these reactions are their general applicability, the commercial availability of the starting materials and, of course, the relevance of the products afforded.

Far from being an exhaustive survey of such a broad topic, the following section is intended to represent an overview of the most recent and relevant literature. The aim also is to provide some fundamental knowledge of the field by highlighting the application of general concepts and interesting examples that can be expanded and studied in greater depth.

10.3.1
Nucleophilic Additions and Substitutions

In the vast and diversified scenario described above, reactions involving nucleophiles comprise the "lion's share" of methods, due to the ease of preparing or obtaining suitable starting materials (e.g., alcohols or halides) from commercial suppliers, and because of the apparently countless families of molecules that these transformations are capable of yielding. Typically, Mitsunobu reactions, lactamizations/lactonizations (see Section 10.2), metal-promoted processes, and base- or acid-catalyzed condensations, such as Michael and Pictet–Spengler, are the most recurrent tools in this field.

10.3.1.1 Alkylations

As noted by Banfi in a recent comprehensive review [59], the placing of a leaving group and a suitable nucleophile in an Ugi product paves the way to several possible subsequent substitution patterns. Consequently, many different scaffolds can be obtained upon variation of the nature and the position of the above-mentioned reactive functionalities. At first sight, the alkyl halides would seem the most convenient electrophilic components to select, but their main problem is that such building blocks are prone to compatibility issues with the amine component of the IMCR and/or the enamine tautomer of the imine formed during the Ugi reaction [59]. Nevertheless, some remarkable examples of this strategy using alkyl halides have been reported by different groups. For example, in 2006 Dai reported the preparation of benzoxazinones **115** in good to excellent yields, by means of an intramolecular etherification between a phenol derived from the amine input and a bromine atom embedded in the carboxylic acid input (Scheme 10.20) [60].

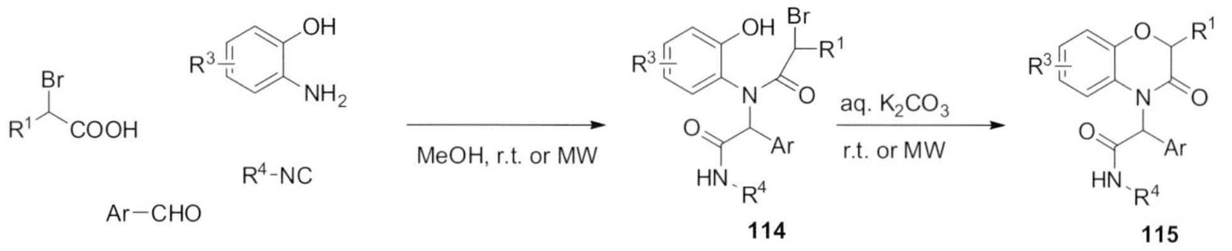

Scheme 10.20

The same approach was exploited by Wessjohann to develop a methodology for the preparation of macrocyclic peptoids [61] bearing the structural motif of biologically relevant natural products, although in very variable yields, ranging from 19% to 90%.

The preparation of piperazine rings with amines participating as the nucleophile has been reported by several groups, and this represents one of the most common targets of IMCR chemistry [62].

Almost 10 years ago, a research group at Procter & Gamble [63] disclosed a methodology for the preparation of peptidomimetics **117** via the Ugi adducts **116**, affording the title compounds in an average 45% yield. In 2001, diketopiperazines **119** were synthesized by Marcaccini [64] from compounds **118**, via an ultrasound-assisted nucleophilic substitution. Although it provided a highly drug-like scaffold with a wide substitution pattern, the method was unfortunately afflicted by limitations such as poor yields when aliphatic aldehydes were employed, and an extensive epimerization during the second step. Finally, acyl-piperazines **120** were obtained in an elegant one-pot fashion by Rossen in 1997 [65]. This early example is still worthy of mention because of its ease and the efficiency of diversity construction; in particular, it was used successfully to build the core of a human immunodeficiency virus protease inhibitor (Scheme 10.21) [66].

10.3.1.2 Mitsunobu Reactions

In more recent times, alcohols have been fruitfully used as "masked" leaving groups that can be activated in mild Mitsunobu processes, and displaced by

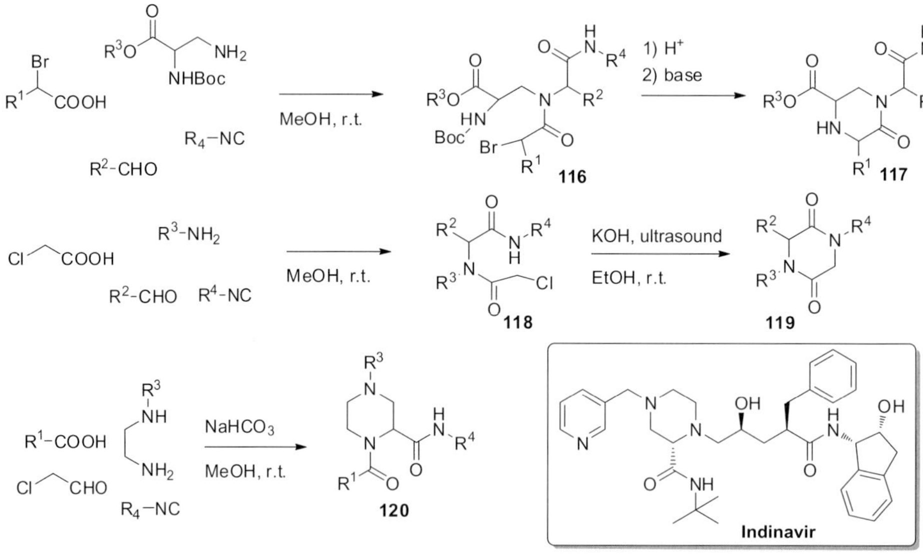

Scheme 10.21

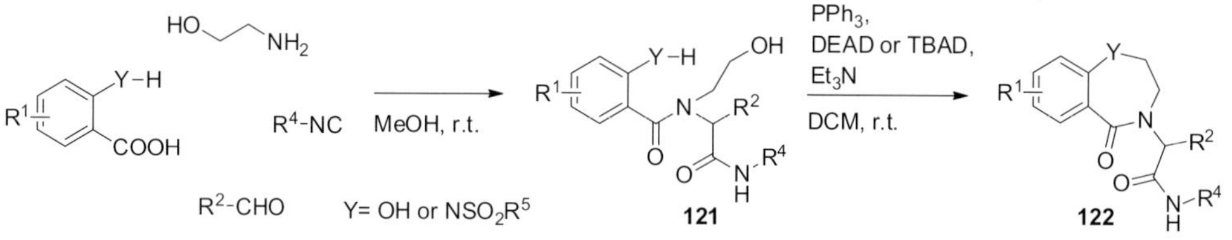

Scheme 10.22

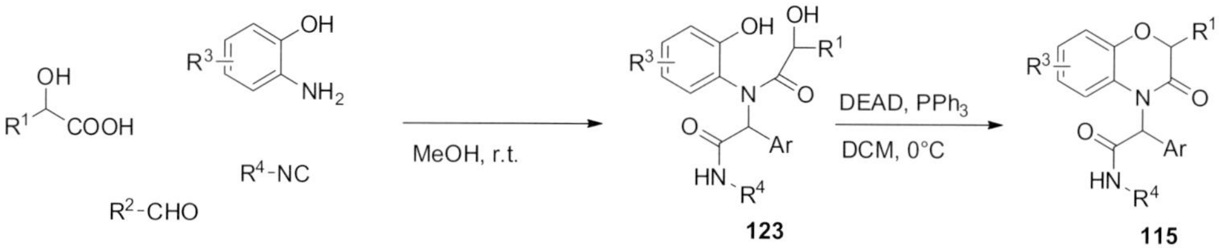

Scheme 10.23

phenols or sulfonamides. These starting materials display almost complete compatibility with other functionalities, and can be easily protected or deprotected in a number of different ways. Furthermore, they may be readily prepared in enantiopure forms, and typically tethered to the amine or carboxylic acid inputs of the condensation [59].

Remarkably, Banfi and coworkers described the use of a combination of ethanolamine along with either salicylic acids or 2-sulfonamide-substituted benzoic acids in the Ugi reaction to produce the intermediate compound **121** which, when reacted under Mitsunobu conditions, would afford functionalized benzodiazepinones and benzoxazepinones **122** (Scheme 10.22). This route represents a straightforward and high-yielding methodology (up to 98%), capable of producing two scaffolds that are regarded as privileged in medicinal chemistry [67, 68].

Notably, optically pure substituted amino-alcohols have also been employed in this pathway [69, 70], and this has resulted in the introduction of an additional diversity point. The same group recently brought to fruition a concise preparation of benzoxazinones **115** by means of another post-Ugi–Mitsunobu strategy, in which o-aminophenols and α-hydroxyacids were used as building blocks in the IMCR condensation (Scheme 10.23) [60, 71].

Although several further examples of this type of post-Ugi transformations have been reported, their complete examination is beyond the scope of this chapter but has been the object of a review [59]. However, it should be noted that the secondary amide generated during the IMCR can also be involved in a post-condensation reaction with hydroxyl groups. In this context, the formation of β-lactam **125** was observed when the Ugi product **124** was treated with diethyl azodicarboxylate and triphenylphosphine (Scheme 10.24) [71].

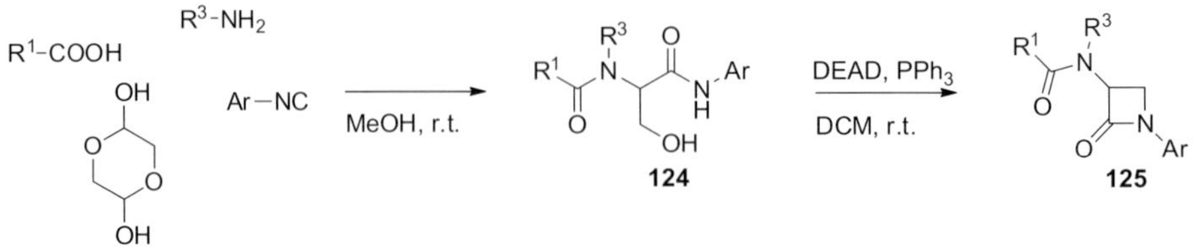

Scheme 10.24

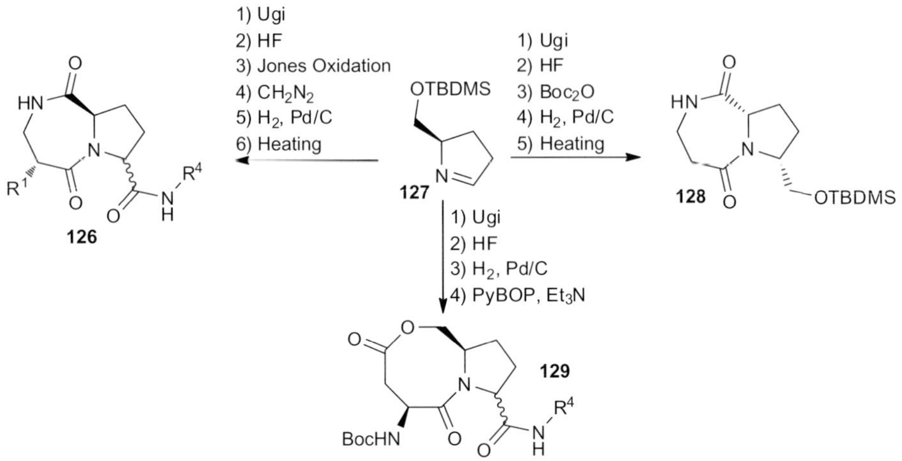

Scheme 10.25

10.3.1.3 Lactonization and Lactamization

The simultaneous presence of an alcohol or an amine and a carboxylic acid on Ugi reagents, which are typically protected or disguised, is one of the most common tactics used to deliver cyclic esters or amides. Even though the majority of the routes to prepare lactams fall within the previously described UDC strategy [6], a few methodologies stand apart. In particular, when enantiomerically pure cyclic imines are employed to induce stereoselection, a longer and more elaborate sequence is required. An elegant example was provided in 2008, again by the group of Banfi [72], for the synthesis of piperidino-fused diazepinediones **126** and **128**, which represent a class of rigidified peptidomimetics mimicking a domain involved in many cell-adhesion and recognition processes. Notably, good (15% to 63%) yields over five or six steps have been reported, and a modified version of the pathway is amenable to the preparation of unusual, eight-membered lactones **129** (Scheme 10.25). These lactones are, however, less attractive structures due to often-encountered and problematic issues of *in vivo* instability. Hence, very few post-Ugi lactonizations have been reported.

10.3.2
Base- or Acid-Promoted Condensations

Within the realm of nucleophile–electrophile interactions, some well-known reactions triggered by basic or acidic conditions have recently met renewed interest to manipulate Ugi products. In particular, Pictet–Spengler-type cyclizations have been exploited by Dömling and Orru and coworkers [73, 74] to prepare complex polycyclic scaffolds containing the piperazine moiety. Both groups have reported medium to excellent yields and high degrees of stereoselectivity, and have illustrated the resemblance of structures **130** and **131** to naturally occurring alkaloids. With regards to Michael additions, at this point it seems appropriate to mention the recent achievements of Andreana [75], who in 2007 described an unprecedented microwave-assisted, one-pot synthesis of diketopiperazines **132** and spirocyclic compounds **133**. Finally, broadly commercially available building blocks with activated methylenes can be exploited to generate molecules that are prone to undergo Knoevenagel condensations. This approach was followed by Torroba in 2004 [76], to design a concise and good-yielding (51-78%) route to quinolinones **134**, that often involved a spontaneous nucleophilic attack onto a carbonyl embedded in the amine component (Scheme 10.26).

10.3.3
Nucleophilic Aromatic Substitutions

Nucleophilic aromatic substitutions have also been widely explored as post-Ugi modifications. Most examples employ palladium catalysis (see Section 10.3.7), although the use of a strong base is often sufficient to synthesize several interesting chemotypes. For example, in 2007 a research group at Priaton described the

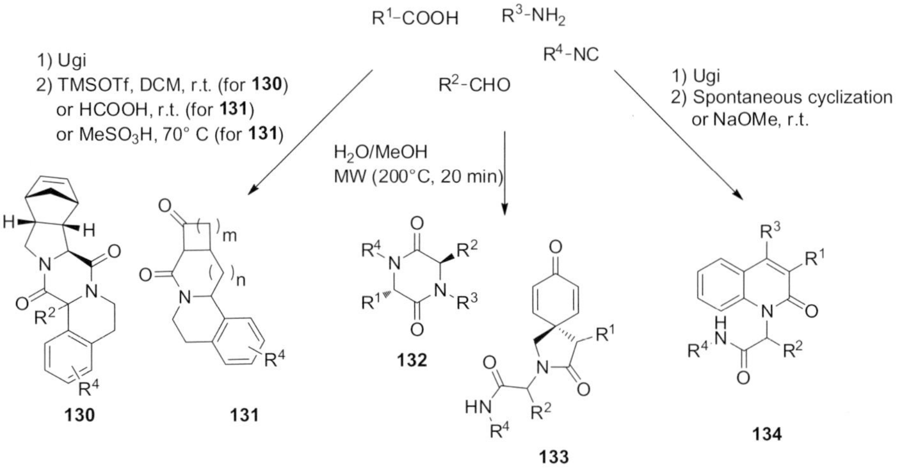

Scheme 10.26

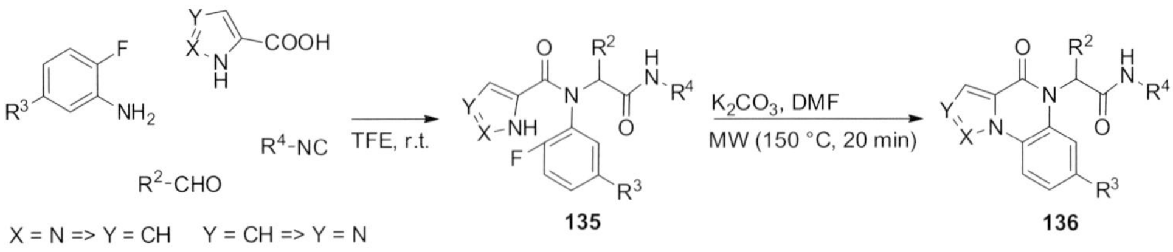

Scheme 10.27

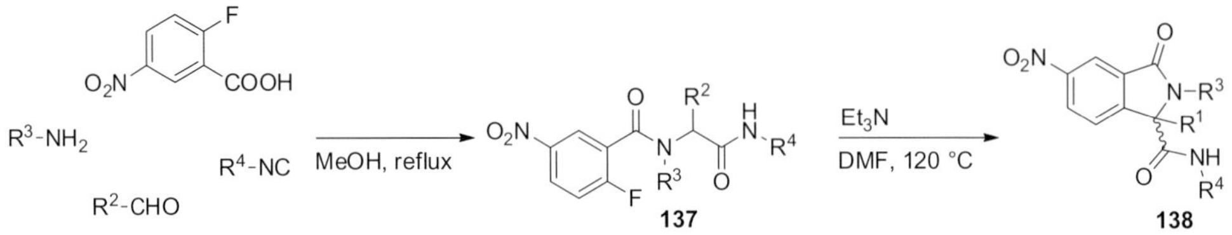

Scheme 10.28

details of a concise procedure [77] which afforded imidazoquinoxalinones and pyrazoloquinoxalinones **136** in an average 45% yield over two steps (Scheme 10.27).

During the same year, Ivachtchenko reported a protocol that enabled access to isoindolinones **138** via the use of 2-fluoro-5-nitrobenzoic acid as the source of the leaving group. Of note in this case was that a carbon nucleophile was generated and target compounds were prepared with no catalysts and in two straightforward synthetic steps (Scheme 10.28) [78].

10.3.4
Palladium-Mediated Reactions

Advances and improvements in the field of palladium chemistry have facilitated the association of several palladium-catalyzed transformations with the Ugi reaction, and this has in turn enabled organic chemists to target a plethora of medicinally relevant heterocycles. Heck and Sonogashira couplings, *N*-arylations, *C*-arylations, C–H functionalizations and Suzuki reactions have in turn opened up new and unexpected molecular diversities when applied to IMCR products. Whilst it is impossible to describe all of the variations and the single possibilities offered by the aforementioned processes, significant examples of recent applications can be provided in order to show typical reactivity patterns and new breakthroughs. Heck couplings, involving double bonds and halides or triflates, have been extensively utilized because of their versatility, and the wide availability of suitable starting materials. In 2008, Dai conceived a Heck-based three-step pathway to assemble biologically attractive arylideneindolinones **139** in an overall yield that ranged from 30% to 57% (Scheme 10.29) [79]. On the other hand, triple bonds have rarely been more employed to perform post-Ugi–Sonogashira couplings.

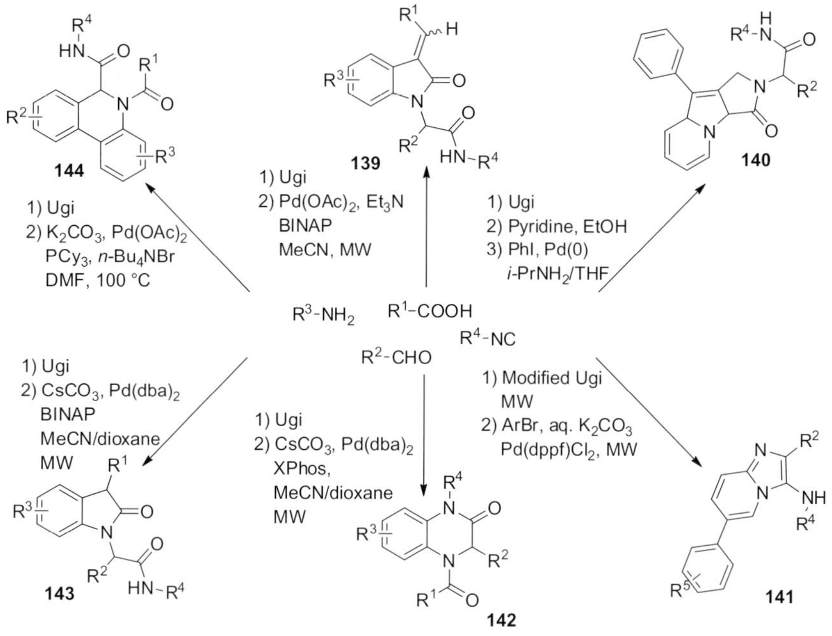

Scheme 10.29

Nevertheless, some creative strategies encompassing this transformation have been tailored to form particular templates, such as indolizines **140**, prepared by El Kaïm by modifying acyclic precursors through a combined Sonogashira-[3+2] cycloaddition protocol (Scheme 10.29) [80]. The target compounds contained a core that is commonly found in alkaloids and therapeutic anti-tumor agents and calcium-channel blockers. Although the use of boronates that are capable of undergoing Suzuki couplings has been only rarely reported, there are some interesting exceptions. The latter include a facile one-pot approach towards functionalized imidazopyridines **141**, which was developed by DiMauro in 2006 (Scheme 10.29) [81]. In this case, 2-aminopyridine-5-boronic acid pinacol ester was an input in a Bienayme condensation, and allowed a subsequent substitution in the 7-position of the ring. Much more commonly, arylation reactions play probably the most important role in this area. Indeed, they do not even require the presence of supplementary reactive groups, as the amidic nitrogen or the aldehyde-derived carbon that is found ubiquitously in any Ugi product serve as suitable "attack points." Of note, it is possible to shift from one reactivity pattern to another by simply changing the reaction conditions. This form of pluripotency has been demonstrated by Zhu [82] who, in 2009, synthesized benzopiperazinones **142** and indolinones **143** from the same set of acyclic precursors by means of a mere ligand shift (Scheme 10.29). Of note, Kalinski had already reported similar products by employing a related methodology in 2006, although in poor yields [83] and starting from two different Ugi adducts. Finally, another palladium-mediated option is the

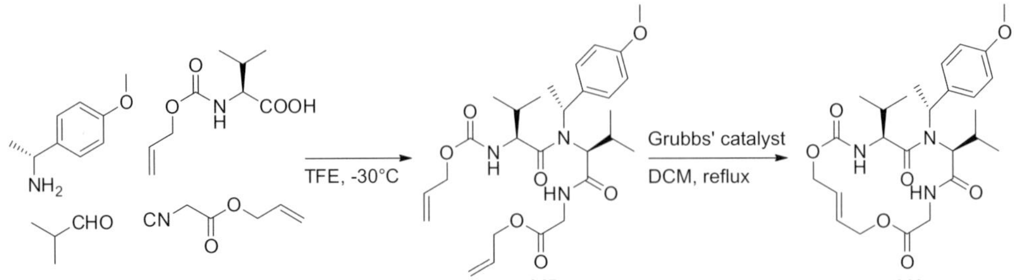

Scheme 10.30

functionalization of an aromatic C–H. In this respect, an outstanding array of diverse quinolines was constructed by Yang, with excellent overall yields that ranged from 78% to 88% (Scheme 10.29) [84]. Significantly, such structures are all potentially biologically relevant, as they display moieties that are present, *inter alia*, in apoptosis-promoting agents and estrogen receptor modulators.

10.3.5
Ring-Closing Metatheses

A straightforward entry into complex natural product-like structures relies on the setting up of a ring-closing metathesis (RCM) after IMCR condensation. This route has proved to be of particular efficacy in the design of challenging peptidomimetics, as it enables the closure of entropically disfavored and large-sized macrocycles. In 2005, Kazmaier carried out an Ugi reaction which involved two inputs bearing terminal double bonds, followed by treatment with a Grubbs' catalyst [85]. In spite of the considerable distance between the two reactive sites, the second transformation afforded cyclic peptoids **146** in a good yield (73%) (Scheme 10.30). Even more charming targets are made affordable via this pathway, provided that further subsequent manipulations are performed after the metathesis step.

For example, in 2007 Banfi disclosed a class of peptidomimetics that was capable of binding and inhibiting a class of integrins with Echistatin [86]. Although this secondary modification may appear unsuitable for library generation, and more fitted to "function-oriented" syntheses, the use of less-sophisticated building blocks in the Ugi step has been reported to prepare collections of compounds. Bridged bicyclic lactams **149** have, in fact, been assembled by a research group at Abbott in three steps, starting from commercial materials and with an olefin metathesis (**147** → **148**) being the key step in the process (Scheme 10.31) [87].

10.3.6
Staudinger–aza-Wittig Reactions

The Staudinger–aza-Wittig reaction has grown increasingly popular during the past few years, due primarily to its ability to construct small and medium-sized nitrogen-containing heterocycles. The sequential coupling of this transformation

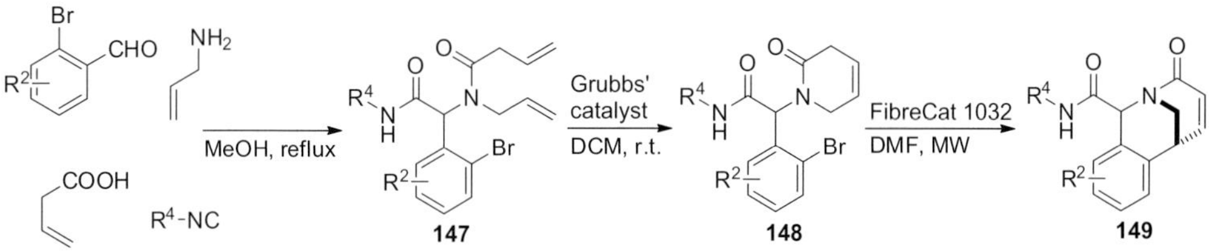

Scheme 10.31

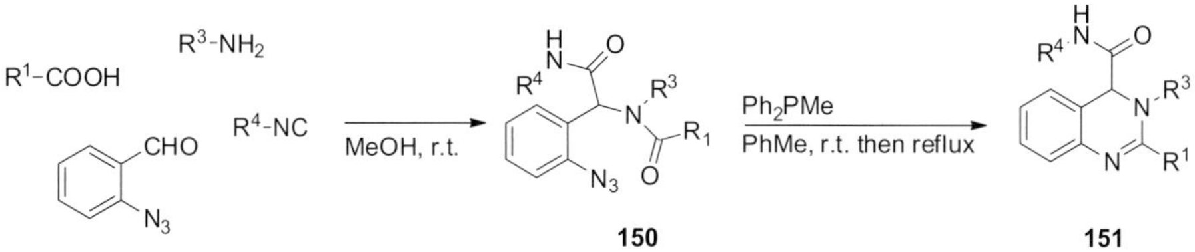

Scheme 10.32

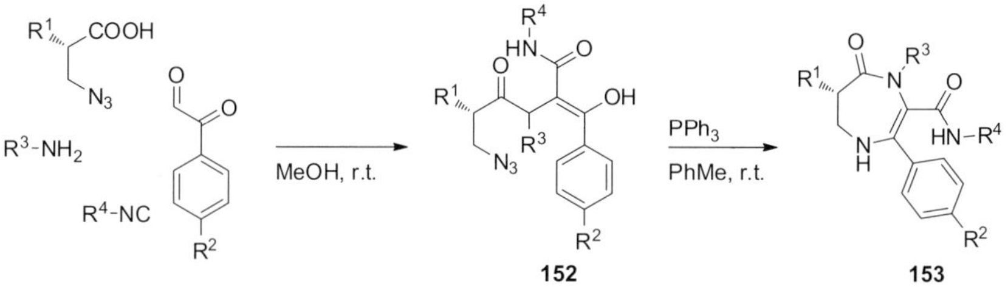

Scheme 10.33

with the Ugi condensation enables a facile access to several diverse and richly decorated chemotypes.

Very recently, Ding disclosed a preparation of trisubstituted dihydroquinazolines **151 by** employing 2-azido-benzaldehyde in an Ugi reaction, followed by an aza-Wittig reaction induced by diphenylmethylphosphine (Scheme 10.32) [88]. Of note, this reaction is one of the few that involve an amide carbonyl, which typically is known for its scarce reactivity towards azides.

In 2010, oxodiazepines **153** were also obtained by Torroba in an average 40% overall yield, via a similar procedure [89], where arylglyoxals provided the carbonyl to be attacked by azide in the second step (Scheme 10.33). Interestingly, the same group was able to access benzodiazepinones by making simple changes to the building blocks employed [89, 90].

10.3.7
Cycloadditions

The use of chemical transformations that result in the formation of cyclic adducts by establishing multiple new connectivities between the atoms upon simple heating, is self-evident. A combination of such methodologies with the high-diversity-generating Ugi reaction can give rise to a legion of densely decorated and unusual chemotypes. Consequently, cycloadditions have undergone intensive investigations to convert the linear Ugi backbone into a series of heterocyclic products. In particular, the Diels–Alder reaction and the [3+2] cycloaddition – which represent the two most popular members of this family – will be examined here.

Since the pioneering studies [91] of Paulvannan in 1999, the Ugi/Diels–Alder sequence has been utilized by numerous groups. In spite of the structural complexity that is achievable, one common feature in this field is the presence of the norbornene or heteronorbornene scaffold in the products. As acutely observed in a review by Basso [92], far from being a coincidence this testifies to a "marriage of convenience" between IMCRs and norbornene chemistry, which is aimed at overcoming the weak points of both. While a detailed overview on this topic has already been provided [92], it seems appropriate at this point to outline some relevant examples. In 2008, glutamate analogues **155** capable of inducing hypoactivity in mice were prepared by Oikawa and colleagues [93] via a pathway that involved a tandem Ugi/Diels–Alder reaction (Scheme 10.34). Quite unusually, the IMCR in this case was not intended to generate diversity, but rather to assemble a scaffold that could be modified afterwards and endowed with different appendages for subsequent functional group transformations.

However, a year later the same group described another protocol to create polycyclic compounds, in which the Ugi/Diels–Alder step contributed towards diversity through the isocyanides and carboxylic acids involved in the IMCR [94]. Indeed, the mere opening of the norbornene core may also afford interesting molecules, as exemplified in 2009 by Xu [95], who performed a three-step, one-pot MCR–cycloaddition–Lewis acid-promoted ring opening that affording isoindolinones in up to 69% yield.

Among other related transformations, worthy of mention at this point is the synthesis of fused azines via a sequential Ugi inverse electron demand Diels–Alder (IEDDA) reaction [96]. This was the first report of an azadiene-containing Ugi adduct that subsequently could perform an IEDDA.

Although the Huisgen reaction has been known for more than a century, it has recently enjoyed a renewed and unprecedented interest following a series of seminal reports by Sharpless on copper catalysis [97]. Consequently, post-Ugi modifications based on azide and alkyne moieties have become quite popular. In

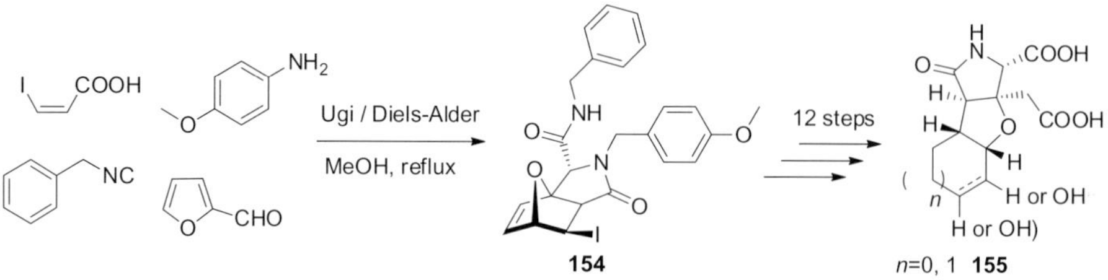

Scheme 10.34

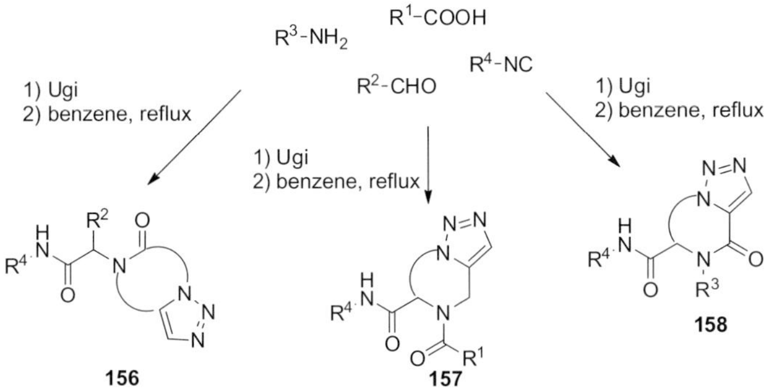

Scheme 10.35

2004, the research group at Abbott [98] developed a general methodology that allowed an efficient entry into different classes of triazolo-fused derivatives **156–158**, employing either commercially available or easily synthesized starting materials (Scheme 10.35).

The same group also reported use of nitrocarboxylic acids, followed by an intramolecular nitrile–oxide cyclization [99] that yielded fused isoxazole and isoxazolines in moderate yields. Remarkably, the cycloaddition-based approach has been also shown to link distant functionalities, as exemplified in the case of the 14- to 16-membered macrocycles assembled by Tron [100], in good conversion. Recently, Nenajdenko and coworkers produced two elegant reports on the preparation of hybrid peptides, by using an Ugi reaction in conjugation with click chemistry. After having introduced the synthesis and use of chiral azidoisocyanide, the Ugi product (carrying an azido group) was subsequently condensed with external alkynes in an intermolecular fashion, to afford triazole-decorated peptides [101, 102].

In short, access to IMCR-derived diversity space has been greatly accelerated during the past 10 years, with the continual discovery of chemo-types from: (i) new IMCRs; and (ii) the functional modification of IMCR products. Although the Ugi reaction contains inherent diversity generating power by assembling four points of diversity in a single reaction, from a drug discovery point of view it is desirable to identify ways in which the highly flexible Ugi product might be constrained. The rigidity of such products are often cited as preferable to more-flexible compounds, as their rigidity helps to diminish the entropic barriers associated with receptor–ligand interactions [103]. The use of bifunctional reagents in the Ugi reaction to achieve this goal is highlighted in the following sections..

10.4
The Bifunctional Approach (BIFA)

One approach that has been adopted by many to obtain constrained Ugi products is the use of a bifunctional reagents; this is a single reagent that contains two

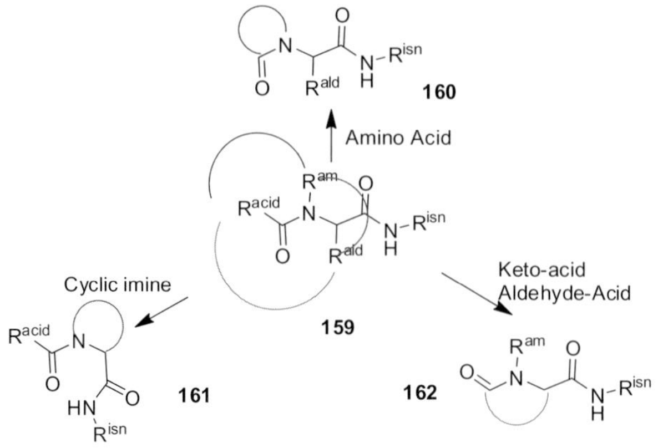

Scheme 10.36

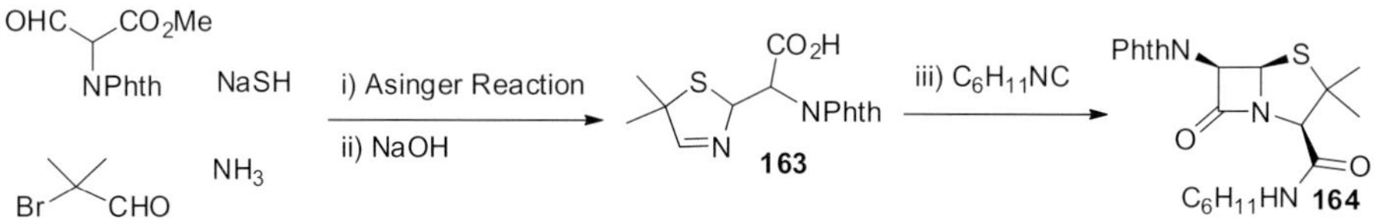

Scheme 10.37

functionalities from the four Ugi components that are tethered together, including amino acids, aldehyde acids, keto acids, cyclic imines, amino isocyanides, and isocyano acids. The possibility of rigidifying the Ugi core chemotype via this approach was first described figuratively by Dömling in 1998 [104], and has also been further described in reviews by Hulme [6, 105]. Here, attention will be focused on the three most widely applied bifunctional reagent classes: the amino acids; cyclic amines; and keto/aldehyde acids (Scheme 10.36).

The earliest reports of constrained Ugi adducts derived from multifunctional precursors appeared during the 1960s, with the preparation of penicillin derivatives (e.g., **164**) that involved sequential Asinger and Ugi four-component reactions (Scheme 10.37). This synthesis represents the shortest preparation of a known penicillin derivative [106], whereby the β-lactam ring is formed after isocyanide addition to the cyclic Schiff base **163**, followed by carboxylate nitrilium ion trapping and acyl transfer to yield the final penicillin core. In this example, a trifunctional precursor is utilized in which the amine, aldehyde, and carboxylic acid inputs may be considered tethered. Although several reports are available on the use of trifunctional reagents, attention here will be focused primarily on the more traditionally used bifunctional reagents.

10.4.1
Applications of Amino Acids

When considering the use of a bifunctional reagent to construct biologically relevant small molecules, it is most logical first to consider using naturally occurring alpha-amino acids in the Ugi reaction. α-Amino acids, which are the building blocks of proteins, contain the shortest possible single carbon linker between the amine and carboxylic acid functionalities. If alpha-amino acids are utilized in a typical Ugi reaction, then the reaction might be expected to yield the corresponding α-lactam **169**; however, as a result of the energy required to form the cyclic aziridinone structure, the Ugi intermediate **166** is intercepted by methanol to yield aminodiacetic adducts, **167**. Ugi designated this transformation the Ugi five-center, four-component reaction (U-5-4CR) [107], and has successfully demonstrated the use of the initial adduct to generate piperazine-2,6-diones **168** (Scheme 10.38) [108]. Interestingly, this condensation product has the same generic core as several commercial angiotensin-converting enzyme inhibitors [109], and further applications will undoubtedly be identified.

If the length of the tether is increased by one carbon and beta-amino acids are employed, however, then the Ugi reaction will proceed as expected via an intramolecular Mumm rearrangement, to yield β-lactams (Scheme 10.39). In fact, this

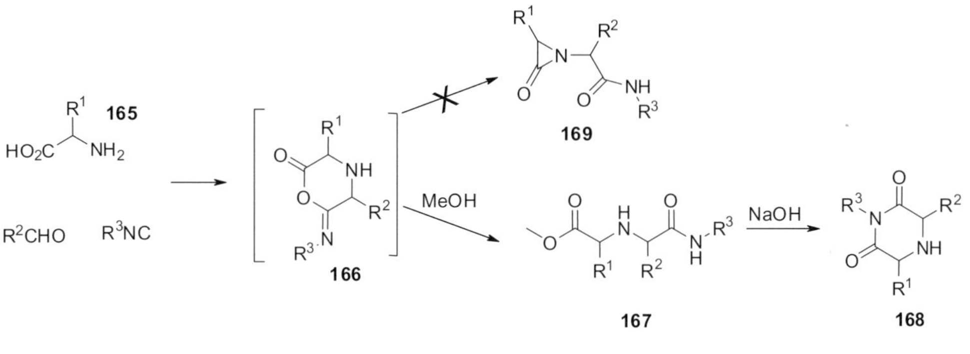

Scheme 10.38

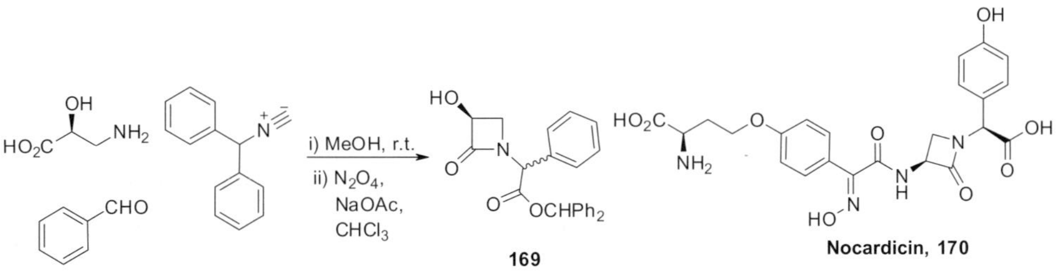

Scheme 10.39

methodology was successfully utilized by Hofheinz and coworkers in 1981 to produce several hundred β-lactam analogues of Nocardicin **170** [110]. Nocardicin A, as isolated in 1975, was the first reported monocyclic β-lactam to demonstrate potentially useful antibacterial activity [111]. In the Hofheinz methodology, the use of enantiopure β-amino acids ensures the correct configuration at the 3-position of the lactam **169** (Scheme 10.39).

At a similar time, Hatanaka and workers [112, 113] employed the same methodology to prepare carbapenem derivatives via the Ugi condensation of 3-aminoglutaric acid mono-*tert*-butyl ester **171**, formaldehyde, and 4-nitrobenzyl isocyanide (R_2-NC). This led to the production of β-lactam **173**, which was formed via a collapse of the seven-membered ring intermediate **172** during an intramolecular Mumm rearrangement. The initial β-lactam formation was followed by a conversion of the 4-nitrobenzyl amide to the 4-nitrobenzyl ester, with subsequent stereoselective Dieckmann condensation in the synthesis of 2-oxocarbapenem **175** (Scheme 10.40).

This methodology can also be utilized when aromatic and nonaromatic beta-amino acids are used. In fact, Feuleop and workers [114] utilized this strategy to facilitate the preparation of bicyclic β-lactams **178** in a high-throughput manner (Scheme 10.41). In addition, further chemotypes were accessible via an acid-promoted cleavage of the lactam ring, producing structures **179** and **181**, respectively (Scheme 10.41).

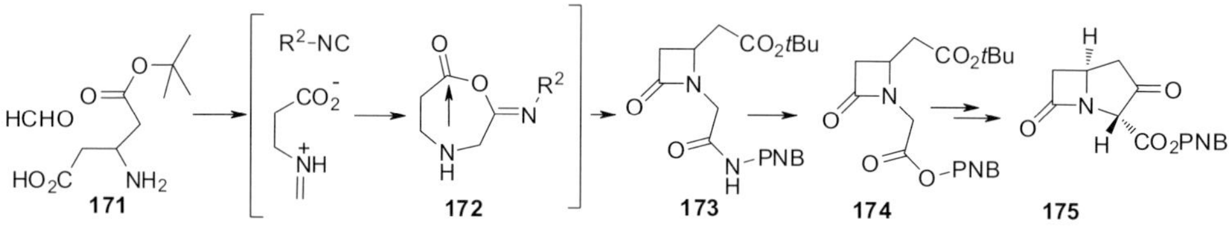

Scheme 10.40

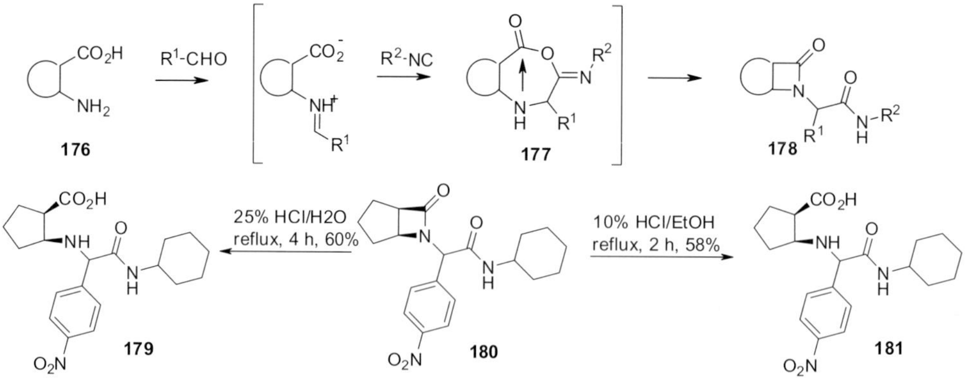

Scheme 10.41

The length of the amine to carboxylic acid tether has been extended by Kim and workers, via the use of a dipeptide [115]. This strategy enabled the elegant preparation of a series of 17 diketopiperazines with the generic formula **186** (Scheme 10.42). A select example (**187**) was produced in good yield, and with a moderate diastereoselectivity ratio; moreover, the yield was optimal when a less-nucleophilic solvent (trifluoroethanol) and microwave irradiation were used in the reaction. Improvements in the yield of kinetically slow Ugi condensations, using trifluoroethanol rather than methanol, have been reported and the mechanistic rational has been discussed [5].

10.4.2
Applications of Cyclic Imines

Cyclic imines are equivalent to tethering an aldehyde and amine input, and have been widely used in IMCRs. Recently, an application of the Ugi reaction with 2-substituted five-, six-, and seven-membered cyclic imines **188** was reported. These reactions have opened up new routes to both substituted proline and homo-proline derivatives **189**, with the method having been shown to be efficient for the one-step preparation of semi-natural dipeptides containing natural amino acid residues, and fragments of substituted proline or homo-proline (Scheme 10.43) [116, 117].

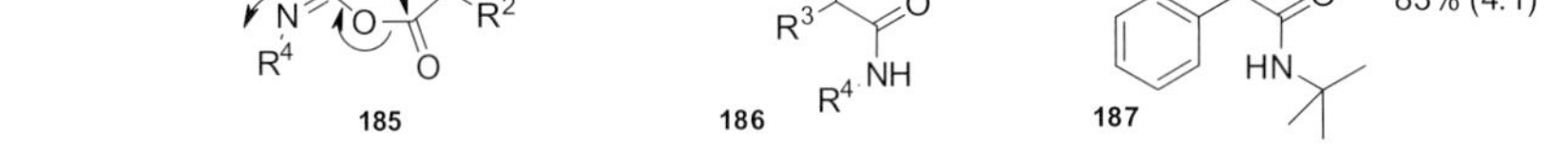

Scheme 10.42

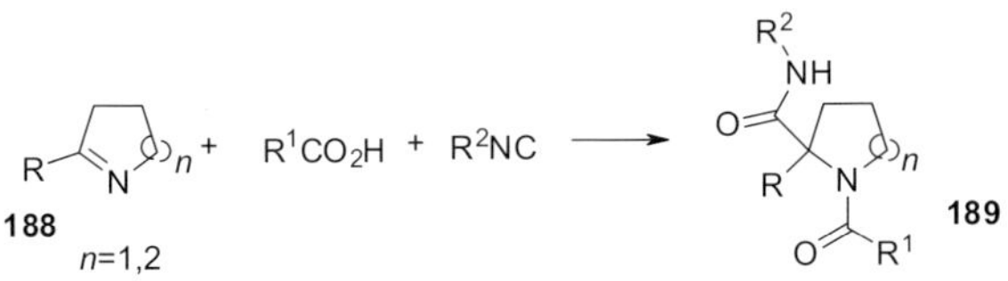

Scheme 10.43

10.4.3
Applications of Tethered Aldehyde and Keto Acids

The most commonly used tethered combination of inputs mines the variety of linker diversity available between the aldehyde and acid components **190** of the Ugi reaction (Scheme 10.44).

Such tethering enables access to lactams of various ring sizes, and was independently reported by three groups, between 1997 and 1998 [118, 119]. For example, Harriman envisioned that compounds containing the constrained peptide-like backbone could potentially be used as modulators of leukocyte trafficking [120]. More specifically, libraries were designed to incorporate integrin recognition and binding motifs for lead generation purposes (i.e., RGD for gpI-IbIIIa, fibrinogen, $\alpha_5\beta_1$, fibronection and LDV for $\alpha_4\beta_1$ and VCAM macromolecular targets of interest) [121]. In each case, the molecules were constructed to have an appropriately positioned t-butyl ester-protected carboxylic acid group that was tethered to the amine input, and simple post-Ugi condensation acid-catalyzed hydrolysis yielded the desired molecules. The syntheses of 2,2-disubstituted pyrrolidinones **195**, piperidinones **196**, azepinones **197**, and azocanones **198** were all reported, although the overall isolated yields were decreased with increasing ring size. Representative examples with yields **195** to **198** are shown in Figure 10.6.

Mjalli et al. described a solid-phase approach to this class of compounds via the use of a solid-supported isocyanide on a Wang resin [122]. Both, Mjalli and Ugi [118] independently extended the above-described methodology to include the formation of benzofused gamma-lactams, **200**. Ugi also employed a subsequent

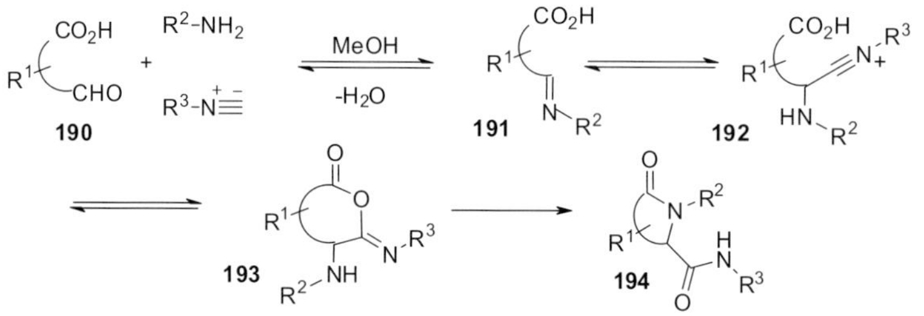

Scheme 10.44

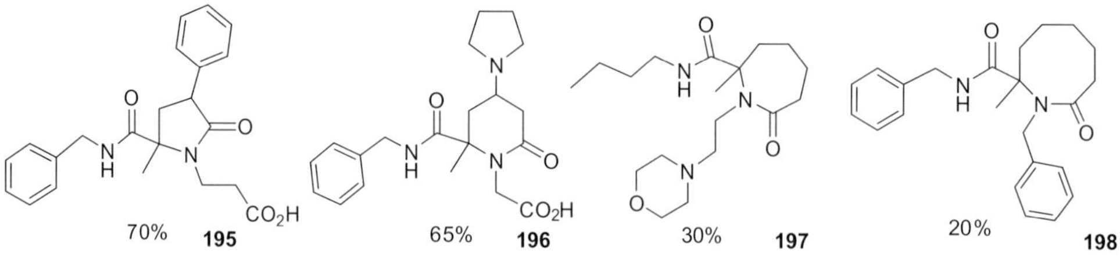

Figure 10.6 Lactams prepared via the UDC approach.

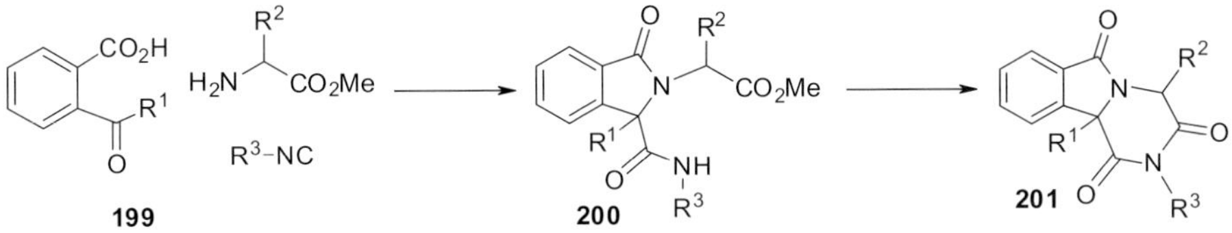

Scheme 10.45

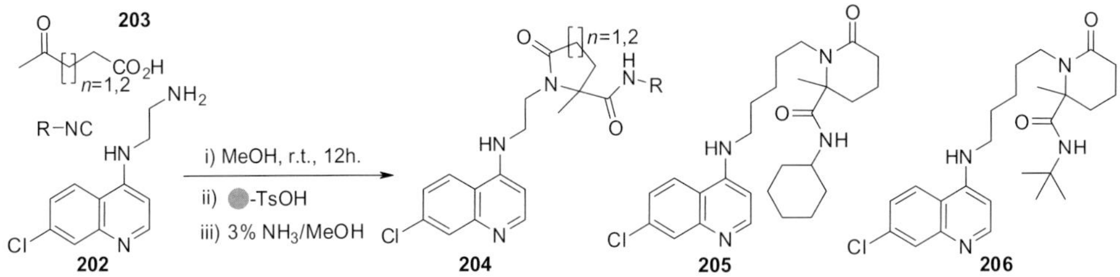

Scheme 10.46

post-condensation modification to produce 1,4-diazabicyclo[4.3.0]nonane-3,5,9-triones, **201** (Scheme 10.45).

Previously, the γ-lactams have demonstrated their properties as modulators of cholesterol absorption, as analgesics, and as bronchodilators. Interestingly, Chibale and coworkers [123, 124] described first the parallel synthesis of a new series of 4-aminoquinolines **204** (Scheme 10.46) via this methodology, and then developed a catch-and-release purification protocol that exploited the affinity of the basic quinoline nitrogen for solid-supported tosic acid. This protocol produced yields that ranged from 60% to 77%, with purities of up to 96%. Compound **205** was the most active agent against a chloroquine-resistant W2 strain of *Plasmodium falciparum* (IC_{50} = 0.096 μM), and also inhibited recombinant falcipain-2 *in vitro* (IC_{50} = 17.6 μM). Compound **206** inhibited the growth of *Trypanosoma brucei* (ED_{50} = 1.44 μM) and exhibited a favorable therapeutic index of 409 against a human KB cell line (Scheme 10.46).

The group of Banfi developed a synthetic route to α-acyloxyaminoamides through the use of keto acids and N-alkylated hydroxylamines in the Ugi reaction [125]. This intramolecular reaction utilized bifunctional keto acids, 4-acetylbutyric acid (**207**, n = 2) or levulinic acid (**207**, n = 1), N-benzylhydroxylamine **208**, and ethyl isocyanoacetate **209**. Subsequently, it was found that when acetylbutyric acid (**207**, n = 2) was employed, the desired oxazepinone **211** was isolated in moderate yield, while the major side product was identified as the corresponding acyclic α-hydroxylaminoamide **212**, into which one molecule of methanol had been incorporated. Alternatively, when levulinic acid (**63**, n = 1) was used, only methyl ester **211** was isolated in good yield (Scheme 10.47).

Apparently, due to the cyclic nature of intermediates **210**, the nucleophilic attack by methanol becomes competitive with an intramolecular migration of the acyl

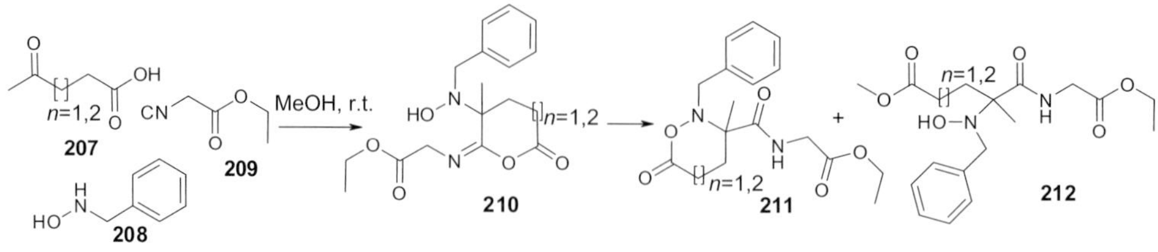

Scheme 10.47

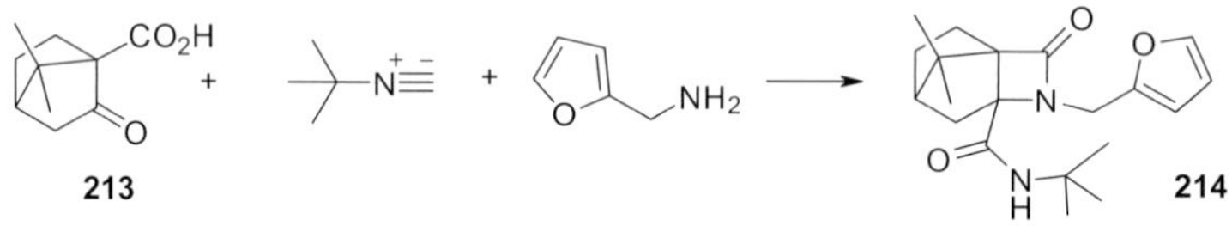

Scheme 10.48

group. More recently, Pirrung et al. described the preparation of β-lactams **214** from β-keto acids **213** by switching from an organic to an aqueous medium (Scheme 10.48) [126]. Subsequent studies of the reactions kinetics revealed a significant rate enhancement in water that enabled a high-throughput and high-yielding product preparation. As such, the enhanced reaction rate of organics in aqueous media has been widely demonstrated, and explained by several phenomena that include the hydrophobic effect, a greater degree of hydrogen bonding in reaction transition states, and also the high cohesive energy density of water. A similar acceleration of the reaction rate in water was observed by Mironov and coworkers [127] when levulinic acids were employed in the production of γ-lactams.

The chemistry research group at Repligen [128] further extended this lactam-forming methodology by utilizing the non-commercially available carboxylic acid bifunctional starting materials **215**, **217**, and **219** to access 1,8-naphtha-δ-lactams **216**, 1,4-benzoxazepin-3-one-5-carboxamides **218**, and 1,5-dibenzooxazocin-4-one-6-carboxamides **220** through the respective 3-CC couplings. The synthetic routes to these unique precursors were described in detail, and the authors also noted existing reports of the synthesis of 3-formylindole-2-carboxylic acid and a variety of other precursors worthy of evaluation for the expeditious preparation of a new "molecular diversity" [129, 130]. Whilst the Repligen group did not make available any data obtained from these studies, it was clearly evident that they had fully recognized the "exploratory power" of this methodology, three examples of which are shown in Scheme 10.49.

In the past, Marcaccini and coworkers have been especially productive in the intramolecular Ugi field, having developed routes to libraries of hexahydro-1,4-thiazepin-5-ones and 1,4-benzothiazepin-5-ones, respectively [131]. Both, 1,4-thiazepines and 1,4-benzothiazepines have been successfully applied as thera-

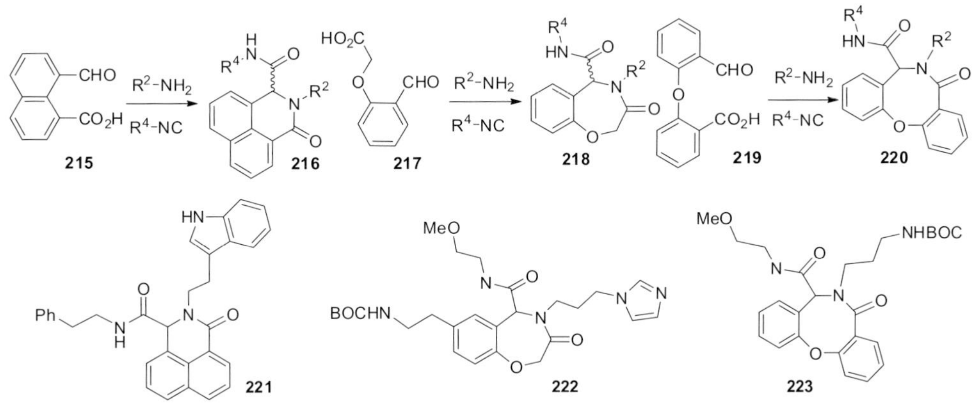

Scheme 10.49

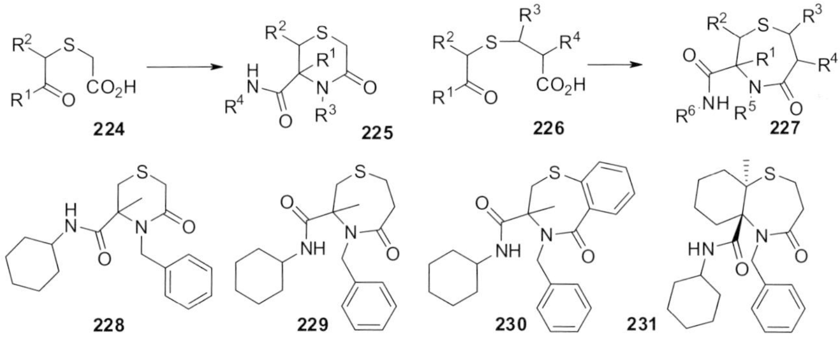

Scheme 10.50

peutics in pharmaceutical research, and the development of atom-economic routes to these chemotypes are of major commercial interest. The general strategy to both six- and seven-membered ring systems is exemplified in Scheme 10.50, where representative examples are shown that, in addition to monocyclic systems **228** and **229**, include both bicyclic and tricyclic chemotypes **230** to **231** (Scheme 10.50).

An interesting caveat to these studies was observed when ammonia was used as a supporting reagent. Single-crystal X-ray diffraction studies indicated a preference for the formation of a tetracyclic structure **233** in 40% yield, as opposed to the expected Ugi product. Subsequently, the yield of the conversion was improved to 91% in the absence of ammonia, but this was explained by an intramolecular Passerini reaction having been invoked to produce **234**. This was followed by amidic NH ring closure to produce the relatively surprising orthoamide functionality, **233** (Scheme 10.51) [132, 133].

Synthetic routes to several heterocyclic chemo-types containing pyrrolo[1,2-a][1,4]diazepine fragments have also been developed, driven by an interest in anti-tumor agents that are known to act against various types of leukemia cell lines,

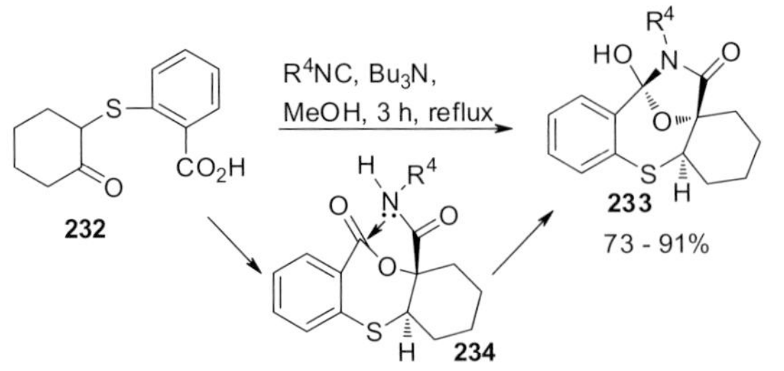

Scheme 10.51

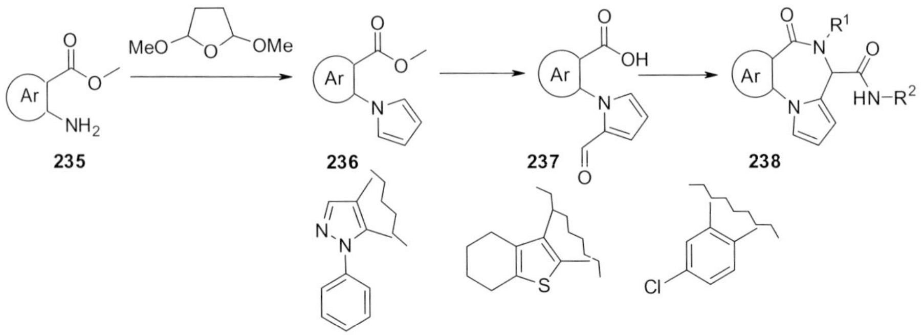

Scheme 10.52 Reagents and conditions: (i) AcOH, 4 h; (ii) HCONMe$_2$/POCl$_3$, 50–60 °C; (iii) 1% aq. NaOH, 40 °C, 8 h; (iv) R^{1}NH$_2$, R^{2}NC, 40 °C, 4–18 h, MeOH.

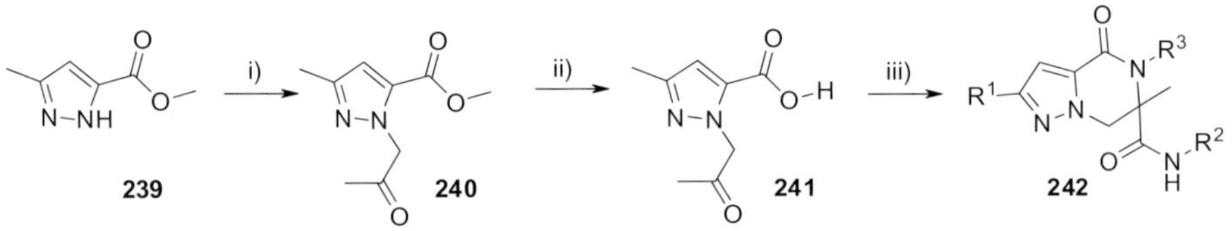

Scheme 10.53 Reagents and conditions: (i) MeCOCH$_2$Cl, K$_2$CO$_3$, 18-crown-6, 1,4-dioxane, reflux, 5 h; (ii) NaOH, H$_2$O, 70 °C 6 h; (iii) R^{2}NC, R^{3}NH$_2$, MeOH, 40 °C.

and also their known physiological profiles as central nervous system-active agents [134, 135]. Bifunctional precursors **237** were prepared (Scheme 10.52) and, after condensation, final tricyclic and tetracyclic products **238** isolated in good yield.

By using alternate attachment points for the bifunctional reactive centers to a central pyrazole ring **241**, the methodology was subsequently and elegantly extended to enable the preparation of arrays of 4-oxo-4,5,6,7-tetrahydropyrazolo [1,5-a] pyrazine-6-carboxamides **242** (Scheme 10.53) [136].

Indole, pyrrole, and imidazole analogues of the aforementioned indazole were also investigated, the aim being to increase the repertoire of available chemotypes from the BIFA methodology even further, producing both pyrrolo- **243**, indolo- **244**, and imidazo-ketopiperazines **245** (Figure 10.7) [137].

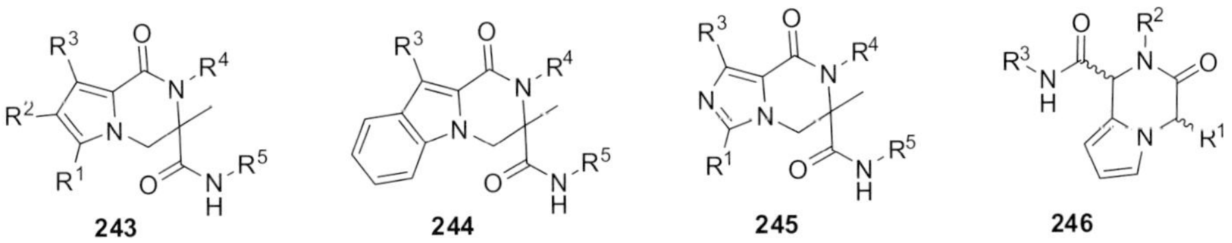

243 **244** **245** **246**

Figure 10.7 Ketopiperazines prepared via the BIFA reaction.

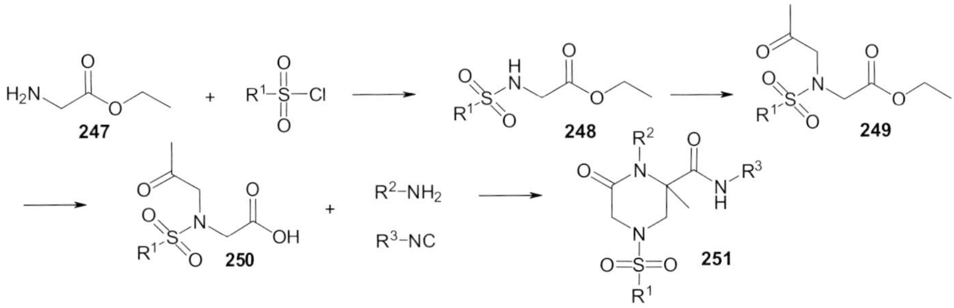

Scheme 10.54 Reagents and conditions: (i) triethylamine; (ii) chloroacetone, K$_2$CO$_3$, 18-crown-6; (iii) KOH, EtOH/H$_2$O; (iv) MeOH.

Recently, Nenajdenko and coworkers reported the application of a novel acid-aldehyde tether enabling the production of fused pyrroloketopiperazines **246** (Figure 10.7) [138]. In an interesting advance, Ivatchenko used the same approach for the one-step construction of peptidomimetic 5-carbomyl-4-sulfonyl-2-piperazinones **251** [139] that have attracted considerable attention as scaffolds in the design of small-molecule modulators for peptidergic receptors [140]. The required starting materials were obtained via a straightforward protocol that started with ethyl glycinate **247** and involved sequential sulfonylation, alkylation, and ester hydrolysis to produce **250**. Of note, the use of 18-crown-6 as a phase-transfer catalyst enabled the use of a relatively mild base (K$_2$CO$_3$), which greatly facilitated the reaction work-up (Scheme 10.54).

10.4.4
Heterocyclic Amidines as a Tethered Ugi Input

Finally, worthy of mention here is an extremely versatile reaction that has been exploited by several groups for the highly efficient synthesis of a variety of "drug-like" amino-3-imidazoles, **131** [8, 141–146]. The α-amino-heterocycle **252** involved in this reaction could be considered as the tethered input and, from a mechanistic standpoint, it is assumed that the isocyanide undergoes a [4+1] cycloaddition to produce the fused imidazo-heterocycle (Scheme 10.55). An extensive review of this chemistry was recently provided by Hulme (see Ref. [147]). The methodology

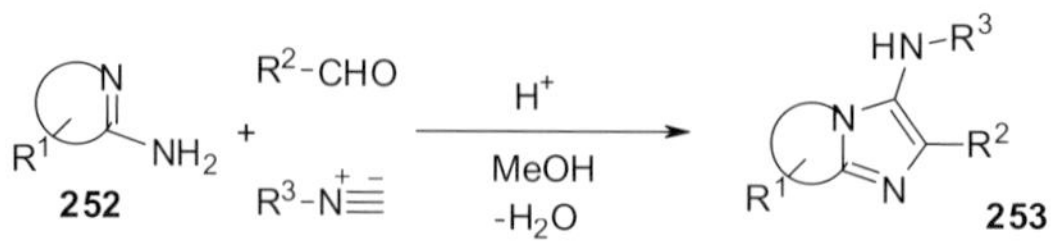

Scheme 10.55

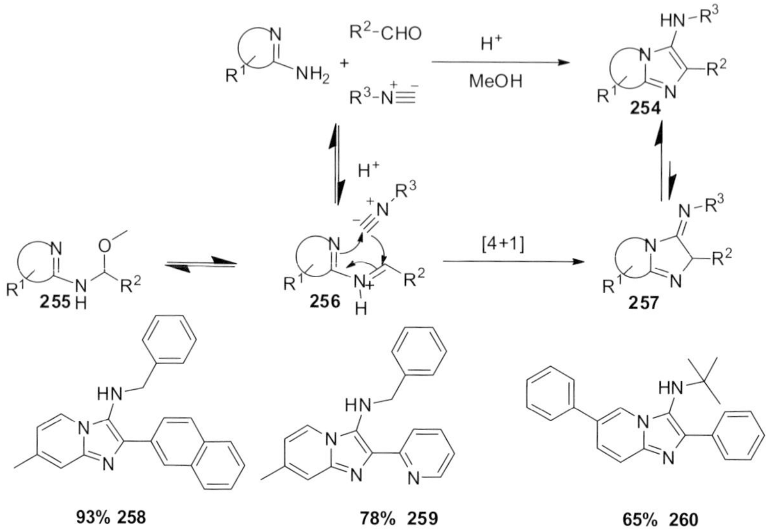

Scheme 10.56

exemplifies the use of intra-molecular "Ugi-like" IMCRs in which one reagent is bifunctional and condensation affords a new ring system, as opposed to a flexible, peptidic-like Ugi product. This methodology proceeds via the reaction of heterocyclic amidines **252**, aldehydes, and isocyanides in the presence of a variety of possible Lewis or Brønsted acids, producing the 3-amino-imidazoheterocycle **253** in one step. The reaction has a broad reactivity domain, and is compatible with a range of amino-heterocycles such as α-amino-thiazoles, α-amino-pyrazines, and α-amino-pyrimidines, to name but a few (Scheme 10.55).

Thus, simply mixing three reagents and a catalyst (usually, the preformation of a Schiff base is necessary) provides the fused imidazo system with the loss only of water, thereby demonstrating a high atom economy and approaching the "ideal synthesis" as originally postulated by Trost [148]. Mechanistically, the 3-CC reaction is believed to proceed via a [4+1] cycloaddition. Finally, the tautomerization of **257** affords the fused 3-amino imidazole **254** (Scheme 10.56).

Intriguingly, this was the first reported example of a [4+1] cycloaddition of a Schiff base, although such compounds have been shown to undergo [4+2] reactions with electron-rich olefins [149]. As noted above, several hetero-aromatic amidines proceed well in this IMCR, although for most electron-poor amidines the reactions are slow. This generally results in an accumulation of undesirable side products that frequently are derived from the addition of methanol to the initial Schiff base, yielding products with the generic structure **255**. Once again, however, the problem can be lessened if a non-nucleophilic compound, such as trifluoroethanol, is used. A small selection of chemotypes accessible via this reaction is shown in Scheme 10.56. The ease of synthesis of these chemo-types is exemplified by the fact that Amgen have reported the preparation of more than 30 000 bicyclic heterocyclic structures via a solution-phase protocol, having performed the chemistry at ambient temperatures in 96-well plates [43]. Interestingly, multiple compound vendors have used this methodology to produce materials for commercial gain, many of which have found their way into corporate screening collections.

10.4.5
Combined Bifunctional and Post-Condensation Modifications

An interesting development was reported by Hulme and coworkers, who combined the BFA with a subsequent post-condensation modification, in which a convertible isocyanide **262** and masked "internal" amine nucleophile **261** were employed [25]. On deprotection, a second ring system was installed, whereupon the methodology represented the first BFA equivalent of the well-documented UDC methodology (Scheme 10.57).

Westermann and coworkers [150, 151] applied the same tethering approach to the preparation of peptide-like bicyclic lactams via a sequential Ugi reaction–olefin metathesis (Scheme 10.58). In this concise synthesis of a fused 6,7-system, the tethered precursor was made available by the ring-opening reaction of pyroglutamic acid derivative **266**, with allylmagnesium bromide, followed by ester hydrolysis to **267**. The lactam formation proceeded well in methanol, while RCM using the Grubb's catalyst gave the bicyclic product **269**, in 81% yield.

In summary, following a hiatus of almost 40 years, the past 10 years have witnessed a resurgence of interest in isocyanide chemistry and, in particular, in the development of new IMCRs and applications. The Ugi reaction and its slight variants fall at the forefront of this field, with its inherent exploratory powers deriving from the plethora of opportunities available to trap out the so-called nitrilium ion

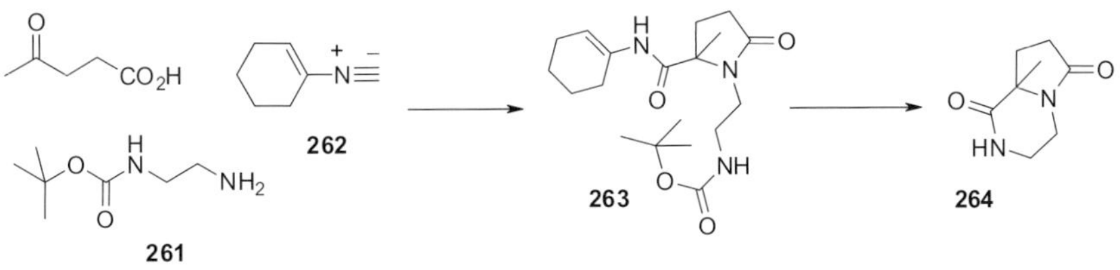

Scheme 10.57

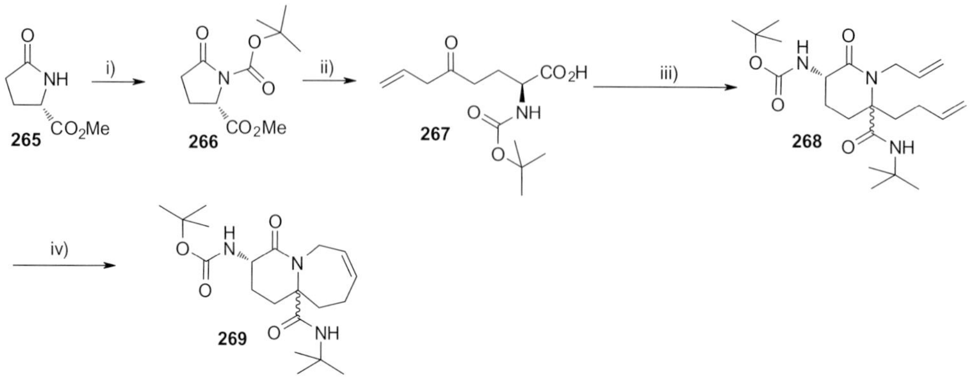

Scheme 10.58 Reagents and conditions: (i) BOC$_2$O, DMAP, 75% yield; (ii) (a) Allylmagnesium bromide; (b) NaOH, 78% yield; (iii) 78% MeOH, ^{t}BuNC, Allylamine, 57% yield; (iv) (Cy$_3$P)$_2$Cl$_2$Ru=CHPh, 5 mol.%, 81% yield.

that is involved in the rather non-obvious mechanism that affords the Ugi condensation product. In this chapter, three main areas have been identified in which this pivotal IMCR is used to access new molecular diversity: (i) from UDC protocols, which often are plate-compatible and so operationally friendly that high-school students may easily be trained to produce libraries of a variety of sizes; (ii) from the most common methodologies available to elaborate the Ugi product from the arsenal of synthetic procedures available to the skilled organic chemist; and (iii) from the tethering strategies of Ugi reagents that afford a variety of conformationally constrained chemotypes. Ultimately, these studies should enable the production of compound libraries with desirable metrics to serve as screening sets in "hit-to-lead" campaigns. The IMCRs often deliver the added dimension of compounds with a built-in "iterative efficiency potential" (IEP), where the speed around the iterative hypothesis–synthesis–screening loop is faster and the total number of required iterations for value creation is reduced. Indeed, IMCRs and compounds with a high IEP are being increasingly recognized as a significant "value addition" as they progress along the drug discovery value chain. This is a key driver for the continued study of enabling methodologies in both academic and industrial research laboratories, and is particularly pertinent to lead-generation exercises. The field of IMCRs in general has an immense "exploratory power," and a plethora of opportunities remain to be discovered and developed to further increase the chemotype "tool box" that is available to those involved in corporate (or academic) file-enhancement strategies. The utilization of IMCR methodologies, to which the isocyanide moiety is central, represents a cost-effective starting point to appropriately balance file-enhancement paradigms from both tactical and strategic perspectives. Hopefully, this chapter will inspire those working in the field of isocyanide-based research to continue making major advances in this area.

Acknowledgments

The authors would like to thank the Office of the Director, NIH, and the National Institute of Mental Health (1RC2MH090878-01) and Nicole Schechter, PSM, for extensive copy-editing.

Abbreviations

3-CC	three-component condensation
4-CC	four-component condensation
ACD	available chemicals directory
ACE	angiotensin-converting enzyme
AMPA	2-amino-3-(5-methyl-3-oxo-1,2- oxazol-4-yl)propanoic acid
BIFA	bifunctional approach
BZD	benzodiazepine
BZDs	1,4-benzodiazepine-2,5-diones
CNS	central nervous system
DIEA	diisopropylethylamine
DMAP	4-dimethylaminopyridine
DMF	dimethylformamide
ED_{50}	effective dose
ELSD	evaporative light-scattering detector
GABA	γ-aminobutyric acid
HATU	2-(7-Aza-1H-benzotriazole-1-yl)-1,1,3,3-tetramethyluronium hexafluorophosphate
HIV	human immunodeficiency virus
IC_{50}	half-maximal inhibition concentration
IEDDA	inverse electron demand Diels–Alder
IEP	iterative efficiency potential
IMCR	isocyanide-based multicomponent reaction
LC/MS	liquid chromatography/mass spectrometry
MCR	multicomponent reaction
MMP	matrix metalloproteinase
PS	polymer-supported
RCM	ring-closing metathesis
SAR	structure–activity relationship
SPE	solid-phase extraction
TBD	7-methyl-1,5,7-triazabicyclo[4.4.0]dec-5-ene
TFA	trifluoroacetic acid
TFE	trifluoroethanol
TMOF	trimethylorthoformate
TMSCN	trimethylsilylcyanide
$TMSN_3$	trimethylsilylazide
UDC	Ugi/deprotect/cyclize
UV	ultraviolet

References

1 Passerini, M. (1921) Isonitriles. I. Compound of *p*-isonitrileazobenzene with acetone and acetic acid. *Gazz. Chim. Ital.*, **51**, 126–129.

2 Ugi, I. and Rosendahl, K. (1963) Isonitriles. XV. Δ1-Cyclohexenylisocyanide. *Justus Liebigs Ann. Chem.*, **666**, 65–67.

3 Viswanadhan, V.N., Balan, C., Hulme, C., Cheetham, J.C., and Sun, Y. (2002) Knowledge-based approaches in the design and selection of compound libraries for drug discovery. *Curr. Opin. Drug. Discov. Dev.*, **5**, 400–406.

4 Ugi, I. (1962) The alpha-addition of immonium ions and anions to isonitriles accompanied by secondary reactions. *Angew. Chem. Int. Ed. Engl.*, **1**, 8–21.

5 Domling, A. (2006) Recent developments in isocyanide based multicomponent reactions in applied chemistry. *Chem. Rev.*, **106**, 17–89.

6 Hulme, C. and Gore, V. (2003) Multi-component reactions: emerging chemistry in drug discovery: from Xylocain to Crixivan. *Curr. Med. Chem.*, **10**, 51–80.

7 Hulme, C. and Nixey, T. (2003) Rapid assembly of molecular diversity via exploitation of isocyanide-based multi-component reactions. *Curr. Opin. Drug Discov. Dev.*, **6**, 921–929.

8 Bienayme, H., Hulme, C., Oddon, G., and Schmitt, P. (2000) Maximizing synthetic efficiency: multi-component transformations lead the way. *Eur. J. Chem.*, **6**, 3321–3329.

9 Keating, T.A. and Armstrong, R.W. (1996) A remarkable two-step synthesis of diverse 1,4-benzodiazepine-2,5-diones using the Ugi four-component condensation. *J. Org. Chem.*, **61**, 8935–8939.

10 Cho, N.S., Song, K.Y., and Parkanyi, C. (1989) Ring closure reactions of methyl *N*-(haloacetyl) anthranilates with ammonia. *J. Heterocycl. Chem.*, **26**, 1807–1810.

11 Bauer, A., Weber, K.H., Danneberg, P., and Kuhn, F.J. (1975) U.S. Patent 3 914 216.

12 Martino, G.D., Massa, S., Corelli, F., Pantaleoni, G., Fanini, D., and Palumbo, G. (1983) C.N.S. agents: neuropsychopharmacological effects of 5H-pyrrolo[2,1-c][1,4]benzodiazepine derivatives. *Eur. J. Med. Chem. Chim. Ther.*, **18**, 347–350.

13 Ananthan, S., Clayton, S.D., Ealick, S.E., Wong, G., Evoniuk, G.E., and Skolnick, P. (1993) Synthesis and structure-activity relationships of 3,5-disubstituted 4,5-dihydro-6H-imidazo[1,5-a][1,4]benzodiazepin-6-ones at diazepam-sensitive and diazepam-insensitive benzodiazepine receptors. *J. Med. Chem.*, **36**, 479–490.

14 Wong, G., Koehler, K.F., Skolnick, P., Gu, Z.Q., Ananthan, S., Schonholzer, P., Hunkeler, W., Zhang, W., and Cook, J.M. (1993) Synthetic and computer-assisted analysis of the structural requirements for selective, high-affinity ligand binding to diazepam-insensitive benzodiazepine receptors. *J. Med. Chem.*, **36**, 1820–1830.

15 Wright, W.B. Jr, Brabander, H.J., Greenblatt, E.N., Day, I.P., and Hardy, R.A. Jr (1978) Derivatives of 1,2,3,11a-tetrahydro-5H-pyrrolo[2,1-c][1,4]benzodiazepine-5,11(10H)-dione as anxiolytic agents. *J. Med. Chem.*, **1978** (21), 1087–1089.

16 Jones, G.B., Davey, C.L., Jenkins, T.C., Kamal, A., Kneale, G.G., Neidle, S., Webster, G.D., and Thurston, D.E. (1990) The noncovalent interaction of pyrrolo[2,1-c][1,4]benzodiazepine-5,11-diones with DNA. *Anti-Cancer Drug Des.*, **5**, 249–264.

17 McDowell, R.S., Blackburn, B.K., Gadek, T.R., McGee, L.R., Rawson, T., Reynolds, M.E., Robarge, K.D., Somers, T.C., Thorsett, E.D., Tischler, M., Webb, R.R., and Venuti, M.C. (1994) From peptide to non-peptide. 2. The *de novo* design of potent, non-peptidal inhibitors of platelet aggregation based on a benzodiazepinedione scaffold.

J. Am. Chem. Soc., **116**, 5077–5083.

18 Keating, T.A. and Armstrong, R.W. (1996) Post-condensation modifications of Ugi four-component condensation products: 1-Isocyanocyclohexene as a convertible isocyanide. Mechanism of conversion, synthesis of diverse structures and demonstration of resin capture. *J. Am. Chem. Soc.*, **118**, 2574–2583.

19 Keating, T.A. and Armstrong, R.W. (1995) Molecular diversity via a convertible isocyanide in the Ugi four-component condensation. *J. Am. Chem. Soc.*, **117**, 7842–7843.

20 Tempest, P., Pettus, L., Gore, V., and Hulme, C. (2003) MCC/S$_N$Ar methodology. Part 2: novel three-step solution phase access to libraries of benzodiazepines. *Tetrahedron Lett.*, **44**, 1947–1950.

21 Hulme, C., Peng, J., Tang, S.Y., Burns, C.J., Morize, I., and Labaudiniere, R. (1998) Improved procedure for the solution phase preparation of 1,4-benzodiazepine-2,5-dione libraries via Armstrong's convertible isonitrile and the Ugi reaction. *J. Org. Chem*, **63**, 8021–8023.

22 Hulme, C., Peng, J., Morton, G., Salvino, J.M., Herpin, T., and Labaudiniere, R. (1998) Novel safety-catch linker and its application with a Ugi/De-Boc/Cyclization (UDC) strategy to access carboxylic acids, 1,4-benzodiazepines, diketopiperazines, ketopiperazines and dihydroquinoxalinones. *Tetrahedron Lett.*, **39**, 7227–7230.

23 Tempest, P., Ma, V., Thomas, S., Hua, Z., Michael, G.K., and Hulme, C. (2001) Two-step solution-phase synthesis of novel benzimidazoles utilizing a UDC (Ugi/de-Boc/cyclize) strategy. *Tetrahedron Lett*, **42**, 4959–4962.

24 Dietrich, J., Kaiser, C., Meurice, N., and Hulme, C. (2010) Concise two-step solution phase syntheses of four novel dihydroquinazoline scaffolds. *Tetrahedron Lett.*, **51**, 3951–3955.

25 Hulme, C., Ma, L., Cherrier, M.P., Romano, J.J., Morton, G., Duquenne, C., Salvino, J., and Labaudiniere, R. (2000) Novel applications of convertible isonitriles for the synthesis of mono and bicyclic γ-lactams via a UDC strategy. *Tetrahedron Lett.*, **41**, 1883–1887.

26 Hulme, C., Morrissette, M., Volz, F., and Burns, C. (1998) The solution phase synthesis of diketopiperazine libraries via the Ugi reaction: novel application of Armstrong's convertible isonitrile. *Tetrahedron Lett.*, **39**, 1113–1116.

27 Nixey, T., Tempest, P., and Hulme, C. (2002) Two-step solution-phase synthesis of novel quinoxalinones utilizing a UDC (Ugi/de-Boc/cyclize) strategy. *Tetrahedron Lett.*, **43**, 1637–1639.

28 Hulme, C., Peng, J., Louridas, B., Menard, P., Krolikowski, P., and Kumar, N.V. (1998) Applications of *N*-BOC-diamines for the solution phase synthesis of ketopiperazine libraries utilizing a Ugi/De-BOC/Cyclization (UDC) strategy. *Tetrahedron Lett.*, **39**, 8047–8050.

29 Miller, J.F., Koch, K., and Piscopio, A.D. (1997) Application of the ester enolate Claisen/ring-closing metathesis manifold to natural product synthesis. 214th ACS National Meeting, Las Vegas, NV, ORGN-232.

30 Lindhorst, T., Bock, H., and Ugi, I. (1999) A new class of convertible isocyanides in the Ugi four-component reaction. *Tetrahedron*, **55**, 7411–7420.

31 Kennedy, A.L., Fryer, A.M., and Josey, J.A. (2002) A new resin-bound universal isonitrile for the Ugi 4CC reaction: preparation and applications to the synthesis of 2,5-diketopiperazines and 1,4-benzodiazepine-2,5-diones. *Org. Lett.*, **4**, 1167–1170.

32 Chen, J.J., Golebiowski, A., McClenaghan, J., Klopfenstein, S.R., and West, L. (2001) Universal Rink–isonitrile resin: application for the traceless synthesis of 3-acylamino imidazo[1,2-*a*] pyridines. *Tetrahedron Lett.*, **42**, 2269–2271.

33 Flynn, D.L., Zelle, R.E., and Grieco, P.A. (1983) A mild two-step method for the hydrolysis of lactams and secondary amides. *J. Org. Chem.*, **48**, 2424–2426.

34 Plunkett, M.J. and Ellman, J.A. (1995) A silicon-based linker for traceless

solid-phase synthesis. *J. Org. Chem.*, **1995** (60), 6006–6007.
35 Hulme, C. and Cherrier, M.P. (1999) Novel applications of ethyl glyoxalate with the Ugi MCR. *Tetrahedron Lett.*, **40**, 5295–5299.
36 Kelder, J., Grootenhuis, P.D.J., Bayada, D.M., Delbressine, L.P.C., and Ploemen, J.P. (1999) Polar molecular surface as a dominating determinant for oral absorption and brain penetration of drugs. *Pharm. Res.*, **16**, 1514–1519.
37 Hulme, C., Ma, L., and Labaudiniere, R. (2000) Novel applications of resin bound α-amino acids for the synthesis of benzodiazepines (via Wang resin) and ketopiperazines (via hydroxymethyl resin). *Tetrahedron Lett.*, **41**, 1509–1514.
38 Szardenings, A.K., Antonenko, V., Campbell, D.A., DeFrancisco, N., Ida, S., Shi, L., Sharkov, N., Tien, D., Wang, Y., and Navre, M. (1999) Identification of highly selective inhibitors of collagenase-1 from combinatorial libraries of diketopiperazines. *J. Med. Chem.*, **42**, 1348–1357.
39 Hulme, C., Ma, L., Romano, J., and Morrissette, M. (1999) Remarkable three-step-one-pot solution phase preparation of novel imidazolines utilizing a UDC (Ugi/de-Boc-Cyclize) strategy. *Tetrahedron Lett.*, **40**, 7925–7928.
40 Zhang, W. and Tempest, P. (2004) Highly efficient microwave-assisted fluorous Ugi and post-condensation reactions for benzimidazoles and quinoxalinones. *Tetrahedron Lett.*, **45**, 6757–6760.
41 Tempest, P., Ma, V., Kelly, M.G., Jones, W.J., and Hulme, C. (2001) MCC/S$_N$Ar methodology. Part 1: novel access to a range of heterocyclic cores. *Tetrahedron Lett.*, **42**, 4963–4968.
42 Hulme, C., Bienayme, H., Nixey, T., Chenera, B., Jones, W., Tempest, P., and Smith, A. (2003) Library generation via postcondensation modifications of isocyanide-based multicomponent reactions. *Methods Enzymol. (Combinatorial Chem., Part B)*, **369**, 469–496.
43 Nixey, T., Kelly, M., Semin, D., and Hulme, C. (2002) Short solution phase preparation of fused azepine-tetrazoles via a UDC (Ugi/de-Boc/cyclize) strategy. *Tetrahedron Lett.*, **43**, 3681–3684.
44 Nixey, T., Kelly, M., and Hulme, C. (2000) The one-pot solution-phase preparation of fused tetrazole-ketopiperazines. *Tetrahedron Lett.*, **41**, 8729–8733.
45 Marcaccini, S., Miliciani, M., and Pepino, R. (2005) A facile synthesis of 1,4-benzodiazepine derivatives via Ugi four-component condensation. *Tetrahedron Lett.*, **46**, 711–713.
46 Hulme, C., Chappeta, S., Griffith, C., Lee, Y.S., and Dietrich, J. (2009) An efficient solution phase synthesis of triazadibenzoazulenones: "designer isonitrile free" methodology enabled by microwaves. *Tetrahedron Lett.*, **50**, 1939–1942.
47 Desaubry, L., Wermuth, C.G., and Bourguignon, J.-J. (1995) Synthesis of a conformationally constrained analogue of BW A78U, an anticonvulsant adenine derivative. *Tetrahedron Lett.*, **36**, 4249–4252.
48 Hulme, C., Chappeta, S., and Dietrich, J. (2009) A simple, cheap alternative to "designer convertible isonitriles" expedited with microwaves. *Tetrahedron Lett.*, **50**, 4054–4057.
49 Xu, Z., Dietrich, J., Shaw, A.Y., and Hulme, C. (2010) Two-step syntheses of fused quinoxaline-benzodiazepines and bis-benzodiazepines. *Tetrahedron Lett.*, **51**, 4566–4569.
50 Hulme, C., Ma, L., Romano, J.J., Morton, G., Tang, S.Y., Cherrier, M.P., Choi, S., Salvino, J., and Labaudiniere, R. (2000) Novel applications of carbon dioxide/MeOH for the synthesis of hydantoins and cyclic ureas via the Ugi reaction. *Tetrahedron Lett.*, **41**, 1889–1893.
51 Hanessian, S. and Yang, R.Y. (1996) Solution- and solid-phase synthesis of 5-alkoxyhydantoin libraries with a three-fold functional diversity. *Tetrahedron Lett.*, **37**, 5835–5838.
52 Short, K.M., Ching, B.W., and Mjalli, A.M.M. (1996) The synthesis of hydantoin 4-imides on solid support. *Tetrahedron Lett.*, **37**, 7489–7492.

53 Gunawan, S., Nichol, G.S., Chappeta, S., Dietrich, J., and Hulme, C. (2010) Concise preparation of novel tricyclic chemotypes: fused hydantoin–benzodiazepines. *Tetrahedron Lett.*, **51**, 4689–4692.

54 Kerwin, R. (2006) Discontinued drugs in 2005: schizophrenia drugs. *Expert Opin. Invest. Drugs*, **15**, 1487–1495.

55 Ishii, A., Yamakawa, M., and Toyomaki, Y. (1991) U.S. Patent 4 985 453.

56 Habashita, H., Kokubo, M., Hamano, S.-I., Hamanaka, N., Toda, M., Shibayama, S., Tada, H., Sagawa, K., Fukushima, D., Maeda, K., and Mitsuya, H. (2006) Design, synthesis, and biological evaluation of the combinatorial library with a new spirodiketopiperazine scaffold. Discovery of novel potent and selective low-molecular-weight CCR5 antagonists. *J. Med. Chem.*, **49**, 4140–4152.

57 Ayaz, M., Dietrich, J., and Hulme, C. (2011) A novel route to synthesize libraries of quinoxalines via Petasis methodology in two synthetic operations. *Tetrahedron Lett.*, **52**, 4821–4823.

58 Sunderhous, J.D. and Martin, S.F. (2009) Application of multicomponent reactions to the synthesis of diverse heterocyclic scaffolds. *Chem. Eur. J.*, **15**, 1300–1308.

59 Banfi, L., Riva, R., and Basso, A. (2010) Coupling isocyanide-based multicomponent reactions with aliphatic or acylic nucleophilic substitution processes. *Synlett*, 23–41.

60 Xing, X.L., Wu, J.L., Feng, G.F., and Dai, W.M. (2006) Microwave-assisted one-pot U-4CR and intramolecular O-alkylation toward heterocyclic scaffolds. *Tetrahedron*, **62**, 6774–6781.

61 de Greef, M., Abeln, S., Belksami, K., Dömling, A., Orru, R.V.A., and Wessjohann, L.A. (2006) Rapid combinatorial access to macrocyclic ansapeptoids and ansapeptides with natural-product-like core structure. *Synthesis*, 3997–4004.

62 Dömling, A. and Huang, Y. (2010) Piperazine scaffolds via isocyanide-based multi component reactions. *Synthesis*, 2859–2883.

63 Golebiowski, A., Jozwik, J., Klopfenstein, S.R., Colson, A.O., Grieb, A.L., Russell, A.F., Rastogi, V.L., Diven, C.F., Portlock, D.E., and Chen, J.J. (2002) Solid support synthesis of putative peptide β-turn mimetics via Ugi reaction for diketopiperazine formation. *J. Comb. Chem.*, **4**, 584–559.

64 Marcaccini, S., Pepino, R., and Pozo, M.C. (2001) A facile synthesis of 2,5 diketopiperazines based on isocyanide chemistry. *Tetrahedron Lett.*, **42**, 2727–2728.

65 Rossen, K., Sager, J., and Di Michele, L.M. (1997) An efficient and versatile synthesis of piperazine-2-carboxamides. *Tetrahedron Lett.*, **38**, 3183–3186.

66 Rossen, K., Pye, P.J., Di Michele, L.M., Volante, R.P., and Reider, P.J. (1998) An efficient asymmetric hydrogenation approach to the synthesis of the Crixivan® piperazine intermediate. *Tetrahedron Lett.*, **39**, 6823–6826.

67 Banfi, L., Basso, A., Guanti, G., Lecinska, P., and Riva, R. (2006) Multicomponent synthesis of dihydrobenzoxazepinones by coupling Ugi and Mitsunobu reactions. *Org. Biomol. Chem.*, **4**, 4236–4240.

68 Banfi, L., Basso, A., Guanti, G., Kielland, N., Repetto, C., and Riva, R. (2007) Ugi multicomponent reaction followed by an intramolecular nucleophilic substitution: convergent multicomponent synthesis of 1-sulfonyl 1,4-diazepan-5-ones and of their benzo-fused derivatives. *J. Org. Chem.*, **72**, 2151–2160.

69 Banfi, L., Basso, A., Guanti, G., Lecinska, P., Riva, R., and Rocca, V. (2007) Multicomponent synthesis of novel 2- and 3-substituted dihydrobenzo[1,4]oxazepinones and tetrahydrobenzo[1,4]diazepin-5-ones and their conformational analysis. *Heterocycles*, **73**, 699–728.

70 Banfi, L., Basso, A., Cerulli, V., Guanti, G., Monfardini, I., and Riva, R. (2010) Multicomponent synthesis of dihydrobenzoxazepinones, bearing four diversity points, as potential α-helix mimics. *Mol. Divers.*, **14**, 425–442.

71 Banfi, L., Basso, A., Giardini, L., Riva, R., Rocca, V., and Guanti, G. (2011) Tandem Ugi MCR/Mitsunobu

cyclization as a short, protecting-group-free route to benzoxazinones with four diversity points. *Eur. J. Org. Chem.*, 100–109.

72 Banfi, L., Basso, A., Guanti, G., Lecinska, P., and Riva, R. (2008) Multicomponent synthesis of benzoxazinones via tandem Ugi/Mitsunobu reactions. An unexpected cine-substitution. *Mol. Divers.*, **12**, 187–190.

73 Wang, W., Ollio, S., Herdtweck, E., and Doemling, A. (2011) Polycyclic compounds by Ugi–Pictet–Spengler sequence. *J. Org. Chem.*, **76**, 637–644.

74 Znabet, A., Zonneveld, J., Janssen, E., De Kanter, F.J.J., Helliwell, M., Turner, N.J., Ruijter, E., and Orru, R.V.A. (2010) Asymmetric synthesis of synthetic alkaloids by a tandem biocatalysis/Ugi/Pictet–Spengler-type cyclization sequence. *Chem. Commun.*, **46**, 7706–7708.

75 Santra, S. and Andreana, P.R. (2007) A one-pot, microwave-influenced synthesis of diverse small molecules by multicomponent reaction cascades. *Org. Lett.*, **9**, 5035–5038.

76 Marcaccini, S., Pepino, R., Cruz Pozo, M., Basurto, S., Garcia Valverde, M., and Torroba, T. (2004) One-pot synthesis of quinolin-2-(1H)-ones via tandem Ugi–Knoevenagel condensations. *Tetrahedron Lett.*, **45**, 3999–4001.

77 Spatz, J.H., Umkehrer, M., Kalinski, C., Ross, G., Burdack, C., Kolb, J., and Bach, T. (2007) Combinatorial synthesis of 4-oxo-4H-imidazo[1,5-a]quinoxalines and 4-oxo-4H-pyrazolo[1,5-a] quinoxalines. *Tetrahedron Lett.*, **48**, 8060–8064.

78 Trifilenkov, A.S., Ilyin, A.P., Kysil, V.M., Sandulenko, Y.B., and Ivachtchenko, A.V. (2007) One-pot tandem complexity-generating reaction based on Ugi four-component condensation and intramolecular cyclization. *Tetrahedron Lett.*, **48**, 2563–2567.

79 Dai, W.M., Shi, J., and Wu, J. (2008) Synthesis of 3-arylideneindolin-2-ones from 2-aminophenols by Ugi four-component reaction and Heck carbocyclization. *Synlett*, 2716–2720.

80 El Kaïm, L., Gizolme, M., and Grimaud, L. (2007) New indolizine template from the Ugi reaction. *Synlett*, 227–230.

81 DiMauro, E.F. and Kennedy, J.M. (2007) Rapid synthesis of 3-amino-imidazopyridines by a microwave-assisted four-component coupling in one-pot. *J. Org. Chem.*, **72**, 1013–1016.

82 Erb, W., Neuville, L., and Zhu, J. (2009) Ugi post-functionalization, from a single set of Ugi adducts to two distinct heterocycles by microwave-assisted palladium-catalyzed cyclizations: tuning the reaction pathway by ligand switch. *J. Org. Chem.*, **74**, 3109–3115.

83 Kalinski, C., Umkehrer, M., Ross, G., Kolb, J., Burdack, C., and Hiller, W. (2006) Highly substituted indol-2-ones, quinoxalin-2-ones and benzodiazepin-2,5-diones via a new Ugi(4CR)-Pd assisted N-aryl amidation strategy. *Tetrahedron Lett.*, **47**, 3423–3426.

84 Ma, Z., Xiang, Z., Luo, T., Lu, K., Xu, Z., Chen, J., and Yang, Z. (2006) Synthesis of functionalized quinolones via Ugi and Pd-catalyzed intramolecular arylation reactions. *J. Comb. Chem.*, **8**, 696–704.

85 Kazmaier, U., Hebach, C., Watzke, A., Maier, S., Mues, H., and Huch, V. (2005) A straightforward approach towards cyclic peptides *via* ring-closing metathesis – scope and limitations. *Org. Biomol. Chem.*, **3**, 136–145.

86 Banfi, L., Basso, A., Damonte, D., De Pellegrini, F., Galatini, A., Guanti, G., Monfardini, I., and Scapolla, C. (2007) Synthesis and biological evaluation of new conformationally biased integrin ligands based on a tetrahydroazoninone scaffold. *Bioorg. Med. Chem. Lett.*, **17**, 1341–1345.

87 Ribelin, T.P., Judd, A.S., Akritopolou-Zanze, I., Henry, R.F., Cross, J.L., Whittern, D.N., and Djuric, S.W. (2007) Concise construction of novel bridged bicyclic lactams by sequenced Ugi/RCM/Heck reactions. *Org. Lett.*, **24**, 5119–5122.

88 Zhong, Y., Wang, L., and Ding, M.W. (2011) New efficient synthesis of 2,3,4-trisubstituted 3,4-dihydroquinazolines by a Ugi 4CC/Staudinger/aza-Wittig

sequence. *Tetrahedron*, **67**, 3714–3723.

89 Lecinska, P., Corres, N., Moreno, D., Valverde, M.G., Marcaccini, S., and Torroba, T. (2010) Synthesis of pseudopeptidic (S)-6-amino-5-oxo-1,4-diazepines and (S)-3-benzyl-2-oxo-1,4-benzodiazepines by an Ugi 4CC Staudinger/aza-Wittig sequence. *Tetrahedron*, **66**, 6783–6788.

90 Sanudo, M., Valverde, M.G., Marcaccini, S., Delgado, J.J., Rojo, J., and Torroba, T. (2009) Synthesis of benzodiazepine β-turn mimetics by an Ugi 4CC/Staudinger/aza-Wittig sequence. Solving the conformational behavior of the Ugi 4CC adducts. *J. Org. Chem.*, **74**, 2189–2192.

91 Paulvannan, K. (1999) Preparation of tricyclic nitrogen heterocycles via tandem four component condensation/intramolecular Diels–Alder reaction. *Tetrahedron Lett.*, **40**, 1851–1854.

92 Basso, A., Banfi, L., and Riva, R. (2010) A marriage of convenience: combining the power of isocyanide-based multicomponent reactions with the versatility of (hetero)norbornene chemistry. *Eur. J. Org. Chem.*, 1831–1841.

93 Ikoma, M., Oikawa, M., Gill, M.B., Swanson, G.T., Sakai, R., Shimamoto, K., and Sasaki, M. (2008) Regioselective domino metathesis of 7-oxanorbornenes and its applications to the synthesis of biologically active glutamate analogues. *Eur. J. Org. Chem.*, 5215–5220.

94 Ikoma, M., Oikawa, M., and Sasaki, M. (2009) Chemospecific allylation and domino metathesis of 7-oxanorbornenes for skeletal and appendage diversity. *Eur. J. Org. Chem.*, 72–84.

95 Huang, X. and Xu, J. (2009) One-pot facile synthesis of substituted isoindolinones via an Ugi four-component condensation/Diels–Alder cycloaddition/deselenization–aromatization sequence. *J. Org. Chem.*, **74**, 8859–8861.

96 Akritopolou-Zanze, I., Wang, Y., Zhao, H., and Djuric, S.W. (2009) Synthesis of substituted fused pyridines, pirazines and pyrimidines by sequential Ugi/inverse electron demand Diels–Alder. *Tetrahedron Lett.*, **50**, 5773–5776.

97 Himo, F., Lovell, T., Hilgraf, R., Rostovtsev, V.V., Noodleman, L., Sharpless, K.B., and Fokin, V.V. (2005) Copper(I)-catalyzed synthesis of azoles. DFT study predicts unprecedented reactivity and intermediates. *J. Am. Chem. Soc.*, **127**, 210–216.

98 Akritopolou-Zanze, I., Gracias, V., and Djuric, S.W. (2004) A versatile synthesis of fused triazolo-derivatives by sequential Ugi/alkyne-azide cycloaddition reactions. *Tetrahedron Lett.*, **45**, 8439–8441.

99 Akritopolou-Zanze, I., Gracias, V., Moore, J.D., and Djuric, S.W. (2004) Synthesis of novel fused isoxazole and isoxazolines by sequential Ugi/INOC reactions. *Tetrahedron Lett.*, **45**, 3421–3423.

100 Pirali, T., Tron, G.C., and Zhu, J. (2006) One-pot synthesis of macrocycles by a tandem three-component reaction and intramolecular [3+2] cycloaddition. *Org. Lett.*, **18**, 4145–4148.

101 Sokolova, N.V., Gulevich, A.V., Sokolova, N.V., Mironov, A.V., and Balenkova, E.S. (2010) Chiral isocyanoazides: efficient bifunctional reagents for bioconjugation. *Eur. J. Org. Chem.*, 1445–1449.

102 Nenajdenko, V.G., Latyshev, G.V., Lukashev, N.V., and Nenajdenko, V.G. (2011) *Org. Biomol. Chem.*, **9**, 4921–4926.

103 Ruben, A.J., Kiso, Y., and Freire, E. (2006) Overcoming roadblocks in lead optimization: a thermodynamic perspective. *Chem. Biol. Drug. Des.*, **67**, 2–4.

104 Dömling, A. (1998) Isocyanide-based multi component reactions in combinatorial chemistry. *Comb. Chem. High Throughput Screening*, **1**, 1–22.

105 Hulme, C. and Dietrich, J. (2009) Emerging molecular diversity from the intra-molecular Ugi reaction: iterative efficiency in medicinal chemistry. *Mol. Divers.*, **13**, 195–207.

106 Ugi, I. (1962) Isonitriles. XI. Synthesis of simple penicillamic acid derivatives. *Chem. Ber.*, **95**, 136–140.

107 Demharter, A., Hoerl, W., Herdtweck, E., and Ugi, I. (1996) Synthesis of chiral

1,1-iminodicarboxylic acid derivatives from α-amino acids, aldehydes, isocyanides, and alcohols by the diastereoselective five-center-four-component reaction. *Angew. Chem. Int. Ed. Engl.*, **35**, 173–175.

108 Ugi, I., Goebel, M., Gruber, B., Heilingbrunner, M., Heib, C., Horl, W., Kern, O., Starnecker, M., and Dömling, A. (1996) Molecular libraries in liquid phase via Ugi-MCR. *Res. Chem. Intermed.*, **22**, 625–644.

109 Hayashi, K., Nunami, K., Kato, J., Yoneda, N., Kubo, M., Ochiai, T., and Ishida, R. (1989) Studies on angiotensin converting enzyme inhibitors. 4. Synthesis and angiotensin converting enzyme inhibitory activities of 3-acyl-1-alkyl-2-oxoimidazolidine-4-carboxylic acid derivatives. *J. Med. Chem.*, **32**, 289–297.

110 Isenring, H.P. and Hofheinz, W. (1981) A simple two-step synthesis of diphenylmethyl esters of 2-oxo-1-azetidineacetic acids. *Synthesis*, 385–387.

111 Kamiya, T., Teraji, T., Hashimoto, M., Nakaguchi, O., and Oku, T. (1976) Studies on β-lactam antibiotics. III. Synthesis of 2-methyl-3-cephem derivatives. *J. Am. Chem. Soc.*, **98**, 2342–2344.

112 Hatanaka, M., Nitta, H., and Ishimar, T. (1984) The synthesis of the 1-carbopenem antibiotic (±)-ps-5 and its 6-epi analogue. *Tetrahedron Lett.*, **25**, 2387–2390.

113 Hatanaka, M., Yamamoto, Y.-I., Nitta, H., and Ishimaru, T. (1981) A novel synthesis of the carbapen-2-em derivatives. *Tetrahedron Lett.*, **22**, 3883–3886.

114 Kanizsai, I., Szakonyi, Z., Sillanpää, R., and Fülöp, F. (2006) A comparative study of the multicomponent Ugi reactions of an oxabicycloheptene-based β-amino acid in water and in methanol. *Tetrahedron Lett.*, **47**, 9113–9116.

115 Cho, S., Keum, G., Kang, S.B., Han, S.Y., and Kim, Y. (2003) An efficient synthesis of 2,5-diketopiperazine derivatives by the Ugi four-center three-component reaction. *Mol. Divers.*, **6**, 283–286.

116 Banfi, L., Basso, A., Guanti, G., and Riva, R. (2004) Enantio- and diastereoselective synthesis of 2,5-disubstituted pyrrolidines through a multicomponent Ugi reaction and their transformation into bicyclic scaffolds. *Tetrahedron Lett.*, **45**, 6637–6640.

117 Nenajdenko, V.G., Gulevich, A.V., and Balenkova, E.S. (2006) The Ugi reaction with 2-substituted cyclic imines. Synthesis of substituted proline and homoproline derivatives. *Tetrahedron*, **62**, 5922–5930.

118 Hanusch-Kompa, C. and Ugi, I. (1998) Multi-component reactions 13: synthesis of γ-lactams as part of a multiring system via Ugi-4-centre-3-component reaction. *Tetrahedron Lett.*, **39**, 2725–2728.

119 Harriman, G.C.B. (1997) Synthesis of small and medium sized 2,2-disubstituted lactams via the "intramolecular" three component Ugi reaction. *Tetrahedron Lett.*, **38**, 5591–5594.

120 Keenan, R.M., Miller, W.H., Chet Kwon, Ali, F.E., et al. (1997) Discovery of potent nonpeptide vitronectin receptor ($\alpha_v\beta_3$) antagonists. *J. Med. Chem.*, **40**, 2289–2292.

121 Samanen, J.M., Ali, F.E., Barton, L.S., Bondinell, W.E., et al. (1996) Potent, selective, orally active 3-oxo-1,4-benzodiazepine GPIIb/IIIa integrin antagonists. *J. Med. Chem.*, **39**, 4867–4870.

122 Short, K.M. and Mjalli, A.M.M. (1997) A solid-phase combinatorial method for the synthesis of novel 5- and 6-membered ring lactams. *Tetrahedron Lett.*, **38**, 359–362.

123 Chibale, K. (2005) Economic drug discovery and rational medicinal chemistry for tropical diseases. *Pure Appl. Chem.*, **77**, 1957–1964.

124 Musonda, C.C., Gut, J., Rosenthal, P.J., Yardley, V., Carvalho de Souzad, R.C., Chibale, K. (2006) Application of multicomponent reactions to antimalarial drug discovery. Part 2: new antiplasmodial and antitrypanosomal 4-aminoquinoline γ- and δ-lactams via a "catch and release" protocol. *Bioorg. Med. Chem.*, **14**, 5605–5615.

125 Basso, A., Banfi, L., Guanti, G., and Riva, R. (2005) One-pot synthesis of α-acyloxyaminoamides via nitrones as imine surrogates in the Ugi MCR. *Tetrahedron Lett.*, **46**, 8003–8006.

126 Pirrung, M. and Sarma, K.D. (2004) β-Lactam synthesis by Ugi reaction of β-keto acids in aqueous solution. *Synlett*, 1425–1427.

127 Mironov, M.A., Ivantsova, M.N., and Mokrushin, V.S. (2003) Ugi reaction in aqueous solutions: a simple protocol for libraries production. *Mol. Divers.*, **6**, 193–197.

128 Zhang, J., Jacobson, A., Rusche, J.R., and Herlihy, W. (1999) Unique structures generated by Ugi 3CC reactions using bifunctional starting materials containing aldehyde and carboxylic acid. *J. Org. Chem.*, **64**, 1074–1076.

129 Bleasdale, D.A. and Jones, D.W. (1991) 2-Benzopyran-3-ones as synthetic building blocks; regioselective Diels–Alder additions with simple olefins leading to aromatic steroids. *J. Chem. Soc., Perkin Trans. 1*, 1683–1692.

130 Moody, C.J. and Rahimtoola, K.F. (1990) Diels–Alder reactivity of pyrano[4,3-b] indol-3-ones, indole 2,3-quinodimethane analogues. *J. Chem. Soc., Perkin Trans. 1*, 673–679.

131 Marcaccini, S., Miguel, D., Torroba, T., and Garcia-Valverde, M. (2003) 1,4-Thiazepines, 1,4-benzothiazepin-5-ones, and 1,4-benzothioxepin orthoamides via multicomponent reactions of isocyanides. *J. Org. Chem.*, **68**, 3315–3318.

132 Arrhenius, G., Bladridge, K.K., Richards-Gross, S., and Siegel, J.S. (1997) Glycolonitrile oligomerization: structure of isolated oxazolines, potential heterocycles on the early earth. *J. Org. Chem.*, **62**, 5522–5525.

133 Laurenti, D., Santelli-Rouvier, C., Pepe, G., and Santelli, M. (2000) Synthesis of cis,cis,cis-tetrasubstituted cyclobutanes. Trapping of tetrahedral intermediates in intramolecular nucleophilic addition. *J. Org. Chem.*, **65**, 6418–6422.

134 Kaneko, T., Wong, H., Doyle, T.W., Rose, W.C., and Bradner, W.T. (1985) Bicyclic and tricyclic analogs of anthramycin. *J. Med. Chem.*, **28**, 388–392.

135 Hara, T., Kayama, Y., Mori, T., Itoh, K., Fujimori, H., Sunami, T., Hashimoto, Y., and Ishimoto, S. (1978) Diazepines. 5. Synthesis and biological action of 6-phenyl-4H-pyrrolo[1,2-a][1,4] benzodiazepines. *J. Med. Chem.*, **21**, 263–268.

136 Ilyin, A.P., Kuzovkova, J., Shkirando, A., and Ivachtchenko, A. (2005) An efficient synthesis of 3-oxo-1,2,3,4-tetrahydropyrrolo[1,2-a]pyrazine-1-carboxamides using novel modification of Ugi condensation. *Heterocycl. Commun.*, **11**, 523–526.

137 Ilyin, A.P., Parchinski, V.Z., Peregudova, J.N., Trifilenkov, A.S., Poutsykina, E.B., Tkachenko, S.E., Kravchenko, D.V., and Ivachtchenko, A.V. (2006) Synthesis of 7,8-dihydrothieno[3′2′:4,5]pyrrolo[1,2-a] pyrazin-5(6H)-ones using a modification of four-component Ugi reaction. *Synth. Comm.*, **36**, 903–910.

138 Nenajdenko, V.G., Reznichenko, A.L., and Balenkova, E.S. (2007) Diastereoselective Ugi reaction without chiral amines: the synthesis of chiral pyrroloketopiperazines. *Tetrahedron*, **63**, 3031–3041.

139 Ilyin, A.P., Trifilenkov, A.S., Kurashvili, I.D., Krasavin, M., and Ivachtchenko, A.V. (2005) One-step construction of peptidomimetic 5-carbamoyl-4-sulfonyl-2-piperazinones. *J. Comb. Chem.*, **7**, 360–363.

140 Athanassios-Giannis, T.K. (1993) Peptidomimetics for receptor ligands – discovery, development, and medical perspectives. *Angew. Chem. Int. Ed. Engl.*, **32**, 1244–1267.

141 Blackburn, C., Guan, B., Fleming, P., Shiosaki, K., and Tsai, S. (1998) Parallel synthesis of 3-aminoimidazo[1,2-a] pyridines and pyrazines by a new three-component condensation. *Tetrahedron Lett.*, **39**, 3635–3638.

142 Groebke, K., Weber, L., and Mehlin, F. (1998) Synthesis of imidazo[1,2-a] annulated pyridines, pyrazines and pyrimidines by a novel three-component condensation. *Synlett*, 661–663.

143 Bienayme, H. and Bouzid, K. (1998) A new heterocyclic multicomponent

144 Masquelin, T., Bui, H., Brickley, B., Stephenson, G., Schwerkoske, J., and Hulme, C. (2006) Sequential Ugi/Strecker reactions via microwave assisted organic synthesis: novel 3-center-4-component and 3-center-5-component multi-component reactions. *Tetrahedron Lett.*, **47**, 2989–2991.

143 reaction for the combinatorial synthesis of fused 3-aminoimidazoles. *Angew. Chem. Int. Ed. Engl.*, **37**, 2234–2237.

145 Schwerkoske, J., Masquelin, T., Perun, T., and Hulme, C. (2005) New multicomponent reaction accessing 3-aminoimidazo[1,2-a]pyridines. *Tetrahedron Lett.*, **46**, 8355–8357.

146 Georges, G.J., Vercauteren, D.P., Evrard, G.H., Durant, F.V., George, P.G., and Wick, A.E. (1993) Characterization of the physico-chemical properties of the imidazopyridine derivative alpidem. Comparison with zolpidem. *Eur. J. Med. Chem.*, **28**, 323–335.

147 Hulme, C. and Lee, Y.-S. (2008) Emerging approaches for the syntheses of bicyclic imidazo[1,2-x]-heterocycles. *Mol. Divers.*, **12**, 1–15.

148 Trost, B.M. (1991) The atom economy: a search for synthetic efficiency. *Science*, **254**, 1471.

149 Grieco, P.A. and Bahsas, A. (1988) Role reversal in the cyclocondensation of cyclopentadiene with heterodienophiles derived from arylamines and aldehydes: synthesis of novel tetrahydroquinolines. *Tetrahedron Lett.*, **29**, 5855–5858.

150 Krelaus, R. and Westermann, B. (2004) Preparation of peptide-like bicyclic lactams via a sequential Ugi reaction-olefin metathesis approach. *Tetrahedron Lett.*, **45**, 5987–5990.

151 Westermann, B., Diedrichs, N., Krelaus, R., Walter, A., and Gedrath, I. (2004) Diastereoselective synthesis of homologous bicyclic lactams-potential building blocks for peptide mimics. *Tetrahedron Lett.*, **45**, 5983–5986.

11
Synthesis of Pyrroles and Their Derivatives from Isocyanides
Noboru Ono and Tetsuo Okujima

11.1
Introduction

The most widely used methods for constructing pyrrole rings are cyclization reactions, referred to as the Knorr and the Parr–Knorr syntheses and cycloaddition reactions, all of which have been extensively reviewed [1]. In addition to the Knorr-type cyclization, isocyanide cyclization is today becoming increasingly important in pyrrole synthesis, where two C–C bonds are formed via nucleophilic addition and isocyanide cyclization. Subsequently, isonitrile cyclization leads to the synthesis of a variety of heterocycles, including pyrroles, oxazoles, thiazole, imidazole, and indoles [2]. In this chapter, pyrrole synthesis based on the van Leusen reaction, using *p*-tolylsulfonylmethyl isocyanide (TosMIC) [3], and also the Barton–Zard reaction, using alkyl isocyanoacetates [4], are described. These reactions offer a useful strategy for the synthesis of pyrrole-containing natural products [5], such as heme, chlorophyll, vitamin B_{12}, lukianol A [6], lamellarin G [7], halitulin [8], and dictyodendrin [9]. The pyrrole alkaloids of prodigiosin – the deep-red-colored pigment that is produced by Gram-negative bacteria – afford a new series of biologically active compounds [10]. Furthermore, pyrroles are important key components in the materials sciences, being employed as conductive polymers [11], as semiconductors for application to organic electronics [12], as synthetic dyes in photodynamic therapy (PDT) [13], as solar cells [14], and in nonlinear optics [15]. Some typical pyrrole derivatives that are found naturally occurring, or in the materials sciences, are shown in Figure 11.1. Porphyrin CP – which is porphyrin fused with bicyclo[2.2.2]octadiene (BCOD) rings at the β,β-positions; see Figure 11.1 – is soluble in organic solvents, and is converted into insoluble tetrabenzoporphyrin (TBP) by heating at 150 °C. This process is employed in the fabrication of a p-type semiconductor that is based on TBP and used in solar cells [16] or field-effect transistors (FETs) [17], via a solution process.

The boron–dipyrromethene (BODIPY) dye, when fused with a phenanthrene ring, demonstrates a bright, long-wavelength fluorescence that is useful for detection purposes in microscopy [18]. In addition, cyclo[8]pyrroles with long alkyl side chains have potential utility as chemosensors for nitroaromatic explosives [19]. In

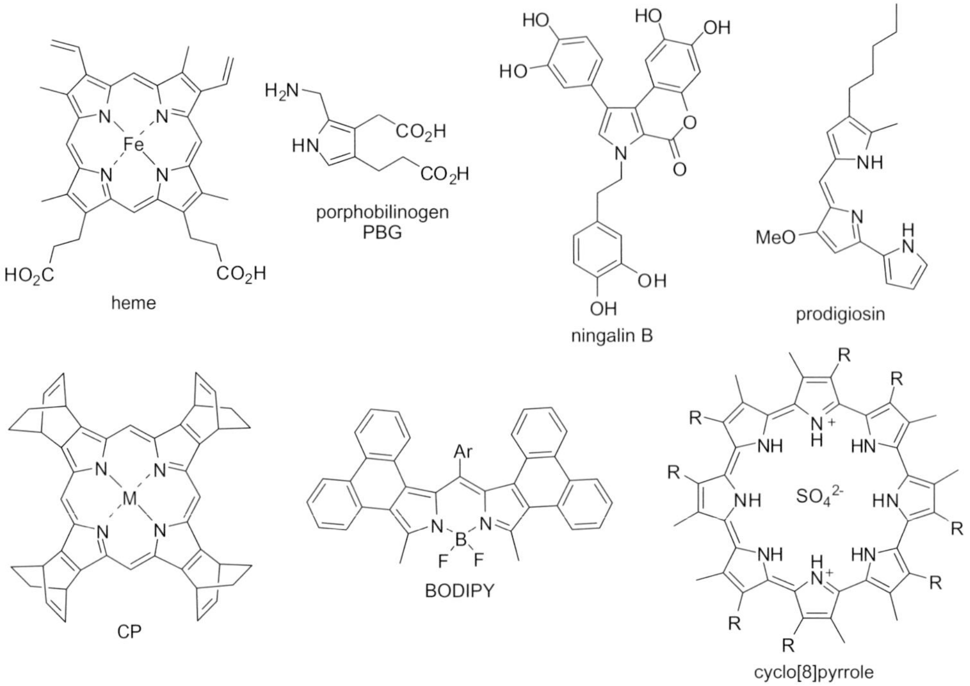

Figure 11.1 Typical pyrrole derivatives found in nature or used in the materials sciences.

order to prepare such materials, pyrroles fused with either a bicyclic ring or an aromatic ring, or substituted with long alkyl groups are required. In this situation, the Barton–Zard reaction is more suitable for the synthesis of such pyrroles, and is far superior to the classical Knorr method.

11.2
Synthesis of Pyrroles Using TosMIC

In 1972, van Leusen and coworkers reported a useful method for the synthesis of pyrroles, by using TosMIC and electron-deficient alkenes. The process is shown in Scheme 11.1, where the isocyano function undergoes typical α-addition reactions while the tosyl group serves two functions, serving as a leaving group and also enhancing the acidity of the α-carbon. Today, TosMIC is better known as van Leusen's reagent, and has been used extensively in the synthesis of a variety of heterocyclic compounds [3]. Thus, pyrroles with carbonyl, cyano, and nitro groups at the β-positions may be readily prepared, as shown in Scheme 11.1. Quinone can be also used for the van Leusen reaction, with fused pyrrole **2** being obtained via reaction with naphthoquinone [20] (Scheme 11.2). Since the scope and limitation of the van Leusen reaction have been well documented [3], its description will be mimimized in this chapter.

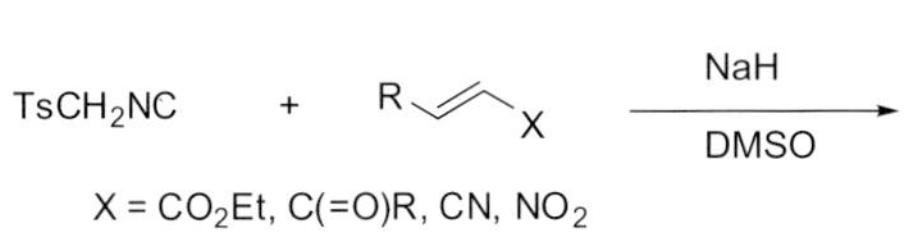

11.2 Synthesis of Pyrroles Using TosMIC

TsCH$_2$NC + R–CH=CH–X $\xrightarrow[\text{DMSO}]{\text{NaH}}$ 3,4-disubstituted pyrrole (R, X on pyrrole)

X = CO$_2$Et, C(=O)R, CN, NO$_2$

Scheme 11.1

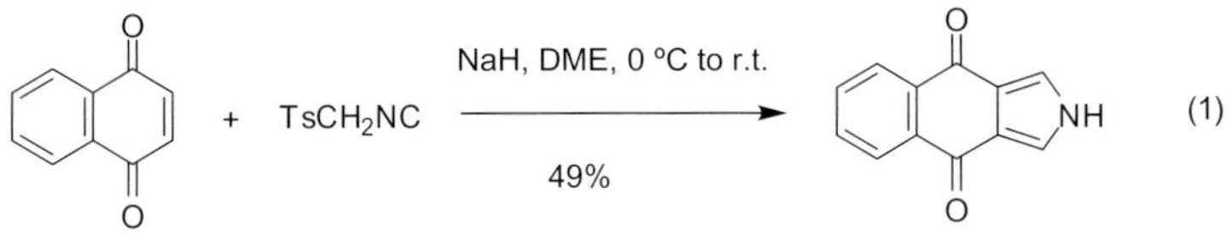

Naphthoquinone (1) + TsCH$_2$NC $\xrightarrow[49\%]{\text{NaH, DME, 0 °C to r.t.}}$ benz[f]isoindole-4,9-dione (2)   (1)

Scheme 11.2

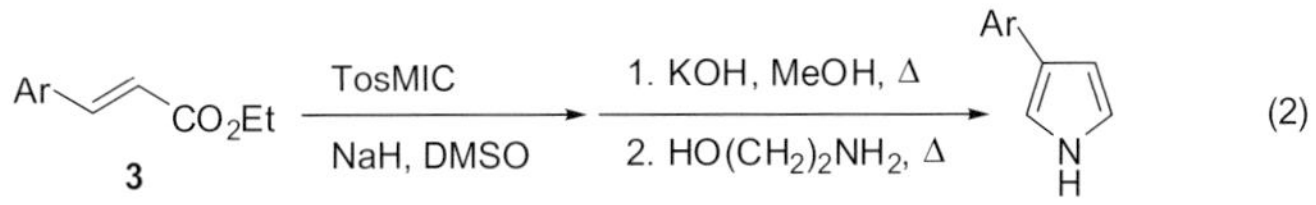

Ar–CH=CH–CO$_2$Et (3) $\xrightarrow[\text{NaH, DMSO}]{\text{TosMIC}}$ → $\xrightarrow[\text{2. HO(CH}_2)_2\text{NH}_2, \Delta]{\text{1. KOH, MeOH, }\Delta}$ 3-aryl pyrrole (4)   (2)

Scheme 11.3

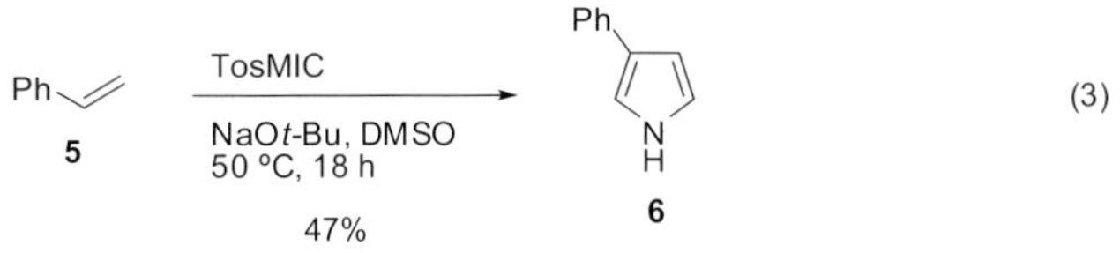

Ph–CH=CH$_2$ (5) $\xrightarrow[\substack{\text{NaO}t\text{-Bu, DMSO}\\ \text{50 °C, 18 h}\\ 47\%}]{\text{TosMIC}}$ 3-phenyl pyrrole (6)   (3)

Scheme 11.4

α-Free pyrroles prepared by this route are useful precursors of polypyrroles or porphyrins. This reaction has been used extensively in the synthesis of a variety of natural or unnatural products, as the functional groups at the β-position are readily transformed into other groups. For example, 3-aryl pyrroles – which are important as functional materials – can be prepared from aromatic aldehydes via the Horner–Wadsworth–Emmons olefination, the van Leusen reaction, and subsequent demethoxycarbonylation in good overall yields [21] (Scheme 11.3). As a more efficient route, however, 3-aryl- and 4,4-diarypyrroles can be directly prepared in good yields by the reaction of arylalkenes with TosMIC in the presence of NaOt-Bu in dimethylsulfoxide (DMSO), as shown in Schemes 11.4 and 11.5. The reaction proceeds rapidly when electron-poor aryl groups are attached to the alkenes, as shown in Scheme 11.5 [22].

Both halogenation and formylation take place at the α-positions of pyrroles selectively, and this is useful for the conversion of α-free pyrroles to porphyrins and other pigments. For example, the van Leusen reaction of diethyl glutaconate

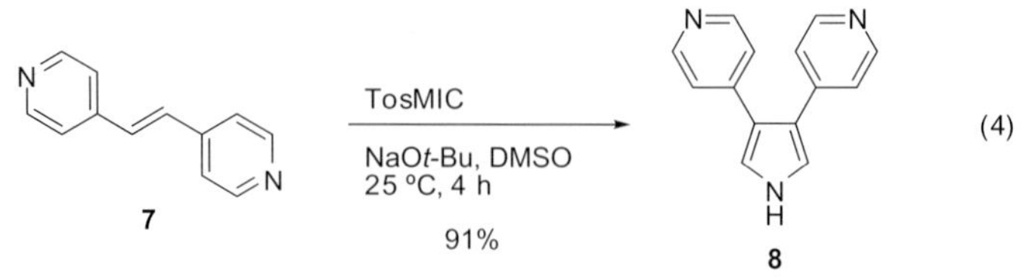

Scheme 11.5

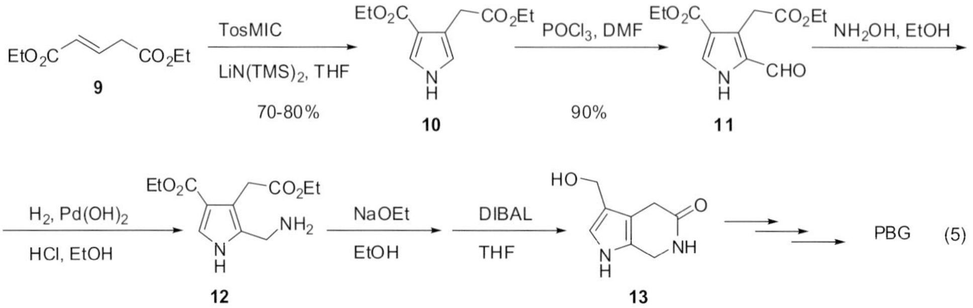

Scheme 11.6

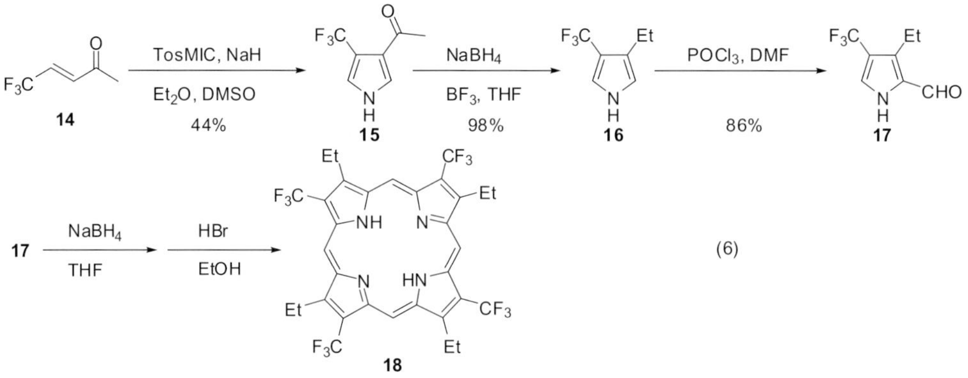

Scheme 11.7

produces pyrrole **10** in approximately 80% yield, and this in turn is converted into α-formylpyrrole **11** in 90% yield on treatment with $POCl_3$ and dimethylformamide (DMF) (Scheme 11.6). The formyl group is further transformed into an aminomethyl group via reduction of the corresponding oxime, and finally is converted into porphobilinogen (PBG in Figure 11.1), which is a key substrate in the biosynthesis of porphyrins [23]. Another example is shown in synthesis of pyrrole-2-carbaldehyde substituted with CF_3 group (Scheme 11.7). Here, the reduction of this pyrrole with $NaBH_4$, followed by tetramerization on treatment with an acid and oxidation, produces porphyrin with CF_3 groups [24].

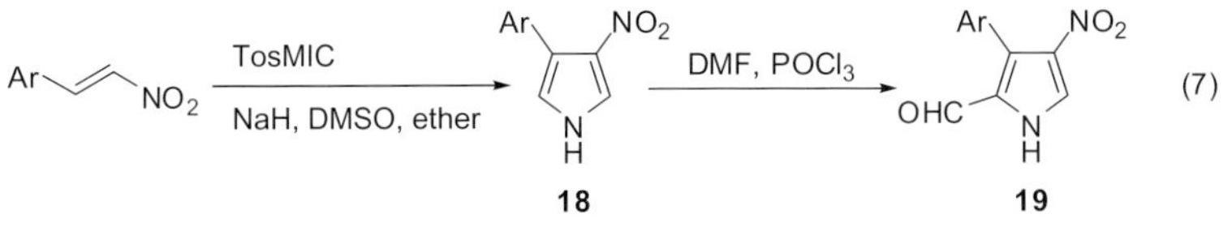

11.2 Synthesis of Pyrroles Using TosMIC

Scheme 11.8

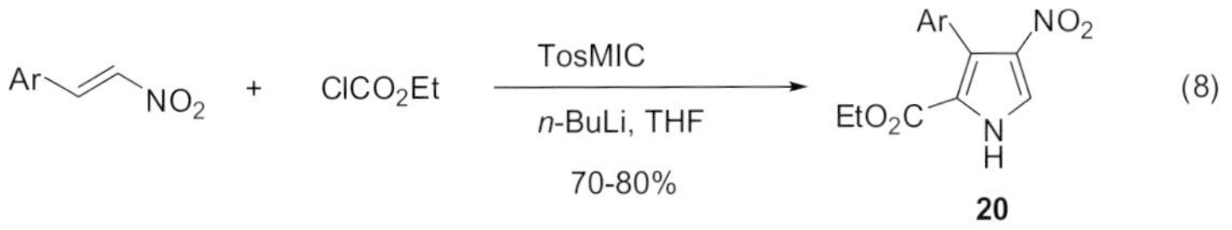

Scheme 11.9

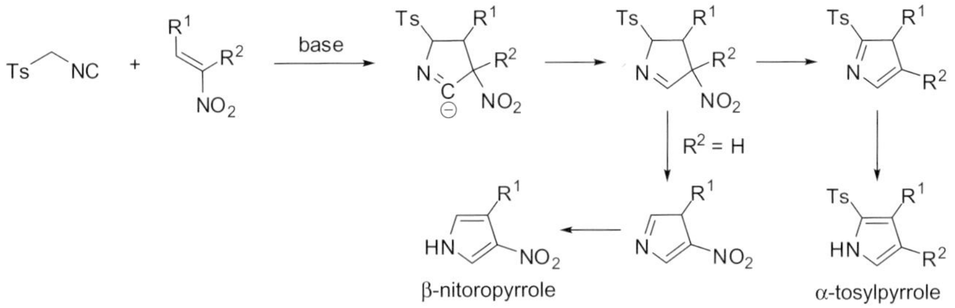

Scheme 11.10

The reaction of nitroalkenes with TosMIC is very useful for the preparation of β-nitropyrroles, which are not available via the usual nitration of pyrroles. Typically, pyrroles are too reactive to produce α-nitropyrroles in good yields; rather, they are often decomposed to provide polymeric compounds under nitration conditions, using HNO_3 and H_2SO_4. β-Nitropyrroles are sufficiently reactive to produce 4-nitropyrrole-2-carbaldehyde with $POCl_3$ and DMF [25] (Scheme 11.8). Recently, a simple synthesis of pyrroles with such substitution patterns was reported, whereby the treatment of a mixture of TosMIC, ethyl chloroformate, and nitrostyrenes with n-BuLi afforded ethyl 4-nitro-3-arylpyrrole-2-carboxylate in good yield [26] (Scheme 11.9).

The reaction of nitroalkenes with TosMIC may proceed in two ways, to produce either β-nitropyrroles or α-tosylpyrroles, depending on the structure of the nitroalkenes employed. The mechanism of pyrrole formation via isocyanide cyclization is shown in Scheme 11.10 where, when R^2 is H, an elimination of sulfenic acid occurs to produce β-nitropyrroles. In contrast, when $R^2 \neq H$, the elimination of nitrous acid predominates to yield α-tosylpyrroles (see Section 11.3.1; Barton–Zard reaction). Although various bases such as NaH, DBU (1,8-diazabicyclo[5.4.0]undec-7-ene), and KOt-Bu are sufficiently effective to induce the van Leusen and

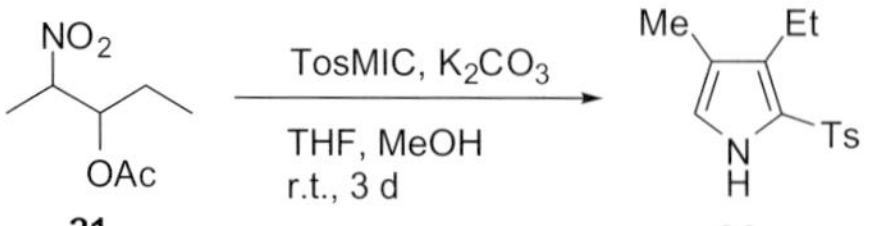

21 22 (9)

Scheme 11.11

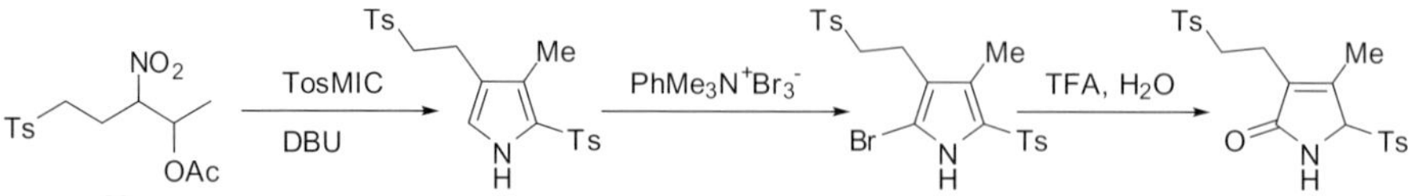

23 24 25 26 (10)

Scheme 11.12

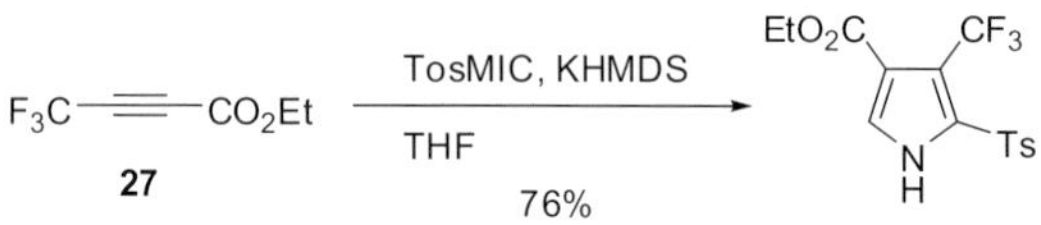

27 28 (11)

Scheme 11.13

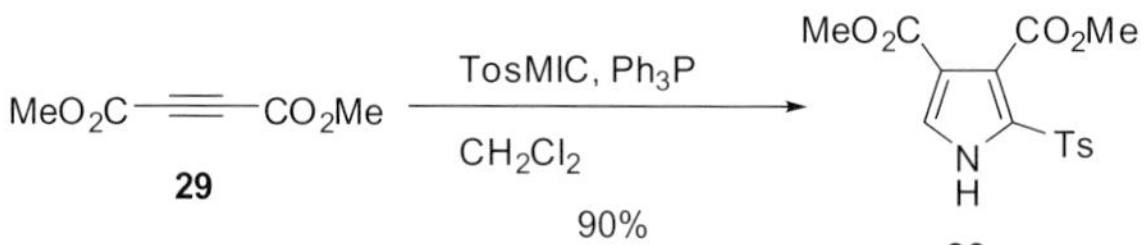

29 30 (12)

Scheme 11.14

Barton–Zard reactions, K_2CO_3 provides a superior yield in some reactions of TosMIC with nitroalkenes, as shown in Scheme 11.11 [27]. 2-Tosylpyrroles prepared by the reaction of nitroalkenes with TosMIC are useful for the synthesis of biologically important pigments, such as chlorins and phytochromes. The bromination of 2-tosylpyrroles, followed by hydrolysis, affords pyrrolin-2-one, which is the D-component of phytochrome (Scheme 11.12). Recently, Inomata and coworkers have employed this strategy very effectively to synthesize model compounds of phytochrome [28].

The Barton–Zard-type of pyrrole synthesis can be accomplished by using alkynes with electron-withdrawing groups (EWGs). The reaction of TosMIC with electron-deficient alkynes, such as CF_3-substituted acetylenecarboxylate or dimethyl acetylenedicarboxylate in the presence of KHMDS [29] or Ph_3P [30], produces 2-tosylpyrroles with CF_3 or ester groups at 3- and 4-positions, in good yields (Schemes 11.13 and 11.14). Similar reactions are also possible by using ethyl isocyanoacetate and alkynes with a base or other catalysts.

11.3
Synthesis of Pyrroles Using Isocyanoacetates

Today, ethyl isocyanoacetate is one of the most important isocyanides used in organic synthesis and, as it can be prepared from glycine, its industrial production is easily achieved. Indeed, the Nippon Synthetic Chemical Industry Co. Ltd, in Osaka, Japan, currently produces ethyl isocyanoacetate on a large scale, and it is available commercially at very low cost. Other esters, such as benzyl or *t*-butyl esters, are also easily prepared from glycine via a simple procedure. The isocyanoacetates serve as very useful regents for the synthesis of a wide variety of heterocycles or amino acids via multicomponent reactions (MCRs). In fact, a recent excellent review of isocynoacetate has been prepared in which the various pyrrole synthetic methods are described [4d–f]. One important application of isocyanoacetates in organic synthesis is the Barton–Zard pyrrole synthesis, which provides pyrrole-2-carboxylate via the reaction with nitroalkenes or α,β-unsaturated sulfones. The pyrroles prepared in this way may easily be converted into porphyrins, polypyrroles, and BODIPY dyes, all of which are important in the materials sciences (see Scheme 11.15). Although the Barton–Zard reaction has emerged only quite recently in porphyrin chemistry, it is now used routinely for porphyrin synthesis [31]. The properties of porphyrins or other related π-conjugated molecules may be finely controlled by the choice of substituents R^1 and R^2, which are derived from nitroalkenes or vinylsulfones. Compared to a traditional synthesis of porphyrins or polypyrroles, based on Knorr reaction, a wide variety of R^1 and R^2 may be readily introduced into such materials via the Barton–Zard reaction.

11.3.1
Synthesis from Nitroalkenes

In 1985, Barton and Zard reported that the base-catalyzed reaction of nitroalkenes or β-nitroacetates with alkyl isocyanoacetate produced pyrrole-2-carboxylates

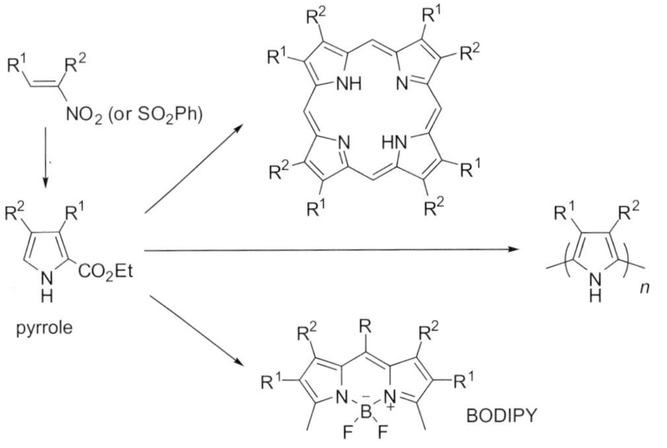

Scheme 11.15

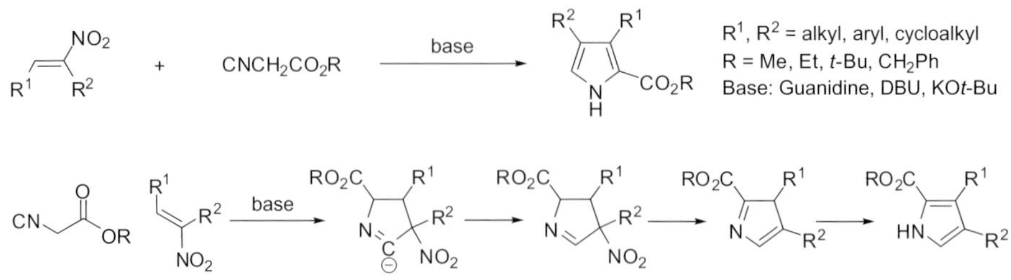

Scheme 11.16

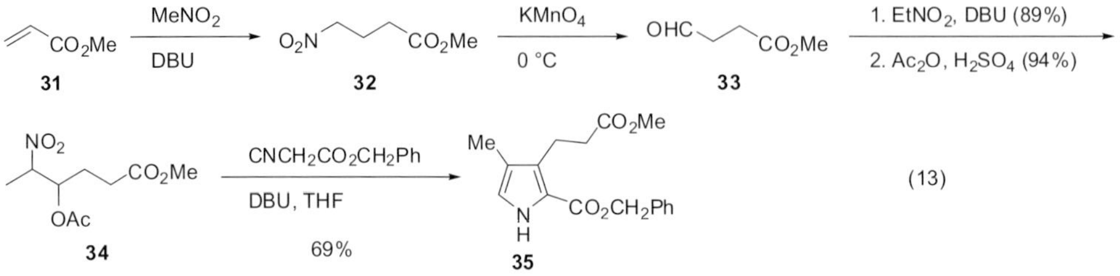

Scheme 11.17

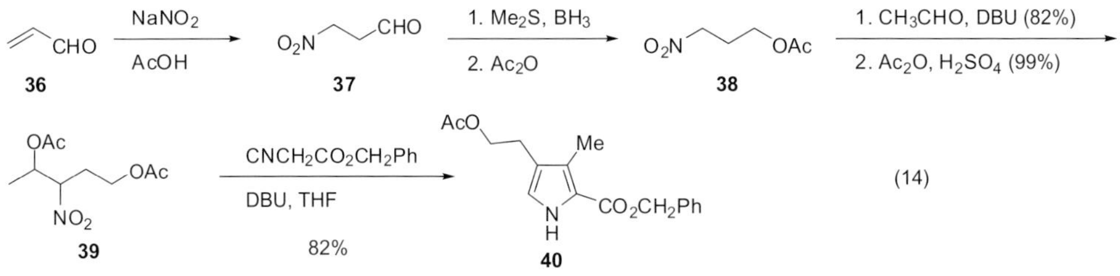

Scheme 11.18

(Scheme 11.16) [4], whereby the nitro group acted as an activating group for the C–C bond-formation process and as a leaving group in a pyrrole-forming step. Today, this type of pyrrole synthesis is widely referred to as the Barton–Zard reaction, and has become an important method for pyrrole synthesis.

The main advantage of the Barton–Zard reaction over other methods is that various groups may be readily introduced into 3- and 4-positions of pyrrole-2-carboxylates by a simple procedure, because the nitroalkenes may be prepared by a variety of methods, including the Henry reaction or the nitration of alkenes [32, 33]. Furthermore, the reaction proceeds very cleanly, such that polymer-supported reagents can be used to generate an array of 1,2,3,4-tetrasubstituted pyrrole derivatives without a need for any chromatographic purification steps [34]. The convenience of the Barton–Zard reaction is shown clearly in Schemes 11.17 and 11.18 [35]. In this case, the Michael addition of nitro compounds, the Nef reaction, and the Henry reaction are employed sequentially to obtain the requisite β-nitroacetates such as **34** and **39** for the Barton–Zard reaction, which is usually carried out using DBU in tetrahydrofuran (THF) as a base-solvent system. Recently, it was reported that THF containing a radical inhibitor (butylated hydroxy toluene; BHT) caused a retardation of the Barton–Zard reactions so as to produce a lower yield. Thus, the use of methyl *tert*-butyl ether (MTBE) was recommend as a solvent, whereby a large-scale Barton–Zard reaction was reported to produce 322 g of ethyl 3,4-diethylpyrrole-2-carboxylate (**43**) in 90% overall yield via a one-pot process (see Scheme 11.19) [36]. This pyrrole is useful as a starting material for the production of octaethylporphyrin (see Section 11.4). To obtain **43** on such a large scale is not easily achieved by using any other method; although typically, most nitroalkenes react with isocyanoacetates in the presence of DBU to produce the desired pyrroles in good yield, the Barton–Zard reaction cannot be induced by DBU when the nitroalkene is sterically hindered (e.g., **44**). Consequently, a much stronger base, such as phosphazene base P_4-t-Bu, was required to afford pyrrole **45** in 35% yield (Scheme 11.20) [37].

The Barton–Zard reaction has been applied extensively to the synthesis of biologically active pyrroles. One such example, porphobilinogen—a key building block in the pigments of living cells, such as heme and vitamin B_{12}—can be prepared via the Barton–Zard reaction, as shown in Scheme 11.21 [38].

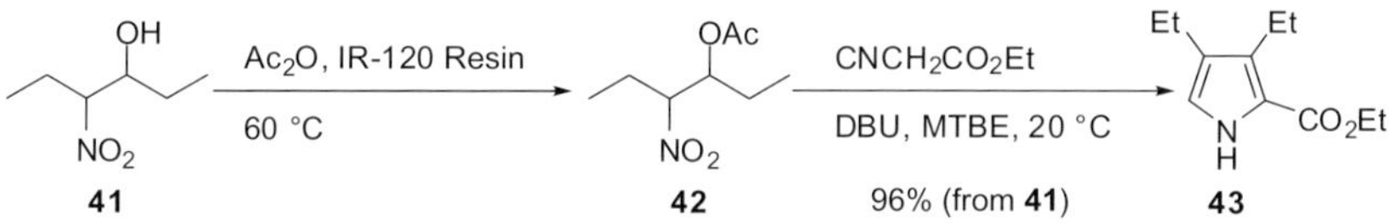

Scheme 11.19

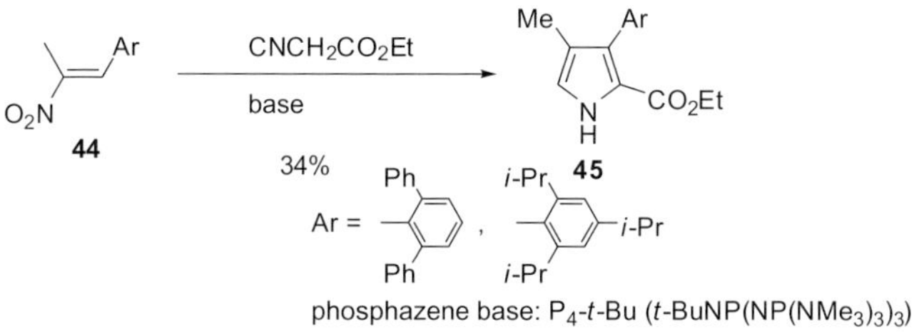

Scheme 11.20

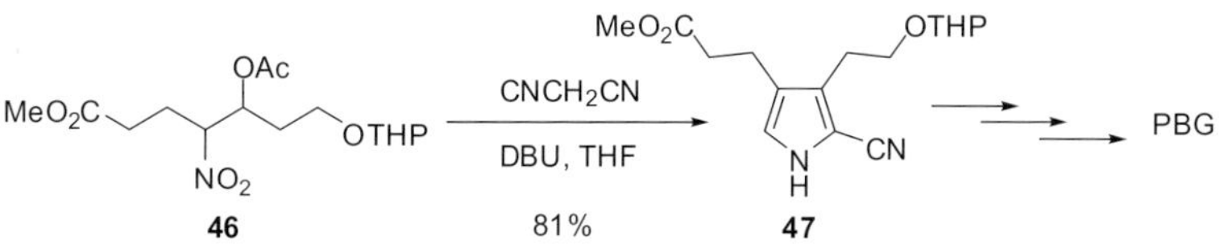

Scheme 11.21

An additional application of the Barton–Zard reaction to create biologically active compounds is shown in Scheme 11.22, where (+)-deoxypyrrololine, a potential biochemical marker for the diagnosis of osteoporosis, is prepared [39].

Pyrrolostatin, a novel inhibitor of lipid peroxidation, consists of a pyrrole-2-carboxylic acid with a geranyl group at the 4 position. Pyrrolostatin can be readily prepared by applying the Barton–Zard reaction, as shown in Scheme 11.23 [40].

Phytochrome, a chromoprotein, is found in a variety of higher plants and is involved with plant growth, development, and morphogenesis. The total synthesis of phytochromobilin and its related compounds (Figure 11.2) was achieved by Inomata and coworkers in an effort to understand the biological activities of these

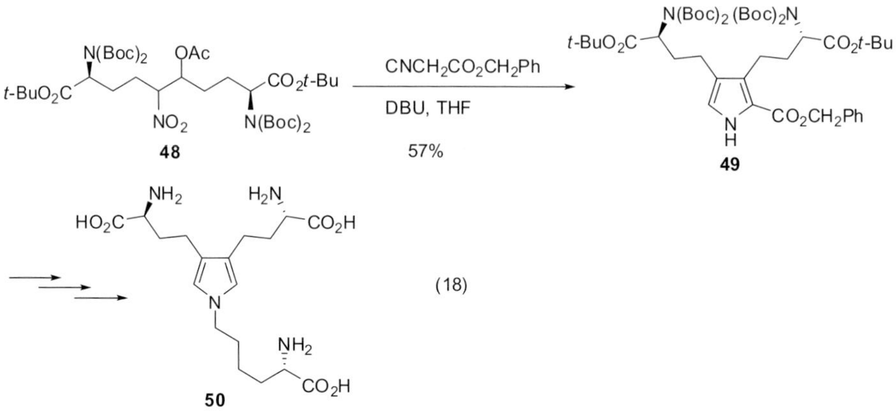

Scheme 11.22

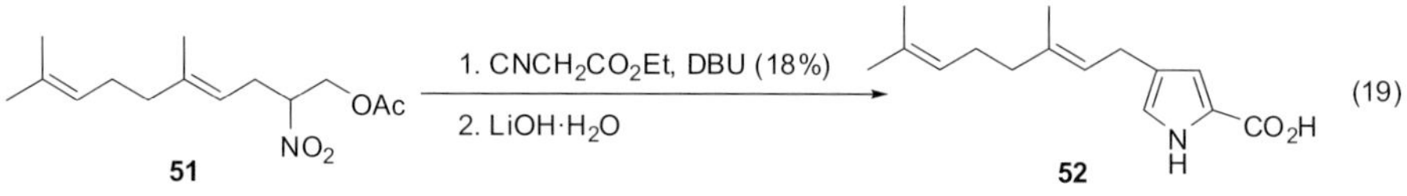

Scheme 11.23

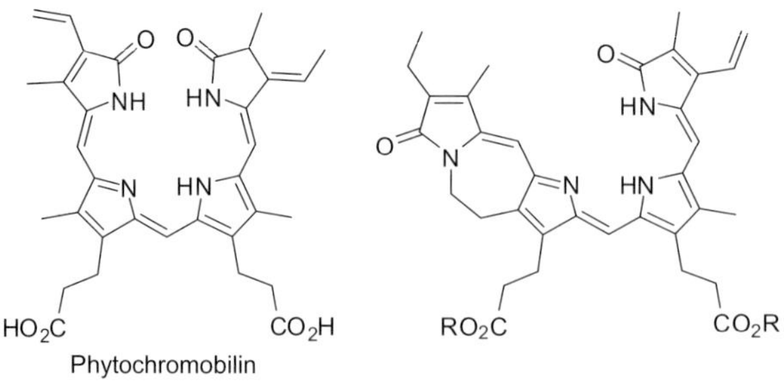

Figure 11.2 Phytochromobilin and related compounds.

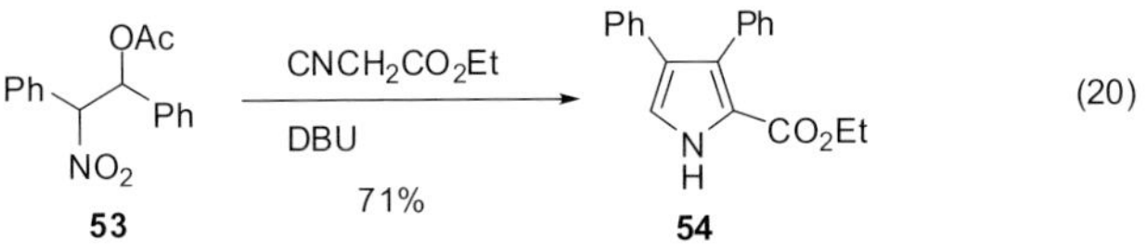

11.3 Synthesis of Pyrroles Using Isocyanoacetates | 395

Scheme 11.24

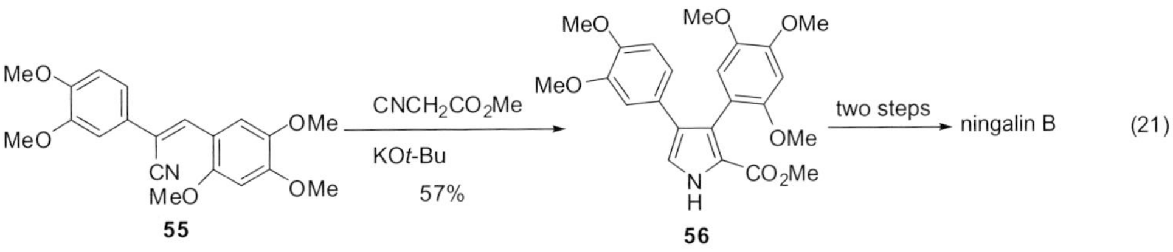

Scheme 11.25

materials. In this case, the Barton–Zard reaction was used to prepare the requisite pyrrole rings [41], with side chains such as methyl, vinyl, ethyl, and propionic acid being readily introduced via an application of nitroaldol reaction and a subsequent Barton–Zard reaction (see Schemes 11.17 and 11.18).

3,4-Diarylpyrroles, which are important in the synthesis of pyrrole alkaloids or porphyrins with altered electronic properties, can be readily prepared via the Barton–Zard reaction, using nitroalkenes (Scheme 11.24) [42]. However, the requisite α-nitrostilbenes, or their equivalents, are more difficult to prepare in some cases. Interestingly, α,β-unsaturated nitriles can be used in place of nitroalkenes for the preparation of 3,4-diarylpyrroles (Scheme 11.25) [43]. Marine natural products such as ningalin B (see Figure 11.1) may be readily prepared using this methodology; indeed, Anderson recently used this reaction to prepare 3,4-diarylpyrrole-2-carboxylate as a starting material for dodecaarylporphyrin, where the aryl group was is 4-iodo-3-alkoxyphenyl [44].

Pyrrole-2-carbaldehydes are important building blocks in the synthesis of oligopyrrolic macrocycles, such as porphyrins and linear oligopyrroles. Although such formylpyrroles are generally prepared via the Vilsmeier–Haack reaction of α-free pyrroles, in an alternative method the ester group of the Barton–Zard pyrrole is converted into a formyl group via reduction and oxidation. Subsequently, the Barton–Zard reaction of isocyanide substituted with Weinreb amide with nitroalkenes, followed by reduction with $LiAlH_4$, provides an excellent route to the preparation of pyrrole-2-carbaldehyde (Scheme 11.26) [45]. 4-Formylpyrrole-2-carboxylates may be prepared via the Barton–Zard reaction, using nitroacetaldehyde dimethylacetal, as shown in Scheme 11.27 [46]. Thus, the formyl group can be introduced at either the 2- or 4-position, in regioselective fashion. The 4-formyl group is then reduced to a 4-hydroxymethyl group with $NaBH_4$, which may be replaced by various nucleophiles such as RSH or ROH. The internal cyclization of 4-hydroxymethylpyrrole leads to bowl-shaped cyclononatripyrroles that adopt a

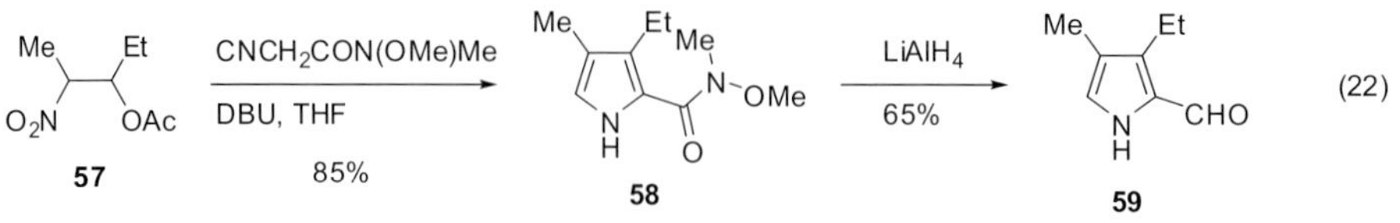

Scheme 11.26

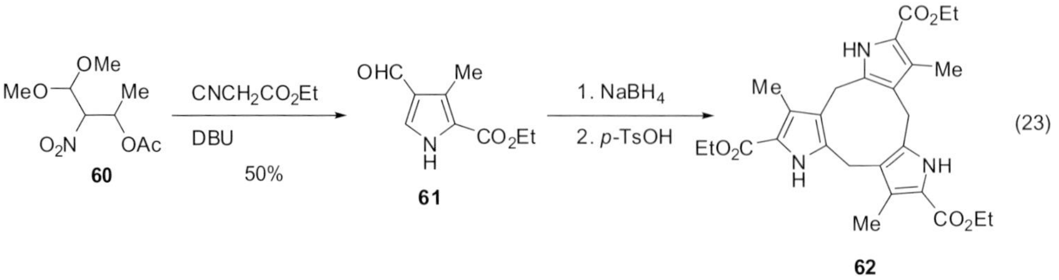

(22)

Scheme 11.27

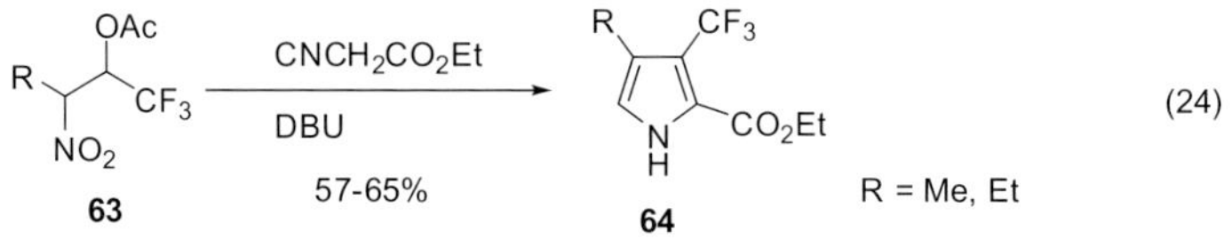

(23)

Scheme 11.28

(24) R = Me, Et

crown conformation in solution [46]. 3-Alkanoyl pyrroles may also be prepared by the similar procedures [47]. Indeed, the Barton–Zard reaction provides a simple route towards 3-trifluoromethyl pyrroles, which are not only interesting as biologically active compounds but also the targets of intensive investigations [48]. Trifluoromethyl-substituted β-nitroalkyl acetates may be readily prepared, starting from trifluoroacetic acid (TFA) and nitroalkanes; the latter pair react with isocyanoacetate in the presence of DBU to produce trifluoromethylpyrrole in good yield (Scheme 11.28) [49]. Other examples of pyrroles prepared via the Barton–Zard reaction include pyrroles with long alkyl groups [50], sugar molecules [51], and/or naphthyl groups [52]. The last of these pyrroles **67** has a chiral nature, and may be separated to generate an axially dissymmetric pyrrole (Figure 11.3). These compounds are very important starting materials for the synthesis of a variety of functional materials, the properties of which are controlled by the choice and presence of long alkyl groups, sugars, and aryl groups.

11.3.2
Synthesis from α,β-Unsaturated Sulfones

In 1984, Magnus reported that vinylsulfone-substituted pyrrole **68** would be converted into bipyrrole **69** on treatment with ethyl isocyanoacetate, in good yield

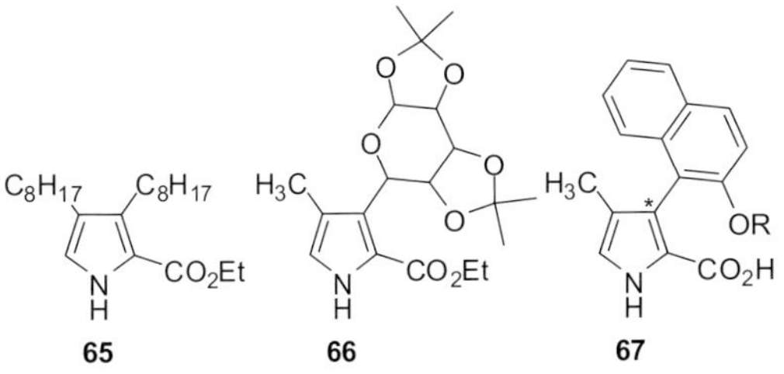

Figure 11.3 Some examples of compounds prepared via the Barton–Zard reaction.

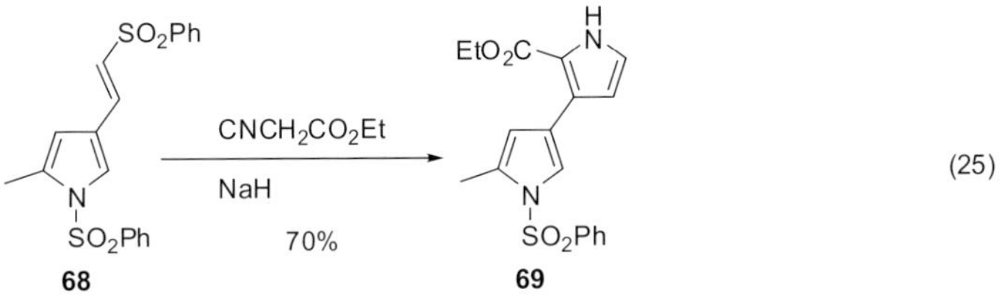

Scheme 11.29

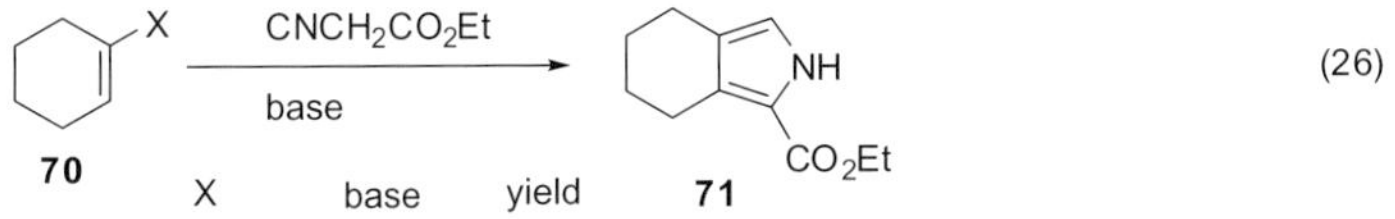

Scheme 11.30

X	base	yield
NO$_2$	DBU	90%
SO$_2$Ph	KO*t*-Bu	48%

(Scheme 11.29) [53]. This was the first description of the Barton–Zard-type pyrrole synthesis, though an earlier report by Arnold showed that the reaction of 1-phenylsulfonylcyclohexene **70** with ethyl isocyanoacetate in the presence of KO*t*-Bu would produce ethyl 4,5,6,7-tetrahydro-2*H*-isoindole-1-carboxylate **71**, in 48% yield (Scheme 11.30), whereas DBU was not effective for this reaction [54]. The same compound (**71**) was obtained in 90% yield by using 1-nitrocyclohexene and DBU as a base [55]. As nitroalkenes are typically more reactive as a Michael acceptor than vinyl sulfones, a Barton–Zard reaction using nitroalkenes would proceed more rapidly than if vinyl sulfones were used. However, the use of cyclic vinyl sulfones in the Barton–Zard reaction provides the advantage of a good availability of the requisite starting materials. For example, the Diels–Alder product **72** is readily converted into α,β-unsaturated sulfone **74** via the addition of PhSCl, followed by oxidation of the sulfides to the sulfones and the elimination of HCl; the Barton–Zard reaction of **74** then provides pyrrole **75** in 43% yield (Scheme 11.31) [56].

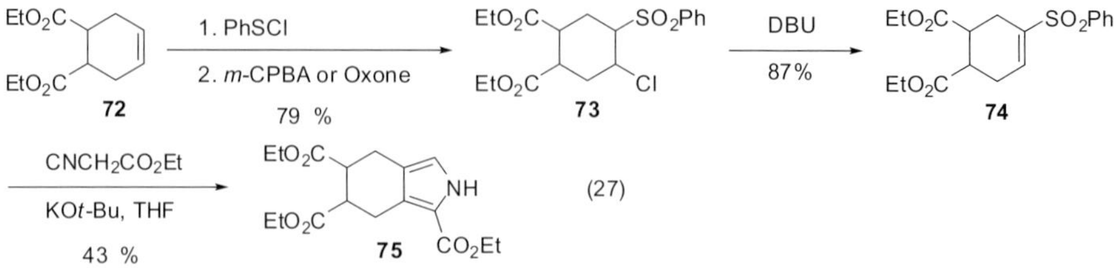

Scheme 11.31

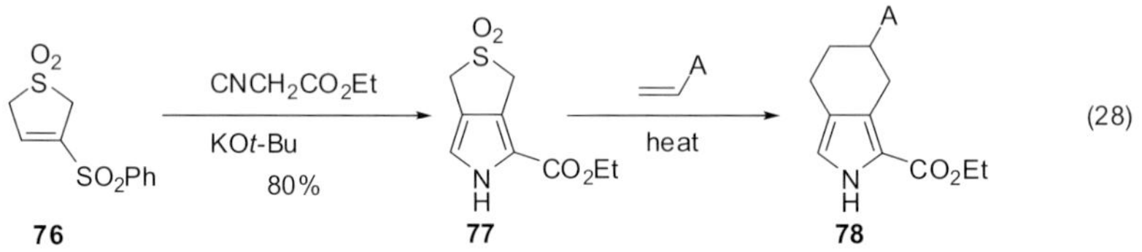

Scheme 11.32

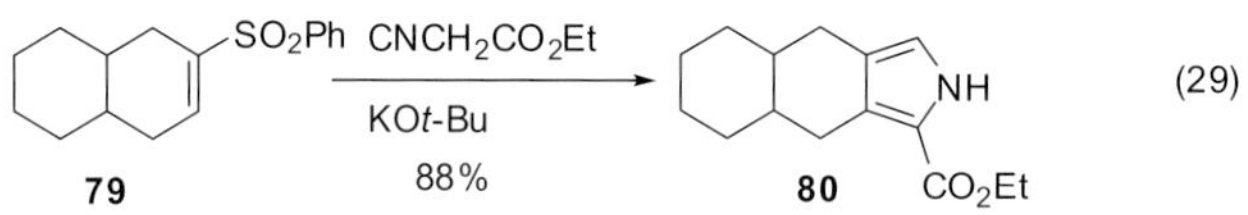

Scheme 11.33

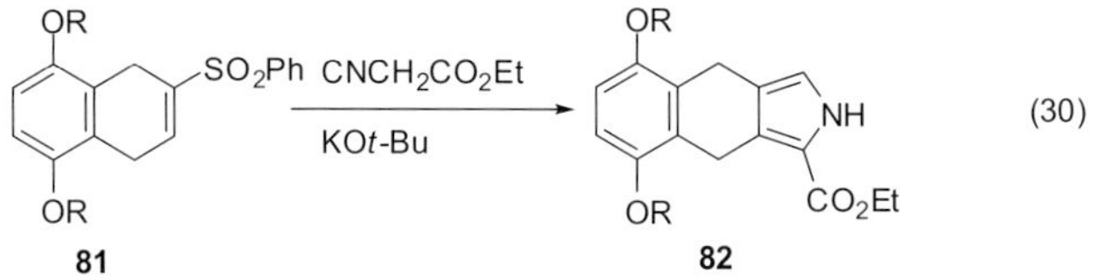

Scheme 11.34

Some other examples of this reaction are shown in Schemes 11.32 to 11.35. For example, the Barton–Zard reaction of 1-phenylsulfonyl-3-sulfolene affords 3,4-fused pyrrolo-3-sulfolene **77** [57], which undergo a Diels–Alder reaction with various dienophiles such as electron-deficient alkenes [58] involving C_{60} [59]. Vinogradov and coworkers have developed a strategy which employs tetrahydroisoindoles such as **71** and **75** as isoindole synthons in the preparation of benzoporphyrins. The pyrroles **80** and **82** may also be prepared via the Barton–Zard reaction, and used in synthesis of naphthoporphyrins [60]. In this case, the requisite starting materials **79** and **81** are prepared from the corresponding cyclohexene rings via a sequence similar to that shown in Scheme 11.31. However, a more useful method for the synthesis of dihydroisoindole was accomplished via the Diels–Alder reaction of tosylacetylene with butadiene, followed by the Barton–Zard reaction (see Scheme 11.35) [61].

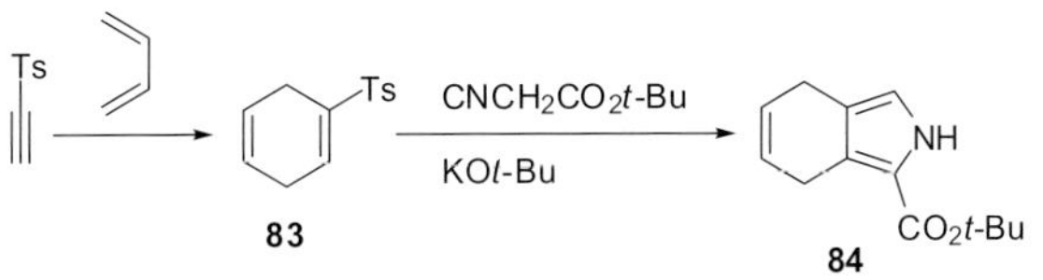

11.3 Synthesis of Pyrroles Using Isocyanoacetates | 399

Scheme 11.35

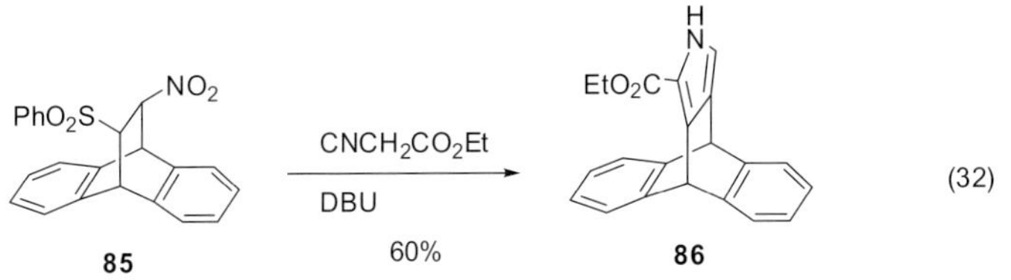

Scheme 11.36

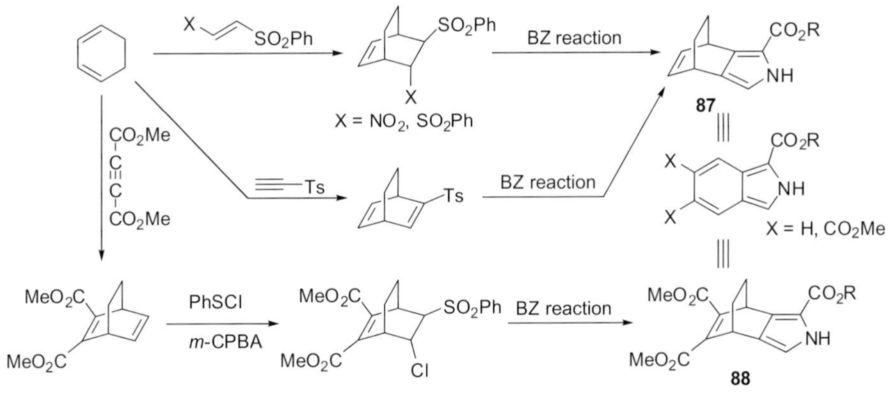

Scheme 11.37

In 1997, an elegant method that could be used to prepare the triptycene type of pyrrole **86** was reported via the Diels–Alder reaction of β-sulfonylnitroethylene with anthracene, followed by a Barton–Zard reaction (Scheme 11.36) [62a]. This process was shown to be very valuable for the synthesis of pyrroles fused with BCOD units, as detailed in Scheme 11.37. The first such synthesis was achieved via a Diels–Alder reaction of 1,3-cyclohexadiene with β-(phenylsulfonyl) nitroethylene [62], but the procedure was subsequently improved by employing (E)-1,2-bis(phenylsulfonyl)ethylene, followed by a Barton–Zard reaction [63]. (Z)-1,2-Bis(phenylsulfonyl)ethylene was also shown to be effective for this purpose. The Diels–Alder reaction of sulfonylacetylene, and the subsequent Barton–Zard reaction, represents the easiest means of obtaining the requisite pyrrole [64]. A combination of 1,3-cyclohexadienes with various dienophiles and sulfonylation affords various types of BCOD-fused pyrroles, which are regarded as masked isoindoles (see Scheme 11.37) [65]. As these pyrroles are converted to isoindoles by heating, they have been used extensively in the synthesis of benzoporphyrins (see Section 11.4). This strategy may be further extended to the synthesis of 4,9-dihydro-4,9-ethano-2H-benz[f]isoindole **91** as a synthon of 2H-benz[f]isoindole [66]. In this case, the requisite starting materials are prepared via a Diels–Alder reaction of benzyne or benzoquinone with 1,3-cyclohexadiene. The reaction of benzyne with 1,3-cyclohexadiene produces **89**, which is in turn converted into **90** by the addition of PhSCl, oxidation, and the elimination of HCl. Sulfone **90** is transformed into pyrrole **91** by the Barton–Zard reaction (Scheme 11.38). The pyrrole **93** can be prepared by a reaction of 1,4-naphthoquinone (**92**) with 1,3-cyclohexadiene, followed by sulfonylation and the Barton–Zard reaction (Scheme 11.39) [67]. The pyrroles **87**, **91**, and **93** are subsequently converted into pyrroles fused with benzene, naphthalene, and anthracene, respectively, via the retro Diels–Alder reaction as discussed in Section 11.4.

The Barton–Zard reaction using α,β-unsaturated sulfones is especially useful for the preparation of pyrrole-2-carboxylates with EWGs at the 4-position. For example, CF_3, CN, and CO_2R may be readily introduced into the β-position of pyrroles in good yields by this process, as shown in Schemes 11.40 and 11.41 [68].

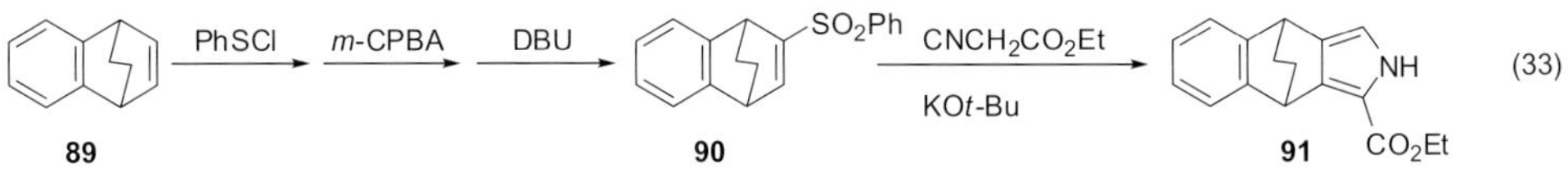

Scheme 11.38

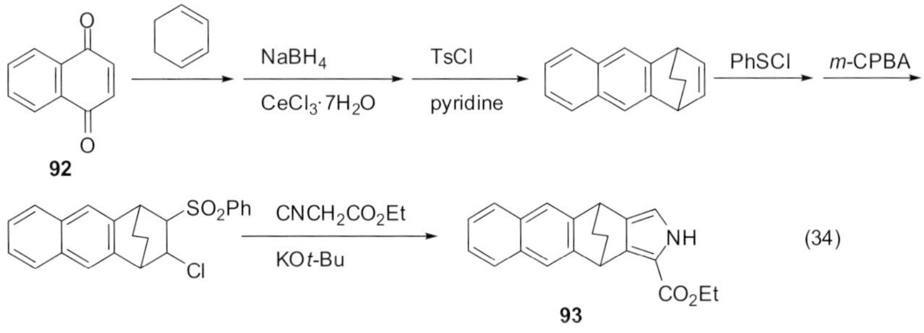

Scheme 11.39

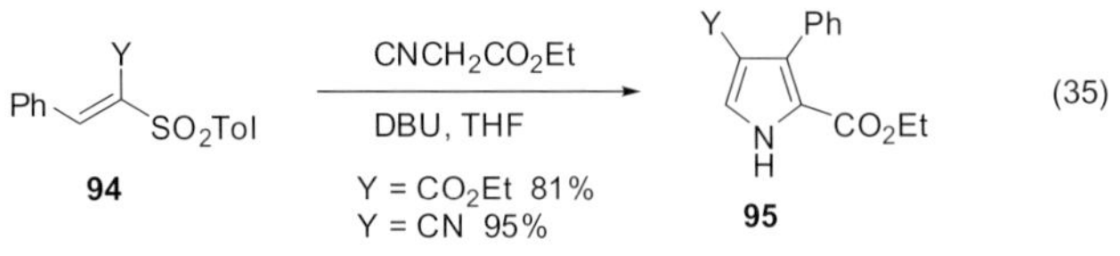

Scheme 11.40

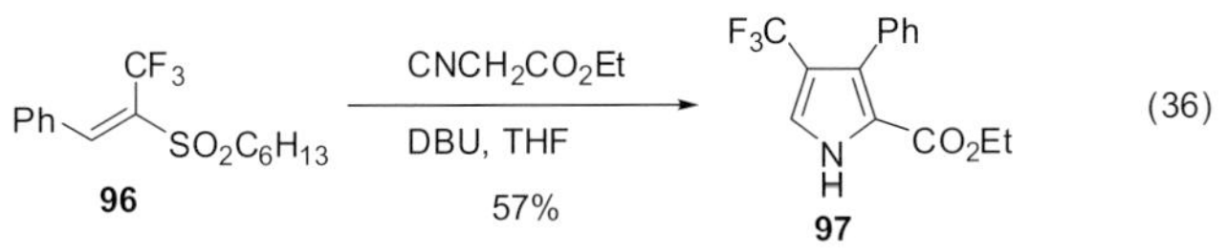

11.3 Synthesis of Pyrroles Using Isocyanoacetates

Scheme 11.41

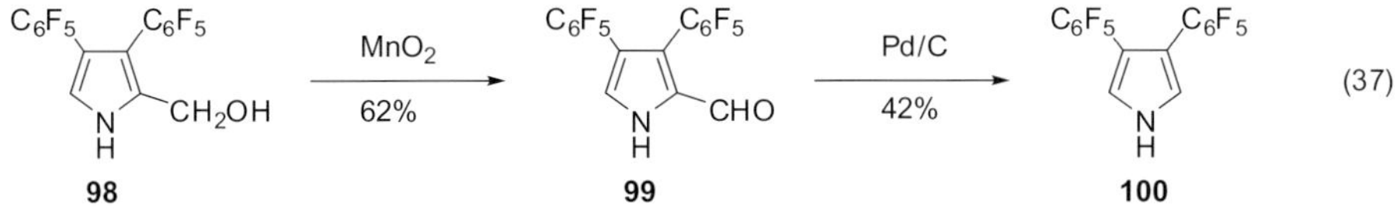

(36): Ph-C(CF$_3$)=C(SO$_2$C$_6$H$_{13}$)(Ph) (**96**) + CNCH$_2$CO$_2$Et, DBU, THF, 57% → 3-(F$_3$C)-4-Ph-pyrrole-2-CO$_2$Et (**97**)

Scheme 11.42

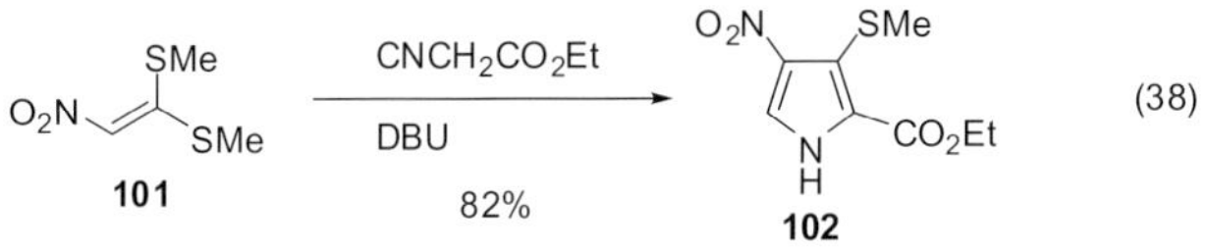

(37): 3,4-bis(C$_6$F$_5$)-2-(CH$_2$OH)pyrrole (**98**) → [MnO$_2$, 62%] → 3,4-bis(C$_6$F$_5$)-2-CHO-pyrrole (**99**) → [Pd/C, 42%] → 3,4-bis(C$_6$F$_5$)pyrrole (**100**)

Scheme 11.43

(38): O$_2$N-C(SMe)=C(SMe) (**101**) + CNCH$_2$CO$_2$Et, DBU, 82% → 3-O$_2$N-4-SMe-pyrrole-2-CO$_2$Et (**102**)

The choice of base for the Barton–Zard reaction depends on the reactivity of the electron-deficient alkenes. Although DBU is sufficiently reactive in the reaction with nitroalkenes, a stronger base such as KO*t*-Bu is required in the reaction of α,β-unsaturated sulfones shown in Scheme 11.30. As two EWGs are present in **98**, the use of DBU leads to the production of pyrroles **99** in good yields, but when a nonionic strong base, such as *t*-butyl-iminotris(pyrrolidino)phosphorane (BTPP), is employed in the reaction of Scheme 11.41, the yield is increased to 79%. However, when Y = F in Scheme 11.40, the yield of 4-fluoropyrrole-2-carboxylate will be poor, due to the formation of byproducts [69]. Pyrroles with C$_6$F$_5$ groups at the β-positions may be prepared via the Barton–Zard reaction of α,β-unsaturated sulfones with C$_6$F$_5$ groups, followed by reduction with LiAlH$_4$, oxidation with MnO$_2$, and decarbonylation with Pd/C (Scheme 11.42) [70].

Nitroketene *S,S*- and *N,S*-acetals with an EWG may also be employed in the Barton–Zard reaction, where the methylthio group is eliminated to form pyrroles with EWGs (NO$_2$, CN, CO$_2$R) at the 4-position, as shown in Scheme 11.43 [71].

11.3.3
Synthesis from Alkynes

The reaction of alkyl isocyanoacetate with electron-deficient alkynes provides an efficient method for the synthesis of pyrrole-2-carboxylates, as shown in Schemes 11.44 to 11.46. In this reaction, which is catalyzed either by a base or copper [72], MTBE is again recommended as the solvent (though it is unclear why MTBE is

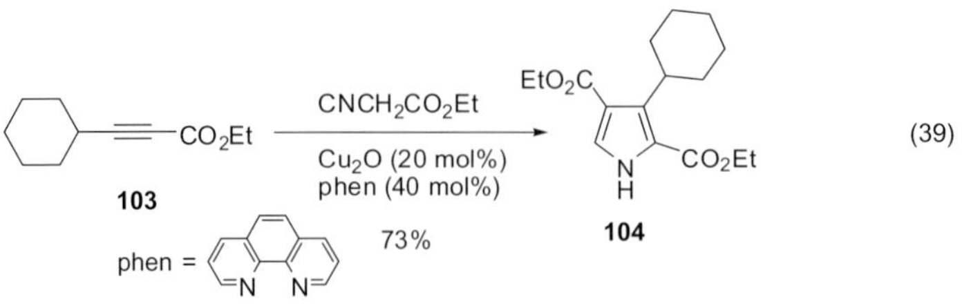

Scheme 11.44

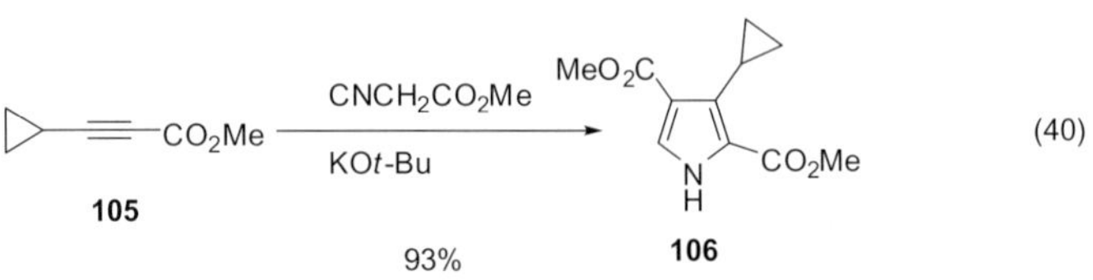

$$\text{103} + \text{CNCH}_2\text{CO}_2\text{Et} \xrightarrow[\text{phen (40 mol\%)}]{\text{Cu}_2\text{O (20 mol\%)}} \text{104} \quad (39)$$

Cy–C≡C–CO$_2$Et (103); product 104: pyrrole with Cy, CO$_2$Et (EtO$_2$C) groups; 73%

phen = 1,10-phenanthroline

Scheme 11.45

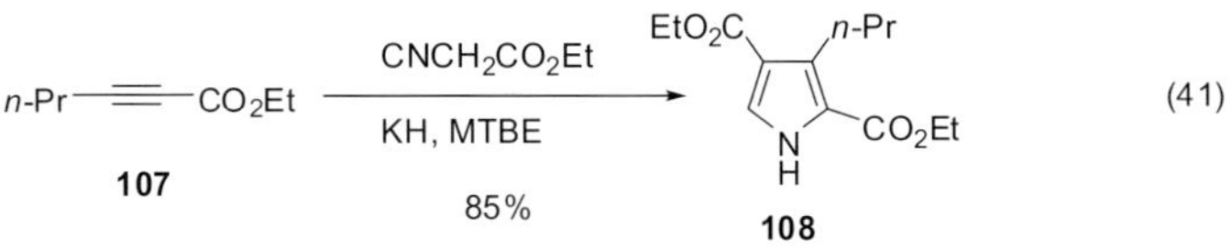

$$\text{105} + \text{CNCH}_2\text{CO}_2\text{Me} \xrightarrow{\text{KO}t\text{-Bu}} \text{106} \quad (40)$$

cyclopropyl–C≡C–CO$_2$Me (105); product 106; 93%

Scheme 11.46

$$\text{107} + \text{CNCH}_2\text{CO}_2\text{Et} \xrightarrow{\text{KH, MTBE}} \text{108} \quad (41)$$

n-Pr–C≡C–CO$_2$Et (107); product 108; 85%

superior to THF) [37]. These pyrroles can be prepared via the Barton–Zard reactions using α,β-unsaturated sulfones (as shown in Scheme 11.19), with the choice of method depending on the availability of the starting materials.

11.3.4
Synthesis from Aromatic Nitro Compounds: Isoindole Derivatives

The most impressive feature of the Barton–Zard reaction is the synthesis of pyrroles fused with aromatic rings from aromatic nitro compounds. In 1991, Ono and coworkers described the synthesis of pyrroles fused with polycyclic aromatic rings, as shown in Scheme 11.47 [73]. In this case, the ethoxycarbonyl group is readily removed by heating with KOH in ethylene glycol to produce the α-free pyrroles that serve as important starting materials for porphyrins or polypyrroles. Lash and coworkers also reported the synthesis of phenanthropyrrole **110** and phenanthrolinopyrrole **111**, respectively, using the same procedure [74]. The method that employs aromatic nitro compounds in the Barton–Zard reaction is very attractive, as these may be readily prepared via the nitration of aromatic compounds.

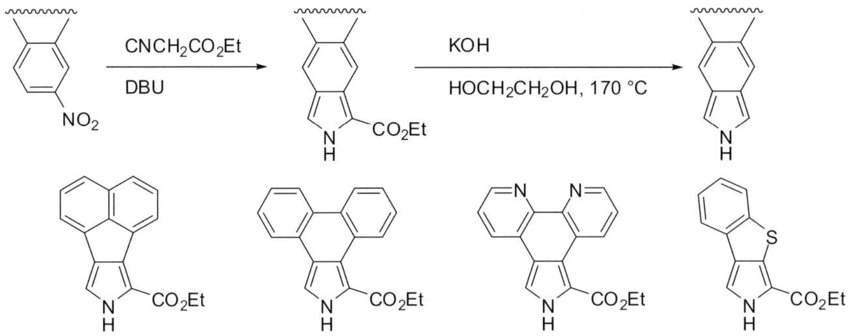

11.3 Synthesis of Pyrroles Using Isocyanoacetates | 403

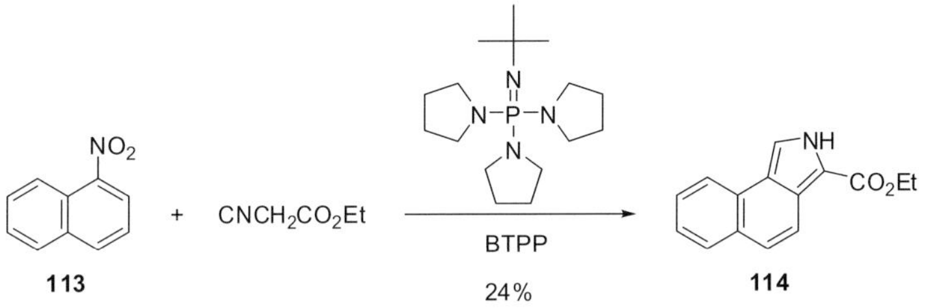

Scheme 11.47

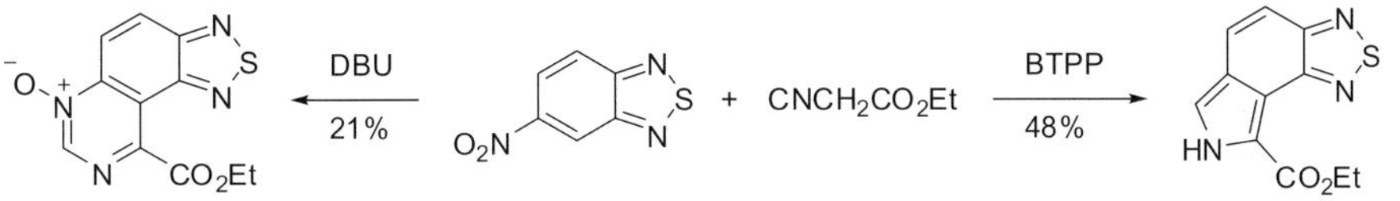

Compounds: **109** 70%, **110** 75%, **111** 87%, **112** 60%

Scheme 11.48

$$113 + CNCH_2CO_2Et \xrightarrow[24\%]{BTPP} 114 \qquad (42)$$

Scheme 11.49

$$115 \xleftarrow[21\%]{DBU} 116 + CNCH_2CO_2Et \xrightarrow[48\%]{BTPP} 117 \qquad (43)$$

In contrast, 1- or 2-nitronaphthalenes are less reactive in the Barton–Zard reaction than are the nitroaromatics used in Scheme 11.47. A super-strong nonionic base, such as phosphazene base BTPP, was effective in inducing the Barton–Zard reaction of 1-nitronaphthalene to produce **114** in 24% yield, whereas the DBU-catalyzed reaction provided the same product in only 2% yield (Scheme 11.48) [75]. In some cases, when DBU either fails or provides nonpyrrolic products, the use of BTPP will lead to the provision of pyrroles, as shown in Scheme 11.49 [75]. Thus, the use of BTPP as a base expands the scope and limitation of the Barton–Zard reaction for aromatic nitro compounds [75, 76]. Recently, both the pyrrenopyrrole **119** and corannulenopyrrole **121** were prepared in good yields by the

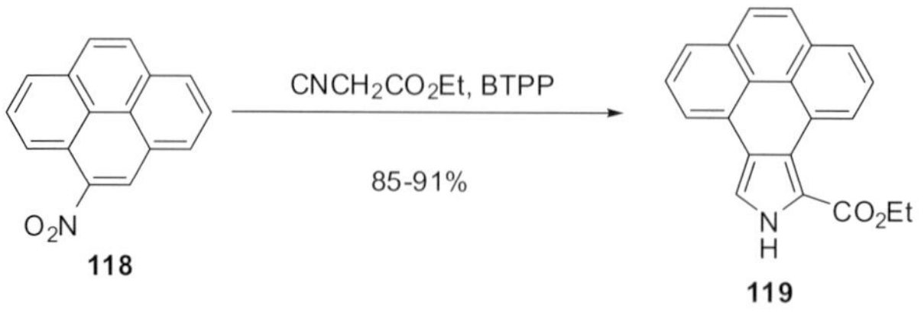
CNCH2CO2Et, BTPP
85-91%
O2N
118
N
H
CO2Et
119

(44)

Scheme 11.50

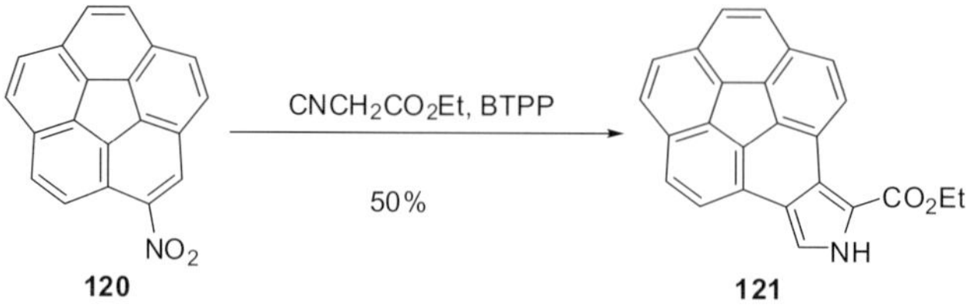
CNCH2CO2Et, BTPP
50%
NO2
120
CO2Et
NH
121

(45)

Scheme 11.51

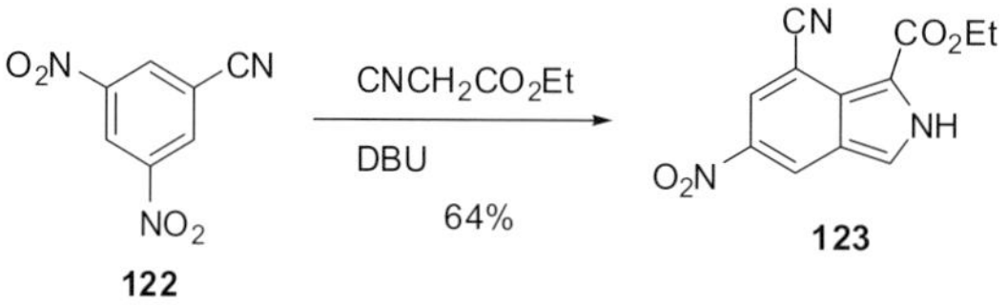
O2N
CN
NO2
122
CNCH2CO2Et
DBU
64%
CN
CO2Et
NH
O2N
123

(46)

Scheme 11.52

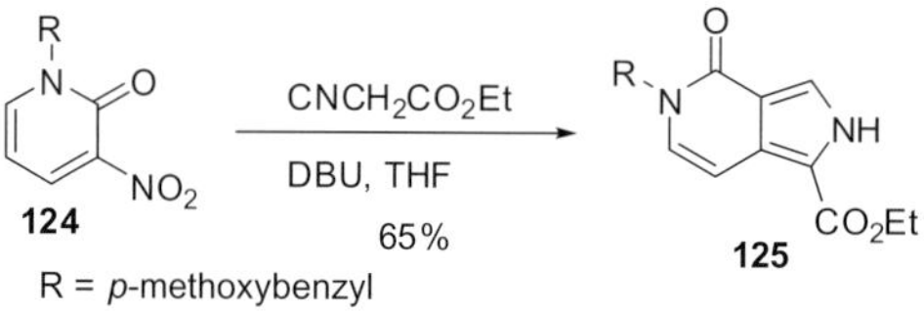
R
N
O
NO2
124
R = p-methoxybenzyl
CNCH2CO2Et
DBU, THF
65%
O
R
N
NH
CO2Et
125

(47)

Scheme 11.53

corresponding nitro compounds via the Barton–Zard reaction, using BTPP as a base (see Schemes 11.50 [77] and 11.51 [78]).

Although nitrobenzene alone is not sufficiently reactive to produce isoindole via the Barton–Zard reaction, *m*-dinitrobenzene or its derivative **122** can provide an isoindole (e.g., **123**) directly via the Barton–Zard reaction (Scheme 11.52) [79].

The Barton–Zard reaction can be applied to various heterocyclic nitro compounds to provide isoindole derivatives, all of which are difficult to prepare via other methods. Some examples are presented in Schemes 11.53 [80], 11.54 [80], 11.55 [81], and 11.56 [81]. The 3-nitroindoles show an interesting reactivity towards

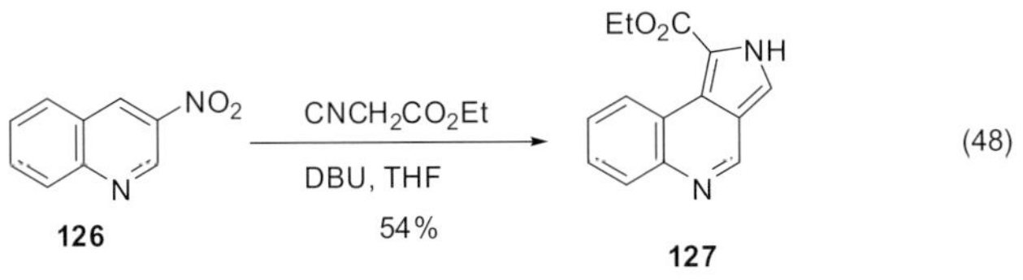

Scheme 11.54

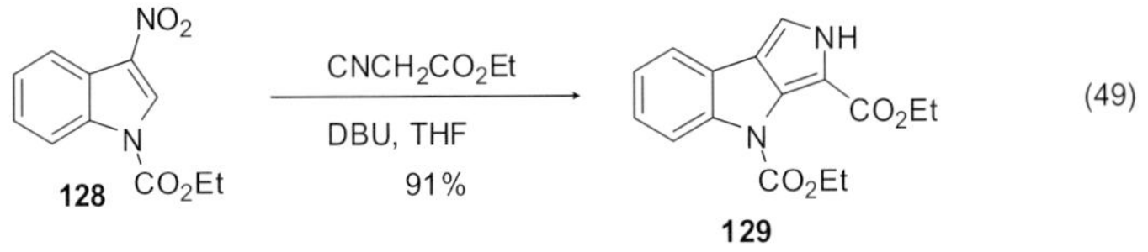

Scheme 11.55

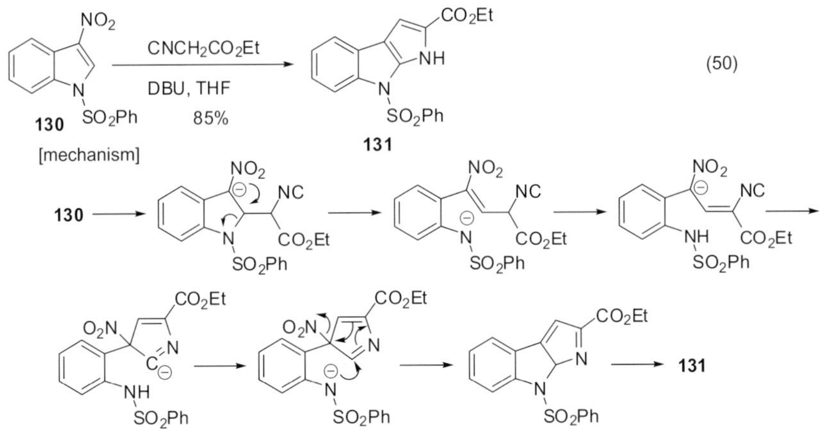

Scheme 11.56

the anion of ethyl isocyanoacetate; N-alkoxycarbonyl derivatives provide the normal product, the pyrrolo[3,4-b]indole ring system (Scheme 11.55), whereas the N-sulfonyl derivatives produce the pyrrolo[2,3-b]indole ring system (Scheme 11.56). β-Nitroporphyrins **132** also undergo the Barton–Zard reaction to afford fused pyrroloporphyrins **133**, as shown in Scheme 11.57, which may serve as useful precursors for highly conjugated porphyrin oligomers [82].

Thus, the Barton–Zard reaction of nitroalkenes or nitro aromatic compounds with ethyl isocyanoacetate has a wide applicability for pyrrole synthesis. However, it does not afford always pyrroles, but in some cases will afford pyrimidine or pyrazoles, as shown in Schemes 11.58 [83], 11.59 [84], and 11.60 [85].

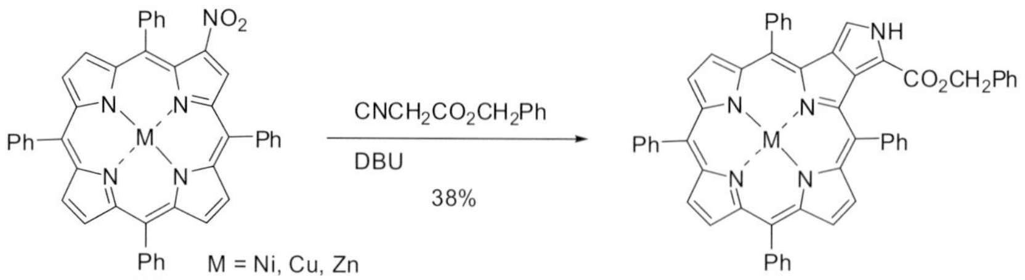

Scheme 11.57

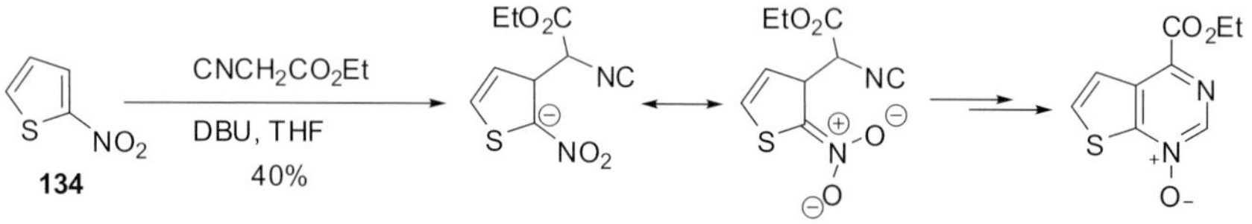

(51)

Scheme 11.58

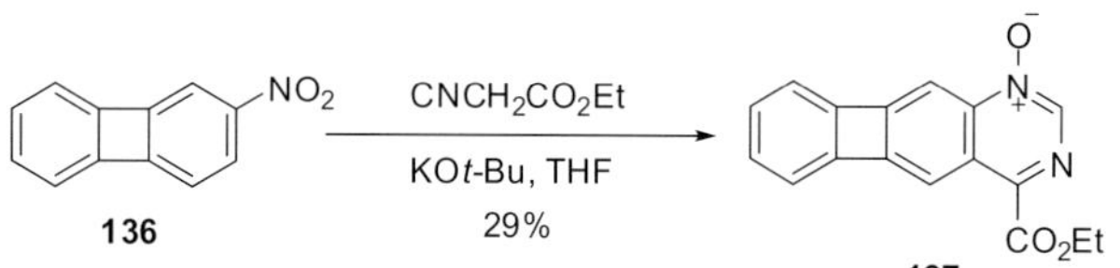

(52)

Scheme 11.59

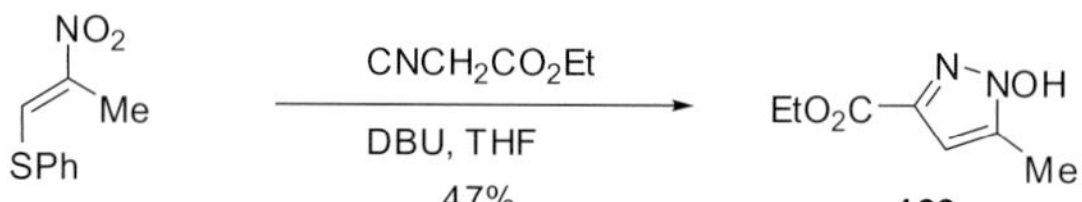

(53)

Scheme 11.60

(54)

An additional limitation is that the Barton–Zard reaction of nitrobenzene or nitronaphthalene does not produce benzo[c]pyrrole (2H-isoindole, **140**) or naphtho[2,3-c]pyrrole (2H-benz[f]isoindole, **141**) directly. However, when using cyclic α,β-unsaturated sulfones, the Barton–Zard reaction will produce their equivalents, as shown in Scheme 11.61. Isoindole frameworks are important as building blocks of the phthalocyanines or benzoporphyrins. Typically, isoindoles are prepared via the oxidation of isoindolines, which usually are obtained from 1,2-disbustituted benzenes [86]. However, isoindole synthons prepared via the Barton–Zard reaction may find their utility in porphyrin chemistry (as discussed in Section 11.4).

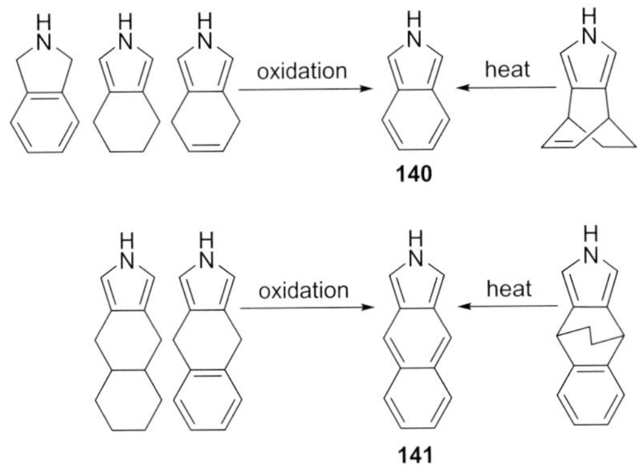

11.4 Synthesis of Porphyrins and Related Compounds

Scheme 11.61

11.4
Synthesis of Porphyrins and Related Compounds

The Barton–Zard pyrroles described in Section 11.3 are converted into porphyrins by several routes. The simplest method is tetramerization of 2-hydroxymethylpyrroles which are readily obtained by the reduction of ester function pyrrole carboxylates, and the subsequent oxidation. This method is useful for synthesis of symmetrically substituted *meso*-free porphyrins (Ono method). The ester function is readily removed to give α-free pyrroles, which react with aldehydes to give *meso*-substituted porphyrins (Lindsey method). Interestingly, the Lindsey procedure affords not only porphyrins but also expanded or contracted porphyrins. Also various types of porphyrin isomers such as sapphyrin, corrole, porphycene, core-modified porphyrins are prepared from the Barton–Zard pyrroles. Furthermore, Lash has developed [2+2] condensation of dipyrromethane or [3+1] condensation of tripyrrane for synthesis of unsymmetrical porphyrins, where the Barton–Zard pyrroles are extensively used.

11.4.1
Tetramerization

In 1985, a simple method was reported for the conversion of Barton–Zard pyrroles into porphyrins, whereby a synthetic sequence of: (i) reduction with $LiAlH_4$; (ii) H^+; and (iii) an oxidizing agent provided porphyrins in 30–50% yield. Today, this is a standard protocol for porphyrin synthesis, with the Barton–Zard reaction readily affording various pyrrole-2-carboxylates, porphyrins (e.g., **142** [55]) or octaethyl-porphyrin (OEP) via this procedure (Scheme 11.62). When the substituents at the 3- and 4-positions are not identical, four possible isomers may be formed. However, the scrambling of the porphyrin isomers can be minimized to produce Type I porphyrins selectively, by the choice of substituents and reaction

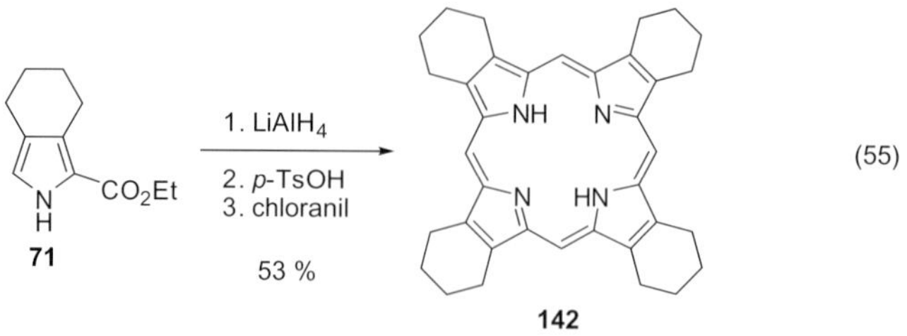

Scheme 11.62

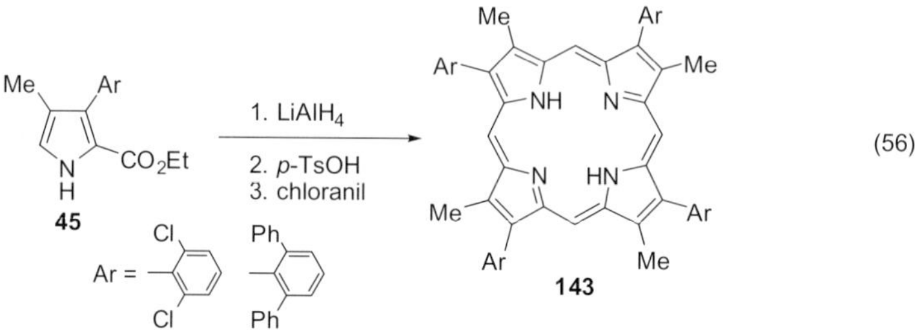

(55)

Scheme 11.63

(56)

conditions. When the substituents are sterically hindered or EWGs, then Type I porphyrins will be obtained as shown in Schemes 11.63 [37] and 11.64 [87]. Oxoiron(IV) porphyrin π-cation radicals are believed to participate in biochemical oxidations, such as horseradish peroxidase and catalase. Iron complexes of porphyrins substituted with a bulky group at the β- or *meso*-positions and their oxidized species provide a good model of such systems [88–90]. The electron paramagnetic resonance (EPR) and ultraviolet-visible spectra of these model compounds, when derived from β-substituted porphyrins, are strikingly similar to those of compound I of natural catalase and peroxidase.

Pyrroles with EWGs prepared by the Barton–Zard reaction are converted into Type I porphyrins via a standard procedure (LiAlH$_4$-reduction, H$^+$-cyclization, and oxidation). As an example, the reduction of 3-aryl-2,4-diethoxycarbonylpyrroles **144** with LiAlH$_4$ produces 3-aryl-2-hydroxymethyl-4-ethoxycarbonylpyrroles selectively, which in turn affords porphyrinogens **145** on treatment with acids. In some cases, hexaphyrinogens **146** may also be isolated; these serve as intermediates to the porphyrins or hexaphyrins, but are not generally isolated under aerobic conditions. Although the porphyrinogens **145** are oxidized to the corresponding porphyrins **147** on treatment with 2,3-dichloro-5,6-dicyanobenzoquinone (DDQ), the hexaphyrinogens **146** are very stable under the conditions using DDQ, as shown in Scheme 11.64 [87]. The strong intramolecular hydrogen bonding between C=O and NH maintains a rigid framework of hexaphyrinogens; this has been confirmed using X-ray analysis, and represents a first case of the isolation

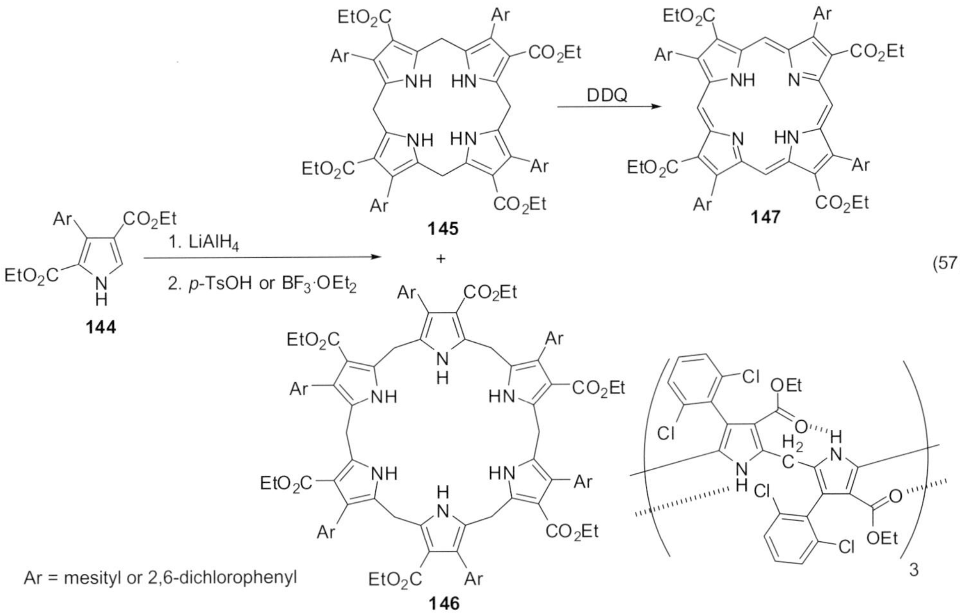

11.4 Synthesis of Porphyrins and Related Compounds

Scheme 11.64

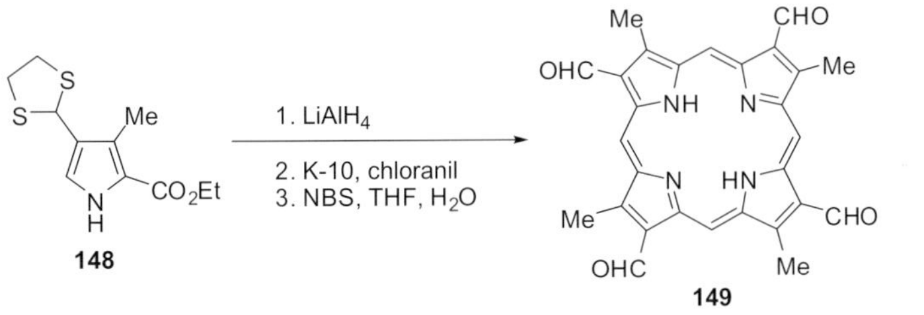

Ar = mesityl or 2,6-dichlorophenyl

(57)

Scheme 11.65

(58)

of stable hexaphyrinogens. As hexaphyrinogens are stabilized by their hydrogen bonding with β-ethoxycarbonyl groups, pyrroles with other EWGs (such as C_6F_5, CF_3) produce – on a selective basis – not hexaphyrinogens but rather porphyrinogens which, ultimately, will be oxidized to the corresponding electron-deficient porphyrins [91].

In order to minimize the scramble of porphyrin isomers, solid acid catalysts such as K-10 are effective, as shown in Scheme 11.65, where β-tetraformylporphyrin is prepared via the Barton–Zard pyrrole [92].

The bicyclo[2.2.2]octadiene (BCOD)-fused pyrrole **87** prepared via the Baton–Zard reaction may be employed as a masked isoindole, as it provides isoindole on

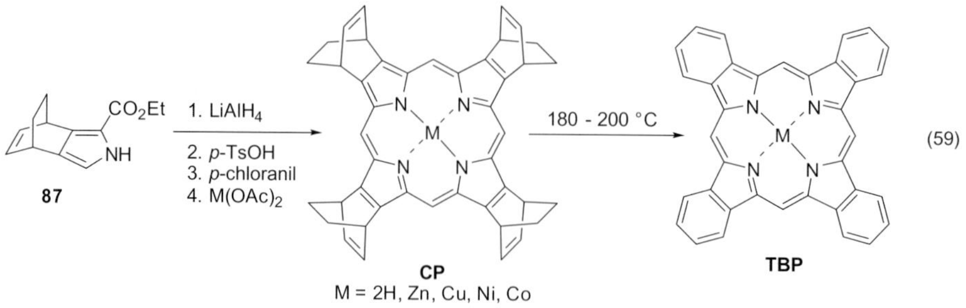

Scheme 11.66

heating via the retro Diels–Alder reaction. As an example, tetrabenzoporphyrin (TBP) and its metal complexes may be prepared using this method (Scheme 11.66) [62]. The CPs are obtained, starting from **87**, in 50–80% yield via a standard method (LiAlH$_4$, H$^+$, oxidizing agent). As CP is soluble in organic solvents, it can be purified by using column chromatography to produce a pure sample (99.9% purity, as assessed with high-performance liquid chromatography) that may then be converted into pure TBP, in quantitative yield, by heating at 180–200 °C. Thus, insoluble TBP can be prepared in a highly pure form via this thermal process [65b, 93]. The extension of TBP synthesis leads to the synthesis of tetranaphtho- and tetraanthraporphyrins from the corresponding BCOD-fused pyrroles **91** and **93** [66, 67].

The oxidation of fused cyclohexene or cyclohexane rings at the β-positions is also useful in the preparation of benzo- and naphtho-porphyrins. For example, TBP may be prepared starting from pyrrole **77**, using the Barton–Zard reaction; this would then be followed by a Diels–Alder reaction with vinylphenylsulfone, porphyrin synthesis, the elimination of sulfonyl groups, and the oxidative aromatization of fused-cyclohexene rings [58].

The Diels–Alder reaction of acenes such as anthracene or pentacene with appropriate dienophiles, followed by the Barton–Zard reaction and conversion into porphyrins, provides a unique porphyrin framework which has a deep cavity (Scheme 11.67) [94]. Such porphyrins with well-defined cavities may find use as enzyme models and for size-selective molecular recognition purposes.

Fluoranthene-fused TBP **154** was also synthesized, using a similar strategy, as a highly conjugated porphyrin [95]. The tetramerization of pyrrolecarboxylate **152** led to the production of porphyrin **153**, which was in turn converted into TBP and fused with four fluoranthene rings via the retro Diels–Alder reaction, as shown in Scheme 11.68. The absorptions of **154** demonstrated a red-shifted Soret band at 464 nm (log ε 5.45) and a strong Q band at 751 nm (log ε 5.65).

The Diels–Alder adduct **155** between pyrrole **77** in Scheme 11.32 and *meso*-tetraphenylporphyrin (TPP) is further converted into TBP fused with four TPPs, as shown in Scheme 11.69, via an oxidative aromatization of the precursor [96].

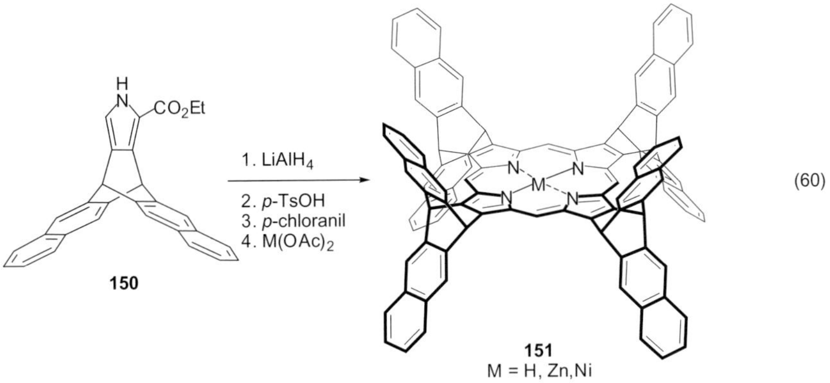

11.4 Synthesis of Porphyrins and Related Compounds

(60)

Scheme 11.67

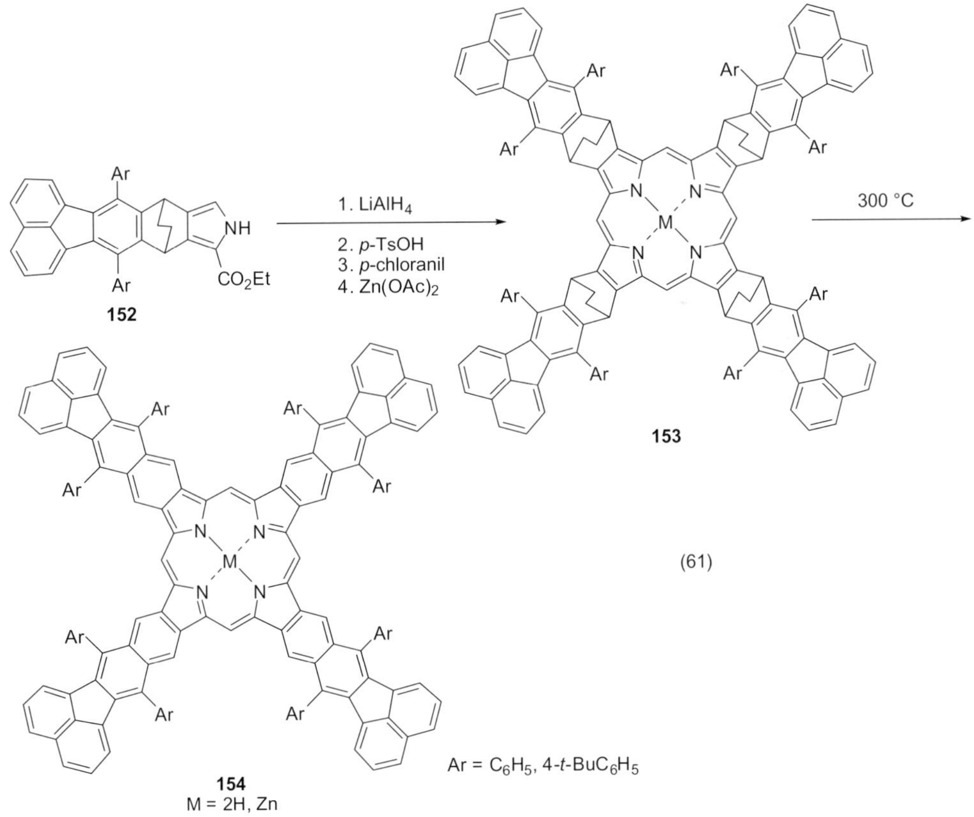

(61)

Ar = C$_6$H$_5$, 4-t-BuC$_6$H$_5$

Scheme 11.68

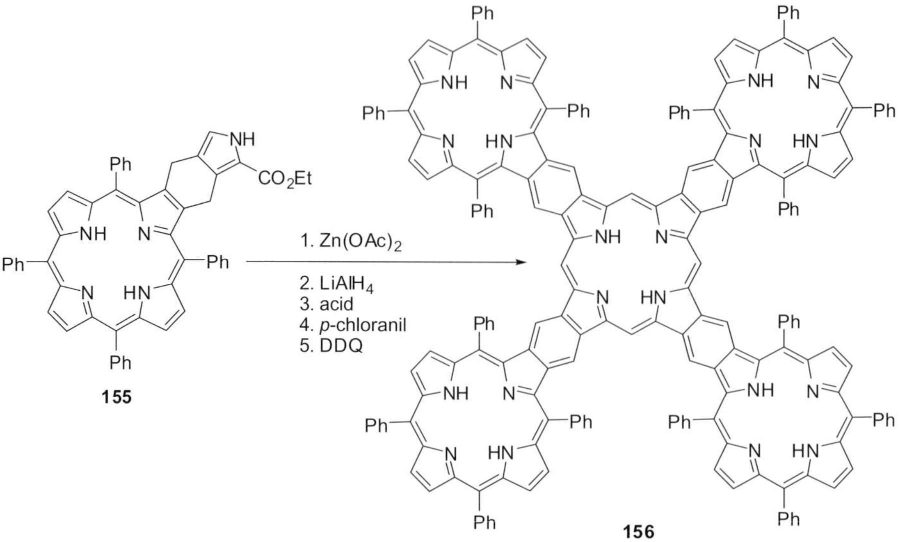

Scheme 11.69

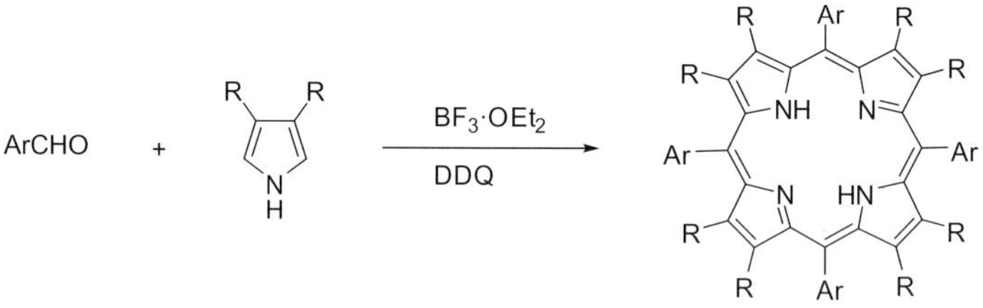

Scheme 11.70

11.4.2
Meso-Tetraarylporphyrins via the Lindsey Procedure

The reaction of α-free pyrroles with aromatic aldehydes in the presence of a Lewis acid, with subsequent oxidation, represents a very important method for porphyrin synthesis. Lindsey and coworkers have examined this process closely to identify the preferred conditions for porphyrin synthesis (Scheme 11.70) [97].

As the extension of porphyrin conjugation induces a strong absorption in the far-red region of the visible spectrum, such highly conjugated porphyrins have during recent years been the subject of intense research, as they are important materials for use as nonlinear optical systems, sensors, and sensitizers of PDT [98]. The use of pyrroles fused with aromatic rings represents a valuable means of extending porphyrin conjugation, with such pyrroles being prepared via the Barton–Zard reaction using aromatic nitro compounds (except for nitrobenzene,

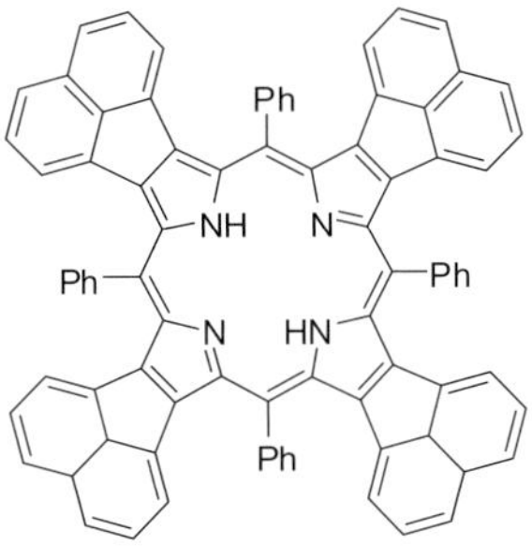

Figure 11.4 Structure and absorption maxima of *meso*-phenylacenaphtho- and phenanthroporphyrins.

157 λ_{max} 556, 638, 705, 790 nm

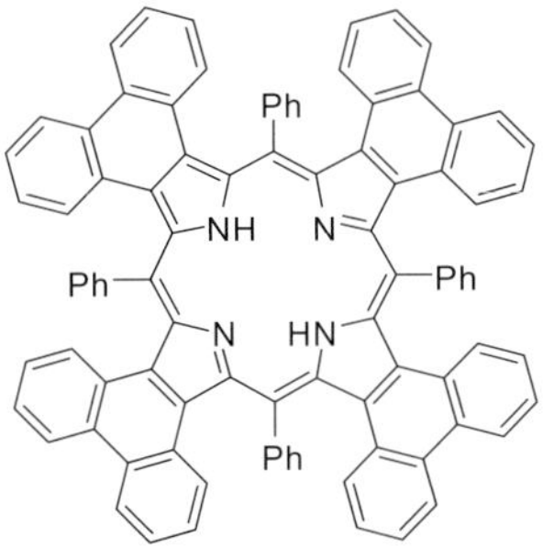

158 λ_{max} 577, 725, 796 nm

as shown in Scheme 11.47). Recently, new series of conjugated porphyrins have been prepared via this strategy, using polycyclic aromatic nitro compounds, by the groups of Ono [83, 99] and Lash [100], in independent fashion (for a review of this topic, see Ref. [31]).

The effectiveness of this approach in inducing the red-shift of absorption is shown in Figure 11.4. In this respect, Lash and coworkers have prepared a series of tetraacenaphthoporphyrins with *meso*-aryl groups, which exhibited highly redshifted absorption spectra [100]. In the case of these materials, the Soret band appeared at 556 nm, and it was claimed that the unusual red-shift was due to a special effect of an acenaphthylene group. Subsequently, Shen and coworkers reported that the absorption of a series of *meso*-tetraaryltetraphenathroporphyrin was more red-shifted than that of the acenaphthoporphyrins [101]. Consequently, there seems to be no special meaning to the acenaphthylene rings in terms of their red-shifted absorption of porphyrins. When the unusual red-shift of these nonplanar porphyrins was recently subjected to theoretical calculations, it became clear that the red-shifted absorption of the porphyrins in Figure 11.4 was due to the nonplanarity of porphyrins in raising the highest occupied molecular orbital (HOMO) energy and conjugation between the phenyl and expanded porphyrin rings [102].

The main limitation of the Barton–Zard pyrroles in the preparation of porphyrins is that nitrobenzene and 2-nitronaphthalene do not provide 2*H*-isoindole or benz[*f*]isoindole, respectively. In general, such isoindole derivatives are difficult to prepare owing to their instability. *Meso*-substituted TBPs, tetranaphtho[2,3] porphyrins (TNPs), and tetraanthra[2,3]porphyrins (TAPs) may be synthesized via retro Diels–Alder or oxidative–aromatization methods from the corresponding precursors prepared by the Lindsey method. The BCOD-fused porphyrins with *meso*-substituents are then converted into the *meso*-substituted TBPs **160** by heating in near-quantitative yields, as shown in Scheme 11.71 [62, 103]. A similar strategy affords *meso*-arylTNPs and *meso*-arylTAPs, by heating the precursors at 290–300 °C; the latter were prepared via the deesterification of **91** and **93**, followed

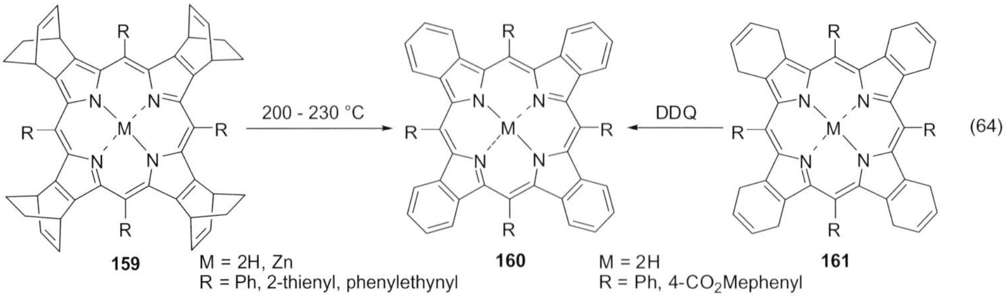

Scheme 11.71

159 M = 2H, Zn
R = Ph, 2-thienyl, phenylethynyl

200 - 230 °C →

160 M = 2H
R = Ph, 4-CO$_2$Mephenyl

← DDQ

161 (64)

by the Lindsey method [66, 67]. Vinogradov and coworkers have reported the details of an oxidative–aromatization method to synthesize *meso*-substituted TBPs, TNPs, and TAPs [60, 61]. In this case, the pyrroles **71**, **75**, **80**, **82**, and **84** were treated with KOH in refluxing ethylene glycol to provide α-free pyrroles that may then be converted into **161** and the corresponding porphyrins (Scheme 11.71).

11.4.3
[3+2] and [2+2] Methods

Lash and coworkers have prepared porphyrins with exocyclic rings by employing the MacDonald [2+2] or [3+1] condensation method. As the numbers, positions, and types of exocyclic ring affect the electronic properties of the porphyrins, this strategy affords a highly effective method for controlling the HOMO and LUMO energy levels of porphyrins. Examples of porphyrins prepared by employing this strategy are shown in Schemes 11.72 to 11.76 [104].

The [3+1] condensation method with BCOD-fused tripyrrane and its derivatives and a subsequent retro Diels–Alder reaction afforded core-modified and π-expanded benzoporphyrins, as shown in Figure 11.5. The core-modified benzoporphyrins **167–171** were prepared from the corresponding BCOD-fused thia-, dithia-, thiaoxa-, carba-, and azuliporphyrins, with or without *meso*-aryl groups [105]. The similar [3+1] porphyrin synthesis with BCOD-fused tripyrrane and phenanthrolinotripyrranes afforded acenaphthobenzoporphyrins and fluoranthobenzoporphyrins with intense Q bands at 600–700 nm [65c], and phenanthrolinoporphyrins **171** [104c] and phenanthrolinobenzoporphyrins, respectively. Phenathrolinoporphyrins with two coordination sites afforded the porphyrin-fused phenanthroline-Ru(III) complexes **175–178**, as shown in Scheme 11.77 and Figure 11.6 [106].

11.4.4
Expanded, Contracted, and Isomeric Porphyrins

Sapphyrins are typical expanded porphyrins, the absorption band of which at 453 nm is extremely narrow and strong. Sapphyrins that are fused with BCOD

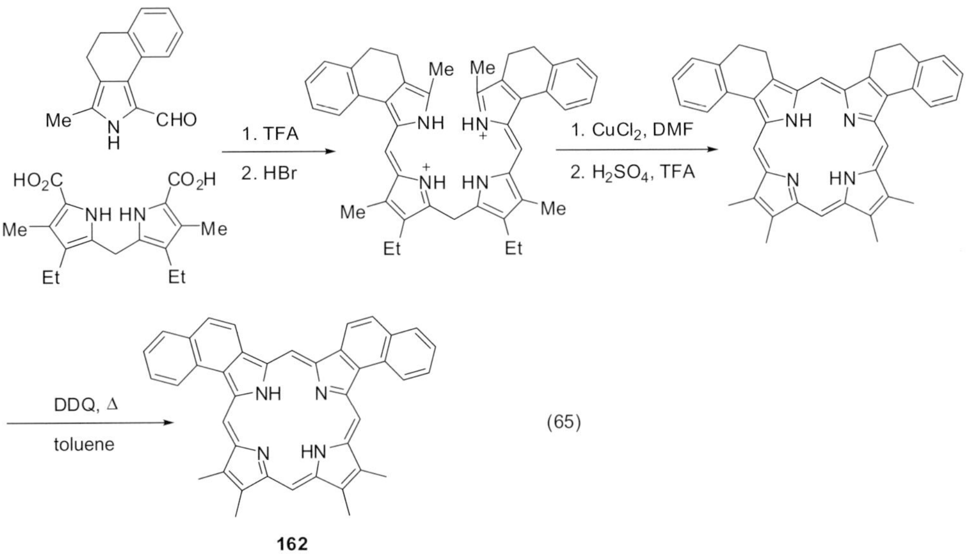

11.4 Synthesis of Porphyrins and Related Compounds

Scheme 11.72

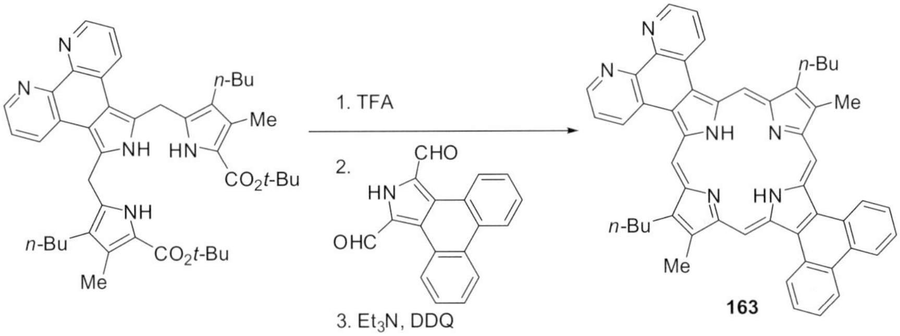

Scheme 11.73

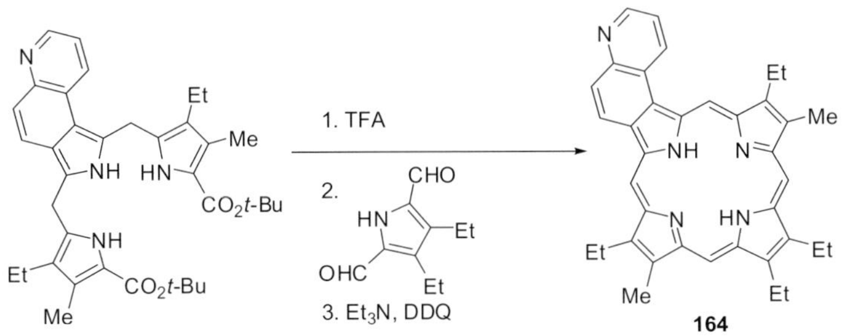

Scheme 11.74

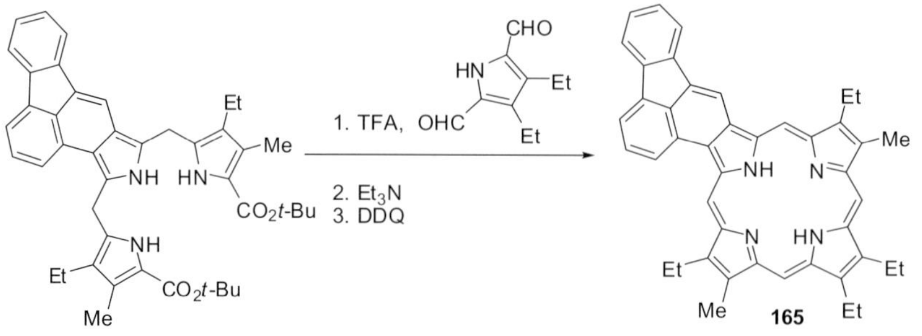

(68)

Scheme 11.75

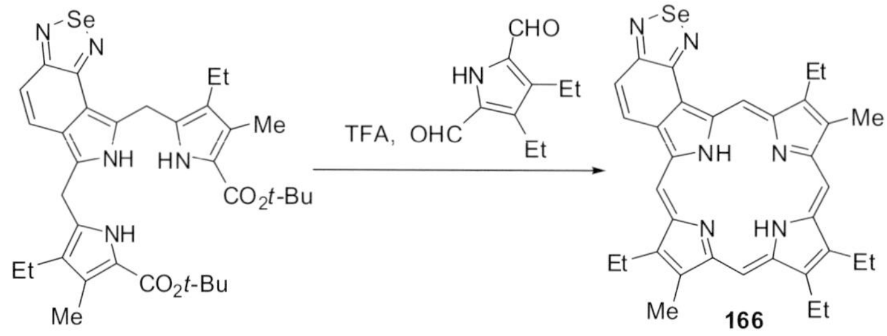

(69)

Scheme 11.76

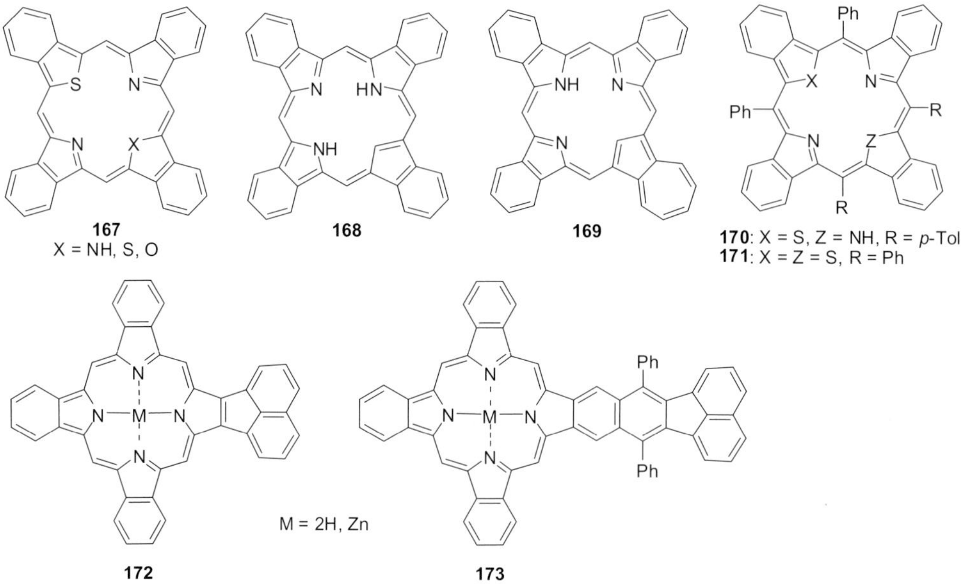

Figure 11.5 Core-modified TBPs, and acenaphtho- and fluorantho-BPs.

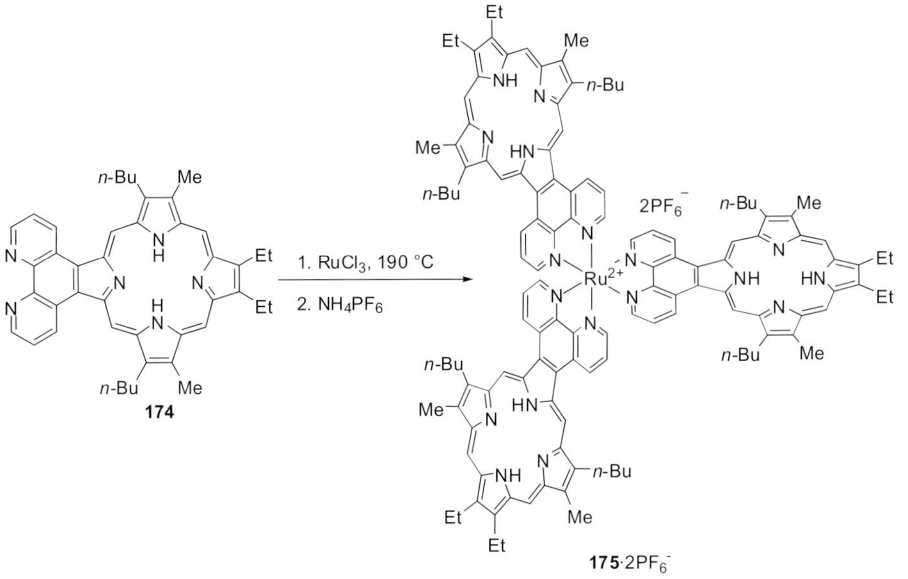

11.4 Synthesis of Porphyrins and Related Compounds | 417

Scheme 11.77

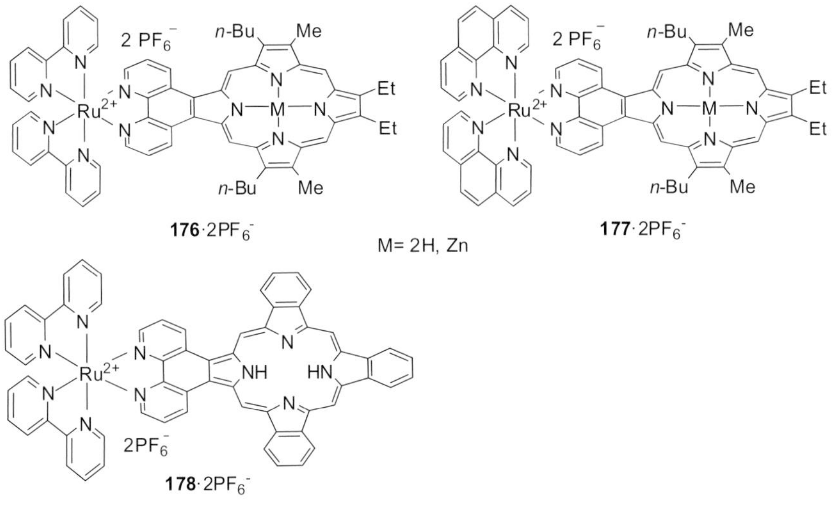

Figure 11.6 Porphyrin-fused phenanthroline–Ru(III) complexes.

rings demonstrate absorptions at 453–456 nm, and are converted on heating at 200 °C into mono-, di-, tri-, and pentabenzosapphyrins (Scheme 11.78) [107]. Lash and coworkers have reported that acenaphthosapphyrin and phenathrosapphyrin, and also core-modified sapphyrins, can be generated via [4+1] sapphyrin synthesis (Figure 11.7) [108].

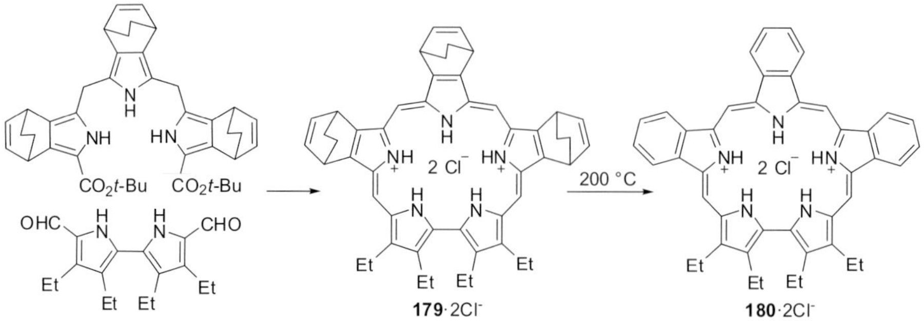

418 | *11 Synthesis of Pyrroles and Their Derivatives from Isocyanides*

(71)

Scheme 11.78

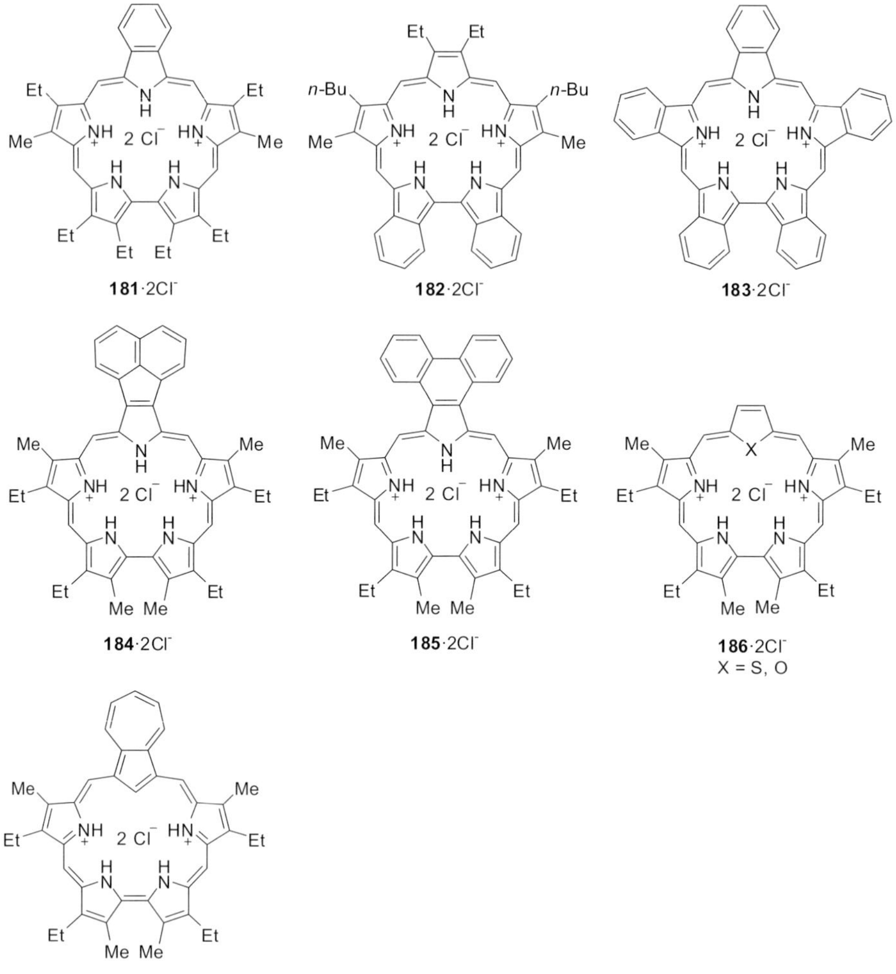

181·2Cl⁻

182·2Cl⁻

183·2Cl⁻

184·2Cl⁻

185·2Cl⁻

186·2Cl⁻
X = S, O

187·2Cl⁻

Figure 11.7 Benzo-, acenaphtho-, and phenanthro-, and core-modified sapphyrins.

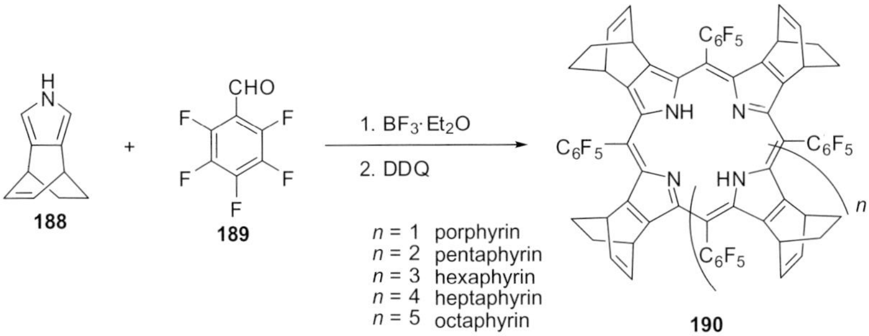

11.4 Synthesis of Porphyrins and Related Compounds

Scheme 11.79

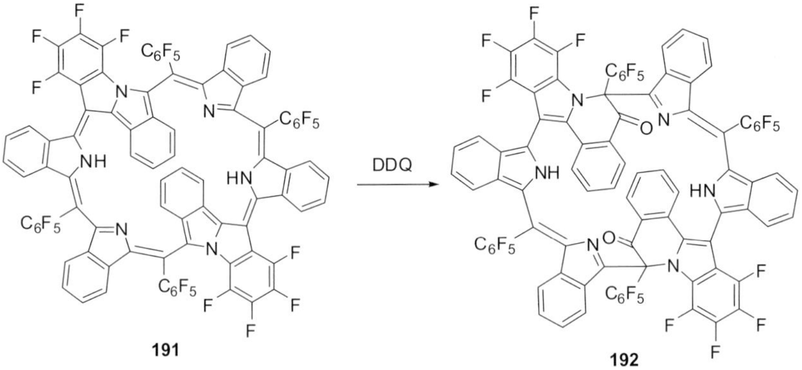

$n = 1$ porphyrin
$n = 2$ pentaphyrin
$n = 3$ hexaphyrin
$n = 4$ heptaphyrin
$n = 5$ octaphyrin

(72)

Scheme 11.80

(73)

Recently, Osuka and coworkers have developed the preparation of *meso*-aryl expanded porphyrins by the reaction of pentafluorobenzaldehyde with pyrrole, under conditions employing $BF_3 \cdot OEt_2$ and DDQ. This strategy has been applied to the reaction of BCOD-fused pyrrole to produce porphyrin, pentaphyrin, and hexaphyrin fused with BCOD (Scheme 11.79). Subsequent heating provided the corresponding benzo-derivatives. Interestingly, a doubly N-fused benzohexaphyrin **191** was formed in the case of a BCOD-fused hexaphyrin; upon DDQ oxidation, this was further rearranged to form a fluorescent macrocycle **192**, as shown in Scheme 11.80 [109].

Cyclo[8]pyrrole **193** is a ring-expanded porphyrin with no *meso*-bridges, which was reported by Sessler and coworkers based on an oxidative coupling of 2,2′-bipyrrole with $FeCl_3$ [110]. The photophysical, anion-binding, and liquid-crystalline properties of β-alkyl cyclo[8]pyrroles have been studied in-depth, along with their electronic structures [19, 111]. In 2011, cyclo[8]isoindole **194** was reported based on the retro Diels–Alder strategy (Figure 11.8) [112].

In the past, porphyrin dimers or further oligomers with highly ordered molecular systems have been subjected to extensive studies of electron and energy transfer

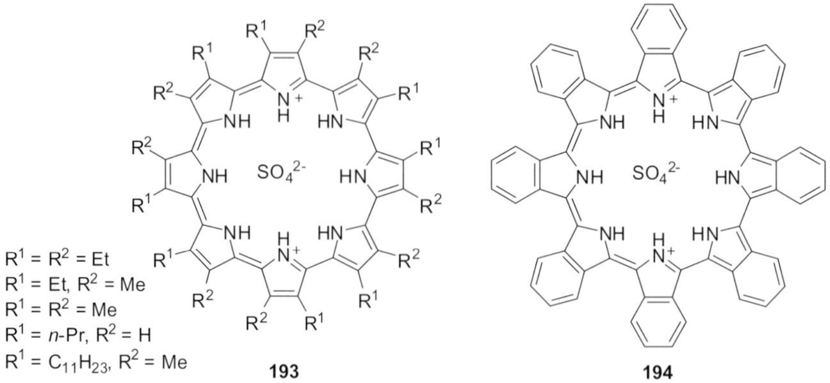

$R^1 = R^2 = Et$
$R^1 = Et, R^2 = Me$
$R^1 = R^2 = Me$
$R^1 = n\text{-Pr}, R^2 = H$
$R^1 = C_{11}H_{23}, R^2 = Me$

Figure 11.8 Cyclo[8]pyrroles.

between the porphyrin units, which are linked with suitable spacers such as aromatic rings or ethylene units [113]. The Barton–Zard reaction provides a novel approach to the construction of porphyrin oligomers, which are difficult to prepare using other methods. For example, *meso*-free porphyrin dimers linked with 1,3- and 1,4-phenylenes may be prepared from the corresponding phenylenedipyrroles via a double [3+1] condensation. The acid-catalyzed condensation of phenylene-linked bis(tripyrrane) with diformylpyrrole, followed by oxidation, then affords the porphyrin dimers **195** [114]. Directly β,β′-linked bisporphyrin **196** may be prepared via a [2+2] MacDonald condensation of β,β′-linked bis(dipyrromethane) with 2 equiv. of diformyldipyrromethane [115a]. Alternatively, such porphyrin dimers may be prepared via a Suzuki coupling, starting from β-bromoporphyrins [115b]. The inverse-type [3+1] porphyrin synthesis of BCOD-linked bispyrrole with diformyltripyrrane affords the gable-type porphyrin dimer linked to a rigid BCOD ring; this may be converted into an insoluble conjugated planar bisporphyrin **197** by heating at 200 °C [116]. Other synthetic approaches to the creation of directly fused porphyrin dimers, trimers, or oligomers have been developed by many other research groups (see Figure 11.9) [117].

Recently, the creation of benzene-fused contracted or isomeric porphyrins via retro Diels–Alder or oxidative-aromatization methods were reported; these included triphyrin(2.1.1) **200** [118], porphycene **201** [119], and corrole **202** [120] (Figure 11.10).

11.4.5
Functional Dyes from Pyrroles

Pyrroles developed from isocyanides provide an interesting range of functional dyes, including polypyrrole, BODIPYs, and porphyrins for a variety of applications such as conductive polymers, fluorescent dyes, organic field effect transistors (OFETs), organic photovoltaic (OPV) devices, nonlinear optics, and sensors.

Figure 11.9 Linked porphyrin oligomers.

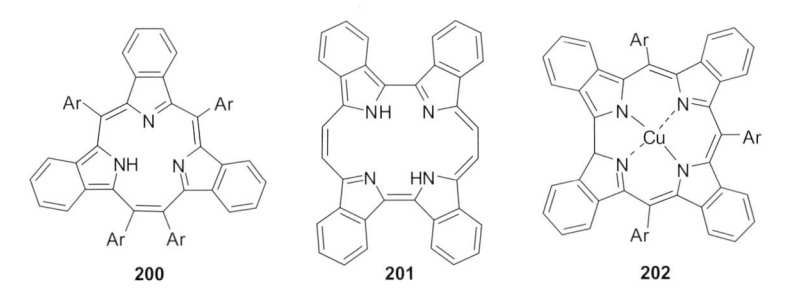

Figure 11.10 Benzene-fused contracted or isomeric porphyrins.

Polypyrroles may be readily obtained by the anodic oxidation or chemical oxidation of α-free pyrroles [11]. The Barton–Zard reaction affords α-free pyrroles with various substituents at the 3- and/or 4-positions, and these serve as important starting materials for polypyrroles. For example, polypyrroles fused with aromatic rings may be obtained by either anodic oxidation or chemical oxidation [73a]. The

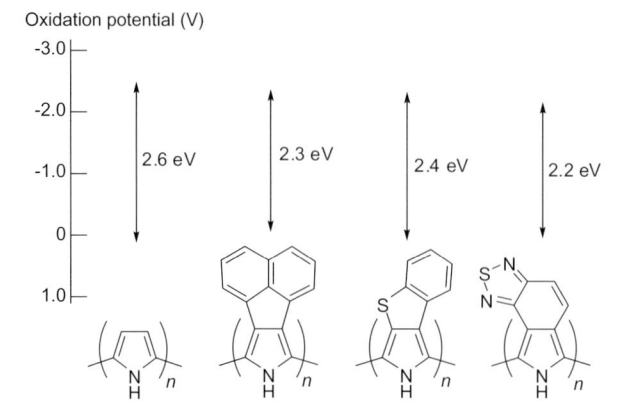

Figure 11.11 Energy levels of conducting polypyrroles.

Figure 11.12 BODIPYs fused with exocyclic rings.

203 λ_{max} 530 nm
204 λ_{max} 560 nm
205 λ_{max} 603 nm
206 λ_{max} 630 nm
207 λ_{max} 657 nm
208 λ_{max} 711 nm
209 λ_{max} 765 nm

band gaps of these polypyrroles may be considerably lower than those of the unsubstituted polypyrroles or of alkyl-substituted polypyrroles. Furthermore, the band gaps and HOMO–LUMO energy levels may be finely controlled by changing the nature of the fused aromatic rings and their substituents (Figure 11.11).

BODIPYs are highly fluorescent materials that have been used widely in a variety of scientific fields. They are especially important for use as laser dyes, as molecular probes for biochemical studies, as fluorescent sensors, and/or in various optoelectronic devices. Details of this subject are available in excellent reviews [18, 121]. The colors of the BODIPYs can be finely tuned by the choice of alkyl or fused rings of pyrroles, as shown in Figure 11.12, where the absorption ranges from 530 nm to 765 nm [122].

Table 11.1 Tetrabenzoporphyrin (TBP)-based organic field effect transistors, by solution process.

Material	Mobility (cm^{2}V^{-1}s^{-1})	V_{Th} (V)	On/off ratio
2HTBP	0.06	4	10^5
CuTBP	0.92	5	2×10^5
NiTBP	0.4	9	3×10^4
ZnTBP	0.013	12	500

BCOD-fused porphyrin CPs prepared from **87** are key compounds for the application of TBPs and their analogues as solution-processed organic semiconductors. The spin-coating of CPs, followed by thermal annealing, affords an insoluble crystalline thin film of TBP. The TBP-based OFETs fabricated via this solution-based process generally demonstrate a high performance, with high charge mobilities of between 10^{-2} and $100\,\text{cm}^2\text{V}^{-1}\text{s}^{-1}$ (see Table 11.1) [123].

TBP/PCBM ([6,6]-phenyl C61 butyric acid methyl ester) thin-films on indium–tin oxide (ITO) electrodes, prepared using a solution-based process, show photocurrent generation with the highest IPCE (incident-photon-to-electron conversion efficiency) value of 6.8% [124]. Thus, TBPs might be used as a p-type semiconductor for OPV devices. Recently, three-layered p-i-n OPV devices composed of TBP and fullerene derivatives fabricated by a solution-based process have been developed, and demonstrated a good performance with a high IPCE of more than 5% [16, 124].

Dye-sensitized solar cells based on porphyrins have also been developed during recent years. In this case, the porphyrins are functionalized as a sensitizer such as TPP–tetracarboxylic acid–ZnO, D–π–A system of porphyrin containing a carbazole-linked triphenylamine as the second electron donor, a trichromophoric sensitizer consisting of covalently linked BODIPY, zinc porphyrin, and squaraine units, and a perylene anhydride fused porphyrin covering the visible and near-infrared regions [125]. These exhibited a high overall conversion efficiency of over 5% due to their electron-donating ability, stability, wide-range, and strong absorptions.

11.5 Conclusion

The synthesis of pyrroles and their related compounds has been described in this chapter, with attention focused on the van Leusen reaction of electron-deficient alkenes or alkynes using TosMIC to afford α-free or α-tosylpyrroles. The Barton–Zard reaction, using alkyl isocyanoacetate, affords 3,4-substituted pyrrole-2-carboxylates, which may then be converted into porphyrinoids, poly/oligopyrroles, and BODIPY dyes. Various substituents are readily introduced into the 3- and

4-positions of pyrrole using these methods. The main advantage of the Barton–Zard reaction over other pyrrole syntheses is that the nitroarenes react with isocyanoacetate in the presence of a base to produce pyrroles that are fused with an aromatic ring. Although nitrobenzene does not afford 2*H*-isoindole via the Barton–Zard reaction, isoindole, benz[*f*]isoindole, and naphtho[2,3-*f*]isoindole are each obtained starting from nitro-, phenylsulfonyl-, or tosyl-substituted cyclohexene or BCOD derivatives via the Barton–Zard reaction, with subsequent oxidative-aromatization or retro Diels–Alder reaction. The Barton–Zard pyrroles thus obtained are suitable for the synthesis of symmetrically β-substituted *meso*-free porphyrins, using the Ono method, and of *meso*-substituted porphyrins using the Lindsey method. Unsymmetrical porphyrins may be obtained via the MacDonald [2+2] condensation of dipyrromethane, or the [3+1] condensation of tripyrrane. Today, a wide variety of porphyrinoids and their related compounds, such as core-modified, isomeric, ring-expanded and contracted porphyrins, BODIPYs, and polypyrroles, may also be synthesized from the Barton–Zard pyrroles. Recently, the development of the synthesis of these π-conjugated molecules, based on van Leusen and Barton–Zard reactions, has provided a series of functional materials that are suitable for applications to organic light-emitting diodes and OPV devices, in OFETs, as photosensitizers for PDT, as chemosensors, and in biological probes.

References

1 (a) Gilchrist, T.L. (1985) *Heterocyclic Chemistry*, Pitman, London; (b) Jones, A. (1990) *Pyrroles*, John Wiley & Sons, Inc., New York; (c) Gossauer, A. (1994) *Houben-Weyl*, vol. E6a (ed. R.P. Kreher), Thieme Verlag, Stuttgart, pp. 556–798; (d) Sundberg, R.J. and Smith, K.M. (1984) *Comprehensive Heterocyclic Chemistry*, vol. 4, Pergamon Press, Oxford, pp. 313–442; (e) Black, D.S. (2001) *Science of Synthesis*, vol. 9 (ed. G. Mass), Thieme Verlag, Stuttgart, pp. 441–552; (f) Yu, M. and Pagenkopf, B.L. (2003) *Org. Lett.*, **5**, 5099–5101 and references therein; (g) Estévez, V., Villacampa, M., and Menéndez, J.C. (2010) *Chem. Soc. Rev.*, **39**, 4402–4421.

2 Campo, J., García-Valverde, M., Marcaccini, S., Rojo, M.J., and Torroba, T. (2006) *Org. Biomol. Chem.*, **4**, 757–765.

3 (a) van Leusen, D. and van Leusen, A.M. (2001) *Org. React.*, **57**, 417–666; (b) Original paper: Van Leusen, A.M., Boerma, G.J.M., Helmholdt, R.B., Siderius, H., and Strating, J. (1972) *Tetrahedron Lett.*, **13**, 2367–2368.

4 (a) Barton, D.H.R. and Zard, S.Z. (1985) *J. Chem. Soc. Chem. Commun.*, 1098–1100; (b) Full paper: Barton, D.H.R., Kervagoret, J., and Zard, S.Z. (1990) *Tetrahedron*, **46**, 7587–7598; (c) Ono, N. (2008) *Heterocycles*, **75**, 243–284; (d) Gulenvich, A.V., Zhdanko, A.G., Orru, R.V.A., and Nenajdenko, V.G. (2010) *Chem. Rev.*, **110**, 5235–5331; (e) Pyrrole synthesis by the reaction of aldehyde with isocyanoacetate: Suzuki, M., Miyoshi, M., and Matsumoto, K. (1974) *J. Org. Chem.*, **39**, 1980; (f) Matsumoto, K., Suzuki, M., Ozaki, Y., and Miyoshi, M. (1976) *Agr. Biol. Chem.*, **40**, 2271–2274.

5 (a) Battersby, A.R. (1987) *Nat. Prod. Rep.*, **4**, 77–87; (b) Falk, H. (1989) *The Chemistry of Linear Oligopyrroles and Bile Pigments*, Springer, Berlin; (c) Franck, B. and Nonn, A. (1995) *Angew. Chem. Int. Ed. Engl.*, **34**, 1795–1811.

6 Fürstner, A., Weintritt, H., and Hupperts, A. (1995) *J. Org. Chem.*, **60**, 6637–6641.

7 Heim, A., Terpin, A., and Steglich, W. (1997) *Angew. Chem. Int. Ed. Engl.*, **36**, 155–156.

8 Kashman, Y., Koren-Goldshlager, G., Gravalos, M.D.G., and Schleyer, M. (1999) *Tetrahedron Lett.*, **40**, 997–1000.
9 Fürstner, A., Domostoj, M.M., and Scheiper, B. (2005) *J. Am. Chem. Soc.*, **127**, 11620–11621.
10 (a) Fürstner, A. (2003) *Angew. Chem. Int. Ed.*, **42**, 3582–3603; (b) Bhaduri, A.P. (1990) *Synlett*, 557–564; (c) Bellina, F. and Rossi, R. (2006) *Tetrahedron*, **62**, 7213–7256.
11 Skotheim, T.A., Elsenbaumer, R.L., and Reynolds, J.R. (eds) (1998) *Handbook of Conducting Polymers*, CRC Press, New York.
12 Klauk, H. (ed.) (2006) *Organic Electronics, Materials, Manufacturing and Application*, Wiley-VCH Verlag GmbH, Weinheim.
13 (a) Zollinger, H. (2003) *Color Chemistry, Synthesis, Properties, and Applications of Organic Dyes and Pigments*, Wiley-VCH, Zurich; (b) Bonnet, R. (2000) *Chemical Aspects of Photodynamic Therapy*, Gordon and Breach Science Publishers, Amsterdam.
14 Review: Martínez-Díaz, M.V., de la Torre, G., Torres, T. (2010) *Chem. Commun.*, **46**, 7090–7108.
15 Calvete, M., Yang, G.Y., and Hanack, M. (2004) *Synth. Met.*, **141**, 231–243.
16 Matsuo, Y., Sato, Y., Niinomi, T., Soga, I., Tanaka, H., and Nakamura, E. (2009) *J. Am. Chem. Soc.*, **131**, 16048–16050.
17 Aramaki, S., Sakai, Y., and Ono, N. (2004) *Appl. Phys. Lett.*, **84**, 2085–2087.
18 Descalzo, A.B., Hu, H.-J., Xue, Z., Hoffmann, K., Shen, Z., Weller, M.G., You, X.-Z., and Rurack, K. (2008) *Org. Lett.*, **10**, 1581–1584.
19 Stępień, M., Donnio, B., and Sessler, J.L. (2007) *Angew. Chem. Int. Ed.*, **46**, 1431–1435.
20 Di Santo, R., Costi, R., Massa, S., and Artico, M. (1996) *Synth. Commun.*, **26**, 1839–1847.
21 (a) Pavri, N.P. and Trudell, M.L. (1997) *J. Org. Chem.*, **62**, 2649–2651; (b) Balasubramanian, T. and Lindsey, J.S. (1999) *Tetrahedron*, **55**, 6771–6784.
22 Smith, N.D., Huang, D., and Cosford, N.D.P. (2002) *Org. Lett.*, **4**, 3537–3539.
23 De Leon, C.Y. and Ganem, B. (1996) *J. Org. Chem.*, **61**, 8730–8731.
24 Aoyagi, K., Haga, T., Toi, H., Aoyama, Y., Mizutani, T., and Ogoshi, H. (1997) *Bull. Chem. Soc. Jpn*, **70**, 937–943.
25 (a) Ono, N., Muratani, E., and Ogawa, T. (1991) *J. Heterocycl. Chem.*, **28**, 2053–2055; (b) Leusink, F.R., ten Have, R., van den Berg, K.J., and van Leusen, A.M. (1992) *J. Chem. Soc. Chem. Commun.*, 1401–1402.
26 Baxendale, I.R., Buckle, C.D., Ley, S.V., and Tamborini, L. (2009) *Synthesis*, 1485–1493.
27 Bobál, F. and Lightner, D.A. (2001) *J. Heterocycl. Chem.*, **38**, 527–530.
28 Inomata, K. (2008) *Bull. Chem. Soc. Jpn*, **81**, 25–59 and references therein.
29 Larionov, O.V. and de Meijere, A. (2005) *Angew. Chem. Int. Ed.*, **44**, 5664–5667.
30 Alizadeh, A., Masrouri, H., Rostamnia, S., and Movahedi, F. (2006) *Helv. Chim. Acta*, **89**, 923–926.
31 (a) Ono, N., Yamada, H., and Okujima, T. (2010) *Handbook of Porphyrin Science*, vol. 2 (eds K.M. Kadish, K.M. Smith, and R. Guilard), World Scientific, Singapore, pp. 1–102; (b) Lash, T.D. (2000) *The Pophyrin Handbook*, vol. 2 (eds K.M. Kadish, K.M. Smith, and R. Guilard), Academic Press, San Diego, pp. 125–200.
32 (a) Ono, N. (2001) *The Nitro Group in Organic Synthesis*, John Wiley & Sons, Inc., New York; (b) Ono, N. (2001) *1-Nitroalkenes. Science of Synthesis*, vol. 33 (ed. C.A. Molander), Thieme Verlag, Stuttgart, pp. 337–370.
33 Review of the nitroaldol reaction: Luzzio, F.A. (2001) *Tetrahedron*, **57**, 915–945.
34 Caldarelli, M., Habermann, J., and Ley, S.V. (1999) *J. Chem. Soc., Perkin Trans. 1*, 107–110.
35 Ono, N., Katayama, H., Nishiyama, S., and Ogawa, T. (1994) *J. Heterocycl. Chem.*, **31**, 707–710.
36 Bhattacharya, A., Cherukuri, S., Plata, R.E., Patel, N., Tamez, V. Jr, Grosso, J.A., Peddicord, M., and Palaniswamy, V.A. (2006) *Tetrahedron Lett.*, **47**, 5481–5484.
37 Bag, N., Chern, S.-S., Peng, S.-M., and Chang, C.K. (1995) *Tetrahedron Lett.*, **36**, 6409–6412.
38 Adamczyk, M., Fishpaugh, J.R., Heuser, K.J., Ramp, J.M., Reddy, R.E., and

Wong, M. (1998) *Tetrahedron*, **54**, 3093–3112.
39 Adamczyk, M., Johnson, D.D., and Reddy, R.E. (2001) *J. Org. Chem.*, **66**, 11–19.
40 Fumoto, Y., Eguchi, T., Uno, H., and Ono, N. (1999) *J. Org. Chem.*, **64**, 6518–6521.
41 (a) Kakiuchi, T., Kinoshita, H., and Inomata, K. (1999) *Synlett*, 901–904; (b) Hammam, M.A.S., Murata, Y., Kinoshita, H., and Inomata, K. (2004) *Chem. Lett.*, **33**, 1258–1259 and references therein.
42 Ono, N., Miyagawa, H., Ueta, T., Ogawa, T., and Tani, H. (1998) *J. Chem. Soc., Perkin Trans. 1*, 1595–1601.
43 Bullington, J.L., Wolf, R.R., and Jackson, P.F. (2002) *J. Org. Chem.*, **67**, 9439–9442.
44 Hoffmann, M., Wilson, C.J., Odell, B., and Anderson, H.L. (2007) *Angew. Chem. Int. Ed.*, **46**, 3122–3125.
45 Coffin, A.R., Roussell, M.A., Tserlin, E., and Pelkey, E.T. (2006) *J. Org. Chem.*, **71**, 6678–6681.
46 (a) Fumoto, Y., Uno, H., Ito, S., Tsugumi, Y., Sasaki, M., Kitawaki, Y., and Ono, N. (2000) *J. Chem. Soc., Perkin Trans. 1*, 2977–2981; (b) Conformation analysis of cyclononatripyrroles: Uno, H., Fumoto, Y., Inoue, K., Ono, N. (2003) *Tetrahedron*, **59**, 601–605.
47 Boëlle, J., Schneider, R., Gérardin, P., and Loubinoux, B. (1997) *Synthesis*, 1451–1456.
48 Review of trifluoromethylpyrroles: Muzalevskiy, V.M., Shastin, A.V., Balenkova, E.S., Haufe, G., Nenajdenko, V.G. (2009) *Synthesis*, 3905–3929.
49 Ono, N., Kawamura, H., and Maruyama, K. (1989) *Bull. Chem. Soc. Jpn*, **62**, 3386–3388.
50 Ono, N. and Maruyama, K. (1988) *Bull. Chem. Soc. Jpn*, **61**, 4470–4472.
51 Ono, N., Bougauchi, M., and Maruyama, K. (1992) *Tetrahedron Lett.*, **33**, 1629–1632.
52 Furusho, Y., Tsunoda, A., and Aida, T. (1996) *J. Chem. Soc., Perkin Trans. 1*, 183–190.
53 Halazy, S. and Magnus, P. (1984) *Tetrahedron Lett.*, **25**, 1421–1424.
54 Arnold, D.P., Burgessdean, L., Hubbard, J., and Rahman, M.A. (1994) *Aus. J. Chem.*, **47**, 969–974.
55 Ono, N. and Maruyama, K. (1988) *Chem. Lett.*, 1511–1514.
56 Haake, G., Struve, D., and Montforts, F.-P. (1994) *Tetrahedron Lett.*, **35**, 9703–9704.
57 Abel, Y., Haake, E., Haake, G., Schmidt, W., Struve, D., Walter, A., and Montforts, F.-P. (1998) *Helv. Chim. Acta*, **81**, 1978–1996.
58 Vincente, M.G.H., Tomé, A.C., Walter, A., and Cavaleiro, J.A.S. (1997) *Tetrahedron Lett.*, **38**, 3639–3642.
59 Ishida, H., Itoh, K., Ito, S., Ono, N., and Ohno, M. (2001) *Synlett*, 296–298.
60 (a) Finikova, O.S., Cheprakov, A.V., Beletskaya, I.P., Carroll, P.J., and Vinogradov, S.A. (2004) *J. Org. Chem.*, **69**, 522–535; (b) Finikova, O., Galkin, A., Rozhkov, V., Cordero, M., Hägerhäll, C., and Vinogradov, S. (2003) *J. Am. Chem. Soc.*, **125**, 4882–4893; (c) Finikova, O.S., Cheprakov, A.V., Carroll, P.J., and Vinogradov, S.A. (2003) *J. Org. Chem.*, **68**, 7517–7520; (d) Finikova, O.S., Cheprakov, A.V., and Vinogradov, S.A. (2005) *J. Org. Chem.*, **70**, 9562–9572.
61 (a) Filatov, M.A., Cheprakov, A.V., and Beletskaya, I.P. (2007) *Eur. J. Org. Chem.*, 3468–3475; (b) Filatov, M.A., Lebedev, A.Y., Vinogradov, S.A., and Cheprakov, A.V. (2008) *J. Org. Chem.*, **73**, 4175–4185.
62 (a) Ito, S., Murashima, T., and Ono, N. (1997) *J. Chem. Soc., Perkin Trans 1*, 3161–3165; (b) Ito, S., Murashima, T., Uno, H., and Ono, N. (1998) *Chem. Commun.*, 1661–1662; (c) Ito, S., Ochi, N., Murashima, T., Uno, H., and Ono, N. (2000) *Heterocycles*, **52**, 399–411.
63 Uno, H., Ito, S., Wada, M., Watanabe, H., Nagai, M., Hayashi, A., Murashima, T., and Ono, N. (2000) *J. Chem. Soc., Perkin Trans. 1*, 4347–4355.
64 Okujima, T., Jin, G., Hashimoto, Y., Yamada, H., Uno, H., and Ono, N. (2006) *Heterocycles*, **70**, 619–626.
65 (a) Ito, S., Uno, H., Murashima, T., and Ono, N. (2001) *Tetrahedron Lett.*, **42**, 45–47; (b) Murashima, T., Tsujimoto, S., Yamada, T., Miyazawa, T., Uno, H., Ono, N., and Sugimoto, N. (2005)

Tetrahedron Lett., **46**, 113–116; (c) Okujima, T., Komobuchi, N., Uno, H., and Ono, N. (2006) Heterocycles, **67**, 255–267; (d) Okujima, T., Jin, G., Hashimoto, Y., Yamada, H., Uno, H., and Ono, N. (2006) Heterocycles, **70**, 619–626.

66 Ito, S., Ochi, N., Uno, H., Murashima, T., and Ono, N. (2000) Chem. Commun., 893–894.

67 Yamada, H., Kuzuhara, D., Takahashi, T., Shimizu, Y., Uota, K., Okujima, T., Uno, H., and Ono, N. (2008) Org. Lett., **10**, 2947–2950.

68 Uno, H., Tanaka, M., Inoue, T., and Ono, N. (1999) Synthesis, 471–474.

69 Uno, H., Sakamoto, K., Tominaga, T., and Ono, N. (1994) Bull. Chem. Soc. Jpn, **67**, 1441–1448.

70 Uno, H., Inoue, K., Inoue, T., Fumoto, Y., and Ono, N. (2001) Synthesis, 2255–2258.

71 Misra, N.C., Panda, K., Ila, H., and Junjappa, H. (2007) J. Org. Chem., **72**, 1246–1251.

72 Kamijo, S., Kanazawa, C., and Yamamoto, Y. (2005) J. Am. Chem. Soc., **127**, 9260–9266.

73 (a) Ono, N., Hironaga, H., Simizu, K., Ono, K., Kuwano, K., and Ogawa, T. (1994) Chem. Commun., 1019–1020; (b) The synthesis of isoindoles from aromatic nitro compounds was originally reported in 1991, see: Maruyama, K., Kawamura, H., and Ono, N. (1993) Abstracts, Chemical Congress of Japan, Yokohama, 1A435, 1991; see also Maruyama, K., Kawamura, H., and Ono, N. (1993) Chem. Abstr., **118**, 80798h.

74 Lash, T.D., Novak, B.H., and Lin, Y. (1994) Tetrahedron Lett., **35**, 2493–2494.

75 Murashima, T., Tamai, R., Fujita, K., Uno, H., and Ono, N. (1996) Tetrahedron Lett., **37**, 8391–8394.

76 Lash, T.D., Thompson, M.L., Werner, T.M., and Spence, J.D. (2000) Synlett, 213–216.

77 Gandhi, V., Thompson, M.L., and Lash, T.D. (2010) Tetrahedron, **66**, 1787–1799.

78 Boedigheimer, H., Ferrence, G.M., and Lash, T.D. (2010) J. Org. Chem., **75**, 2518–2527.

79 Murashima, T., Tamai, R., Nishi, K., Nomura, K., Fujita, K., Uno, H., and Ono, N. (2000) J. Chem. Soc., Perkin Trans. 1, 995–998.

80 Murashima, T., Nishi, K., Nakamoto, K., Kato, A., Tamai, R., Uno, H., and Ono, N. (2002) Heterocycles, **58**, 301–310.

81 (a) Pelkey, E.T. and Gribble, G.W. (1997) Chem. Commun., 1873–1874; (b) Pelkey, E.T., Chang, L., and Gribble, G.W. (1996) Chem. Commun., 1909–1910.

82 Jaquinod, L., Gros, C., Olmstead, M.M., Antolovich, M., and Smith, K.M. (1996) Chem. Commun., 1475–1476.

83 Ono, N., Hironaga, H., Ono, K., Kaneko, S., Murashima, T., Ueda, T., Tsukamura, C., and Ogawa, T. (1996) J. Chem. Soc., Perkin Trans. 1, 417–423.

84 Uoyama, H., Ono, N., and Uno, H. (2007) Heterocycles, **72**, 363–372.

85 Uno, H., Kinoshita, T., Matsumoto, K., Murashima, T., Ogawa, T., and Ono, N. (1996) J. Chem. Res. (S), 76–77.

86 (a) Donohoe, T.J. (2000) Science of Synthesis, vol. 10 (ed. E.J. Thomas), Thieme Verlag, Stuttgart, p. 658; (b) Ohmura, T., Kijima, A., and Suginome, M. (2011) Org. Lett., **13**, 1238–1241 and references therein.

87 Uno, H., Inoue, T., Fumoto, Y., Shiro, M., and Ono, N. (2000) J. Am. Chem. Soc., **122**, 6773–6774.

88 Fujii, M. (1993) J. Am. Chem. Soc., **115**, 4641–4648.

89 Czarnecki, K., Proniewicz, L.M., Fujii, H., and Kincaid, J.R. (1996) J. Am. Chem. Soc., **118**, 4680–4685.

90 Ayoygou, K., Mandon, D., Fischer, J., Weiss, R., Müther, M., Schünemann, V., Trautwein, A.X., Bill, E., Terner, J., Jayaraj, K., Gold, A., and Austin, R.N. (1996) Chem. Eur. J., **2**, 1159–1163.

91 Uno, H., Inoue, K., Inoue, T., and Ono, N. (2003) Org. Biomol. Chem., **1**, 3857–3865.

92 Fumoto, Y., Uno, H., Murashima, T., and Ono, N. (2001) Heterocycles, **54**, 705–720.

93 Okujima, T., Hashimoto, Y., Jin, G., Yamada, H., Uno, H., and Ono, N. (2008) Tetrahedron, **64**, 2405–2411.

94 (a) Uno, H., Watanabe, H., Yamashita, Y., and Ono, N. (2005) Org. Biomol.

Chem., **3**, 448–453; (b) Schlögl, J. and Kräutler, B. (1999) *Synlett*, 969–971.
95 Nakamura, J., Okujima, T., Tomimori, Y., Komobuchi, N., Yamada, H., Uno, H., and Ono, N. (2010) *Heterocycles*, **80**, 1165–1175.
96 Vicente, M.G.H., Cancilla, M.T., Lebrilla, C.B., and Smith, K.M. (1998) *Chem. Commun.*, 2355–2356.
97 (a) Lindsey, J.S., Schreiman, I.C., Hsu, H.C., Kearney, P.C., and Marguerettaz, A.M. (1987) *J. Org. Chem.*, **52**, 827–836; (b) Lindsey, J.S. (2000) *The Porphyrin Handbook*, vol. 1 (eds K.M. Kaddish, K.M. Smith, and R. Guilard), Academic Press, San Diego, pp. 45–118.
98 Bonnett, R. (1995) *Chem. Soc. Rev.*, **24**, 19–33.
99 Murashima, T., Fujita, K., Ono, K., Ogawa, T., Uno, H., and Ono, N. (1996) *J. Chem. Soc., Perkin Trans. 1*, 1403–1407.
100 (a) Lash, T.D. and Chandrasekar, P. (1996) *J. Am. Chem. Soc.*, **118**, 8767–8768; (b) Spence, J.D. and Lash, T.D. (2000) *J. Org. Chem.*, **65**, 1530–1539.
101 (a) Xu, H.-J., Shen, Z., Okujima, T., Ono, N., and You, X.-Z. (2006) *Tetrahedron Lett.*, **47**, 931–934; (b) Xu, H.-J., Mack, J., Descalzo, A.B., Shen, Z., Kobayashi, N., You, X.-Z., and Rurack, K. (2011) *Chem. Eur. J.*, **17**, 8965–8983.
102 Mack, J., Asano, Y., Kobayashi, N., and Stillman, M.J. (2005) *J. Am. Chem. Soc.*, **127**, 17697–17711.
103 Shen, Z., Uno, H., Shimizu, Y., and Ono, N. (2004) *Org. Biomol. Chem.*, **2**, 3442–3447.
104 (a) Manley, J.M., Roper, T.J., and Lash, T.D. (2005) *J. Org. Chem.*, **70**, 874–891; (b) Lash, T.D., Werner, T.M., Thompson, M.L., and Manley, J.M. (2001) *J. Org. Chem.*, **66**, 3152–3159; (c) Lash, T.D., Lin, Y., Novak, B.H., and Parikh, M.D. (2005) *Tetrahedron*, **61**, 11601–11614; (d) Cillo, C.M. and Lash, T.D. (2005) *Tetrahedron*, **61**, 11615–11627; (e) Lash, T.D. and Gandhi, V. (2000) *J. Org. Chem.*, **65**, 8020–8026.
105 (a) Shimizu, Y., Shen, Z., Okujima, T., Uno, H., and Ono, N. (2004) *Chem. Commun.*, 374–375; (b) Okujima, T., Komobuchi, N., Shimizu, Y., Uno, H., and Ono, N. (2004) *Tetrahedron Lett.*, **45**, 5461–5464; (c) Mack, J., Bunya, M., Shimizu, Y., Uoyama, H., Komobuchi, N., Okujima, T., Uno, H., Ito, S., Stillman, M.J., Ono, N., and Kobayashi, N. (2008) *Chem. Eur. J.*, **14**, 5001–5020; (d) Uno, H., Shimizu, Y., Uoyama, H., Tanaka, Y., Okujima, T., and Ono, N. (2008) *Eur. J. Org. Chem.*, 87–98.
106 Okujima, T., Mifuji, A., Nakamura, J., Yamada, H., Uno, H., and Ono, N. (2009) *Org. Lett.*, **11**, 4088–4091.
107 Ono, N., Kuroki, K., Watanabe, E., Ochi, N., and Uno, H. (2004) *Heterocycles*, **62**, 365–373.
108 Richter, D.T. and Lash, T.D. (2004) *J. Org. Chem.*, **69**, 8842–8850.
109 Inokuma, Y., Matsunari, T., Ono, N., Uno, H., and Osuka, A. (2005) *Angew. Chem. Int. Ed.*, **44**, 1856–1860.
110 (a) Seidel, D., Lynch, V., and Sessler, J.L. (2002) *Angew. Chem. Int. Ed.*, **41**, 1422–1425; (b) Köhler, T., Seidel, D., Lynch, V., Arp, F.O., Ou, Z., Kadish, K.M., and Sessler, J.L. (2003) *J. Am. Chem. Soc.*, **125**, 6872–6873.
111 (a) Eller, L.R., Stępień, M., Fowler, C.J., Lee, J.T., Sessler, J.L., and Moyer, B.A. (2007) *J. Am. Chem. Soc.*, **129**, 11020–11021; (b) Sessler, J.L., Karnas, E., Kim, S.K., Ou, Z., Zhanf, M., Kadish, K.M., Ohkubo, K., and Fukuzumi, S. (2008) *J. Am. Chem. Soc.*, **13**, 15256–15257; (c) Gorski, A., Köhler, T., Seidel, D., Lee, J.T., Orzanowska, G., Sessler, J.L., and Waluk, J. (2005) *Chem. Eur. J.*, **11**, 4179–4184; (d) Lim, J.M., Yoon, Z.S., Shin, J.-Y., Kim, K.S., Yoon, M.-C., and Kim, D. (2009) *Chem. Commun.*, 261–273.
112 Okujima, T., Jin, G., Matsumoto, N., Mack, J., Mori, S., Ohara, K., Kuzuhara, D., Ando, C., Ono, N., Yamada, H., Uno, H., and Kobayashi, N. (2011) *Angew. Chem. Int. Ed.*, **50**, 5699–5703.
113 Aratani, N. and Osuka, A. (2001) *Bull. Chem. Soc. Jpn*, **74**, 1361–1379.
114 Fumoto, Y., Uno, H., Tanaka, K., Tanaka, M., Murashima, T., and Ono, N. (2001) *Synthesis*, 399–402.
115 (a) Uno, H., Kitawaki, Y., and Ono, N. (2002) *Chem. Commun.*, 116–117; (b) Bringmann, G., Rüdenauer, S., Götz, D.C.G., Gulder, T.A.M., and Reichert,

M. (2006) *Org. Lett.*, **8**, 4743–4746.

116 Ito, S., Nakamoto, K., Uno, H., Murashima, T., and Ono, N. (2001) *Chem. Commun.*, 2696–2697.

117 (a) Inokuma, Y., Ono, N., Uno, H., Kim, D.Y., Noh, S.B., Kim, D., and Osuka, A. (2005) *Chem. Commun.*, 3782–3784; (b) Jaquinod, L., Siri, O., Khoury, R.G., and Smith, K.M. (1998) *Chem. Commun.*, 1261–1262.

118 (a) Xue, Z.-L., Shen, Z., Mack, J., Kuzuhara, D., Yamada, H., Okujima, T., Ono, N., You, X.-Z., and Kobayashi, N. (2008) *J. Am. Chem. Soc.*, **130**, 16478–16479; (b) Kuzuhara, D., Yamada, H., Xue, Z., Okujima, T., Mori, S., Shen, Z., and Uno, H. (2011) *Chem. Commun.*, **47**, 722–724.

119 Kuzuhara, D., Mack, J., Yamada, H., Okujima, T., Ono, N., and Kobayashi, N. (2009) *Chem. Eur. J.*, **15**, 10060–10069.

120 Pomarico, G., Nardis, S., Paolesse, R., Ongayi, O.C., Courtney, B.H., Fronczek, F.R., and Vicente, M.G. (2011) *J. Org. Chem.*, **76**, 3765–3773.

121 (a) Wood, T.E. and Thompson, A. (2007) *Chem. Rev.*, **107**, 1831–1861; (b) Loudet, A. and Burgess, K. (2007) *Chem. Rev.*, **107**, 4891–4932; (c) Ulrich, G., Ziessel, R., and Harriman, A. (2008) *Angew. Chem. Int. Ed.*, **47**, 1184–1201.

122 (a) Wada, M., Ito, S., Uno, H., Murashima, T., Ono, N., Urano, T., and Urano, Y. (2001) *Tetrahedron Lett.*, **42**, 6711–6713; (b) Ono, N., Yamamoto, T., Shimada, N., Kuroki, K., Wada, M., Utsunomiya, R., Yano, T., Uno, H., and Murashima, T. (2003) *Heterocycles*, **61**, 433–447; (c) Shen, Z., Röhr, H., Rurack, K., Uno, H., Spieles, M., Schulz, B., Reck, G., and Ono, N. (2004) *Chem. Eur. J.*, **10**, 4853–4871; (d) Okujima, T., Tomimori, Y., Nakamura, J., Yamada, H., Uno, H., and Ono, N. (2010) *Tetrahedron*, **66**, 6895–6900; (e) Tomimori, Y., Okujima, T., Yano, T., Mori, S., Ono, N., Yamada, H., and Uno, H. (2011) *Tetrahedron*, **67**, 3187–3193.

123 (a) Aramaki, S., Sakai, Y., and Ono, N. (2004) *Appl. Phys. Lett.*, **84**, 2085–2087; (b) Shea, P.B., Kanicki, J., Pattison, L.R., Petroff, P., Kawano, M., Yamada, H., and Ono, N. (2006) *J. Appl. Phys.*, **100**, 034502/1–034502/7; (c) Shea, P.B., Pattison, L.R., Kawano, M., Chen, C., Chen, J., Petroff, P., Martin, D.C., Yamada, H., Ono, N., and Kanicki, J. (2007) *Synth. Met.*, **157**, 190–197; (d) Shea, P.B., Chen, C., Kanicki, J., Pattison, L.R., Petroff, P., Yamada, H., and Ono, N. (2007) *Appl. Phys. Lett.*, **90**, 233107/1–233107/3.

124 (a) Ku, S.-Y., Liman, C.D., Cochran, J.E., Toney, M.F., Chabinyc, M.L., and Hawker, C.J. (2011) *Adv. Mater.*, **23**, 2289–2293; (b) Guide, M., Dang, X.-D., and Nguyen, T.-Q. (2011) *Adv. Mater.*, **23**, 2313–2319.

125 (a) Tu, W., Lei, J., Wang, P., and Ju, H. (2011) *Chem. Eur. J.*, **17**, 9440–9447; (b) Lee, M.J., Seo, K.D., Song, H.M., Kang, M.S., Eom, Y.K., Kang, H.S., and Kim, H.K. (2011) *Tetrahedron Lett.*, **52**, 3879–3882; (c) Warnan, J., Buchet, F., Pellegrin, Y., Blart, E., and Odobel, F. (2011) *Org. Lett.*, **13**, 3944–3947; (d) Jiao, C., Zu, N., Huang, K.-W., Wang, P., and Wu, J. (2011) *Org. Lett.*, **13**, 3652–3655.

12
Isocyanide-Based Multicomponent Reactions towards Benzodiazepines
Yijun Huang and Alexander Dömling

12.1
Introduction

The benzodiazepines, a family of drugs that are used to relieve insomnia and anxiety, as well as to treat muscle spasms and prevent seizures [1], are one of the most widely prescribed medications for the central nervous system. In fact, during the past 40 years diazepam has been one of the most frequently prescribed drugs worldwide, with over 40 medications highlighting the benzodiazepines as classic privileged structures with a broad range of therapeutic treatments [2, 3]. Following the discovery of the benzodiazepine family, many synthetic derivatives with a wide pharmacological spectrum have been created [4]. Indeed, the development of new synthetic approaches to the benzodiazepines has attracted considerable attention with regards to the discovery of biologically active compounds.

Although the chemical syntheses of benzodiazepines have been investigated extensively since the 1960s [5], it was only during the early 1990s that Ellman *et al.* developed a general method for the solid-phase synthesis of 1,4-benzodiazepine libraries [6]. As shown in Scheme 12.1, 2-aminobenzophenone derivatives **1** were first attached to the polystyrene solid support, and then transformed to 1,4-benzodiazepine derivatives **5** by constitute linear syntheses over five steps. The benzodiazepine products **6** were cleaved from the support in very high overall yields (85–100% conversion from support-bound starting materials **2**). The library of 1,4-benzodiazepine derivatives was evaluated by using a screening assay against the cholecystokinin A receptor [7].

In contrast to constitute linear syntheses, multicomponent reaction (MCR) chemistry allows for the synthesis of arrays of compounds in a highly efficient and diverse manner [8, 9]. Typically, MCRs allow the resource- and cost-effective, rapid, and convergent synthesis of diverse compound libraries, and also greatly improve the efficiency to explore the chemical space with limited synthetic effort [10]. Isocyanide-based MCRs (IMCRs), such as the Ugi and Passerini reactions, provide powerful tools for the production of arrays of drug-like compounds, and with a high atom economy [11]. The post-condensation reactions of MCR products have

Isocyanide Chemistry: Applications in Synthesis and Material Science, First Edition. Edited by Valentine Nenajdenko.
© 2012 Wiley-VCH Verlag GmbH & Co. KGaA. Published 2012 by Wiley-VCH Verlag GmbH & Co. KGaA.

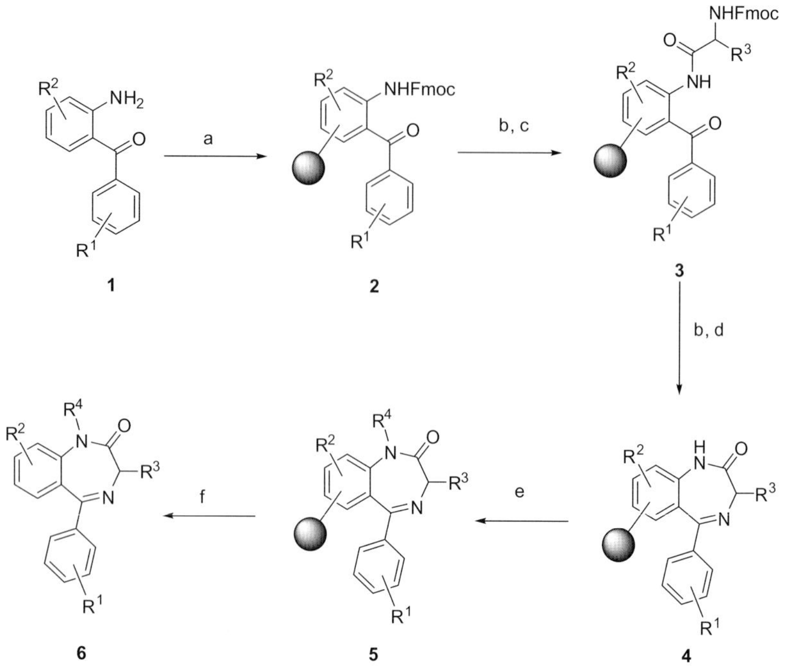

Scheme 12.1 Ellman's 1,4-benzodiazepine synthesis. Reaction conditions: (a) Attachment of 2-aminobenzophenones to an acid-cleavable linker; (b) 20% piperidine in DMF; (c) N-FMOC-amino acid fluoride, 4-methyl-2,6-di-*tert*-butylpyridine; (d) 5% acetic acid in DMF, 60 °C; (e) Lithiated 5-(phenylmethyl)-2-oxazolidinone in THF, −78 °C, followed by alkylating agent and DMF; (f) TFA/H$_2$O/Me$_2$S (95:5:10).

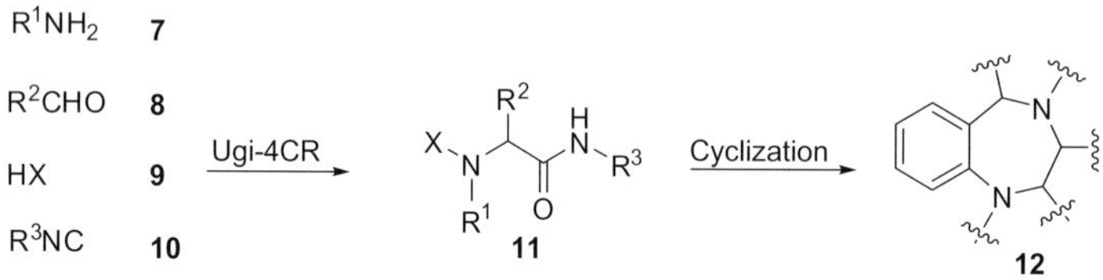

Figure 12.1 The general strategy for the synthesis of benzodiazepines via Ugi-multicomponent reactions.

been investigated extensively for the construction of heterocycles [12]. Indeed, during the past few decades IMCR chemistry has been applied to the synthesis of diverse libraries based on benzodiazepine scaffolds as drug-like compounds, an example being the modulators of germ cell nuclear factor (GCNF) to regulate stem cell differentiation [13]. In general, benzodiazepine heterocyclic cores **12** can be achieved by the Ugi reaction of bifunctional starting materials in conjunction with post-condensation cyclization (Figure 12.1).

12.2
1,4-Benzodiazepine Scaffolds Assembled via IMCR Chemistry

12.2.1
Two-Ring Systems

Armstrong et al. reported a dramatically improved route to 1,4-benzodiazepine-2,5-diones by using the Ugi four-component reaction (Ugi-4CR) [14]. As shown in Scheme 12.2, the Ugi reaction employs a variety of anthranilic acids **13**, amines **14**, aldehydes **15**, and 1-isocyanocyclohexene **16** as a convertible isocyanide, which is cleaved under acidic conditions to form the products **17b** with a seven-membered diazepine ring. This example shows the advantage of the MCR for the synthesis of a benzodiazepine scaffold via such a concise route. Another feature is the higher input diversity offered by the Ugi-4CR, compared to traditional syntheses that rely on the coupling of amino acids with anthranilic acids [15].

Subsequently, Hulme et al. developed an improved high-yielding solution-phase synthesis of 1,4-benzodiazepine-2,5-diones [16]. In this case, N-Boc-protected anthranilic acids **1** were employed for the Ugi reaction using Armstrong's convertible isocyanide **16** (Scheme 12.3). Following a one-pot deprotection and cyclization, the 1,4-benzodiazepine-2,5-diones **21** were obtained from both a 96-well plate format and a scale-up procedure. Later, n-butylisonitrile was used as a cheaper and more atom-economic alternative to "designer convertible isonitriles" for the synthesis of 1,4-benzodiazepine-2,5-diones [17].

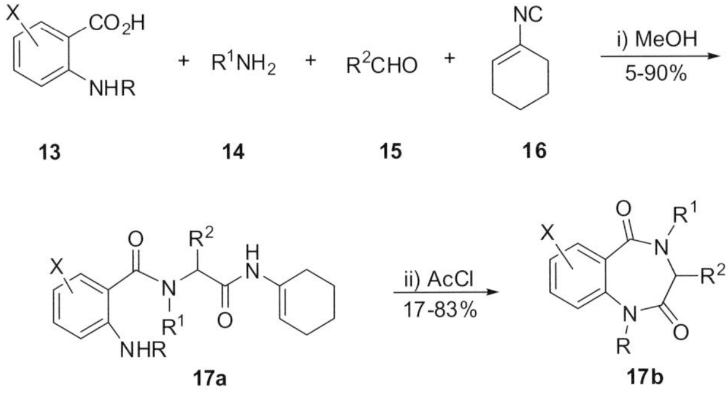

Scheme 12.2 Armstrong's 1,4-benzodiazepine-2,5-dione synthesis.

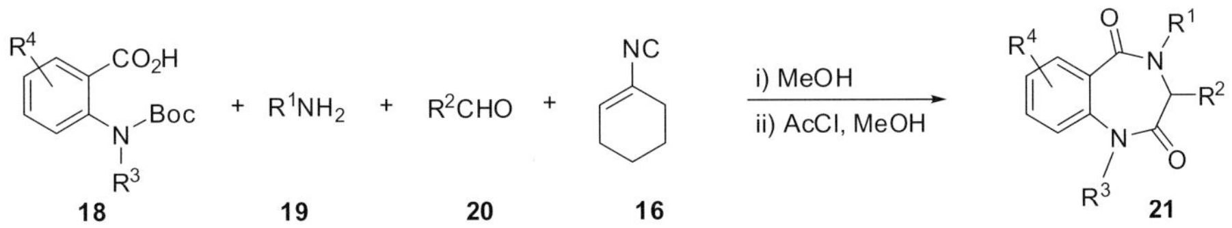

Scheme 12.3 Hulme's 1,4-benzodiazepine-2,5-dione synthesis.

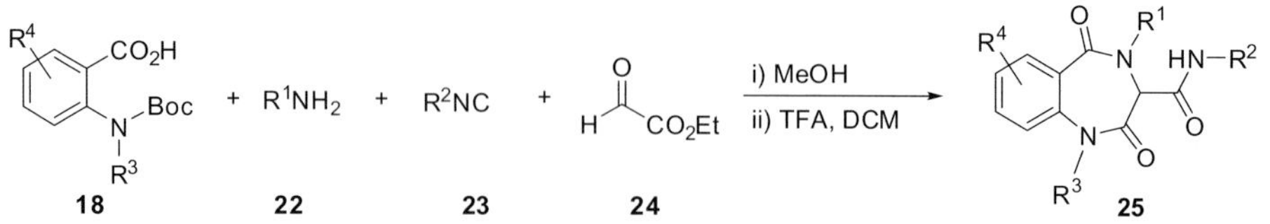

434 | *12 Isocyanide-Based Multicomponent Reactions towards Benzodiazepines*

Scheme 12.4 Synthesis of 1,4-benzodiazepine-2,5-diones.

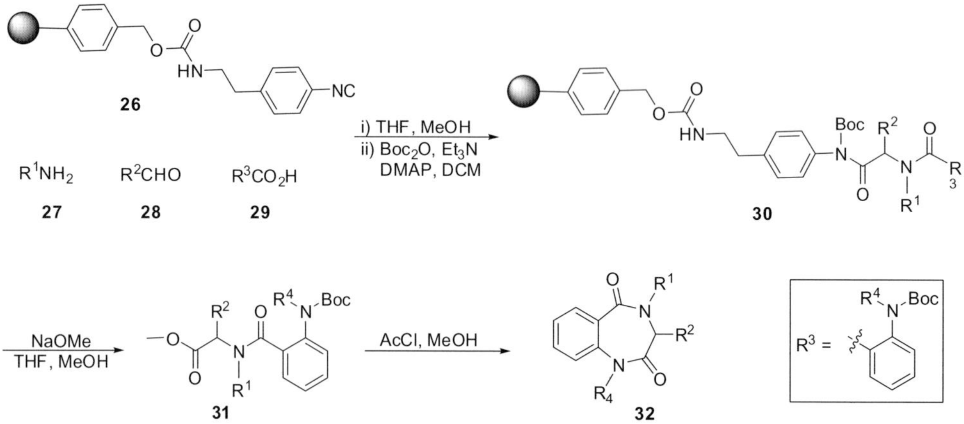

Scheme 12.5 UDC approach using resin-bound isocyanide **1**.

A UDC (Ugi/deprotection/cyclization) strategy was further explored for the preparation of 1,4-benzodiazepine-2,5-diones [18]. For this, ethyl glyoxylate **24** serves as a bifunctional starting material, while the aldehyde functional group is employed in the Ugi-4CR and the glyoxal ester functionality is cyclized with the deprotected intramolecular amino nucleophile (Scheme 12.4). This versatile "three-step, one-pot" procedure allows access to diverse arrays of 1,4-benzodiazepine-2,5-diones **25** in high yield and purity.

Polymer-supported reagents were investigated for the automated synthesis of diverse 1,4-benzodiazepine-2,5-diones, in order to avoid the tedious and costly parallel purification of the solution-phase products. The resin-bound isocyanide **26** is an example of a safety-catch linker, which can be resin-cleaved upon Boc-activation with a variety of nucleophiles [19]. The Ugi reaction of polymer-supported isocyanide **26** was followed by a methoxide safety-catch clipping strategy. The subsequent solution-phase cyclization allowed access to 1,4-benzodiazepine-2,5-diones **32** (Scheme 12.5). Later, Chen *et al.* introduced the Rink-isonitrile resin as a universal platform for applying the UDC strategy [20].

Hulme *et al.* further applied a UDC strategy with resin-bound α-amino acids **33** for the synthesis of 1,4-benzodiazepine-2,5-diones [21]. The Ugi reaction of N-Boc-protected anthranilic acids, with subsequent acid treatment, allowed the preparation of diverse arrays of products (Scheme 12.6). This two-step protocol proved to

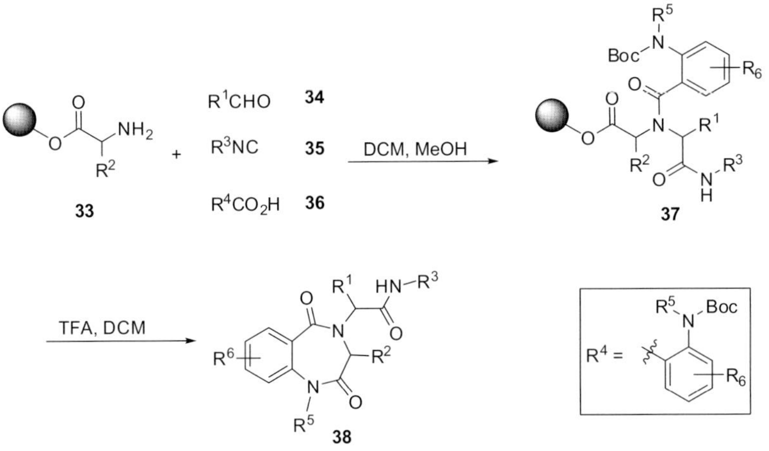

Scheme 12.6 UDC approach using resin-bound amino acid **1**.

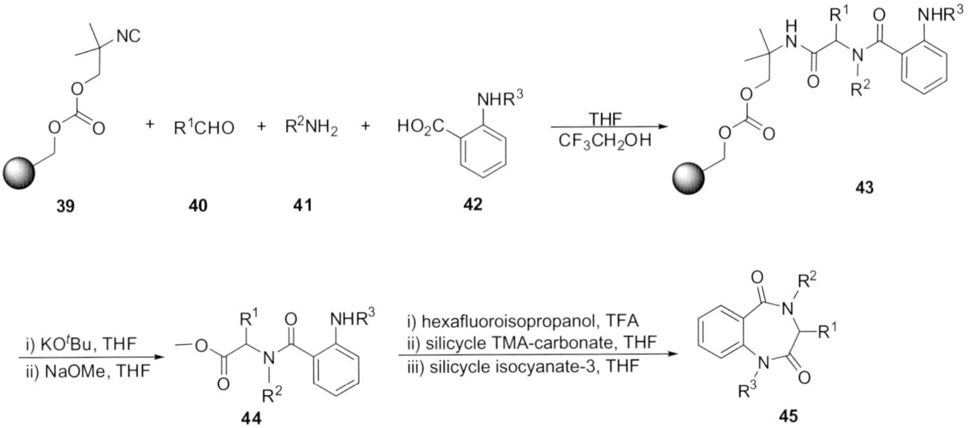

Scheme 12.7 Kennedy's 1,4-benzodiazepine-2,5-dione synthesis. Silicycle TMA-carbonate and isocyanate-3 were used to free the base amine and scavenge any uncyclized product, respectively.

be capable of generating 1,4-benzodiazepines **38** with five potential points of diversity. This library was also advanced to production in a 96-well format, using an automation procedure.

Kennedy et al. developed a resin-bound isocyanide **39** (as an extension of a class of convertible isonitriles reported by Ugi et al. [22]) for the synthesis of 1,4-benzodiazepine-2,5-diones [23]. The convertible isonitrile was utilized in the Ugi-4CR, whereupon the resin-bound Ugi product went cleavage to form **44** (Scheme 12.7). The intramolecular cyclization between the carboxylic acid ester and the secondary amine yielded the 1,4-diazepine ring. By following this

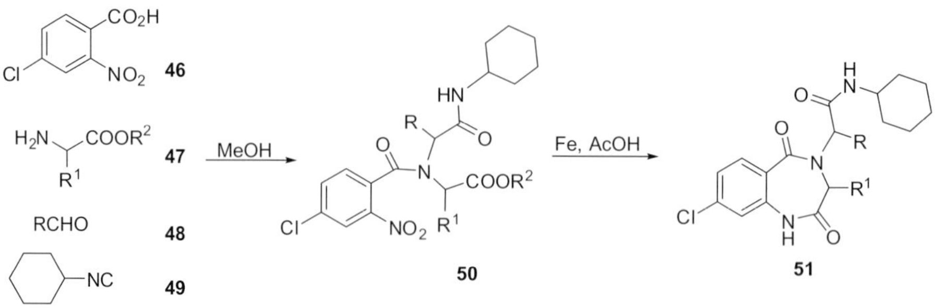

Scheme 12.8 Marcaccini's 1,4-benzodiazepine-2,5-dione synthesis.

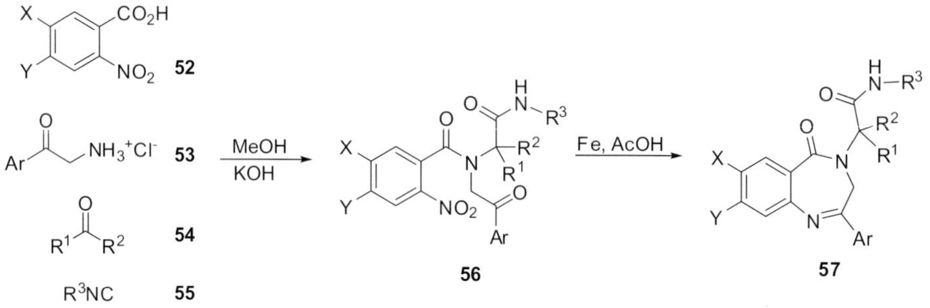

Scheme 12.9 Synthesis of 4,5-dihydro-3H-1,4-benzodiazepin-5-ones.

procedure, it was possible to prepare 80 compounds of a 1,4-benzodiazepine-2,5-dione plate in parallel format.

As an alternative to the synthetic routes using anthranilic acid derivatives, o-nitrobenzoic acid derivatives were employed as bifunctional starting materials for the Ugi-4CR and subsequent reductive cyclization. Marcaccini et al. described a convenient two-step synthesis of 1,4-benzodiazepine-2,5-diones starting from 4-chloro-2-nitrobenzoic acid 46 [24]. In this case, the reaction of **46**, α-amino esters **47**, aldehydes **48**, and cyclohexyl isocyanide **49** afforded the Ugi products **50**, in which the nitro group was reduced to the amino group, followed by an immediate cyclization to the final products **51** (Scheme 12.8).

Following this strategy, 2-nitrobenzoic acids **52** were applied for the synthesis of 4,5-dihydro-3H-1,4-benzodiazepin-5-ones [25]. The reaction of **52**, phenacylamine hydrochloride **53**, aldehyde or ketone **54**, and isocyanide **55** led to the generation of the Ugi products **5** (Scheme 12.9). The reductive cyclization of compounds **56** then yielded benzodiazepines **57** in good overall yields.

Hulme et al. developed a strategy which combined the MCR with S$_N$Ar cyclization to access arrays of biologically relevant benzodiazepines [26, 27]. For this, N-Boc-protected α-aminoaldehydes **59** were employed in the Ugi-4CR, with the scavenging resins polystyrene (PS)-tosylhydrazine (PS-TsNHNH$_2$) and PS-diisopropylethylamine (PS-DIEA) being used to remove any excess aldehyde **59**

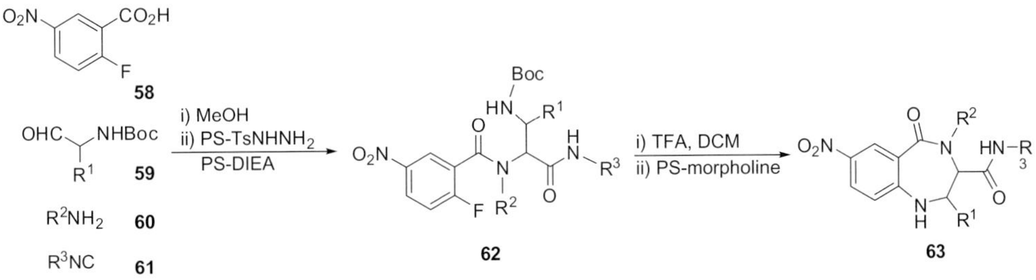

Scheme 12.10 Synthesis of 1,4-benzodiazepine-5-ones.

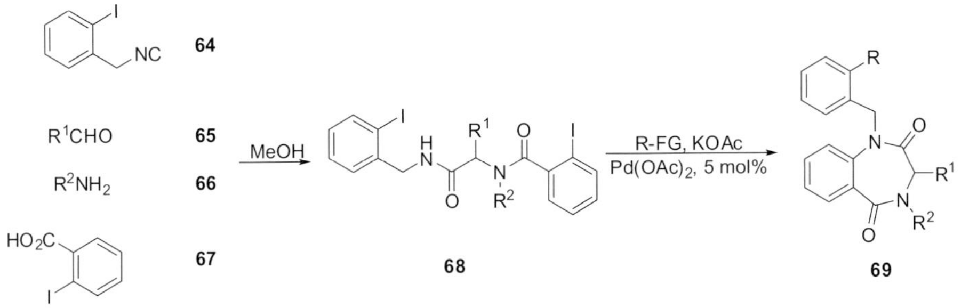

Scheme 12.11 Synthesis of functionalized benzodiazepinediones. R-FG: dihydrofuran, methyl acrylate, dimethylacrylamide, acrylonitrile.

and unreacted acid **58**. In this reaction, the intermediate **62** underwent deprotection and subsequent S_NAr cyclization to yield 1,4-benzodiazepine-5-ones **63** (Scheme 12.10).

Zhu et al. developed a metal-catalyzed process for the synthesis of functionalized benzodiazepinediones [28, 29]. Here, o-iodobenzyl isonitrile **64**, o-iodobenzoic acid **67**, amines, and aldehydes were employed as inputs for the Ugi reaction (Scheme 12.11). As a consequence, the Ugi product **68** was transformed to **69** via a palladium-catalyzed domino process that involved intramolecular N-arylation and intermolecular C–C bond formation. A major feature of this procedure was the facile formation of 1,4-diazepine ring and the site-selective functionalization.

Kalinski et al. reported a straightforward two-step synthesis of 1,4-benzodiazepine-2,5-diones using Ugi-4CR coupled with a palladium-assisted intramolecular N-aryl amidation [30]. In this case, 2-bromobenzoic acids **72** derivates were employed in the Ugi reaction, and further involved in the intramolecular N-aryl amidation reaction (Scheme 12.12). The products **74b** were generated with four points of diversity from readily available starting materials.

Andreana et al. developed two MCR approaches for the synthesis of 1,4-benzodiazepine-3-ones [31], with both routes employing a bifunctional starting material **75** as substrate for the Ugi reaction and the Michael addition (Scheme 12.13). The nitro group of either o-nitrobenzaldehyde **77** or o-nitrobenzylamine

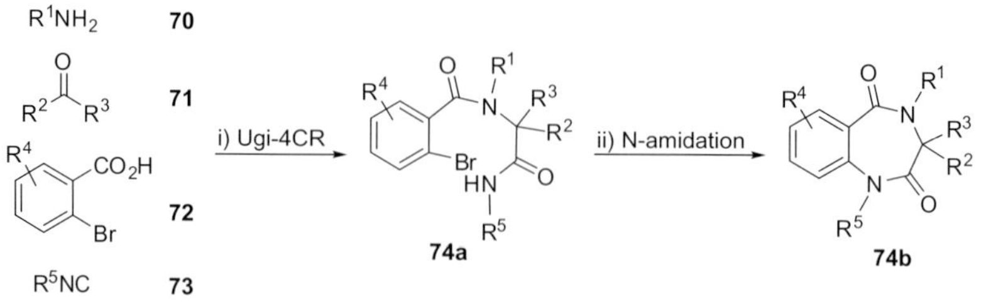

Scheme 12.12 Ugi-4CR/N-aryl amidation strategy.

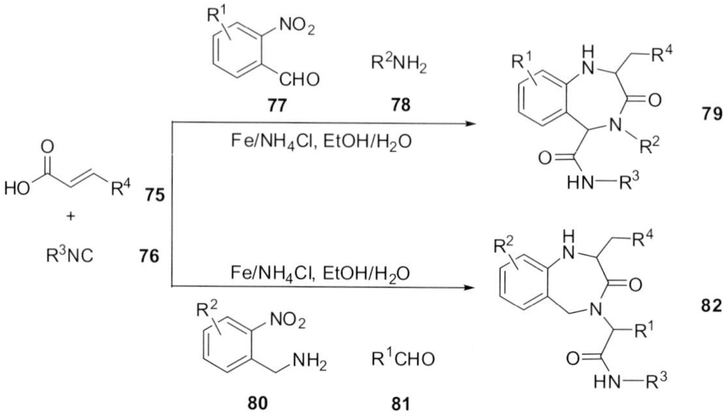

Scheme 12.13 Synthesis of 1,4-benzodiazepine-3-ones.

80 was reduced to the amino functionality, which can facilitate an aza-Michael cyclization to form the 1,4-diazepine ring. Subsequently, the regiochemically differentiated products **79** and **82** were obtained via the one-pot, two-step reaction protocol.

Yan et al. developed fluorous displaceable linker-facilitated synthesis of 1,4-benzodiazepine-2,5-dione libraries [32]. Here, perfluorooctanesulfonyl-protected 4-hydroxy benzaldehydes **84** were used as the limiting agent for the Ugi-4CR to form condensed products **86** (Scheme 12.14). Two approaches were investigated for the synthesis of **87** via an intramolecular cyclization between the ester and the amino component, which was introduced by 2-nitrobenzoic acids or Boc-protected anthranilic acids. The products with a fluorous tag were purified using fluorous solid-phase extraction (F-SPE), while the subsequent microwave-assisted Suzuki coupling reactions introduced a biaryl functionality to replace the fluorous tag.

This strategy was extended to the synthesis of biaryl-substituted 1,4-benzodiazepine-2,5-diones using a Ugi–cyclization–Suzuki protocol [33]. In this case, cyclohexylisocyanide and methyl 2-isocyanoacetate were used as convertible isocyanides **91** (Scheme 12.15), while microwave-promoted deprotection and

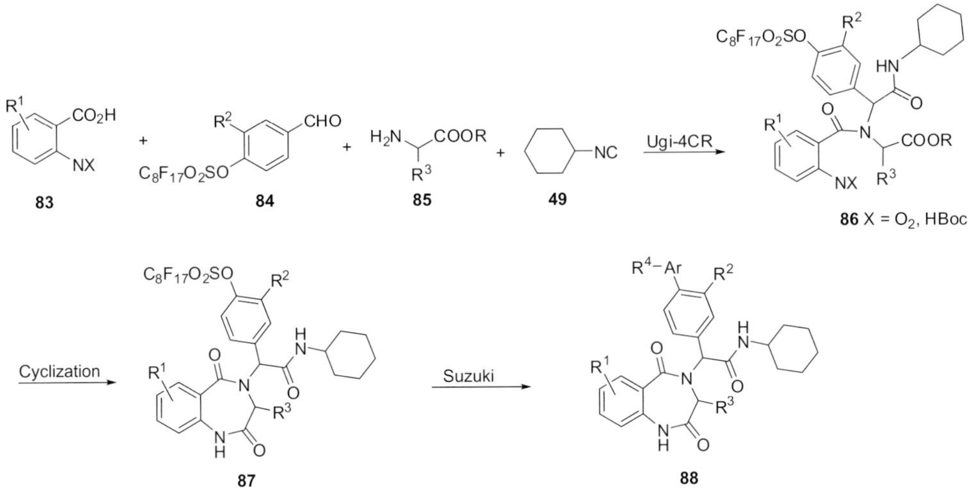

12.2 1,4-Benzodiazepine Scaffolds Assembled via IMCR Chemistry

Scheme 12.14 Fluorous synthesis of 1,4-benzodiazepine-2,5-diones.

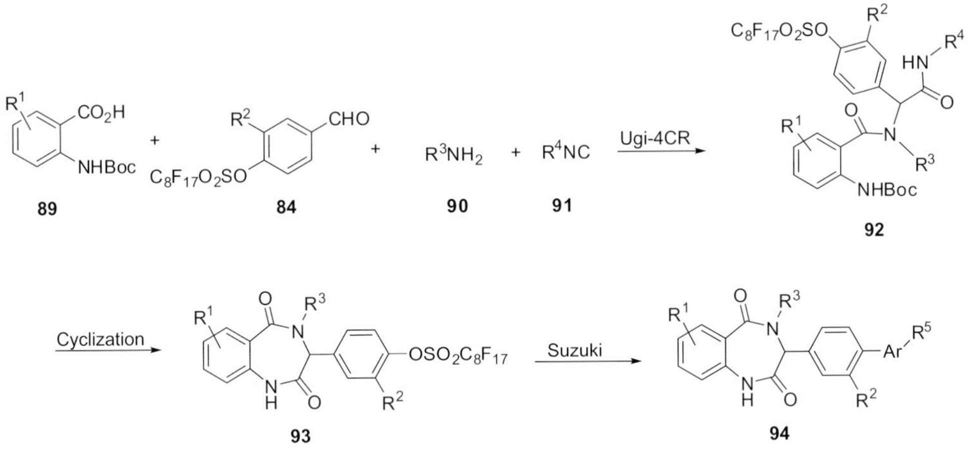

Scheme 12.15 The Ugi–cyclization–Suzuki protocol.

cyclization reactions yielded the 1,4-benzodiazepine-2,5-dione scaffold **93**. Suzuki coupling reactions further derivatized a biaryl group on the position, thus introducing the fluorous tag.

Torroba et al. described the synthesis of 5-oxobenzo[e][1,4]diazepine-3-carboxamides via a sequential Ugi reaction–Staudinger/aza-Wittig cyclization [34]. The intermediate Ugi products **98** were characterized as two opposite conformers of the enol form (Scheme 12.16), and then treated with triphenylphosphine to form a seven-membered ring of the benzodiazepine scaffold **99**. The pseudopeptidic backbone was superimposed well with type I, I′, II, and II′ β-turn motifs.

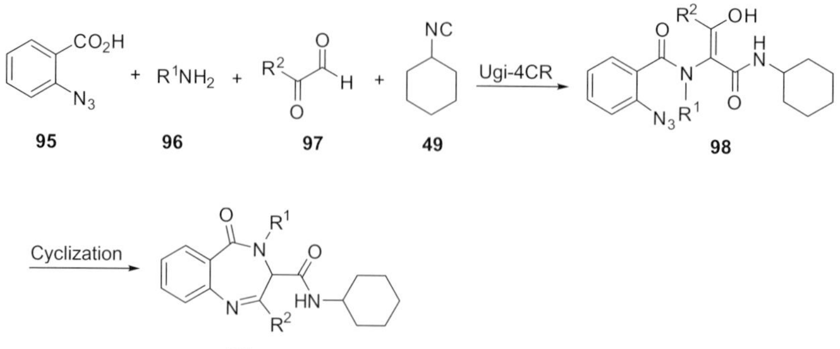

Scheme 12.16 Torroba's 1,4-benzodiazepine synthesis.

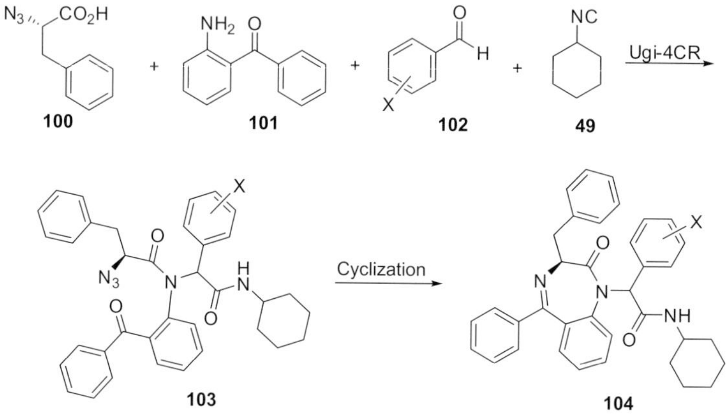

Scheme 12.17 The Ugi-4CR/Staudinger/aza-Wittig protocol.

By applying the same strategy, (S)-3-phenyl-2-azidopropionic acid **100**, 2-aminobenzophenone **101**, p-substituted benzaldehydes **102**, and cyclohexyl isocyanide **49** were employed for the Ugi-4CR [35], and the Ugi products **103** were isolated by filtration and recrystallization (Scheme 12.17). After treatment with triphenylphosphine, (S)-3-benzyl-2-oxo-1,4-benzodiazepines were obtained in good yields after chromatography purification. The products **103** and **104** were identified as single S isomers in the phenylalanine moiety, while **104** was isolated as an equimolecular mixture of the two diastereomers.

12.2.2
Fused-Ring Systems

In combination with the Ugi-4CR and the subsequent intramolecular cyclization, Zhu et al. reported a two-step synthesis of dihydroazaphenanthrene-fused benzodiazepinediones [28]. Here, the Ugi-4CR yielded the linear amide **109**, which then underwent a palladium-catalyzed intramolecular α-arylation to produce **110**

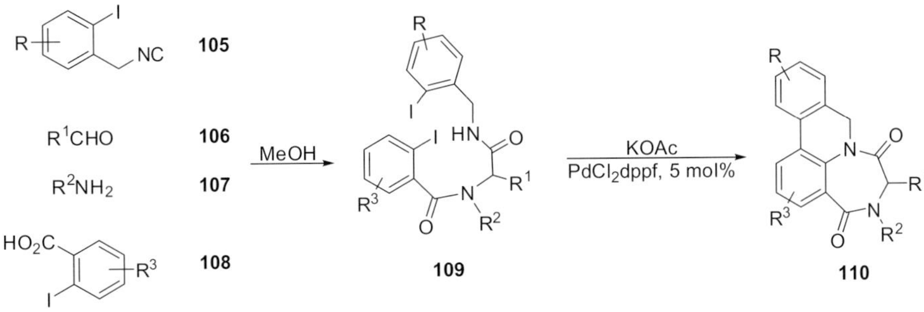

Scheme 12.18 Synthesis of dihydroazaphenanthrene-fused benzodiazepinediones.

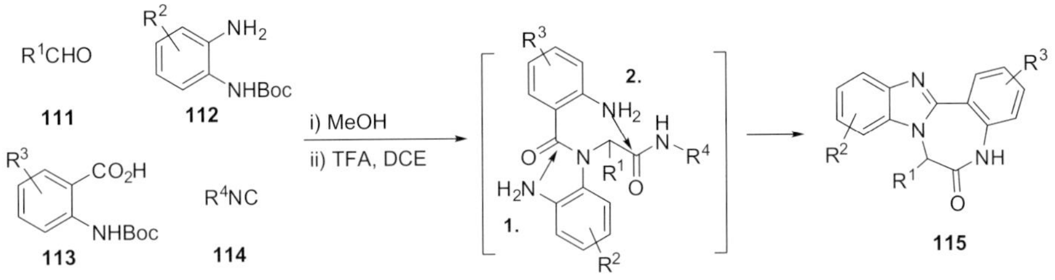

Scheme 12.19 Synthesis of triazadibenzoazulenones. 1, Benzimidazole formation; 2, benzodiazepine formation.

(Scheme 12.18). The scope and reliability of this process indicated the potential in the diversity-oriented synthesis of this class of compound.

Hulme et al. developed a two-step solution-phase protocol for the synthesis of arrays of triazadibenzoazulenones [36]. In this case, the Ugi reaction employed aldehydes **111**, mono-Boc-protected phenylene diamines **112**, Boc-protected anthranilic acids **113**, and a designer "universal isonitrile" **114** (Scheme 12.19). The Ugi product was treated with trifluoroacetic acid (TFA) to unmask any internal amino nucleophiles, which in turn initiated two tandem ring-forming reactions. The isocyanide component (n-butylisonitrile) was selected to enable the formation of 1,4-diazepine ring-fused products **115**.

Similarly, a double UDC strategy was adopted to synthesize fused quinoxalinone–benzodiazepines and bis-benzodiazepines [37]. In this reaction, N-Boc-1,2-phenylenediamine **118** and ethyl glyoxylate **24** were employed to pre-form the Schiff base, which was then incorporated into the Ugi reaction. In an alternative route, tert-butyl 2-aminomethyl phenylcarbamate **120** was used as the amine source for the Ugi reaction (Scheme 12.20). The Ugi products were cyclized under microwave irradiation to afford the tetracyclic scaffolds **119** and **121**, with control of the desired cyclization mode being achieved by employing 4-tert-butyl cyclohexen-1-yl isocyanide **117**.

Voskressensky et al. developed an efficient method for the synthesis of heteroannulated 1,4-benzodiazepines [38], in which the Ugi five-center four-component

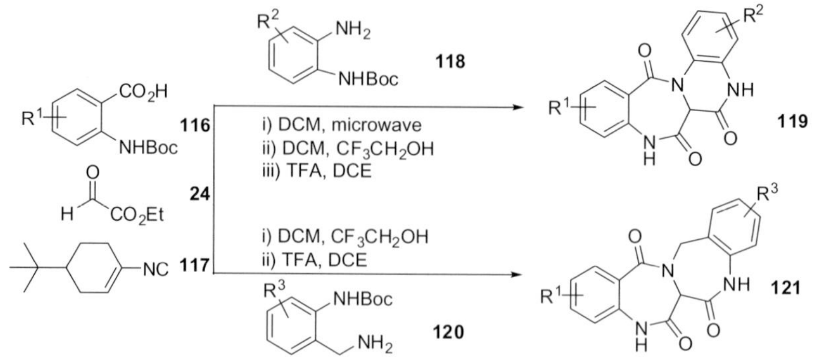

Scheme 12.20 Synthesis of fused benzodiazepinediones.

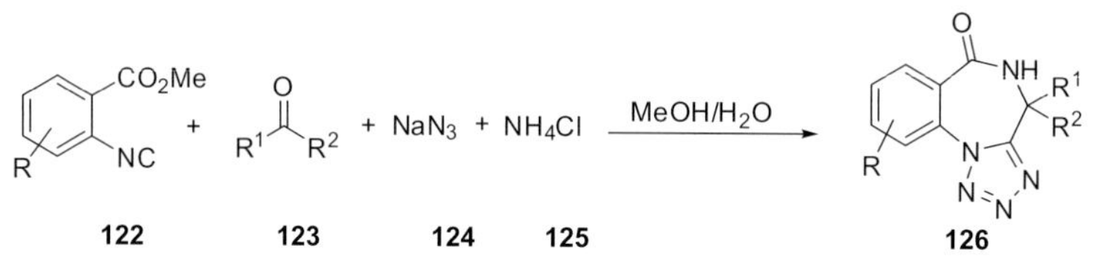

Scheme 12.21 Synthesis of tetrazolo[1,5-*a*][1,4]benzodiazepines.

reaction (U-5C-4CR) employed isocyanides **122**, ketones **123**, sodium azide **124**, and ammonium chloride **125** (Scheme 12.21). The tetrazolodiazepines **126** produced were precipitated from the reaction mixture in aqueous methanol. Compounds derived from this scaffold have demonstrated both platelet aggregation inhibitory and cholecystokinin agonist activities.

Hulme et al. reported two-step, solution-phase protocols for the synthesis of fused dihydroquinazoline–benzodiazepine tetracycles [39], where *N*-Boc-protected anthranilic acid **127** and mono-protected 2-aminobenzylamine (**130** or **132**) were employed for the Ugi-4CR. The acid treatment enabled two sequential ring-closing transformations to form the tetracyclic scaffolds **131** and **133** (Scheme 12.22). This method was facilitated by the use of microwave irradiation and *n*-butyl isocyanide to control the rate of each ring-forming transformation.

Ivachtchenko et al. developed an efficient method for the synthesis of pyrrolo[1,2-*a*][1,4]-diazepines [40]. In this case, the bifunctional starting material **134** was designed with aldehyde and acid functional groups, which were involved in the Ugi reaction with amines **135** and isocyanides **136** (Scheme 12.23). Based on the modification of the Ugi-4CR, the products **137** with 1,4-diazepine ring were formed in excellent yields. This route was shown to provide a valuable strategy towards accessing pyrrolo[1,2-*a*][1,4]diazepines and their bioisosteric analogues.

When Akritopoulou-Zanze et al. applied sequential Ugi/alkyne–azide cycloaddition reactions to the synthesis of triazolobenzodiazepines [41], the azide function-

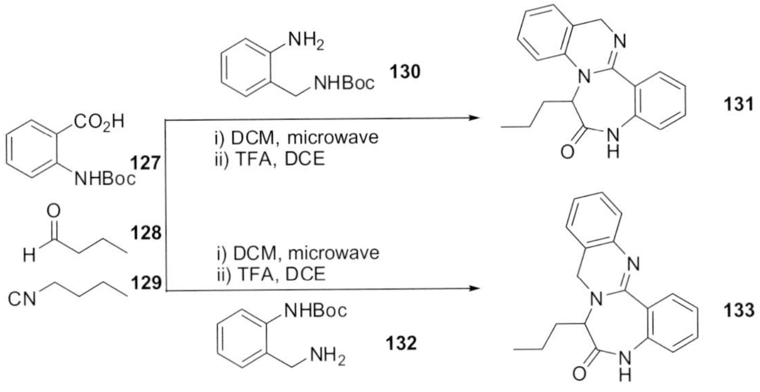

Scheme 12.22 Synthesis of dihydroquinazoline–benzodiazepine tetracycles.

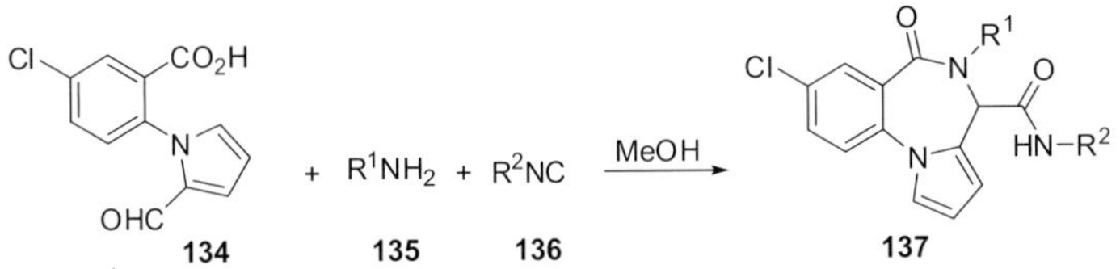

Scheme 12.23 Synthesis of pyrrolo[1,2-a][1,4]-diazepines.

ality could be introduced on the carboxylic acid or aldehyde inputs, while an acetylenic functionality could be incorporated on the amine or carboxylic acid inputs (Scheme 12.24). The Ugi reactions proceeded smoothly to provide the desired intermediates, which were heated in benzene to afford the cyclized products, in excellent yields.

12.3
1,5-Benzodiazepine Scaffolds Assembled via IMCR Chemistry

Shaabani et al. discovered an efficient MCR approach for the synthesis of 1,5-benzodiazepine derivatives [42], which utilized a one-pot reaction of aromatic diamines **151**, linear or cyclic ketones **152**, and isocyanides **153** in the presence of water, to yield the 1,5-benzodiazepine scaffold **154** (Scheme 12.25). The products were isolated in high yields, using only a catalytic quantity of p-toluenesulfonic acid at room temperature. This main features of this methodology were the easy work-up procedure, the high atom economy, combinatorial diversity, and a lack of undesirable side reactions.

Subsequently, Shaabani et al. developed an alternative method for the synthesis of 1,5-benzodiazepine-2,4-diones, starting from diamines [43]. In this one-pot reaction of aromatic diamines **155**, Meldrum's acid **156**, and isocyanides **157**, the

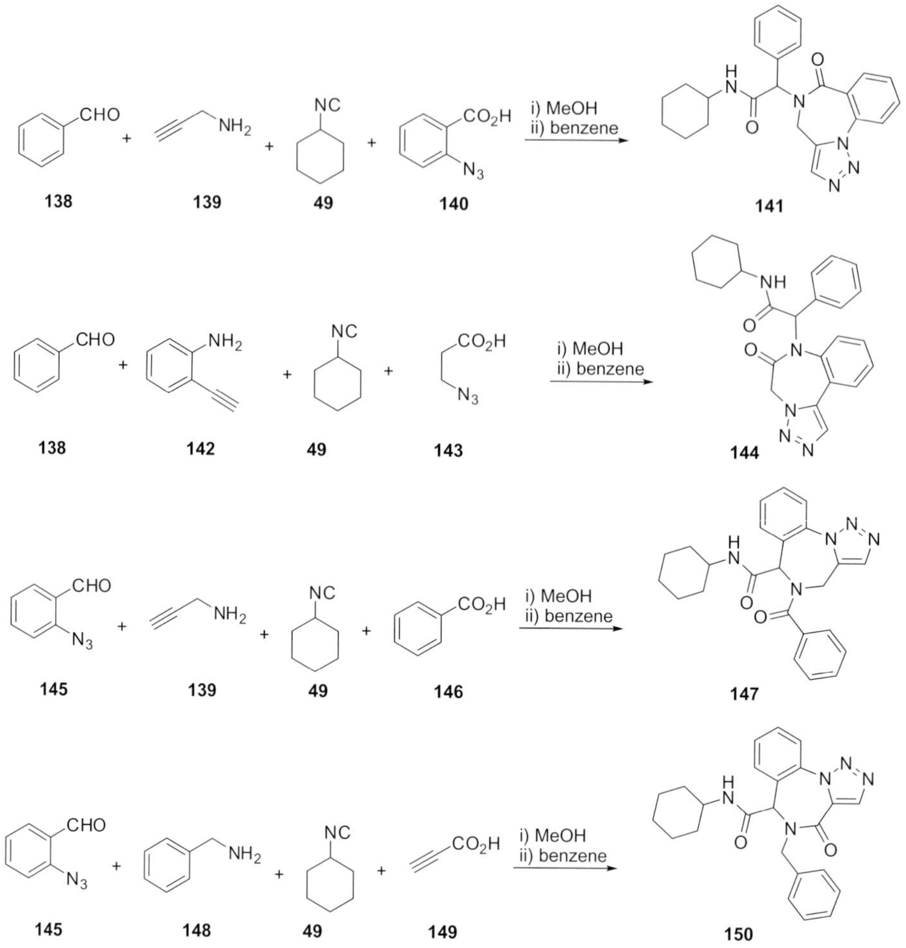

Scheme 12.24 The Ugi/alkyne–azide cycloaddition strategy.

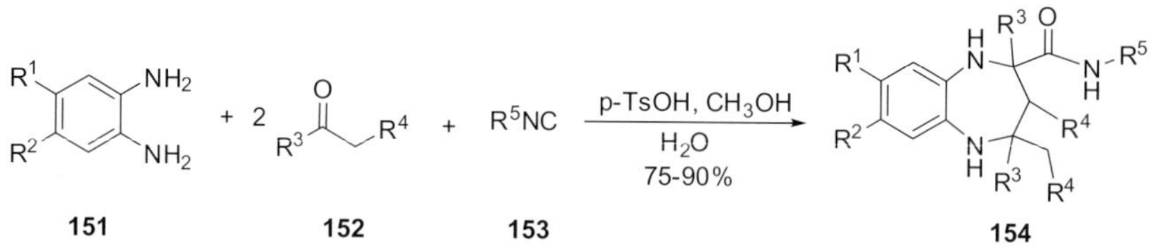

Scheme 12.25 Shaabani's synthesis of 1,5-benzodiazepines.

desirable products **158** were provided in high yields (Scheme 12.26). In this case, the Meldrum's acid was first converted to three components (1,3-dicarbonyl moiety, acetone, and water), and these were then recombined in the reaction sequence. The reactions were set up under ambient temperature, without the need for either a catalyst or activation.

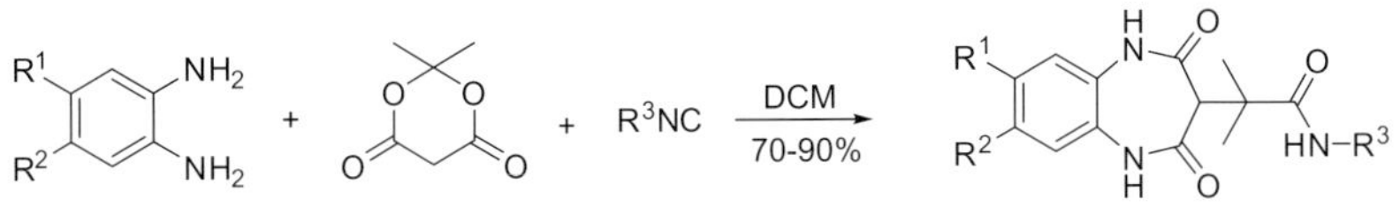

155 156 157 158

Scheme 12.26 Synthesis of 1,5-benzodiazepine-2,4-diones.

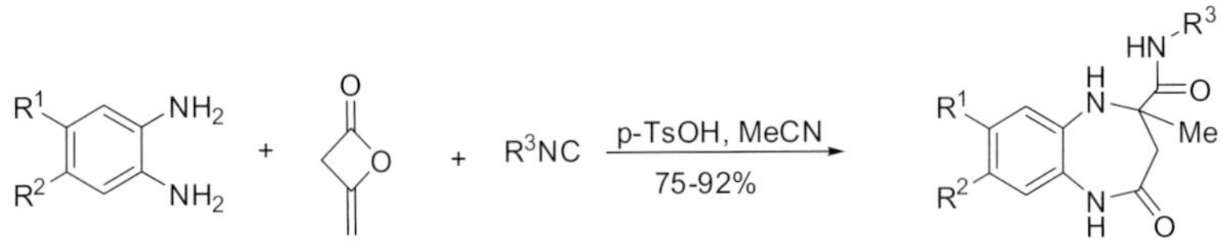

159 160 161 162

Scheme 12.27 Synthesis of 1,5-benzodiazepine-2-ones **162**.

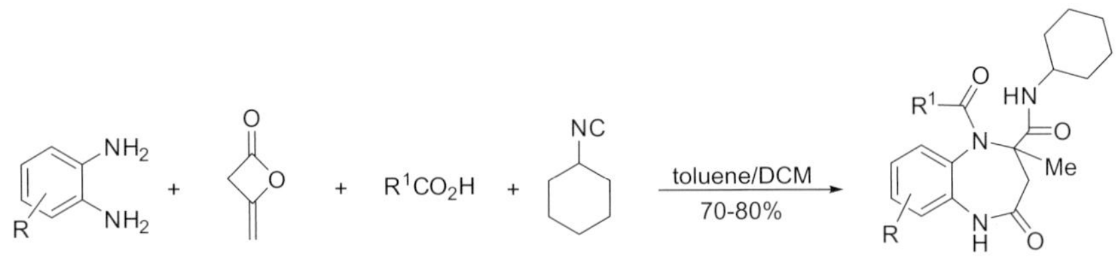

163 160 164 49 165

Scheme 12.28 Synthesis of 1,5-benzodiazepine-2-ones **165**.

A diketene-based MCR was investigated for the synthesis of 1,5-benzodiazepine-2-ones [44]. In this case, a one-pot reaction of aromatic 1,2-diamines **159**, diketene **160**, and various isocyanides **161** yielded the 1,5-diazepine products **162**, with regiochemical control (Scheme 12.27). In this reaction, three points of diversity could be introduced into the products, with good to excellent yields.

Alizadeh *et al.* developed a concise route for the synthesis of substituted 1,5-benzodiazepine-2-ones [45]. As an alternative to the classical Ugi reaction, *o*-phenylenediamine and diketene were considerably extended as the amine and oxo components, respectively (Scheme 12.28). The products were generated *in situ* from the four-component reaction of *o*-phenylenediamine **163**, diketene **160**, aromatic or aliphatic carboxylic acid **164**, and isocyanide **49**. Conformational isomerism was observed in the solution phase, because the free rotation around amide bonds was sterically restricted by the bulk substitutions.

Hulme *et al.* reported the synthesis of a 1,5-benzodiazepine scaffold using Ugi-4CR/S_NAr methodology [27, 46]. In this case, the Ugi reaction employed the bifunctional starting materials 2-fluoro-5-nitro benzoic acid **166** and *N*-Boc-*ortho*-phenylene diamine **167** (Scheme 12.29). The Ugi product was treated with TFA to remove the protection group, after which a PS-supported, base-catalyzed cyclization led to the dibenzoazepinone **170** in 54% yield.

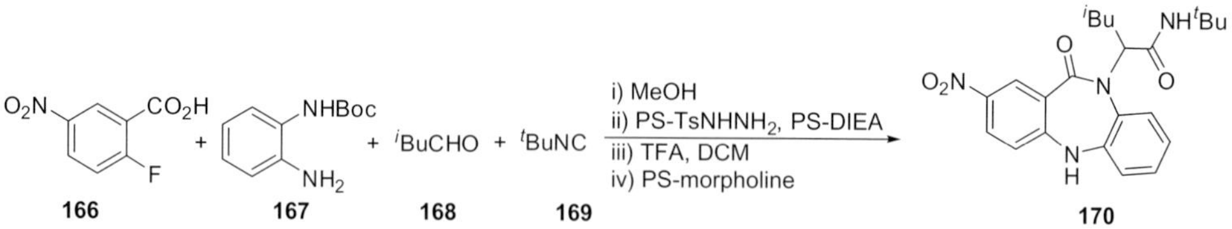

Scheme 12.29 The Ugi-4CR/S$_N$Ar methodology.

12.4
Outlook

During recent years, IMCRs have been successfully applied to the synthesis of benzodiazepines, in both an efficient and a diverse manner. In fact, over several decades these innovative approaches have become increasingly popular as a means of rapidly generating drug-like benzodiazepine libraries. Moreover, in taking advantage of these concise and powerful synthetic methodologies, this strategy should fulfill drug discovery efforts on the privilege scaffold. Undoubtedly, in the near future the rational design of novel MCRs will become increasingly important in revealing the as-yet uncharted chemical space of benzodiazepines. Clearly, the generation of diverse benzodiazepine libraries, representing enormous chemical space, will provide a valuable toolbox for structure-based drug design and drug discovery.

References

1 Sternbach, L.H. (1979) The benzodiazepine story. *J. Med. Chem.*, **22** (1), 1–7.
2 Duarte, C.D., Barreiro, E.J., and Fraga, C.A.M. (2007) Privileged structures: a useful concept for the rational design of new lead drug candidates. *Mini-Rev. Med. Chem.*, **7** (11), 1108–1119.
3 Welsch, M.E., Snyder, S.A., and Stockwell, B.R. (2010) Privileged scaffolds for library design and drug discovery. *Curr. Opin. Chem. Biol.*, **14** (3), 347–361.
4 Hadjipavlou-Litina, D. and Hansch, C. (1994) Quantitative structure-activity relationships of the benzodiazepines. A review and reevaluation. *Chem. Rev.*, **94** (6), 1483–1505.
5 Archer, G.A. and Sternbach, L.H. (1968) Chemistry of benzodiazepines. *Chem. Rev.*, **68** (6), 747–784.
6 Bunin, B.A. and Ellman, J.A. (1992) A general and expedient method for the solid-phase synthesis of 1,4-benzodiazepine derivatives. *J. Am. Chem. Soc.*, **114** (27), 10997–10998.
7 Bunin, B.A., Plunkett, M.J., and Ellman, J.A. (1994) The combinatorial synthesis and chemical and biological evaluation of a 1,4-benzodiazepine library. *Proc. Natl Acad. Sci. USA*, **91** (11), 4708–4712.
8 Armstrong, R.W., Combs, A.P., Tempest, P.A., Brown, S.D., and Keating, T.A. (1996) Multiple-component condensation strategies for combinatorial library synthesis. *Acc. Chem. Res.*, **29** (3), 123–131.
9 Domling, A., Wang, K., and Wang, W. (2012) Chemistry and biology of multicomponent reactions. *Chem. Rev.*, in press.
10 Hulme, C. (2005) Applications of multicomponent reactions in drug discovery–lead generation to process development, in *Multicomponent Reactions*

(eds J. Zhu and H. Bienaymé), John Wiley & Sons Inc., New York, pp. 311–341.

11 Domling, A. (2006) Recent developments in isocyanide based multicomponent reactions in applied chemistry. *Chem. Rev.*, **106** (1), 17–89.

12 Marcaccini, S. and Torroba, T. (2005) Post-condensation modifications of the Passerini and Ugi reactions, in *Multicomponent Reactions* (eds J. Zhu and H. Bienaymé), John Wiley & Sons Inc., New York, pp. 33–75.

13 Roughten, A., Rong, Y., Quintero, J., Ohlmeyer, M., Kultgen, S., Kingsbury, C., and Ho, K.-K. (2007) Preparation of benzodiazepines as modulators of germ cell nuclear factor to regulate of stem cell differentiation. 2007-US61984 2007095495.

14 Keating, T.A. and Armstrong, R.W. (1996) A remarkable two-step synthesis of diverse 1,4-benzodiazepine-2,5-diones using the Ugi four-component condensation. *J. Org. Chem.*, **61** (25), 8935–8939.

15 Moroder, L., Lutz, J., Grams, F., RudolphBohner, S., Osapay, G., Goodman, M., and Kolbeck, W. (1996) A new efficient method for the synthesis of 1,4-benzodiazepine-2,5-dione diversomers. *Biopolymers*, **38** (3), 295–300.

16 Hulme, C., Peng, J., Tang, S.Y., Burns, C.J., Morize, I., and Labaudiniere, R. (1998) Improved procedure for the solution phase preparation of 1,4-benzodiazepine-2,5-dione libraries via Armstrong's convertible isonitrile and the Ugi reaction. *J. Org. Chem.*, **63** (22), 8021–8023.

17 Hulme, C., Chappeta, S., and Dietrich, J. (2009) A simple, cheap alternative to "designer convertible isonitriles" expedited with microwaves. *Tetrahedron Lett.*, **50** (28), 4054–4057.

18 Hulme, C. and Cherrier, M.P. (1999) Novel applications of ethyl glyoxalate with the Ugi MCR. *Tetrahedron Lett.*, **40** (29), 5295–5299.

19 Hulme, C., Peng, J., Morton, G., Salvino, J.M., Herpin, T., and Labaudiniere, R. (1998) Novel safety-catch linker and its application with a Ugi/De-BOC/cyclization (UDC) strategy to access carboxylic acids, 1,4-benzodiazepines, diketopiperazines, ketopiperazines and dihydroquinoxalinones. *Tetrahedron Lett.*, **39** (40), 7227–7230.

20 Chen, J.J., Golebiowski, A., Klopfenstein, S.R., and West, L. (2002) The universal Rink-isonitrile resin: applications in Ugi reactions. *Tetrahedron Lett.*, **43** (22), 4083–4085.

21 Hulme, C., Ma, L., Kumar, N.V., Krolikowski, P.H., Allen, A.C., and Labaudiniere, R. (2000) Novel applications of resin bound alpha-amino acids for the synthesis of benzodiazepines (via Wang resin) and ketopiperazines (via hydroxymethyl resin). *Tetrahedron Lett.*, **41** (10), 1509–1514.

22 Lindhorst, T., Bock, H., and Ugi, I. (1999) A new class of convertible isocyanides in the Ugi four-component reaction. *Tetrahedron*, **55** (24), 7411–7420.

23 Kennedy, A.L., Fryer, A.M., and Josey, J.A. (2002) A new resin-bound universal isonitrile for the Ugi 4CC reaction: preparation and applications to the synthesis of 2,5-diketopiperazines and 1,4-benzodiazepine-2,5-diones. *Org. Lett.*, **4** (7), 1167–1170.

24 Faggi, C., Marcaccini, S., Pepino, R., and Pozo, M.C. (2002) Studies on isocyanides and related compounds: synthesis of 1,4-benzodiazepine-2,5-diones via Ugi four-component condensation. *Synthesis*, **18**, 2756–2760.

25 Marcaccini, S., Miliciani, M., and Pepino, R. (2005) A facile synthesis of 1,4-benzodiazepine derivatives via Ugi four-component condensation. *Tetrahedron Lett.*, **46** (4), 711–713.

26 Tempest, P., Pettus, L., Gore, V., and Hulme, C. (2003) MCC/SNAr methodology. Part 2: novel three-step solution phase access to libraries of benzodiazepines. *Tetrahedron Lett.*, **44** (9), 1947–1950.

27 Tempest, P., Ma, V., Kelly, M.G., Jones, W., and Hulme, C. (2001) MCC/SNAr methodology. Part 1: novel access to a range of heterocyclic cores. *Tetrahedron Lett.*, **42** (30), 4963–4968.

28 Cuny, G., Bois-Choussy, M., and Zhu, J.P. (2004) Palladium- and copper-catalyzed synthesis of medium- and

large-sized ring-fused dihydroazaphenanthrenes and 1,4-benzodiazepine-2,5-diones. Control of reaction pathway by metal-switching. *J. Am. Chem. Soc.*, **126** (44), 14475–14484.

29 Salcedo, A., Neuville, L., Rondot, C., Retailleau, P., and Zhu, J. (2008) Palladium-catalyzed domino intramolecular N-arylation/intermolecular C-C bond formation for the synthesis of functionalized benzodiazepinediones. *Org. Lett.*, **10** (5), 857–860.

30 Kalinski, C., Umkehrer, M., Ross, G., Kolb, J., Burdack, C., and Hiller, W. (2006) Highly substituted indol-2-ones, quinoxalin-2-ones and benzodiazepin-2,5-diones via a new Ugi(4CR)-Pd assisted N-aryl amidation strategy. *Tetrahedron Lett.*, **47** (20), 3423–3426.

31 De Silva, R.A., Santra, S., and Andreana, P.R. (2008) A tandem one-pot, microwave-assisted synthesis of regiochemically differentiated 1,2,4,5-tetrahydro-1,4-benzodiazepin-3-ones. *Org. Lett.*, **10** (20), 4541–4544.

32 Liu, A., Zhou, H., Su, G., Zhang, W., and Yan, B. (2009) Microwave-assisted fluorous synthesis of a 1,4-benzodiazepine-2,5-dione library. *J. Comb. Chem.*, **11** (6), 1083–1093.

33 Zhou, H., Zhang, W., and Yan, B. (2010) Use of cyclohexylisocyanide and methyl 2-isocyanoacetate as convertible isocyanides for microwave-assisted fluorous synthesis of 1,4-benzodiazepine-2,5-dione library. *J. Comb. Chem.*, **12** (1), 206–214.

34 Sanudo, M., Garcia-Valverde, M., Marcaccini, S., Delgado, J.J., Rojo, J., and Torroba, T. (2009) Synthesis of benzodiazepine beta-turn mimetics by an Ugi 4CC/Staudinger/aza-Wittig sequence. Solving the conformational behavior of the Ugi 4CC adducts. *J. Org. Chem.*, **74** (5), 2189–2192.

35 Lecinska, P., Corres, N., Moreno, D., Garcia-Valverde, M., Marcaccini, S., and Torroba, T. (2010) Synthesis of pseudopeptidic (S)-6-amino-5-oxo-1,4-diazepines and (S)-3-benzyl-2-oxo-1,4-benzodiazepines by an Ugi 4CC Staudinger/aza-Wittig sequence. *Tetrahedron*, **66** (34), 6783–6788.

36 Hulme, C., Chappeta, S., Griffith, C., Lee, Y.-S., and Dietrich, J. (2009) An efficient solution phase synthesis of triazadibenzoazulenones: "designer isonitrile free" methodology enabled by microwaves. *Tetrahedron Lett.*, **50** (17), 1939–1942.

37 Xu, Z., Dietrich, J., Shaw, A.Y., and Hulme, C. (2010) Two-step syntheses of fused quinoxaline-benzodiazepines and bis-benzodiazepines. *Tetrahedron Lett.*, **51** (34), 4566–4569.

38 Borisov, R.S., Polyakov, A.I., Medvedeva, L.A., Khrustalev, V.N., Guranova, N.I., and Voskressensky, L.G. (2010) Concise approach toward tetrazolo[1,5-a][1,4]benzodiazepines via a novel multicomponent isocyanide-based condensation. *Org. Lett.*, **12** (17), 3894–3897.

39 Dietrich, J., Kaiser, C., Meurice, N., and Hulme, C. (2010) Concise two-step solution phase syntheses of four novel dihydroquinazoline scaffolds. *Tetrahedron Lett.*, **51** (30), 3951–3955.

40 Ilyn, A.P., Trifilenkov, A.S., Kuzovkova, J.A., Kutepov, S.A., Nikitin, A.V., and Ivachtchenko, A.V. (2005) New four-component Ugi-type reaction. Synthesis of heterocyclic structures containing a pyrrolo[1,2-a][1,4]diazepine fragment. *J. Org. Chem.*, **70** (4), 1478–1481.

41 Akritopoulou-Zanze, I., Gracias, V., and Djuric, S.W. (2004) A versatile synthesis of fused triazolo derivatives by sequential Ugi/alkyne-azide cycloaddition reactions. *Tetrahedron Lett.*, **45** (46), 8439–8441.

42 Shaabani, A., Maleki, A., and Mofakham, H. (2008) Novel multicomponent one-pot synthesis of tetrahydro-1H-1,5-benzodiazepine-2-carboxamide derivatives. *J. Comb. Chem.*, **10** (4), 595–598.

43 Shaabani, A., Rezayan, A.H., Keshipour, S., Sarvary, A., and Ng, S.W. (2009) A novel one-pot three-(in situ five-)component condensation reaction: an unexpected approach for the synthesis of tetrahydro-2,4-dioxo-1H-benzo[b][1,5]diazepine-3-yl-2-methylpropanamide derivatives. *Org. Lett.*, **11** (15), 3342–3345.

44 Shaabani, A., Maleki, A., Hajishaabanha, F., Mofakham, H., Seyyedhamzeh, M., Mahyari, M., and Ng, S.W. (2009) Novel syntheses of tetrahydrobenzodiazepines and dihydropyrazines via isocyanide-based multicomponent reactions of

diamines. *J. Comb. Chem.*, **12** (1), 186–190.

45 Zohreh, N., Alizadeh, A., Bijanzadeh, H.R., and Zhu, L.-G. (2010) Novel approach to 1,5-benzodiazepine-2-ones containing peptoid backbone via one-pot diketene-based Ugi-4CR. *J. Comb. Chem.*, **12** (4), 497–502.

46 Tempest, P., Ma, V., Thomas, S., Hua, Z., Kelly, M.G., and Hulme, C. (2001) Two-step solution-phase synthesis of novel benzimidazoles utilizing a UDC (Ugi/de-Boc/cyclize) strategy. *Tetrahedron Lett.*, **42** (30), 4959–4962.

13
Applications of Isocyanides in the Synthesis of Heterocycles
Irini Akritopoulou-Zanze

13.1
Introduction

In the past, isocyanides have been used as starting materials in the preparation of numerous diverse structures, and have found many applications in the generation of heterocyclic structures. Recently, several excellent reviews have detailed the synthesis of heterocycles [1] and nitrogen heterocycles [2], and also of the multi-component reactions (MCRs) of isocyanides to produce heterocycles [3–5], nitrogen heterocycles [6], and piperazines [7].

In this chapter, details are provided of the synthesis of the most common aromatic heterocyclic rings, organized in a tabular format. Also highlighted are the major reactions in which an isocyanide is employed as a starting material to produce heterocyclic products. Only those reactions in which the corresponding heterocycle is prepared in a one-pot procedure or in one step, followed by straightforward secondary manipulations, are considered. The "Reactant" column contains details of all starting materials that contribute atoms to the final molecule, regardless of the sequence of addition. In cases of multiple step reactions, the reader is encouraged to consult with the primary literature to identify the exact reaction conditions.

13.2
Furans

Few reports exist on the synthesis of furans starting from isocyanides. The majority of these employ the reactions of aldehydes with zwitterionic intermediates, generated from dialkyl acetylenedicarboxylate (DMAD) and isocyanides, to yield tetra-substituted 2-aminofurans (Table 13.1). The reactions are versatile, and are equally effective with aliphatic and aromatic aldehydes. "Greener" versions using water [11, 12] or PEG400 [14] as the solvent have also been developed. Cinnamyl aldehydes [14, 17], tosyl imines [18], tricarbonyl compounds [19], as well as benzoyl chlorides [20], also participate in this reaction at room temperature to provide the

Isocyanide Chemistry: Applications in Synthesis and Material Science, First Edition. Edited by Valentine Nenajdenko.
© 2012 Wiley-VCH Verlag GmbH & Co. KGaA. Published 2012 by Wiley-VCH Verlag GmbH & Co. KGaA.

Table 13.1 Synthesis of 2-aminofurans.

Products	Reactants	Conditions (yields %)	Ref.
(2,3-diester-5-R3-4-R1NH-furan)	alkyne diester + isocyanide + aldehyde R3CHO	Dry benzene, 80°C, 2–48 h (54–68)	[8–15]
		or H_2O, r.t., 1.5–5 h (73–95)	
		or H_2O, PTC, 80°C, 1–2 h (76–82)	
		or CH_2Cl_2, r.t., 24 h (82–90)	
		or PEG400, r.t., 3–4 h (75–90)	
		or DMF, 180 W, 60 μl flow, 1180 μm capillary (76–79)	
(furan with indole substituent)	alkyne diester + isocyanide + N-Boc indole-3-carbaldehyde	Dry benzene, 25°C, 4–6 h (75–94)	[16]
(furan with vinyl-Ar)	alkyne diester + isocyanide + cinnamaldehyde	CH_2Cl_2, r.t., 2 days (58–72)	[14, 17]
		or PEG400, r.t., 3–4 h (75–96)	
(dimethyl ester furan with R2)	alkyne diester + isocyanide + N-Ts imine	Benzene, argon, r.t., 18 h (72–96)	[18]
(dimethyl ester furan with OR2)	alkyne diester + isocyanide + mandelate	CH_2Cl_2, r.t., 12 h (37–60)	[19]
(dimethyl ester furan with aryl-X, X = Cl, NO_2)	alkyne diester + isocyanide + 4-X-benzoyl chloride (X = Cl, NO_2)	CH_2Cl_2, r.t., 24 h (70–80)	[20]

Table 13.1 (Continued)

Products	Reactants	Conditions (yields %)	Ref.
Ph-acetyl-methyl-furan with R₁-NH	Ph enone with R₁-N⁺≡C⁻	CH₂Cl₂, reflux, 12 h (73–76)	[21]
R₂, CN, R₃-indole furan with R₁-NH	R₂-CHO, CN-acetyl-indole, R₁-N⁺≡C⁻	CH₃CO₂NH₄, EtOH, 78 °C, 1 h (61–91)	[22]
t-Bu-NH, Ar, CF₃ furan with t-Bu-NH	t-Bu-N⁺≡C⁻, Ar-CO-CH₂-CO-CF₃	CH₂Cl₂, r.t., 24 h (72–80)	[23]
Tetrasubstituted furan with R₁, R₂, R₃	R₂-O-CO-C≡C-CO-O-R₂, ⁻C≡N⁺-R₁, R₃-CO-O-CO-R₃	CH₂Cl₂, r.t., 24 h (10–85)	[24]
Chromone-fused furan with R₁, R₂, R₃	R₂-O-CO-C≡C-CO-O-R₂, R₁-N⁺≡C⁻, ⁻C≡N⁺-R₁, R₃-salicylaldehyde, Meldrum's acid	H₂O, r.t., 12 h (83–96)	[25]

furan products in good yields. Several other methods have been developed for the synthesis of 2-aminofurans, the majority of which are multicomponent in nature and allow for the synthesis of highly functionalized furan derivatives.

13.3
Pyrroles

Isocyanides have been used extensively for the preparation of disubstituted (Table 13.2), trisubstituted (Table 13.3), and tetrasubstituted (Table 13.4) pyrroles. The

Table 13.2 Synthesis of 2,3- and 3,4-disubstituted pyrroles.

Product	Reactants	Conditions (yields %)	Ref.
3-acyl pyrrole, R_1, R_2(C=O)	R_1CH=CH-C(O)R_2; Tos-N$^+\equiv$C$^-$	NaH, DMSO/Et$_2$O, r.t., 15 min–3 h (10–88)	[26, 27]
3-nitro pyrrole with R	R-CH=CH-NO$_2$; Tos-N$^+\equiv$C$^-$	[bmIm]Br, KOH, 20 °C, 1.5–12 h (38–64)	[28]
3-sulfonyl pyrrole, R_1, S(O)$_2$$R_2$	R_1CH=CH-S(O)$_2$$R_2$; Tos-N$^+\equiv$C$^-$	NaH, DMSO/Et$_2$O, r.t., 6 h (63–72)	[29, 30]
3,4-diaryl pyrrole Ar$_1$, Ar$_2$/H	Ar$_1$CH=CH-Ar$_2$/H; Tos-N$^+\equiv$C$^-$	t-BuONa, DMSO, 25–100 °C, 1–18 h (44–91)	[31]
2,3-disubstituted pyrrole, R, X; X = CO$_2$Et, CO$_2$t-Bu, 4-NO$_2$Ph	R-C≡C-H; X-N$^+\equiv$C$^-$; X = CO$_2$Et, CO$_2$t-Bu, 4-NO$_2$Ph	CuBr, Cs$_2$CO$_3$, DMF, 120 °C, 3 h (5–88)	[32]
3-nitro pyrrole with R	R-C(=CH-N$^+\equiv$C$^-$Tos)-NO$_2$	t-BuOK, DME, 0 °C to 20 °C, 1 h (14–94)	[33]
2,3-disubstituted pyrrole R, X; X = CN, CO$_2$Et, COPh, Ph, Me	R-C(=CH-N$^+\equiv$C$^-$Tos)-CH(X)(C(O)R_1); X = CN, CO$_2$Et, COPh, Ph, Me	t-BuOK, or Na, or NaOH, DME, or MeOH or EtOH, 0 °C to 20 °C, 1 h (0–99)	[34]

Table 13.3 Synthesis of 2,3,4-trisubstituted pyrroles.

Product	Reactants	Conditions (yields %)	Ref.
R_2, X on pyrrole (R_1 at 2, NH); X = CO_2Me, COR, CN, NO_2	R_2–CH=CH–X; Tos–CH(R_1)–$N^+\equiv C^-$; X = CO_2Me, COR, CN, NO_2	NaH, DMSO/Et_2O, r.t. (10–91); or t-BuOK, THF, −80 °C, 1 h (71–99); or n-BuLi, THF, −78 °C, 1 h (40–71)	[26, 33, 35–41]
R_2, X on pyrrole (R_1 at 2, NH); X = CO_2Me, COR, CN	R_2–CH=CH–X; Bt–CH(R_1)–$N^+\equiv C^-$; X = CO_2Me, COR, CN	t-BuOK, THF, reflux (0–92)	[42]
Pyrrole with R_1O–C(O)– at 2, R_2 at 3, –C(O)O–R_1 at 4	R_1O–C(O)–CH_2–$N^+\equiv C^-$; R_2–C(O)–H; R_1O–C(O)–CH_2–$N^+\equiv C^-$	n-BuLi, THF, −60 °C, then r.t. (68–91); or DBU, THF, 15–50 °C, 2 h (43–71)	[43–45]
Pyrrole with R_1O–C(O)– at 2, R_2 at 3, R_3 at 4	R_2–CH=C(R_3)–NO_2; R_1O–C(O)–CH_2–$N^+\equiv C^-$	DBU, THF, 75 °C, 16 h (55–95); or DBU, t-BuOMe, 20 °C, 4–20 h (78–91); or superstrong noionic base, THF −20 to −15 °C, 2 h (100–100)	[46–49]
Pyrrole with R_1O–C(O)– at 2, R_2 at 3, R_3 at 4	R_2–CH=C(R_3)–S(O)–Ph; R_1O–C(O)–CH_2–$N^+\equiv C^-$	t-BuOK, NaH or KH, THF, 0 °C to r.t., 3–12 h (0–63)	[50–54]
Pyrrole with R_1O–C(O)– at 2, R_2 at 3, R_3 at 4	R_2–CH(OAc)–CH(R_3)–NO_2; R_1O–C(O)–CH_2–$N^+\equiv C^-$	DBU, THF, 25 °C, 10–12 h (41–74)	[55, 56]

(Continued)

Table 13.3 (Continued)

Product	Reactants	Conditions (yields %)	Ref.
Pyrrole with R, Me, Tos, NH	R-CH(OAc)-CH(Me)-NO$_2$; Tos-CH$_2$-N$^+$≡C$^-$	[bmIm]Br, KOH, 20 °C, 2 h (80–93)	[28]
Pyrrole: X = NO$_2$, CN, COPh, COMe, CO$_2$Et; Z = SMe, NR$_1$R$_2$; R = CO$_2$Et, Ts, Ph, 4-Cl-Ph	Z-C(X)=C(Y)-SMe ; R-CH$_2$-N$^+$≡C$^-$; X = NO$_2$, CN, COPh, COMe, CO$_2$Et; Y = H, CO$_2$Et, COPh, COMe; Z = SMe, NR$_1$R$_2$; R = CO$_2$Et, Ts, Ph, 4-Cl-Ph	DBU, DMF, 120 °C, 5–8 h or t-BuOK, THF, −78 °C to r.t., 2 h (65–86)	[57]
Pyrrole with Ar$_1$, Ar$_2$, CO$_2$R$_1$	Ar$_1$-C(CN)=CH-Ar$_2$; R$_1$O-CO-CH$_2$-N$^+$≡C$^-$	t-BuOK, THF, 0 °C, 1 h (51–60)	[58]
major: R, NO$_2$, CO$_2$Et pyrrole (NH); minor: R, CO$_2$Et, Tos pyrrole	R-CH=CH-NO$_2$; Cl-CO-O-Et ; Tos-CH$_2$-N$^+$≡C$^-$	n-BuLi, THF, −78 °C to r.t., 2–4 days (Major 63–88, Minor 3–6)	[59]
R, CO$_2$Me, Tos pyrrole	R-C≡C-CO-CH$_3$; Tos-CH$_2$-N$^+$≡C$^-$	DBU, DMF, −50 °C, 0.5 h then −10 °C, 1 h (7–12)	[60]

Table 13.3 (Continued)

Product	Reactants	Conditions (yields %)	Ref.
Pyrrole with R-O-C(O)- groups at 3,4-positions, Tos at 2-position	R-O-C(O)-C≡C-C(O)-O-R; Tos-CH₂-N⁺≡C⁻	Cat. 1-Methylimidazole, CH₂Cl₂, r.t., 2 h (90–95)	[61]
Pyrrole with X at 2-position, R at 3-position, Y at 4-position X = CO₂Me, CO₂Et, Ph, CO₂t-Bu, Tos, CN, CONEt₂, P(O)(OEt)₂ Y = CO₂Et, CO₂t-Bu, COR, CONEt₂, CN, SO₂Ph, P(O)(OEt)₂	R≡Y; X-CH₂-N⁺≡C⁻ X = CO₂Me, CO₂Et, Ph, CO₂t-Bu, Tos, CN, CONEt₂, P(O)(OEt)₂ Y = CO₂Et, CO₂t-Bu, COR, CONEt₂, CN, SO₂Ph, P(O)(OEt)₂	Cu₂O, dioxane, 1,10-phenanthroline, 100 °C, 1–14 h (11–79) or KH, t-BuOMe, 20 °C, 4–20 h (78–91) or Cs₂CO₃, DMF, 90 °C (81) or t-BuOK, or KHMDS or CuSPh or Cu-NP, DMF, 85 °C, 12 h (45–97)	[32, 48, 62–64]
Pyrrole with CO₂Me at 2, R at 3, SO₂-aryl-CH₂OH at 4	R-C≡C-SO₂-aryl-CH₂-O-C(O)-(resin); MeO-C(O)-CH₂-N⁺≡C⁻	1. Cu₂O, DMF, 95 °C, 16 h 2. LiOH, THF (50–57)	[65]
Pyrrole with X at 2, R at 3, Y at 4 X = CO₂t-Bu, CONEt₂, P(O)(OEt)₂, Tos Y = CO₂Et, COMe, CONEt₂, CN	Y≡R; X-CH₂-N⁺≡C⁻ X = CO₂t-Bu, CONEt₂, P(O)(OEt)₂, Tos Y = CO₂Et, COMe, CONEt₂, CN	Dppf, dioxane, 100 °C, 0.5–24 h (17–73)	[62]

Table 13.4 Synthesis of 2,3,4,5-tetrasubstituted pyrroles.

Product	Reactants	Conditions (yields %)	Ref.
2-amino-pyrrole with X, R_3, R_2, aryl-S, R_1; X = CN, CO_2Et	alkene with X, R_3, CN; thiophenol with R_2; isocyanide $R_1-N^+\equiv C^-$; X = CN, CO_2Et	H_2O/Py, MeCN, r.t., 4–13 h (35–90)	[66]
pyrrole with R_3, R_2, R_1, ester $R_1-O-C(=O)-$, NH	β-ketoester $R_1-C(=O)-CH(R_3)-C(=O)-R_2$... ; $R_1-O-C(=O)-CH_2-N^+\equiv C^-$	Cat. $Rh_4(CO)_{12}$, toluene, 80 °C, 4 h (52–84)	[67]
pyrrole with R_2, R_3, R_1N, R_4, CN	alkene R_2, R_3, imine with R_4, CN; $R_1-N^+\equiv C^-$	$AlCl_3$, toluene, 90 °C, 15–24 h (67–88)	[68]
major: pyrrole NC, R_1, R_2, NHt-Bu, t-Bu; minor: pyrrole NC, R_2, R_1, NHt-Bu, t-Bu	alkyne R_1≡R_2; t-Bu-$N^+\equiv C^-$; t-Bu-$N^+\equiv C^-$	Cat. [Pd], DCE, 80 °C, 16–18 h (Major 49–95, Minor 5–51)	[69]

majority of the pyrroles are prepared via the addition of α-acidic isocyanides to Michael acceptors, under basic conditions. The reactions are quite versatile, and give rise to a variety of substitution patterns. Van Leusen was the first to explore these reactions, using TosMIC (p-toluenesulfonylmethyl isocyanide) [26]. Various other α-acidic isocyanides have been employed, leading to di substituted and trisubstituted pyrroles. One very useful variation of this reaction is the Barton–Zard synthesis [46], which utilizes nitroalkenes as the Michael acceptors. The yields of this reaction is improved substantially when t-BuOMe is used as the solvent [48], or when superstrong nonionic bases are employed [49].

The addition of α-acidic isocyanides to activated acetylenes under basic conditions and low temperatures has also been reported [60]. The yields of these reactions are improved in the presence of catalytic amounts of copper [62, 64], phosphine [62], and 1-methylimidazole [61].

13.4
Oxazoles

One of the main routes for the preparation of oxazoles is the reaction of α-acidic isocyanides with aldehydes, under basic conditions. When the isocyanide is TosMIC, the reaction is known as the van Leusen reaction [70], and this has been used extensively for the preparation of 5-monosubstituted and 4,5-disubstituted oxazoles (Tables 13.5 and 13.6). The use of ionic liquids as solvents greatly

Table 13.5 Synthesis of 5-substituted oxazoles.

Product	Reactants	Conditions (yields %)	Ref.
R-oxazole	Tos-CH$_2$-N$^+$≡C$^-$; R-CHO	K$_2$CO$_3$, MeOH, reflux, 2 h (57–71)	[70, 71]
R-oxazole	Bt-CH$_2$-N$^+$≡C$^-$; R-CHO	t-BuONa, THF, 0 °C, 2 h (35–69)	[72]

Table 13.6 Synthesis of 4,5-disubstituted oxazoles.

Product	Reactants	Conditions (yields %)	Ref.
R$_1$, R$_2$-oxazole	Tos-CHR$_1$-N$^+$≡C$^-$; R$_2$-CHO	t-BuOK, MeOH, reflux, 1 h (62–75) or K$_2$CO$_3$, MeOH, reflux, 0.5–3 h (47–88) or K$_2$CO$_3$, [bmim]Br, r.t., 4–24 h (0–95)	[35, 73, 74]
Tos, R-oxazole	Tos-CH$_2$-N$^+$≡C$^-$; R-CHO	K$_2$CO$_3$, MeOH, 20 °C, 10 min–2.5 h (53–80)	[70]

(Continued)

Table 13.6 (Continued)

Product	Reactants	Conditions (yields %)	Ref.
Tos-oxazole with R	Tos-CH2-N+≡C−; R-C(=O)-O-Me	2 equiv. n-BuLi, THF, −70 °C, 0.3–2 h (25–70)	[41]
R1-O-C(=O)-oxazole with R2	R1-O-C(=O)-CH2-N+≡C−; R2-C(=O)-Cl	Et3N, THF, r.t., 48 h (70–92) or n-BuLi, THF, −55 °C to −20 °C, 2 h (11–92) or superstrong nonionic base, THF, r.t., 0.5 h (91–100)	[49, 75–77]
R1-O-C(=O)-oxazole with R2	R1-O-C(=O)-CH2-N+≡C−; (R2-C(=O))2O	DBU, THF, r.t., 10 h (60–85)	[78]
R1-S-oxazole with R2	R1-S-CH2-N+≡C−; (R2-C(=O))2O	n-BuLi, THF, −70 °C to 20 °C, 2 h (30)	[79]
R1R2N-C(=O)-oxazole with R3	R1R2N-CH2-N+≡C−; R3-C(=O)-Cl	t-BuOK, THF, 0 °C to r.t., 1 h, (59)	[76, 80]
oxazole with R1, R2N, R3	R3-CH(N+≡C−)-N(R1)-C(=O)-R2	90–130 °C, neat, (44–71)	[80]
complex pyrrole-oxazole product (R2HN, SH, CO2Et, Et, R1)	R1-O-C(=O)-CH2-N+≡C−; R2HN-C(=O)-C(=C(S-CH2CH2-S))-C(=O)-CH=CH-R1	DBU, MeCN, r.t. −80 °C, 2–24 h (50–89)	[81]

Table 13.7 Synthesis of 2,5-disubstituted oxazoles.

Product	Reactants	Conditions (yields, %)	Ref.
R_2–oxazole–R_1	R_1–$N^+{\equiv}C^-$; Cl–C(=O)–R_2	2,6-lutidine, toluene, 80°C, 24 h (40–81)	[82, 83]
Et–O–oxazole(R_1)–C(R_2)(OSiMe$_3$)	$N^+{\equiv}C^-$–TMS–Cl ; Et–O–C(=O)–O–C(=O)–R_2 with R_1	Cat. Zn(OTf)$_2$, NEM, CH$_2$Cl$_2$, 0°C to r.t., 24 h (45–64)	[84]
R_1–O–oxazole–S–R_2	$N^+{\equiv}C^-$; R_1–O–C(=O)– ; Cl–S–R_2	CH$_2$Cl$_2$, −50°C to r.t., 1 h then Et$_3$N, −20°C to r.t., 1 h (quant.)	[85]
R_2(R_1)N–oxazole(R_3)–C(R_4)(OSiR$_5$)	$N^+{\equiv}C^-$; R_5Si–Cl ; R_2(R_1)N–C(=O)– ; R_3–C(=O)–R_4	Cat. Zn(OTf)$_2$, NEM, CH$_2$Cl$_2$, r.t., 12 h (20–84)	[86]
R_2(R_1)N–oxazole(R_3)–C(R_4)(N(R_5)$_2$)	$N^+{\equiv}C^-$; R_5–N(H)–R_5 ; R_2(R_1)N–C(=O)– ; R_3–C(=O)–R_4	Et$_3$N·HCl, MeOH, r.t., 12 h (73–92)	[86]

enhances the yields and also the ease of operation of this reaction [74]. The coupling of isocyanides with acyl halides under strong basic conditions also resulted in 4,5-disubstituted oxazoles [75–77], whereas under mild basic conditions 2,5-disubstituted oxazoles are obtained (Table 13.7) [82, 83].

Variants of the Passerini reaction, employing α-isocyanoacetamides, aldehydes and amines [87], or truncated versions of this reaction, employing α-isocyanoacetamides and aldehydes [88], have also been used extensively for the preparation of polysubstituted oxazoles (Table 13.8). These reactions were also developed to be enantioselective [89–91].

13.5 Isoxazoles

To date, very few reports have been made on the synthesis of isoxazoles from isocyanides (Table 13.9). One such synthesis employs the [4+1] cycloaddition of

Table 13.8 Synthesis of 2,4,5-trisubstituted oxazoles.

Product	Reactants	Conditions (yields, %)	Ref.
		Toluene, LiBr, 60 °C, 4 h (38–89)	[88]
		Cat. chiral (salen)Al(III)Cl, toluene, −20 °C, 24–48 h, 53–80% e.e. (30–90) or cat. binol-derived phosphoric acid, toluene, Et$_2$AlCl, −20 to 40 °C, 24 h, 54–81% e.e. (57–95)	[89, 90]
R1 = 4-NO$_2$Ph, 4-pyridyl	R1 = 4-NO$_2$Ph, 4-pyridyl	Toluene, r.t. (23–97)	[92, 93]
		MeOH, r.t. to 60 °C, 4 h (60–96) or MgSO$_4$, DMF, r.t., 17 h (0–69)	[87, 94–96]
		Cat. chiral phosphoric acid, toluene, −20 °C, 20–24 h, 66–90% e.e. (80–95)	[91]
		Benzene, r.t.-reflux, 0.5–24 h (30–75)	[97]
		Et$_3$N, CH$_2$Cl$_2$, r.t., 1 h (50–83)	[98]

13.5 Isoxazoles

Table 13.8 (Continued)

Product	Reactants	Conditions (yields, %)	Ref.
		Xylene, 140 °C, 4 h (60–88)	[99]
		BF$_3$·Et$_2$O, Et$_2$O, r.t., 2.5 h (52–65)	[100]
		THF, r.t., Me$_2$AlCl or ZnCl$_2$, r.t., 24 h (52–85)	[101]
		Cat. Zn(OTf)$_2$, NEM, CH$_2$Cl$_2$, r.t., 2 h (35–72)	[102]
		Cat. Zn(OTf)$_2$, pyridine, CH$_2$Cl$_2$/THF, r.t., 12 h (35–79)	[102]
		Cat. toluenesulfonic acid, acetone, r.t., 1 h (25–32)	[103]

Table 13.9 Synthesis of isoxazoles.

Product	Reactants	Conditions (yields, %)	Ref.
isoxazole with R_1-NH, R_2	R_1-N$^+$≡C$^-$ X-C(R$_2$)=N-OH, X = Br, Cl	Na$_2$CO$_3$, CH$_2$Cl$_2$ (0–93)	[104, 105]
Ar-C(O)-NH-R / R-N(H)-isoxazole	Ar-C≡N$^+$-R, Ar-CH=CH-NO$_2$, R-N$^+$≡C$^-$	H$_2$O, 80 °C, 2.5 h (80–93)	[106]

Table 13.10 Synthesis of monosubstituted thiazoles.

Product	Reactants	Conditions (yields, %)	Ref.
thiazole, X = CN, CO$_2$Et	X-CH$_2$-N$^+$≡C$^-$, H-C(=S)-O-Et, X = CN, CO$_2$Et	NaCN, EtOH, r.t. or 45–50 °C, 30 min (23–87)	[107, 108]
thiazole-CH(R$_2$)-N(R$_1$)-C(O)-R$_3$	MeO-CH(OMe)-N$^+$≡C$^-$, R$_1$-NH$_2$, R$_2$-CHO, R$_3$-C(O)-SH	1. MeOH, r.t. (31–92) 2. TMSCl, NaI, MW, MeCN (66–91)	[115]

in-situ-generated nitrosoalkenes, deriving from α-bromoketone oximes and isocyanides [104], and was used for the preparation of CC chemokine receptor-4 (CCR4) antagonists [105]. The other synthesis involved a novel MCR of 2 equiv. of isocyanide with nitrostyrenes in water, under heating [106].

13.6
Thiazoles

Thiazoles are prepared from α-acidic isocyanides upon reaction with thionoesters [107, 108] (Table 13.10), thiocarbonyl compounds [109], or isothiocyanates [110–114] (Table 13.11). Thiazoles may also be prepared via a two-step process using an Ugi reaction, followed by ring formation [115].

Table 13.11 Synthesis of disubstituted thiazoles.

Product	Reactants	Conditions (yields, %)	Ref.
		NaCN, EtOH, 45–50 °C, 0.5 h (22–82)	[107, 108]
		KOH, t-BuOH, 6–12 h (48–79)	[109]
		t-BuOK, THF (53–77) or PS-BEMP, MeCN, flow, 55 °C, 45 min (48–97)	[110–113]
		NaH, DME 0 °C or NaH, DMSO, 20 °C (9–90)	[114]
		Et$_3$N, DMF (71–89)	[121]
		CH$_2$Cl$_2$/MeOH (8–84)	[117, 118]
		1. CH$_2$Cl$_2$/MeOH 2. LiOH, H$_2$O/THF (26–85)	[119]

Table 13.12 Synthesis of trisubstituted thiazoles.

Product	Reactants	Conditions (yields, %)	Ref.
(thiazole structure with R, R₁, R₂, R₃, R₄)	R₁–NH₂; methyl 3-dimethylamino-2-isocyano-acrylate; R–Br; R₄–C(O)SH	CH₂Cl₂/MeOH (25–57)	[120]
(thiazole structure with R, R₁, R₂, R₃)	R–O-acrylate isocyanide with NMe₂; R₁–C(O)–R₂; R₃–C(O)SH	BF₃·OEt₂	[122]

3-Dimethylamino-2-isocyano-acrylates are useful reagents that have been used reliably for the synthesis of thiazoles and imidazoles [116]. In the case of thiazoles, 3-dimethylamino-2-isocyano-acrylates participate in a new MCR with aldehydes, amines, and thiocarboxylic acids to yield 2,4-disubstituted thiazoles (Table 13.12) [117–120].

13.7
Imidazoles

The van Leusen reaction of TosMIC with imines (Tables 13.13, 13.5 and 13.16) [123], under basic conditions, is one of the major schemes used to synthesize imidazoles. The reaction of 3-dimethylamino-2-isocyano-acrylate (Table 13.14) [124], and also 3-bromo-2-isocyano-acrylate (Table 13.16) [125, 126], with amines has also been employed successfully for the synthesis of imidazole derivatives. In the case of reactions with hydroxylamines and hydrazines, 1-hydroxy- and 1-amino-imidazoles are obtained [127].

13.8
Pyrazoles

Isocyanides have been used primarily for the preparation of 5-aminopyrazoles via reactions with hydrazones (Table 13.17) [143–146]. Furthermore, a novel one-pot procedure using isocyanides, activated acetylenes and hydrazine carboxamides, has also provided 2-aminopyrazoles in good yields [147]. A titanium-catalyzed MCR that incorporates only the carbon of the isocyano moiety has also been reported [148].

Table 13.13 Synthesis of monosubstituted imidazoles.

Product	Reactants	Conditions (yields, %)	Ref.
R-imidazole (4-substituted, NH)	Tos–N⁺≡C⁻; R–CH=N–X, X = SO₂NMe₂, Tos	K_2CO_3, MeOH/DME, reflux, 1–2 h (6–75), then (for X = SO₂NMe₂) 30% aq. HBr, reflux, 90 min (49–94)	[128]
2-(N-acetyl-NR₁-aminomethyl)imidazole with R₂	(MeO)₂CH–CH(N⁺≡C⁻); R_1–NH₂; R_2–CHO; AcSH	1. MeOH, r.t. (45–83) 2. NH₄OAc/EtOH, reflux, then HCl, reflux (59–72)	[129]

Table 13.14 Synthesis of 1,4-disubstituted imidazoles.

Product	Reactants	Conditions (yields, %)	Ref.
1-(CH₂CO₂R)-4-(CO₂R)-imidazole	R–O–CO–CH₂–N⁺≡C⁻ (×2)	Cat. AgOAc, MeCN, r.t., 2–20 h (88–90)	[130]
1-Ar-4-(CO₂Et)-imidazole	Et–O–CO–CH₂–N⁺≡C⁻; ⁻C≡N⁺–Ar	Cat. Cu₂O, 1,10-phenanthroline, THF, 25–80 °C, 0.75–5 h (10–98)	[131, 132]
1-R-4-(CO₂Et)-imidazole	EtO₂C–C(=CH–NMe)–N⁺≡C⁻; R–NH₂	Neat or n-BuOH, 70–150 °C, 1.5–72 h (31–89)	[124]
1-R-4-(CO₂H)-imidazole	(resin)–O–CO–C(=CH–NMe)–N⁺≡C⁻; R–NH₂	1. MW, 220 °C, 15 min 2. 50% TFA/CH₂Cl₂, 60 min (20–90)	[133]

Table 13.15 Synthesis of 4,5-, 1,5-, and 1,2-disubstituted imidazoles.

Product	Reactants	Conditions (yields, %)	Ref.
Tos, R 4,5-imidazole	Tos–N≡C; R–≡N	2 equiv. n-BuLi, THF, −65 °C 0.3 h (33)	[41]
R_1, R_2 imidazole	Tos–CR$_1$–N≡C; R_2CHO; NH$_4$OH	Piperazine, THF, r.t. 18 h (23–81)	[73, 134]
MeO, HO-aryl imidazole	Tos–N≡C; resin-bound aldehyde	DMF, 130 °C, 20 min then 50% TFA/CH$_2$Cl$_2$, r.t., 1.5 h (82–92)	[135]
2-Tos-amino imidazole (Ar)	EtO–CH(OEt)–CH$_2$–N≡C; ArNH$_2$; TosNNaCl	Cat. TEBA, CH$_2$Cl$_2$, 20 °C, 2 days (40–75)	[136]

Table 13.16 Synthesis of 1,4,5-trisubstituted and 1,2,4,5-tetrasubstituted imidazoles.

Product	Reactants	Conditions (yields, %)	Ref.
R_1, R_2, R_3-imidazole	Tos–CR$_1$–N≡C; R_2CH=N–R$_3$	NaH, DME, 20 °C, 1 h (60–88) or K$_2$CO$_3$ or t-BuNH$_2$, MeOH/DME, −20 to −76 °C, 1–16 h (10–96) or t-BuOK, DMSO (62–78) or pyridine, r.t. 5 days (40) or cat. t-BuNH$_2$, Bi(OTf)$_3$, MeOH (60–73)	[35, 123, 137–141]
Tos, R_1, R_2-imidazole	Tos–N≡C; R$_1$–CCl=N–R$_2$	DME/DMSO, r.t., 1 h (60–85)	[137]

Table 13.16 (Continued)

Product	Reactants	Conditions (yields, %)	Ref.
Imidazole with R₁, R₂, R₃ substituents	Tos-CH(R₁)-N⁺≡C⁻; R₂-CHO; H₂N-R₃	K₂CO₃, DMF, r.t., 3–5 h (62–87)	[73, 134]
Imidazole-carboxamide linked to COOH via (CH₂)ₙ, n=2,10, with Ar, R₂, R on ring	Resin-O-C(O)-(CH₂)ₙ-N⁺≡C⁻, n=2,10; Ar-C(O)-CHO; H₂N-R₁; R₂-COOH	CHCl₃/MeOH, 23 °C or 65 °C, then NH₄OAc, AcOH, 100 °C, 20 h then 10% TFA/CH₂Cl₂ (16–49)	[142]
4-Tos-5-HS-imidazole (N-R)	Tos-CH₂-N⁺≡C⁻; R-N=C=S	n-BuLi, THF, −65 °C (20–54)	[114]
Et-O-C(O)-imidazole with R₂S, N-R₁	MeO-C(O)-CH₂-N⁺≡C⁻; R₁-N=C=S; R₂-X; X = Cl, Br, I	PS-BEMP, MeCN, flow, 55 °C, 45 min (4–48)	[113]
MeO-C(O)-imidazole with R₁, N-R₂	MeO-C(O)-C(=CR₁Br)-N⁺≡C⁻; R₂-NH₂	Et₃N, DMF, 25 °C, 6–48 h (34–85)	[125–127]
(EtO)₂P(O)-imidazole with R₁, N-R₂, R₃	(EtO)₂P(O)-C(=CCl₂)-N⁺≡C⁻; R₁R₂NH; R₃NH₂	Et₂O, 0 °C, 2–3 h (38–73)	[103]

Table 13.17 Synthesis of pyrazoles.

Product	Reactants	Conditions (yields, %)	Ref.
Pyrazole with R_2, R_3, R_4, R_5 substituents	R_2≡R_3; R_1-N^+≡C^-; R_4-NH_2; NH_2-HN-R_5	Titanium catalyst, toluene, 100–150 °C, 24 h (24–50)	[148]
Pyrazole diester (Et, Et, R_1, Ar)	R_2-O—alkyne—$O-R_2$; R_1-N^+≡C^-; Ph-C(O)-NH-C(O)-NH-Ar	Acetone, r.t., 36 h (60–72)	[147]
Pyrazole with R_1, R_2, R_3	X-C(R_2)=N-HN-R_3 (X = Br, Cl); R_1-N^+≡C^-	Na_2CO_3, CH_2Cl_2, r.t., 24 h (0–96)	[105, 143]
Pyrazole with R_1, Ar, R_2, R_5	R_3R_4N-C(O)-C(R_2)=C(R_5)-NH-N-Ar; R_1-N^+≡C^-	Toluene, 1,2-dibromo-ethane, reflux, 12 h (0–80)	[144, 145]
Pyrazole with CN, R_1, Ar	NC-C=N-HN-Ar; R_1-N^+≡C^-	Toluene, reflux (50–52)	[146]

13.9
Oxadiazoles and Triazoles

Reports on the use of isocyanides for the synthesis of oxadiazoles and triazoles are very rare, and involve unique MCRs (Table 13.18). In the case of oxadiazoles, an iminophorphorane is reacted with carboxylic acids [149]. The coupling of TosMIC with diazonium salts resulted in 1,2,4-triazoles [150], while the reaction of isocyanides with acylchlorides and tetrazoles under $ZnCl_2$ activation produced 1,3,4-triazoles [151]. Finally 1,2,5-triazoles may be obtained from *in-situ*-formed nitrilimines by incorporation of the isocyano moiety only [152].

Table 13.18 Synthesis of oxadiazoles and triazoles.

Product	Reactants	Conditions (yields, %)	Ref.
(aryl-oxadiazole with R on phenyl)	Ph₃P=N-N⁺≡C⁻ ; R-C₆H₄-C(=O)OH	CHCl₃, r.t., 6 h (73–94)	[149]
(Tos-triazole with Ar)	Tos-N⁺≡C⁻ ; ⁻XN≡N⁺-Ar, X⁻ = BF₄, Cl	K₂CO₃, DMSO/ MeOH, H₂O, 0 °C, 10 min (15–94)	[150]
(triazole with O=C-R₂, R₁, R₃)	R₁-N⁺≡C⁻ ; R₂-C(=O)Cl ; tetrazole-R₃ (N-H)	Heat neat the first two reagents at 60 °C, 1 h, then ZnCl₂, toluene, 80 °C, 24 h (23–79)	[151]
(triazole Ar₁, Ar₂)	t-Bu-N⁺≡C⁻ ; Ar₁-C(=O)Cl ; HN(Ar₂)	Benzene, Et₃N, reflux, 1 h or 20 °C, 24 h (20–88)	[152]

13.10
Tetrazoles

A plethora of reactions exists that employ isocyanides to produce tetrazoles (Table 13.19). The direct reaction of isocyanides with hydrazoic acid under acidic conditions is an old and well-established reaction [153]; more modern versions include the use of trimethylsilyl azide [154] and acid-catalyzed conditions [155]. The Passerini reaction has also found some new applications to the synthesis of tetrazoles, by employing catalytic ZnI conditions to improve the yields of the reaction [156], in addition to catalytic [(salen)AlIIIMe] to prepare enantiomerically pure products [157]. The Ugi reaction has been widely used for the preparation of tetrazoles, by employing either hydrazoic acid [158–160] or trimethylsilyl azide. The latter approach was used to prepare antitubulin chalcones [161] and H₃-receptor antagonists [162]. In addition, by combining this method with subsequent post-Ugi transformations, complex tetrazole-containing structures were obtained [163–165]. The Ugi synthesis of tetrazoles has also been developed for solid-phase reactions [166]. The one-pot reaction of isocyanides with N-fluoropyridinium fluorides and

Table 13.19 Synthesis of tetrazoles.

Product	Reactants	Conditions (yields, %)	Ref.
1-R-tetrazole	$R-N^+{\equiv}C^-$; HN_3	Toluene, r.t.-reflux, 3 h (9–63)	[153]
1-R-tetrazole	$R-N^+{\equiv}C^-$; $TMSN_3$	cat. HCl, Et_2O, 60 °C, 4–24 h (57–92) or $ZnBr_2$, MeOH, reflux, 3–4 h (70–92)	[154, 155]
Boc-NH-CH(R$_2$)-C(OH)(R$_1$)-tetrazole	$R_1-N^+{\equiv}C^-$; Boc-NH-CH(R$_2$)-CHO; HN_3	10 mol%. ZnI, CH_2Cl_2, 18–69 h (35–90)	[156]
R$_2$-CH(OH)-tetrazole(R$_1$)	$R_1-N^+{\equiv}C^-$; R_2-CHO; HN_3	Cat. [(salen)AlIIIMe], toluene, −40 °C, 48 h, 51–97% e.e. (45–99)	[157]
R_4R_5N-CR$_2R_3$-tetrazole(R$_1$)	$R_1-N^+{\equiv}C^-$; R_2-CO-R_3; R_4-NH-R_5; HN_3	Benzene/MeOH, 0–20 °C, 0.15–500 h (60–93)	[158–160]
R_4R_5N-CR$_2R_3$-tetrazole(R$_1$)	$R_1-N^+{\equiv}C^-$; R_2-CO-R_3; R_4-NH-R_5; $TMSN_3$	MeOH, r.t., 3–72 h (11–96)	[161–165]
R_2R_3N-CHR$_1$-tetrazole(CH-)	(polymer-bound isocyanide); R_1-CHO; R_2-NH-R_3; $TMSN_3$	1. THF/MeOH, r.t., 16 h 2. 15% TFA/CH_2Cl_2, r.t., 30 min (25–89)	[166]
major: 2-(pyridyl)tetrazole (R_1, R_2); minor: R_2-pyridyl-C(O)-NH-R_1	N-fluoropyridinium (R$_2$) F^-; $R_1-N^+{\equiv}C^-$; $TMSN_3$	$CHCl_3$, −50 °C, 1 h (major 34–84; minor 9–36). Preparation of N-fluoropyridinium fluoride is by fluorine gas, pyridine, $CHCl_3$, −78 °C to −50 °C, removal of fluorine	[167]
5-Ar-1-R-tetrazole	$R-N^+{\equiv}C^-$; NaN_3; $ArB(OH)_2$	Br_2, MeCN, r.t., then r.t. or 65 °C, then K_2CO_3, Pd(PPh$_3$)$_4$, toluene, 110 °C (23–98)	[168]

Table 13.20 Synthesis of benzofurans and benzimidazoles.

Product	Reactants	Conditions (yields, %)	Ref.
Benzofuran with R_1, R_2, R_3, NH	Phenol with R_1, piperazine-CH_2Ph with R_2, OH, $^-C{\equiv}N^+{-}R_3$	Toluene, 110 °C, 1–6 h, 1,2-dibromoethane (60–91)	[169]
Benzofuran with R, R_1, $N{-}R_2$, R_3, NH	Salicylaldehyde with R, $R_1{-}N(H){-}R_2$, OH, $^-C{\equiv}N^+{-}R_3$	CH_2Cl_2, −10 °C, then silica gel, r.t. (80–94) or CAN, EtOH, r.t., 2 h (65–85)	[170, 171]
Benzofuran with R, $HN{-}Ar_1$, Ar_2, NH	Aryl imine with R, $N{-}Ar_1$, OH, $^-C{\equiv}N^+{-}Ar_2$	cat. $BF_3{\cdot}Et_2O$, CH_2Cl_2, 0 °C, 1.5 h (60–74)	[25]
Benzimidazole with R	2-bromophenyl isocyanide ($N^+{\equiv}C^-$, Br), $H_2N{-}R$	Cat Cu^I, cat. 1,10-phenanthroline, Cs_2CO_3, DMF, 20–90 °C, 16 h (38–70)	[172]

trimethylsilyl azide yielded tetrazolyl pyridines, in good yields [167]. A one-pot, three-step transformation of isocyanides to tetrazoles has also been reported [168] which involves the *in-situ* formation of isocyanide dibromides, reactions with sodium azide, and Suzuki couplings of the corresponding 5-bromotetrazoles.

13.11
Benzofurans and Benzimidazoles

Benzofurans are prepared from suitably substituted phenols and isocyanides (Table 13.20) [25, 169–171], while benzimidazoles are prepared in a two-step, one pot procedure from 2-bromophenyl isocyanide and amines in the presence of a copper catalyst [172].

13.12
Indoles

A variety of isocyanide-based reactions yielding indoles have been developed, and the subject has been extensively reviewed [173]. A large majority of these reactions involve α-additions, where the isocyano moiety acts either as a nucleophile [174] or as an electrophile (Table 13.21) [175–180].

Ruthenium- [181, 182], palladium- [183, 184], and zirconium- [185] mediated indole formations have also been reported.

Another major class of transformations for the synthesis of indoles are the radical reactions. *o*-Alkenyl-isocyanobenzenes undergo tin-mediated radical 5-*exo-trig* cyclizations to produce 3-substituted indoles [186–188], and these reactions have been utilized for the synthesis of tryptostatins A and B [189] and (–)vindoline [186]. The 2-stannyl intermediates [187] of this reaction can be further exploited to generate additional substitutions at the 2-position. The 2-thio derivatives are also prepared via radical cyclization reactions [190–192], while 2-boryl and 2-silyl indoles are prepared from *o*-alkenyl-isocyanobenzenes via copper-catalyzed cyclizations [193].

Table 13.21 Synthesis of indoles.

Product	Reactants	Conditions (yields, %)	Ref.
Indole with R-CH(OH)-C(=O)- substituent	o-isocyano benzyl with OCH(OMe)O and R-CHO	Cat. D, L-10-CSA, CH_2Cl_2, r.t., 20 h (65–74)	[195]
2,5,6-trisubstituted indole (R_1, R_2, R_3)	o-isocyano benzyl chloride with R_1, R_2 and R_3Li	THF, –78 °C, 2 h (49–74)	[175]
3-substituted indole (R_1, R)	o-isocyano methylbenzene (R_1, R)	2 equiv. LDA or LTMP, –78 °C, diglyme (42–100)	[176, 177]
3-acyl indole (R, R_1)	o-isocyano benzyl ketone (R, R_1)	2 equiv. LDA or LTMP, –78 °C, diglyme (55–68) or Cu_2O, benzene, reflux, 2 h (60–85)	[178, 179]
3-X indole, X = CN, CO_2Me	o-isocyano benzyl-X, X = CN, CO_2Me	Cu_2O, benzene, 55 °C, 6 h (86–92)	[180]

Table 13.21 (Continued)

Product	Reactants	Conditions (yields, %)	Ref.
indole with R_1 at 3, R_2 at 7	2-isocyanobenzyl with R_1, R_2	Cat. Ru(dmpe)$_2$H(naphthyl), C$_6$D$_6$, 140 °C, 12 h–24 h	[181, 182]
3-R indole	2-isocyano styrene with R	n-Bu$_3$SnH, cat. AIBN, MeCN, 100 °C, 1 h then H$_3$O$^+$ (18–91)	[187–189]
		or R$_1$SH, cat. AIBN, MeCN, 100 °C, 15 min then Raney Ni, EtOH, r.t. (29–76)	
3-R indole; 3-R quinoline	2-isocyano phenylacetylene with R	n-Bu$_3$SnH, cat. AIBN, MeCN, reflux, 0.25 h (11–82)	[190, 194]
2,3-di-R indole with R substituent	2-isocyano aryl ketone	6% HCl, MeOH/H$_2$O, r.t., 30 min then 10% aq. NaOH, r.t., 30 min (43–90)	[177, 179]
2-Me-3-R indole	2-isocyano styrene with R	Pd(dppp)MeCl, CH$_2$Cl$_2$, r.t. (46) then HCl (quant.)	[183]
3-(CH(NEt$_2$))-2-(4-Y-aryl) indole; Y = H, OMe, NO$_2$	vinyl isocyanide + Et$_2$NH + 4-X-Y-aryl; Y = H, OMe, NO$_2$; X = I, OTf	Cat. Pd(OAc)$_2$, cat. dppp, THF, 40 °C, 10–42)	[184]
3-R-2-SnBu$_3$ indole	2-isocyano styrene with R, n-BuSnH	Cat. AIBN, MeCN, 100 °C, 1 h	[187]
3-(CH$_2$CO$_2$Me)-2-B(pin) indole	methyl 2-isocyanocinnamate, B$_2$(pin)$_2$	Cat. CuOAc, PPh$_3$, MeOH, THF, 25 °C, 5 h (57–98)	[193]

(Continued)

Table 13.21 (Continued)

Product	Reactants	Conditions (yields, %)	Ref.
(indole with 2-SiMe₂Ph and 3-CH₂CO₂Me, R substituent)	methyl (2-isocyanophenyl)acrylate; (pin)B-SiMe₂Ph	Cat. CuOAc, PPh₃, MeOH, THF, 0 °C, 5 h (75–89)	[193]
(indole with 2-SR and 3-CH(RS)TMS)	(2-isocyanophenyl)acetylene-TMS; RSH, RSH	Cat. AIBN, MeCN, 110 °C, 0.25 h (49–94)	[190]
(indole with R₁, R₂, 2-SR₄, 3-C(O)R₃)	(2-isocyanophenyl) with C(O)R₃, S, R₄X, R₁, R₂	1. S, cat. Selenium, Et₃N, THF, r.t., 1.5 h then NaH, r.t., 30 min then R₄X, r.t. 30 min (64–91)	[191]
(indole with 2-SPh and 3-CH(PhS)R)	(2-isocyanophenyl)vinyl R; Ph-S-S-Ph	(PhTe)₂, hν (>400 nM), CDCl₃, 40 °C, 5–31 h (42–50)	[192]
(indole with R₂ at 3, 2-C(O)NH-R₁, N-R₁)	(2-acylphenyl) with two ⁻C≡N⁺R₁	BF₃·Et₂O	[174]
(EDG-indole with 3-NHR₁, 2-R₂)	(EDG-aryl) ⁻C≡N⁺R₁, N=R₂	Diphenyl trifluoromethylsulfonylphosphoramidate, CH₂Cl₂, r.t., 2 h (12–96)	[197]
(indole with 3-CH₂Ar(R₃), 2-C(O)NHR₁, R₂)	(2-Br, NH₂-aryl R₂) cinnamaldehyde R₃; ⁻C≡N⁺R₁; HCO-OH	2,2,2-Trifluoroethanol, r.t., 1–3 days (15–38)	[196]
(indole with Ar at 4, 3-Ar, 2-R, 7-R)	Si(Ar)≡, ≡Ar; ⁻C≡N⁺R, ⁻C≡N⁺R	"Cp₂Zr", then H₂O	[185]

When o-alkynyl-isocyanobenzenes are subjected to radical conditions, a competitive 6-*endo-dig* cyclization takes place to provide substituted quinolines [190, 194].

Previously, MCRs have also been used in the preparation of indoles. For example, a Passerini-type condensation of o-isocyanobenzene acetaldehyde dimethylacetals with aldehydes yielded 1-substituted indoles, in good yields [195]. The Ugi/Heck reaction of acrylic aldehydes, bromoanilines, acids and isocyanides yielded 2-carboxamide-substituted indoles [196], while an interrupted Ugi reaction enabled the preparation of 3-amino-substituted indoles [197].

13.13
Quinolines

Quinolines are prepared primarily from o-alkenyl or o-alkynyl-isocyanobenzenes (Table 13.22). These substrates undergo 6-*endo-dig* cyclizations under either

Table 13.22 Synthesis of quinolines.

Product	Reactants	Conditions (yields, %)	Ref.
		MBDA, ether, reflux, 2 h (0–87)	[206]
		LDA, diglyme, −78 °C, 1 h (47–97)	[207]
		n-Bu$_3$SnH, cat. AIBN, MeCN, reflux, 0.25 h (11–82)	[190, 194]
		MeOH, 50 °C, 20 h (66–86)	[198]
		DABCO, CH$_2$Cl$_2$, 0 °C, 30 min (31–90)	[199]

(*Continued*)

Table 13.22 (Continued)

Product	Reactants	Conditions (yields, %)	Ref.
quinoline with R at 3, NEt₂ at 2	2-isocyanophenyl alkyne with R, Et₂NH	Et₂NH, r.t., 24 h (65–94)	[198]
quinoline with R₁, R₂, Cl at 2	2-isocyanophenyl alkyne with R₁, R₂, n-BuNCl	CH₂Cl₂, 40–60 °C, 1.5–24 h (0–99)	[200]
quinoline with R₁, I at 4, R₂ at 3, I at 2	2-isocyanophenyl alkyne with R₁, R₂, I₂	hν (>300 nM), CHCl₃, r.t., 4 h (0–86)	[205]
quinoline with R₁, R₂, R₃	2-isocyanophenyl vinyl OMe with R₁, R₂, R₃Li	DME, −78 °C to r.t., 1 h (55–91)	[201]
2,3-difluoroquinoline with R₁, R₂	2-isocyanophenyl CF=CF with R₁, R₂M	Toluene, HMPA, −78 or 0 °C to r.t., 2 h (60–78)	[202]
M = MgBr, MgCl, Li, GeLi			
3-fluoroquinoline with R	2-isocyanophenyl CF=CHF with R	n-Bu₃SnLi, THF, −78 °C 0.25–2 h (65–80)	[203]
3-fluoroquinoline with R, E at 2	2-isocyanophenyl CF=CF with R, E⁺	n-Bu₃SnLi, THF, −78 °C to r.t., 2 h (52–87)	[203]
quinoline with n-Bu, F at 3, X at 2; X = H, I	2-isocyanophenyl CF=CF with n-Bu	n-Bu₃SnH, cat. AIBN, toluene, 80 °C, 1 h then DBU, H₃O⁺ or I₂ (43–60)	[204]
3-methyl-2-SPh-quinoline and 3-CH₂SPh-2-SPh-quinoline	2-isocyanostyrene, Ph-S-S-Ph	hν (>400 nM), (PhTe)₂ or (PhSe)₂, CDCl₃, 40 °C, 7–30 h (14–39)	[192]

Table 13.23 Synthesis of quinoxalines.

Product	Reactants	Conditions (yields, %)	Ref.
(quinoxaline with R, NH-cyclohexyl)	o-phenylenediamine, ArCHO, cyclohexyl isocyanide	Cat. Fe(ClO$_4$)$_3$, MeCN, reflux, 2 h (91–93)	[210]
(dihydroquinoxaline with R$_1$, R$_2$, R$_3$, R$_4$)	substituted o-phenylenediamine, R$_4$CHO, R$_1$-NC	con. HCl, MeOH, r.t., 18 h then DDQ, benzene, r.t., 1–3 h (33–54)	[208, 209]
(quinoxaline with R$_1$, R$_2$, Ar)	substituted o-phenylenediamine, ArCHO, TosMIC	DBU or DABCO, toluene, 80 °C, 4 h (trace-91)	[211]

nucleophile-induced [198–203] or radical conditions [190, 194, 204], and photochemical cyclizations have also been reported [192, 205]. Alternatively, o-ketoenolates may cyclize in the presence of bivalent magnesium ions to provide 4-hydroxy-3-quinolinecarboxylic acid derivatives [206]. o-Acetaldehyde-dimethyl acetals, on the other hand, cyclize in an excess of lithium diisopropylamide (LDA) to produce 2-methoxyquinolines [207].

13.14
Quinoxaline

Quinoxalines can be prepared from isocyanides, aldehydes, and o-phenylenediamines in two sequential reactions, under acidic conditions, to first form 1,4-dihydroquinoxalines, followed by oxidation with 2,3-dichloro-5,6-dicyanobenzoquinone (DDQ) (Table 13.23) [208, 209]. When the reaction is performed in the presence of catalytic ferric perchlorate, either under neutral conditions [210] or in the presence of a base [211], the corresponding quinoxalines are obtained in a one-pot procedure, in good yields.

Abbreviations

AIBN	azobisisobutyronitrile
Bt	benzotriazol
[bmIm]Br	1-butyl-3-methylimidazolium bromide
CSA	camphorsulfonic acid
Cu-NP	preactivated nanosize copper powder
DCE	dichloroethane
dmpe	1,2-bis(dimethylphosphino)ethane
dppf	1,3-bis(diphenylphosphino)butane
dppp	bis(diphenylphosphino)propane
KHMDS	potassium hexamethyldisilazide
LDA	lithium diisopropylamide
LTMP	2,2,6,6-tetramethylpiperidine
MBDA	magnesium bis(diisopropylamide)
NEM	N-ethylmorpholine
PCT	benzyltriethyl ammonium chloride
PEG	poly(ethylene glycol)
PS-BEMP	polymer-supported 2-tert-butylimino-2-diethylamino-1,3-dimethyl-perhydro-1,3,2-diaza-phosphine
TEBA	triethylbenzylammonium chloride
Tos	p-CH$_3$C$_6$H$_4$SO$_2$

References

1 Sadjadi, S. and Heravi, M.M. (2011) Recent application of isocyanides in synthesis of heterocycles. *Tetrahedron*, **67**, 2707–2752.

2 Lygin, A.V. and de Meijere, A. (2010) Isocyanides in the synthesis of nitrogen heterocycles. *Angew. Chem. Int. Ed.*, **49**, 9094–9124.

3 Ivachtchenko, A.V., Ivanenkov, Y.A., Kysil, V.M., Krasavin, M.Y., and Ilyin, A.P. (2010) Multicomponent reactions of isocyanides in the synthesis of heterocycles. *Russian Chem. Rev.*, **79**, 787–817.

4 Elders, N., Ruijter, E., Nenajdenko, V.G., and Orru, R.V.A. (2010) α-Acidic isocyanides in multicomponent chemistry, in *Topics in Heterocyclic Chemistry 23. Synthesis of Heterocycles via Multicomponent Reactions I* (eds R. Orru and E. Ruijter), Springer, Heidelberg, pp. 129–159.

5 Banfi, L., Basso, A., and Renata, R. (2010) Synthesis of heterocycles through classical Ugi and Passerini reactions followed by secondary transformations involving one or two additional functional groups, in *Topics in Heterocyclic Chemistry, 23. Synthesis of Heterocycles via Multicomponent Reactions I* (eds R. Orru and E. Ruijter), Springer, Heidelberg, pp. 1–39.

6 Dömling, A. (2008) Isocyanide-based multicomponent reactions (IMCRs) as a valuable tool with which to synthesize nitrogen-containing compounds, in *Amino Group Chemistry. From Synthesis to the Life Sciences* (ed. A. Ricci), Wiley-VCH, Weinheim, pp. 149–183.

7 Huang, Y., Khoury, K., and Dömling, A. (2010) The piperazine space in isocyanide-based MCR chemistry, in *Topics in Heterocyclic Chemistry 23. Synthesis of Heterocycles via*

Multicomponent Reactions I (eds R. Orru and E. Ruijter), Springer, Heidelberg, pp. 85–127.

8 Nair, V. and Vinod, A.U. (2000) The reaction of cyclohexyl isocyanide and dimethyl acetylenedicarboxylate with aldehydes: a novel synthesis of 2-aminofuran derivatives. *Chem. Commun.*, 1019–1020.

9 Nair, V., Vinod, A.U., Abhilash, N., Menon, R.S., Santhi, V., Varma, R.L., Viji, S., Mathew, S., and Srinivas, R. (2003) Multicomponent reactions involving zwitterionic intermediates for the construction of heterocyclic systems; one pot synthesis of aminofurans and iminolactones. *Tetrahedron*, **59**, 10279–10286.

10 Hazeri, N., Maghsoodlou, M.T., Habibi-Khorassani, S.M., Marandi, G., Khandan-Barani, K., Ziyaadini, M., and Aminkhani, A. (2007) Synthesis of novel 2-pyridyl-substituted 2,5-dihydro-2-imino- and 2-amino-furan derivatives *via* a three component condensation of alkyl isocyanides and acetylenic esters with di-(2-pyridyl) ketone or 2-pyridinecarboxaldehyde. *ARKIVOC*, (i), 173–179.

11 Azizian, J., Mohammadizadeh, M.R., Mohammadi, A.A., and Karimi, A.R. (2005) A modified and green methodology for preparation of polysubstituted furans. *Heteroatom Chem.*, **16**, 259–262.

12 Yadav, J.S., Reddy, B.V.S., Subashree, S., Sadashiv, K., and Rao, D.K. (2007) Organic synthesis in water: green protocol for the synthesis of 2-amino furan derivatives. *J. Mol. Catal. A: Chem.*, **272**, 128–131.

13 Asghari, S. and Qandalee, M. (2010) A facile and efficient reaction of aromatic aldehydes with the isocyanide-dialkylacetylenedicarboxylate zwitter ions: formation of 2-amino-5-aryl furan derivatives. *Asian J. Chem.*, **22**, 3435–3438.

14 Reddy, B.V.S., Somashekar, D., Reddy, A.M., Yadav, J.S., and Sridhar, B. (2010) PEG 400 as a reusable solvent for 1,4-dipolar cycloadditions via a three-component reaction. *Synthesis*, 2069–2074.

15 Bremmer, W.S. and Organ, M.G. (2007) Multicomponent reactions to form heterocycles by microwave-assisted continuous flow organic synthesis. *J. Comb. Chem.*, **9**, 14–16.

16 Reddy, B.V.S., Reddy, A.M., Somashekar, D., Yadav, J.S., and Sridhar, B. (2010) Three-component one-pot synthesis of 3-(2-furanyl)indoles from acetylenedicarboxylate, isocyanide, and 3-formylindole. *Synthesis*, 2571–2576.

17 Asghari, S., and Qandalee, M. (2007) A facile one-pot synthesis of amino furans using *trans*-cinnamaldehyde in the presence of nucleophilic isocyanides. *Acta Chim. Slov.*, **54**, 638–641.

18 Nair, V., Vinod, A.U., and Rajesh, C. (2001) A novel synthesis of 2-aminopyrroles using a three-component reaction. *J. Org. Chem.*, **66**, 4427–4429.

19 Nair, V. and Deepthi, A. (2006) A novel reaction of vicinal tricarbonyl compounds with the isocyanide-DMAD zwitterion: formation of highly substituted furan derivatives. *Tetrahedron Lett.*, **47**, 2037–2039.

20 Yavari, I., Mokhtarporyani-Sanandaj, A., Moradi, L., and Mirzaei, A. (2008) Reaction of benzoyl chlorides with Huisgen's zwitterion: synthesis of functionalized 2,5-dihydro-1*H*-pyrroles and tetrasubstituted furans. *Tetrahedron*, **64**, 5221–5225.

21 Yavari, I., Shaabani, A., and Maghsoodlou, M.T. (1997) On the reaction between alkyl isocyanides and 3-benzylidene-2,4-pentanedione. A convenient synthetic route to densely functionalized furans. *Monatsh. Chem.*, **128**, 697–700.

22 Sun, C., Ji, S.-J., and Liu, Y. (2007) Facile synthesis of 3-(2-furanyl)indoles via a multicomponent reaction. *Tetrahedron Lett.*, **48**, 8987–8989.

23 Mosslemin, M.H., Yavari, I., Anary-Abbasinejad, M., and Nateghi, M.R. (2004) Reaction between *tert*-butyl isocyanide and 1,1,1-trifluoro-4-aryl-butane-2,4-diones. Synthesis of new trifluoromethylated furan derivatives.

J. Fluorine Chem., **125**, 1497–1500.

24 Bayat, M., Imanieh, H., Shiraz, N.Z., and Qavidel, M.S. (2010) One-pot, three-component reaction of isocyanides, dialkyl acetylenedicarboxylates and non-cyclic anhydrides: synthesis of 2,5-diaminofuran derivatives and dialkyl (*E*)-2-[(*N*-acyl-*N*-alkylamino)carbonyl]-2-butenedioates. *Monatsh. Chem.*, **141**, 333–338.

25 Kobayashi, K., Shirai, Y., Fukamachi, S., and Konishi, H. (2010) Synthesis of 2,3-bis(arylamino)benzofurans and 2,3-bis(arylimino)-2,3-dihydrobenzofurans by a Lewis acid-catalyzed reaction of 2-aryliminophenols with aryl isocyanides. *Synthesis*, 666–670.

26 van Leusen, A.M., Siderius, H., Hoogenboom, B.E., and van Leusen, D. (1972) A new and simple synthesis of the pyrrole ring system from Michael acceptors and tosylmethylisocyanides. *Tetrahedron Lett.*, 5337–5340.

27 Hormaza, A. and Perez, O.F.A. (2009) Synthesis of a new series of pyrroles via cycloaddition. *Rev. Soc. Quim. Perú*, **75**, 12–16.

28 Qin, J., Zhang, J., Wu, B., Zheng, Z., Yang, M., and Yu, X. (2009) Efficient and mild protocol for the synthesis of 4(3)-substituted 3(4)-nitro-1*H*-pyrroles and 3-substituted 4-methyl-2-tosyl-1*H*-pyrroles from nitroolefins and tosylmethyl isocyanide in ionic liquids. *Chin. J. Chem.*, **27**, 1782–1788.

29 Padmavathi, V., Reddy, B.J.M., Sarma, M.R., and Thriveni, P. (2004) A simple strategy for the synthesis of 3,4-disubstituted pyrroles. *J. Chem. Res.*, 79–80.

30 Padmavathi, V., Reddy, B.J.M., Mahesh, K., Thriveni, P., and Padmaja, A. (2010) *E,E*-Bis(styryl)sulfones – synthons for a new class of bis(heterocycles). *J. Heterocycl. Chem.*, **47**, 825–830.

31 Smith, N.D., Huang, D., and Cosford, N.D.P. (2002) One step synthesis of 3-aryl- and 3,4-diaryl-(1*H*)-pyrroles using tosylmethyl isocyanide (TOSMIC). *Org. Lett.*, **4**, 3537–3539.

32 Lygin, A.V., Larionov, O.V., Korotkov, V.S., and de Meijere, A. (2009) Oligosubstituted pyrroles directly from substituted methyl isocyanides and acetylenes. *Chem. Eur. J.*, **15**, 227–236.

33 van Leusen, D., Flentge, E., and van Leusen, A.M. (1991) Chemistry of sulfonylmethyl isocyanides. 35. An efficient synthesis of 3-nitropyrroles. *Tetrahedron*, **47**, 4639–4644.

34 van Leusen, D., van Echten, E., and van Leusen, A.M. (1992) Synthesis of 3,4-disubstituted pyrroles bearing substituents of electron-withdrawing and/or electron-donating nature. *J. Org. Chem.*, **57**, 2245–2249.

35 Possel, O. and van Leusen, A.M. (1977) Synthesis of oxazoles, imidazoles and pyrroles with the use of mono-substituted tosylmethyl isocyanides. *Heterocycles*, **7**, 77–80.

36 Pavri, N.P. and Trudell, M.L. (1997) An efficient method for the synthesis of 3-arylpyrroles. *J. Org. Chem.*, **62**, 2649–2651.

37 Reyes-Gutiérrez, P.E., Camacho, J.R., Ramirez-Apan, M.T., Osornio, Y.M., and Martinez, R. (2010) Synthesis of 5,6-dihydropyrrolo[2,1-*a*]isoquinolines featuring an intramolecular radical-oxidative cyclization of polysubstituted pyrroles, and evaluation of their cytotoxic activity. *Org. Biomol. Chem.*, **8**, 4374–4382.

38 Barton, D.H.R., Kervagoret, J., and Zard, S.Z. (1990) A useful synthesis of pyrroles from nitroolefins. *Tetrahedron*, **46**, 7587–7598.

39 ten Have, R., Leusink, F.R., and van Leusen, A.M. (1996) Chemistry of sulfonylmethyl isocyanides. 42. An efficient synthesis of substituted 3(4)-nitropyrroles from nitroalkenes and tosylmethyl isocyanides. *Synthesis*, 871–876.

40 Poulard, C., Cornet, J., Legoupy, S., Dujardin, G., Dhal, R., and Huet, F. (2009) Synthesis of polysubstituted pyrroles. *Lett. Org. Chem.*, **6**, 359–361.

41 van Nispen, S.P.J.M., Mensink, C., and van Leusen, A.M. (1980) Chemistry of sulfonylmethyl isocyanides. 21. Use of dilithio-tosylmethyl isocyanide in the synthesis of oxazoles and imidazoles. *Tetrahedron Lett.*, **21**, 3723–3726.

42 Katritzky, A.R., Cheng, D., and Musgrave, R.P. (1997) Synthesis of imidazoles and pyrroles: BETMIC and TOSMIC as complementary reagents. *Heterocycles*, **44**, 67–70.

43 Schöllkopf, U. and Meyer, R. (1977) Syntheses with 2-metalated isocyanides, XXXVIII. Trialkylmethyl-substituted glycines and pyrrole-2,4-dicarboxylic esters from 2-isocyanoacrylic esters and carbanions. *Liebigs Ann. Chem.*, 1174–1182.

44 Suzuki, M., Miyoshi, M., and Matsumoto, K. (1974) A convenient synthesis of 3-substituted pyrrole-2,4-dicarboxylic acid esters. *J. Org. Chem.*, **39**, 1980.

45 Matsumoto, K., Suzuki, M., Ozaki, Y., and Miyoshi, M. (1976) The synthesis of 3-substituted pyrrole-2,4-dicarboxylic acid esters: reaction of methyl isocyanoacetate with aldehydes. *Agr. Biol. Chem.*, **40**, 2271–2274.

46 Barton, D.H.R. and Zard, S.Z. (1985) A new synthesis of pyrroles from nitroalkenes. *J. Chem. Soc., Chem. Commun.*, 1098–1100.

47 Lash, T.D., Bellettini, J.R., Bastian, J.A., and Couch, K.B. (1994) Synthesis of pyrroles from benzyl isocyanoacetate. *Synthesis*, 170–172.

48 Bhattacharya, A., Cherukuri, S., Plata, R.E., Patel, N., Tamez, V. Jr, Grosso, J.A., Peddicord, M., and Palaniswamy, V.A. (2006) Remarkable solvent effect in Barton–Zard pyrrole synthesis: application in an efficient one-step synthesis of pyrrole derivatives. *Tetrahedron Lett.*, **47**, 5481–5484.

49 Tang, J. and Verkade, J.G. (1994) Nonionic superbase-promoted synthesis of oxazoles and pyrroles: facile synthesis of porphyrins and α-C-acyl amino acid esters. *J. Org. Chem.*, **59**, 7793–7802.

50 Halazy, S. and Magnus, P. (1984) 1-Phenylsulfonyl-1,3-butadiene. An electrophilic equivalent to 1,3-butadiene for the synthesis of 3,3^1-bipyrroles. *Tetrahedron Lett.*, **25**, 1421–1424.

51 Haake, G., Struve, D., and Montforts, F.-P. (1994) A useful preparation of pyrroles from α,β-unsaturated sulfones. *Tetrahedron Lett.*, **35**, 9703–9704.

52 Uno, H., Sakamoto, K., Tominaga, T., and Ono, N. (1994) A new approach to β-fluoropyrroles based on the Michael addition of isocyanomethylide anions to α-fluoroalkenyl sulfones and sulfoxides. *Bull. Chem. Soc. Jpn*, **67**, 1441–1448.

53 Abel, Y., Haake, E., Haake, G., Schmidt, W., Struve, D., Walter, A., and Montforts, F. (1998) A simple and flexible synthesis of pyrroles from α,β-unsaturated sulfones. *Helv. Chim. Acta*, **81**, 1978–1996.

54 Vicente, M.G.H., Tome, A.C., Walter, A., and Cavaleiro, J.A.S. (1997) Synthesis and cycloaddition reactions of pyrrole-fused 3-sulfolenes: a new versatile route to tetrabenzoporphyrins. *Tetrahedron Lett.*, **38**, 3639–3642.

55 Ono, N., Katayama, H., Nisyiyama, S., and Ogawa, T. (1994) Regioselective synthesis of 5-unsubstituted benzyl pyrrole-2-carboxylates from benzyl isocyanoacetate. *J. Heterocycl. Chem.*, **31**, 707–710.

56 Ono, N., Kawamura, H., and Maruyama, K. (1989) A convenient synthesis of trifluoromethylated pyrroles and porphyrins. *Bull. Chem. Soc. Jpn*, **62**, 3386–3388.

57 Misra, N.C., Panda, K., Ila, H., and Junjappa, H. (2007) An efficient highly regioselective synthesis of 2,3,4-trisubstituted pyrroles by cycloaddition of polarized ketene S,S- and N,S-acetals with activated methylene isocyanides. *J. Org. Chem.*, **72**, 1246–1251.

58 Bullington, J.L., Wolff, R.R., and Jackson, P.F. (2002) Regioselective preparation of 2-substituted 3,4-diaryl pyrroles: a concise total synthesis of Ningalin B. *J. Org. Chem.*, **67**, 9439–9442.

59 Baxendale, I.R., Buckle, C.D., Ley, S.V., and Tamborini, L. (2009) A base-catalysed one-pot three-component coupling reaction leading to nitrosubstituted pyrroles. *Synthesis*, 1485–1493.

60 Saikachi, H., Kitagawa, T., and Sasaki, H. (1979) Synthesis of furan derivatives. LXXXVI. Reaction of tosylmethyl isocyanide with methyl 3-substituted propiolates as acetylenic Michael

acceptors. *Chem. Pharm. Bull.*, **27**, 2857–2861.

61 Adib, A., Bagher, M., Sheikhi, E., and Bijanzadeh, H.R. (2011) 1-Methylimidazole-catalyzed reaction between tosylmethyl ioscyanide and dialkyl acetylenedicarboxylates: an efficient synthesis of functionalized pyrroles. *Chin. Chem. Lett.*, **22**, 314–317.

62 Kamijo, S., Kanazawa, C., and Yamamoto, Y. (2005) Copper- or phosphine-catalyzed reaction of alkynes with isocyanides. Regioselective synthesis of substituted pyrroles controlled by the catalyst. *J. Am. Chem. Soc.*, **127**, 9260–9266.

63 Cai, Q., Zhou, F., Xu, T., Fu, L., and Ding, K. (2011) Copper-catalyzed tandem reactions of 1-(2-iodoary)-2-yn-1-ones with isocyanides for the synthesis of 4-oxo-indeno[1,2-b]pyrroles. *Org. Lett.*, **13**, 340–343.

64 Larionov, O.V. and de Meijere, A. (2005) Versatile direct synthesis of oligosubstituted pyrroles by cycloaddition of α-metalated isocyanides to acetylenes. *Angew. Chem. Int. Ed.*, **44**, 5664–5667.

65 Gao, D., Zhai, H., Parvez, M., and Back, T.G. (2008) 1,3-Dipolar cycloadditions of acetylenic sulfones in solution and on solid supports. *J. Org. Chem.*, **73**, 8057–8068.

66 Kolontsova, A.N., Ivantsova, M.N., Tokareva, M.I., and Mironov, M.A. (2010) Reaction of isocyanides with thiophenols and gem-diactivated olefins: a one-pot synthesis of substituted 2-aminopyrroles. *Mol. Divers.*, **14**, 543–550.

67 Takaya, H., Kojima, S., and Murahashi, S.-I. (2001) Rhodium complex-catalyzed reaction of isonitriles with carbonyl compounds: catalytic synthesis of pyrroles. *Org. Lett.*, **3**, 421–424.

68 Fontaine, P., Masson, G., and Zhu, J. (2009) Synthesis of pyrroles by consecutive multicomponent reaction/ [4+1] cycloaddition of α-iminonitriles with isocyanides. *Org. Lett.*, **11**, 1555–1558.

69 Tsukada, N., Wada, M., Takahashi, N., and Inoue, Y. (2009) Synthesis of dinuclear palladium complexes having two parallel isocyanide ligands, and their application as catalysts to pyrrole formation from tert-butylisocyanide and alkynes. *J. Organomet. Chem.*, **694**, 1333–1338.

70 van Leusen, A.M., Hoogenboom, B.E., and Siderius, H. (1972) A novel and efficient synthesis of oxazoles from tosylmethylisocyanide and carbonyl compounds. *Tetrahedron Lett.*, 2369–2372.

71 Saikachi, H., Kitagawa, T., Sasaki, H., and van Leusen, A.M. (1979) Synthesis of furan derivatives. LXXXV. Condensation of heteroaromatic aldehydes with tosylmethyl isocyanide. *Chem. Pharm. Bull.*, **27**, 793–796.

72 Katritzky, A.R., Chen, Y.-X., Yannakopoulou, K., and Lue, P. (1989) Benzotriazol-1-ylmethyl isocyanide, a new synthon for CH-N=C transfer. Syntheses of α-hydroxyaldehydes, 4-ethoxy-2-oxazolines and oxazoles. *Tetrahedron Lett.*, **30**, 6657–6660.

73 Sisko, J., Kassick, A.J., Mellinger, M., Filan, J.J., Allen, A., and Olsen, M.A. (2000) An investigation of imidazole and oxazole syntheses using aryl-substituted TosMIC reagents. *J. Org. Chem.*, **65**, 1516–1524.

74 Wu, B., Wen, J., Zhang, J., Li, J., Xiang, Y.-Z., and Yu, X.-Q. (2009) One-pot van Leusen synthesis of 4,5-disubstituted oxazoles in ionic liquid. *Synlett*, 500–504.

75 Suzuki, M., Iwasaki, T., Matsumoto, K., and Okumura, K. (1972) Convenient syntheses of aroylamino acids and α-amino ketones. *Synth. Commun.*, **2**, 237–242.

76 Ozaki, Y., Maeda, S., Iwasaki, T., Matsumoto, K., Odawara, A., Sasaki, Y., and Morita, T. (1983) Syntheses of 5-substituted oxazole-4-carboxylic acid derivatives with inhibitory activity on blood platelet aggregation. *Chem. Pharm. Bull.*, **31**, 4417–4424.

77 Henneke, K.-W., Schöllkopf, U., and Neudecker, T. (1979) Syntheses with α-metalated isocyanides. XLII. Bi-, ter- and quateroxazoles from α-anionized isocyanides and acylating agents; α-amino ketones and α,α'-

diamino ketones. *Liebigs Ann. Chem.*, 1370–1387.
78 Suzuki, M., Iwasaki, T., Miyoshi, M., Okumura, K., and Matsumoto, K. (1973) New convenient syntheses of α-C-acylamino acids and α-amino ketones. *J. Org. Chem.*, **38**, 3571–3575.
79 van Leusen, A.M., and van Gennep, H.E. (1973) Preparation of thiomethylisocyanides and their use in heterocyclic syntheses. *Tetrahedron Lett.*, 627–628.
80 Chupp, J.P. and Leschinsky, K.L. (1980) Heterocycles from substituted amides. VII (1,2). Oxazoles from 2-isocyanoacetamides. *J. Heterocycl. Chem.*, **17**, 705–709.
81 Li, Y., Xu, X., Tan, J., Xia, C., Zhang, D., and Liu, Q. (2011) Double isocyanide cyclization: a synthetic strategy for two-carbon-tethered pyrrole/oxazole pairs. *J. Am. Chem. Soc.*, **133**, 1775–1777.
82 dos Santos, A., El Kaïm, L., Grimaud, L., and Ronsseray, C. (2009) Unconventional oxazole formation from isocyanides. *Chem. Commun.*, 3907–3909.
83 El Kaïm, L., Grimaud, L., and Schiltz, A. (2009) One-pot synthesis of oxazoles using isocyanide surrogates. *Tetrahedron Lett.*, **50**, 5235–5237.
84 Xia, Q. and Ganem, B. (2002) Metal-promoted variants of the Passerini reaction leading to functionalized heterocycles. *Org. Lett.*, **4**, 1631–1634.
85 Bossio, R., Marcaccini, S., and Pepino, R. (2003) 1986) A novel synthetic route to oxazoles: one-pot synthesis of 2-arylthio-5-alkoxyoxazoles. *Heterocycles*, **24**, 2005.
86 Wang, Q., Xia, Q., and Ganem, B. (2003) A general synthesis of 2-substituted-5-aminooxazoles: building blocks for multifunctional heterocycles. *Tetrahedron Lett.*, **44**, 6825–6827.
87 Sun, X., Janvier, P., Zhao, G., Bienaymé, H., and Zhu, J. (2001) A novel multicomponent synthesis of polysubstituted 5-aminooxazole and its new scaffold-generating reaction to pyrrolo[3,4-b]pyridine. *Org. Lett.*, **3**, 877–880.
88 Cuny, G., Gámez-Montaño, R., and Zhu, J. (2004) Truncated diastereoselective Passerini reaction, a rapid construction of polysubstituted oxazole and peptides having an α-hydroxy-β-amino acid component. *Tetrahedron*, **60**, 4879–4885.
89 Wang, S.-X., Wang, M.-X., Wang, D.-X., and Zhu, J. (2007) Chiral salen-aluminum complex as a catalyst for enantioselective α-addition of isocyanides to aldehydes: asymmetric synthesis of 2-(1-hydroxyalkyl)-5-aminooxazoles. *Org. Lett.*, **9**, 3615–3618.
90 Yue, T., Wang, M.-X., Wang, D.-X., Masson, G., and Zhu, J. (2009) Catalytic asymmetric Passerini-type reaction: chiral aluminum-organophosphate-catalyzed enantioselective α-addition of isocyanides to aldehydes. *J. Org. Chem.*, **74**, 8396–8399.
91 Yue, T., Wang, M.-X., Wang, D.-X., Masson, G., and Zhu, J. (2009) Brønsted acid-catalyzed enantioselective three-component reaction involving the α addition of isocyanides to imines. *Angew. Chem. Int. Ed.*, **48**, 6717–6721.
92 Bonne, D., Dekhane, M., and Zhu, J. (2007) Modulating the reactivity of α-isocyanoacetates: multicomponent synthesis of 5-methoxyoxazoles and furopyrrolones. *Angew. Chem. Int. Ed.*, **46**, 2485–2488.
93 Lalli, C., Bouma, M.J., Bonne, D., Masson, G., and Zhu, J. (2011) Exploiting the divergent reactivity of α-isocyanoacetate: multicomponent synthesis of 5-alkoxyoxazoles and related heterocycles. *Chem. Eur. J.*, **17**, 880–889.
94 Fayol, A., Housseman, C., Sun, X., Janvier, P., Bienaymé, H., and Zhu, J. (2005) Synthesis of α-isocyano-α-alkyl(aryl)acetamides and their use in the multicomponent synthesis of 5-aminooxazole, pyrrolo[3,4-b]pyridin-5-one and 4,5,6,7-tetrahydrofuro[2,3-c] pyridine. *Synthesis*, 161–165.
95 Janvier, P., Sun, X., Bienaymé, H., and Zhu, J. (2002) Ammonium chloride-promoted four-component synthesis of pyrrolo[3,4-b]pyridin-5-one. *J. Am. Chem. Soc.*, **124**, 2560–2567.
96 Elders, N., Ruijter, E., de Kanter, F.J.J., Groen, M.B., and Orru, R.V.A. (2008)

Selective formation of 2-imidazolines and 2-substituted oxazoles by using a three-component reaction. *Chem. Eur. J.*, **14**, 4961–4973.

97 Chupp, J.P. and Leschinsky, K.L. (1980) Heterocycles from substituted amides. VIII. (1,2) Oxazole derivatives from reaction of isocyanates with 2-isocyanoacetamides. *J. Heterocycl. Chem.*, **17**, 711–715.

98 Mossetti, R., Pirali, T., Tron, G.C., and Zhu, J. (2010) Efficient synthesis of α-ketoamides via 2-acyl-5-aminooxazoles by reacting acyl chlorides and α-isocyanoacetamides. *Org. Lett.*, **12**, 820–823.

99 Wu, J., Chen, W., Hu, M., Zou, H., and Yu, Y. (2010) Synthesis of polysubstituted 5-aminooxazoles from α-diazocarbonyl esters and α-isocyanoacetamides. *Org. Lett.*, **12**, 616–618.

100 Deyrup, J.A. and Killion, K.K. (1972) The reaction of N-acyl imines with t-butyl isocyanide. *J. Heterocycl. Chem.*, **9**, 1045–1048.

101 Zhang, J., Coqueron, P.-Y., Vors, J., and Ciufolini, M.A. (2010) Synthesis of 5-amino-oxazole-4-carboxylates from α-chloroglycinates. *Org. Lett.*, **12**, 3942–3945.

102 Wang, Q. and Ganem, B. (2003) New four-component condensations leading to 2,4,5-trisubstituted oxazoles. *Tetrahedron Lett.*, **44**, 6829–6832.

103 Schnell, M., Ramm, M., and Köckritz, A. (1994) α-Substituted phosphonates. 64. Phosphono-substituted imidazoles and other heterocycles from diethyl [(2,2-dichloro-1-isocyano)-ethenyl] phosphonate. *J. Prakt. Chem./Chemiker-Zeitung*, **336**, 29–37.

104 Buron, C., El Kaïm, L., and Uslu, A. (1997) A new straightforward formation of aminoisoxazoles from isocyanides. *Tetrahedron Lett.*, **38**, 8027–8030.

105 Wang, X., Xu, F., Xu, Q., Mahmud, H., Houze, J., Zhu, L., Akerman, M., Tonn, G., Tang, L., McMaster, B.E., Dairaghi, D.J., Schall, T.J., Collins, T.L., and Medina, J.C. (2006) Optimization of 2-aminothiazole derivatives as CCR4 antagonists. *Bioorg. Med. Chem. Lett.*, **16**, 2800–2803.

106 Adib, M., Mahdavi, M., Ansari, S., Malihi, F., Zhu, L.-G., and Bijanzadeh, H.R. (2009) Reaction between isocyanides and nitrostyrenes in water: a novel and efficient synthesis of 5-(alkylamino)-4-aryl-3-isoxazolecarboxamides. *Tetrahedron Lett.*, **50**, 7246–7248.

107 Hartman, G.D. and Weinstock, L.M. (1976) A novel 1,3-thiazole synthesis via α-metallated isocyanides and thiono esters. *Synthesis*, 681–682.

108 Hartman, G.D. and Weinstock, L.M. (1979) Thiazoles from ethyl isocyanoacetate and thiono esters: ethyl thiazole-4-carboxylate. *Org. Syntheses*, **59**, 183–190.

109 Oldenziel, O.H. and van Leusen, A.M. (1972) Chemistry of sulfonylmethylisocyanides. 4. A new synthesis of thiazoles from tosylmethylisocyanide and carboxymethyl dithioates. *Tetrahedron Lett.*, 2777–2778.

110 Suzuki, M., Moriya, T., Matsumoto, K., and Miyoshi, M. (1982) A new convenient synthesis of 5-amino-1,3-thiazole-4-carboxylic acids. *Synthesis*, 874–875.

111 Solomon, D.M., Rizvi, R.K., and Kaminski, J.J. (1987) Observations on the reactions of isocyanoacetate esters with isothiocyanates and isocyanates. *Heterocycles*, **26**, 651–674.

112 Boros, E.E., Johns, B.A., Garvey, E.P., Koble, C.S., and Miller, W.H. (2006) Synthesis and HIV-integrase strand transfer inhibition activity of 7-hydroxy[1,3]thiazolo[5,4-b]pyridin-5(4H)-ones. *Bioorg. Med. Chem. Lett.*, **16**, 5668–5672.

113 Baxendale, I.R., Ley, S.V., Smith, C.D., Tamborini, L., and Voica, A.-F. (2008) A bifurcated pathway to thiazoles and imidazoles using a modular flow microreactor. *J. Comb. Chem.*, **10**, 851–857.

114 van Nispen, S.P.J.M., Bregman, J.H., van Engen, D.G., van Leusen, A.M., Saikachi, H., Kitagawa, T., and Sasaki, H. (1982) Chemistry of sulfonylmethyl isocyanide. Part 23. Synthesis of thiazoles and imidazoles from isothiocyanates and tosylmethyl

isocyanide. Base-induced ring transformation of 5-amino-1,3-thiazoles to imidazole-5-thiols. *J. Royal Netherlands Chem. Soc.*, **101**, 28–34.

115 Kazmaier, U. and Ackermann, S. (2005) A straightforward approach towards thiazoles and endothiopeptides *via* Ugi reaction. *Org. Biomol. Chem.*, **3**, 3184–3187.

116 Dömling, A. and Illgen, K. (2005) 1-Isocyano-2-dimethylamino-alkenes: versatile reagents in diversity-oriented organic synthesis. *Synthesis*, 662–667.

117 Heck, S. and Dömling, A. (2000) A versatile multi-component one-pot thiazole synthesis. *Synlett*, 424–426.

118 Kolb, J., Beck, B., Almstetter, M., Heck, S., Herdtweck, E., and Dömling, A. (2003) New MCRs: The first 4-component reaction leading to 2,4-disubstituted thiazoles. *Mol. Diversity*, **6**, 297–313.

119 Henkel, B., Westner, B., and Dömling, A. (2003) Polymer-bound 3-N,N-(dimethylamino)-2-isocyanoacrylate for the synthesis of thiazoles via a multicomponent reaction. *Synlett*, 2410–2412.

120 Umkehrer, M., Kolb, J., Burdack, C., and Hiller, W. (2005) 2,4,5-Trisubstituted thiazole building blocks by a novel multi-component reaction. *Synlett*, 79–82.

121 Yamada, M., Fukui, T., and Nunami, K. (1995) A novel synthesis of methyl 5-substituted thiazole-4-carboxylates using 3-bromo-2-isocyanoacrylates (BICA). *Tetrahedron Lett.*, **36**, 257–260.

122 Dömling, A., Henkel, B., and Beck, B. (2004) WO 2004005269.

123 van Leusen, A.M. and Oldenziel, O.H. (1972) Chemistry of sulfonylmethyl isocyanides. 3. Synthesis of tosyl-substituted imidazoles from tosylmethylisocyanide and imidoyl chlorides. *Tetrahedron Lett.*, 2373–2374.

124 Helal, C.J. and Lucas, J.C. (2002) A concise and regioselective synthesis of 1-alkyl-4-imidazolecarboxylates. *Org. Lett.*, **4**, 4133–4134.

125 Hiramatsu, K., Nunami, K., Hayashi, K., and Matsumoto, K. (1990) Synthesis of amino acids and related compounds. 39. A facile synthesis of methyl 1,5-disubstituted imidazole-4-carboxylates. *Synthesis*, 781–782.

126 Nunami, K., Yamada, M., Fukui, T., and Matsumoto, K. (1994) A novel synthesis of methyl 1,5-disubstituted imidazole-4-carboxylates using 3-bromo-2-isocyanoacrylates (BICA). *J. Org. Chem.*, **59**, 7635–7642.

127 Yamada, M., Fukui, T., and Nanami, K. (1995) Synthesis of 5-substituted methyl-1-hydroxyimidazole-4-carboxylates and 5-substituted methyl-1-aminoimidazole-4-carboxylates using 3-bromo-2-isocyanoacrylates. *Synthesis*, 1365–1367.

128 ten Have, R., Huisman, M., Meetsma, A., and van Leusen, A.M. (1997) Chemistry of sulfonylmethyl isocyanides. 43. Novel synthesis of 4(5)-monosubstituted imidazoles via cycloaddition of tosylmethyl isocyanide to aldimines. *Tetrahedron*, **53**, 11355–11368.

129 Gulevich, A.V., Balenkova, E.S., and Nenajdenko, V.G. (2007) The first example of a diastereoselective thio-Ugi reaction: a new synthetic approach to chiral imidazole derivatives. *J. Org. Chem.*, **72**, 7878–7885.

130 Grigg, R., Lansdell, M.I., and Thornton-Pett, M. (1999) Silver acetate catalyzed cycloadditions of isocyanoacetates. *Tetrahedron*, **55**, 2025–2044.

131 Kanazawa, C., Kamijo, S., and Yamamoto, Y. (2006) Synthesis of imidazoles through the copper-catalyzed cross-cycloaddition between two different isocyanides. *J. Am. Chem. Soc.*, **128**, 10662–10663.

132 Bonin, M.-A., Giguère, D., and Roy, R. (2007) *N*-Arylimidazole synthesis by cross-cycloaddition of isocyanides using a novel catalytic system. *Tetrahedron*, **63**, 4912–4917.

133 Henkel, B. (2004) Synthesis of imidazole-4-carboxylic acids via solid-phase bound 3-N,N-(dimethylamino)-2-isocyanoacrylate. *Tetrahedron Lett.*, **45**, 2219–2221.

134 Sisko, J. and Mellinger, M. (2002) Development of a general process for the synthesis of highly substituted

imidazoles. *Pure Appl. Chem.*, **74**, 1349–1357.

135 Samanta, S.K., Kylänlahti, I., and Yli-Kauhaluoma, J. (2005) Microwave-assisted synthesis of imidazoles: reaction of *p*-toluenesulfonylmethyl isocyanide and polymer-bound imines. *Bioorg. Med. Chem. Lett.*, **15**, 3717–3719.

136 Bossio, R., Marcaccini, S., Pepino, R., and Torroba, T. (1996) Studies on isocyanides and related compounds. Synthesis of 1-aryl-2-(tosylamino)-1*H*-imidazoles, a novel class of imidazole derivatives. *J. Org. Chem.*, **61**, 2202–2203.

137 van Leusen, A.M., Wildeman, J., and Oldenziel, O.H. (1977) Chemistry of sulfonylmethyl isocyanides. 12. Base-induced cycloaddition of tosylmethyl isocyanides to C,N double bonds. Synthesis of 1,5-disubstituted and 1,4,5-trisubstituted imidazoles from aldimines and imidoyl chlorides. *J. Org. Chem.*, **42**, 1153–1159.

138 Che, H., Tuyen, T.N., Kim, H.P., and Park, H. (2010) 1,5-Diarylimidazoles with strong inhibitory activity against COX-2 catalyzed PGE2 production from LPS-induced RAW 264.7 cells. *Bioorg. Med. Chem. Lett.*, **20**, 4035–4037.

139 Boulos, J. and Schulman, J. (1998) Synthesis of oxazolyl- and furanyl-substituted imidazole hydrochlorides and methiodides. *J. Heterocycl. Chem.*, **35**, 859–863.

140 Sisko, J. (1998) A one-pot synthesis of 1-(2,2,6,6-tetramethyl-4-piperidinyl)-4-(4-fluorophenyl)-5-(2-amino-4-pyrimidinyl)-imidazole: a potent inhibitor of p38 MAP kinase. *J. Org. Chem.*, **63**, 4529–4531.

141 Fodili, M., Nedjar-Kolli, B., Garrigues, B., Lherbet, C., and Hoffmann, P. (2009) Synthesis of imidazoles from ketimines using tosylmethyl isocyanide (TosMIC) catalyzed by bismuth triflate. *Lett. Org. Chem.*, **6**, 354–358.

142 Zhang, C., Moran, E.J., Woiwode, T.F., Short, K.M., and Mjalli, A.M.M. (1996) Synthesis of tetrasubstituted imidazoles via α-(*N*-acyl-*N*-alkylamino)-β-ketoamides on Wang resin. *Tetrahedron Lett.*, **37**, 751–754.

143 Atlan, V., Buron, C., and El Kaïm, L. (2000) A new straightforward formation of aminopyrazoles from isocyanides. *Synlett*, 489–490.

144 Atlan, V., El Kaïm, L., Grimaud, L., Jana, N.K., and Majee, A. (2002) The Mannich reaction of hydrazones amenable to solid phase synthesis: a powerful tool for heterocycle preparation. *Synlett*, 352–354.

145 El Kaïm, L., Gautier, L., Grimaud, L., and Michaut, V. (2003) New insight into the azaenamine behaviour of *N*-arylhydrazones: first aldol and improved Mannich reactions with unactivated aldehydes. *Synlett*, 1844–1846.

146 Ancel, J.E., El Kaïm, L., Gadras, A., Grimaud, L., and Jana, N.K. (2002) Studies towards the synthesis of Fipronil analogues: improved decarboxylation of α-hydrazonoacid derivatives. *Tetrahedron Lett.*, **43**, 8319–8321.

147 Adib, M., Mohammadi, B., and Bijanzadeh, H.R. (2008) A novel one-pot, three-component synthesis of dialkyl 5-(alkylamino)-1-aryl-1*H*-pyrazole-3,4-dicarboxylates. *Synlett*, 3180–3182.

148 Majumder, S., Gipson, K.R., Staples, R.J., and Odom, A.L. (2009) Pyrazole synthesis using a titanium-catalyzed multicomponent coupling reaction and synthesis of withasomnine. *Adv. Synth. Catal.*, **351**, 2013–2023.

149 Ramazani, A. and Souldozi, A. (2008) Iminophosphorane-mediated one-pot synthesis of 1,3,4-oxadiazole derivatives. *ARKIVOC*, (xvi), 235–242.

150 van Leusen, A.M., Hoogenboom, B.E., and Houwing, H.A. (1976) Chemistry of sulfonylmethyl isocyanides. 11. Synthesis of 1,2,4-triazoles from tosylmethyl isocyanide and aryldiazonium compounds. *J. Org. Chem.*, **41**, 711–713.

151 El Kaïm, L., Grimaud, L., and Wagschal, S. (2009) Three-component Nef-Huisgen access to 1,2,4-triazoles. *Synlett*, 1315–1317.

152 Moderhack, D. and Daoud, A. (2003) 1,2,3- and 1,2,4-triazolium salts, pyrazoles, and quinoxalines from diarylnitrilimines and isocyanides: a

study of the scope. *J. Heterocycl. Chem.*, **40**, 625–637.

153 Fallon, F.G. and Herbst, R.M. (1957) Synthesis of 1-substituted tetrazoles. *J. Org. Chem.*, **22**, 933–936.

154 Jin, T., Kamijo, S., and Yamamoto, Y. (2004) Synthesis of 1-substituted tetrazoles via the acid-catalyzed [3+2] cycloaddition between isocyanides and trimethylsilyl azide. *Tetrahedron Lett.*, **45**, 9435–9437.

155 Sureshbabu, V.V., Narendra, N., and Nagendra, G. (2009) Chiral *N*-Fmoc-β-amino alkyl isonitriles derived from amino acids: first synthesis and application in 1-substituted tetrazole synthesis. *J. Org. Chem.*, **74**, 153–157.

156 Schremmer, E.S. and Wanner, K.T. (2007) Zinc iodide as an efficient catalyst in the TMS-azide modified Passerini reaction. *Heterocycles*, **74**, 661–671.

157 Yue, T., Wang, M.-X., Wang, D.-X., and Zhu, J. (2008) Asymmetric synthesis of 5-(1-hydroxyalkyl)tetrazoles by catalytic enantioselective Passerini-type reactions. *Angew. Chem. Int. Ed.*, **47**, 9454–9457.

158 Ugi, I., Beck, F., Bodesheim, F., Fetzer, U., Meyr, R., Rosendahl, K., Steinbrückner, C., and Wischhöfer, E. (1960) Neues über isonitrile. *Angew. Chem.*, **72**, 639.

159 Ugi, I. (1962) The α-addition of immonium ions and anions to isonitriles accompanied with secondary reactions. *Angew. Chem. Int. Ed.*, **1**, 8–21.

160 Mironov, M.A., Tokareva, M.I., Ivantsova, M.N., and Mokrushin, V.S. (2004) Ugi reaction with isocyanoindoles. *Russ. J. Org. Chem.*, **40**, 847–853.

161 Mesenzani, O., Massarotti, A., Giustiniano, M., Pirali, T., Bevilacqua, V., Caldarelli, A., Canonico, P., Sorba, G., Novellino, E., Genazzani, A.A., and Tron, G.C. (2011) Replacement of the double bond of antitubulin chalcones with triazoles and tetrazoles: synthesis and biological evaluation. *Bioorg. Med. Chem. Lett.*, **21**, 764–768.

162 Davenport, A.J., Stimson, C.C., Corsi, M., Vaidya, D., Glenn, E., Jones, T.D., Bailey, S., Gemkow, M.J., Fritz, U., and Hallett, D.J. (2010) Discovery of substituted benzyl tetrazoles as histamine H3 receptor antagonists. *Bioorg. Med. Chem. Lett.*, **20**, 5165–5169.

163 Kalinski, C., Umkehrer, M., Gonnard, S., Jäger, N., Ross, G., and Hiller, W. (2006) A new and versatile Ugi/ S_NAr synthesis of fused 4,5-dihydrotetrazolo[1,5-*a*]quinoxalines. *Tetrahedron Lett.*, **47**, 2041–2044.

164 Marcos, C.F., Marcaccini, S., Menchi, G., Pepino, R., and Torroba, T. (2008) Studies on isocyanides: synthesis of tetrazolyl-isoindolinones via tandem Ugi four-component condensation/intramolecular amidation. *Tetrahedron Lett.*, **49**, 149–152.

165 Nayak, M. and Batra, S. (2010) Isonitriles from the Baylis–Hillman adducts of acrylates: viable precursor to tetrazolo-fused diazepinones via post-Ugi cyclization. *Tetrahedron Lett.*, **51**, 510–516.

166 Chen, J.J., Golebiowski, A., Klopfenstein, S.R., and West, L. (2002) The universal Rink-isonitrile resin: applications in Ugi reactions. *Tetrahedron Lett.*, **43**, 4083–4085.

167 Kiselyov, A.S. (2005) Reaction of *N*-fluoropyridinium fluoride with isonitriles and $TMSN_3$: a convenient one-pot synthesis of tetrazol-5-yl pyridines. *Tetrahedron Lett.*, **46**, 4851–4854.

168 El Kaïm, L., Grimaud, L., and Patil, P. (2011) Three-component strategy toward 5-membered heterocycles from isocyanide dibromides. *Org. Lett.*, **13**, 1261–1263.

169 El Kaïm, L., Grimaud, L., and Oble, J. (2006) New *ortho*-quinone methide formation: application to three-component coupling of isocyanides, aldehydes and phenols. *Org. Biomol. Chem.*, **4**, 3410–3413.

170 Ramazani, A., Mahyari, A.T., Rouhani, M., and Rezaei, A. (2009) A novel three-component reaction of a secondary amine and a 2-hydroxybenzaldehyde derivative with an isocyanide in the presence of silica gel: an efficient one-pot synthesis of benzo[*b*]furan

derivatives. *Tetrahedron Lett.*, **50**, 5625–5627.
171 Mitra, S., Hota, S.K., and Chattopadhyay, P. (2010) Lewis acid catalyzed one-pot selective synthesis of aminobenzofurans and N-alkyl-2-aryl-2-(arylimino)acetamides: product dependence on the nature of the aniline. *Synthesis*, 3899–3905.
172 Lygin, A.V. and de Meijere, A. (2009) Synthesis of 1-substituted benzimidazoles from o-bromophenyl isocyanide and amines. *Eur. J. Org. Chem.*, 5138–5141.
173 Campo, J., García-Valverde, M., Marcaccini, S., Rojo, M.J., and Torroba, T. (2006) Synthesis of indole derivatives via isocyanides. *Org. Biomol. Chem.*, **4**, 757–765.
174 Zeeh, B. (1969) Heterocycles from isocyanides. VI. Synthesis of indole derivatives from aromatic ketones and tert-butyl isocyanide. *Chem. Ber.*, 678–685.
175 Kobayashi, K., Iitsuka, D., Fukamachi, S., and Konishi, H. (2009) A convenient synthesis of 2-substituted indoles by the reaction of 2-(chloromethyl)phenyl isocyanides with organolithiums. *Tetrahedron*, **65**, 7523–7526.
176 Ito, Y., Kobayashi, K., and Saegusa, T. (1977) An efficient synthesis of indole. *J. Am. Chem. Soc.*, **99**, 3532–3534.
177 Ito, Y., Kobayashi, K., Seko, N., and Saegusa, T. (1984) Indole syntheses utilizing o-methylphenyl isocyanides. *Bull. Chem. Soc. Jpn*, **57**, 73–84.
178 Ito, Y., Kobayashi, K., and Saegusa, T. (1979) Reactions of o-tolyl isocyanide with isocyanate and isothiocyanate. Syntheses of N-substituted indole-3-carboxamides and indole-3-thiocarboxamides. *Tetrahedron Lett.*, 1039–1042.
179 Ito, Y., Kobayashi, K., and Saegusa, T. (1979) Indole syntheses with o-tolyl isocyanide. 3-Acylindoles and 2-substituted indoles. *J. Org. Chem.*, **44**, 2030–2032.
180 Ito, Y., Inubushi, Y., Sugaya, T., Kobayashi, K., and Saegusa, T. (1978) Synthesis of indole derivatives by Cu_2O-catalyzed cyclization of o-(α-cyanoalkyl)phenyl isocyanides and o-[α-(methoxycarbonyl)alkyl]phenyl isocyanides. *Bull. Chem. Soc. Jpn*, **51**, 1186–1188.
181 Jones, W.D. and Kosar, W.P. (1986) Carbon–hydrogen bond activation by ruthenium for the catalytic synthesis of indoles. *J. Am. Chem. Soc.*, **108**, 5640–5641.
182 Hsu, G.C., Kosar, W.P., and Jones, W.D. (1994) Functionalization of benzylic carbon–hydrogen bonds. Mechanism and scope of the catalytic synthesis of indoles with [Ru(dmpe)$_2$]. *Organometallics*, **13**, 385–396.
183 Onitsuka, K., Yamamoto, M., Suzuki, S., and Takahashi, S. (2002) Structure and reactivity of (η^3-indolylmethyl)palladium complexes generated by the reaction of organopalladium complexes with o-alkenylphenyl isocyanide. *Organometallics*, **21**, 581–583.
184 Onitsuka, K., Suzuki, S., and Takahashi, S. (2002) A novel route to 2,3-disubstituted indoles via palladium-catalyzed three-component coupling of aryl iodide, o-alkenylphenyl isocyanide and amine. *Tetrahedron Lett.*, **43**, 6197–6199.
185 Zhang, S., Zhang, W.-X., and Xi, Z. (2010) Efficient one-pot synthesis of N-containing heterocycles by multicomponent coupling of silicon-tethered diynes, nitriles, and isocyanides through intramolecular cyclization of iminoacyl-Zr intermediates. *Chem. Eur. J.*, **16**, 8419–8426.
186 Tokuyama, H. and Fukuyama, T. (2002) Indole synthesis by radical cyclization of o-alkenylphenyl isocyanides and its application to the total synthesis of natural products. *Chem. Record*, **2**, 37–45.
187 Fukuyama, T., Chen, X., and Peng, G. (1994) A novel tin-mediated indole synthesis. *J. Am. Chem. Soc.*, **116**, 3127–3128.
188 Tokuyama, H., Watanabe, M., Hayashi, Y., Kurokawa, T., Peng, G., and Fukuyama, T. (2001) A novel protocol for construction of indolylmethyl group at aldehydes and ketones. *Synlett*, 1403–1406.

189 Yamakawa, T., Ideue, E., Shimokawa, J., and Fukuyama, T. (2010) Total synthesis of tryprostatins A and B. *Angew. Chem. Int. Ed.*, **49**, 9262–9265.

190 Rainier, J.D. and Kennedy, A.R. (2000) Cascades to substituted indoles. *J. Org. Chem.*, **65**, 6213–6216.

191 Fukamachi, S., Konishi, H., and Kobayashi, K. (2009) One-pot synthesis of 3-acyl-2-(alkylsulfanyl)indoles and 2-(alkylsulfanyl)-indole-3-carboxylates from (2-isocyanophenyl)methyl ketones or (2-isocyanophenyl)acetates. *Synthesis*, 1786–1790.

192 Mitamura, T., Tsuboi, Y., Iwata, K., Tsuchii, K., Nomoto, A., Sonoda, M., and Ogawa, A. (2007) Photoinduced thiotelluration of isocyanides by using a $(PhS)_2$-$(PhTe)_2$ mixed system, and its application to bisthiolation via radical cyclization. *Tetrahedron Lett.*, **48**, 5953–5957.

193 Tobisu, M., Fujihara, H., Koh, K., and Chatani, N. (2010) Synthesis of 2-boryl- and silylindoles by copper-catalyzed borylative and silylative cyclization of 2-alkenylaryl isocyanides. *J. Org. Chem.*, **75**, 4841–4847.

194 Rainier, J.D., Kennedy, A.R., and Chase, E. (1999) An isonitrile-alkyne cascade to di-substituted indoles. *Tetrahedron Lett.*, **40**, 6325–6327.

195 Rubinshtein, M., James, C.R., Young, J.L., Ma, Y.J., Kobayashi, Y., Gianneschi, N.C., and Yang, J. (2010) Facile procedure for generating side chain functionalized poly(α-hydroxy acid) copolymers from aldehydes via a versatile Passerini-type condensation. *Org. Lett.*, **12**, 3560–3563.

196 Kalinski, C., Umkehrer, M., Schmidt, J., Ross, G., Kolb, J., Burdack, C., Hiller, W., and Hoffmann, S.D. (2006) A novel one-pot synthesis of highly diverse indole scaffolds by the Ugi/Heck reaction. *Tetrahedron Lett.*, **47**, 4683–4686.

197 Schneekloth, J.S., Kim, J. Jr, and Sorensen, E.J. (2009) An interrupted Ugi reaction enables the preparation of substituted indoxyls and aminoindoles. *Tetrahedron*, **65**, 3096–3101.

198 Suginome, M., Fukuda, T., and Ito, Y. (1999) New access to 2,3-disubstituted quinolines through cyclization of o-alkynylisocyanobenzenes. *Org. Lett.*, **1**, 1977–1979.

199 Zhao, J.-J., Peng, C.-L., Liu, L.-Y., Wang, Y., and Zhu, Q. (2010) Synthesis of 2-alkoxy(aroxy)-3-substituted quinolines by DABCO-promoted cyclization of o-alkynylaryl isocyanides. *J. Org. Chem.*, **75**, 7502–7504.

200 Liu, L., Wang, Y., Wang, H., Peng, C., Zhao, J., and Zhu, Q. (2009) Tetrabutylammonium chloride-triggered 6-*endo* cyclization of o-alkynylisocyanobenzenes: an efficient synthesis of 2-chloro-3-substituted quinolines. *Tetrahedron Lett.*, **50**, 6715–6719.

201 Kobayashi, K., Yoneda, K., Miyamoto, K., Morikawa, O., and Konishi, H. (2004) A convenient synthesis of quinolines by reactions of o-isocyano-β-methoxystyrenes with nucleophiles. *Tetrahedron*, **60**, 11639–11645.

202 Ichikawa, J., Wada, Y., Miyazaki, H., Mori, T., and Kuroki, H. (2003) Ring-fluorinated isoquinoline and quinoline synthesis: intramolecular cyclization of o-cyano- and o-isocyano-β,β-difluorostyrenes. *Org. Lett.*, **5**, 1455–1458.

203 Ichikawa, J., Mori, T., Miyazaki, H., and Wada, Y. (2004) C-C bond formation between isocyanide and β,β-difluoroalkene moieties via electron transfer: fluorinated quinoline and biquinoline syntheses. *Synlett*, 1219–1222.

204 Mori, T. and Ichikawa, J. (2007) Radical 6-*endo-trig* cyclization of β,β-difluoro-o-isocyanostyrenes: a facile synthesis of 3-fluoroquinolines and their application to the synthesis of 11-alkylated cryptolepines. *Synlett*, 1169–1171.

205 Mitamura, T. and Ogawa, A. (2011) Synthesis of 2,4-diiodoquinolines via the photochemical cyclization of o-alkynylaryl isocyanides with iodine. *J. Org. Chem.*, **76**, 1163–1166.

206 Kobayashi, K., Nakashima, T., Mano, M., Morikawa, O., and Konishi, H. (2001) Synthesis of 4-hydroxy-3-quinolinecarboxylic acid derivatives by a

condensation/cyclization sequence between o-isocyanobenzoates and magnesium enolates. *Chem. Lett.*, 602–603.

207 Kobayashi, K., Yoneda, K., Mizumoto, T., Umakoshi, H., Morikawa, O., and Konishi, H. (2003) Synthesis of 3-methoxyquinolines via cyclization of 1-isocyano-2-(2-lithio-2-methoxyethenyl)benzenes. *Tetrahedron Lett.*, **44**, 4733–4736.

208 Krasavin, M. and Parchinsky, V. (2008) Expedient entry into 1,4-dihydroquinoxalines and quinoxalines via a novel variant of isocyanide-based MCR. *Synlett*, 645–648.

209 Krasavin, M., Shkavrov, S., Parchinsky, V., and Bukhryakov, K. (2009) Imidazo[1,2-a]quinoxalines accessed via two sequential isocyanide-based multicomponent reactions. *J. Org. Chem.*, **74**, 2627–2629.

210 Heravi, M.M., Baghernejad, B., and Oskooie, H.A. (2009) A novel three-component reaction for the synthesis of N-cyclohexyl-3-aryl-quinoxaline-2-amines. *Tetrahedron Lett.*, **50**, 767–769.

211 Neochoritis, C., Stephanidou-Stephanatou, J., and Tsoleridis, C.A. (2009) Heterocyclizations via TosMIC-based multicomponent reactions: a new approach to one-pot facile synthesis of substituted quinoxaline derivatives. *Synlett*, 302–305.

14
Renaissance of Isocyanoarenes as Ligands in Low-Valent Organometallics

Mikhail V. Barybin, John J. Meyers, Jr, and Brad M. Neal

14.1
Historical Perspective

The birth of isocyanide coordination chemistry can be traced back to 1869 (Alfred Werner was only three years old at that time!), when Gautier isolated several compounds of the composition "Ag(CNR)(CN)" [1]. Gautier recognized correctly that the extraordinarily repulsive odor generated when silver cyanide was treated with organohalides was not associated with the formation of organonitriles, as had been considered previously by Meyer [2] and Lieke [3], but rather with the production of species that were isomeric to nitriles. Some 90 years later, an impressively large number of homoleptic (i.e., binary) isocyanide complexes involving metals of the Cr, Mn, Fe, Co, Ni, and Cu triads were featured in Malatesta's monograph *Isocyanide Complexes of Metals* [4]. The analogy between carbon monoxide (:C≡O) and isocyanides (:C≡N–R) as ligands became apparent well before the introduction of Hoffmann's concept of isolobal relationships [5]. Efforts toward accessing isocyanide congeners of various metal carbonyls, and particularly those containing highly electron-rich metal centers such as carbonylmetalates [6], continue to this date.

In his 1959 review [4], Malatesta emphasized the complete lack of isocyanide-metalates (i.e., isocyanide complexes of metals in sub-zero oxidation states) that would be analogous to carbonylmetalates such as $[Fe(CO_4)]^{2-}$, $[Co(CO)_4]^-$, and $[Rh(CO_4)]^-$ that were already known at that time [6]. Malatesta also commented on the fact that Group 4 and 5 transition metals ". . . did not show any tendency to form complexes with isocyanides," and pointed at the absence of zero-valent isocyanide complexes of the elements with odd atomic numbers. It took many years to begin filling the above gaps in the isocyanide coordination chemistry. Indeed, $Co_2(CNR)_8$ (R = ^{t}Bu, 2,6-xylyl, mesityl), the first zero-valent isocyanide complexes of an odd-numbered transition metal, were obtained in 1977 [7], the first binary isocyanide adduct of a Group 5 metal, $[V(CN^tBu)_6]^{2+}$, was reported in 1980 [8], and the initial homoleptic isocyanidemetalate, $[Co(CNXyl)_4]^-$ (Xyl = 2,6-xylyl or

Isocyanide Chemistry: Applications in Synthesis and Material Science, First Edition. Edited by Valentine Nenajdenko.
© 2012 Wiley-VCH Verlag GmbH & Co. KGaA. Published 2012 by Wiley-VCH Verlag GmbH & Co. KGaA.

2,6-dimethylphenyl), was isolated in 1989 [9]. While no homoleptic isocyanide complexes of Group 4 transition metals are yet known, several Group 4 heteroleptic species – including CpTi(CNXyl)$_4$E (E = I, SnPh$_3$, and SnMe$_3$) – have been described [10].

The up-to-date list of isolable binary isocyanide complexes of transition metals is compiled in Table 14.1. Here, the entries in italic print correspond to a few species [11] that had been known, but were omitted from a similar table of neutral and cationic binary isocyanide complexes published in 1998 [12]; in contrast, those entries in bold print reflect new types of complexes that have been established since 1998. In Table 14.1, no distinction is provided between mono- and polydentate, or terminal versus bridging, isocyanide ligands. For example, a mononuclear complex featuring three η^2-bound diisocyanide ligands would be represented as M(CNR)$_6$, while a dinuclear complex with six terminal and two μ_2-bridging isocyanide ligands would be shown as M$_2$(CNR)$_8$. As is clearly shown in Table 14.1, the old paradigm that isocyanide ligands are greatly inferior to CO in accommodating zero-valent – and, especially, sub-valent – metals has been shattered.

Unlike CO, which has a dipole moment of only 0.12 Debye [13], organic isocyanides are quite polar substances; for instance, the dipole moment of CNPh is 3.44 Debye [13]. Given that the dominant resonance form of a typical isocyanide with a hydrocarbon substituent R places the negative charge on the terminal carbon atom ($^-$:C≡N$^+$–R), isocyanide ligands are considered stronger σ-donors and weaker π-acceptors compared to CO. In other words, the so-called σ-donor/π-acceptor ratio of an isocyanide ligand is significantly higher than that of CO [14]. This argument has been used frequently to explain why isocyanides may be less suitable for stabilizing electron-rich metal ions than CO. The two reactions shown in Scheme 14.1 are certainly consistent with the above reasoning [8, 15].

In general, aryl isocyanides (CNAr) are better at accommodating low-valent metals than alkyl isocyanides, presumably due to the possibility of a delocalization of back-donated electron density into the aromatic ring, as described by the resonance contribution M=C=N$^+$=Ar$^-$. However, Lentz et al. provided not only convincing X-ray structural evidence, but also ^{13}C nuclear magnetic resonance (NMR) and infrared (IR) spectroscopic data, that certain isocyanide ligands with fluorinated substituents are very strong π-acceptors [16]. In particular, CNCF$_3$ – which, unfortunately, has very poor thermal stability in the condensed phase – appears to be on par with (if not more powerful than) CO in terms of its π-accepting ability.

The remarkable versatility of isocyanide ligands compared to CO can be nicely illustrated by the recently crystallographically characterized series of complexes [Ta(CNXyl)$_7$]$^+$, [Ta(CNXyl)$_6$I], [Ta(CNXyl)$_5$(NO)], cis-[Ta(CNXyl)$_4$(NO)$_2$]$^+$, and [Ta(CNXyl)$_6$]$^-$ [17]. While [Ta(CO)$_6$]$^-$ is well known [18], the carbonyl analogues [Ta(CO)$_7$]$^+$, [Ta(CO)$_6$I], [Ta(CO)$_5$(NO)] and [Ta(CO)$_4$(NO)$_2$]$^+$ do not exist. To be fair, however, [Ta(CNR)$_5$]$^{3-}$, which is an isocyanide congener of [Ta(CO)$_5$]$^{3-}$ [6, 19], is not (yet) known either. A quantitative testimony to the electronic flexibility of 2,6-xylyl isocyanide is provided in Figure 14.1, which compares the ^{13}C NMR and vibrational signatures of the CNXyl ligands in four octahedral, formally Ta(–I) complexes to those of free CNXyl. The isocyanide ligands in [Ta(CNXyl)$_4$(NO)$_2$]$^+$

Table 14.1 Homoleptic isocyanide complexes of transition metals isolated to date.

Group of Periodic Table

5	6	7	8	9	10	11
[V(CNR)$_6$]$^-$	Cr(CNR)$_6$	[Mn(CNR)$_5$]$^-$	[Fe(CNR)$_4$]$^{2-}$	[Co(CNR)$_4$]$^-$	**Ni(CNR)$_3$**	[Cu(CNR)$_4$]$^+$
V(CNR)$_6$	[Cr(CNR)$_6$]$^+$	[Mn(CNR)$_6$]$^+$	Fe(CNR)$_5$	**Co(CNR)$_4$**	Ni(CNR)$_4$	**[Cu$_2$(CNRNC)$_3$]$^{2+}$**
[V(CNR)$_6$]$^+$	[Cr(CNR)$_6$]$^{2+}$	[Mn$_2$(CNR)$_6$]$^{2+}$	Fe$_2$(CNR)$_9$	Co$_2$(CNR)$_8$	Ni$_4$(CNR)$_7$	
[V(CNR)$_6$]$^{2+}$	[Cr(CNR)$_6$]$^{3+}$		[Fe(CNR)$_6$]$^{2+}$	**[Co(CNR)$_4$]$^+$**	[Ni(CNR)$_4$]$^{2+}$	
	[Cr(CNR)$_7$]$^{2+}$			[Co(CNR)$_5$]$^+$	Ni$_4$(CNR)$_6$	
				[Co(CNR)$_5$]$^{2+}$		
				[Co$_2$(CNR)$_{10}$]$^{2+}$		
[Nb(CNR)$_6$]$^-$	[Mo(CNR)$_6$]	[Tc(CNR)$_6$]$^+$	Ru(CNR)$_5$	Rh$_2$(CNR)$_8$	**Pd(CNR)$_2$**	[Ag(CNR)$_2$]$^+$
[Nb(CNR)$_7$]$^+$	[Mo(CNR)$_7$]$^{2+}$		Ru$_2$(CNR)$_9$	[Rh(CNR)$_4$]$^+$	Pd$_3$(CNR)$_6$	[Ag(CNR)$_4$]$^+$
			[Ru(CNR)$_6$]$^{2+}$	[Rh$_2$(CNR)$_8$]$^{2+}$	[Pd(CNR)$_4$]$^{2+}$	
			[Ru$_2$(CNR)$_{10}$]$^{2+}$	[Rh$_2$(CNRNC)$_4$]$^{2+}$	[Pd$_2$(CNR)$_6$]$^{2+}$	
					[Pd$_3$(CNR)$_8$]$^{2+}$	
[Ta(CNR)$_6$]$^-$	W(CNR)$_6$	[Re(CNR)$_6$]$^+$	Os(CNR)$_5$	[Ir(CNR)$_4$]$^+$	Pt$_3$(CNR)$_6$	[Au(CNR)$_2$]$^+$
[Ta(CNR)$_7$]$^+$	[W(CNR)$_7$]$^{2+}$		[Os(CNR)$_6$]$^{2+}$		Pt$_7$(CNR)$_{12}$	
			[Os$_2$(CNR)$_{10}$]$^{2+}$		[Pt(CNR)$_4$]$^{2+}$	
					[Pt$_2$(CNR)$_6$]$^{2+}$	

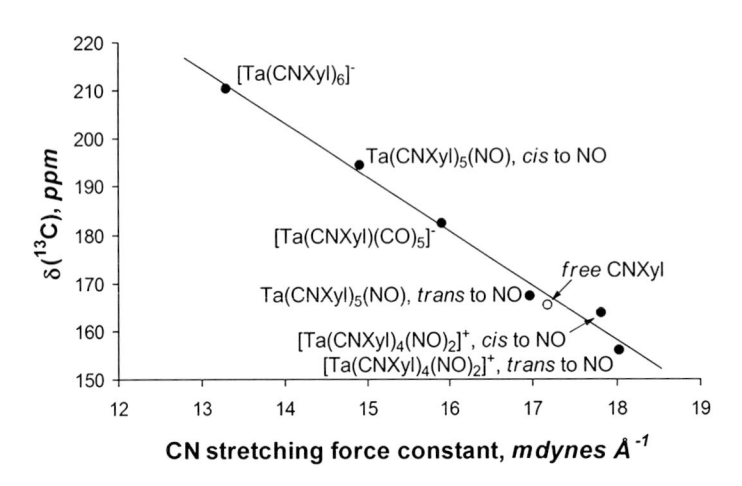

14 Renaissance of Isocyanoarenes as Ligands in Low-Valent Organometallics

$$Co_2(CO)_8 + 5\ CNMe \xrightarrow{Et_2O} [Co(CNMe)_5][Co(CO)_4] + 4\ CO$$

$$3\ [Et_4N]^+[V(CO)_6]^- + 6\ CN^tBu \xrightarrow{1.7\ PhICl_2} [V(CN^tBu)_6][V(CO)_6]_2 + 6\ CO$$

Scheme 14.1

Figure 14.1 ^{13}C NMR chemical shifts for the ligating carbon atom of CNXyl plotted against the corresponding C–N stretching force constants for [Ta(CNXyl)$_6$]$^-$, Ta(CNXyl)$_5$(NO), [Ta(CNXyl)(CO)$_5$]$^-$, and [Ta(CNXyl)$_4$(NO)$_2$]$^-$. The open circle refers to the properties of the uncomplexed CNXyl ligand. Xyl = 2,6-dimethylphenyl. Adapted with permission from Ref. [17]; © 2007, American Chemical Society.

serve essentially as σ-donors, whereas those in [Ta(CNXyl)$_6$]$^-$ clearly behave as powerful π-acceptors [17].

During the past decade, the number of commercially available isocyanides has doubled to reach approximately 25, of which only five are classified as aryl isocyanides (2,6-dimethylphenyl isocyanide; 4-methoxyphenyl isocyanide; 2-naphthyl isocyanide; 2-chloro-6-methylphenyl isocyanide; and 1,4-phenylene diisocyanide: all from Sigma-Aldrich). This surge has undoubtedly been driven by the growing demands for new substrates in combinatorial drug discovery research that employs isocyanide-based multicomponent reactions (MCR) [20].

The coordination chemistry of isocyanides, including polydentate versions thereof, has been at the heart of numerous fundamental and practical advances in organic and organometallic syntheses, catalysis, and diagnostic medicine, as well as in surface, polymer, and materials sciences. Whilst several excellent reviews describing these topics are available [14, 20–32], in this chapter attention will be focused on recent breakthroughs in the chemistry of isocyanidemetalates and related electron-rich complexes. In addition, the coordination and surface chemistries of the recently identified families of unusual isocyanoarene ligands that feature either nonbenzenoid substituents or extremely sterically encumbered benzenoid substituents, will be highlighted.

14.2 Isocyanidemetalates and Related Low-Valent Complexes

14.2.1 Introduction

This section highlights the chemistry of binary isocyanidemetalates documented to date, which include species containing Co(-I), Fe(-II), Mn(-I), V(-I), Nb(-I), and Ta(-I) metal centers. Redox-related, higher-valent congeners thereof, as well as a few currently known nonhomoleptic isocyanidemetalates, are also accounted for. Notably, all isolable isocyanidemetalates contain *aryl* isocyanide ligands, and can be viewed as the isocyanide analogues of the long-established carbonylmetalates $[Co(CO)_4]^-$, $[Fe(CO)_4]^{2-}$, $[Mn(CO)_5]^-$, and $[M(CO)_6]^-$ (M = V, Nb, Ta) [6]. The isocyanidemetalate counterparts of the highly reduced anions $[Ti(CO)_6]^{2-}$, $[Cr(CO)_5]^{2-}$, $[V(CO)_5]^{3-}$, $[Co(CO)_3]^{3-}$, $[Mn(CO)_4]^{3-}$, $[Cr(CO)_4]^{4-}$, and their second and third row equivalents [6] have yet to be discovered, if they can be accessed at all.

14.2.2 Four-Coordinate Isocyanidemetalates and Redox-Related Complexes

In 1989, Cooper and Warnock reported the synthesis of the initial binary isocyanidemetalate, namely $[Co(CNXyl)_4]^-$ (Figure 14.2), through the substitution of all four ethylene ligands in the labile anion $[Co(C_2H_4)_4]^-$ [9]; later, the same authors developed additional elegant routes to $[Co(CNXyl)_4]^-$ that involved reductions of Cp_2Co, $Co_2(CNXyl)_8$, or $[Co(CNXyl)_5]^+$ (Scheme 14.2) [33]. Complex $[Co(CNXyl)_4]^-$ undergoes an electrophilic addition upon treatment with Ph_3SnCl to form five-coordinate $Co(CNXyl)_4(SnPh_3)$. The nature of the "Co(-I)" intermediate(s) generated via a potassium naphthalenide reduction of Cp_2Co remains unclear, although it might be related (at least to some degree) to that of polyarenecobaltates(I-), as described by Ellis and coworkers (Figure 14.3) [34]. Notably, $[Co(\eta^4\text{-anthracene})_2]^-$ reacts cleanly with CNXyl to afford $[Co(CNXyl)_4]^-$ (Scheme 14.3) [34b].

Recently, Figueroa *et al.* introduced the coordination chemistry of two remarkably bulky aryl isocyanide ligands $CNAr^{Mes2}$ and $CNAr^{Dipp2}$, that are based on the *m*-terphenyl scaffold (Figure 14.4) [35]. The reduction of CoI_2 with sodium amalgam in the presence of $CNAr^{Mes2}$ afforded $[Co(CNAr^{Mes2})_4]^-$, an extremely sterically encumbered congener of $[Co(CNXyl)_4]^-$ (Scheme 14.4) [36]. The $[Co(CNAr^{Mes2})_4]^-$ and $[Co(CNXyl)_4]^-$ anions exhibit ν_{NC} bands at 1821 and 1815 cm^{-1}, respectively, in their IR spectra, which is consistent with a strong backbonding interaction between the formally Co(-I) centers and the isocyanide ligands in these complexes. The crystallographically characterized salts $[Ph_3PNPPh_3][Co(CNAr^{Mes2})_4]$

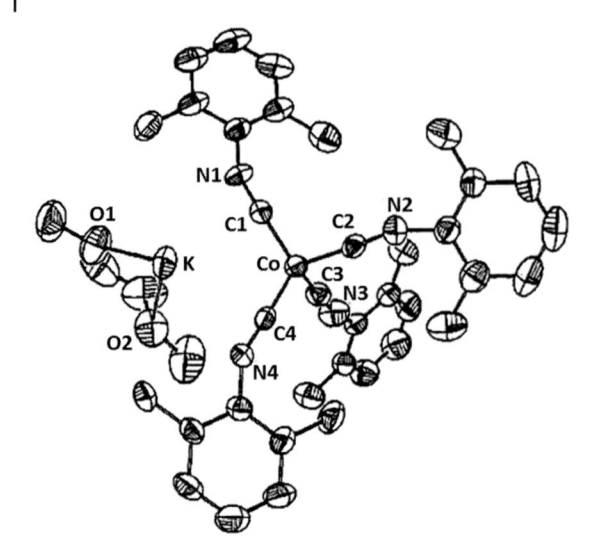

Figure 14.2 ORTEP diagram for [K(DME)][Co(CNXyl)$_4$]. The hydrogen atoms have been omitted for clarity. Adapted with permission from Ref. [9]; © 1994, American Chemical Society.

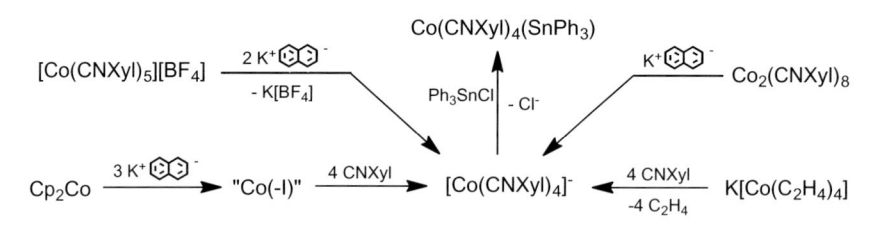

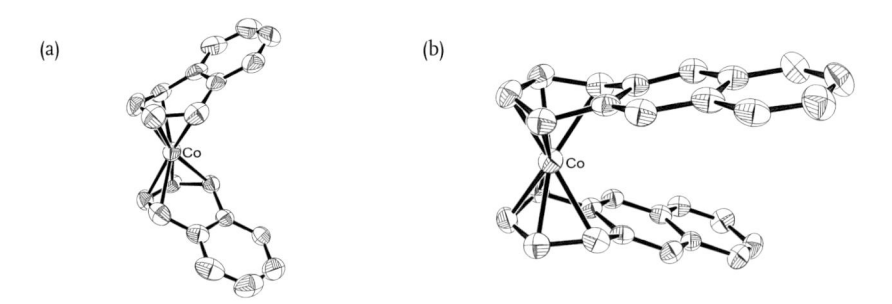

Scheme 14.2

Figure 14.3 ORTEP diagrams. (a) [Co(η^4-naphthalene)$_2$]$^-$ anion in "triple salt" [K(18-crown-6)]$_3$[Co(η^4–C$_{10}$H$_8$)$_2$][Co(η^4–C$_{10}$H$_8$)$_2$(η^2–C$_2$H$_4$)$_2$]; (b) [Co(η^4-anthracene)$_2$]$^-$ anion in [K(18-crown-6)(THF)$_2$][Co(η^4–C$_{14}$H$_{10}$)] generated using the ORTEP-3 program from CIF data in Ref. [34]. The hydrogen atoms have been omitted for clarity.

Scheme 14.3

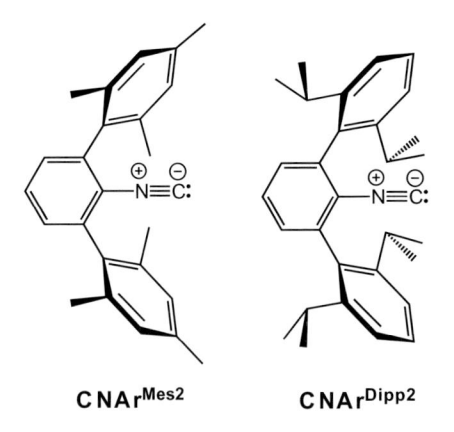

CoBr$_2$ $\xrightarrow[\text{−2 KBr}]{\text{1) 3 K}^+\text{[2.2.2]cryptand or 18-crown-6}}$ [Co(η^4-anthracene)$_2$]$^-$ $\xrightarrow[\text{−2 C}_{14}\text{H}_{10}]{\text{4 CNXyl}}$ [Co(CNXyl)$_4$]$^-$

CNAr^{Mes2}   **CNAr^{Dipp2}**

Figure 14.4 Bulky *m*-terphenyl isocyanides CNAr^{Mes2} and CNAr^{Dipp2}, as synthesized by Figureoa et al. [35].

Scheme 14.4

CoCl$_2$ + 4 CNAr^{Mes2} $\xrightarrow[\text{−2 NaCl}]{\text{ex. Na/Hg}}$ Na[Co(CNAr^{Mes2})$_4$] $\xrightarrow[\text{− Na[OTf]}]{\text{[Cp}_2\text{Fe][OTf]}}$ Co(CNAr^{Mes2})$_4$

[Ph$_3$P=N=PPh$_3$][Co(CNAr^{Mes2})$_4$] $\xleftarrow[\text{− NaCl}]{\text{[PPN]Cl}}$ Na[Co(CNAr^{Mes2})$_4$] $\xrightarrow[\text{− Na[OTf]}]{\text{1) 2 [Cp}_2\text{Fe][OTf], toluene}\\\text{2) Na[BAr}^F_4\text{]}}$ [Co(CNAr^{Mes2})$_4$][BAr$^F_4$]

(Figure 14.5a) and [K(DME)][Co(CNXyl)$_4$] (Figure 14.2) have essentially identical average Co–C and C–NR bond distances, as well as very similar average C–N–C angles (Table 14.2). However, substantial variability in C–N–C bending occurs among the individual ligands in [K(DME)][Co(CNXyl)$_4$], due to a perturbation of the anion's geometry by close cation–anion interactions [33].

The similarity between the two aforementioned isocyanidecobaltates(-I) ends when their oxidation chemistry is considered. Indeed, oxidation of the [Co(CNXyl)$_4$]$^-$ anion produces the dimer Co$_2$(CNXyl)$_8$, while its monovalent homoleptic analogue is the five-coordinate cation [Co(CNXyl)$_5$]$^+$ [33]. However, one-electron oxidation of [Co(CNAr^{Mes2})$_4$]$^-$ affords zero-valent, mononuclear, paramagnetic Co(CNAr^{Mes2})$_4$ (Figure 14.5b; Scheme 14.4) [36]. This stable analogue of the non-isolable "Co(CO)$_4$" radical exhibits intriguing temperature-dependent changes in its X-ray structural and electron paramagnetic resonance (EPR) spectroscopic characteristics. Moreover, two-electron oxidation of [Co(CNAr^{Mes2})$_4$]$^-$ in a non-coordinating

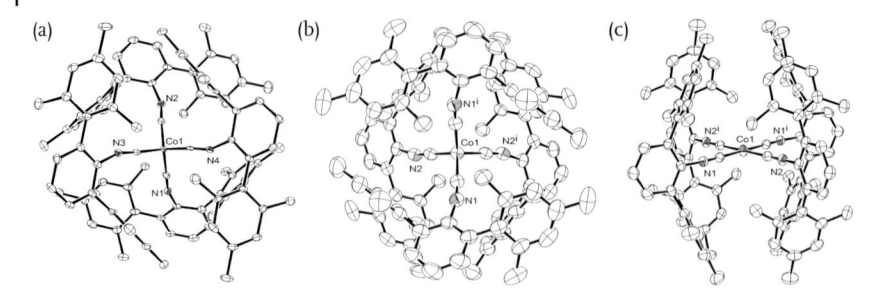

Figure 14.5 ORTEP diagrams. (a) The [Co(CNAr^{Mes2})$_4$]$^-$ anion in [Ph$_3$PNPPh$_3$][Co(CNAr^{Mes2})$_4$]; (b) Co(CNAr^{Mes2})$_4$; (c) The cation [Co(CNAr^{Mes2})$_4$]$^+$ in [Co(CNAr^{Mes2})$_4$][B([3,5-(CF$_3$)$_2$C$_6$H$_3$)$_4$] generated using the ORTEP-3 program from CIF data in Ref. [36]. The hydrogen atoms have been omitted for clarity.

Table 14.2 Selected IR spectroscopic and X-ray structural data for CNXyl, CNAr^{Mes2} and series of binary isocyanide complexes of Co, Fe, Mn, V, Nb, and Ta containing metal centers in various oxidation states.

Compound	ν_{NC} (cm^{-1})	M–C bond length (Å)	C–NR bond length (Å)	C–N–C angle (°)	Reference
CNXyl	2117	–	1.160 (3)	179.4 (2)	[40]
CNAr^{Mes2}	2120	–	1.158 (3)	178.1 (3)	[35a]
[Co(CNAr^{Mes2})$_4$]$^+$	2085	1.863 (0)	1.168 (0)	170.8 (3)	[36]
[Co(CNAr^{Mes2})$_4$]	1941	1.828 (2)	1.180 (4)	166 (1)	[36]
[Co(CNAr^{Mes2})$_4$]$^-$	1821	1.784 (5)	1.207 (8)	158 (3)	[36]
[Co(CNXyl)$_4$]$^-$	1815	1.79 (2)	1.20 (2)	156 (7)	[33]
[Fe(CNXyl)$_4$]$^{2-}$	1675	1.765 (3)	1.237 (7)	144 (3)	[38]
[Mn(CNXyl)$_5$]$^-$	1920, 1710	1.80 (5)	1.21 (2)	151 (8)	[11a]
[V(CN^{t}Bu)$_6$]$^{2+}$	2197	2.10 (1)	1.14 (2)	178 (1)	[8b]
[V(CNXyl)$_6$]$^+$	2033	2.07 (2)	1.169 (6)	173 (2)	[53]
[V(CNXyl)$_6$]	1939	2.026 (7)	1.186 (5)	163 (4)	[53]
[V(CNXyl)$_6$]$^-$	1823	1.98 (3)	1.20 (2)	158 (10)	[53]
[Nb(CNXyl)$_6$]$^-$	1820	2.141 (3)	1.191 (4)	157.0 (3)	[17]
[Ta(CNXyl)$_7$]$^+$	2029	2.15 (1)	1.17 (1)	173 (5)	[17]
[Ta(CNXyl)$_6$]$^-$	1824	2.127 (3)	1.199 (4)	155.7 (3)	[17]

Numbers in parentheses are the standard deviations from the mean.

solvent medium produces the 16-electron, diamagnetic, square-planar cation [Co(CNAr^{Mes2})$_4$]$^+$ (Figure 14.5c). This is in marked contrast to the related [Co(CO)$_4$]$^+$ cation that is detected in the gas phase, and which is believed to be C$_{2v}$-symmetric and paramagnetic [37]. Within the series [Co(CNAr^{Mes2})$_4$]^z, the average Co–C bond shortens, the average C–NR bond lengthens, and the average C–N–C angle becomes more bent upon reducing the charge z from +1 to −1, due to the increase in the extent of backbonding (Table 14.2).

Soon after describing [Co(CNXyl)$_4$]$^-$, Cooper et al. presented convincing, albeit circumstantial, evidence that supported the possibility of accessing binary isocyanidemetalate dianions [11e]. A four-electron reduction of Ru(CNR)$_4$Cl$_2$ (R = Xyl, ^{t}Bu) with potassium naphthalenide at −78 °C in tetrahydrofuran (THF) produced highly thermally unstable species that were formulated as isocyanideruthenates (-II), [Ru(CNR)$_4$]$^{2-}$, based on their oxidative addition reactivity with several electrophiles, including Ph$_3$SnCl [11e]. Some 15 years later, Ellis and Brennessel reduced an *in-situ*-generated Fe(CNXyl)$_4$Br$_2$ by following the above reaction protocol to isolate the complex K$_2$[Fe(CNXyl)$_4$], which afforded *trans*-[Fe(CNXyl)$_4$(SnPh$_3$)$_2$] upon treatment with Ph$_3$SnCl (Scheme 14.5; Figure 14.6) [38]. *Trans*-[Fe(CN^{t}Bu)$_4$(SnPh$_3$)$_2$] was also prepared from the non-isolable intermediate [Fe(CN^{t}Bu)$_4$]$^{2-}$. The dianion [Fe(CNXyl)$_4$]$^{2-}$ was characterized crystallographically as a salt with two [K[2.2.2.]cryptand]$^+$ cations (Figure 14.6a), and remains the sole isolable isocyanidemetalate(-II) known to date. It constitutes the isocyanide analogue of the oldest carbonylmetalate [Fe(CO)$_4$]$^{2-}$ (often referred to as "Collman's reagent") that initially was prepared by Hieber and Leutert in 1931 [39].

The CNXyl ligands in [Fe(CNXyl)$_4$]$^{2-}$ feature the most bent average C–N–C angle (144(3)°), the longest average isocyanide C–N bond distance (1.237 Å), and the lowest energy ν_{NC} IR signature (1675 cm^{-1} in THF) documented for any terminal isocyanide ligand in a binary metal complex (Table 14.2). For "free" CNXyl, the corresponding parameters are: <(C–N–C) = 179.4°, d(C≡NXyl) = 1.160(3) Å, and ν_{NC} = 2117 cm^{-1} (in THF) [40]. Also interesting is the fact that the average Fe–C

$$\text{FeBr}_2 \xrightarrow[-2\text{ KBr, } -4\text{ C}_{10}\text{H}_8]{\begin{array}{l}1)\ 4\text{ CNXyl} \\ 2)\ 4\text{ K}^+[\text{cryptand}]^-\end{array}} \text{K}_2[\text{Fe(CNXyl)}_4] \xrightarrow[-2\text{ KCl}]{2\text{ Ph}_3\text{SnCl}} \textit{trans}\text{-[Fe(CNXyl)}_4(\text{SnPh}_3)_2]$$

Scheme 14.5

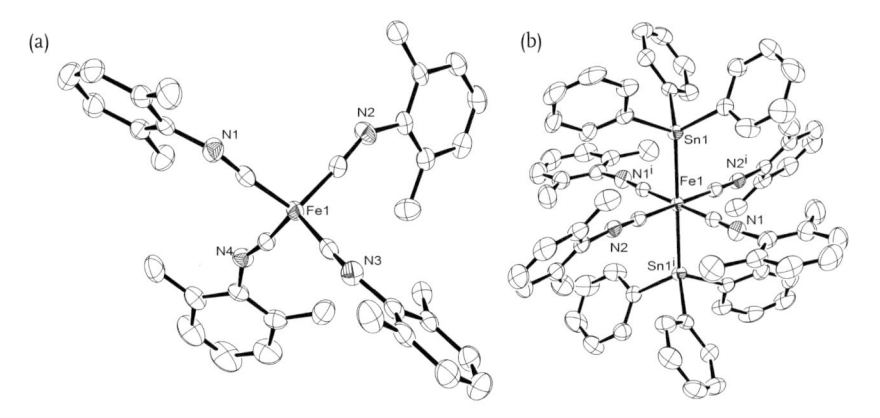

Figure 14.6 ORTEP diagrams. (a) The [Fe(CNXyl)$_4$]$^{2-}$ dianion in [K([2.2.2]cryptand)]$_2$[Fe(CNXyl)$_4$]; (b) *Trans*-[Fe(CNXyl)$_4$(SnPh$_3$)$_2$] generated using the ORTEP-3 program from CIF data in Ref. [38]. The hydrogen atoms have been omitted for clarity.

bond distance in [Fe(CNXyl)$_4$]$^{2-}$ is only 0.02 Å longer than that in [Fe(CO)$_4$]$^{2-}$ [41]. Thus, the isocyanide ligands in [Fe(CNXyl)$_4$]$^{2-}$ behave as very powerful π-acceptors and experience an extremely large extent of backbonding, as may be represented by the resonance contribution "Fe=C=N–Xyl."

14.2.3
Five-Coordinate Isocyanidemetalates

In contrast to a variety of five-coordinate carbonylmetalates [M(CO)$_5$]$^-$, [M(CO)$_5$]$^{2-}$, and [M(CO)$_5$]$^{3-}$ that contain Group 7, 6, or 5 metal ions, respectively [6], only one homoleptic isocyanidemetalate, namely [Mn(CNXyl)$_5$]$^-$, has been reported to date [11a]. As summarized in Scheme 14.6, Cooper and coworkers accessed the quite thermally sensitive K[Mn(CNXyl)$_5$] by employing a potassium naphthalenide reduction of ClMn(CNXyl)$_5$, and initially established its nature as a binary isocyanide complex of Mn(-I) through its reaction with Ph$_3$SnCl. The successful isolation of the crystallographically characterized salt [K(18-crown-6)(DME)][Mn(CNXyl)$_5$] must be credited to the remarkable technical skills of the research group, as samples had to be maintained at temperatures below −30 °C (including any manipulations) during the reaction work-up. The anion [Mn(CNXyl)$_5$]$^-$ features a trigonal bipyramidal (tbp) geometry (Figure 14.7), with the axial CNXyl ligands being on average 16° less bent than the CNXyl ligands in the equatorial positions. The latter fact is consistent with the expectation that π-acceptor ligands in the equatorial sites of a tbp complex should engage in more effective backbonding than those in the axial sites.

Despite the straightforward reactivity of [Mn(CNXyl)$_5$]$^-$ toward Ph$_3$SnCl, its interactions with MeI, EtI, and MeOSO$_2$CF$_3$ proved to be quite different from what might had been expected based on the known reactions of [Mn(CO)$_5$]$^-$ with such organic electrophiles. The Mn(I) alkylation products were isolated in low to mediocre yields, and featured new bidentate ligands formed via double isocyanide/alkyl insertion [42].

Very recently, Figueroa et al. discovered the golden mean of the robust reactivity characteristics of [Mn(CO)$_5$]$^-$ and the steric/electronic tunability at the metal center that can be provided by isocyanide ligands. The treatment of BrMn(CO)$_5$ with the bulky m-terphenyl isocyanides CNAr^{Mes2} or CNAr^{Dipp2}, as shown in Figure 14.4, followed by a two-electron reduction of the corresponding Mn(I) substitution products gave the mixed carbonyl/isocyanidemanganates(-I) [Mn(CO)$_2$(CNAr^{Mes2})$_3$]$^-$ or [Mn(CO)$_3$(CNAr^{Dipp2})$_2$]$^-$ in 20% and 47% yields, respectively (Scheme 14.7) [43].

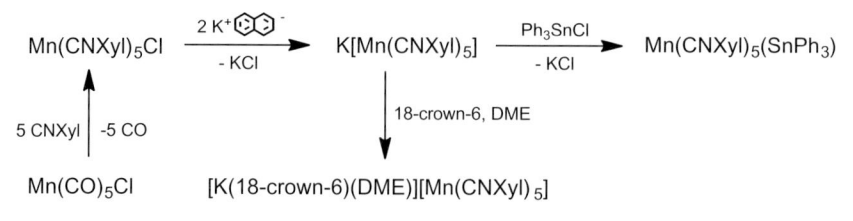

Mn(CNXyl)$_5$Cl $\xrightarrow[\text{− KCl}]{2\text{ K}^+\text{○○}^-}$ K[Mn(CNXyl)$_5$] $\xrightarrow[\text{− KCl}]{\text{Ph}_3\text{SnCl}}$ Mn(CNXyl)$_5$(SnPh$_3$)

5 CNXyl ↑ −5 CO

18-crown-6, DME ↓

Mn(CO)$_5$Cl   [K(18-crown-6)(DME)][Mn(CNXyl)$_5$]

Scheme 14.6

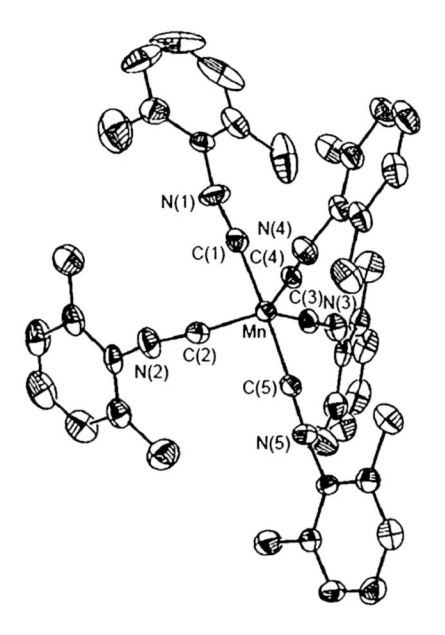

Figure 14.7 ORTEP diagram for the anion in [K(DME)][Mn(CNXyl)$_5$]. The hydrogen atoms have been omitted for clarity. Adapted from Ref. [11a] with permission from The Royal Society of Chemistry.

Scheme 14.7

$$3\ CNAr^{Mes2},\ -3\ CO:\ BrMn(CO)_5 \rightarrow BrMn(CO)_2(CNAr^{Mes2})_3 \xrightarrow[-KBr]{1)\ 2\ K^+\text{(18-crown-6)}^-;\ 2)\ 18\text{-crown-6}} [K(18\text{-crown-6})][Mn(CO)_2(CNAr^{Mes2})_3]$$

$$2\ CNAr^{Dipp2},\ -2\ CO:\ BrMn(CO)_5 \rightarrow BrMn(CO)_3(CNAr^{Dipp2})_2 \xrightarrow[-NaBr]{ex.\ Na/Hg} Na[Mn(CO)_3(CNAr^{Dipp2})_2]$$

Given that CO is certainly a stronger π-acid than the CNAr^{Mes2} ligand, the axial placement of both carbonyl ligands in [Mn(CO)$_2$(CNAr^{Mes2})$_3$]$^-$ (Figure 14.8a) must be forced by the steric requirements of the three bulky Ar^{Mes2} substituents in the complex. In contrast, the three CO ligands in the tbp anion [Mn(CO)$_3$(CNAr^{Dipp2})$_2$]$^-$ occupy equatorial sites (Figure 14.8b), which is favorable both sterically and electronically [44].

The treatment of [Mn(CO)$_3$(CNAr^{Dipp2})$_2$]$^-$ with HCl and MeI cleanly afforded mer,trans-HMn(CO)$_3$(CNAr^{Dipp2})$_2$ and mer,trans-MeMn(CO)$_3$(CNAr^{Dipp2})$_2$, respectively. This reactivity profile is similar to that of [Mn(CO)$_5$]$^-$, but not of [Mn(CNXyl)$_5$]$^-$ [43]. The reactions of [Mn(CO)$_3$(CNAr^{Dipp2})$_2$]$^-$ with MeSiCl$_3$ and SnCl$_2$ produced mer,trans-(Cl$_2${Me}Si)Mn(CO)$_3$(CNAr^{Dipp2})$_2$ and mer,trans-ClSnMn(CO)$_3$(CNAr^{Dipp2})$_2$, respectively. The latter metallostannylene complex is particularly intriguing, as its Sn center does not carry a bulky substituent [43].

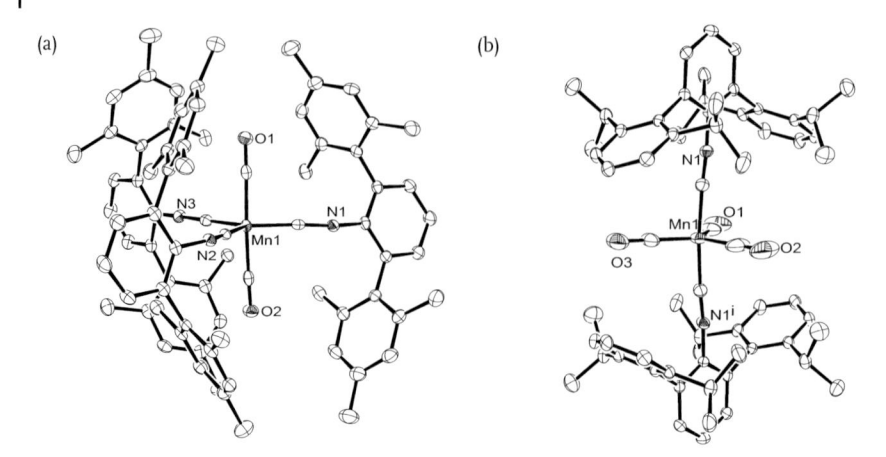

Figure 14.8 ORTEP diagrams. (a) The [Mn(CNAr^{Mes2})$_3$(CO)$_2$]$^-$ anion in [K(18-crown-6)(DME)][Mn(CO)$_2$(CNAr^{Dipp2})$_3$]; (b) The [Mn(CO)$_3$(CNAr^{Dipp2})$_2$]$^-$ anion in [K(NCMe)$_2$][Mn(CO)$_3$(CNAr^{Dipp2})$_2$] generated using the ORTEP-3 program from CIF data in Ref. [43]. The hydrogen atoms have been omitted for clarity.

14.2.4
Six-Coordinate Isocyanidemetalates and Redox-Related Complexes

Binary hexacoordination in mononuclear isocyanidemetalate complexes with an 18-electron count at the metal center can, in principle, be accommodated in two scenarios: [M(CNR)$_6$]$^-$ (M = V, Nb, Ta) or [M(CNR)$_6$]$^{2-}$ (M = Ti, Zr, Hf). These would be analogous to the well-established Group 5 monoanionic [M(CO)$_6$]$^-$ (M = V, Nb, Ta) or Group 4 dianionic [M(CO)$_6$]$^{2-}$ (M = Ti, Zr, Hf) species, respectively [6]. No Group 4 transition metal complexes with the metal center having a formally negative oxidation number and containing at least one isocyanide ligand are currently known, although a few higher-valent Group 4 metal complexes featuring up to four discrete isocyanide ligands have been described [10]. Earlier examples of isolable, isocyanide-containing Group 5 metalates included [V(CO)$_5$(CNR)]$^-$ (R = Me, C$_6$H$_{11}$, ^{t}Bu, ^{n}Bu, Ph) [45], [M(CO)$_5$(CN^{t}Bu)]$^-$ (M = Nb, Ta) [19, 46], and [M(CO)(CNMe)(dmpe)$_2$]$^-$ (M = Nb, Ta) [47], all of which contain only one isocyanide ligand. More recently, the synthesis of the anion [Ta(CO)$_5$(CNXyl)]$^-$ has also been reported (Scheme 14.8) [17b]. In addition, the complexes cis-[V(NO)$_2$(CN^{t}Bu)$_4$]$^+$ [48], CpV(NO)$_2$(CN^{t}Bu) [49] and VX(NO)$_2$(CNR)$_3$ (R = ^{t}Bu, X = Cl; R = ^{i}Pr, X = Br, Cl; R = C$_6$H$_{11}$, X = Br, I) [50] that have vanadium centers in −1 formal oxidation state were reported during the 1980s.

The scarcity of Group 4 and 5 metal isocyanides may be related to the fact that early transition metal complexes usually exhibit relatively high coordination numbers, and can induce a reductive C–C coupling of isocyanide ligands [29, 51]. While almost all reported cases of such reductive coupling reactions involve alkyl isocyanides, at least one example of reductively coupled aryl isocyanide ligands (CNXyl) has been documented (Figure 14.9).

14.2 Isocyanidemetalates and Related Low-Valent Complexes

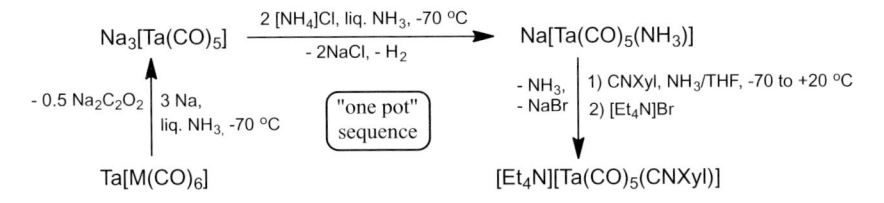

$$Na_3[Ta(CO)_5] \xrightarrow[-2NaCl, -H_2]{2\,[NH_4]Cl,\ liq.\ NH_3,\ -70\,°C} Na[Ta(CO)_5(NH_3)]$$

- 0.5 Na$_2$C$_2$O$_2$ | 3 Na, liq. NH$_3$, -70 °C

"one pot" sequence

- NH$_3$, - NaBr | 1) CNXyl, NH$_3$/THF, -70 to +20 °C 2) [Et$_4$N]Br

Ta[M(CO)$_6$]

[Et$_4$N][Ta(CO)$_5$(CNXyl)]

Scheme 14.8

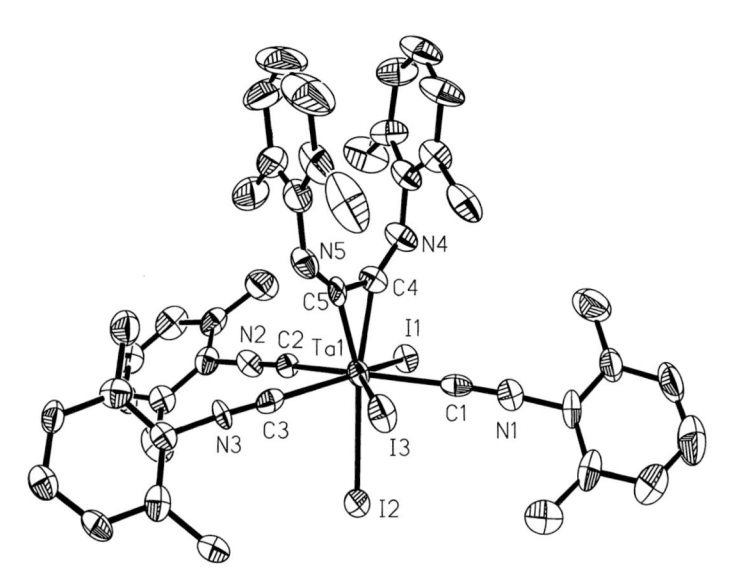

Figure 14.9 ORTEP diagram for Ta(CNXyl)$_3$(Xyl(H)NC≡CN(H)Xyl)I$_3$. The hydrogen atoms have been omitted for clarity. Adapted with permission from Ref. [52].

At the turn of the 20th century, Ellis and Barybin isolated the first homoleptic isocyanide complex of vanadium(0), V(CNXyl)$_6$ (Figure 14.10a), by substituting both naphthalene rings in labile V(η^6-C$_{10}$H$_7$R)$_2$ (R = H, Me) (Scheme 14.9) [53]. This thermally stable isocyanide analogue of very sensitive V(CO)$_6$ [54] undergoes either one-electron reduction to form binary isocyanovanadate(-I) [V(CNXyl)$_6$]$^-$ (Figure 14.10b) or one-electron oxidation to afford [V(CNXyl)$_6$]$^+$ (Figure 14.10c). The homoleptic series [V(CNXyl)$_6$]$^{1+,0,1-}$ can also be accessed from V(CO)$_6$, as summarized in Scheme 14.10 [53b]. Unlike Co(CNAr^{Mes2})$_4$ [36], the paramagnetic, 17-electron species V(CNXyl)$_6$ is EPR-silent, but exhibits sharp resonances in its ^{1}H and ^{13}C NMR spectra. The paramagnetic, low-spin d^4 complex [V(CNXyl)$_6$]$^+$ also produces sharp ^{1}H and ^{13}C NMR resonances [53b]. The ^{1}H paramagnetic shifts for both V(CNXyl)$_6$ and [V(CNXyl)$_6$]$^+$ are essentially "contact" in origin, and indicate an unpaired spin delocalization into the π-systems of the 2,6-xylyl substituents.

The energies of the ν_{NC} bands in the IR spectra of [V(CNXyl)$_6$]^z (z = +1, 0, −1) are well below that for the free CNXyl ligand, which indicates a substantial extent of backbonding, even in the cation [V(CNXyl)$_6$]$^+$ (Table 14.2). The trends in V–C

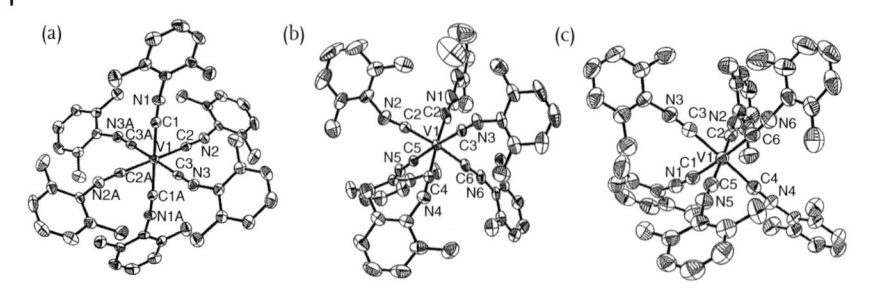

Figure 14.10 ORTEP diagrams. (a) V(CNXyl)$_6$; (b) The [V(CNXyl)$_6$]$^-$ anion in Cs[V(CNXyl)$_6$]; (c) The [V(CNXyl)$_6$]$^+$ cation in [V(CNXyl)$_6$][PF$_6$]. The hydrogen atoms have been omitted for clarity. Adapted with permission from Ref. [53b]; © 2000, American Chemical Society.

Scheme 14.9

Scheme 14.10

and C–NXyl bond distances, C–N–C angles, as well as the ν_{NC} energies for [V(CNXyl)$_6$]^z (z = +1, 0, −1) and [V(CN^{t}Bu)]$^{2+}$, are apparent from the corresponding data in Table 14.2, and can be explained by increasing the extent of V(dπ)→CNR(π^*) backbonding upon decreasing the oxidation number of the metal ion.

The niobium and tantalum congeners of the isocyanidevanadate(-I) [V(CNXyl)$_6$]$^-$ were accessed by Ellis, Barybin and coworkers via the sequential treatment of solutions of [M(CO)$_6$]$^-$ (M = Nb, Ta) with I$_2$ and CNXyl, followed by a two-electron reduction of the resultant seven-coordinate complex M(CNXyl)$_6$I with excess CsC$_8$ (Scheme 14.11) to yield Cs[M(CNXyl)$_6$] (Figure 14.11a) [17]. Oxidative decarbonylation of [M(CO)$_6$]$^-$ with 2 equiv. of Ag[BF$_4$] in the presence of CNXyl cleanly afforded the homoleptic cationic complexes [M(CNXyl)$_7$]$^+$ (Scheme 14.11). Unlike paramagnetic six-coordinate [V(CNXyl)$_6$]$^+$, the Nb and Ta cations [M(CNXyl)$_7$]$^+$ are diamagnetic and contain seven discrete CNXyl ligands (Figure 14.11b). Interestingly, the CNXyl ligands support the direct nitrosylation of [M(CO)$_6$]$^-$ (M = Nb, Ta) with 2 equiv. of [NO][BF$_4$] to produce *cis*-[M(NO)$_2$(CNXyl)$_4$]$^+$ (Scheme 14.11) [17]. Even though *cis*-[M(NO)$_2$(CNXyl)$_4$]$^+$ formally contain Nb(-I) and Ta(-I) centers, the CNXyl ligands in these complexes behave primarily as σ-donors to accommodate

14.2 Isocyanidemetalates and Related Low-Valent Complexes

$$[Et_4N][M(CO)_6] \xrightarrow[- 6 CO, - [Et_4N][BF_4]]{4\ CNXyl,\ 2\ [NO][BF_4]} cis\text{-}[M(NO)_2(CNXyl)_4][BF_4]_2$$
M = Nb, Ta

$$[Et_4N][M(CO)_6] \xrightarrow[- 6 CO, - [Et_4N][BF_4]]{7\ CNXyl,\ 2\ Ag[BF_4]} [M(CNXyl)_7][BF_4]$$
M = Nb, Ta   M = Nb, Ta

- [Et₄N]I 1) I₂
- 6 CO   2) 6 CNXyl

$$\xrightarrow{[Bu_4N]I,\ -CNXyl,\ -[Bu_4N][BF_4]}$$

MI(CNXyl)₆ $\xrightarrow{ex.\ CsC_8}$ Cs[M(CNXyl)₆]
M = Nb, Ta   M = Nb, Ta

1) ex. KC₈
2) RSO₂N(NO)Me
$\xrightarrow{- KI,\ - K[RSO_2NMe]}$ M(NO)(CNXyl)₅
R = p-tolyl   M = Ta

Scheme 14.11

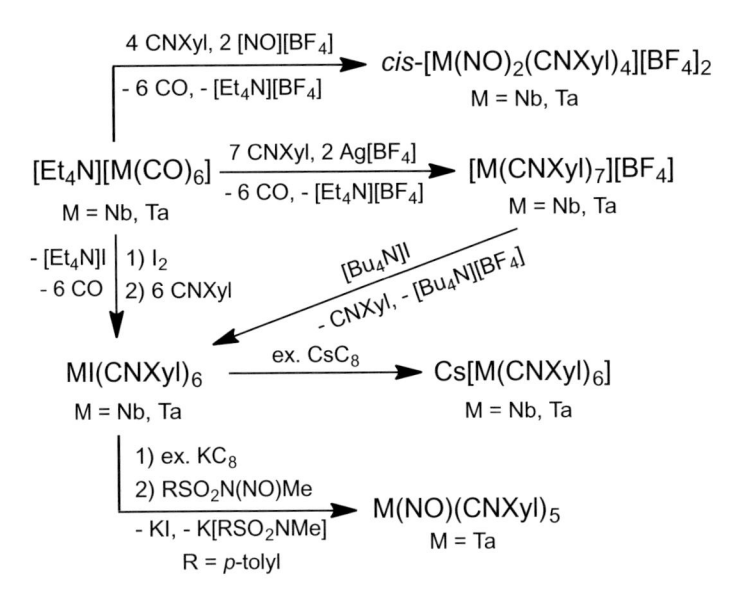

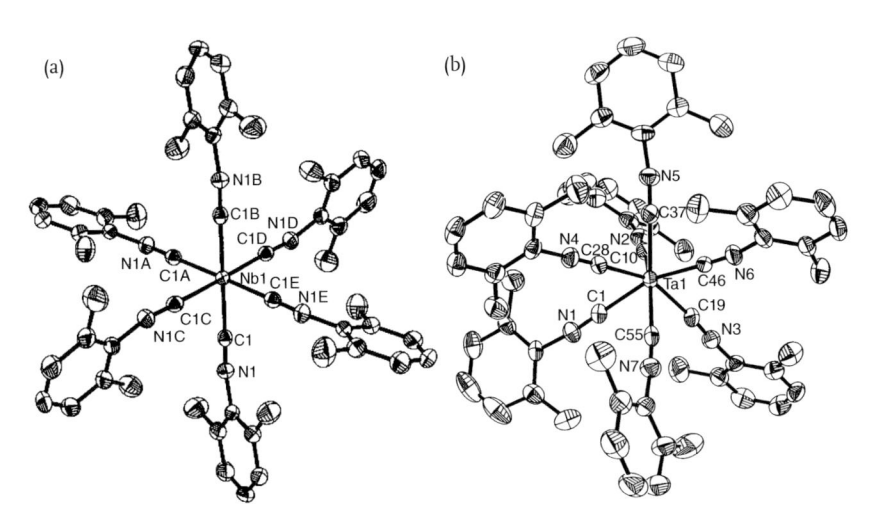

Figure 14.11 ORTEP diagrams. (a) The [Nb(CNXyl)₆]⁻ anion in Cs[Nb(CNXyl)₆]; (b) The [Ta(CNXyl)₇]⁺ cation in [Ta(CNXyl)₇][BF₄]. The hydrogen atoms have been omitted for clarity. Adapted with permission from Ref. [17b]; © 2007, American Chemical Society.

two linear NO ligands, which are extremely powerful π-acceptors. The treatment of Ta(CNXyl)₆I with Diazald®, a mild source of NO⁺, afforded the mono-nitrosyl complex Ta(CNXyl)₆(NO) (Scheme 14.11) [17].

While isocyanidemetalate(-I) analogues of long-established [M(CO)₆]⁻ (M = V, Nb, Ta) [6] have finally been discovered, the existence of thermally stable

[V(CNXyl)$_6$]$^+$, [M(CNXyl)$_7$]$^+$ (M = Nb, Ta), M(CNXyl)$_6$I (M = Nb, Ta), cis-[M(NO)$_2$(CNXyl)$_4$]$^+$ (M = Mb, Ta), cis-[V(NO)$_2$(CN^{t}Bu)$_4$]$^+$ and Ta(NO)(CNXyl)$_5$, for which no carbonyl equivalents can be obtained, clearly underscores the remarkable electronic versatility of 2,6-xylylisocyanide compared to CO as a ligand.

14.3
Coordination and Surface Chemistry of Nonbenzenoid Isocyanoarenes

14.3.1
Isocyanoazulenes

With the exception of isocyanoferrocene, all of the isocyanoarenes that were known up until the early 2000s possessed benzenoid aryl substituents [21]. Recently, however, a new class of isocyanoarenes has been established incorporating the nonbenzenoid aromatic system of azulene [21]. The latter is an unusual hydrocarbon which is composed of fused five- and seven-membered rings (Figure 14.12). The isocyanoazulene compounds, together with their corresponding abbreviations, that have been reported by the present authors' group to date are illustrated in Figure 14.13 [55–59]. Given the polar nature of the azulenic scaffold (Figure 14.12), these species can be considered as stable derivatives of the hypothetical isocyanocyclopentadienide anion or isocyanotropylium cation [21]. The isocyanoazulenes shown in Figure 14.13 are all crystalline substances that exhibit a good thermal stability (except for CN^{6}Az and CN^{2}Az^{CN}), and can for practical purposes be considered air-stable.

As shown in Figure 14.14a, the frontier molecular orbitals of azulene have complementary density distributions; consequently, the general reactivity profile of the azulenic scaffold is governed by the rules summarized in Figure 14.14b. Notably, the preparation of 2-substituted azulenes almost always requires an appropriate substituent to be in place before the azulenic ring is closed; this requirement is due to the quite unreactive nature of the C–H bond at the 2-position of the azulenic framework. The "parent" isocyanoazulenes CN^{1}Az, CN^{4}Az, CN^{6}Az, as well as the di-*t*-butyl derivative CN^{5}Az^{tBu}, can all be accessed regioselectively from azulene itself [21, 56].

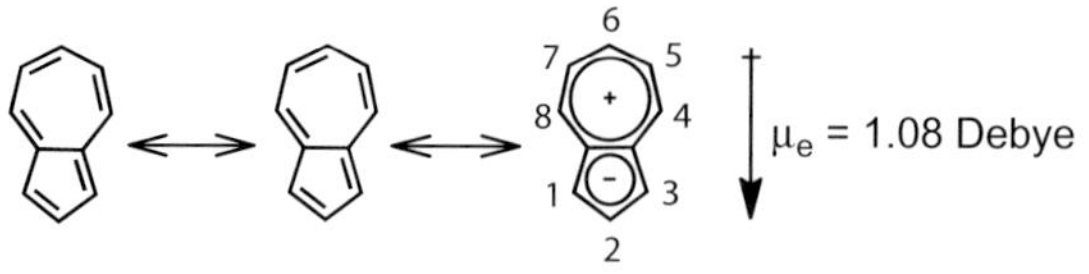

μ_e = 1.08 Debye

Figure 14.12 The resonance forms of azulene, emphasizing the polar nature and numbering scheme of the azulenic scaffold. Adapted with permission from Ref. [21].

14.3 Coordination and Surface Chemistry of Nonbenzenoid Isocyanoarenes

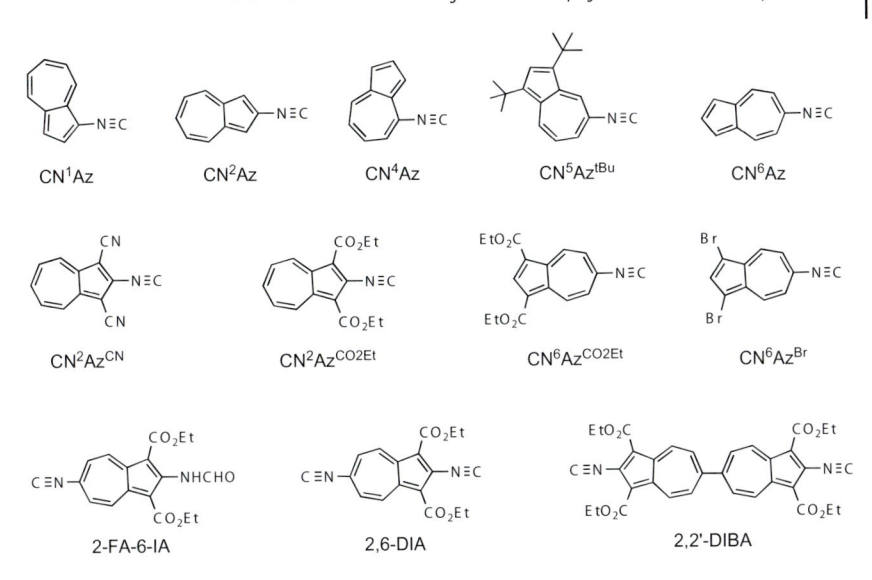

Figure 14.13 Structures and the corresponding abbreviations of the isocyanoazulene derivatives known to date.

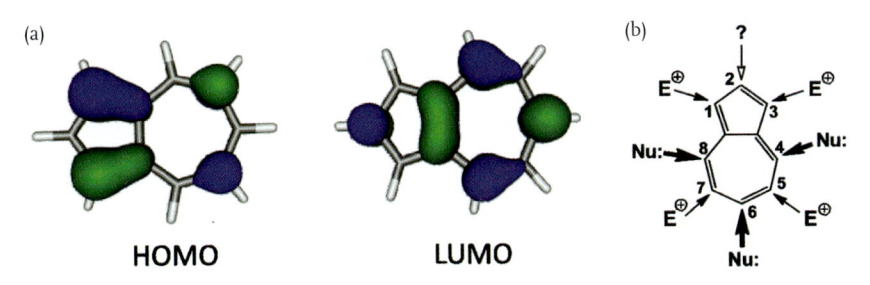

Figure 14.14 (a) Lowest unoccupied molecular orbital (LUMO) and highest occupied molecular orbital (HOMO) of azulene; (b) Sites of nucleophilic (Nu:) and electrophilic (E$^+$) attacks of the azulenic scaffold as general strategies for its functionalization. Adapted with permission from Ref. [21].

14.3.2
Organometallic η^5-Isocyanocyclopentadienides

The currently known transition metal-stabilized η^5-isocyanocyclopentadienide ligands, along with their corresponding abbreviations used in this chapter, are shown in Figure 14.15. While the first reports on the synthesis of isocyanoferrocene and two of its complexes (CNFc)Cr(CO)$_5$ and cis-[(CNFc)$_2$Fe(CO)$_4$] appeared during the late 1980s, the chemistry of this redox-active, nonbenzenoid isocyanoarene ligand had remained idle until 2002 [60], presumably due to the rather tedious and inefficient procedures for the preparation of its precursor

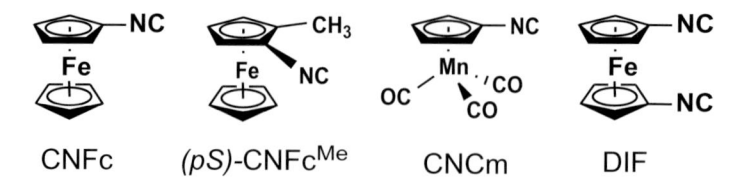

Figure 14.15 Structures and the corresponding abbreviations of the η^5-bound isocyanocyclopentadienide ligands reported to date.

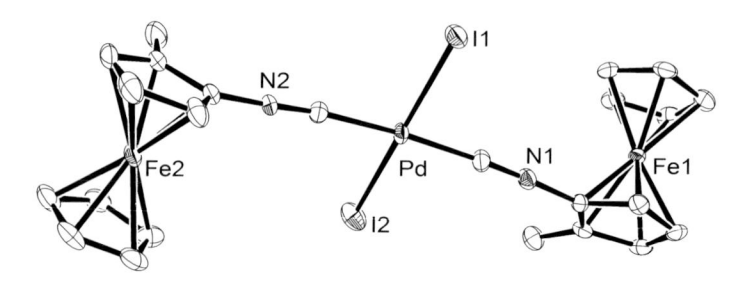

Figure 14.16 ORTEP diagram for trans-[PdI$_2${($_p$S)-CNFc^{Me}}$_2$] generated using the ORTEP-3 program from CIF data in Ref. [65]. The hydrogen atoms have been omitted for clarity.

H$_2$NFc available at the time [21]. Indeed, depending on the synthetic route chosen, CNFc could be obtained in only 0.6–5% overall yields, starting from ferrocene [21].

Following reports of greatly improved syntheses of aminoferrocene at the turn of the 20th century [61–63], the present authors' group developed a highly reproducible procedure for the preparation of isocyanoferrocene that would afford CNFc in a 38% overall yield, based on the starting ferrocene [60]. Previously, an unknown isocyanocymantrene, CNCm, was synthesized from commercial cymantrene, CpMn(CO)$_3$, in a 45% overall yield through a convenient three-step reaction sequence [64]. In addition, a planar–chiral version of isocyanoferrocene, namely ($_p$S)-1-isocyano-2-methylferrocene, was recently isolated in an enantiomerically pure form and characterized crystallographically as an adduct with PdI$_2$ (Figure 14.16) [65]. Finally, an efficient synthesis of 1,1'-diisocyanoferrocene (DIF) from 1,1'-ferrocenedicarboxylic acid was reported by Hessen and van Leusen in 2001 [63].

14.3.3
Homoleptic Complexes of Nonbenzenoid Isocyanoarenes

To date, the coordination chemistry of azulene and its derivatives has been dominated by multi-hapto interactions of the azulenic scaffold with transition metal centers that disrupt the aromaticity of the azulenic nucleus [21, 66]. Interest in the family of isocyanoazulenes originated from the quest for metal–organic systems in which the azulenic core was electronically coupled to an electron-rich metal atom or ion while retaining its aromatic character. Because of the cylindrical sym-

14.3 Coordination and Surface Chemistry of Nonbenzenoid Isocyanoarenes

metry of its π-system, the isocyanide group appeared to be an attractive choice as a π-conducting junction. The treatment of $Cr(\eta^6\text{-naphthalene})_2$, a storable source of atomic chromium, with 6 equiv. of various isocyanoazulenes, including CN^1Az, CN^2Az, CN^4Az, CN^5Az^{tBu}, and C^6NAz, afforded the corresponding binary complexes $Cr(CNR)_6$ that featured six discrete isocyanoazulene ligands. The monocationic versions thereof were prepared via oxidation of the neutral complexes with, for example, AgX (X = BF_4, SbF_6).

The isomeric, low-spin d⁵ complexes $[Cr(CN^xAz)_6]^+$ (x = 1, 2, 4, or 6) (Figure 14.17) provide interesting ¹H and ¹³C NMR patterns that suggest unpaired electron spin delocalization from the Cr(I) center into the π*-systems of the azulenyl substituents by means of backbonding to the isocyanide junction [56]. For instance, Figure 14.18a displays the ¹H NMR spectrum of $[Cr(CN^1Az)_6]^+$ that features substantial paramagnetic shifts of the ¹H resonances with respect to its diamagnetic, low-spin d⁶ analogue $Cr(CN^1Az)_6$. These paramagnetic shifts are essentially

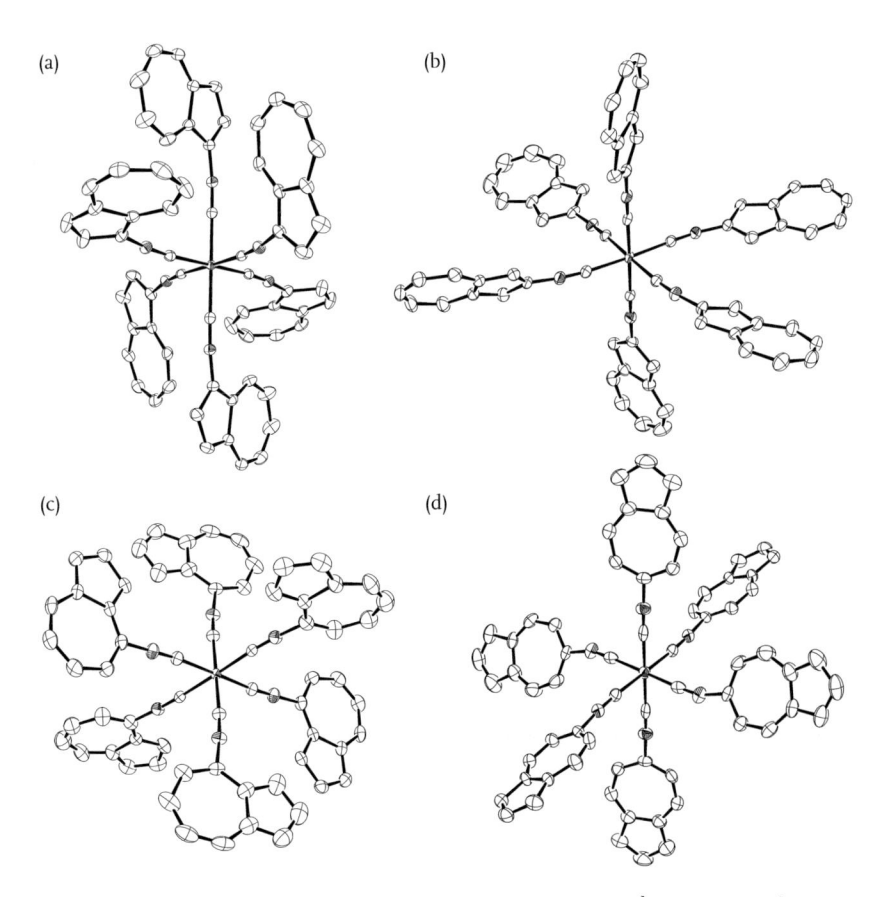

Figure 14.17 ORTEP diagrams for the isomeric cations in (a) $[Cr(CN^1Az)_6][V(CO)_6]$, (b) $[Cr(CN^2Az)_6][BF_4]$, (c) $[Cr(CN^4Az)_6][SbF_6]$, and (d) $[Cr(CN^6Az)_6][BF_4]$. The hydrogen atoms have been omitted for clarity. Adapted with permission from Ref. [21].

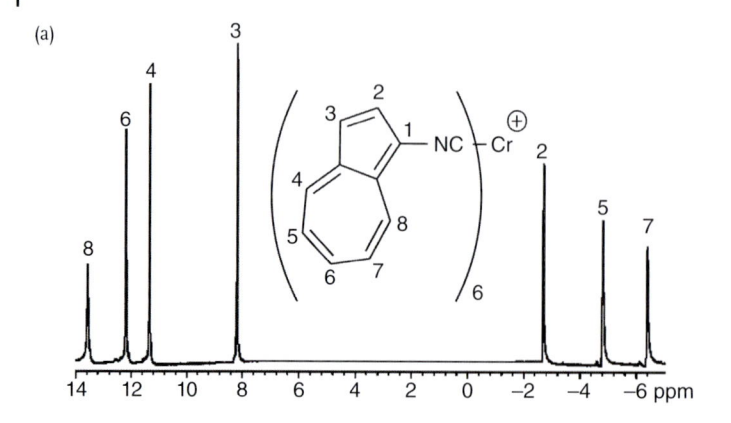

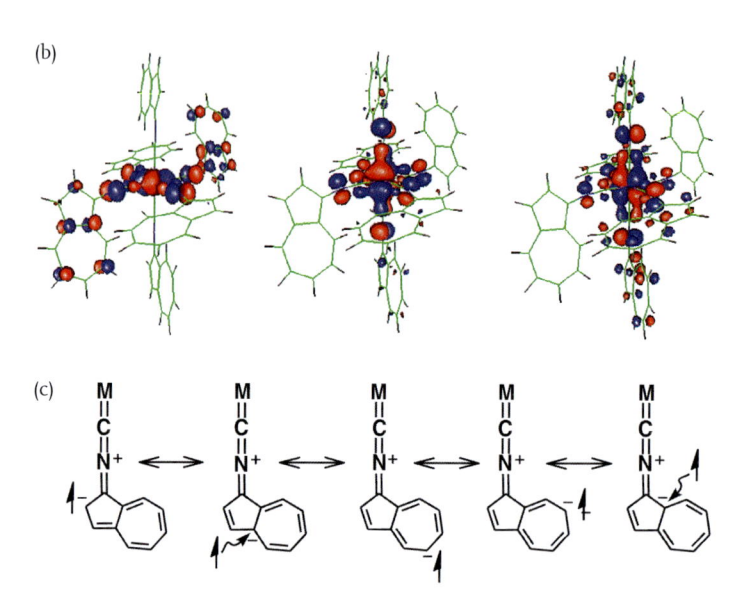

Figure 14.18 (a) ^{1}H NMR spectrum of [Cr(CN^{1}Az)$_6$][V(CO)$_6$] in CD$_2$Cl$_2$ at 25 °C; (b) The "t$_{2g}$"-like set of the highest occupied MOs for the low-spin d^5 complex [Cr(CN^{1}Az)$_6$]$^+$. Adapted with permission from Ref. [56]; © 2005, American Chemical Society; (c) Resonance description of unpaired spin delocalization in [Cr(CN^{1}Az)$_6$]$^+$ through Cr(dπ)→CN^{1}Az(pπ*) backbonding.

"contact" in origin because of the octahedral symmetry of the cation. The ^{1}H NMR peaks for the hydrogen atoms 2, 5, and 7 of the azulenic frameworks in [Cr(CN^{1}Az)$_6$]$^+$ are shifted upfield, which suggests the presence of unpaired electron spin density in the p-orbitals of the corresponding carbon atoms. This phenomenon is nicely illustrated by the density functional theory (DFT)-generated pictures of the "t$_{2g}$"-like molecular orbitals of [Cr(CN^{1}Az)$_6$]$^+$ (Figure 14.18b) that clearly show contributions from the p-orbitals of the carbon atoms 2, 5, and 7, as well as Cr(dπ)→CN^{1}Az(pπ*) backbonding interactions between the Cr(I)-center and the

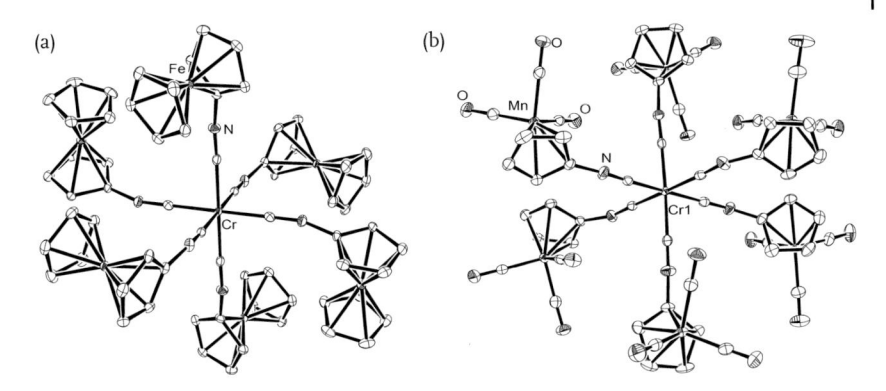

Figure 14.19 ORTEP diagrams for (a) Cr(CNFc)$_6$ and (b) Cr(CNCm)$_6$ generated using the ORTEP-3 program from CIF data in Ref. [64]. The hydrogen atoms have been omitted for clarity.

Table 14.3 Selected IR spectroscopic and X-ray structural data for binary isocyanide complexes of chromium that contain CNPh, CNFc, or CNCm ligands.

Complex	v_{NC} (cm^{-1})	Cr–C bond length (Å)	C–NR bond length (Å)	C–N–C angle (°)	Reference
[Cr(CNPh)$_6$]$^{3+}$	2205	2.07 (1)	1.139 (5)	172 (3)	[67]
[Cr(CNPh)$_6$]$^{2+}$	2139	2.014 (5)	1.158 (5)	177 (1)	[67]
[Cr(CNPh)$_6$]$^{+}$	2056	1.98 (1)	1.159 (5)	177 (1)	[67]
Cr(CNPh)$_6$	1950	1.938 (3)	1.176 (4)	173.7 (2)	[68]
[Cr(CNFc)$_6$]$^{2+}$	2131	2.019 (17)	1.150 (5)	175 (3)	[64]
[Cr(CNFc)$_6$]$^{+}$	2053	1.972 (13)	1.160 (3)	171 (4)	[64]
Cr(CNFc)$_6$	1971	1.937 (7)	1.178 (5)	162 (4)	[64]
Cr(CNCm)$_6$	1947	1.93 (2)	1.18 (1)	162 (14)	[64]

Numbers in parentheses are the standard deviations from the mean.

isocyanide junctions. The ^{1}H NMR pattern in Figure 14.18a can also be rationalized by considering a simple valence-bond description of backbonding in [Cr(CN^{1}Az)$_6$]$^+$, as illustrated in Figure 14.18c.

The air-stable homoleptic, heptanuclear complexes of CNFc and CNCm were isolated for chromium in three different oxidation states: [Cr(CNR)$_6$]$^{0,1+,2+}$ (Figure 14.19) [64]. Akin to the series [Cr(CNR)$_6$]$^{0,1+,2+,3+}$ [67, 68], [V(CNXyl)$_6$]$^{1-,0,1+}$ [53b], and [Co(CNAr^{Mes2})$_6$]$^{1-,0,1+}$ [36], for [Cr(CNFc)$_6$]$^{0,1+,2+}$ the average M–C bond elongates, the average C–NR distance decreases, and the average C–N–C angle widens upon successive oxidation of the metal center (Tables 14.2 and 14.3) [64].

It is instructive to consider the extensive data in Table 14.4, which contains the values of the half-wave potentials for the reversible redox couples [Cr(CNR$_6$)]$^{0/1+}$ and [Cr(CNR$_6$)]$^{1+/2+}$, and essentially constitutes a quantitative electrochemical

Table 14.4 Half-wave $E_{1/2}$ redox potentials (in volts versus Cp_2Fe/Cp_2Fe^+) for $[Cr(CNR)_6]^{z/z+1}$ couples in $CH_2Cl_2/[^nBu_4N][PF_6]$.

Couple	Substituent R										
	naphthyl	azulenyl	Cp(CO)₂Mn-C₅H₄	EtO₂C-azulenyl-CO₂Et	azulenyl	phenyl	xylyl	indenyl	Fc(η⁵)	azulenyl	alkyl
$[Cr(CNR)_6]^{0/1+}$	−0.44	−0.49	−0.53	−0.67	−0.69	−0.83	−0.87	−0.97	−0.98	−1.03	−1.54
$[Cr(CNR)_6]^{1+/2+}$	+0.02	−0.03	−0.04	−0.14	−0.16	−0.21	−0.29	−0.35	−0.42	−0.32	−0.77
Reference	[56]	[56]	[64]	[71]	[56]	[70]	[56]	[64]	[56]	[71]	[69]
	σ-donor/π-acceptor ratio increases →										

measure of the σ-donor/π-acceptor ratios of various isocyanide ligands. Based on the data shown in Table 14.4, it is evident that the electron richness of the metal center in zero-valent $Cr(CN^4Az)_6$ is practically the same as that in the monovalent $[Cr(CN^1Az)_6]^+$. According to the data in Table 14.4, the isocyanoazulenes CN^4Az and CN^6Az appear to have the lowest σ-donor/π-acceptor ratios among any known isocyanide with a purely hydrocarbon substituent [21].

Even though the electronic characteristics of isocyanoferrocene as a ligand have been thought to be comparable to those of CNMe, based on similarities of the electrochemical behavior of $(OC)_5Cr(CNFc)$ and $(OC)_5Cr(CNMe)$ [72, 73], and on the fact that the Hammett constants for the Fc and Me substituents are identical [74], the σ-donor/π-acceptor ratio of CNFc is clearly closer to that of CNPh rather than an alkyl isocyanide. The DFT analysis of isocyanoferrocene suggested that its LUMO and LUMO+2 are well-suited for delocalizing π-backbonded electron density beyond the isocyanide junction upon coordination to an electron-rich metal center [60, 64]. In addition, as the CNCm ligand is a substantially more powerful π-acceptor than CNFc, the σ-donor/π-acceptor behavior of an η⁵-isocyclopentadienide can indeed be tuned by varying the nature of the metal unit coordinated to its five-membered ring.

14.3.4
Bridging Nonbenzenoid Isocyanoarenes

In 2005, Chisholm, Barybin, Dalal and coworkers demonstrated through a series of UV-visible-near infrared, EPR, cyclic voltammetry, and DFT studies of the 2,6-azulene-dicarboxylate bridge linking M–M quadruply bonded units (M = Mo, W), that the π*-system of the 2,6-azulenic scaffold is exceptionally well-suited to supporting the charge delocalization between electron reservoirs [75]. Shortly thereafter, the first mononuclear and dinuclear complexes of 2,6-diisocyanoazulene (2,6-DIA) with $[M(CO)_5]$ (M = Cr, W) fragments were reported [57]. The M(dπ)→ 2,6-DIA(pπ*) charge transfer in these complexes occurs in the visible range due to the quite low energy of 2,6-DIA's LUMO.

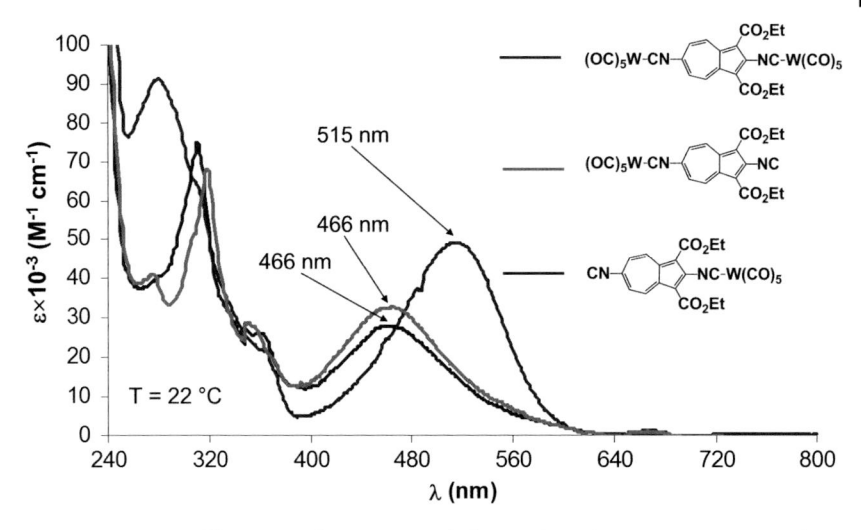

Figure 14.20 UV-visible spectra of [(OC)$_5$W]$_2$(μ-η^1:η^1-2,6-DIA) and two isomeric mononuclear complexes [(OC)$_5$W](η^1-2,6-DIA) in CH$_2$Cl$_2$. Adapted with permission from Ref. [57]; © 2006, American Chemical Society.

Interestingly, the energy of the M(dπ)→2,6-DIA(pπ*) charge transfer band observed for the mononuclear complexes (OC)$_5$M(η^1-2,6-DIA) (M = Cr, W) does not depend on which end of the asymmetric 2,6-DIA ligand is coordinated to the metal pentacarbonyl moiety (Figure 14.20).

In addition, the binucleation of mononuclear species (OC)$_5$M(η^1-2,6-DIA) to form [(OC)$_5$M]$_2$(μ-η^1:η^1-2,6-DIA) (M = Cr, W) is accompanied by a substantial redshift in the M(dπ)→2,6-DIA(pπ*) charge transfer band (Figure 14.20). This indicates that the entire π*-system of 2,6-DIA, and not just its isocyanide junctions, is involved in the M(dπ)→2,6-DIA(pπ*) charge transfer. In the case of the tungsten system, this red shift is much greater than that observed earlier by Bennett et al. [76] for the analogous 1,4-diisocyanobenzene-bridged system (Table 14.5).

While optimizing the synthesis of 2,6-DIA, it was noted that a dehydration of the 2-formamido terminus in 2,6-diformamido-1,3-diethoxycarbonylazulene proceeded significantly more slowly than that of the 6-formamido end, and this allowed for a preparation of the 2-formamido-6-isocyanoazulene derivative 2-FA-6-IA (Figure 14.13) in a highly regioselective manner (Scheme 14.12). This, in turn, permitted a regioselective heterobimetallic complexation of the 2,6-DIA linker via a stepwise installation and coordination of its isocyanide termini, as illustrated in Scheme 14.12. Switching the order of addition of the two M(CO)$_5$(THF) reactants in Scheme 14.12 afforded the isomeric binuclear product featuring interchanged Cr/W sites (Figure 14.21) [21, 57].

In 2010, the synthesis of the first linear 2,2'-diisocyano-6,6'-biazulenyl (2,2'-DIBA) linker (Figure 14.13) was accomplished via an unexpected "one-pot" homocoupling of 2-amino-6-bromo-1,3-diethoxycarbonylazulene, followed by standard

Table 14.5 Red-shifts in energy of the metal-to-ligand charge transfer (MLCT) that occur upon binucleation of 2,6-diisocyano-1,3-diethoxycarbonylazulene (2,6-DIA) or 1,4-diisocyanobenzene (DIB) ligands.

Binucleation process	ΔE_{MLCT} (cm^{-1})	Reference
$(OC)_5Cr(\eta^1\text{-}2,6\text{-DIA}) \rightarrow$ $[(OC)_5Cr]_2(\mu\text{-}\eta^1\text{:}\eta^1\text{-}2,6\text{-DIA})$	1650	[57]
$(OC)_5W(\eta^1\text{-}2,6\text{-DIA}) \rightarrow$ $[(OC)_5W]_2(\mu\text{-}\eta^1\text{:}\eta^1\text{-}2,6\text{-DIA})$	2042	[57]
$(OC)_5W(\eta^1\text{-DIB}) \rightarrow$ $[(OC)_5W]_2(\mu\text{-}\eta^1\text{:}\eta^1\text{-DIB})$	1221	[76]

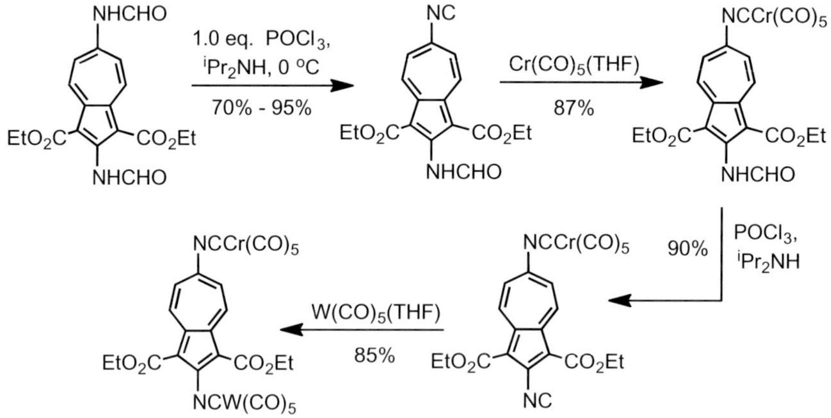

Scheme 14.12

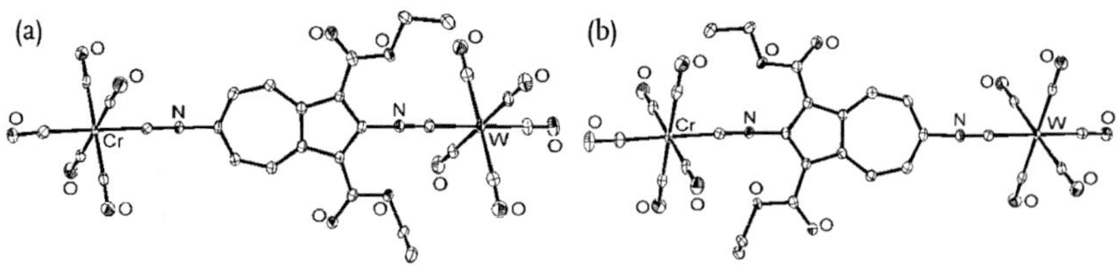

Figure 14.21 ORTEP diagrams for two isomeric heterobimetallic complexes [(OC)$_5$W] $(\mu\text{-}\eta^1\text{:}\eta^1\text{-}2,6\text{-DIA})[Cr(CO)_5]$. The hydrogen atoms have been omitted for clarity. Adapted with permission from Ref. [21].

formylation and dehydration procedures [58]. This molecule exhibits a reversible two-electron reduction to form a singlet dianion. The long axis of 2,2′-DIBA, which is the only structurally characterized biazulenyl derivative known, spans more than 17 Å (Figure 14.22). Similar to 2,6-DIA, 2,2′-DIBA readily coordinates to metal fragments, such as [W(CO)$_5$], via its isocyanide termini.

Figure 14.22 ORTEP diagram for 2,2'-DIBA generated using the ORTEP-3 program from CIF data in Ref. [58]. The hydrogen atoms have been omitted for clarity.

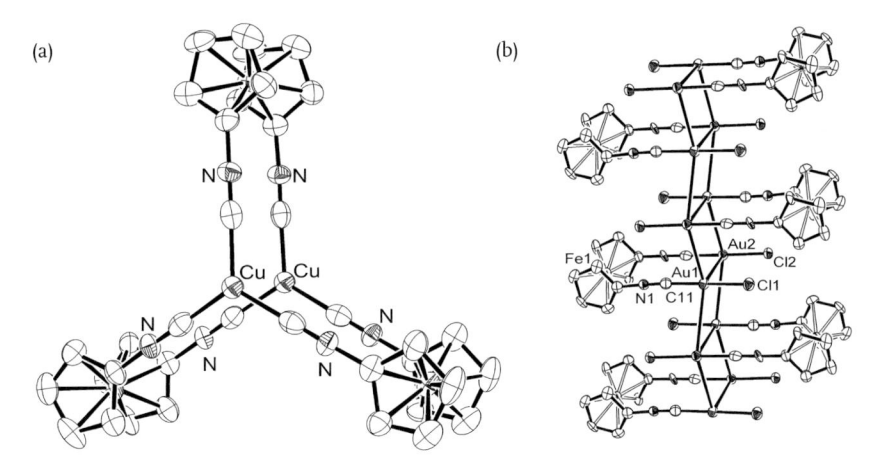

Figure 14.23 ORTEP diagrams for (a) [Cu$_2$(μ-DIF)$_3$]$^{2+}$ and (b) [(ClAu)$_2$(μ-DIF)]$_\infty$ generated using the ORTEP-3 program from CIF data in Refs [77, 80]. The hydrogen atoms have been omitted for clarity.

The coordination chemistry of DIF has recently been established by Siemeling and coworkers. The structurally characterized complexes of this redox-active ditopic isocyanide ligand included [(OC)$_5$Cr]$_2$(μ-DIF) [77, 78], [Ag$_2$(μ-DIF)]$^{2+}$ [79], paddle-wheel-like [Cu$_2$(μ-DIF)$_3$]$^{2+}$ (Figure 14.23a) [80], and [(ClAu)$_2$(μ-DIF)]$_\infty$ (Figure 14.23b) [77, 78]. The latter structure constitutes an intriguing model for understanding the self-assembly of DIF on gold surfaces [77, 78]. Adducts of DIF with a few Au(I) acetylides have also been reported [78, 81].

14.3.5
Self-Assembled Monolayer Films of Nonbenzenoid Isocyano- and Diisocyanoarenes on Gold Surfaces

The chemisorption of isocyanoarenes onto metal surfaces to yield the corresponding self-assembled monolayer (SAM) films provides valuable initial platforms for a variety of potential nanotechnological applications, including organic electronics [27, 82–88]. A recent review by Angelici and Lasar [23] provides a detailed account of the adsorption of organic isocyanides onto gold, silver, copper, platinum, palladium, nickel, rhodium, and chromium metal surfaces [23]. The adsorption of isocyanides onto various gold surfaces, including Au(111) films, gold powder, and gold nanoparticles, is of particular practical interest due to the oxidative stability of gold in ambient atmospheres. Indeed, a large body of experimental and theoretical studies have indicated that organic isocyanides adsorb onto gold in the terminal upright (η^1) fashion, with only the isocyanide carbon atom interacting with the surface [23]. Upon the adsorption of a benzenoid isocyanide molecule to a gold surface, the energy of the v_{NC} band is shifted to a higher energy, typically by 50–80 cm^{-1}, compared to that of free isocyanide (Table 14.6) [89]. This is a consequence of the isocyanide carbon's lone pair, which interacts with Au, being somewhat antibonding with respect to the C≡NR π-bond.

The self-assembly of many benzenoid isocyanoarenes on gold surfaces requires sample handling and storage under an inert atmosphere in order to avoid any marked deterioration of the isocyanide groups to isocyanates, RN=C=O, which give a characteristic stretching band at around 2270 cm^{-1} in their IR spectra [90, 91]. Polydentate isocyanoarenes that can adsorb to the gold surface via multiple isocyanide junctions constitute a notable exception in this regard. For example, SAMs of tetraisocyano-*meta*-cyclophane (TIMC), the structure of which is shown in Figure 14.24a, exhibit an enhanced kinetic stability toward gold-promoted oxidation. The reflection–absorption IR spectra of the SAMs of this tetraisocyanide, which are prepared and stored in air without exclusion of ambient lighting, indicate the absorption of TIMC to the Au(111) surface via all four of its isocyanide groups (Figure 14.24b) and lack any features that could be attributed to isocyanate formation [92].

The isocyanoazulenes CN^{2}Az, CN^{2}Az^{CN}, CN^{2}Az^{CO2Et}, CN^{6}Az, CN^{6}Az^{CO2Et}, 2-FA-6-IA, 2,6-DIA, both isomers of (OC)$_5$Cr(η^1-2,6-DIA), and 2,2′-DIBA each formed well-defined SAM films on the Au(111) surface [58, 59, 93]. These films were prepared without precautions to exclude air and ambient lighting. The grazing angle reflection–absorption IR spectra of these films, along with ellipsometric film thickness measurements, suggest an approximately terminal upright coordination of the molecules in all of these SAMs (Table 14.6).

The IR spectrum of CN^{2}Az^{CN} in CHCl$_3$ solution is shown in Figure 14.25a. This species is particularly interesting because it features two types of IR spectroscopic reporters: an isocyanide group and nitrile groups. Upon adsorption to Au(111), the v_{NC} band corresponding to the isocyanide junction undergoes a 42 cm^{-1} blue shift that signifies a coordination of the isocyanide group to the gold surface (Figure 14.25b). The v_{CN} of the ancillary nitrile reporters then experiences a small (10 cm^{-1}) red shift upon absorption of CN^{2}Az^{CN}, which may be explained by the fact that the Au–CNR interaction slightly depletes the substituent ^{2}Az^{CN} of its π-electron density [59]. The exposure of a SAM of CN^{2}Az^{CN} on Au(111) to a solution of the related thiol, HS^{2}Az^{CN}, results in complete displacement of the isocyanide monolayer and formation of the new thiolate SAM, which can be conveniently

Table 14.6 Infrared vibrational data for selected isocyanoarenes before and after adsorption on gold surfaces.

Molecule	v_{NC} of free isocyanoarene (cm^{-1})	v_{NC} of surface bound isocyanide (cm^{-1})	v_{XY} of ancillary reporter (cm^{-1})	Observed (theoretical) thickness (Å)	Reference
DIB	2125	2181	2120 (isocyanide)	10 (10.3)	[89]
DIB	2127	2170	2122 (isocyanide)	Not reported	[88]
DIB	2129	2172	2121 (isocyanide)	12.7 (10.3)	[90]
DIBP	2126	2121	2191 (isocyanide)	14 (14.6)	[89]
DIBP	2127	2121	2190 (isocyanide)	13.9 (14.7)	[90]
TIMC	2119	2175	–	5.8 ± 5 (6.4)	[92]
CN^{2}Az	2127	2174	–	Not reported	[93]
CN^{6}Az	2117	2176	–	Not reported	[93]
CN^{6}Az^{Br}	2115	2179	–	Not reported	[93]
2-FA-6-IA	2116	2170	–	Not reported	[93]
CN^{2}Az^{CO2Et}	2127	2169	–	8 ± 1 (10.5)	[93]
CN^{6}Az^{CO2Et}	2115	2178	–	11 ± 3 (12.5)	[93]
2,6-DIA	2116, 2125	2163	2117 (isocyanide)	14 ± 1 (12.5)	[93]
2,2'-DIBA	2130	2170	2119 (isocyanide)	20.5 ± 2.4 (19.1)	[58]
CN^{2}Az^{CN}	2114	2156	2214 (nitrile)	11.7 ± 0.6 (10.7)	[59]
DIF	2118	2181	–	14.1 (8.8)	[77, 79]

See Figures 14.13, 14.15, and 14.24 for structures corresponding to the abbreviations.
DIB = 1,4-diisocyanobenzene; DIBP = 4,4'-diisocyanobiphenyl.

identified by the ancillary nitrile reporters (Figure 14.25c) [59]. This observation is consistent with the approximately twofold higher affinity of thiolate for metallic gold compared to that of isocyanide [23, 94].

The IR solution spectrum of the (OC)$_5$Cr(η^1-2,6-DIA) isomer shown in Figure 14.26a features four IR-active bands in v_{NC} and v_{CO} stretching regions [21, 93]. The v_{NC} bands at 2115 and 2135 cm^{-1} correspond to the free isocyanide terminus and the Cr-bound isocyanide group, respectively (Figure 14.26b). The other two more

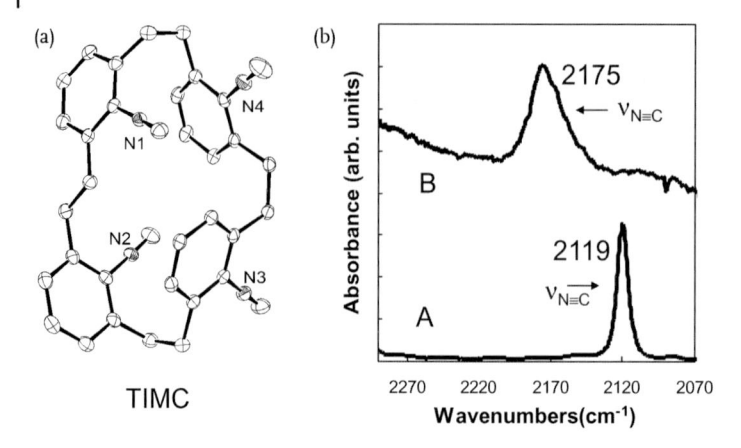

Figure 14.24 (a) ORTEP diagram for TIMC generated using the ORTEP-3 program from CIF data in Ref. [92]. The hydrogen atoms have been omitted for clarity; (b) IR spectrum of TIMC in CH$_2$Cl$_2$ solution (bottom) and reflection–absorption IR spectrum of a SAM film of TIMC on Au(111). Adapted with permission from Ref. [92]; © 2008, American Chemical Society.

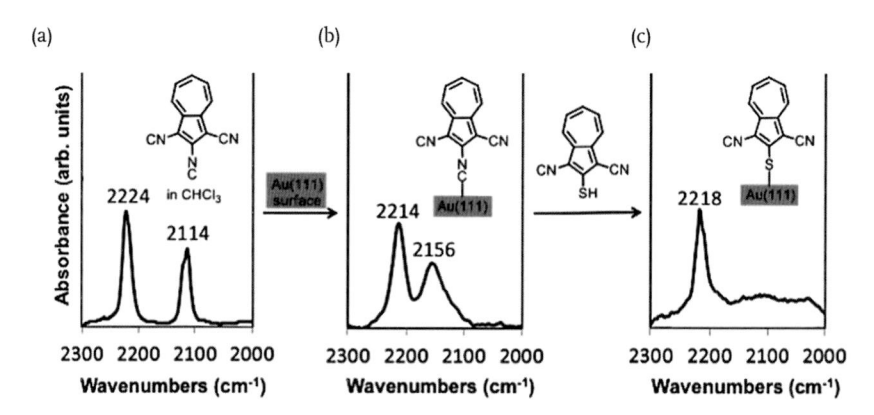

Figure 14.25 (a) IR spectrum of CN^{2}Az^{CN} in CHCl$_3$; (b) Reflection–absorption IR spectrum of CN^{2}Az^{CN} adsorbed onto Au(111); (c) Reflection–absorption IR spectrum of the sample in (b) after exposure to a solution of 2-mercapto-1,3-dicyanoazulene in THF. Reproduced from Ref. [59] with permission from The Royal Society of Chemistry.

intense bands at 2043 and 1962 cm^{-1} are due to $v_{CO}(A_1)$ and $v_{CO}(E)$ vibrations associated with the [Cr(CO)$_5$] moiety. Upon SAM film formation on Au(111), the band at 2115 cm^{-1} moves to 2174 cm^{-1}, thereby indicating coordination of the free isocyanide group of (OC)$_5$Cr(η^1-2,6-DIA) to the gold surface (Figure 14.26c). In addition, the relative intensities of the $v_{CO}(A_1)$ and $v_{CO}(E)$ bands change dramatically compared to the solution spectrum, which suggests an approximately parallel orientation of the equatorial carbonyl ligands to the metal surface and, thus, an upright coordination of the entire molecule on the surface. The measured

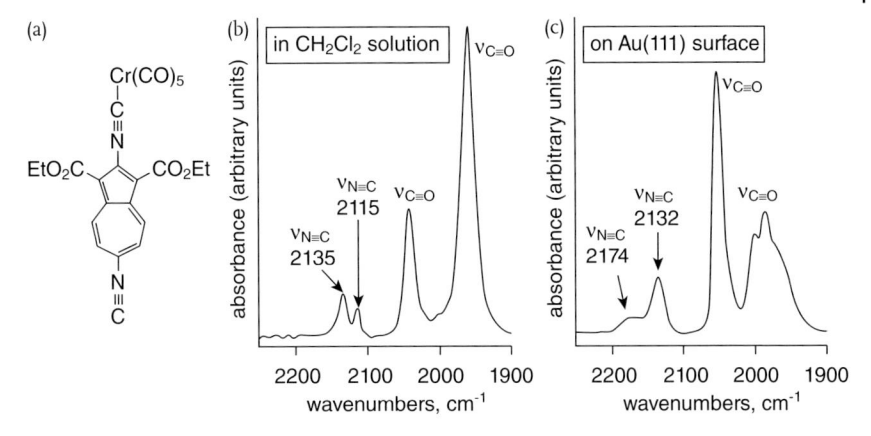

Figure 14.26 (a) IR spectrum of the mononuclear complex [(OC)$_5$Cr](η^1-2,6-DIA) in CH$_2$Cl$_2$ solution (the [(OC)$_5$Cr] fragment is bound to the 2-isocyanide terminus of 2,6-DIA); (b) Reflection–absorption IR spectrum of [(OC)$_5$Cr](η^1-2,6-DIA) adsorbed on Au(111). Adapted with permission from Ref. [21].

ellipsometric thickness of the SAM of this (OC)$_5$Cr(η^1-2,6-DIA) complex on Au(111) is 18(2) Å, which is practically identical to the expected thickness value of 17 Å for the perfectly upright orientation on the gold surface.

A comparison of the reflection–absorption IR spectra for the SAMs of 2,6-DIA with those for the SAMs of CN^{2}Az^{CO2Et} and CN^{6}Az^{CO2Et} on Au(111) suggests that the diisocyanoazulene 2,6-DIA binds preferentially to the gold surface via its 2-isocyano group, which is a slightly better electron donor than the 6-isocyano terminus (Table 14.6) [93]. SAM films of the diisocyanobiazulene 2,2′-DIBA also feature a terminal upright coordination of this nearly 2 nm-long molecule (Figure 14.27) [58]. These SAM films of 2,6-DIA and 2,2′-DIBA pave the way for the opportunity to probe the conductivity properties of linear molecular wires based on the 2,6-azulenic scaffold.

Finally, Siemeling et al. reported that DIF forms well-ordered SAM films on Au(111), and coordinates via both of its isocyanide groups to the metal surface. The ν_{NC} value of 2118 cm^{-1} blue shifts to 2181 cm^{-1} upon adsorption. In addition, the dipodal adsorption mode of DIF Au(111) was indicated by X-ray photoelectron spectroscopy measurements, which showed single N(1s) emission [77, 79].

14.4
Conclusions and Outlook

In this chapter, several relatively recent developments in the transition metal chemistry of isocyanoarenes have been addressed. The discovery of isolable isocyanidemetalates and related low-valent complexes has significantly narrowed a longstanding fundamental gap in the coordination chemistry of CNR and CO ligands at the electron-rich extreme. It is now apparent that isocyanoarene ligands

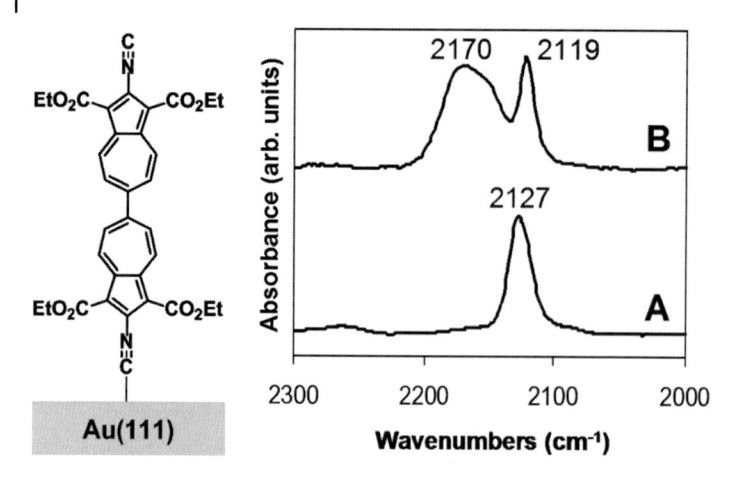

Figure 14.27 Left: Schematic representation of the terminal upright coordination of 2,2′-DIBA to the gold surface. Right: IR spectrum (A) of 2,2′-DIBA in CH_2Cl_2 and reflection–absorption IR spectrum (B) of 2,2′-DIBA adsorbed on Au(111). Adapted with permission from Ref. [58]; © 2010, American Chemical Society.

with hydrocarbon substituents R can behave not only as powerful π-acceptors, but also adjust their σ-donor/π-acceptor characteristics in a wide continuous range to accommodate electronic requirements at the metal center. The recently emerged bulky isocyanoarenes, such as $CNAr^{Mes2}$ and $CNAr^{Dipp2}$, are destined to bring revolutionary changes to the landscape of modern coordination and organometallic chemistry of isocyanides, including applications in catalysis. The unprecedented low-valent, low-coordinate complexes, such as $[Co(CNAr^{Mes2})_4]^z$ [36], $Ni(CNAr^{Dipp2})_3$ [95], and $Pd(CNAr^{Dipp2})_2$ [96] exhibit unique reactivity patterns and are the most profound examples of capitalizing on the steric variability of the substituent R in a CNR ligand. The families of isocyanoazulenes and η^5-bound isocyanocyclopentadienides (including planar–chiral versions thereof) provide yet another dimension for expanding electronic and steric versatilities of an isocyanoarene ligand in coordination and surface chemistry. The linear, isocyanide-terminated 2,6-azulenic and biazulenic frameworks constitute intriguing new platforms for developing azulene-based nanoarchitectures for applications in functional organic electronics.

Acknowledgments

M.V.B. gratefully acknowledges the US National Science Foundation (CAREER Award CHE-0548212), DuPont (Young Professor Award), Kansas NSF EPSCoR, Kansas Technology Enterprise Corporation, and the University of Kansas for the financial support of the research featured in Section 14.3. The authors are also indebted to Professor Cindy L. Berrie for her important role in developing the surface chemistry of isocyanoazulenes.

References

1 Gautier, A. (1869) Ueber die einwirkung der säuren auf die carbylamine. *Justus Liebigs Ann. Chem.*, **151** (2), 239–244.

2 Meyer, E. (1856) Ueber cyanäthyl und eine neue bildung des aethylamin. *J. Prakt. Chem.*, **68** (1), 279–295.

3 Lieke, W. (1859) Ueber das cyanallyl. *Justus Liebigs Ann. Chem.*, **112** (3), 316–321.

4 Malatesta, L. (1959) Isocyanide complexes of metals, in *Progress in Inorganic Chemistry*, vol. 1 (ed. F.A. Cotton), John Wiley & Sons, Inc, Hoboken, NJ, pp. 283–379.

5 Hoffmann, R. (1982) Building bridges between inorganic and organic chemistry. *Angew. Chem. Int. Ed. Engl.*, **21** (10), 711–724.

6 (a) Ellis, J.E. (2003) Metal carbonyl anions: from $[Fe(CO)_4]^{2-}$ to $[Hf(CO)_6]^{2-}$ and beyond. *Organometallics*, **22** (17), 3322–3338; (b) Ellis, J.E. (2006) Adventures with substances containing metals in negative oxidation states. *Inorg. Chem.*, **45** (8), 3167–3186; (c) Ellis, J.E. (2006) Adventures with substances containing metals in negative oxidation states (addition/correction). *Inorg. Chem.*, **45** (14), 5710–5710.

7 (a) Barker, G.K., Galas, A.M.R., Green, M., Howard, J.A.K., Stone, F.G.A., Turney, T.W., Welch, A.J., and Woodward, P. (1977) Synthesis and reactions of octakis(*t*-butyl isocyanide) dicobalt and pentakis(*t*-butyl isocyanide) ruthenium; X-ray crystal and molecular structures of $[Co_2(Bu^tNC)_8]$ and $[Ru(Ph_3P)(Bu^tNC)_4]$. *J. Chem. Soc. Chem. Commun.*, 256–258; (b) Yamamoto, Y., and Yamazaki, H. (1978) Studies on interaction of isocyanide with transition-metal complexes. 18. Synthesis and reactions of dicobalt octaisocyanide. *Inorg. Chem.*, **17** (11), 3111–3114.

8 (a) Silverman, L.D., Dewan, J.C., Giandomenico, C.M., and Lippard, S.J. (1980) Molecular structure and ligand-exchange reactions of trichlorotris(*tert*-butyl isocyanide)vanadium(III). Synthesis of the hexakis(*tert*-butyl isocyanide) vanadium(II) cation. *Inorg. Chem.*, **19** (11), 3379–3383; (b) Silverman, L.D., Corfield, P.W.R., and Lippard, S.J. (1981) Synthesis and structure of hexakis(*tert*-butyl isocyanide)vanadium(II) hexacarbonylvanadate(-I). *Inorg. Chem.*, **20** (9), 3106–3109.

9 Warnock, G.F. and Cooper, N.J. (1989) The First Transition-Metal Isonitrilate: synthesis and characterization of $K[Co(2,6-Me_2C_6H_3NC)_4]$. *Organometallics*, **8** (7), 1826–1827.

10 Allen, J.M. and Ellis, J.E. (2008) Synthesis and characterization of titanium tetraisocyanide complexes, $[CpTi(CNXyl)_4E]$, E = I, $SnPh_3$, and $SnMe_3$. *J. Organometallic Chem.*, **693** (8–9), 1536–1542.

11 (a) $[Mn(CNR)_5]^-$: Utz, T.L., Leach, P.A., Geib, S.J., and Cooper, N.J. (1997) Synthesis, derivatization, and structural characterization of $[Mn(CNC_6H_3Me_2-2,6)_5]^-$, a five-coordinate isonitrilate complex containing Mn^{-1}. *Chem. Commun.*, 847–848; (b) $[Mn(CNR)_6]^{2+}$: Mann, K.R., Cimolino, M., Geoffroy, G.L., Hammond, G.S., Orio, A.A., Albertin, G., and Gray, H.B. (1976) Electronic structures and spectra of hexakisphenylisocyanide complexes of Cr(0), Mo(0), W(0), Mn(I), and Mn(II). *Inorg. Chim. Acta*, **16**, 97–101; (c) $[Tc(CNR)_6]^+$: Abrams, M.J., Davison, A., Jones, A.G., Costello, C.E., and Pang, H. (1983) Synthesis and characterization of hexakis(alkyl isocyanide) and hexakis(aryl isocyanide) complexes of technetium(I). *Inorg. Chem.*, **22** (20), 2798–2800; (d) $[Fe(CNR)_6]^{2+}$: Bonati, F. and Minghetti, G. (1970) New isocyanide complexes of iron(II). *J. Organometallic Chem.*, **22** (1), 183–194; (e) $[Ru(CNR)_4]^{2-}$: Corella, J.A., Thompson, R.L., and Cooper, N.J. (1992) (Isocyanide)ruthenate analogues of tetracarbonylferrate. *Angew. Chem. Int. Ed. Engl.*, **31** (1), 83–84; (f) $[Ru(CNR)_6]^{2+}$ and $[Os(CNR)_6]^{2+}$: Tetrick, S.M. and Walton, R.A. (1985) Homoleptic isocyanide complexes of ruthenium(II) and osmium(II). *Inorg. Chem.*, **24** (21),

3363–3366; (g) Rh$_2$(CNR)$_8$: Yamamoto, Y. and Yamazaki, H. (1984) Studies of the interaction of isocyanides with transition-metal complexes. XXVI. The preparation of octakis(aryl isocyanide)dirhodium. *Bull. Chem. Soc. Jpn*, **57** (1), 297–298; (h) Ni$_4$(CNR)$_6$: Muetterties, E.L., Band, E., Kokorin, A., Pretzer, W.R., and Thomas, M.G. (1980) Metal clusters. 23. Tetranuclear nickel alkyl isocyanide clusters. *Inorg. Chem.*, **19** (6), 1552–1560; (i) [Ag(CNR)$_4$]$^+$: Pierce, J.L., Wigley, D.E., and Walton, R.A. (1982) Homoleptic isocyanide and mixed nitrosyl-isocyanide complexes of chromium and molybdenum: secondary ion mass spectrometry studies. *Organometallics*, **1** (10), 1328–1331.

12 Weber, L. (1998) Homoleptic isocyanide metalates. *Angew. Chem. Int. Ed. Engl.*, **37** (11), 1515–1517.

13 Hammick, D.L., New, R.C.A., Sidgwick, N.V., and Sutton, L.E. (1930) CCXLIV – Structure of the isocyanides and other compounds of bivalent carbon. *J. Chem. Soc.*, 1876–1887.

14 Triechel, P.M. (1973) Transition-metal isocyanide complexes. *Adv. Organomet. Chem.*, **11**, 21–86.

15 Sacco, A. (1953) Complex cobaltocarbonyl isocyanide salts. *Gazz. Chim. Ital.*, **83**, 632–636.

16 (a) Lentz, D. (1994) Organometallic chemistry of fluorinated isocyanides, in *Inorganic Fluorine Chemistry: Toward the 21st Century* (eds J.S. Thrasher and S.H. Strauss), American Chemical Society, Washington, D.C, pp. 265–285; (b) Lentz, D. (1994) Fluorinated isocyanides – more than ligands with unusual properties. *Angew. Chem. Int. Ed. Engl.*, **33** (13), 1315–1331.

17 (a) Barybin, M.V., Young, V.G. Jr, and Ellis, J.E. (1999) First homoleptic isocyanides of niobium and tantalum. *J. Am. Chem. Soc.*, **121** (39), 9237–9238; (b) Barybin, M.V., Brennessel, W.W., Kucera, B.E., Minyaev, M.E., Sussman, V.J., Young, V.G., and Ellis, J.E. (2007) Homoleptic isocyanidemetalates of 4d- and 5d-transition metals: [Nb(CNXyl)$_6$]$^-$, [Ta(CNXyl)$_6$]$^-$, and derivatives thereof. *J. Am. Chem. Soc.*, **129** (5), 1141–1150.

18 Ellis, J.E., Warnock, G.F., Barybin, M.V., and Pomije, M.K. (1995) New PF$_3$ and carbonyl chemistry of tantalum. *Chem. Eur. J.*, **1** (8), 521–527.

19 Warnock, G.F.P., Sprague, J., Fjare, K.L., and Ellis, J.E. (1983) Highly reduced organometallics. 10. Synthesis and chemistry of the pentacarbonylmetallate(3-) ions of niobium and tantalum, M(CO)$_5$$^{3-}$. *J. Am. Chem. Soc.*, **105** (3), 672–672.

20 Dömling, A. (2006) Recent developments in isocyanide based multicomponent reactions in applied chemistry. *Chem. Rev.*, **106** (1), 17–89.

21 Barybin, M.V. (2010) Nonbenzenoid aromatic isocyanides: new coordination building blocks for organometallic and surface chemistry. *Coord. Chem. Rev.*, **254** (11–12), 1240–1252.

22 Harvey, P.D., Clément, S., Knorr, M., and Husson, J. (2010) Luminescent organometallic coordination polymers built on isocyanide bridging ligands, in *Macromolecules Containing Metal and Metal-like Elements*, Volume 10: Photophysics and Photochemistry of Metal-Containing Polymers (eds A.S.A.-E. Aziz, C.E. Carraher, Jr, P.D. Harvey, C.U. Pittman, and M. Zeldin), 1st edn John Wiley & Sons, Inc, Hoboken, pp. 45–87.

23 Lazar, M. and Angelici, R.J. (2009) Isocyanide binding modes on metal surfaces and in metal complexes, in *Modern Surface Organometallic Chemistry* (eds J.-M. Basset, R. Psaro, D. Roberto, and R. Ugo), Wiley-VCH Verlag GmbH, Weinheim, pp. 513–556.

24 Heimel, G., Romaner, L., Zojer, E., and Bredas, J.-L. (2008) The interface energetics of self-assembled monolayers on metals. *Acc. Chem. Res.*, **41** (6), 721–729.

25 Puddephatt, R.J. (2001) Coordination polymers: polymers, rings, oligomers containing gold(I) centres. *Coord. Chem. Rev.*, **216–217**, 313–332.

26 Nakano, T. and Okamoto, Y. (2001) Synthetic helical polymers: conformation and function. *Chem. Rev.*, **101** (12), 4013–4038.

27 Hong, S., Reifenberger, R., Tian, W., Datta, S., Henderson, J., and Kubiak, C.P. (2000) Molecular conductance

spectroscopy of conjugated, phenyl-based molecules on Au(111): the effect of end groups on molecular conduction. *Superlattices Microstruct.*, **28** (4), 289–303.

28 Sharma, V. and Piwnica-Worms, D. (1999) Metal complexes for therapy and diagnosis of drug resistance. *Chem. Rev.*, **99** (9), 2545–2560.

29 Lippard, S. and Carnahan, E. (1993) 15 years of reductive coupling: what have we learned? *Acc. Chem. Res.*, **26** (3), 90–97.

30 Hahn, F.E. (1993) The coordination chemistry of multidentate isocyanide ligands. *Angew. Chem. Int. Ed. Engl.*, **32** (5), 650–665.

31 Singleton, E. and Oosthuizen, H.E. (1983) Metal isocyanide complexes. *Adv. Organometallic Chem.*, **22**, 209–310.

32 Yamamoto, Y. (1980) Zerovalent transition metal complexes of organic isocyanides. *Coord. Chem. Rev.*, **32** (3), 193–233.

33 Leach, P.A., Geib, S.J., Corella, J.A., Warnock, G.F., and Cooper, N.J. (1994) Synthesis and structural characterization of [Co(CN(2,6-$C_6H_3Me_2$))$_4$]$^-$, the first transition metal isonitrilate. *J. Am. Chem. Soc.*, **116** (9), 8566–8574.

34 (a) Brennessel, W.W., Young, V.G., and Ellis, J.E. (2006) Towards homoleptic naphthalenemetalates of the later transition metals: isolation and characterization of naphthalenecobaltates(1-). *Angew. Chem. Int. Ed. Engl.*, **45** (43), 7268–7271; (b) Brennessel, W.W., Young, V.G., and Ellis, J.E. (2002) Bis(1,2,3,4-η^4-anthracene)cobaltate(1-). *Angew. Chem. Int. Ed. Engl.*, **41** (7), 1211–1215.

35 (a) Fox, B.J., Sun, Q.Y., DiPasquale, A.G., Fox, A.R., Rheingold, A.L., and Figueroa, J.S. (2008) Solution behavior and structural properties of Cu(I) complexes featuring *m*-terphenyl isocyanides. *Inorg. Chem.*, **47** (19), 9010–9020; (b) Ditri, T.B., Fox, B.J., Moore, C.E., Rheingold, A.L., and Figueroa, J.S. (2009) Effective control of ligation and geometric isomerism: direct comparison of steric properties associated with bis-mesityl and bis-diisopropylphenyl *m*-terphenyl isocyanides. *Inorg. Chem.*, **48** (17), 8362–8375.

36 Margulieux, G.W., Weidemann, N., Lacy, D.C., Moore, C.E., Rheingold, A.L., and Figueroa, J.S. (2010) Isocyano analogues of [Co(CO)$_4$]^n: a tetraisocyanide of cobalt isolated in three states of charge. *J. Am. Chem. Soc.*, **132** (14), 5033–5035.

37 Ricks, A.M., Bakker, J.M., Douberly, G.E., and Duncan, M. (2009) An infrared spectroscopy and structures of cobalt carbonyl cations, Co(CO)$^{n+}$ ($n = 1$–9). *J. Phys. Chem. A*, **113** (16), 4701–4708.

38 Brennessel, W.W. and Ellis, J.E. (2007) Fe(CNXyl)$_4$]$^{2-}$: an isolable and structurally characterized homoleptic isocyanidemetalate dianion. *Angew. Chem. Int. Ed. Engl.*, **46** (4), 598–600.

39 Hieber, W. and Leutert, F. (1931) Äthylendiamin-substituierte eisencarboyle und eine neue bildungsweise von eisencarbonylwasserstoff (XI. mitteil. über metallcarbonyle). *Ber. Dtsch Chem. Ges.*, **64** (11), 2832–2839.

40 Mathieson, T., Schier, A., and Schmidbaur, H. (2001) Supramolecular chemistry of gold(I) thiocyanate complexes with thiophene, phosphine and isocyanide ligands, and the structure of 2,6-dimethylphenyl isocyanide. *J. Chem. Soc. Dalton Trans.*, 1196–1200.

41 Chin, H.B. and Bau, R. (1976) The crystal structure of disodium tetracarbonylferrate. Distortion of the tetracarbonylferrate(2-) anion in the solid state. *J. Am. Chem. Soc.*, **98** (9), 2434–2439.

42 Utz, T.L., Leach, P.A., Geib, S.J., and Cooper, N.J. (1997) Formation of the 1,4-diazabutadien-2-yl complex [Mn(CNPh*)$_4${C(=NPh*)C(CH$_3$)=N(Ph*)}] through methylation of a manganese(-I) isonitrilate. *Organometallics*, **16** (19), 4109–4114.

43 Stewart, M.A., Moore, C.E., Ditri, T.B., Labios, L.A., Rheingold, A.L., and Figueroa, J.S. (2011) Electrophilic functionalization of well-behaved manganese monoanions supported by *m*-terphenyl isocyanides. *Chem. Commun.*, **47** (1), 406–408.

44 Rossi, A.R. and Hoffmann, R. (1975) Transition metal pentacoordination. *Inorg. Chem.*, **14** (2), 365–374.

45 (a) Ellis, J.E. and Fjare, K.L. (1982) Highly reduced organometallics. 7. The synthesis of alkyl and phenyl isocyanide and related monosubstituted vanadium carbonyl anions, V(CO)$_5$L$^-$, by the thermal substitution of (amine) pentacarbonylvanadate(1-), V(CO)$_5$NH$_3^-$. *Organometallics*, **1** (7), 898–903; (b) Ihmels, K. and Rehder, D. (1985) Pentacarbonylvanadates(-I) containing carbon, nitrogen, and group 16 ligands. The relation between IR and vanadium-51 NMR shift parameters. *Organometallics*, **4** (8), 1340–1347.

46 (a) Ellis, J.E., Fjare, K.L., and Warnock, G.F. (1995) Synthesis and ligand substitution reactions of [Ta(CO)$_5$NH$_3$]$^{-1}$. *Inorg. Chim. Acta*, **240** (1-2), 379–384; (b) Barybin, M.V., Ellis, J.E., Pomije, M.K., Tinkham, M.L., and Warnock, G.F. (1998) Syntheses and properties of homoleptic carbonyl and trifluorophosphane niobates: [Nb(CO)$_6$]$^-$, [Nb(PF$_3$)$_6$]$^-$ and [Nb(CO)$_5$]$^{3-}$. *Inorg. Chem.*, **37** (25), 6518–6527.

47 Carnahan, E.M. and Lippard, S.J. (1992) Formation of highly functionalized metal-bound acetylenes by reductive coupling of carbon monoxide and methyl isocyanide ligands. *J. Am. Chem. Soc.*, **114** (11), 4166–4174.

48 Herberhold, M. and Trampisch, H. (1983) Dinitrosyl vanadium complexes. *Inorg. Chim. Acta*, **70** (2), 143–146.

49 Herberhold, M. and Trampisch, H. (1982) Spectroscopic studies of cyclopentadienyldinitrosylvanadium complexes CpV(NO)$_2$L (L = Lewis base). *Z. Naturforsch.*, **37b** (5), 614–619.

50 Naeumann, F. and Rehder, D. (1984) Dinitrosylvanadium complexes: [VL$_4$(NO)$_2$]Br (L = N, P, As, Sb, O and S donor/acceptor) and [VX(CNR)$_3$(NO)$_2$] (X = Cl, Br, I). *Z. Naturforsch.*, **39b** (12), 1654–1661.

51 Collazo, C., Rodewald, D., Schmidt, H., and Rehder, D. (1996) Niobium-centered C-C coupling of isonitriles. *Organometallics*, **15** (22), 4884–4887.

52 Barybin, M.V. (1999) Novel chemistry of low-valent early transition metal complexes. Dissertation, University of Minnesota.

53 (a) Barybin, M.V., Young, V.G., and Ellis, J.E. (1998) Syntheses and structural characterizations of the first 16-, 17-, and 18-electron homoleptic isocyanide complexes of vanadium: hexakis(2,6-dimethylphenyl isocyanide)vanadium(I, 0, -I). *J. Am. Chem. Soc.*, **120** (2), 429–430; (b) Barybin, M.V., Young, V.G., and Ellis, J.E. (2000) First Paramagnetic zerovalent transition metal isocyanides. Syntheses, structural characterizations, and magnetic properties of novel low-valent isocyanide complexes of vanadium. *J. Am. Chem. Soc.*, **122** (19), 4678–4691.

54 Natta, G., Ercoli, R., Calderazzo, F., Alberola, A., Corradini, P., and Allegra, G. (1959) Properties and structure of a new metal carbonyl: V(CO)$_6$. *Atti Accad. Naz. Lincei Cl. Sci. Fis. Mat. Nat.*, **27**, 107–112.

55 Robinson, R.E., Holovics, T.C., Deplazes, S.F., Lushington, G.H., Powell, D.R., and Barybin, M.V. (2003) First isocyanoazulene and its homoleptic complexes. *J. Am. Chem. Soc.*, **125** (15), 4432–4433.

56 Robinson, R.E., Holovics, T.C., Deplazes, S.F., Powell, D.R., Lushington, G.H., Thompson, W.H., and Barybin, M.V. (2005) Five possible isocyanoazulenes and electron-rich complexes thereof: a quantitative organometallic approach for probing electronic inhomogeneity of the azulenic framework. *Organometallics*, **24** (10), 2386–2397.

57 Holovics, T.C., Robinson, R.E., Weintrob, E., Toriyama, M., Lushington, G.H., and Barybin, M.V. (2006) The 2,6-diisocyanoazulene motif: synthesis and efficient mono- and heterobimetallic complexation with controlled orientation of the azulenic dipole. *J. Am. Chem. Soc.*, **128** (7), 2300–2309.

58 Maher, T.R., Spaeth, A.D., Neal, B.M., Berrie, C.L., Thompson, W.H., Day, V.W., and Barybin, M.V. (2010) Linear 6,6′-biazulenyl framework featuring isocyanide termini: synthesis, structure, redox behavior, complexation, and self-assembly on Au(111). *J. Am. Chem. Soc.*, **132** (45), 15924–15926.

59 Neal, B.M., Vorushilov, A.S., DeLaRosa, A.M., Robinson, R.E., Berrie, C.L., and Barybin, M.V. (2011) Ancillary nitrile

substituents as convenient IR spectroscopic reporters for self-assembly of mercapto- and isocyanoazulenes on Au(111). *Chem. Commun.*, **47** (38), 10803–10805.

60 Barybin, M.V., Holovics, T.C., Deplazes, S.F., Lushington, G.H., Powell, D.R., and Toriyama, M. (2002) First homoleptic complexes of isocyanoferrocene. *J. Am. Chem. Soc.*, **124** (46), 13668–13669.

61 Bildstein, B., Malaun, M., Kopacka, H., Wurst, K., Mitterböck, M., Ongania, K.-H., Opromolla, G., and Zanello, P. (1999) N,N′-Diferrocenyl-N-heterocyclic carbenes and their derivatives. *Organometallics*, **18** (21), 4325–4336.

62 Kavallieratos, K., Hwang, S., and Crabtree, R.H. (1999) Aminoferrocene derivatives in chloride recognition and electrochemical sensing. *Inorg. Chem.*, **38** (22), 5184–5186.

63 van Leusen, D. and Hessen, B. (2001) 1,1′-Diisocyanferrocene and a convenient synthesis of ferrocenylamine. *Organometallics*, **20** (1), 224–226.

64 Holovics, T.C., Deplazes, S.F., Toriyama, M., Powell, D.R., Lushington, G.H., and Barybin, M.V. (2004) Organometallic isocyanocyclopentadienides: a combined synthetic, spectroscopic, structural, electrochemical, and theoretical investigation. *Organometallics*, **23** (12), 2927–2938.

65 McGinnis, D.M., Deplazes, S.F., and Barybin, M.V. (2011) Synthesis, properties and complexation of ($_p$S)-1-isocyano-2-methylferrocene, the first planar-chiral isocyanide ligand. *J. Organomet. Chem.*, **696** (25), 3939–3944.

66 Churchill, M.R. (1970) Transition metal complexes of azulene and related ligands, in *Progress in Inorganic Chemistry*, vol. 11 (ed. S.J. Lippard), John Wiley & Sons, Inc, Hoboken, NJ, pp. 53–98.

67 Bohling, D.A. and Mann, K.R. (1984) X-ray structural characterization of [Cr(CNPh)$_6$][CF$_3$SO$_3$], [Cr(CNPh)$_6$][PF$_6$]$_2$, and [Cr(CNPh)$_6$][SbCl$_6$]$_3$·CH$_2$Cl$_2$. Completion of a unique series of complexes in which the metal attains four different oxidation states while maintaining identical ligation. *Inorg. Chem.*, **23** (10), 1426–1432.

68 Ljungström, E. (1978) The crystal structure of hexakis(phenyl isocyanide) chromium(0), Cr(CNC$_6$H$_5$)$_6$. *Acta Chem. Scand.*, **32A** (1), 47–50.

69 Mialki, W.S., Wigley, D.E., Wood, T.E., and Walton, R.A. (1982) Homoleptic seven- and six-coordinate alkyl isocyanide complexes of chromium: synthesis, characterization, and redox and substitution chemistry. *Inorg. Chem.*, **21** (2), 480–485.

70 (a) Bullock, J.P. and Mann, K.R. (1989) UV-Visible IR thin-layer spectroelectrochemical studies of hexakis(aryl isocyanide)chromium complexes. In situ generation and characterization of four oxidation states. *Inorg. Chem.*, **28** (21), 4006–4011; (b) Treichel, P.M. and Essenmacher, G.J. (1976) Oxidations of hexakis(aryl isocyanide)chromium(0) complexes. *Inorg. Chem.*, **15** (1), 146–150.

71 Deplazes, S.F. (2008) Ligand design, coordination, and electrochemistry of nonbenzenoid aryl isocyanides. Dissertation, University of Kansas.

72 El-Shihi, T., Siglmüller, F., Herrmann, R., Fernanda, M., Carvalho, N.N., and Pombeiro, A.J.L. (1987) Isocyanide derivatives of ferrocene. Preparation, complexation and redox properties. *J. Organomet. Chem.*, **335** (2), 239–247.

73 El-Shihi, T., Siglmüller, F., Herrmann, R., Fernanda, M., Carvalho, N.N., and Pombeiro, A.J.L. (1987) Electrochemical study of isocyanide derivatives of ferrocene and of their complexes with the chromium pentacarbonyl {Cr(CO)$_5$} center. *Port. Electrochim. Acta*, **5**, 179–184.

74 Nesmeyanov, A.N., Perovalova, E.G., Gubin, S.P., Grandberg, K.I., and Kozlovsky, A.G. (1966) Electronic properties of the ferrocenyl as a substituent. *Tetrahedron Lett.*, **22** (7), 2381–2387.

75 Barybin, M.V., Chisholm, M.H., Dalal, N., Holovics, T.H., Patmore, N.J., Robinson, R.E., and Zipse, D.J. (2005) Long-range electronic coupling of MM quadruple bonds (M = Mo or W) via a 2,6-azulenedicarboxylate bridge. *J. Am. Chem. Soc.*, **127** (43), 15182–15190.

76 Grubisha, D.S., Rommel, J.S., Lane, T.M., Tysoe, W.T., and Bennett, D.W.

(1992) Communication between metal centers in tungsten(0)-tungsten(II) complexes bridged by 1,4-diisocyanobenzene: is the ligand pi system involved? *Inorg. Chem.*, **31** (24), 5022–5027.

77 Siemeling, U., Rother, D., Bruhn, C., Fink, H., Weidner, T., Träger, F., Rothenberger, A., Fenske, D., Priebe, A., Maurer, J., and Winter, R. (2005) The interaction of 1,1′-diisocyanoferrocene with gold: formation of monolayers and supramolecular polymerization of an aurophilic ferrocenophane. *J. Am. Chem. Soc.*, **127** (4), 1102–1103.

78 Siemeling, U., Rother, D., and Bruhn, C. (2008) Reactions of gold(I) acetylides with 1,1′-diisocyanoferrocene: from orthodox to unorthodox behavior. *Organometallics*, **27** (24), 6419–6426.

79 Weidner, T., Ballav, N., Zharnikov, M., Priebe, A., Long, N.J., Winter, R., Rothenberger, A., Fenske, D., Rother, D., Bruhn, C., Fink, H., and Siemeling, U. (2008) Dipodal ferrocene-based adsorbate molecules for self-assembled monolayers on gold. *Chem. Eur. J.*, **14** (14), 4346–4360.

80 Siemeling, U., Klapp, L.R.R., and Bruhn, C. (2010) Tri- and tetracoordinate copper(I) complexes of 1,1′-diisocyanoferrocene. *Z. Anorg. Allg. Chem.*, **636** (3-4), 539–542.

81 Siemeling, U., Rother, D., and Bruhn, C. (2007) "Schizoid" reactivity of 1,1′-diisocyanoferrocene. *Chem. Commun.*, **41**, 4227–4229.

82 Chen, J., Wang, W., Klemic, J., Reed, M.A., Axelrod, B.W., Kaschak, D.M., Rawlett, A.M., Price, D.W., Dirk, S.M., Tour, J.M., Grubisha, D.S., and Bennett, D.W. (2002) Molecular wires, switches, and memories. *Ann. N. Y. Acad. Sci.*, **960**, 69–99.

83 McCreery, R. (2004) Molecular electronic junctions. *Chem. Mater.*, **16** (23), 4477–4496.

84 Choi, S.H., Kim, B., and Frisbie, C.D. (2008) Electrical resistance of long conjugated molecular wires. *Science*, **320** (5882), 1482–1486.

85 Samori, P. and Cacialli, F. (eds) (2011) *Functional Supramolecular Architectures: For Organic Electronics and Nanotechnology*, vols 1–2, Wiley-VCH Verlag GmbH, Weinheim.

86 Chu, C., Ayers, J.A., Stefanescu, D.M., Walker, B.R., Gorman, C.B., and Parsons, G.N. (2007) Enhanced conduction through isocyanide terminal groups in alkane and biphenylene molecules measured in molecule/nanoparticle/molecule junctions. *J. Phys. Chem. C*, **111** (22), 8080–8085.

87 Kim, B., Beebe, J.M., Jun, Y., Zhu, X.-Y., and Frisbie, C.D. (2006) Correlation between HOMO alignment and contact resistance in molecular junctions: aromatic thiols versus aromatic isocyanides. *J. Am. Chem. Soc.*, **128** (15), 4970–4971.

88 Murphy, K.L., Tysoe, W.T., and Bennett, D.W. (2004) A comparative investigation of aryl isocyanides chemisorbed to palladium and gold: an ATR-IR spectroscopic study. *Langmuir*, **20** (5), 1732–1738.

89 Henderson, J.I., Feng, S., Bein, T., and Kubiak, C.P. (2000) Adsorption of diisocyanides on gold. *Langmuir*, **16** (15), 6183–6187.

90 Swanson, S.A., McClain, R., Lovejoy, K.S., Alamdari, N.B., Hamilton, J.S., and Scott, J.C. (2005) Self-assembled diisocyanide monolayer films on gold and palladium. *Langmuir*, **21** (11), 5034–5039.

91 Stapleton, J.J., Daniel, T.A., Uppili, S., Cabarcos, O.M., Naciri, J., Shashidhar, R., and Allara, D.L. (2005) Self-assembly, characterization, and chemical stability of isocyanide-bound molecular wire monolayers on gold and palladium surfaces. *Langmuir*, **21** (24), 11061–11070.

92 Toriyama, M., Maher, T.R., Holovics, T.C., Vanka, K., Day, V.W., Berrie, C.L., Thompson, W.H., and Barybin, M.V. (2008) Multipoint anchoring of the [2.2.2.2]metacyclophane motif to a gold surface via self-assembly: coordination chemistry of a cyclic tetraisocyanide revisited. *Inorg. Chem.*, **47** (8), 3284–3291.

93 DuBose, D.L., Robinson, R.E., Holovics, T.C., Moody, D.R., Weintrob, E.C., Berrie, C.L., and Barybin, M.V. (2006) Interaction of mono- and diisocyanoazulenes with gold surfaces: first examples of self-assembled

94 monolayer films involving azulenic scaffolds. *Langmuir*, **22** (10), 4599–4606.
94 Love, J.C., Estroff, L.A., Kriebel, J.K., Nuzzo, R.G., and Whitesides, G.M. (2005) Self-assembled monolayers of thiolates on metals as a form of nanotechnology. *Chem. Rev.*, **105** (4), 1103–1169.
95 Emerich, B.M., Moore, C.E., Fox, B.J., Rheingold, A.L., and Figueroa, J.S. (2011) Protecting-group-free access to a three-coordinate nickel(0) tris-isocyanide. *Organometallics*, **30** (9), 2598–2608.
96 Labios, L.A., Millard, M.D., Rheingold, A.L., and Figueroa, J.S. (2009) Bond activation, substrate addition and catalysis by an isolable two-coordinate Pd(0) bis-isocyanide monomer. *J. Am. Chem. Soc.*, **131** (32), 11318–11319.

15
Carbene Complexes Derived from Metal-Bound Isocyanides: Recent Advances

Konstantin V. Luzyanin and Armando J.L. Pombeiro

15.1
Introduction

Isocyanides (RN≡C) constitute one of the most versatile types of reagent in organic chemistry [1]. Their transformations, for example via the so-called multicomponent reactions (MCRs) of the Ugi and Passerini types, allow the synthesis of a wide range of products, including bis-amides, heterocycles, and peptides [1–5]. Isocyanides have also attracted the attention of coordination chemists due to their versatile σ-electron donor and π-electron acceptor properties [6–10]. Hence, they behave as strong π-acceptors when binding a suitable electron-rich metal center, yet can be activated towards an electrophilic attack to produce aminocarbyne species (e.g., Route I in Scheme 15.1) [10–12]. However, at a common Lewis acid metal center, they act as soft Pearson's bases and often undergo nucleophilic attack to form various types of product (e.g., Routes II–IV in Scheme 15.1) [9].

Among the complexes obtained via the transformation of metal-bound isocyanides, the metallocarbene species – and, in particular, those with a heterocyclic skeleton (*N*-heterocyclic carbenes, NHCs) [13–17] – represent one of the most extensively studied types of compound in organometallic chemistry and catalysis of the past few decades. The attractive properties of the NHCs – notably their high chemical and thermal stabilities, low toxicity and good availability – have led to the production of complexes with these species that can challenge the dominance of the commonly used tertiary phosphine-based catalysts over a wide range of cross-coupling systems, including the Suzuki–Miyaura, Heck, and Sonogashira reactions [16]. The NHCs and their related structural acylic analogues, the acyclic diaminocarbenes (ADCs) [18], display similar electronic stabilizations and exhibit comparable net electron-donor and steric properties [14]. Metallacarbenes of both types can be prepared either via a complexation of the pre-prepared free carbene (typically generated *in situ* from its precursors) to a metal center, or via the transformation of activated metal-bound isontriles, which is a highly atom-efficient procedure.

Isocyanide Chemistry: Applications in Synthesis and Material Science, First Edition. Edited by Valentine Nenajdenko.
© 2012 Wiley-VCH Verlag GmbH & Co. KGaA. Published 2012 by Wiley-VCH Verlag GmbH & Co. KGaA.

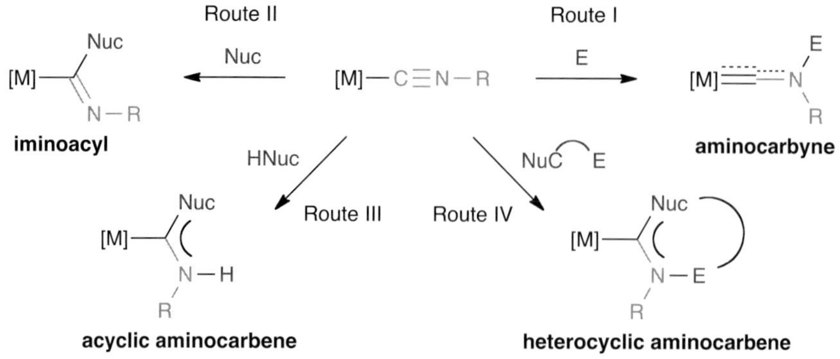

Scheme 15.1 Reactivity modes for metal-bound isocyanides.

Whilst carbene complexes derived from corresponding isocyanide species were initially reviewed in 2001 by Michelin, Pombeiro et al. [9], over the past decade many new examples have appeared. Consequently, the aim of this chapter is to summarize recent results relating to the metal-mediated generation of aminocarbene ligands from isocyanides, with the available data organized on the basis of the types of reaction that lead to metal–carbene species.

15.2
Coupling of the Isocyanide Ligand with Simple Amines or Alcohols

The cationic isocyanide complexes fac-[Mn(CNR)(CO)$_3$(bipy)]$^+$ react with NH$_2$Me to produce the acyclic diaminocarbene compounds fac-[Mn(ADC)(CO)$_3$(bipy)]$^+$ (Scheme 15.2), as reported by Ruiz et al. [19]. In these products, the carbene ligands are deprotonated when treated with KOH to yield the corresponding neutral formamidinyl (ADC-H) derivatives fac-[Mn(ADC-H)(CO)$_3$(bipy)]; this deprotonation has been shown to be reversible.

Fehlhammer et al. have shown that, under aprotic conditions [20], [Cr(CO)$_5$(CNCCl$_3$)] reacts with triphenylphosphine in the presence of aromatic aldehydes or ketones to produce the α-chloroalkenylisocyanide complexes $cis(Z)$- and $trans(E)$-[Cr(CO)$_5$(CNCCl=CR^{1}R^2)] [R^1 = H, R^2 = 4-C$_6$H$_4$F, 4-C$_6$H$_4$–CH=C(Cl)NCCr(CO)$_5$]. The reactions of these species with pyrrolidine afford, upon the

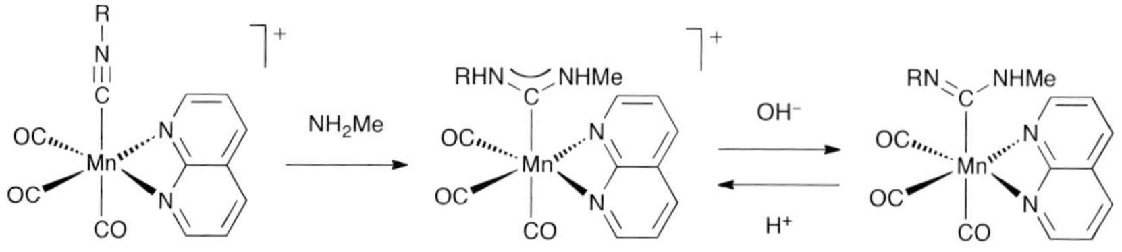

Scheme 15.2

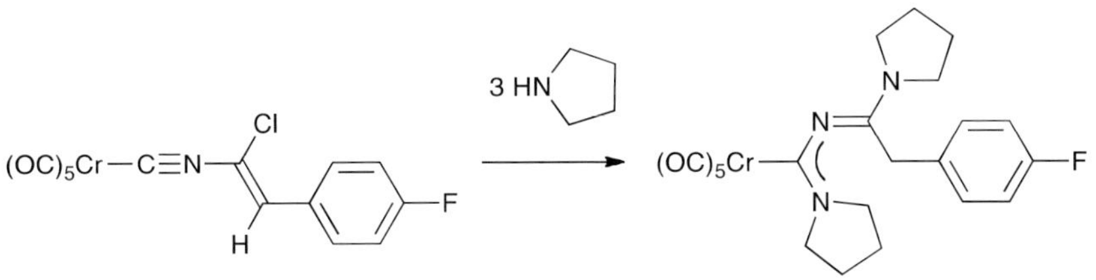

15.2 Coupling of the Isocyanide Ligand with Simple Amines or Alcohols | 533

Scheme 15.3

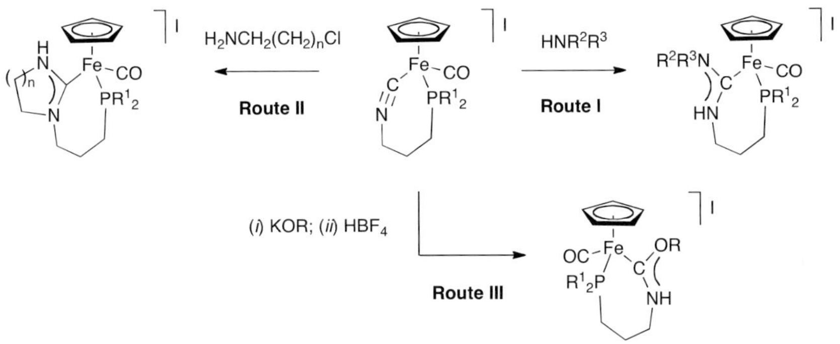

Scheme 15.4

expected nucleophilic attack of this amine towards the isocyanide moiety, complexes that contain unusual amino(imino)carbene ligands (Scheme 15.3).

Mehrkhodavandi et al. [21] showed that the reactions of the phosphine-tethered isocyanide iron(II) complex [(Cp)Fe(CO)(PNC)]I (Cp = η^5-C$_5$H$_5$, PNC = CNCH$_2$CH$_2$PR$^1_2$, R^1 = tBu, Ph) with primary and secondary amines (Scheme 15.4, Route I) afford the corresponding acyclic (diamino)carbene complexes [(Cp)Fe(CO)(PNHCX)]I. In this way, five- and six-membered cyclic (diamino)carbene complexes [(Cp)Fe(CO)(PNCHN-nCy)]I (nCy = five- or six-membered ring) were generated in two steps from the reaction of [CpFe(CO)(PNC)]I with 2-chloroethyl and 3-chloropropylamine, followed by an intramolecular cyclization (Scheme 15.4, Route II). The addition of alkoxides to [(Cp)Fe(CO)(PNC)]I yields the respective ylidene complexes [(Cp)Fe(CO)(PNCO(X))] which, upon protonation by HBF$_4$, form the acyclic oxyamino carbene species [(Cp)Fe(CO)(PNHCO(X))][BF$_4$] (X = Me, Et, iPr; Scheme 15.4, Route III). The related five- and six-membered cyclic oxyamino carbene complexes [(Cp)Fe(CO)(PNCO-nCy)]Cl are formed by the concerted reaction of [(Cp)Fe(CO)(PNC)]I with 2-chloroethoxide and 3-chloropropoxide, followed by intramolecular cyclization [21].

Vicente et al. [22] showed that the reaction of [Tl$_2${S$_2$C=C{C(O)Me}$_2$}]$_n$ with [MCl$_2$(NCPh)$_2$] and the isocyanide (CNR) (1:1:2) provides the isocyanide complexes [M{η^2-S$_2$C=C{C(O)Me}$_2$}(CNR)$_2$] [M = Pd, Pt; R = tBu, xylyl (Xyl)]. Coupling of the latter species with diethylamine then produces the mixed isocyanide carbene complexes [M{η^2-S$_2$C=C{C(O)Me}$_2$}(CNR){C(NEt$_2$)(NHR)}], regardless of

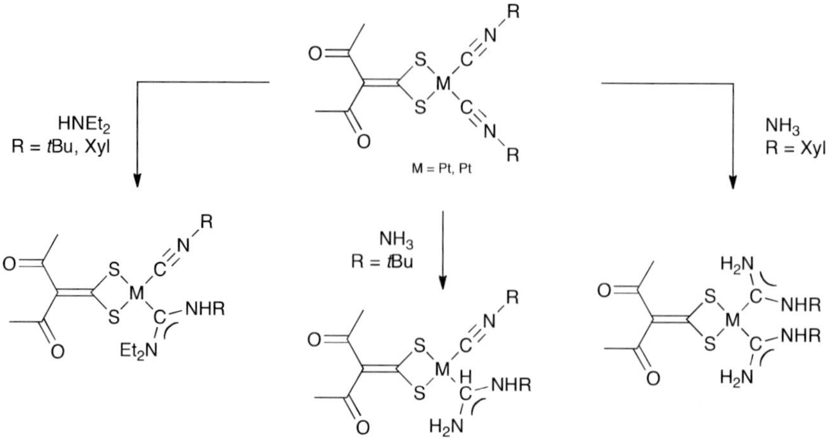

Scheme 15.5

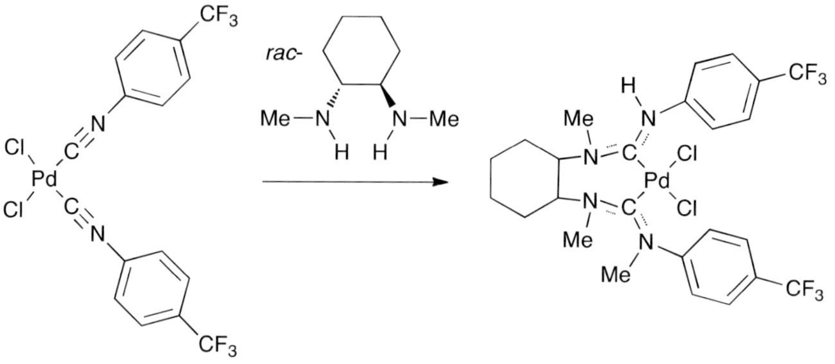

Scheme 15.6

the molar ratio of the reagents, while the addition of an excess of ammonia provides [M{η^2-(S,S')S$_2$ C=C{C(O)Me}$_2$}(CNtBu){C(NH$_2$)(NHtBu)}] or [M{η^2-(S,S')S$_2$ C=C{C(O)Me}$_2$}{C(NH$_2$)(NHXyl)}$_2$] in accord with the nature of the substituent, R [22] (Scheme 15.5).

Slaughter et al. [23] reported that the addition of trans-N,N'-dimethyl-1,2-diaminocyclohexane to a palladium bis(arylisocyanide) complex leads to the one-step formation of the first chiral bis(acyclic diaminocarbene) complex (Scheme 15.6). This is thermally stable under dinitrogen, but undergoes a slow oxidation to form a bis(amidine) complex under air. In a related study [24], the reaction of cis-dichlorobis(p-trifluoromethylphenylisocyanide)palladium(II) with N,N'-bis[(R)-1-phenylethyl]-1,3-diaminopropane afforded an enantiomerically pure, C1-symmetric bis(acyclic diaminocarbene)PdCl$_2$ complex.

Reaction of the bromo-bridged dimeric monocarbene complex [PdBr$_2$(iPr$_2$-bimy)]$_2$ (iPr$_2$-bimy = 1,3-diisopropylbenzimidazolin-2-ylidene) with various isocyanides afforded a series of mixed carbene–isocyanide complexes [PdBr$_2$(iPr$_2$-bimy)

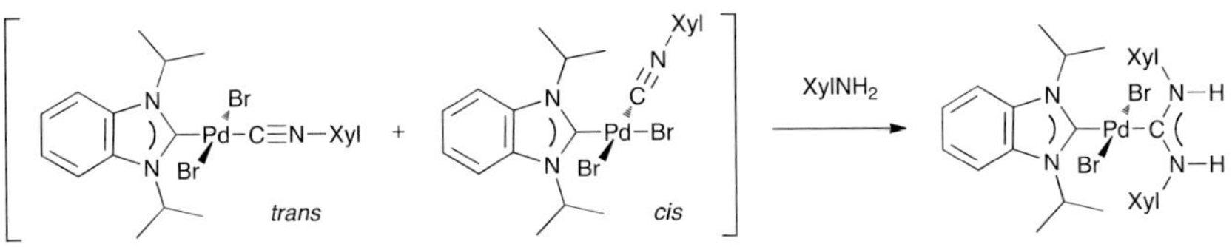

15.2 Coupling of the Isocyanide Ligand with Simple Amines or Alcohols

Scheme 15.7

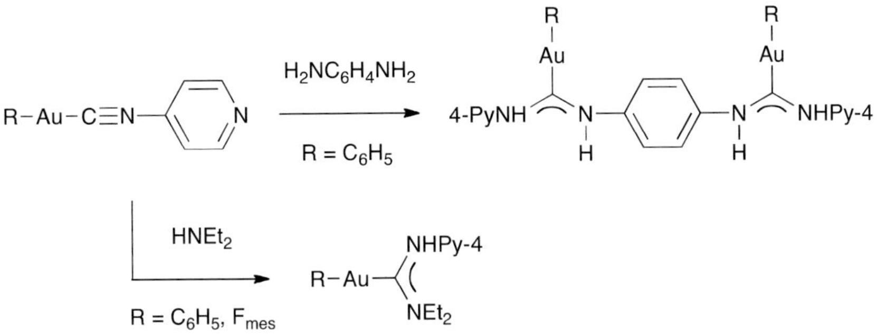

Scheme 15.8

(CNR)], as shown by Huynh et al. [25]. The reaction of the latter with the 2,6-dimethylaniline afforded a new mixed NHC–ADC Pd(II) complex in low yield (Scheme 15.7) [25].

Vicente et al. [26] reported the preparation of the arenediethynylgold(I) complex [{(tBuNH)(Et$_2$N)}CAuC≡C(mes)C≡CAuC{(NHtBu)(NEt$_2$)}] containing aminocarbene ligands starting from [(NCtBu)AuC≡C(mes)C≡CAu(NCtBu)] and diethylamine.

Howell et al. [27] reported that the gold(I)–isocyanide complexes [XAu(CNPh)] (X = Br, Cl) and [(NCS)Au(CNMe)] would react with HNMe(CH$_2$CH$_2$O)$_n$Me (n = 1–11) or HNEt$_2$ to provide the gold–aminocarbene species [XAu{C(NHPh) (MeN(CH$_2$CH$_2$O)$_n$Me)}] and [(NCS)Au{C(NHMe)(NEt$_2$)}]$_2$, respectively. The metal–aminocarbenes thus obtained were further employed as substrates for the direct laser writing of gold decoration onto ceramics [27].

Monomeric gold(I) carbenes of the type [AuR{C(NR^{1}R^2)(NHPy-4)}] [Py-4 = 4-pyridyl; R = C$_6$F$_5$, F$_{mes}$ = 2,4,6-tris(trifluoromethyl)phenyl] have been obtained by reaction of the corresponding isocyanide compounds [AuR(CNPy-4)] with primary (H$_2$NMe) or secondary (HNEt$_2$) amines (e.g., Scheme 15.8), as reported by Espinet et al. [28].

The results of single crystal X-ray diffraction studies have shown that, in some of the complexes, the NHPy-4 moiety forms supramolecular macrocycles supported by hydrogen-bond interactions, either with the N–H groups of other molecules or with water. In addition, dimeric gold(I) carbenes were obtained using a diamine as a nucleophile (Scheme 15.8). Notably, most of the complexes described here display luminescent properties [28].

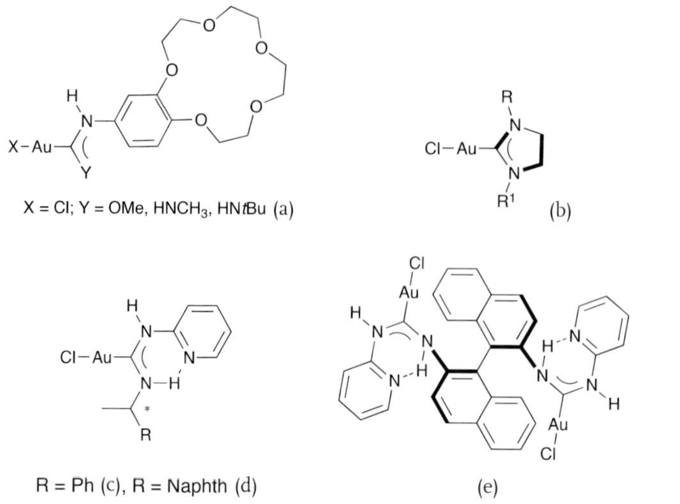

Figure 15.1 Gold(I)–aminocarbene complexes.

In a related study conducted by Espinet et al. [29], the isocyanide in complexes [AuX(CNPy-2)] (X = Cl, C$_6$F$_5$, F$_{mes}$) were shown to react with H$_2$NMe or HNEt$_2$ to produce the aminocarbene species [AuX{C(NR^{1}R^2)(NHPy-2)}]. The NHPy-2 moiety generated in this reaction gives rise to either intramolecular (for primary amines) or intermolecular (for secondary amines) hydrogen bonds. Except for the fluoromesityl derivatives, the carbene complexes display luminescent properties [29].

The new crown ether isocyanide CNR (R = benzo-15-crown-5) was synthesized and the gold(I) derivatives [AuX(CNR)] (X = Cl, C$_6$F$_5$, Br, I), [Au(C$_6$F$_4$OCH$_2$C$_6$H$_4$OC$_n$H$_{2n+1}$-p)(CNR)] (n = 4, 8, 10, 12), and [Au(C$_6$F$_4$OCH$_2$C$_6$H$_2$-3,4,5-(OC$_n$H$_{2n+1}$)$_3$(CNR)] (n = 4, 8, 12) were obtained, starting from the appropriate gold(I) precursors, as shown by Espinet et al. [30]. A nucleophilic attack at the coordinated isocyanide in [AuCl(CNR)] by either methanol or a primary amine (HY) produced the carbene derivatives [AuCl{C(NHR)(OMe)}] and [AuCl{C(NHR^1)(NHR)}] (R^1 = Me, nBu; Figure 15.1, *a*). All of the thus-obtained derivatives are luminescent at room temperature, both in the solid state and in solution [30].

Echavarren, Espinet et al. [31] prepared the series of diaminocarbene complexes [AuCl{C(NHR)(NHR^1)}] and [AuCl{C(NHR)(NEt$_2$)}] (R = tBu, 4-Tol, Xyl, 4-C$_6$H$_4$COOH, 4-C$_6$H$_4$COOEt, R^1 = Me, nBu, iPr, nheptyl, 4-Tol) by reaction of the corresponding gold–isocyanide complexes [AuCl(CNR)] with several primary amines or diethylamine. All of the carbene species were shown to be highly selective catalysts for skeletal rearrangement, the methoxycyclization of 1,6-enynes, and for other related gold-catalyzed transformations.

A series of gold(I)–isocyanide complexes was accomplished, as described by Hashmi et al. [32], via the replacement of tht (tetrahydrothiophene) in [AuCl(tht)] with R^{1}NC, and further conversion to the corresponding diaminocarbene gold(I) complexes upon reaction with primary and symmetrical secondary amines. Subsequent nuclear magnetic resonance (NMR) studies permitted an analysis of the different diastereomers present in solution. On production, the metal–carbenes were employed as catalysts in the gold-catalyzed phenol synthesis (via the intramolecular cyclization of furanyne), reaching an unprecedented turnover number (TON) of approximately 3000 moles of product per mole of catalyst with this substrate. Good conversions for the hydration of phenylacetylene were also obtained [32].

In another study conducted by Hashmi et al. [33], gold(I) complexes bearing unsymmetrically substituted NHCs were obtained from the corresponding metal–isocyanides upon nucleophilic attack of 2-chloroethylammonium chlorides (in the presence of an Et_3N), followed by intramolecular cyclization (Figure 15.1, *b*).

Both, mononuclear and dinuclear chiral gold(I) complexes containing acyclic diaminocarbene ligands were prepared by the reactions of isocyanide gold(I) complexes with chiral amines or diamines (Figure 15.1, *c–e*), as demonstrated by Espinet et al. [34]. The reactions proceeded without racemization, and the absolute configuration of the chiral centers was maintained intact. All of these complexes may serve as catalysts in the cyclopropanation of vinyl arenes, and also in the intramolecular hydroalkoxylation of allenes, providing good yields but only modest or poor enantioselectivities. The diaminocarbene ligands prepared in this way are compatible with different functions and reaction conditions, and are worth considering as alternative systems to NHCs or phosphines in gold-catalyzed reactions.

15.3
Coupling of the Isocyanide Ligand with Functionalized Amines or Alcohols

A variety of NHC complexes of general formula *fac*-[Mn(NHC)(CO)$_3$(bipy)]$^+$ were prepared by Ruiz et al. [35] by the reaction of manganese(I)–isocyanide complexes *fac*-[Mn(CNR)(CO)$_3$(bipy)]$^+$ with propargylamines and propargylic alcohols (Scheme 15.9).

Formation of the cyclic aminocarbene complexes was shown to proceed through an initial nucleophilic attack of the amine or the alcohol on the isocyanide, followed by an intramolecular cyclization process that implies a formal hydroamination reaction of the alkyne residue. This approach allows the synthesis of Mn(I) complexes that contain a variety of carbene ligands of imidazoline-2-ylidene and imidazolidine-2-ylidene types, as well as their analogous N,O-heterocyclic carbenes [35].

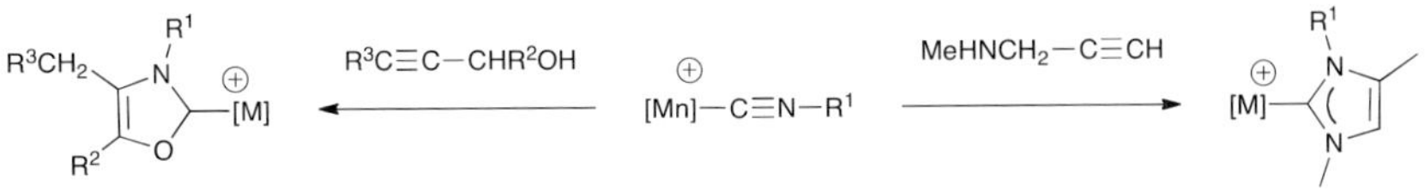

Scheme 15.9

15.4
Coupling of the Isocyanide Ligand with a Hydrazine or Hydrazone

Slaughter et al. have described the palladium-templated addition of phenylhydrazine to methylisocyanide, leading to a Chugaev-type chelating dicarbene–palladium

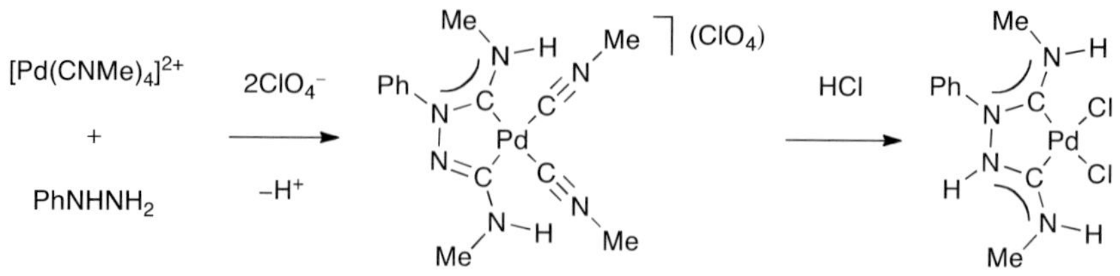

Scheme 15.10

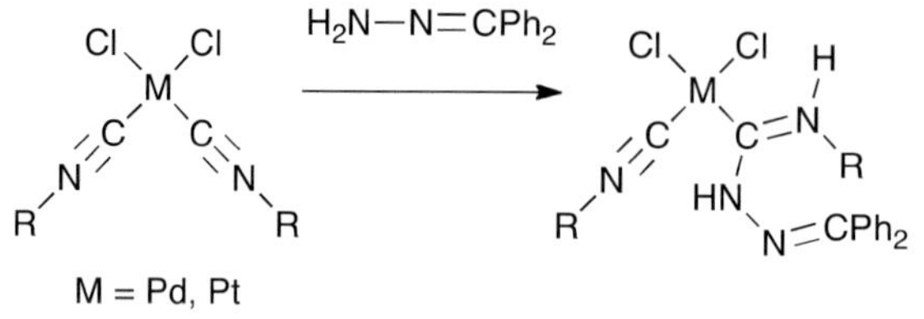

Scheme 15.11

complex (Scheme 15.10) [36, 37]. In this case, when isopropylisocyanide was used as a starting material, the reaction afforded an unprecedented complex that contained aminohydrazinocarbene ligands. In related studies [37, 38], a series of comparable dicarbene–palladium complexes derived from the addition of unsubstituted and substituted hydrazines to Pd–isocyanides was described, and their catalytic activities in the Suzuki–Miyaura cross-coupling reaction were evaluated. Slaughter et al. [39] also showed that the reaction of cis-dichlorobis(p-trifluoromethylphenylisocyanide)palladium(II) with hydrazobenzene would afford a complex with a monodentate aminohydrazino carbene ligand and an unreacted arylisocyanide. Upon heating, this complex was partially converted into a chelating bis(acyclic diaminocarbene) complex. In solution, an equilibrium between the monocarbene and bis(carbene) complexes was observed [39].

Luzyanin, Pombeiro, Kukushkin et al. showed that the metal-mediated reaction between cis-[MCl$_2$(C≡NR)$_2$] (M = Pd, Pt; R = cyclohexyl (Cy), tBu, Xyl, 4-MeOC$_6$H$_4$) and benzophenone hydrazone, H$_2$N–N=CPh$_2$, affords the carbene species (Scheme 15.11), which displays an excellent catalytic activity for the Suzuki–Miyaura cross-coupling under mild conditions [40].

15.5
Coupling of the Isocyanide Ligand with an Imine or Amidine

The first example of coupling between metal-bound isocyanides and imines was recently described by the present authors [41]. Thus, the interplay between cis-[PtCl$_2$(CNXyl)$_2$] and Ph$_2$C=NH resulted in an addition of the benzophenone imine to one isocyanide ligand to yield the aminoimino-carbene cis-[PtCl$_2$(CNXyl){C(N=CPh$_2$)N(H)Xyl}]. This adduct is unstable in solution, even at room tempera-

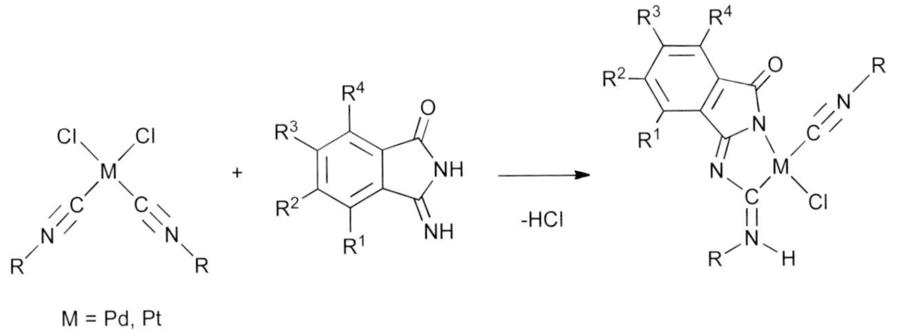

15.5 Coupling of the Isocyanide Ligand with an Imine or Amidine

Scheme 15.12

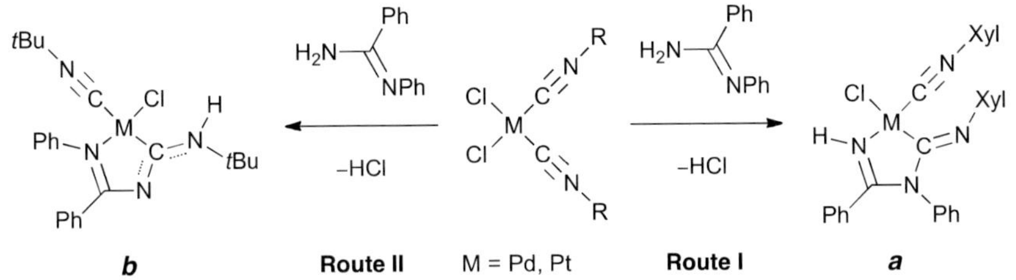

Scheme 15.13

ture, and leads to the diaminocarbene cis-[PtCl$_2$(CNXyl){C(NH$_2$)N(H)Xyl}], which is formally the product of the addition of ammonia to one isocyanide ligand in cis-[PtCl$_2$(CNXyl)$_2$] [41].

The coupling between isocyanide in cis-[MCl$_2$(C≡NR)$_2$] (M = Pd, Pt; R = Cy, tBu, Xyl) and various unsubstituted or substituted 3-iminoisoindolin-1-ones HN=CC$_6$R^{1}R^{2}R^{3}R^{4}CONH (R^1–R^4 = H; R^1, R^3, R^4 = H, R^2 = Me; R^1, R^4 = H; R^2, R^3 = Cl) affords complexes [MCl{C(N–C(C$_6$R^{1}R^{2}R^{3}R^{4}CON))=N(H)R}(C≡NR)] bearing a novel type of carbene ligands (Scheme 15.12), as also shown by the present authors [42].

Kukushkin et al. have described the metal-mediated coupling between one or two isocyanide ligands in cis-[MCl$_2$(C≡NR)$_2$] (M = Pd, Pt) and N-phenylbenzamidine, HN=C(Ph)NHPh (Scheme 15.13) [43]. The coupling proceeds with different regioselectivities upon varying the R group. When an aromatic (R = Xyl) isocyanide is used, N-phenylbenzamidine coordinates to the metal by the amino moiety, and the nucleophilic attack proceeds via the NPh center of the benzamidine (Scheme 15.13, Route I), producing [MCl{C(N(Ph)C(Ph)=NH)=NXyl}(C≡NXyl)] (Scheme 15.13, *a*). For R = tBu, HN=C(Ph)NHPh coordinates to the metal by the NPh moiety, and the addition occurs via the HN=C center of the nucleophile (Scheme 15.13, Route II) to afford [MCl{C(NC(Ph)=NPh)=NHtBu}(C≡NtBu)] (Scheme 15.12, *b*). When R = Cy, a mixture of two products derived from the addition of N-phenylbenzamidine by the two nucleophilic centers was detected. The obtained aminocarbene species showed a high catalytic activity towards the Sonogashira cross-coupling [43].

15.6
Intramolecular Cyclizations of Functionalized Isocyanide Ligands

The reaction of 2-azidophenyl isocyanide with [M(CO)$_5$(THF)] (M = Cr, W; THF = tetrahydrofuran) yields the corresponding isocyanide complex [M(CO)$_5$(CNR)] (R = 2-azidophenyl) which, upon further reaction of triphenylphosphine with the azido function of the isocyanide ligand, produces the corresponding 2-triphenylphosphiniminophenyl isocyanide species (Scheme 15.14), as reported by Hahn et al. [44]. The latter compound undergoes hydrolysis with H$_2$O/HBr to afford triphenylphosphine oxide and the complex containing the unstable 2-aminophenyl isocyanide ligand. This ligand cyclizes spontaneously via an intramolecular nucleophilic attack of the primary amine at the isocyanide carbon atom to yield the 2,3-dihydro-1H-benzimidazol-2-ylidene complexes (NH,NH–NHC species). Double deprotonation of the thus-generated NHC ligand with KOtBu and reaction with 2 equiv. of allyl bromide forms the corresponding N,N-dialkylated benzannulated NHC complex (Scheme 15.14). In another investigation conducted by Hahn et al. [45], the tungsten complex [W(CO)$_5$(CNR)] (R = 2-azidophenyl) reacted with PMe$_3$ at the azido function which, upon subsequent hydrolysis, produced the complex with the corresponding NHC ligand; the latter compound was arylated using the same methodology.

In a related study carried out by Michelin et al. [46], chromium and molybdenum complexes with 2-(azidomethyl)phenyl isocyanide 2-(CH$_2$N$_3$)C$_6$H$_4$NC (AziNC), viz. [M(CO)$_5$(AziNC)], undergo the Staudinger reaction with PPh$_3$ followed by hydrolysis with H$_2$O to produce NHC complexes. Within the same study, the tungsten–isocyanide complex [W(CO)$_5$(CNC$_6$H$_4$-2-CH$_2$I)] reacts with MeNH$_2$ to afford in low yield the NHC species [W(CO)$_5${CN(H)C$_6$H$_4$-2-CH$_2$N(Me)}].

Hahn et al. [47] showed that the reaction of 2-nitrophenyl isocyanide with [M(CO)$_5$(THF)] (M = Cr, Mo, W) yields the expected isocyanide complexes

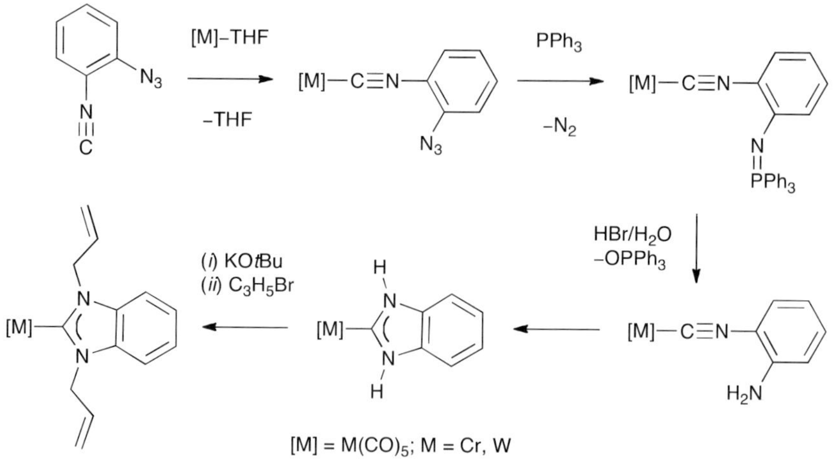

Scheme 15.14

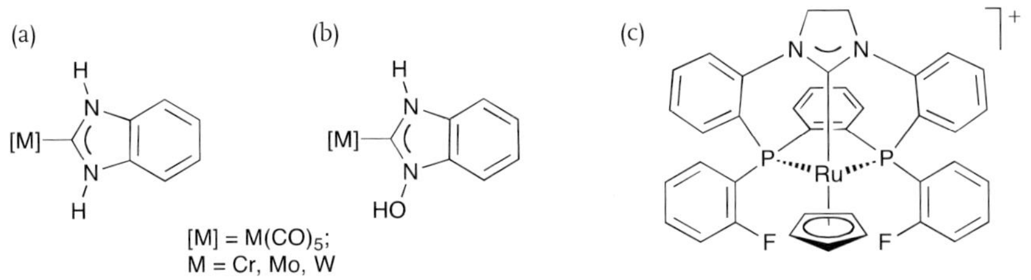

15.6 Intramolecular Cyclizations of Functionalized Isocyanide Ligands | 541

Figure 15.2 Types of diaminocarbene complexes of Cr, Mo, W (*a*, *b*) and Ru (*c*).

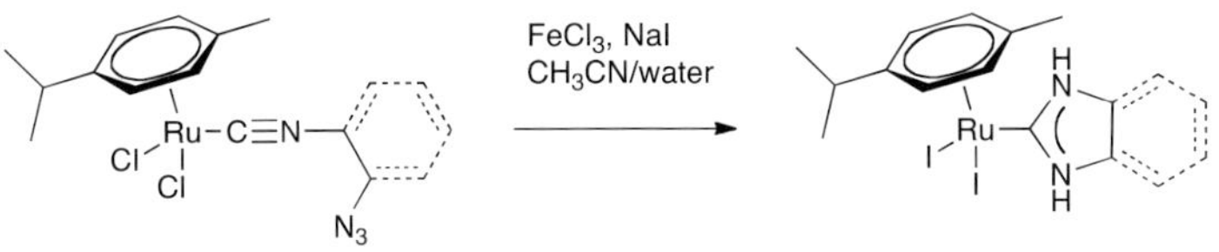

Scheme 15.15

[M(CO)$_5$(CNR)] which are reduced by elemental tin under protic conditions to produce the corresponding NH,NH–NHC complexes upon spontaneous cyclization (Figure 15.2, *a*). When hydrazine hydrate is used instead of tin, an incomplete reduction of the nitro group in the starting isocyanide complex is observed, followed by an intramolecular cyclization to produce the NH,NOH–NHC complexes (Figure 15.2, *b*).

The reaction of [RuCl$_2$(*p*-cymene)]$_2$ with 2-azidoethyl isocyanide or 2-azidophenyl isocyanide leads to the corresponding isocyanide complex [RuCl$_2$(*p*-cymene)(CNR)] [CNR = CN–CH$_2$CH$_2$–N$_3$, CN-(*o*-C$_6$H$_4$)–N$_3$] (Scheme 15.15). Reduction of the azido group using FeCl$_3$/NaI in acetonitrile/water affords complexes with the 2-amino-substituted isocyanide ligands, which undergo intramolecular ring closure to yield [RuI$_2$(*p*-cymene)(NHC)] [48]. Similarly, reduction of the azide moiety in the ruthenium isocyanide/phosphine complexes [(Cp)Ru(P–P)(CNR)] [P–P = 2PPh$_3$, 2PMe$_3$, dppe, dppp; CNR = CNCH$_2$CH$_2$N$_3$, CN(*o*-C$_6$H$_4$)N$_3$] with Zn/NH$_4$Cl/H$_2$O in methanol affords the corresponding [(Cp)Ru(P–P)(NHC)] species [49].

Hahn et al. reported that replacement of the chloride ligand in [(Cp)RuCl(L)] [L = bis[di(2-fluorophenyl)phosphino]benzene] with CNCH$_2$CH$_2$N$_3$, 2-azidoethyl isocyanide, affords [(Cp)Ru(L)(CNCH$_2$CH$_2$N$_3$)]Cl, or [(Cp)Ru(L)(CNCH$_2$CH$_2$N$_3$)] [BF$_4$] upon treatment with [NH$_4$][BF$_4$]. Reduction of the azido group of the coordinated isocyanide with Zn/NH$_4$Cl/H$_2$O generates the 2-aminoethyl isocyanide ligand that cyclizes intramolecularly to produce a metal-bound NH,NH–NHC species. Deprotonation of the thus-obtained NHC ligands results in an intramolecular nucleophilic attack of the amido nitrogen atoms at the fluorinated phenyl groups of the diphosphine ligand to afford a complex with the facially coordinated macrocyclic [11]ane-P$_2$C^{NHC} species (Figure 15.2, *c*) [50].

The same authors reported the preparation of a structurally related iron(II) complex containing 1,2-bis[bis-(2-fluorophenyl)phosphanyl]benzene and

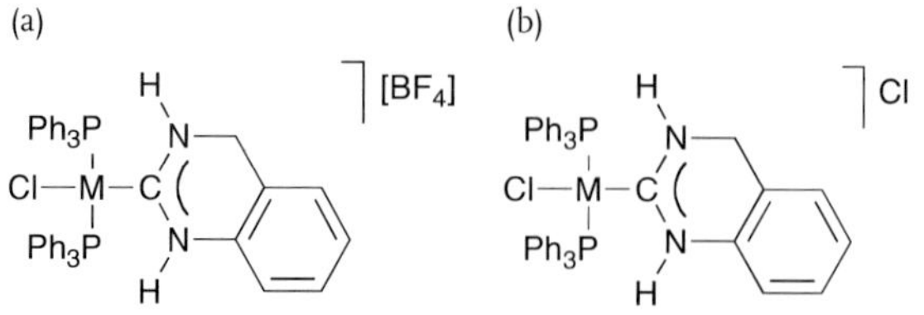

Figure 15.3 Platinum(II)–NHC and palladium(II)–NHC complexes.

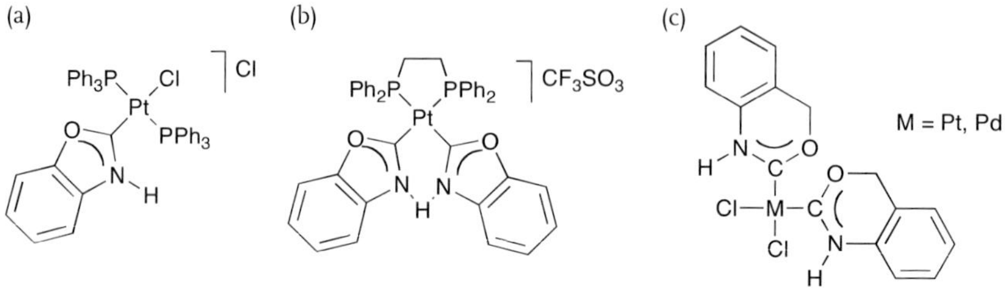

Figure 15.4 Platinum(II) and palladium(II) complexes with heterocyclic oxyaminocarbene ligands.

2-azidoethyl isocyanide ligands [51]. The Staudinger reaction with PPh$_3$ at the azido function, followed by hydrolysis of the iminophosphorane with HBr, yields a novel iron(II) complex containing the above-mentioned NHC ligand. The latter, in the presence of a base, affords the corresponding complex with the above-described facially coordinated macrocyclic [11]ane-P$_2$C^{NHC} ligand [51].

Michelin et al. [52] showed that the coordination of 2-(azidomethyl)phenyl isocyanide, 2-(CH$_2$N$_3$)C$_6$H$_4$N≡C (AziNC), to Pt(II) and Pd(II) centers affords isocyanide species of the types trans-[MCl(AziNC)(PPh$_3$)$_2$][BF$_4$] and [MCl$_2$(AziNC)$_2$]. The complexes trans-[MCl(AziNC)(PPh$_3$)$_2$][BF$_4$] then react with PPh$_3$ and H$_2$O to afford the heterocyclic carbene species trans-[MCl{CN(H)C$_6$H$_4$-2-CH$_2$N(H)}(PPh$_3$)$_2$][BF$_4$] (Figure 15.3, *a*). The interplay of [MCl$_2$(AziNC)$_2$] with PPh$_3$ results in the displacement of one isocyanide with the formation of cis-[MCl$_2$(AziNC)(PPh$_3$)]. The latter, upon reaction with 2 equiv. of PPh$_3$ and H$_2$O, leads to the cationic NHC complex trans-[MCl{CN(H)C$_6$H$_4$-2-CH$_2$N(H)}(PPh$_3$)$_2$]Cl (Figure 15.3, *b*) [52, 53].

In another study, the coordination of novel arsonium-substituted isocyanides to Pt(II) centers provided metal–isocyanide species which could be further converted into the corresponding indolidin-2-ylidene derivatives in the presence of NEt$_3$ [54].

Hahn et al. [55] demonstrated that 2-(trimethylsiloxy)phenyl isocyanide, upon reaction with cis-[PtCl$_2$(PPh$_3$)$_2$], followed by a subsequent hydrolysis of the O–SiMe$_3$ bond, yielded the cyclic carbene species trans-[Pt(benzoxazolin-2-ylidene)Cl(PPh$_3$)$_2$]Cl (Figure 15.4, *a*). Reaction of the dinuclear complex [Pt(μ-Cl)(dppe)]$_2$(CF$_3$SO$_3$)$_2$ [dppe = 1,2-bis(diphenylphosphanyl)ethane] with 2 equiv. of 2-(trimethylsiloxy) phenyl isocyanide did not yield the monocarbene complex but rather the dicarbene species [Pt(benzoxazolinato-2-ylidene)(benzoxazolin-2-ylidene)(dppe)](CF$_3$SO$_3$) (Figure 15.4, *b*).

The coordination of 2-(trimethylsiloxymethyl)phenyl isocyanide, 2-(CH$_2$OSiMe$_3$)C$_6$H$_4$N≡C to Pt(II) and Pd(II) centers affords compounds *cis*-[MCl$_2$(CNC$_6$H$_4$-2-CH$_2$OSiMe$_3$)$_2$] and *cis*-[PdCl$_2$(CNC$_6$H$_4$-2-CH$_2$OSiMe$_3$)(PPh$_3$)], as reported by Michelin *et al.* [56]. The isocyanide complexes were transformed to the corresponding cyclic aminooxycarbene derivatives in the presence of a catalytic amount of fluoride ions (TBAF = tetrabutylammonium fluoride) in MeOH (Figure 15.4, *c*).

Vicente *et al.* [57] described the base-promoted transformation of the iminoacyl isocyanide complex *trans*-[Pd{C(=NXyl)C$_6$H$_4$NHC(O)NHTol-2}I(CNXyl)$_2$] to yield the iminoacyl amido carbene C,N,C pincer compound [Pd{κ^{3}C,N,C–C(=NXyl)C$_6$H$_4$NC(O)NTolC(NHXy)-2}(CNXyl)]. The latter species was shown to be an efficient precatalyst in the Heck and Suzuki coupling reactions [57].

15.7
Coupling of Isocyanides with Dipoles

Reactions of the azide gold(I) complex [Au(N$_3$)(PPh$_3$)] with the isocyanides R^{1}NC (R^1 = Xyl, *t*Bu, Cy) yielded the corresponding neutral tetrazolyl(phosphine) complexes [Au(tetrazolyl)(PPh$_3$)] (Scheme 15.16, *a–c*), as shown by Raubenheimer *et al.* [58]. Alkylation of the species *a* (R^1 = Xyl) with methyl triflate on N4 allowed the isolation of the corresponding carbene species *d*. For R^1 = Cy, the complex of type *c* was not isolated in the pure form but was converted, upon triphenylphosphine replacement by isocyanide, into the mixed tetrazolyl-isocyanide complex [Au(tetrazolyl)(CNCy)]. The latter reacted with CyNH$_2$ to produce the corresponding acyclic diaminocarbene [Au(tetrazolyl){C(NHCy)$_2$}] [58].

The reaction between *cis*-[PdCl$_2$(C≡NR)$_2$] (R = Cy, *t*Bu, Xyl) and the acyclic nitrones O$^+$N(R^2)=C(H)R^3 (R^2 = Me, CH$_2$Ph; R^3 = 4-MeC$_6$H$_4$) provides the carbene complexes [PdCl$_2${C(ONR^{2}C^{c}HR^3)=N^{d}R}(C≡NR)(C^c–N^d)] (Scheme 15.17), as shown by Pombeiro, Kukushkin *et al.* [59]. The latter species are originated from the previously unreported metal-mediated [2+3] cycloaddition of nitrones to coordinated isocyanides. The coupling of *cis*-[PdCl$_2$(C≡NR)$_2$] with the nonaromatic cyclic nitrone $^-$O$^+$N^a=CHCH$_2$CH$_2$C^{b}Me$_2$(N^a–C^b) leads to the corresponding carbene species [PdCl$_2${C(ON^{a}CMe$_2$CH$_2$CH$_2$C^{b}H)=N^{e}R}(C≡NR)(N^a–C^b)(C^b–N^e)] [59].

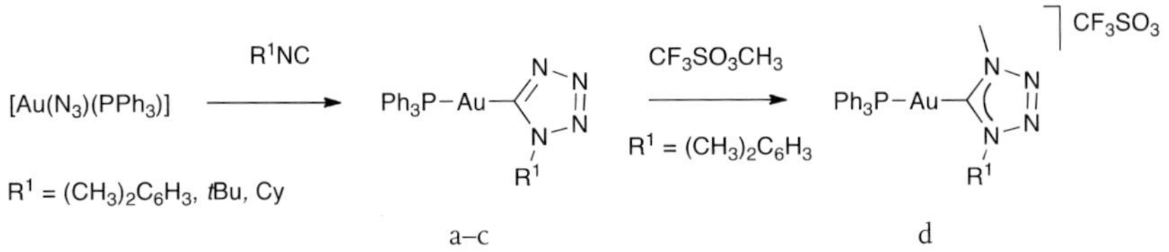

Scheme 15.16

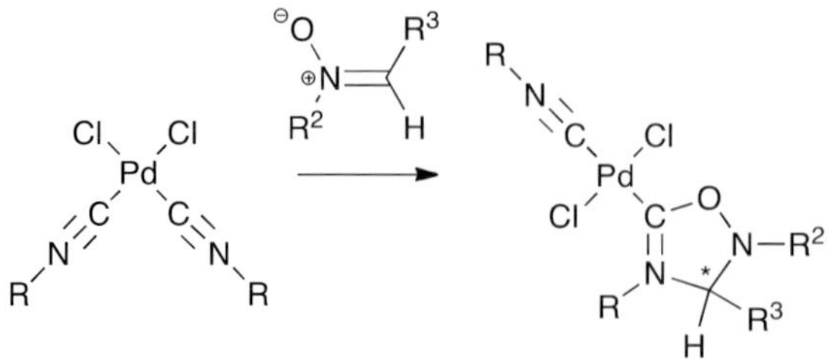

Scheme 15.17

15.8
Other Reactions

The isocyanide complex [(Cp*)Mo(CNtBu)$_3$(PhC≡CPh)][BF$_4$] (Cp* = η^5-C$_5$Me$_5$) reacts with one additional equivalent of CNtBu to yield [(Cp*)Mo{=C(NHtBu)C(Ph)=C(Ph)CN}(CNtBu)$_2$][BF$_4$], containing an η^3-vinylcarbene ligand that is formed from the coupling of two isocyanide ligands with diphenylacetylene, with the concomitant protonation of one of these isocyanide fragments and dealkylation of the other one (Scheme 15.18, Route I), as shown by Connelly et al. [60]. Protonation of the former complex with HBF$_4$ generates the aminovinylcarbene complex [(Cp*)Mo{=C(Ph)C(Ph)C=NHtBu}(CNtBu)$_2$][BF$_4$]$_2$, by inducing the coupling of a formally protonated isocyanide ligand with diphenylacetylene (Scheme 15.18, Route II). The latter reacts with CNtBu to give the above-mentioned η^3-vinylcarbene complex (Scheme 15.18, Route III). A similar reaction of the vinylcarbene compound [(Cp*)Mo{=C(Ph)C(Ph)C=NHtBu}(CNtBu)$_2$][BF$_4$]$_2$ with P(OMe)$_3$ generates the respective η^3-vinylcarbene [(Cp*)Mo{=C(NHtBu)C(Ph)=C(Ph)CN}(CNtBu){P(OMe)$_3$}][BF$_4$] involving the same coupling and elimination pattern.

Pétillon, Schollhammer et al. reported [61] that the reductive coupling between isocyanide ligands attached to adjacent molybdenum atoms in the bis(arylisocyanide) complex [(Cp)$_2$Mo$_2$(μ-SMe)$_3$(XylNC)$_2$][BF$_4$], promoted by hydrosulfide anion, affords the dimetallaimidoyl(amino)carbene derivative [(Cp)$_2$Mo$_2$(μ-SMe)$_3$(μ-η^1(C):

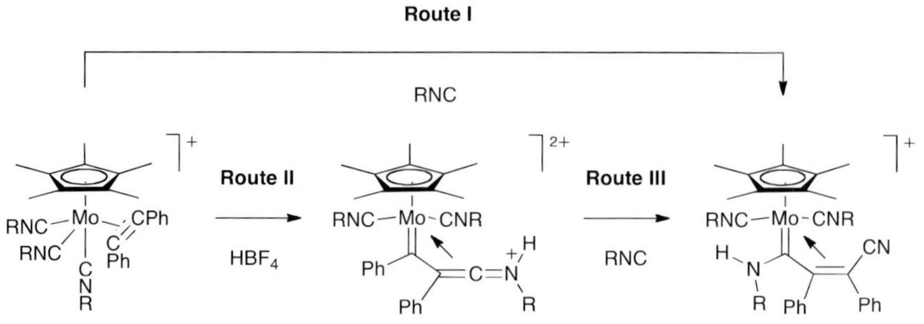

Scheme 15.18

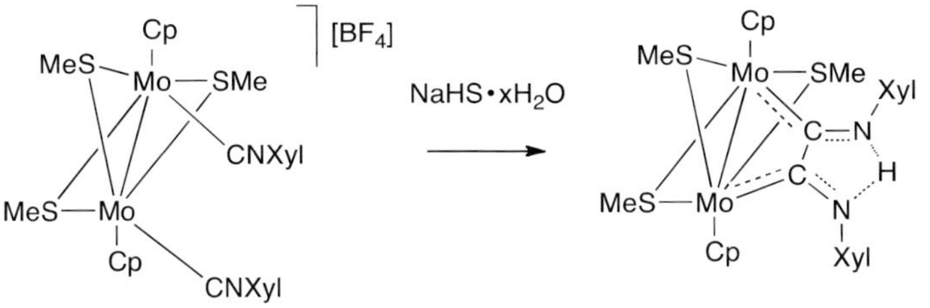

Scheme 15.19

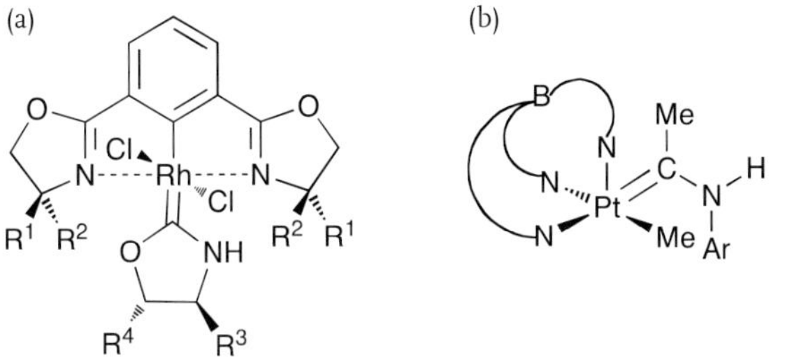

Figure 15.5 Rhodium(III)–carbene and platinum(II)–carbene complexes.

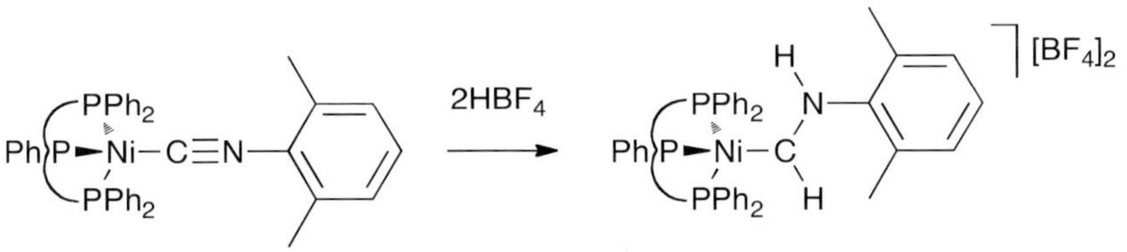

Scheme 15.20

$\eta^1(C)$–C(NHXyl)C(NXyl)}], in which both isocyanide groups of the starting isocyanide complex are now linked by a new C–C bond (Scheme 15.19).

Nishiyama et al. [62] showed that compounds [(Phebox)RhCl$_2$(CNCH$_2$R^3)] [Phebox = 2,6-bis(oxazolinyl)phenyl; CNCH$_2$R^3 = methyl isocyanoacetate, p-(tolylsulfonyl)methyl isocyanide], bearing a ligated isocyanide with an acidic methylene group, readily react with aldehydes (R^{4}CHO) in the presence of tBuOK to produce the aldol adducts as diastereomeric mixtures of chiral Fischer carbene complexes (Figure 15.5, *a*).

The nickel isocyanide complex [Ni(triphos)(CNXyl)] [triphos = *bis*(2-diphenylphosphinoethyl)phenylphosphine], prepared by reaction of [Ni(COD)$_2$] with triphos and CNXyl, adds 2 equiv. of HBF$_4$ to afford the stable dicationic nickel carbene complex [Ni(triphos){C(H)N(H)Xyl}][BF$_4$]$_2$ (Scheme 15.20), as demonstrated by Kubiak et al. [63]. The proposed mechanism for this reaction is based on the initial metal protonation, resulting in a nickel hydride complex, followed by insertion of the isocyanide to produce a nickel imino formyl species which, upon protonation, results in the final carbene compound.

In another investigation conducted by Templeton et al. [64], platinum(II) aminocarbene complexes were synthesized upon nucleophilic attack by metal alkyl reagents at the isocyanide ligand of the corresponding isocyanide/tris(3,5-dimethylpyrazolyl)borate platinum(II) complexes, followed by the addition of acid (Figure 15.5, *b*).

15.9
Final Remarks

In the above-described studies, the most recent investigations of the preparation of metal–carbene species from corresponding metal-bound isocyanides have been grouped together according to the main types of reactions involved. From this, a variety of conclusions can be drawn, as follows.

First, although a wide range of transition metals has been used in the successful generation of ligated carbenes from isocyanides, metals from Groups 4 and 5 – as well as a few from other groups (e.g., Tc, Re, Os, Co, Ir, Cu, and Ag) – have been neglected or studied to a much lesser degree during the period covered by these investigations. Clearly, a further extension of studies involving these metals is to be expected in the future.

Second, various types of carbene ligand may be obtained via elegant and frequently atom-efficient synthetic procedures, starting from CNR substrates. The typical processes leading to ADC and NHC species represent stoichiometric addition reactions that generate well-defined complexes.

Third, despite the fact that ADC ligands have been explored much less as catalysts compared to NHC derivatives, the previously reported results in this field [24, 31, 32, 34, 37, 38, 40, 43, 65] clearly indicate a high potential for ADC complexes in catalysis. Taking into account that the generation of M-ADCs via other routes, as distinct from the use of metal-ligated isocyanides, is often associated with experimental difficulties, it is to be expected that the development of a metal-mediated approach to complexes with ADC ligands from isocyanide precursors will play a major role in the further involvement of these compounds in catalysis.

Acknowledgments

These studies have been partially supported by the Fundação para a Ciência e a Tecnologia (FCT), Portugal (including FCT projects PTDC/QUI-QUI/098760/2008, PTDC/QUI-QUI/109846/2009 and PEst-OE/QUI/UI0100/2011). K.V.L. would like to thank the FCT and the Instituto Superior Técnico (IST) for his research contract (Ciência 2008 program).

References

1 Ugi, I. (1971) *Isonitrile Chemistry*, Academic Press, New York.
2 Ugi, I., Lohberger, S., and Karl, R. (1991) The Passerini and Ugi reactions, in *Comprehensive Organic Synthesis: Selectivity for Synthetic Efficiency* (eds B.M. Trost and C.H. Heathcock), Pergamon, Oxford, pp. 1083–1109.
3 Ugi, I., Marquarding, D., and Urban, R. (1982) Syntheses of peptides by four-component condensation, in *Chemistry and Biochemistry of Amino Acids, Peptides and Proteins* (ed. B. Weinstein), Marcel Dekker, New York, pp. 245–289.
4 Dömling, A. (2006) Recent developments in isocyanide based multicomponent reactions in applied chemistry. *Chem. Rev.*, **106** (1), 17–89.
5 Dömling, A. and Ugi, I. (2000) Multicomponent reactions with isocyanides, *Angew. Chem. Int. Ed.*, **39** (18), 3168–3210.
6 Pombeiro, A.J.L. (2007) Characterization of coordination compounds by electrochemical parameters. *Eur. J. Inorg. Chem.*, 1473–1482.
7 Pombeiro, A.J.L. (2005) Electron-donor/acceptor properties of carbynes, carbenes, vinylidenes, allenylidenes and alkynyls as measured by electrochemical ligand parameters. *J. Organomet. Chem.*, **690** (24–25), 6021–6040.
8 Pombeiro, A.J.L. (1997) Molecular electrochemistry in coordination chemistry: metal-ligand bonds and their activation by electron transfer. *New J. Chem.*, **21** (6–7), 649–660.
9 Michelin, R.A., Pombeiro, A.J.L., and Guedes da Silva, M.F.C. (2001) Aminocarbene complexes derived from nucleophilic addition to isocyanide ligands. *Coord. Chem. Rev.*, **218**, 75–112.
10 Pombeiro, A.J.L., Guedes da Silva, M.F.C., and Michelin, R.A. (2001) Aminocarbyne complexes derived from isocyanides activated toward electrophilic addition. *Coord. Chem. Rev.*, **218**, 43–74.
11 Pombeiro, A.J.L. and Guedes da Silva, M.F.C. (2001) Coordination chemistry of CNH_2, the simplest aminocarbyne. *J. Organomet. Chem.*, **617–618**, 65–69.
12 Pombeiro, A.J.L. (2001) Coordination chemistry of CNH, the simplest isocyanide. *Inorg. Chem. Commun.*, **4** (10), 585–597.
13 Herrmann, W.A. (2002) N-heterocyclic carbenes: a new concept in organometallic catalysis, *Angew. Chem. Int. Ed.*, **41** (8), 1290–1309.
14 Hahn, F.E. and Jahnke, M.C. (2008) Heterocyclic carbenes: synthesis and coordination chemistry. *Angew. Chem. Int. Ed.*, **47** (17), 3122–3172.
15 Glorius, F. (2007) N-heterocyclic carbenes in catalysis – an introduction. *Top. Organomet. Chem.*, **21**, 1–20.
16 Nolan, S.P. and Navarro, O. (2007) C–C bond formation by cross-coupling, in *Comprehensive Organometallic Chemistry III* (ed. A. Canty), Elsevier, Oxford, pp. 1–38.
17 Díez-González, S., Marion, N., and Nolan, S.P. (2009) N-heterocyclic carbenes in late transition metal catalysis. *Chem. Rev.*, **109** (8), 3612–3676.
18 Vignolle, J., Catton, X., and Bourissou, D. (2009) Stable noncyclic singlet carbenes. *Chem. Rev.*, **109** (8), 3333–3384.
19 Ruiz, J. and Perandones, B.F. (2009) Acyclic diamino carbene complexes of manganese(I): synthesis, deprotonation, and subsequent multiple insertion reaction of alkynes. *Organometallics*, **28** (3), 830–836.
20 Langenhahn, V., Beck, G., Zinner, G., Lentz, D., Herrschaft, B., and Fehlhammer, W.P. (2007) Reactions at the coordinated trichloromethyl isocyanide. Part VII. α-Chloroalkenylisocyanide versus oxazolin-2-ylidene(ato) complex formation. *J. Organomet. Chem.*, **692**, 2936–2948.
21 Yu, I., Wallis, C.J., Patrick, B.O., Diaconescu, P.L., and Mehrkhodavandi, P. (2010) Phosphine-tethered carbene ligands: template synthesis and reactivity of cyclic and acyclic functionalized carbenes. *Organometallics*, **29** (22), 6065–6076.
22 Vicente, J., Chicote, M.T., Huertas, S., and Jones, P.G. (2003) 1,1-ethylenedithiolato complexes of palladium(II) and platinum(II) with

isocyanide and carbene ligands. *Inorg. Chem.*, **42** (14), 4268–4274.

23 Wanniarachchi, Y. and Slaughter, L.M. (2007) One-step assembly of a chiral palladium bis(acyclic diaminocarbene) complex and its unexpected oxidation to a bis(amidine) complex. *Chem. Commun.*, 3294–3296.

24 Wanniarachchi, Y.A., Subramanium, S.S., and Slaughter, L.M. (2009) Palladium complexes of bis(acyclic diaminocarbene) ligands with chiral N-substituents and 8-membered chelate rings. *J. Organomet. Chem.*, **694**, 3297–3305.

25 Han, Y. and Huynh, H.V. (2009) Mixed carbene-isocyanide Pd(II) complexes: synthesis, structures and reactivity towards nucleophiles. *Dalton Trans.*, 2201–2209.

26 Vicente, J., Chicote, M.T., Alvarez-Falcón, M.M., Abrisqueta, M.-A., Hernández, F.J., and Jones, P.G. (2003) New arenediethynylgold(I) complexes. Crystal structures of [Ph₃PAuC°C(phenylendiyl-1,3)C°CAuPPh₃] and [Ph₃PAuC°C(mesitylendiyl-1,3)C°CAuPPh₃]. *Inorg. Chim. Acta*, **347**, 67–74.

27 Heathcote, R., Howell, J.A.S., Jennings, N., Cartlidge, D., Cobden, L., Coles, S., and Hursthouse, M. (2007) Gold(I)-isocyanide and gold(I)-carbene complexes as substrates for the laser decoration of gold onto ceramic surfaces. *Dalton Trans.*, 1309–1315.

28 Bartolomé, C., Carrasco-Rando, M., Coco, S., Cordovilla, C., Espinet, P., and Martín-Alvarez, J.M. (2007) Gold(I)–carbenes derived from 4-pyridylisocyanide complexes: supramolecular macrocycles supported by hydrogen bonds, and luminescent behavior. *Dalton Trans.*, 5339–5345.

29 Bartolomé, C., Carrasco-Rando, M., Coco, S., Cordovilla, C., Martín-Alvarez, J.M., and Espinet, P. (2008) Luminescent gold (I) carbenes from 2-pyridylisocyanide complexes: structural consequences of intramolecular versus intermolecular hydrogen-bonding interactions. *Inorg. Chem.*, **47** (5), 1616–1624.

30 Arias, J., Bardají, M., and Espinet, P. (2008) Luminescence and mesogenic properties in crown-ether-isocyanide or carbene gold(I) complexes: luminescence in solution, in the solid, in the mesophase, and in the isotropic liquid state. *Inorg. Chem.*, **47** (9), 3559–3567.

31 Bartolomé, C., Ramiro, Z., García-Cuadrado, D., Pérez-Galán, P., Raducan, M., Bour, C., Echavarren, A.M., and Espinet, P. (2010) Nitrogen acyclic gold(I) carbenes: excellent and easily accessible catalysts in reactions of 1,6-enynes. *Organometallics*, **29** (4), 951–956.

32 Hashmi, S.K., Hengst, T., Lothschütz, C., and Rominger, F. (2010) New and easily accessible nitrogen acyclic gold(I) carbenes: structure and application in the gold-catalyzed phenol synthesis as well as the hydration of alkynes. *Adv. Synth. Catal.*, **352** (8), 1315–1337.

33 Hashmi, S.K., Lothschütz, C., Böhling, C., Hengst, T., Hubbert, C., and Rominger, F. (2010) Carbenes made easy: formation of unsymmetrically substituted N-heterocyclic carbene complexes of palladium(II), platinum(II) and gold(I) from coordinated isonitriles and their catalytic activity. *Adv. Synth. Catal.*, **352** (17), 3001–3012.

34 Bartolomé, C., García-Cuadrado, D., Ramiro, Z., and Espinet, P. (2010) Synthesis and catalytic activity of gold chiral nitrogen acyclic carbenes and gold hydrogen bonded heterocyclic carbenes in cyclopropanation of vinyl arenes and in intramolecular hydroalkoxylation of allenes. *Inorg. Chem.*, **49** (21), 9758–9764.

35 Ruiz, J., Perandones, B.F., García, G., and Mosquera, M.E.G. (2007) Synthesis of N-heterocyclic carbene complexes of manganese(I) by coupling isocyanide ligands with propargylamines and propargylic alcohols. *Organometallics*, **26** (23), 5687–5695.

36 Moncada, A., Tanski, J., and Slaughter, L. (2005) Sterically controlled formation of monodentate versus chelating carbene ligands from phenylhydrazine. *J. Organomet. Chem.*, **690**, 6247–6251.

37 Slaughter, L.M. (2008) "Covalent self-assembly" of acyclic diaminocarbene ligands at metal centers. *Commun. Inorg. Chem.*, **29**, 46–72.

38 Moncada, A.I., Manne, S., Tanski, J.M., and Slaughter, L.M. (2006) Modular chelated palladium diaminocarbene

complexes: synthesis, characterization, and optimization of catalytic Suzuki–Miyaura cross-coupling activity by ligand modification. *Organometallics*, **25** (2), 491–505.

39 Wanniarachchi, Y. and Slaughter, L.M. (2008) Reversible chelate ring opening of a sterically crowded palladium bis(acyclic diaminocarbene) complex. *Organometallics*, **27** (6), 1055–1062.

40 Luzyanin, K.V., Tskhovrebov, A.G., Carias, M.C., Guedes da Silva, M.F.C., Pombeiro, A.J.L., and Kukushkin, V. Yu. (2009) Novel metal-mediated (M = Pd, Pt) coupling between isonitriles and benzophenone hydrazone as a route to aminocarbene complexes exhibiting high catalytic activity (M = Pd) in the Suzuki–Miyaura reaction. *Organometallics*, **28** (22), 6559–6566.

41 Luzyanin, K.V., Guedes da Silva, M.F.C., Kukushkin, V. Yu., and Pombeiro, A.J.L. (2009) First example of an imine addition to coordinated isonitrile. *Inorg. Chim. Acta*, **362** (3), 833–838.

42 Luzyanin, K.V., Guedes da Silva, M.F.C., Kukushkin, V. Yu., and Pombeiro, A.J.L. (2008) Coupling between 3-iminoisoindolin-1-ones and complexed isonitriles as a metal-mediated route to a novel type of palladium and platinum iminocarbene species. *Organometallics*, **27** (20), 833–888.

43 Tskhovrebov, A.G., Luzyanin, K.V., Kuznetsov, M.L., Sorokoumov, V.N., Balova, I.A., Haukka, M., and Kukushkin, V. Yu. (2011) Substituent R-dependent regioselectivity switch in nucleophilic addition of N-phenylbenzamidine to Pd II- and Pt II-complexed isonitrile RNC giving aminocarbene-like species. *Organometallics*, **30** (4), 863–874.

44 Hahn, F.E., Langenhahn, V., Meier, N., Lügger, T., and Fehlhammer, W.P. (2003) Template synthesis of benzannulated N-heterocyclic carbene ligands. *Chem. Eur. J.*, **9** (3), 704–712.

45 Hahn, F.E., Langenhahn, V., and Pape, T. (2005) Template synthesis of tungsten complexes with saturated N-heterocyclic carbene ligands. *Chem. Commun.*, 5390–5392.

46 Basato, M., Facchin, G., Michelin, R.A., Mozzon, M., Pugliese, S., Sgarbossa, P., and Tassan, A. (2003) Transition metal coordination and reactivity of 2-(azidomethyl)-, 2-(chloromethyl)-, and 2-(iodomethyl)phenyl isocyanides. *Inorg. Chim. Acta*, **356**, 349–356.

47 Hahn, F.E., García Plumed, C., Münder, M., and Lügger, T. (2004) Synthesis of benzannulated N-heterocyclic carbene ligands by a template synthesis from 2-nitrophenyl isocyanide. *Chem. Eur. J.*, **10** (24), 6285–6293.

48 Kaufhold, O., Flores-Figueroa, A., Pape, T., and Hahn, F.E. (2009) Template synthesis of ruthenium complexes with saturated and benzannulated NH,NH-stabilized N-heterocyclic carbene ligands. *Organometallics*, **28** (3), 896–901.

49 Flores-Figueroa, A., Kaufhold, O., Feldmann, K.-O., and Hahn, F.E. (2009) Synthesis of NHC complexes by template controlled cyclization of beta-functionalized isocyanides. *Dalton Trans.*, 9334–9342.

50 Flores-Figueroa, A., Kaufhold, O., Hepp, A., Fröhlich, R., and Hahn, F.E. (2009) Synthesis of a ruthenium(II) complex containing an [11]ane-P2C-NHC (NHC = imidazolidin-2-ylidene) macrocycle. *Organometallics*, **28** (21), 6362–6369.

51 Flores-Figueroa, A., Pape, T., Weigand, J.J., and Hahn, F.E. (2010) Template-controlled formation of an [11]ane-P2C-NHC macrocyclic ligand at an iron(II) template. *Eur. J. Inorg. Chem.*, 2907–2910.

52 Basato, M., Benetollo, F., Facchin, G., Michelin, R.A., Mozzon, M., Pugliese, S., Sgarbossa, P., Sbovata, S.M., and Tassan, A. (2004) The Staudinger reaction of platinum(II)- and palladium(II)-coordinated 2-(azidomethyl)phenyl isocyanide. X-ray structure of trans-[PtCl{(H)}(PPh$_3$)$_2$][BF$_4$]·CDCl$_3$·H$_2$O. *J. Organomet. Chem.*, **689**, 454–462.

53 Facchin, G., Michelin, R., Mozzon, M., Pugliese, S., Sgarbossa, P., and Tassan, A. (2002) Synthesis and transition metal coordination of 2-(azidomethyl) phenyl isocyanide. *Inorg. Chem. Commun.*, **5**, 915–918.

54 Facchin, G., Michelin, R., Mozzon, M., Sgarbossa, P., and Tassan, A. (2004) Synthesis and cyclization reactions of platinum(II)-coordinated

arsonium-substituted phenyl isocyanides, o-(IR$_3$As–CH$_2$)C$_6$H$_4$NC. *Inorg. Chim. Acta*, **357**, 3385–3389.

55 Hahn, F.E., Klusmann, D., and Pape, T. (2008) Template synthesis of platinum complexes with benzoxazolin-2-ylidene ligands. *Eur. J. Inorg. Chem.*, 4420–4424.

56 Facchin, G., Michelin, R.A., Mozzon, M., and Tassan, A. (2002) Synthesis, coordination and reactivity of 2-isocyanides. *J. Organomet. Chem.*, **662**, 70–76.

57 Vicente, J., Abad, J.-A., López-Serrano, J., Jones, P.G., Nájera, C., and Botella-Segura, L. (2005) Synthesis and reactivity of ortho-palladated arylureas. Synthesis and catalytic activity of a C,N,C pincer complex. Stoichiometric syntheses of some N-heterocycles. *Organometallics*, **24** (21), 5044–5057.

58 Gabrielli, W.F., Nogai, S.D., McKenzie, J.M., Cronje, S., and Raubenheimer, H.G. (2009) Tetrazolyl and tetrazolylidene complexes of gold: a synthetic and structural study. *New J. Chem.*, **33** (11), 2208–2218.

59 Luzyanin, K.V., Tskhovrebov, A.G., Guedes da Silva, M.F.C., Haukka, M., Pombeiro, A.J.L., and Kukushkin, V. Yu. (2009) Metal-mediated [2+3] cycloaddition of nitrones to palladium-bound isonitriles. *Chem. Eur. J.*, **15**, 5969–5978.

60 Adams, C.J., Anderson, K.M., Bartlett, I.M., Connelly, N.G., Orpen, G., and Paget, T.J. (2002) Ligand-induced and reductively induced alkyne–isocyanide coupling reactions of [Mo(CNBut)$_3$ (PhCCPh)(η^5-C$_5$Me$_5$)][BF$_4$]. *Organometallics*, **21** (16), 3454–3463.

61 Ojo, W.-S., Pétillon, F.Y., Schollhammer, P., and Talarmin, J. (2008) C-C, C-S, and C-N coupling versus dealkylation processes in the cationic tris(thiolato) dimolybdenum(III) complexes [Mo$_2$Cp$_2$ (μ-SMe)$_3$L$_2$]$^+$ (L = xylNC, *t*-BuNC, CO, MeCN). *Organometallics*, **27** (16), 4207–4222.

62 Motoyama, Y., Shimozono, K., Aoki, K., and Nishiyama, H. (2002) Formation of Fischer carbene complexes in asymmetric aldol-type condensation of an isocyanide component on bis(oxazolinyl) phenylrhodium(III) complexes with aldehydes: stereochemistry, structural characterization, and mechanistic studies. *Organometallics*, **21** (8), 1684–1696.

63 Hou, H., Gantzel, P.K., and Kubiak, C.P. (2003) Efficient synthesis of N-aryl nickel(II) carbenes by protonation of nickel(0) isocyanide complexes. *Organometallics*, **22** (14), 2817–2819.

64 Engelman, K.L., White, P.S., and Templeton, J.L. (2010) Synthesis of isonitrile, iminoacyl, and aminocarbene Tp/Pt complexes. *Organometallics*, **29** (21), 4943–4949.

65 Dhudshia, B. and Thadani, A.N. (2006) Acyclic diaminocarbenes: simple, versatile ligands for cross-coupling reactions. *Chem. Commun.*, 668–670.

16
Polyisocyanides
Niels Akeroyd, Roeland J.M. Nolte, and Alan E. Rowan

16.1
Introduction

16.1.1
Chiral Polymers

The polyisocyanides constitute a unique class of synthetic polymers, one of their defining features being the chiral and highly ordered conformation of the polymeric backbone. Highly ordered macromolecular architectures are also found throughout Nature, where they are employed in a wide range of applications, including the storage of information, as catalysts, and as construction units [1]. The best-known examples are the proteins, DNA, and polysaccharides, which themselves are chiral and optically active [2].

(Bio)macromolecules may be chiral in different ways. The first form of chirality relates to the well-known *R* and *S* isomers of the monomeric building blocks, which will in turn lead to optical activity, especially when the macromolecule consists of optically pure monomers. Second, chirality may be a result of the secondary and higher order structures of a (macro)molecule. The most famous secondary structures found in Nature are the α-helix and the β-sheet [3], and the double-helical structure of DNA [4]. When (bio)macromolecules form secondary structures this often leads to a loss of entropy, and therefore a compensation for such loss is required. This compensation can be realized by an increase in enthalpic interactions (e.g., hydrogen-bonding, hydrophobic, and electrostatic interactions), or by an increase in the entropy of the environment. The latter can be achieved, among other ways, by the release of water molecules that are bound to the macromolecule, for instance via hydrogen bonding.

The helix, as found in Nature is an intriguing chiral object. It may adopt either a left-handed (*M*) or right-handed (*P*) form, which are non-superimposable mirror images of each other. Helices in Nature are built up from chiral monomeric units, although a helical structure may also be formed even if the macromolecule does not contain any classical *R* or *S* stereo-centers in its main chain or side chain [6].

Isocyanide Chemistry: Applications in Synthesis and Material Science, First Edition. Edited by Valentine Nenajdenko.
© 2012 Wiley-VCH Verlag GmbH & Co. KGaA. Published 2012 by Wiley-VCH Verlag GmbH & Co. KGaA.

16 Polyisocyanides

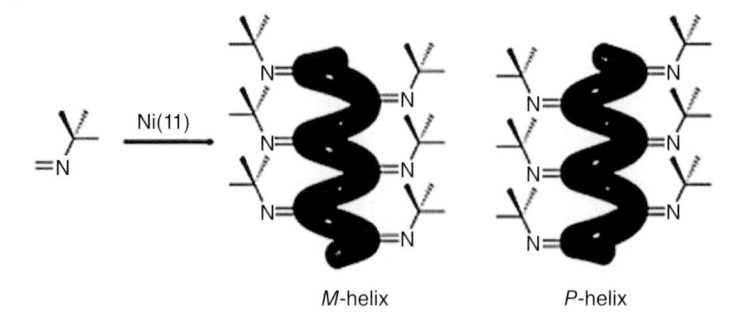

Figure 16.1 Left-handed (*M*) and right-handed (*P*) forms of poly(*tert*-butylisocyanide) prepared from *tert*-butylisocyanide using Ni(II) catalysts [5].

In 1955, Nata reported the first synthetic polymer exhibiting a helical structure, namely, stereoregular isotactic polypropylene in the solid state [7]. The first example of a polymer displaying helicity in solution was reported by Pino and Lorenzi in 1960 [8]. The polymers prepared by Pino *et al.* were isotactic and generated only short-range helices, but these were extremely dynamic in solution. The first polymer found to form a stable helical structure in solution was poly(*tert*-butylisocyanide) [9], which was prepared from achiral *tert*-butylisocyanide by the catalytic action of a nickel salt (Figure 16.1) and could be resolved into (*P*) and (*M*) helices, both of which demonstrated optical activity in solution. Later, Okamoto and coworkers described the preparation of stable optically active vinyl polymers in solution, namely poly(triphenylmethyl methacrylate), which was synthesized via an anionic polymerization reaction using a chiral catalyst [10].

Since their discovery, the details of several different synthetic helical polymers have been reported. Apart from the above-mentioned examples, polypeptides [11–16], poly(quinoxaline-2,3-diyl)s [17, 18], polyguanidines [19–21], polyacetylenes bearing chiral side chains [22–27], polychlorals [28–31], poly(methacrylate ester)s [32], and polyisocyanides [1, 5, 9, 33–38] are all known to form helical structures.

These different polymers may be allocated to two classes:

- Polymers that have a low helix inversion barrier; their optical activity can be changed by interactions of solvent or other chiral molecules with the polymer backbone. Polyacetylenes and some polyisocyanides belong to this group.

- Polymers with a high helix inversion barrier; their optical activity can be influenced by introducing chiral sides into the side chain of the polymer and/or by using a chiral catalyst. Examples of polymers having a high helix inversion barrier include bulky or peptidic polyisocyanides, polychlorals, polyguanidines, and poly(methacrylate ester)s [5].

In this chapter, attention will be focused on the latter class of polyisocyanides, in which the helices are so stable that even at high temperatures racemization does not occur [9, 33]. With the initial details of polyisocyanides having been provided in a book, *New Methods for Polymer Synthesis* [39] and also in reviews [1, 18, 40–42], the present text will concentrate on the newer developments of these intriguing macromolecules.

16.1.2
Polyisocyanides and Their Monomers

The monomeric isocyanides, from which polyisocyanides are prepared, were first reported by Lieke in 1859 [43]. Later, in 1867, Gautier described the synthesis of isocyanides from silver cyanide and alkyl iodides [44], while Hofmann showed that isocyanides could be prepared via the reaction of an amine with CCl_2 [45]. One of the easiest indications of a successful synthesis of a (liquid) isocyanide is, unmistakably, the characteristic pungent and repulsive odor of the compound [46]. The above-mentioned methods are not really suitable for the large-scale production of isocyanides, and this initially limited the investigations into these compounds. The situation was changed dramatically during the late 1950s, however, when Hagedorn and Ugi reported the synthesis of isocyanides from formamides, and this led to a rapid increase in the number of investigations being carried out [47–49].

The synthesis of isocyanides from formamides is achieved using a dehydration reaction. Typically, diphosgene or triphosgene is used as the dehydration agent, although other reagents may also be applied. For example, McCarthy et al. [50] and Mal and coworkers [51] each reported the use of Burgess reagent, while Böhme and Fuchs [52] used phosphoryl trichloride as the dehydration agent. Other reported dehydration agents have included thionyl chloride [53, 54] and cyanuric chloride [55, 56]. A more detailed description of the synthesis of isocyanides is provided in other chapters of this book.

Shortly after details of an improved route to isocyanides was reported, Millich developed the first polymerization procedure for preparing these compounds. In this case, acid-coated glass was used to obtain the first isocyanide macromolecules, which were also referred to as polyisonitriles or polyiminomethylenes [33, 57]. Some 30 years later, it was reported by Nolte et al. that by using acids as initiators, extremely long polymers could be synthesized, with molecular lengths of up to 14 μm [58].

Unlike most polymers (e.g., polyethylene, polystyrene, polymethylmethacrylate, and polyacrylates), polyisocyanides have a unique polymeric architecture, as every carbon atom in the main chain has a substituent. Indeed, it is this architecture that provides these molecules with their unique properties.

16.2
The Polymerization Mechanism

The use of a wide variety of catalysts has been reported for the polymerization of isocyanides, with numerous complexes of the metals in Group 10 of the Periodic Table being the most successful. Nolte and coworkers showed that isocyanides could be polymerized very easily by using Ni(II) complexes or salts, for example:

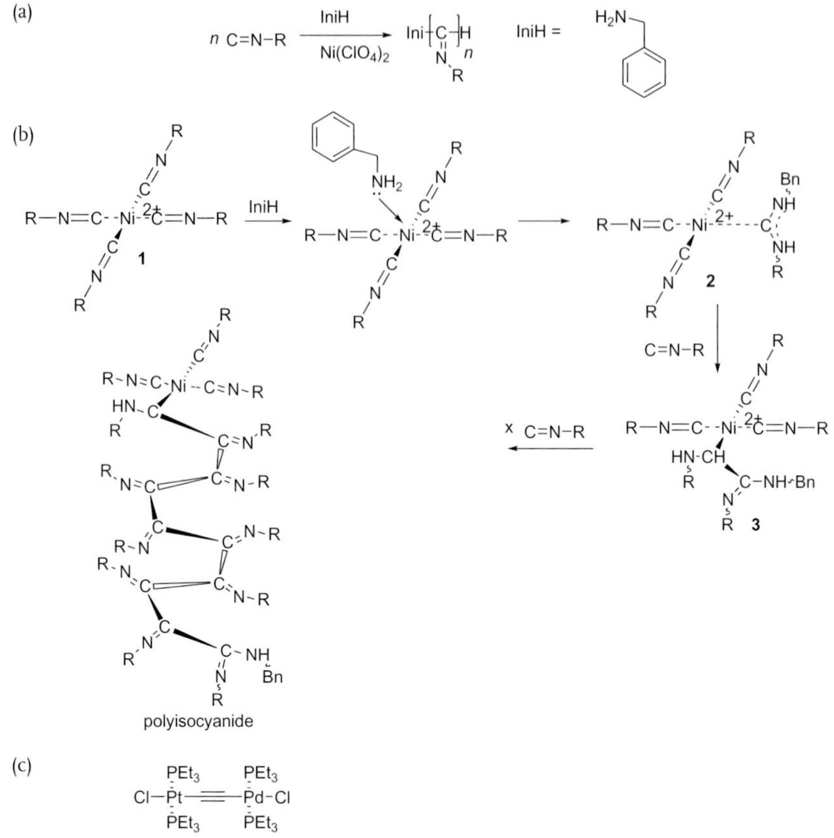

Scheme 16.1 (a,b) The "merry-go-round" mechanism proposed for the polymerization of isocyanides by Ni(II) catalysts [68]; (c) The structure of the Pd–Pt catalyst.

Ni(acac)$_2$ (acac = acetylacetonate anion) [59], NiCl$_2$, and Ni(ClO$_4$)$_2$ [1, 40], to yield M or P helices or a racemic mixture of helices, depending on the chirality of the initiator, the nucleophile, or the solvent [9, 60]. Nowadays, several metal complexes, other than Ni(II), may be used in the polymerization of isocyanides, with details of Pd(II)–Pt(II) (Scheme 16.1), Pd(II)–Pd(II) [61, 62], and Rh(II) complexes all having been reported [5, 63, 64]. Nonetheless, Ni salts remain the most popular catalyst at present.

For this reaction, an initiator (IniH) is required – for example an alcohol or an amine – which provides the end groups of the polymer (Scheme 16.1a). The proposed mechanism for the polymerization reaction using the Ni catalyst is a "merry-go-round" mechanism [1] (see Scheme 16.1b). The reaction starts with the formation of a Ni(II)(isocyanide)$_4$ complex (**1**), which then reacts with the nucleophile (in Scheme 16.1b benzylamine is used as an example) to produce the activated complex (**2**). The activated carbene-like ligand subsequently can attack a neighboring coordinated isocyanide monomer. This is followed by the coordination of a new monomer unit on the empty position in the nickel coordination sphere to yield **3**, which then reacts with further monomers in a circular fashion around the Ni center to produce the polymer. This mechanism provides a deft explanation as to why the tightly packed polymer chain is so easily formed – that is, the nickel center places the monomers in the correct position for each reaction, such that very little movement of the monomers is required. The fourfold coordination symmetry of the nickel complex matches the 4_1 helical architecture of the polymer (*vide infra*).

Nolte and Drenth have used this mechanism to predict the screw sense (i.e., left- or right-handed helicity) of polymers that are formed from optically active isocyanides. Evidence was presented that the screw sense is determined at the nickel center, and is not the result of the polymer chain being folded in the thermodynamically most stable helix conformation after its formation [65, 66].

This mechanism was also used to develop a polymerization procedure in which optically active polymers are generated from achiral isocyanides, whereby one of the screws that is formed at the catalytic nickel center is retarded by the incorporation of a very bulky optically active isocyanide. The other helical screw continues to grow on, eventually consuming all of the monomer [67]. Although this "merry-go-round" mechanism fits with many of the observed polymerization characteristics and the polymer's properties, it remains the subject of much debate However, the crystal structure of complex **2** was recently reported, providing a strong indication that the proposed mechanism is the most plausible route [68].

The mechanism of the polymerization reaction is quite likely more complex than that presented in Scheme 16.1 [69–72]. In fact, investigations conducted by Novak and coworkers, utilizing electron spin resonance (ESR), cyclic voltammetry (CV) and magnetic susceptibility, have revealed that during the reaction a Ni(I) species might also be present [73]; the Ni(I) species could then be oxidized back to the Ni(II) species, simply by purging the reaction mixture with oxygen. Typically, if a polymerization experiment is performed in an oxygen-rich atmosphere, the rates of polymerization will be significantly higher. Typically, first-order kinetics are observed when the reaction is carried out under a nitrogen atmosphere, but this changes to zero-order kinetics under an oxygen atmosphere. When creating a polymer from isocyanides, the heat of polymerization was estimated at about $81\,kJ\,mol^{-1}$, which reflects the energy required to convert a divalent carbon atom in the monomer to a tetravalent carbon atom in the polymer [40].

The polymerization of isocyanides using Ni catalysts also exhibits excellent "living" polymerization behavior [73, 74], which allows for the synthesis of more advanced macromolecular architectures, such as block copolymers. The simplest block copolymer – of just two different isocyanides – can be synthesized by first polymerizing one monomer, and introducing a second monomer only when all of the first monomer has been consumed [74]. An alternative method would be to use end group-functionalized polymers as initiators, for example polystyrene end-capped by an amine that subsequently would be used as the initiator for the nickel complex. Block copolymers with polystyrene, polybutadiene, and polypeptides have all been readily obtained using this approach [75–80].

The above-mentioned μ-ethynediyl Pd(II)–Pt(II) (Scheme 16.1c) and Pd(II)–Pd(II) complexes (the latter compounds being less reactive) are used under slightly different conditions – that is, in refluxing tetrahydrofuran (THF) – and polymerize only aryl isocyanides. In these systems, the monomer inserts only in the Pd–C bond, although when mononuclear complexes of Pd where used only a single monomer unit was found to be inserted. Consequently, the second Pd or Pt metal atom is deemed to play a crucial role, with the Pd–Pt system being superior to the Pd–Pd catalyst.

Polymerizations performed with the Pd–Pt complex have also proved to be of "living" character. This was illustrated by the recent synthesis of photovoltaic block-*co*-polymers [81], and the fact that the polymers formed displayed a low polydispersity index (PDI). In fact, PDI values as low as 1.01 have been reported, which were significantly lower than were obtained with Ni-based systems [82]. Multi-armed star polymers may be synthesized when multiple μ-ethynediyl Pd–Pt catalytic units were used [83, 84]. Polyphenylisocyanides containing dendronized side chains have been synthesized by Iyoda and coworkers using the Pd–Pt catalytic system. In this case, the fact that polyisocyanides with polymeric side chains can be generated proves that the Pd–Pt catalytic system is capable of handling even extremely bulky isocyanide monomers [85, 86]. The size tolerance of this catalyst was further highlighted by Onitsuka *et al.*, who prepared a first-generation dendritic polymerization initiator, allowing for the simultaneous synthesis of four polymers to create a star-like product [87].

As shown by Millich, isocyanides can be polymerized also by acid, although the reaction does not proceed for all types of monomer. When the mechanism of the acid-catalyzed polymerization of (L)-ala-(D)-ala–isocyanide (ala = alanine) was examined in detail by Nolte and coworkers [88], the polymerization was shown – after an initial induction period – to follow first-order kinetics with respect to the monomer. The entropy of activation ($\Delta S^{\ddagger}$) for this reaction was found to be $-170\,\text{J}\,\text{mol}^{-1}\,\text{K}^{-1}$, which was threefold more negative than for the Ni(II)-catalyzed system ($\Delta S^{\ddagger} = -54\,\text{J}\,\text{mol}^{-1}\,\text{K}^{-1}$; a large negative $\Delta S^{\ddagger}$ provides evidence of an exceedingly organized transition state). The activation enthalpy ($\Delta H^{\ddagger}$) for the acid-catalyzed polymerization was $42\,\text{kJ}\,\text{mol}^{-1}$, leading to an overall free energy of activation ($\Delta G^{\ddagger}_{293K}$) of $92\,\text{kJ}\,\text{mol}^{-1}$. The reaction was shown to be extremely stereoselective; typically, the addition of (D)-ala-(L)-ala–isocyanide or (D)-ala-(D)-ala–isocyanide to the growing L–D polymer chain led to a complete blockade of the polymerization reaction. The reaction was initiated by the binding of a proton to an isocyanide monomer, after which new isocyanide molecules would be added to the cationic center. After the reaction of about eight monomers, the oligomer would fold into a helix that could then serve as a template for further growth of the polymer. In this way, polymers with lengths of up to $14\,\mu\text{m}$ could be prepared, corresponding to a molecular weight of approximately 20 000 000 Da [89].

16.3
Conformation of the Polymeric Backbone

When optically active α-phenylethyl isocyanide was polymerized by Millich *et al.* [33], an optically active polymer was obtained which had a higher optical rotation than the monomer. Based on Debye–Scherrer X-ray patterns and space-filling models [90], Millich postulated that the polyisocyanide chain had a helical conformation with four monomers per turn and a helical pitch of 4.1–4.2 Å [91]. This hypothesis was confirmed by Nolte and coworkers, who polymerized the optically inactive *tert*-butyl isocyanide to provide a polymer that could be separated, using chromatography with poly((*S*)-*sec*-butyl isocyanide) as the stationary phase, into (+)- and (−)- rotating fractions. On the basis of circular dichroism (CD) spectroscopy, the (+)- and (−)-rotating fractions were assigned to the (*M*)- and (*P*)- polymer helices, respectively [9, 92].

As indicated above, the polyisocyanides adopt a nonplanar conformation due to the steric hindrance of the side chains, and this leads to a hindered rotation around the bonds that connect the main chain carbon atoms, and hence to stereoisomerism. When two stereoisomers are formed, because of hindered rotation around a single bond, this phenomenon is termed "atropisomerism" (from the Greek *atropos*, meaning no rotation). In the case of polyisocyanides, the barrier around the main-chain carbon bonds is so high that the polymer chain folds into an atropisomeric helical structure during its process of formation. When a nonchiral isocyanide monomer is polymerized with the nickel catalyst and a nonchiral initiator, no driving force for forming either a left-handed or right-handed helix is present, and consequently both helices are obtained in equal amounts. However, the stereoselective formation of one particular helix can be achieved when a chiral initiator [60] or a chiral monomer [41, 67] is used.

During recent years, the helical conformation of a polyisocyanide has been the subject of intense investigation. The initial results reported by Millich [33, 90], Nolte and Drenth [93] revealed a helix with four monomer units per turn (also known as a four-over-one helix or 4_1), separated by 4.1–4.2 Å. The results of subsequent theory-based studies of *tert*-butylisocyanide oligomers indicated that a helical conformation was preferred when the degree of polymerization was increased. The helical structure identified for low-molecular-weight *tert*-butyl isocyanide oligomers showed an average N=C–C=N dihedral angle of 78.6°, which corresponded to 3.75 monomer units per turn. With an increasing degree of polymerization, however, this value changed to 3.6 monomer units per turn, corresponding to a dihedral angle of 84.3°. Calculations performed on ethyl and isopropyl isocyanide oligomers revealed a smaller N=C–C=N dihedral angle, and hence more monomer units per turn. When the structure of poly(methylisocyanide) was calculated, it was found that the methyl group was too small to produce a fixed dihedral angle and, as a result, no atropisomerism was identified [94]. Three other polyisocyanides were also examined by using molecular orbital calculations, namely $(HNC)_n$, $(CH_3NC)_n$, and $(C(CH_3)_3NC)_n$ [95, 96]. It was proposed that, in the hypothetical $(HNC)_n$ polymer, electronic repulsion between the nitrogen lone pairs was the driving force for the main chain to assume a nonplanar conformation. However, in the case of more bulky side chains, as in $(C(CH_3)_3NC)_n$, the driving force for the nonplanar conformation was found to derive from the steric hindrance of the side chains (Figure 16.2a). For the $(CH_3NC)_n$ polymer, a combination of electronic repulsion and steric hindrance was found to be the structure-determining factor.

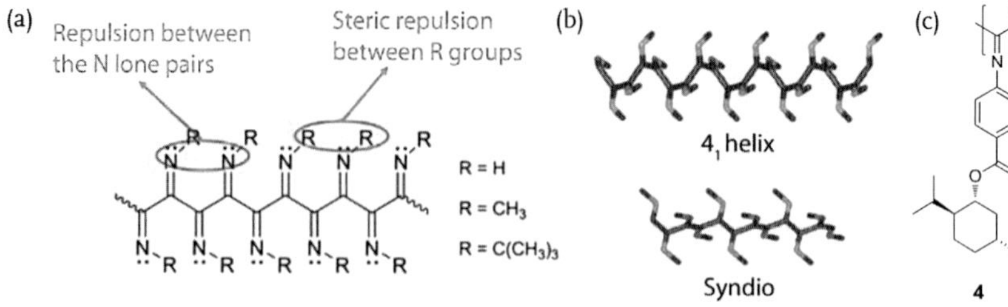

Figure 16.2 (a) The two different driving forces for the nonplanar conformation of a polyisocyanide; (b) The two different conformations proposed for a polyisocyanide [5]; (c) Structure of poly-(1R, 2S, 5R)-2-isopropyl-methylcyclohexyl 4-isocyanobenzoate [118] (4).

The N=C–C=N dihedral angle in the $(HNC)_n$ system was calculated to have a broad variation, but for those polyisocyanides where the nonplanar conformation was the result of steric effects of the side chains, a fourfold helix was found.

Recently, Schwartz et al. reported on the use of vibrational circular dichroism (VCD) for the determination of the screw sense of two enantiomeric polyisocyanides [97]. The C=N-stretch vibration in the VDC spectrum allowed a determination of the screw sense of the polyisocyanide directly, without the use of any further calculations. To determine the helical sense directly by using scanning probe microscopy is also possible, but challenging [98–100], as the resolution of soft organic materials is often difficult when using such a technique. These problems can be overcome, however, by tightly packing the polymers in two dimensions on atomically flat surfaces, as achieved by Yasima et al. for both polyacetylenes and polyisocyanides [101–105]. Subsequently, monolayers of poly(phenylisocyanides) bearing L-lactic acid and L-alanine groups substituted with n-decyl chains were studied, using atomic force microscopy (AFM), by Yashima and coworkers [106]. Long polymers were observed that proved to be one-handed helices, as could be concluded from the appearance of periodic oblique stripes on the polymers [107]. Moreover, left-handed polymers were found to have formed predominantly, with a helical pitch of 1.58 ± 0.09 nm, a value that was slightly larger than had been reported previously (1.30 ± 0.03 nm) for the same polymer when deposited on a highly oriented pyrolytic graphite (HOPG) [104]. These different results were explained by differences in the hydrophobicities of the surfaces used, which led to slight alterations in the polymer architecture.

The helical sense of polyisocyanides has frequently been determined using CD spectroscopy. In fact, Nolte, Drenth and coworkers were able to calculate the helical sense of a polyisocyanide from its CD-spectrum [93]. Porphyrins are well-known spectator groups in CD spectroscopy [108–110], and by utilizing a porphyrin spectator group Takahashi et al. were able to employ exciton-coupled CD spectroscopy to determine the helical sense of poly(aryl isocyanides). This would not have been possible without the porphyrin group, because the aryl and imino moieties would make interpretation of the CD signal below $\lambda = 400$ nm extremely difficult [111]. Because the porphyrin chromophores are stacked so closely within the polymer, the chirality of the backbone is transferred onto the porphyrin mol-

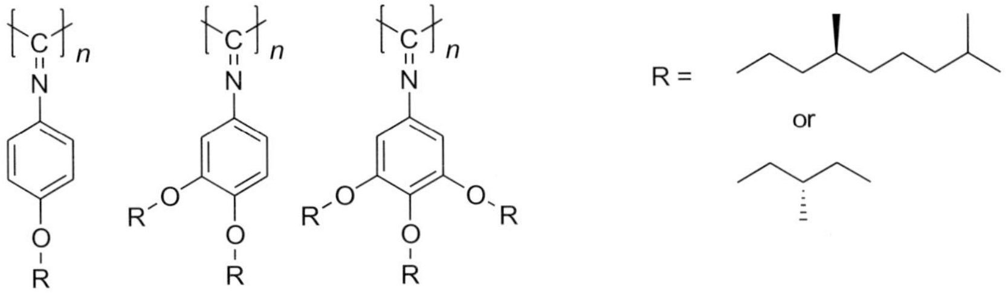

16.3 Conformation of the Polymeric Backbone

Figure 16.3 The chiral poly(phenylisocyanides) synthesized by Yashima and coworkers to study the effect of phenyl group substitution on the Cotton effect in CD spectra [113].

ecules. This led to an exciton-coupled bisignate Cotton effect being identified in the Soret band region of the CD spectrum, and allowed for the screw sense to be determined. The helical sense of polyisocyanides functionalized with diazo groups was also determined by Nolte et al., using a similar approach [112]. Yashima and coworkers have prepared polyphenylisocyanides substituted with one or more chiral chains on the phenyl ring, to examine the effect of the position and the number of chiral groups on the Cotton effects in CD spectra (Figure 16.3). Subsequently, polyisocyanides with only one chiral alkoxy chain on the phenyl ring were found to show no Cotton effect at all, though when two chiral substituents were introduced a small Cotton effect was observed. When three substituents were attached to the phenyl ring, however, a large Cotton effect was identified [113].

The chiral architecture of the polyisocyanide helix can be transferred to other reactions. For example, when a catalyst for direct aldol reactions was prepared based on helical polyisocyanides, the reactions proved to be enantioselective. This polymeric catalyst was one of the first to demonstrate the transfer of chiral information from the polymeric helix to the reaction product [114], and confirmed that the catalytic reaction was carried out close to the helical polymer backbone.

Clericuzio et al. [96] proposed that polyisocyanides, apart from adopting a helical conformation, also can adopt a second–*syndio*–conformation (Figure 16.2b). This conformation was proposed because, according to theoretical calculations, the helical conformation of polyisocyanides is stable, but not at the absolute minimum energy level. In the *syndio* conformation, two neighboring N=C–C=N functions have a flat orientation, and the subsequent sets of these functions alternate by $180 \pm 90°$; this makes the overall polymer structure not helical, but instead highly symmetrical. This theory-based proposal was confirmed experimentally by Yamada and coworkers [115], who showed that the helical stability of a polyisocyanide was increased by the addition of bulky side chains. The *syndio* conformation was observed in the ^{13}C nuclear magnetic resonance (NMR) spectrum of oligomeric compounds derived from phenylisocyanide [116], although when high-molecular-weight polymers of the same monomer were investigated a random coil conformation was found [117]. Green and coworkers noted that some polyisocyanides, when examined using NMR spectroscopy, light scattering, and viscosity measurements [34], could adopt irregular conformations that were related to *syn–anti* isomerism of the >C=N–R groups. This observation was in agreement with the findings of

Takahashi et al., who also identified an irregular conformation in poly(1R, 2S, 5R)-2-isopropyl-methylcyclohexyl 4-isocyanobenzoate (4) (Figure 16.2c).

It was shown, however, that when the polymer was annealed in THF, a *syn–anti* isomerization of the imino groups took place, leading to the energetically most favorable stereoregular conformation of the polymer. This conformational change was concluded from the specific optical rotation of the polymer, which before annealing gave an optical rotation of $[\alpha]_D = +354$, but after annealing a value of $[\alpha]_D = 1038$, and was suggestive of a helical conformation. The same authors also reported that, when a Pd–Pt catalyst was used in refluxing THF instead of a nickel catalyst at room temperature, the irregular conformation was not formed [118]. Taken together, these results indicate that polyisocyanides are more flexible and possess more than one optimum conformation. Yet, despite this geometric flexibility, the polyisocyanides are very stiff when compared to other polymers.

Sugar-functionalized polyphenylisocyanides have been synthesized to investigate the aforementioned molecular recognition phenomena, using acetylated galactose, glucose, and lactose as functional groups. In this case, upon deprotection of the alcohol moieties of the sugar substituents, optically active polymers were obtained. Binding studies with lectin demonstrated a difference between the helical and stiff polyisocyanides and more flexible polyacrylamides, which were functionalized with the same sugar substituents and used for comparison. This difference was attributed to the fact that the sugar units attached to the polyisocyanide have significantly less rotational freedom [119, 120].

A preferred helical handedness of the polyisocyanide chain can also be introduced after the polymerization has taken place. For example, Yashima et al. described the induction of a preferred handedness in optically inactive poly(4-carboxyphenyl isocyanide) by ionic interactions with chiral amines in water. For this purpose, (R)-2-amino-2-phenylethanol, (R)-1-phenylethanamine, and (1R,2S)-1-amino-2,3-dihydro-1H-inden-2-ol were used [121]. The induced chirality was found to be unexpectedly stable, with the induced helix conformation remaining after modification of the side chains of the polyisocyanide and removal of the chiral amines [114, 122]. This behavior was, nevertheless, highly dependent on the solvent system being used; typically, when mixtures of organic solvent and water were used (even up to >50% water) the chiral amine could be removed without any loss of the induced conformation. However, the conformation "memory" was lost when dimethyl sulfoxide (DMSO) [123] or DMSO–water mixtures (<30 vol% water) [124] were applied as the solvent system (Figure 16.4).

Not only the solvent, but also the temperature, had an influence on this "memory" effect. Following removal of the chiral amines, the polymer helices were stable at room temperature but unfolded rapidly upon heating. It was hypothesized that such "memory" behavior was not only due to chiral ionic interactions but may also involve hydrophobic interactions, as the addition of DMSO eliminated this effect [121]. The mechanism of helical induction was studied in detail using CD, infrared (IR), and NMR spectroscopies, and also X-ray diffraction (XRD) [125]. The analytical data subsequently acquired suggested that the unmemorized polymer had a 9_5 helix with a persistence length of 43 nm (measured on the methylated derivative),

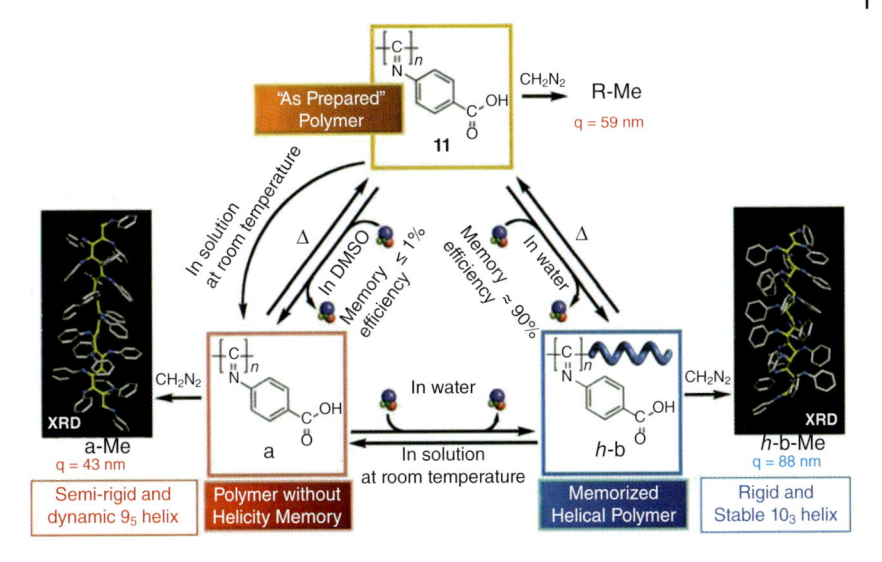

Figure 16.4 Schematic illustration of the induction of helicity in poly(4-carboxyphenyl isocyanide) [5].

while the memorized polymer had a 10_3 helix with a persistence length of 88 nm (also measured on the methylated derivative; see Figure 16.4). Again, these results indicated that the helical "spring-like" architecture of polyisocyanides could readily be tightened and loosened, thus altering the pitch and persistence length.

16.4
Polyisocyanopeptides

As discussed above, the bulkiness of the side groups of a polyisocyanide has a strong influence on the helical stability of the polymer. When side groups that have interactions with each other (e.g., via hydrogen bonds) are used, the helical structure of a polyisocyanide can be stabilized further. Although peptide-functionalized polyisocyanides were first introduced by Nolte and Drenth during the 1980s [126, 127], it was realized only much later that hydrogen bonds as are found in peptide chains may be used to stabilize the helical conformation of a polyisocyanide, and thus obtain ultra-stiff polymer rods [89, 128] (see Figure 16.5).

Circular dichroism studies performed to investigate the polymerization of isocyano-alanine–alanine peptides showed that a minimum of eight monomers was required to observe a CD signal. In other words, as there are approximately four monomer units per helical turn, two helical turns are required to form a stable helical polyisocyanide structure. At this same critical point of eight monomer units, a shift in the ^{1}H NMR spectrum was observed for the amide protons, indicating the formation of a hydrogen-bonding network similar to that seen in

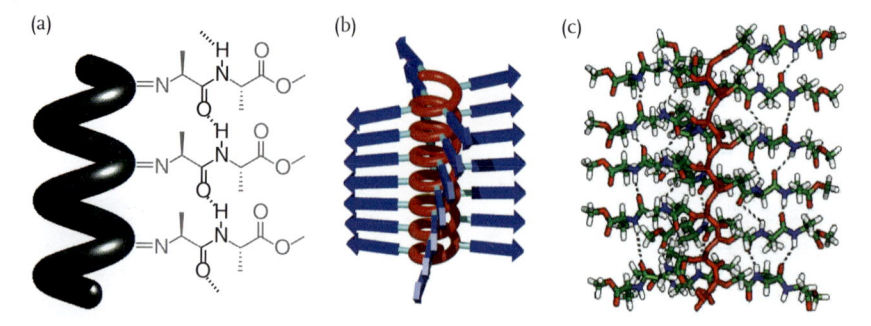

Figure 16.5 (a) The hydrogen-bonding network, which stabilizes the helical structure of a polyisocyanopeptide; (b) Schematic representation of the β-sheet helical structure formed by a polyisocyanopeptide; (c) Hydrogen-bonding network in polyisocyano (L)-ala-(L)-ala-(L)-ala [5].

α-helices [88]. Subsequently, the alanine-based monomers were found to have formed a β-sheet helix with four sets of hydrogen-bonding arrays running parallel to the backbone of the polyisocyanide (Figure 16.5). Architecturally, this structure mimics the β-sheet helices found in Nature [129, 130]. By using IR and ^{1}H NMR spectroscopies, it could be shown clearly that hydrogen-bonding networks were present throughout the polymer chain. Moreover, these networks remained stable for a significant period of time, even when the polymers were dissolved in water [131], and could unfold when the temperature was increased or acid was added [121]. More recently, the hydrogen-bonding arrays in these peptide-functionalized polyisocyanides were investigated using pump-probe spectroscopy, when it was found that in these arrays self-trapped vibrational states could be generated. It is believed that self-trapping states (Davydov solitons) can play an important role in the way in which enzymes and proteins transport energy, that is, via dispersionless wavepackets [132, 133]. The observation of self-trapping served as additional proof that the polyisocyanopeptides possess a "spring-like" character that is stabilized by well-defined hydrogen-bonding networks along the side chains. The "springiness" of these helices can be disrupted by disturbing the hydrogen-bonding networks by the inclusion of an ester unit in the backbone which cannot participate in the hydrogen bonding. However, a consequence of this change was that the self-trapping properties of the polymer were immediately lost [134, 135].

The order of the amino acids in the side chains of the polyisocyanopeptides plays a significant role in the stability of the helical structure. Hydrogen-bonding arrays are formed in the case of the side chain combination glycine–alanine, but not for the combination alanine–glycine. This helical architecture behaves like a natural protein, and in the case of the latter denaturation occurs when the protein is exposed to heat, acids, bases, or hydrogen-bond-forming solvents such as DMSO. In the case of the peptide-based polyisocyanides, it was found that extreme conditions – for example, strong acids such as trifluoroacetic acid (TFA) – are required to break the hydrogen-bonding networks – a fact that highlights the robust nature of the hydrogen-bonding arrays along the polyisocyanopeptide helix. This denaturation can be observed using NMR and CD spectroscopies, and also XRD. Pseudo-hexagonal arrangements of the polymer chains were identified in powder X-ray diffraction (PXRD) experiments on the polyisocyanopeptides, but these were lost upon the addition of TFA in the solid state. Intriguingly, TFA can be used to initiate the polymerization, so as to produce extremely long polymers. Although the rigid nature of the polyisocyanopeptide chains severely hinders the use of size-exclusion chromatography (SEC) [136] measurements, AFM was found to be an ideal analytical tool. In this case, by measuring a large number of chain curvatures and the contour lengths of single polymer molecules, the molecular weights and molecular weight distributions of the polyisocyanopeptides could be determined [137]. In general, AFM also allowed the determination of persistence length, which was found to be 76 nm for polyisocyano-(L)alanyl-(D)alanine. (Somewhat surprisingly, this indicates that the polymers are more rigid than double-stranded DNA, which has a persistence length of 53 nm [138].) The height of the polymers was determined (using AFM under chloroform vapors) as 1.6 nm [139], and slightly less than the expected thickness of 4–5 nm. Intriguingly, the polymers formed by either acid initiation or Ni catalysis both possessed the same architecture as determined with IR and CD spectroscopies, and AFM.

Whilst numerous amino acid combinations have been investigated, somewhat intriguingly the most stable peptidic polyisocyanides were those containing β amino acids. In the case of α amino acids, upon heating or the addition of TFA the helices unfolded but never fully refolded. In the case of the β amino acid derivatives, however, upon warming the kinetic helix was modified to an extremely stable thermodynamic architecture. This latter helical macromolecule could be fully unfolded by the addition of TFA, and refolded by the addition of triethylamine on numerous occasions without the loss of CD signal, thus mimicking many protein architectures [140].

16.5
Polyisocyanides as Scaffolds for the Anchoring of Chromophoric Molecules

The introduction of peptide-based polyisocyanides opened the door to the use of these polymers as macromolecular scaffolds for the anchoring of chromophoric molecules (Figure 16.6) [141–146]. During the 1980s, several attempts had been made to functionalize polyisocyanides with chromophores via the "grafting-on" strategy. However, these reactions were incomplete and caused inhomogeneous polymer stacks; furthermore, the chromophore-functionalized polyisocyanides obtained lacked structural stability [147]. The new amino acid polyisocyanide peptides are accessible via a flexible monomer synthesis strategy, and serve as an ideal scaffold upon which to add chromophores. The rigidity of the polyisocyanopeptide chain (persistence length ≥76 nm) and the long polymer lengths obtained (chains of up to 20 μm can be synthesized) are additional benefits of this system. These unique properties have been utilized in the synthesis of a porphyrin-functionalized polyisocyanide (**6**) via polymerization of the chromophore-functionalized monomer

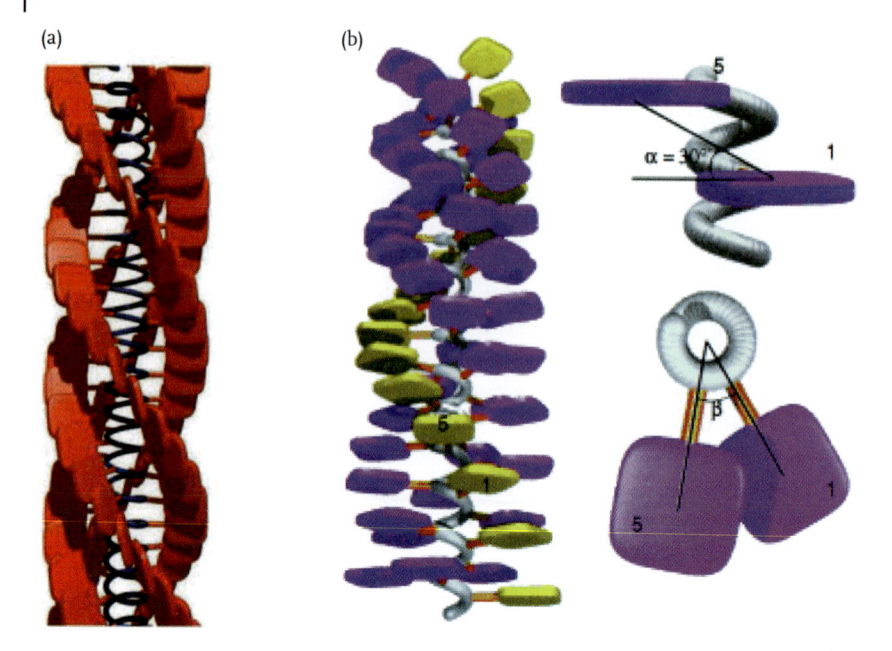

Figure 16.6 (a) The "helter-skelter" arrangement of perylene units along the polymer backbone [144]; (b) Schematic drawing of a porphyrin-functionalized polyisocyanide, showing the stacking of the dye molecules.

Scheme 16.2 Polymerization of a porphyrin-functionalized isocyano-(L)-alanine.

(5) (Scheme 16.2) [148]. The UV/visible spectrum of the polymer was significantly different from that of the monomer, with the latter displaying one Soret band at 421 nm, and the polymer two Soret bands at 413 and 437 nm.

The two Soret bands in the UV/visible spectrum of **6** can be explained by the different interactions that occur between the porphyrin chromophores along the polymer chain. The Soret band at 437 nm indicates that the porphyrin molecules are arranged as J-aggregates, and is attributed to an offset stacking of the porphyrins at positions n and $(n+5)$ – that is, with porphyrins placed one on top of another. The second Soret band is explained by a blue shift caused by interactions between the porphyrin chromophores located at positions n and $(n+1)$, as well as interactions between the chromophores located at positions n and $(n+4)$. Subsequent CD measurements recorded in chloroform showed a strong bisignate Cotton effect at 437 nm, which reversibly decreased and increased in intensity as the sample was heated and cooled. The CD spectrum indicated that chiral interactions were present between the nth and $(n+4)^{th}$ monomer units along the polymer chain, in a slipped conformation. The delocalization of excitons over relatively long distances in the polymer chain causes energy transfer, which can explain the strong signals observed in the CD spectra. When resonance light scattering (RLS) studies were conducted to confirm this conclusion it was found that, upon excitation, the excited state was delocalized over a distance of 100 Å, which corresponded to a stack of about 25 monomeric units. The helical structure of the polymer was studied in more detail using depolarized RLS studies, which showed that the slip angle between the first and fifth porphyrin units was 30° (Figure 16.6b). This corresponds to a helical twist angle of 22°, and an overall helical pitch of 68–71 Å. Fluorescence anisotropy studies then revealed a 25° angle between the helical axis of the polymer and the porphyrin units. The combined results of the fluorescence anisotropy measurements and the RLS studies suggested that the porphyrins form a fourfold "helter-skelter" arrangement along the polymer backbone (Figure 16.6a).

For further studies, zinc ions were inserted in the polymer-anchored porphyrins [149]. In this case, the zinc porphyrin acts as an energy quencher for the free base (H_2) porphyrins. Subsequently, it was shown that a ratio as low as 1:200 of zinc to free base porphyrin was required for full quenching of the free base porphyrin. This indicated an extremely efficient energy transfer over long distances of about 20–30 nm, due to the defined nature of the helical backbone.

More recently, polyisocyanide block copolymers containing free-base porphyrins and Zn-coordinated porphyrins have been prepared by Takahashi and coworkers, using the Pd–Pt polymerization catalyst. By inserting a second metal ion into the free-base porphyrin after preparation of the polymer, these block copolymers could be further converted into block copolymers with two different metal ions coordinating to the porphyrin. The defined architecture of this intriguing compound has interesting applications in the field of light-harvesting systems and molecular photonic devices [150].

Another example in which polyisocyanides are used as a defined scaffold for the anchoring of dye molecules involves perylenes. Perylene diimides are a promising class of compounds in the field of organic photovoltaic cells [151], due largely to their unique photochemical properties. Perylenes absorb strongly in the visible light region of the UV-visible spectrum, and possess a high photochemical stability. Isocyanide monomers bearing perylene side groups were synthesized, and polymerized, and the properties of the polymers then determined [141–146, 152–154]. As in the case of the porphyrins, these perylene-functionalized polyisocyanides possessed an extremely stable helical conformation, due to the

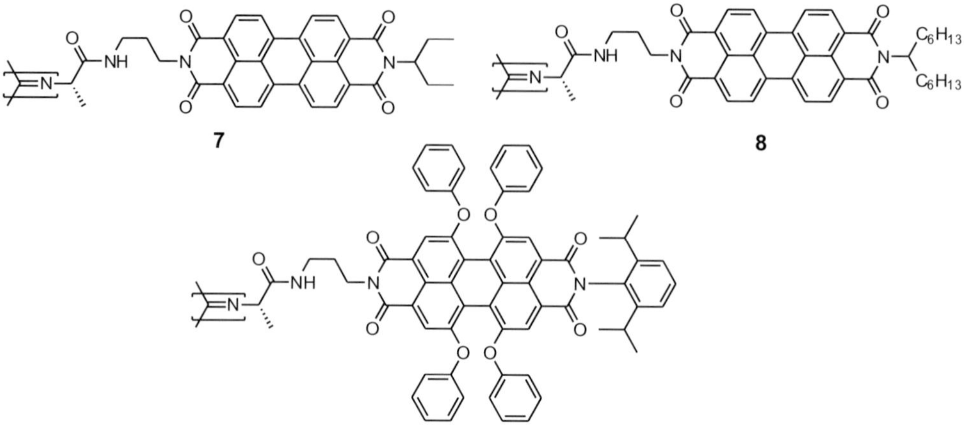

Figure 16.7 Three different PDI-functionalized polyisocyanides.

hydrogen-bonding interactions between the side chains and the π-π stacking interactions between the dye molecules. In fact, the helix conformation was so stable that the helical arrangement remained intact even under prolonged heating up to 180 °C.

A series of three different perylene diimide-functionalized polyisocyanides (**7**, **8**, **9**; Figure 16.7) was synthesized and the properties of the polymers were compared. In particular, the effect of the bulkiness of the perylene diimide on chromophore–chromophore interactions was examined with UV/visible and fluorescence spectroscopies. The results showed that, compared to the respective monomers, the absorption bands in the UV/visible spectra of polymers **7** and **8** were broadened, red-shifted, and also had different relative intensities. These features were all indicative of the presence of strong π–π interactions between the chromophores in these polymers. The fluorescence studies confirmed these interactions because broad, structureless, and red-shifted bands were found, signifying an aggregation of the dye molecules. However, the concentration-independent nature of the spectra was indicative of intramolecular stacking. Fluorescence decay measurements revealed a lifetime of 19.9 ns for polymer **7** compared to 4 ns for the monomeric perylene, thus supporting the hypothesis that emission in the polymer is through an excimer species. In contrast to polymers **7** and **8**, the absorption spectrum of polymer **9**, which contains the bulky perylene diimide, showed hardly any difference when compared to its monomer, while the emission spectrum also presented no evidence for the formation of excimers. These findings can be explained by the fact that the bulkiness of the phenoxy substituents on the perylene diimide in **9** prevents good π–π stacking of the dye molecules, and also indicates that the chromophore–chromophore interactions can be regulated by controlling the bulkiness of the perylene diimide substituents.

One of the challenges in the materials sciences—and, more specifically, in the field of photovoltaic devices—is to obtain control over the position and ordering of chromophores. A perfectly ordered system would allow for high energy and electron transfers over long distances. A photovoltaic device consist of two major components: an electron-accepting material, and an electron-donating material, with a defined architecture as required. The perylene diimides are well known for their electron-accepting properties, and several methods for obtaining well-defined organized structures of these materials have been reported, including perylene crystals [155], aggregates [156–158], and liquid crystals [159]. Polymers **7**, **8**, and **9** (Figure 16.7) represent an interesting option for ordering perylene diimides over distances of 100–200 nm (the average length of polymers obtained via AFM). Subsequent calculations and spectroscopic methods have predicted that these systems would serve as excellent systems for electron transport, and this was confirmed by transient absorption spectroscopy studies that highlighted particularly high exciton migration rates [144].

Photovoltaic devices have also been produced with a number of semiconducting (electron-donating) materials, including poly(3-hexylthiophene) (P3HT), poly(9,9′-dioctylfluorene-*co*-bis(*N*,*N*′-(4-butylphenyl))-bis(*N*,*N*′-phenyl)-1,4-phenyldiamine) (PFB), and poly(9,9-dioctylfluorene-*co*-benzothidiazole) (F8BT) (Figure 16.8). In order to illustrate the effect of the architecture and perylene diimide-coated polyisocyanides, these photovoltaic devices where compared with the free perylene diimide monomer. For the polyisocyanide perylene-based photovoltaic devices, an order of magnitude improvement in power conversion efficiency was found, when compared to identical systems using crystalline perylene diimide. This was due to the strong repellent forces between the polyisocyanide and the F8BT, since if there was a too-great phase separation no improvement was found (Figure 16.8) [143]. The reason for this enhanced performance was a combination of an increased interface between the donor and acceptor material, and a rapid directional migration of the excited electron along the perylene polymers.

Finlayson *et al.* utilized polymer **8** and **9** to form a thin-film transistor (TFT) [146]. In this case, polymer **8** was found to be far superior to polymer **9** and the control (a polyisocyanide without perylene moieties); typically, when polymer **8** was deposited on a surface (silicone or graphite), a network of bundles was formed. Both fiber formation and n-type electroactivity were found in the thus-obtained thin films, with carrier mobilities on the order of $10^{-3}\,cm^2\,V^{-1}\,s^{-1}$ at 350 K, which is extremely high for a single wire-like polymer. The carrier mobility was found to be limited by inter-chain transport processes (e.g., contact resistance and charge injection), but not by migration along the polymer chain.

In addition to photovoltaic and TFT devices, polyisocyanides have also been studied as potential nonlinear optical (NLO) materials. Nolte and coworkers have synthesized a polyisocyano-(*L*)-alaninol derivative with side chains that possess NLO properties (**10**; Figure 16.9). In the polymer, the chromophoric molecules are nicely aligned, leading to very high β-values ($β = 5150 \times 10^{-30}$ esu) [160].

A major drawbacks with polymer **7** was that it became highly insoluble when obtained in high-molecular-weight form. This contrasted with polymer **8**, which

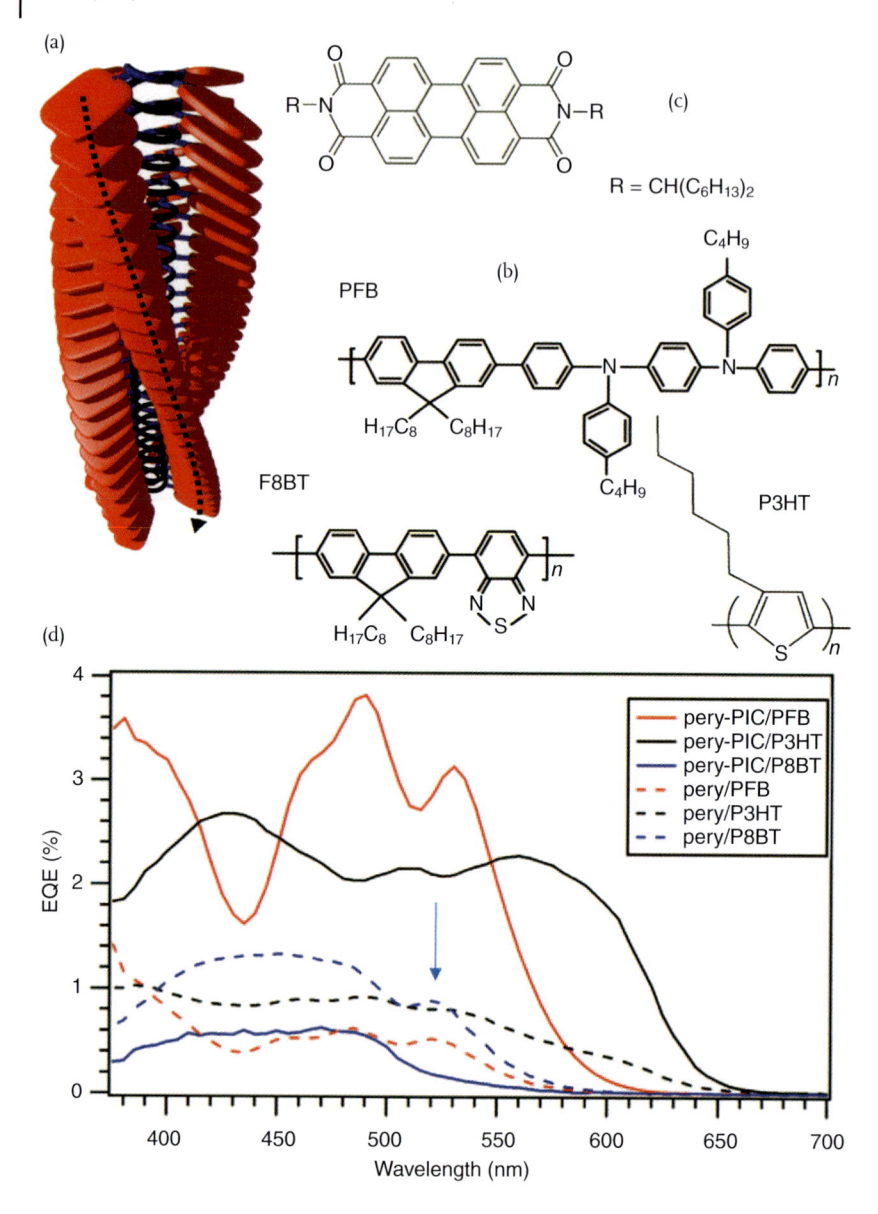

Figure 16.8 (a) A schematic overview of the perylene functional polyisocyanide; (b) Semiconductors; (c) Perylene monomer; (d) Photovoltaic device external quantum efficiency [143].

did not suffer such unfavorable properties due to its long alkane tails; moreover, it demonstrated good processiblity while still displaying the desired electronic effects.

As the synthesis of polymers **7–9** proved to be very laborious (as is typical of functionalized polyisocyanides), an improved grafting method was developed to

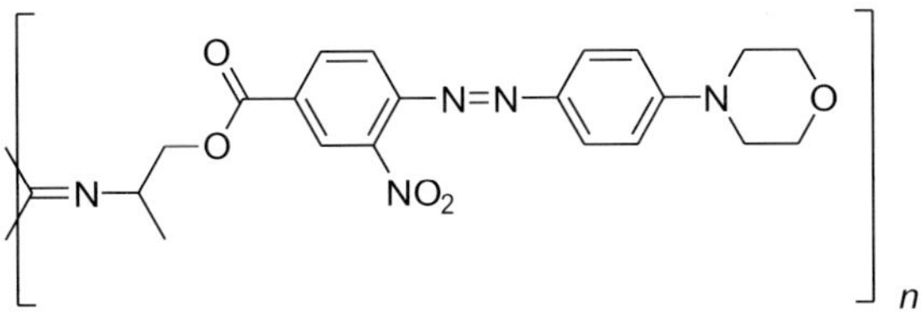

Figure 16.9 Structure of the polyisocyano-(L)-alaninol derivative.

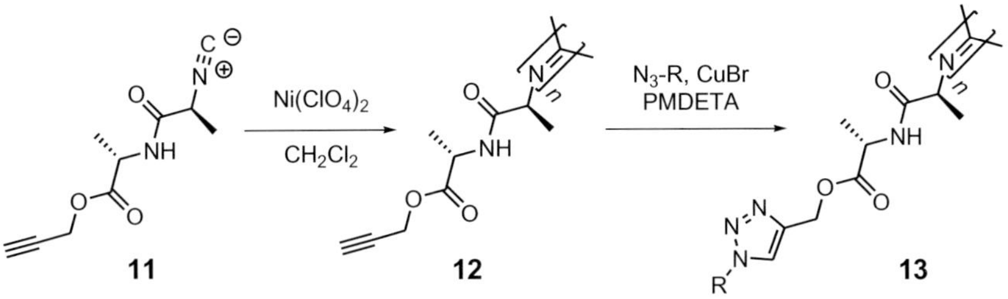

Scheme 16.3 Functionalization of a polyisocyanide via the Cu(I)-catalyzed Huisgen 1,3-dipolar cycloaddition reaction.

prepare perylene-functionalized polyisocyanides. For this purpose, the recently reported copper(I)-catalyzed Huisgen 1,3-dipolar cycloaddition reaction was used [161, 162]. This reaction is known to provide a very efficient means of coupling an acetylene and an azide functionality, to result in a triazole group. To this end, a new alkyne-functionalized monomer (Scheme 16.3; **11**) was synthesized, polymerized (**12**), and subsequently reacted with a perylene-functionalized azide (**13**) [163].

The clicked perylene polymer was analyzed using UV/visible, fluorescence, and CD spectroscopies, and its properties compared to those of the nonclicked polymer **8** (see Figure 16.10). The absorption spectra (Figure 16.10a) of the perylene-clicked polymer **13** and polymer **8** were almost identical. Likewise, hardly any difference was visible in the fluorescence spectra (Figure 16.10b), with only the relative intensity of the clicked polymer being lower than that of polymer **8**. However, a major difference was observed in the CD analysis (Figure 16.10c), where the spectrum of polymer **8** showed a strong signal between 450–500 nm that was not present in the clicked polymer. This lack of chiral expression might have been caused by the longer distance between the chiral helical backbone of polymer **13** and the perylene units in its side chains. With this larger distance, the perylenes might stack in a less chiral manner, and the transfer of chirality from the backbone would be reduced. This weakening of order would also explain the lower intensity of the fluorescence spectrum of the clicked perylene polymer.

The major advantage of polymer grafting via click chemistry is that a whole library of compounds can be synthesized by employing one basic backbone. In addition, the synthesis of polymers with more than one type of functionality can

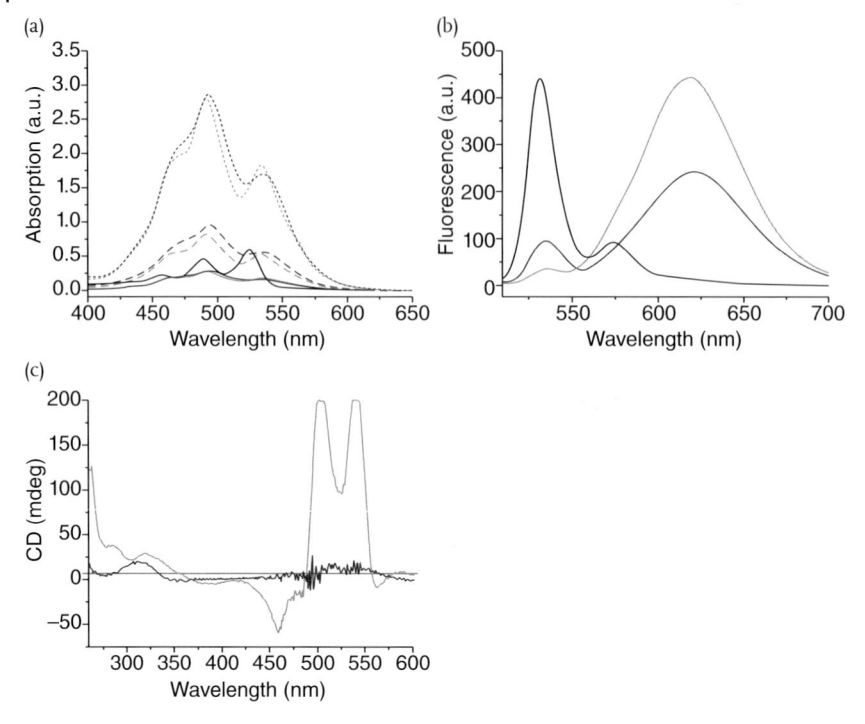

Figure 16.10 UV/visible and fluorescence spectra of polymer **8**, polymer **12**, and the monomer of **8** as reference. (a) UV/visible spectra measured at different concentrations in dichloromethane (8.7 µM, solid lines; 31 µM, dashed lines; 87 µM, dotted lines). For the assignment of the different curves, see original paper; (b) Fluorescence spectra of 0.87 µM solutions of polymers **8** (upper curve at 625 nm) and **12** (middle curve at 625 nm), with monomer of **8** (lower curve at 625 nm) as reference; (c) CD spectra of the two polymers **8** (upper curve at 525 nm) and **12** (lower curve at 525 nm) (87 µM) in dichloromethane [164].

be achieved easily by clicking different azides simultaneously. For example, when azide-functionalized coumarin and perylene functions were clicked onto the same polyisocyanide [163, 164], the reaction product clearly showed an interaction between the two chromophores; for example, a quenched and blue-shifted emission was found for the coumarin signal.

16.6
Functional Polyisocyanides

The above-described method of post-polymerization functionalization of polyisocyanides using click chemistry is, of course, not limited to chromophoric molecules. Rather, the procedure was also used to prepare water-soluble polyisocyanides, by clicking azide-functionalized tetraethylene glycol onto polymer **11** (Scheme 16.3) [164]. In order to make this method even more versatile, polymers from the azide-containing monomers isocyano-(L)-alanyl-(L)-alanine-azidopropyl ester and (isocyano-(D)-alanyl-(L)-alanine-azidopropyl ester were also synthesized, allowing for the functionalization with alkynes, at least in principle. Unfortunately, the first-mentioned monomer was difficult to polymerize into well-defined polymers, while the second monomer led to polymers that were insoluble. However, when these monomers were used in a random copolymerization reaction with the corresponding isocyano-(L or D) alanyl-(L)-alanine methyl ester, well-defined and soluble polymers were obtained to which alkyne functional groups could be clicked [165].

Another post-polymerization modification method involves the introduction of moieties onto a thiolated polyisocyanide via a Michael addition reaction, by dynamic covalent approaches [166]. In this case, pyrenes were attached using a maleimide, iodoacetamide, and thioester coupling (all of these reactions were so-called "click reactions") (Scheme 16.4). These same polymers could also be used to produce biohybrid materials; this was achieved by clicking a biotin molecule onto the thiol functionality via an iodoacetamide coupling. For analytical purposes,

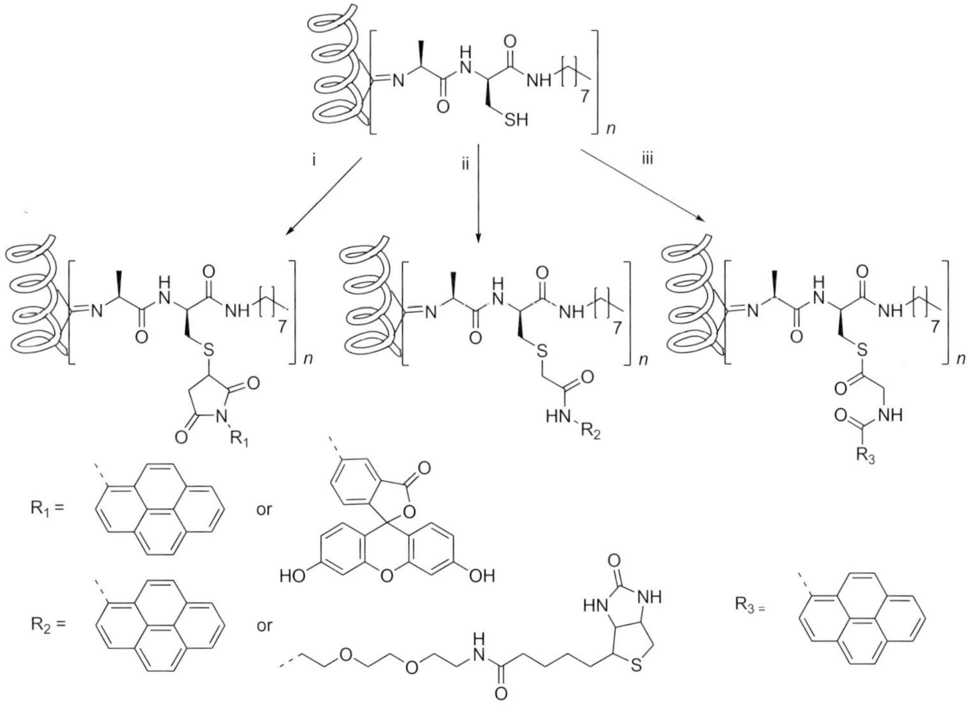

Scheme 16.4 Thiol click reactions used by Le Gac et al. (i) N,N-diisopropylethylamine (DIPEA), CHCl$_3$, room temperature; (ii) DIPEA, CHCl$_3$, room temperature, in the case of R$_2$ = biotin 0.1 M phosphate-buffered saline buffer, pH 8, room temperature; (iii) DIPEA, CHCl$_3$, room temperature.

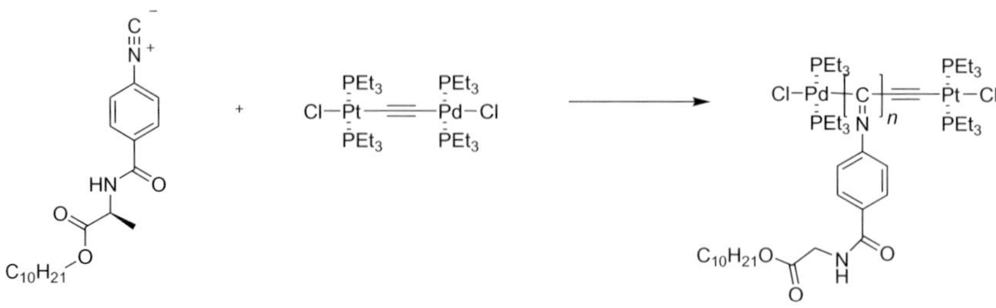

Scheme 16.5 Polymerization of a 4-isocyanobenzoic derivative bearing an L-alanine substituent with a n-decyl chain to form a polyisocyanide that displays LC properties.

a maleimide functional fluorescein derivative was also added, to yield a fluorescent-*co*-biotin random copolymer. When tetravalent streptavidin was added to the biotinylated polymers a precipitate was formed; this proved to be the insoluble polymeric network formed by biotin–streptavidin interaction. These thiol-functionalized polyisocyanides may offer new opportunities in the materials sciences and in biohybrid systems, with multichromophoric scaffolding and protein assembly being among the anticipated applications for these novel polymers.

The helical nature of polyisocyanides has also been exploited in liquid crystal (LC) systems. For example, Yashima *et al.* polymerized an enantiomerically pure 4-isocyanobenzoic acid derivative bearing an L-alanine group substituted with a *n*-decyl chain, with the aim of generating polymeric LCs [104]. When this monomer was polymerized with a Ni catalyst, a polymer with a broad molecular weight distribution (MWD) was obtained that subsequently was shown to produce lyotropic cholesteric LC phases in concentrated solution. When the same monomer was polymerized with the μ-ethynediyl Pt–Pd catalyst, however, a polymer with a low MWD was formed (Scheme 16.5). The use of this Pt–Pd catalyst led to the formation of both left- and right-handed helices which, because they were diastereomeric, could easily be separated on the basis on their different solubilities in acetone. The polymers thus obtained formed thermotropic LC phases with well-defined two- and three- dimensional smectic ordering, and were examined using XRD and AFM [82].

Polyisocyanides derived from 4-isocyanobenzoic acid bearing chiral ester groups have also been synthesized and studied in detail (Figure 16.11) [167]. These were used to prepare chiral stationary phases (CSPs), for liquid chromatography, via an immobilization of the polymers on silica particles. The CSPs obtained, which contained either *P* or *M* helices, were used to separate racemic mixtures of chiral cyclic ethers, for example [168].

An addition of the bulky isocyanide monomer **16** to the polymerization reaction of **14** or **15** (Figure 16.11) was found to inhibit kinetically the growth of the helix that normally was formed, and led to the formation of diastereomers. Compounds **14** and **15** are promesogenic compounds that may interact with liquid crystalline compounds. For example, isocyanide **14** was shown capable of inducing choles-

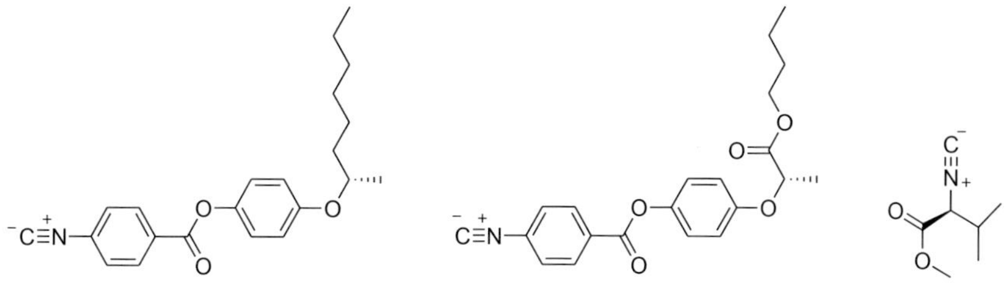

16.6 Functional Polyisocyanides

Figure 16.11 Chemical structures of the chiral monomers **14** (S), **15** (S), and **16** (S).

teric phases in nematic LCs, whereas compound **15** was shown to interact with the smectic C phases in smectic C LCs [169, 170]. When these monomers were polymerized in the presence of LCs, the handedness of the polymer was directly influenced by the handedness of the liquid crystalline phase, which in turn was induced by the chiral monomer. It appeared that, even when the chiral centers in **14** and **15** were quite far away from the polymerizable isocyano group, they still had an effect on the handedness of the polymer. This phenomenon was explained by the monomer having a stereoselective interaction with the growing polymer chain, in similar fashion to its interaction with mesogenic molecules in the liquid crystalline phase. It was shown, using by CD spectroscopy, that the polymer of **14** has a stable conformation at up to 55 °C; however, the introduction of a nitro group on the phenyl ring of **14** led to a less-stable conformation, as indicated by an irreversible loss of optical activity upon heating [171]. A range of isocyanides similar to **14** and **15**, all of which bore the benzoate spacer but with extra added phenyl rings, was synthesized to investigate the transfer of chirality through rigid spacers. The results obtained suggested that the spacer must adopt a twisted conformation in order to effectively transfer the chiral information to the polymer backbone. These results also confirmed that chiral information could, indeed, be transmitted through spacers with lengths up to 21 Å [172, 173].

The use of electroactive components to create chiral devices was elegantly developed by Amablino *et al.*, who synthesized polyisocyanides that were functionalized with tetrathiafulvalene units in their side chains. These polymers were able to adopt three extreme univalent states and two wide mixed valence states, and could be used as multistate redox-switches in organic molecular devices [174, 175]. This concept was further developed by Takahashi and coworkers, who constructed polyisocyanides containing redox-active ferrocenyl groups [176]. In this case it was shown that, upon electrochemical oxidation of the ferrocenyl groups in the polyisocyanide, the Cotton effect in the CD spectrum was changed dramatically, pointing to an unfolding of the polymer chain. Moreover, new absorption bands that were attributed to the formation of ferrocenium chromophores were identified in the UV/visible spectrum. Similar results were obtained when the polymer was

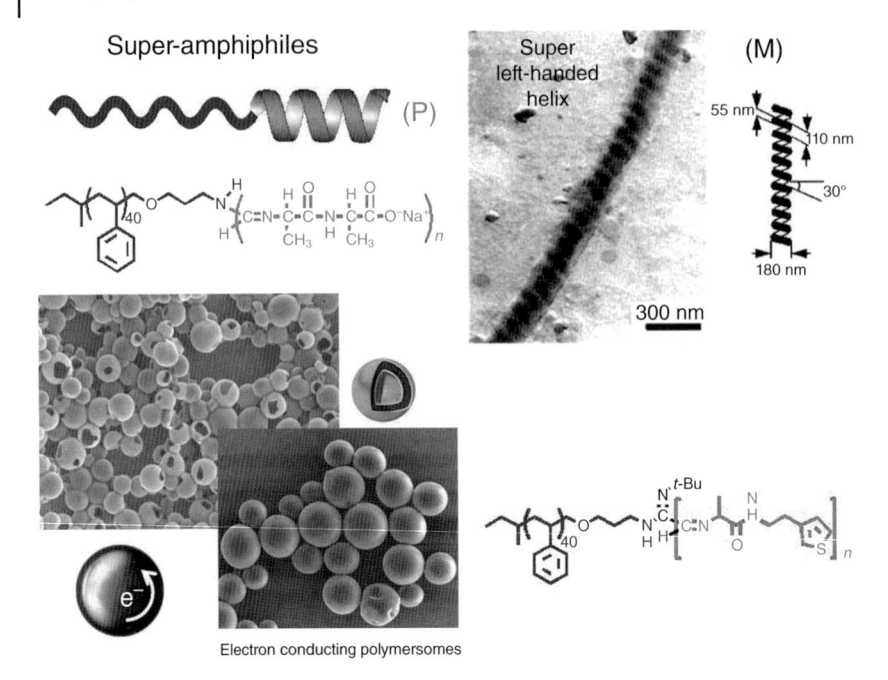

Figure 16.12 Super-amphiphiles formed from polyisocyanides and polystyrene, which form helical super structures and semi-conducting polymersomes in water (for details, see the text).

oxidized chemically; upon reduction of the ferrocenium units, the polymer was seen to refold into a helical conformation, whereupon the original signals in the CD spectrum returned.

In spite of the unique "spring-like" architecture of polyisocyanides, the self-assembly of these macromolecules has been largely unexplored. Although assembly leads to LC phases, very few polyisocyanide nanoarchitectures have been studied, however. The Nolte group has also prepared amphiphilic block copolymers composed of hydrophilic polyisocyanopeptides and hydrophobic polystyrene segments (the so-called "super-amphiphiles") which, on dispersal in water, formed all types of aggregates, including super-helices with a helical sense (left-handed helix) opposite to that of the constituting polyisocyanopeptide (right-handed helix) (Figure 16.12) [76]. When one of the amphiphilic block copolymers was provided with thiophene functions, the super-amphiphile formed polymersomes (bilayer vesicles), both in water and in organic solvents. Following the formation of these supramolecular structures, the thiophene groups in the skin of the polymersomes became crosslinked, which resulted in the creation of semi-conducting particles (Figure 16.12) [177]. The thiophene-containing polymersomes were found to be permeable to low-molecular-weight compounds, but not to high-molecular-weight polymers, including enzymes. This allowed for their use as nanoreactors, and to this end two different enzymes – glucose oxidase and Cal B lipase – have been incorporated into the inner aqueous compartment and the bilayer of the polymersomes, respectively. A third enzyme–horseradish peroxidase (HRP)–was then attached to the surface of the polymersomes. The three-enzyme system created was then applied to a three-step cascade reaction that involved the hydrolysis (by Cal B) and subsequent oxidation (by glucose oxidase) of acetyl glucose; the hydrogen peroxide produced in this reaction was utilized by HRP to oxidize the dye, 2,2-azinobis(3-ethylbenzothiazoline-6-sulfonate (ABTS). This nanoreactor, in which the three enzymes were located at fixed and well-defined positions, proved to be active for prolonged periods, in contrast to a mixture of the three enzymes, which rapidly decomposed and failed to demonstrate any activity [178].

16.7
Conclusions and Outlook

Since the introduction during the 1950s of efficient routes for the synthesis of isocyanides, the number of reports on the polymerization of these compounds has grown rapidly. Isocyanides have proved to be ideal reactive monomers that can be easily polymerized by several metal catalysts, among which nickel is greatly preferred. The polyisocyanides possess a unique structure, which results from the fact that each polymer main-chain carbon atom carries a side chain, which leads to steric hindrance and causes the polymer backbone to fold into a helix. This helical structure can be further stabilized by supramolecular interactions in the side chains, such as hydrogen-bonding and π–π stacking interactions.

The mechanism of the polymerization of isocyanides by nickel catalysts has been studied in detail, and the reaction is thought to take place via a series of insertion reactions that proceed in circular fashion around the nickel center (the "merry-go-round" mechanism). In order for the polymerization to start, a nucleophile is required (usually an alcohol or amine) that ultimately appears as an end group in the polymer. Isocyanides can be easily prepared from peptides; indeed, the resulting polyisocyanopeptides are among the stiffest polymers worldwide, with a persistence length that exceeds that of DNA. Moreover, they possess a β-sheet helical structure, which is stabilized by four arrays of hydrogen bonds that run parallel to the polymer helix axis. These polymers can be used as ideal scaffolds for the anchoring of dye molecules, such as porphyrins and perylenes, leading to extended stacks of organized chromophores, that can be used to transport excitons. Further control over the morphology of these polymers can be obtained by growing them from a surface, which results in interesting NLO materials.

One major future challenge for the polyisocyanides will be their application in electronic and photonic devices. However, this will require their morphology to be controlled over larger distances than the nanometer scale reported to date. The fact that polyisocyanides can be easily modified by post-polymerization reactions (e.g., click chemistry) will undoubtedly open interesting possibilities for the attachment of functional molecules to the polymers. This, in turn, should allow for a broad range of applications for these materials, including drug-delivery systems, sensors, and organic voltaic cells.

References

1 Cornelissen, J.J.L.M., Rowan, A.E., Nolte, R.J.M., and Sommerdijk, N.A.J.M. (2001) Chiral architectures from macromolecular building blocks. *Chem. Rev.*, **101**, 4039–4070.
2 Okamoto, Y. (2000) "Trends in polymer science" chiral polymers. *Prog. Polym. Sci.*, **25**, 159–162.
3 Pauling, L., Corey, R.B., and Branson, H.R. (1951) The structure of proteins. *Proc. Natl Acad. Sci. USA*, **37** (4), 205–211.
4 Watson, J.D. and Crick, F.H.C. (1953) Molecular structure of nucleic acids: a structure for deoxyribose nucleic acid. *Nature*, **171**, 737–738.
5 Schwartz, E., Koepf, M., Kitto, H.J., Nolte, R.J.M., and Rowan, A.E. (2011) Helical poly(isocyanides): past, present and future. *Polym. Chem.*, **2** (1), 33–47.
6 Nakano, T. and Okamoto, Y. (2001) Synthetic helical polymers: conformation and function. *Chem. Rev.*, **101**, 4013–4038.
7 Natta, G., Pino, P., Corradini, P., Danusso, F., Mantica, E., Mazzanti, G., and Moraglio, G. (1955) Crystalline high polymers of α-olefins. *J. Am. Chem. Soc.*, **77** (6), 1708–1710.
8 Pino, P. and Lorenzi, G.P. (1960) Optically active vinyl polymers. II. The optical activity of isotactic and block polymers of optically active α-olefins in dilute hydrocarbon solution. *J. Am. Chem. Soc.*, **82** (17), 4745–4747.
9 Nolte, R.J.M., Van Beijnen, A.J.M., and Drenth, W. (1974) Chirality in polyisocyanides. *J. Am. Chem. Soc.*, **96** (18), 5932–5933.
10 Okamoto, Y., Suzuki, K., Ohta, K., Hatada, K., and Yuki, H. (1979) Optically active poly(triphenylmethyl methacrylate) with one-handed helical conformation. *J. Am. Chem. Soc.*, **101** (16), 4763–4765.
11 Doty, P., Bradbury, J.H., and Holtzer, A.M. (1956) Polypeptides. IV. The molecular weight, configuration and association of poly-γ-benzyl-L-glutamate in various solvents. *J. Am. Chem. Soc.*, **78** (5), 947–954.
12 Lundberg, R.D. and Doty, P. (1957) Polypeptides. XVII. A study of the kinetics of the primary amine-initiated polymerization of N-carboxy-anhydrides with special reference to configurational and stereochemical effects. *J. Am. Chem. Soc.*, **79** (15), 3961–3972.
13 Doty, P. and Lundberg, R.D. (1957) Polypeptides. Xa. Additional comments of the amine-initiated polymerization. *J. Am. Chem. Soc.*, **79** (9), 2338–2339.
14 Chen, F., Lepore, G., and Goodman, M. (1974) Conformational studies of poly[(S)-β-aminobutyric acid]. *Macromolecules*, **7** (6), 779–783.
15 Fernandez-Santin, J.M., Aymami, J., Rodriguez-Galan, A., Munoz-Guerra, S., and Subirana, J.A. (1984) A pseudo [alpha]-helix from poly([alpha]-isobutyl-L-aspartate), a nylon-3 derivative. *Nature*, **311** (5981), 53–54.
16 Appella, D.H., Christianson, L.A., Karle, I.L., Powell, D.R., and Gellman, S.H. (1996) β-Peptide foldamers: robust helix formation in a new family of β-amino acid oligomers. *J. Am. Chem. Soc.*, **118** (51), 13071–13072.
17 Yamada, T., Nagata, Y., and Suginome, M. (2010) Non-hydrogen-bonding-based, solvent-dependent helix inversion between pure P-helix and pure M-helix in poly(quinoxaline-2,3-diyl)s bearing chiral side chains. *Chem. Commun. (Camb.)*, **46**, 4914–4916.
18 Yashima, E., Maeda, K., Iida, H., Furusho, Y., and Nagai, K. (2009) Helical polymers: synthesis, structures, and functions. *Chem. Rev.*, **109** (11), 6102–6211.
19 Tang, H.Z., Novak, B.M., He, J., and Polavarapu, P.L. (2005) A thermal and solvocontrollable cylindrical nanoshutter based on a single screw-sense helical

20 Tang, H.-Z., Lu, Y., Tian, G., Capracotta, M.D., and Novak, B.M. (2004) Stable helical polyguanidines: poly{N-(1-anthryl)-N'-[(R)- and/or (S)-3,7-dimethyloctyl]guanidines}. *J. Am. Chem. Soc.*, **126** (12), 3722–3723.

21 Hill, D.J., Mio, M.J., Prince, R.B., Hughes, T.S., and Moore, J.S. (2001) A field guide to foldamers. *Chem. Rev.*, **101** (12), 3893–4012.

22 Moore, J.S., Gorman, C.B., and Grubbs, R.H. (1991) Soluble, chiral polyacetylenes: syntheses and investigation of their solution conformation. *J. Am. Chem. Soc.*, **113** (5), 1704–1712.

23 Yashima, E., Matsushima, T., and Okamoto, Y. (1995) Poly((4-carboxyphenyl)acetylene) as a probe for chirality assignment of amines by circular dichroism. *J. Am. Chem. Soc.*, **117** (46), 11596–11597.

24 Ciardelli, F., Lanzillo, S., and Pieroni, O. (1974) Optically active polymers of 1-alkynes. *Macromolecules*, **7** (2), 174–179.

25 Simionescu, C.I. and Percec, V. (1982) Progress in polyacetylene chemistry. *Prog. Polym. Sci.*, **8** (1–2), 133–214.

26 Yashima, E. and Maeda, K. (2007) in *Foldamers: Structure, Properties, and Applications* (eds S. Hecht and I. Huc), Wiley-VCH Verlag GmbH, Weinheim, pp. 331–366.

27 Lam, J.W.Y. and Tang, B.Z. (2005) Functional polyacetylenes. *Acc. Chem. Res.*, **38** (9), 745–754.

28 Ute, K., Hirose, K., Kashimoto, H., Hatada, K., and Vogl, O. (1991) Haloaldehyde polymers. 51. Helix-sense reversal of isotactic chloral oligomers in solution. *J. Am. Chem. Soc.*, **113** (16), 6305–6306.

29 Vogl, O. (1994) The rigid single polymer helix. *Prog. Polym. Sci.*, **19**, 1055–1065.

30 Fujiki, M. (1994) Ideal exciton spectra in single- and double-screw-sense helical polysilanes. *J. Am. Chem. Soc.*, **116** (13), 6017–6018.

31 Frey, H., Moeller, M., and Matyjaszewski, K. (1994) Chiral polyguanidine. *Angew. Chem. Int. Ed. Engl.*, **44** (44), 7298–7301.

poly(dipentylsilylene) copolymers. *Macromolecules*, **27** (7), 1814–1818.

32 Okamoto, Y. and Yashima, E. (1990) Asymmetric polymerization of methacrylates. *Prog. Polym. Sci.*, **15** (2), 263–298.

33 Millich, F. and Baker, G.K. (1969) Polyisonitriles. III. Synthesis and racemization of optically active poly(α-phenylethylisonitrile). *Macromolecules*, **2** (2), 122–128.

34 Green, M.M., Gross, R.A., Schilling, F.C., Zero, K., and Crosby, C. (1988) Macromolecular stereochemistry: effect of pendant group structure on the conformational properties of polyisocyanides. *Macromolecules*, **21** (6), 1839–1846.

35 Green, M.M., Gross, R.A., Crosby, C., and Schilling, F.C. (1987) Macromolecular stereochemistry: the effect of pendant group structure on the axial dimension of polyisocyanates. *Macromolecules*, **20** (5), 992–999.

36 Green, M.M., Peterson, N.C., Sato, T., Teramoto, A., Cook, R., and Lifson, S. (1995) A helical polymer with a cooperative response to chiral information. *Science*, **268**, 1860–1866.

37 Green, M.M., Park, J.-W., Sato, T., Teramoto, A., Lifson, S., Selinger, R.L.B., and Selinger, J.V. (1999) The macromolecular route to chiral amplification. *Angew. Chem. Int. Ed. Engl.*, **38** (21), 3138–3154.

38 Suginome, M. and Ito, Y. (eds) (2004) *Polymer Synthesis*. Springer-Verlag, Berlin.

39 Nolte, R.J.M. and Drenth, W. (1992) Synthesis of polymers of isocyanides, in *New Methods for Polymer Synthesis* (ed. W.J. Mijs), Plenum Press, New York, pp. 273–310.

40 Nolte, R.J.M. (1994) Helical poly(isocyanides). *Chem. Soc. Rev.*, **23**, 11–19.

41 Drenth, W. and Nolte, R.J.M. (1979) Poly(iminomethylenes): rigid rod helical polymers. *Acc. Chem. Res.*, **12** (1), 30–35.

42 Vriezema, D.M., Comellas Aragonès, M., Elemans, J.A.A.W., Cornelissen, J.J.L.M., Rowan, A.E., and Nolte, R.J.M. (2005) Self-assembled nanoreactors. *Chem. Rev.*, **105** (4), 1445–1490.

43 Lieke, W. (1859) Ueber das cyanallyl. *Ann. Chem. Pharm.*, **112**, 316–321.

44 Gautier, A. (1867) Ueber die einwirkung des chlorwasserstoffs u. a. auf das Aethyl- und Methylcyanur. *Ann. Chem. Pharm.*, **142**, 289–294.

45 Hofmann, A.W. (1870) Observations on mixed contents. *Ber. Dtsch Chem. Ges.*, **3**, 63.

46 Pirrung, M.C. and Ghorai, S. (2006) Versatile, fragrant, convertible isonitriles. *J. Am. Chem. Soc.*, **128** (36), 11772–11773.

47 Hagedorn, I., Eholzer, U., and Lüttringhaus, A. (1960) Beitrage zur konstitutionsermittlung des antibiotikums xanthocillin. *Chem. Ber.*, **93**, 1584–1590.

48 Ugi, I. and Meyr, R. (1958) New method of preparing isonitriles. *Angew. Chem. Int. Ed. Engl.*, **70**, 702–703.

49 Ugi, I. and Fetzer, U. (1961) Isonitriles. VII. The reaction of cyclohexyl isocyanide with phenylmagnesium bromide. *Chem. Ber.*, **94**, 2239–2243.

50 Creedon, S., M., Kevin Crowley, H., and McCarthy, D., G. (1998) Dehydration of formamides using the Burgess reagent: a new route to isocyanides. *J. Chem. Soc., Perkin. Trans. I*, (6), 1015–1018.

51 Khalpi, S., Dey, S., and Mal, D. (2001) Burgess reagent in organic synthesis. *J. Indian Inst. Sci.*, **81**, 461–476.

52 Böhme, H. and Fuchs, G. (1970) Über Darstellung und Umsetzungen von Formamidomethyl-aminen, -sulfiden und -sulfonen. *Chem. Ber.*, **103** (9), 2775–2779.

53 Walborsky, H.M. and Niznik, G.E. (1972) Synthesis of isonitriles. *J. Org. Chem.*, **37** (2), 187–190.

54 Van der Eijk, J.M., Nolte, R.J.M., and Drenth, W. (1978) Polyisocyanides. 5. Synthesis and polymerization of carbylhistidine and carbylhistamine. *Recl. Trav. Chim. Pays-Bas*, **97**, 46–49.

55 Wittmann, R. (1961) Neue synthese von isonitrilen. *Angew. Chem.*, **73** (6), 219–220.

56 Porcheddu, A., Giacomelli, G., and Salaris, M. (2005) Microwave-assisted synthesis of isonitriles: a general simple methodology. *J. Org. Chem.*, **70** (6), 2361–2363.

57 Millich, F. (1972) Polymerization of isocyanides. *Chem. Rev.*, **72** (2), 101–113.

58 Otten, M.B.J., Ecker, C., Metselaar, G.A., Rowan, A.E., Nolte, R.J.M., Samorì, P., and Rabe, J.P. (2004) Alignment of extremely long single polymer chains by exploiting hydrodynamic flow. *ChemPhysChem*, **5** (1), 128–130.

59 Nolte, R.J.M., Stephany, W., and Drenth, W. (1973) Polyisocyanides. Synthesis and isomerization to polycyanides. *Recl. Trav. Chim. Pays-Bas*, **92**, 83–91.

60 Kamer, P.C.J., Nolte, R.J.M., and Drenth, W. (1988) Screw sense selective polymerization of achiral isocyanides catalyzed by optically active nickel(II) complexes. *J. Am. Chem. Soc.*, **110** (20), 6818–6825.

61 Onitsuka, K., Joh, T., and Takahashi, S. (1992) Reactions of heterobinuclear μ-ethynediyl complexes of palladium and platinum: multiple and successive insertion of isocyanides. *Angew. Chem. Int. Ed. Engl.*, **31**, 851–852.

62 Onitsuka, K., Yanai, K., Takei, F., Joh, T., and Takahashi, S. (1994) Reactions of heterodinuclear μ-ethynediyl palladium-platinum complexes with isocyanides: living polymerization of aryl isocyanides. *Organometallics*, **13**, 3862–3867.

63 Onitsuka, K., Mori, T., Yamamoto, M., Takei, F., and Takahashi, S. (2006) Helical sense selective polymerization of bulky aryl isocyanide possessing chiral ester or amide groups initiated by arylrhodium complexes. *Macromolecules*, **39** (21), 7224–7231.

64 Onitsuka, K., Yamamoto, M., Mori, T., Takei, F., and Takahashi, S. (2006) Living polymerization of bulky aryl isocyanide with arylrhodium complexes. *Organometallics*, **25** (5), 1270–1278.

65 van Beijnen, A.J.M., Nolte, R.J.M., Drenth, W., Hezemans, A.M.F., and van de Coolwijk, P.J.F.M. (1980) Helical configuration of poly(iminomethylenes). Screw sense of polymers derived from optically active alkyl isocyanides. *Macromolecules*, **13** (6), 1386–1391.

66 Van Beijnen, A.J.M., Nolte, R.J.M., Naaktgeboren, A.J., Zwikker, J.W., Drenth, W., and Hezemans, A.M.F. (1983) Helical configuration of poly(iminomethylenes). Synthesis and

CD spectra of polymers derived from optically active isocyanides. *Macromolecules*, **16** (11), 1679–1689.
67 Kamer, P.C.J., Cleij, M.C., Nolte, R.J.M., Harada, T., Hezemans, A.M.F., and Drenth, W. (1988) Atropisomerism in polymers. Screw-sense-selective polymerization of isocyanides by inhibiting the growth of one enantiomer of a racemic pair of helices. *J. Am. Chem. Soc.*, **110** (5), 1581–1587.
68 Metselaar, G.A., Schwartz, E., de Gelder, R., Feiters, M.C., Nikitenko, S., Smolentsev, G., Yalovega, G.E., Soldatov, A.V., Cornelissen, J.J.L., Rowan, A.E., et al. (2007) X-ray spectroscopic and diffraction study of the structure of the active species in the NiII-catalyzed polymerization of isocyanides. *ChemPhysChem*, **8** (12), 1850–1856.
69 Deming, T.J. and Novak, B.M. (1993) Mechanistic studies on the nickel-catalyzed polymerization of isocyanides. *J. Am. Chem. Soc.*, **115**, 9101–9111.
70 Deming, T.J. and Novak, B.M. (1993) Use of copolymerization phenomena in mechanistic studies – monomer substituent effects in nickel-catalyzed isocyanide polymerizations. *Macromolecules*, **26**, 7092–7094.
71 Deming, T.J. and Novak, B.M. (1992) Enantioselective polymerizations of achiral isocyanides. Preparation of optically active helical polymers using chiral nickel catalysts. *Polym. Prepr.*, **33**, 1231–1232.
72 Deming, T.J. and Novak, B.M. (1992) Enantioselective polymerizations of achiral isocyanides – preparation of optically-active helical polymers using chiral nickel-catalysts. *J. Am. Chem. Soc.*, **114**, 7926–7927.
73 Deming, T.J. and Novak, B.M. (1991) Organometallic catalysis in air and water: oxygen-enhanced, nickel-catalyzed polymerizations of isocyanides. *Macromolecules*, **24** (1), 326–328.
74 Deming, T.J. and Novak, B.M. (1991) Polyisocyanides using [(ETA-3-C_3H_5)Ni(OC(O)CF_3)]$_2$ – rational design and implementation of a living polymerization catalyst. *Macromolecules*, **24**, 6043–6045.
75 Deming, T.J. and Novak, B.M. (1991) Change of mechanism block copolymerizations – formation of block copolymers containing helical polyisocyanide and elastomeric polybutadiene segments. *Macromolecules*, **24**, 5478–5480.
76 Deming, T.J., Novak, B.M., and Ziller, J.W. (1994) Living polymerization of butadiene at both chain-ends via a bimetallic nickel initiator – preparation of hydroxytelechelic poly(butadiene) and symmetrical poly(isocyanide-b-butadiene-b-isocyanide) elastomeric triblock copolymers. *J. Am. Chem. Soc.*, **116**, 2366–2374.
77 Cornelissen, J.J.L.M., Van Heerbeek, R., Kamer, P.C.J., Reek, J.N.H., Sommerdijk, N.A.J.M., and Nolte, R.J.M. (2002) Silver nanoarrays templated by block copolymers of carbosilane dendrimers and polyisocyanopeptides. *Adv. Mater.*, **14**, 489–492.
78 Cornelissen, J.J.L.M., Fischer, M., Van Waes, R., Van Heerbeek, R., Kamer, P.C.J., Reek, J.N.H., and Nolte, R.J.M. (2004) Synthesis, characterization and aggregation behavior of block copolymers containing a polyisocyanopeptide segment. *Polymer*, **45** (22), 7417–7430.
79 Cornelissen, J.J.L.M., Fischer, M., Sommerdijk, N.A.J.M., and Nolte, R.J.M. (1998) Helical superstructures from charged poly(styrene)-poly(isocyanodipeptide) block copolymers. *Science*, **280**, 1427–1430.
80 Kros, A., Jesse, W., Metselaar, G.A., and Cornelissen, J.J.L. (2005) Synthesis and self-assembly of rod-rod hybrid poly(γ-benzyl L-glutamate)-block-polyisocyanide copolymers. *Angew. Chem. Int. Ed. Engl.*, **44**, 4349–4352.
81 Wu, Z.-Q., Nagai, K., Banno, M., Okoshi, K., Onitsuka, K., and Yashima, E. (2009) Enantiomer-selective and helix-sense-selective living block copolymerization of isocyanide enantiomers initiated by single-handed helical poly(phenyl isocyanide)s. *J. Am. Chem. Soc.*, **131** (19), 6708–6718.
82 Onouchi, H., Okoshi, K., Kajitani, T., Sakurai, S., Nagai, K., Kumaki, J., Onitsuka, K., and Yashima, E. (2007)

Two- and three-dimensional smectic ordering of single-handed helical polymers. *J. Am. Chem. Soc.*, **130** (1), 229–236.

83 Ohshiro, N., Shimizu, A., Okumura, R., Takei, F., Onitsuka, K., and Takahashi, S. (2000) Living polymerization of aryl isocyanides by multifunctional initiators containing Pd–Pt μ-ethynediyl units. *Chem. Lett.*, **29** (7), 786–787.

84 Onitsuka, K., Yabe, K.-I., Ohshiro, N., Shimizu, A., Okumura, R., Takei, F., and Takahashi, S. (2004) Di- and trifunctional initiators containing Pt-Pd μ-ethynediyl units for living polymerization of aryl isocyanides. *Macromolecules*, **37**, 8204–3867.

85 Tian, Y., Li, Y., and Iyoda, T. (2003) Densely grafted polyisocyanides synthesized by two types of polymerization techniques. *J. Polym. Sci. Part A: Polym. Chem.*, **41** (13), 1871–1880.

86 Tian, Y., Kamata, K., Yoshida, H., and Iyoda, T. (2006) Synthesis, liquid-crystalline properties, and supramolecular nanostructures of dendronized poly(isocyanide)s and their precursors. *Chem. Eur. J.*, **12** (2), 584–591.

87 Onitsuka, K., Shimizu, A., Takei, F., and Takahashi, S. (2009) Multifunctional initiator with platinum-acetylide dendritic core for living polymerization of isocyanides. *J. Inorg. Organomet. Polym. Mater.*, **19** (1), 98–103.

88 Metselaar, G.A., Cornelissen, J.J.L.M., Rowan, A.E., and Nolte, R.J.M. (2005) Acid-initiated stereospecific polymerization of isocyanopeptides. *Angew. Chem. Int. Ed. Engl.*, **44** (13), 1990–1993.

89 Cornelissen, J.J.L.M., Donners, J.J.J.M., de Gelder, R., Graswinckel, W.S., Metselaar, G.A., Rowan, A.E., Sommerdijk, N.A.J.M., and Nolte, R.J.M. (2001) beta-Helical polymers from isocyanopeptides. *Science*, **293** (5530), 676–680.

90 Millich, F. and Sinclair, R.G.I. (1968) Polyisonitriles. I. Original structure and possible rearrangement of poly(α-phenylethylisonitrile). *J. Polym. Sci. Polym. Symp.*, **22**, 33–43.

91 Huang, S.Y. and Hellmuth, E.W. (1974) Conformation of poly(α-phenylethyl isocyanide) in solution. II. Hydrodynamic properties. *Polym. Prepr.*, **15**, 505–508.

92 Van Beijnen, A.J.M., Nolte, R.J.M., and Drenth, W. (1980) Poly(iminomethylenes). Part 13. Helical configuration of poly(tert-butyliminomethylene). Complete resolution and maximum optical rotation. *Recl. Trav. Chim. Pays-Bas*, **99**, 121–123.

93 Van Beijnen, A.J.M., Nolte, R.J.M., Drenth, W., and Hezemans, A.M.F. (1976) Screw sense of polyisocyanides. *Tetrahedron*, **32**, 2017–2019.

94 Huige, C.J.M., Hezemans, A.M.F., Nolte, R.J.M., and Drenth, W. (1993) Molecular-mechanics calculation on oligomers of *tert*-butyl isocyanide. *Recl. Trav. Chim. Pays-Bas*, **112**, 33–37.

95 Kollmar, C. and Hoffmann, R. (1990) Polyisocyanides: electronic or steric reasons for their presumed helical structure? *J. Am. Chem. Soc.*, **112**, 8230–8238.

96 Clericuzio, M., Alagona, G., Ghio, C., and Salvadori, P. (1997) Theoretical investigations on the structure of poly(iminomethylenes) with aliphatic side chains. Conformational studies and comparison with experimental spectroscopic data. *J. Am. Chem. Soc.*, **119**, 1059–1071.

97 Schwartz, E., Domingos, S.R.R., Vdovin, A., Koepf, M., Buma, W.J., Cornelissen, J.J.L.M., Rowan, A.E., Nolte, R.J.M., and Woutersen, S. (2010) Direct access to polyisocyanide screw sense using vibrational circular dichroism. *Macromolecules*, **43** (19), 7931–7935.

98 Schlüter, A.D. and Rabe, J.P. (2000) Dendronized polymers: synthesis, characterization, assembly at interfaces, and manipulation. *Angew. Chem. Int. Ed. Engl.*, **39** (5), 864–883.

99 Sheiko, S.S. and Möller, M. (2001) Visualization of macromolecules: a first step to manipulation and controlled response. *Chem. Rev.*, **101** (12), 4099–4124.

100 Kumaki, J., Sakurai, S.-I., and Yashima, E. (2009) Visualization of synthetic

helical polymers by high-resolution atomic force microscopy. *Chem. Soc. Rev.*, **38** (3), 737–746.

101 Sakurai, S.-I., Ohsawa, S., Nagai, K., Okoshi, K., Kumaki, J., and Yashima, E. (2007) Two-dimensional helix-bundle formation of a dynamic helical poly(phenylacetylene) with achiral pendant groups on graphite. *Angew. Chem. Int. Ed. Engl.*, **46** (40), 7605–7608.

102 Sakurai, S.-I., Okoshi, K., Kumaki, J., and Yashima, E. (2006) Two-dimensional hierarchical self-assembly of one-handed helical polymers on graphite. *Angew. Chem. Int. Ed. Engl.*, **45** (8), 1245–1248.

103 Sakurai, S.-I., Okoshi, K., Kumaki, J., and Yashima, E. (2006) Two-dimensional surface chirality control by solvent-induced helicity inversion of a helical polyacetylene on graphite. *J. Am. Chem. Soc.*, **128** (17), 5650–5651.

104 Kajitani, T., Okoshi, K., Sakurai, S.-I., Kumaki, J., and Yashima, E. (2005) Helix-sense controlled polymerization of a single phenyl isocyanide enantiomer leading to diastereomeric helical polyisocyanides with opposite helix-sense and cholesteric liquid crystals with opposite twist-sense. *J. Am. Chem. Soc.*, **128** (3), 708–709.

105 Kajitani, T., Okoshi, K., and Yashima, E. (2008) Helix-sense-controlled polymerization of optically active phenyl isocyanides. *Macromolecules*, **41** (5), 1601–1611.

106 Kumaki, J., Kajitani, T., Nagai, K., Okoshi, K., and Yashima, E. (2010) Visualization of polymer chain conformations in amorphous polyisocyanide Langmuir–Blodgett films by atomic force microscopy. *J. Am. Chem. Soc.*, **132** (16), 5604–5606.

107 Yashima, E., Maeda, K., and Furusho, Y. (2008) Single- and double-stranded helical polymers: synthesis, structures, and functions. *Acc. Chem. Res.*, **41** (9), 1166–1180.

108 Matile, S., Berova, N., Nakanishi, K., Novkova, S., Philipova, I., and Blagoev, B. (1995) Porphyrins: powerful chromophores for structural studies by exciton-coupled circular dichroism. *J. Am. Chem. Soc.*, **117** (26), 7021–7022.

109 Matile, S., Berova, N., Nakanishi, K., Fleischhauer, J., and Woody, R.W. (1996) Structural studies by exciton coupled circular dichroism over a large distance: porphyrin derivatives of steroids, dimeric steroids, and brevetoxin B. *J. Am. Chem. Soc.*, **118** (22), 5198–5206.

110 Berova, N., Nakanishi, K., and Woody, R.W. (2000) *Circular Dichroism: Principles and Applications*, 2nd edn John Wiley & Sons, Inc., New York.

111 Takei, F., Hayashi, H., Onitsuka, K., Kobayashi, N., and Takahashi, S. (2001) Helical chiral polyisocyanides possessing porphyrin pendants: determination of helicity by exciton-coupled circular dichroism. *Angew. Chem. Int. Ed. Engl.*, **40** (21), 4092–4094.

112 Cornelissen, J.J.L.M., Sommerdijk, N.A.J.M., and Nolte, R.J.M. (2002) Determination of the helical sense in alanine based polyisocyanides. *Macromol. Chem. Phys.*, **203** (10-11), 1625–1630.

113 Kajitani, T., Lin, H., Nagai, K., Okoshi, K., Onouchi, H., and Yashima, E. (2009) Helical polyisocyanides with fan-shaped pendants. *Macromolecules*, **42** (2), 560–567.

114 Miyabe, T., Hase, Y., Iida, H., Maeda, K., and Yashima, E. (2009) Synthesis of functional poly(phenyl isocyanide)s with macromolecular helicity memory and their use as asymmetric organocatalysts. *Chirality*, **21** (1), 44–50.

115 Yamada, Y., Kawai, T., Abe, J., and Iyoda, T. (2001) Synthesis of polyisocyanide derived from phenylalanine and its temperature-dependent helical conformation. *J. Polym. Sci. Part A: Polym. Chem.*, **40**, 399–408.

116 Spencer, L., Euler, W.B., Tranficante, D.D., Kim, M., and Rosen, W. (1998) Complete solution NMR analysis of three oligoimine model compounds. *Magn. Reson. Chem.*, **36**, 398–402.

117 Huang, J.-T., Sun, J., Euler, W.B., and Rosen, W. (1997) Aggregation kinetics and precipitation phenomena in poly(phenylisocyanide). *J. Polym. Sci. Part A: Polym. Chem.*, **35** (3), 439–446.

118 Takei, F., Onitsuka, K., and Takahashi, S. (2005) Thermally induced helical conformational change in poly(aryl isocyanide)s with optically active ester groups. *Macromolecules*, **38**, 1513–1516.

119 Hasegawa, T., Kondoh, S., Matsuura, K., and Kobayashi, K. (1999) Rigid helical poly(glycosyl phenyl isocyanide)s: synthesis, conformational analysis, and recognition by lectins. *Macromolecules*, **32** (20), 6595–6603.

120 Hasegawa, T., Matsuura, K., Ariga, K., and Kobayashi, K. (2000) Multilayer adsorption and molecular organization of rigid cylindrical glycoconjugate poly(phenylisocyanide) on hydrophilic surfaces. *Macromolecules*, **33** (7), 2772–2775.

121 Ishikawa, M., Maeda, K., Mitsutsuji, Y., and Yashima, E. (2003) An unprecedented memory of macromolecular helicity induced in an achiral polyisocyanide in water. *J. Am. Chem. Soc.*, **126** (3), 732–733.

122 Hase, Y., Mitsutsuji, Y., Ishikawa, M., Maeda, K., and Yashima, E. (2007) Unexpected thermally stable, cholesteric liquid-crystalline helical polyisocyanides with memory of macromolecular helicity. *Chem. Asian J.*, **2**, 755–763.

123 Ishikawa, M., Maeda, K., and Yashima, E. (2002) Macromolecular chirality induction on optically inactive poly(4-carboxyphenyl isocyanide) with chiral amines: a dynamic conformational transition of poly(phenyl isocyanide) derivatives. *J. Am. Chem. Soc.*, **124** (25), 7448–7458.

124 Hase, Y., Ishikawa, M., Muraki, R., Maeda, K., and Yashima, E. (2006) Helicity induction in a poly(4-carboxyphenyl isocyanide) with chiral amines and memory of the macromolecular helicity in aqueous solution. *Macromolecules*, **39** (18), 6003–6008.

125 Hase, Y., Nagai, K., Iida, H., Maeda, K., Ochi, N., Sawabe, K., Sakajiri, K., Okoshi, K., and Yashima, E. (2009) Mechanism of helix induction in poly(4-carboxyphenyl isocyanide) with chiral amines and memory of the macromolecular helicity and its helical structures. *J. Am. Chem. Soc.*, **131** (30), 10719–10732.

126 van der Eijk, J.M., Nolte, R.J.M., Drenth, W., and Hezemans, A.M.F. (1980) Optically active polyampholytes derived from L- and D-carbylanayl-L-histidine. *Macromolecules*, **13** (6), 1391–1397.

127 Visser, H.G.J., Nolte, R.J.M., and Drenth, W. (1985) Polymers and copolymers of imidazole-containing isocyanides. Esterolytic activity and enantioselectivity. *Macromolecules*, **18** (10), 1818–1825.

128 Cornelissen, J.J.L., Graswinckel, W.S., Rowan, A.E., Sommerdijk, N.A.J.M., and Nolte, R.J.M. (2003) Conformational analysis of dipeptide-derived polyisocyanides. *J. Polym. Sci. Part A: Polym. Chem.*, **41** (11), 1725–1736.

129 Branden, C. and Tooze, J. (1999) *Introduction to Protein Structure*, vol. 84, 2nd edn, Garland Publishing, New York.

130 Metselaar, G.A., Adams, P.J.H.M., Nolte, R.J.M., Cornelissen, J.J.L.M., and Rowan, A.E. (2007) Polyisocyanides derived from tripeptides of alanine. *Chem. Eur. J.*, **13** (3), 950–960.

131 Donners, J.J.J.M., Nolte, R.J.M., and Sommerdijk, N.A.J.M. (2002) A shape-persistent polymeric crystallization template for $CaCO_3$. *J. Am. Chem. Soc.*, **124** (33), 9700–9701.

132 Davydov, A.S. (1973) The theory of contraction of proteins under their excitation. *J. Theor. Biol.*, **38** (3), 559–569.

133 Scott, A. (1992) Davydov's soliton. *Phys. Rep.*, **217** (1), 1–67.

134 Bodis, P., Schwartz, E., Koepf, M., Cornelissen, J.J.L.M., Rowan, A.E., Nolte, R.J.M., and Woutersen, S. (2009) Vibrational self-trapping in beta-sheet structures observed with femtosecond nonlinear infrared spectroscopy. *J. Chem. Phys.*, **131** (12), 124503.

135 Schwartz, E., Bodis, P., Koepf, M., Cornelissen, J.J.L.M., Rowan, A.E., Woutersen, S., and Nolte, R.J.M. (2009) Self-trapped vibrational states in synthetic [small beta]-sheet helices. *Chem. Commun.*, (31), 4675–4677.

136 Lathe, G.H. and Ruthven, C.R. (1956) The separation of substances and estimation of their relative molecular

sizes by the use of columns of starch in water. *Biochem. J.*, **62** (4), 665–674.

137 Samorí, P., Ecker, C., Gössl, I., de Witte, P.A.J., Cornelissen, J.J.L.M., Metselaar, G.A., Otten, M.B.J., Rowan, A.E., Nolte, R.J.M., and Rabe, J.P. (2002) High shape persistence in single polymer chains rigidified with lateral hydrogen bonded networks. *Macromolecules*, **35** (13), 5290–5294.

138 Rivetti, C., Guthold, M., and Bustamante, C. (1996) Scanning force microscopy of DNA deposited onto mica: equilibration versus kinetic trapping studied by statistical polymer chain analysis. *J. Mol. Biol.*, **264** (5), 919–932.

139 Zhuang, W., Ecker, C., Metselaar, G.A., Rowan, A.E., Nolte, R.J.M., Samorí, P., and Rabe, J.P. (2005) SFM characterization of poly(isocyanodipeptide) single polymer chains in controlled environments: effect of tip adhesion and chain swelling. *Macromolecules*, **38** (2), 473–480.

140 Wezenberg, S.J., Metselaar, G.A., Rowan, A.E., Cornelissen, J.J.L.M., Seebach, D., and Nolte, R.J.M. (2006) Synthesis, characterization, and folding behavior of β-amino acid derived polyisocyanides. *Chem. Eur. J.*, **12** (10), 2778–2786.

141 De Witte, P.A.J., Hernando, J., Neuteboom, E.E., van Dijk, E.M.H.P., Meskers, S.C.J., Janssen, R.A.J., van Hulst, N.F., Nolte, R.J.M., García-Parajó, M.F., and Rowan, A.E. (2006) Synthesis and characterization of long perylenediimide polymer fibers: from bulk to the single-molecule level. *J. Phys. Chem. B*, **110** (15), 7803–7812.

142 Palermo, V., Otten, M.B.J., Liscio, A., Schwartz, E., de Witte, P.A.J., Castriciano, M.A., Wienk, M.M., Nolde, F., De Luca, G., Cornelissen, J.J.L.M., *et al.* (2008) The relationship between nanoscale architecture and function in photovoltaic multichromophoric arrays as visualized by Kelvin probe force microscopy. *J. Am. Chem. Soc.*, **130** (44), 14605–14614.

143 Foster, S., Finlayson, C.E., Keivanidis, P.E., Huang, Y.-S., Hwang, I., Friend, R.H., Otten, M.B.J., Lu, L.-P., Schwartz, E., Nolte, R.J.M., *et al.* (2009) Improved performance of perylene-based photovoltaic cells using polyisocyanopeptide arrays. *Macromolecules*, **42** (6), 2023–2030.

144 Schwartz, E., Palermo, V., Finlayson, C.E., Huang, Y.-S., Otten, M.B.J., Liscio, A., Trapani, S., González-Valls, I., Brocorens, P., Cornelissen, J.J.L.M., *et al.* (2009) "Helter-Skelter-Like" perylene polyisocyanopeptides. *Chem. Eur. J.*, **15** (11), 2536–2547.

145 Hernando, J., de Witte, P.A.J., van Dijk, E.M.H.P., Korterik, J., Nolte, R.J.M., Rowan, A.E., García-Parajó, M.F., and van Hulst, N.F. (2004) Investigation of perylene photonic wires by combined single-molecule fluorescence and atomic force microscopy. *Angew. Chem. Int. Ed. Engl.*, **43** (31), 4045–4049.

146 Finlayson, C.E., Friend, R.H., Otten, M.B.J., Schwartz, E., Cornelissen, J.J.L.M., Nolte, R.J.M., Rowan, A.E., Samorì, P., Palermo, V., Liscio, A., *et al.* (2008) Electronic transport properties of ensembles of perylene-substituted poly-isocyanopeptide arrays. *Adv. Funct. Mater.*, **18** (24), 3947–3955.

147 Razenberg, J.A.S., van der Made, A.W., Smeets, J.W.H., and Nolte, R.J.M. (1985) Cyclohexene epoxidation by the mono-oxygenase model (tetraphenylporphyrinato)manganese(III) acetate-sodium hypochlorite. *J. Mol. Catal.*, **31**, 271–287.

148 De Witte, P.A.J., Castriciano, M., Cornelissen, J.J.L.M., Monsù Scolaro, L., Nolte, R.J.M., and Rowan, A.E. (2003) Helical polymer-anchored porphyrin nanorods. *Chem. Eur. J.*, **9** (8), 1775–1781.

149 De Witte, P.A.J. (2004) *Helical Chromophoric Nanowires*, Radboud University Nijmegen, Nijmegen.

150 Takei, F., Kodama, D., Nakamura, S., Onitsuka, K., and Takahashi, S. (2006) Precise synthesis of porphyrin array scaffolding polyisocyanides. *J. Polym. Sci. Part A: Polym. Chem.*, **44** (1), 585–595.

151 Noma, N., Tsuzuki, T., and Shirota, Y. (1995) Alpha-thiophene octamer as a new class of photoactive material for

photoelectrical conversion. *Adv. Mat.*, **7**, 647–648.

152 Dabirian, R., Palermo, V., Liscio, A., Schwartz, E., Otten, M.B.J., Finlayson, C.E., Treossi, E., Friend, R.H., Calestani, G., Müllen, K., *et al.* (2009) The relationship between nanoscale architecture and charge transport in conjugated nanocrystals bridged by multichromophoric polymers. *J. Am. Chem. Soc.*, **131** (20), 7055–7063.

153 Palermo, V., Schwartz, E., Finlayson, C.E., Liscio, A., Otten, M.B.J., Trapani, S., Müllen, K., Beljonne, D., Friend, R.H., Nolte, R.J.M., *et al.* (2010) Macromolecular scaffolding: the relationship between nanoscale architecture and function in multichromophoric arrays for organic electronics. *Adv. Mater.*, **22** (8), E81–E88.

154 Schwartz, E., Le Gac, S., Cornelissen, J.J.L.M., Nolte, R.J.M., and Rowan, A.E. (2010) Macromolecular multi-chromophoric scaffolding. *Chem. Soc. Rev.*, **39** (5), 1576–1599.

155 Chen, Z., Debije, M.G., Debaerdemaeker, T., Osswald, P., and Würthner, F. (2004) Tetrachloro-substituted perylene bisimide dyes as promising n-type organic semiconductors: studies on structural, electrochemical and charge transport properties. *ChemPhysChem*, **5** (1), 137–140.

156 Liscio, A., De Luca, G., Nolde, F., Palermo, V., Müllen, K., and Samorì, P. (2007) Photovoltaic charge generation visualized at the nanoscale: a proof of principle. *J. Am. Chem. Soc.*, **130** (3), 780–781.

157 Ahrens, M.J., Sinks, L.E., Rybtchinski, B., Liu, W., Jones, B.A., Giaimo, J.M., Gusev, A.V., Goshe, A.J., Tiede, D.M., and Wasielewski, M.R. (2004) Self-assembly of supramolecular light-harvesting arrays from covalent multi-chromophore perylene-3,4:9,10-bis(dicarboximide) building blocks. *J. Am. Chem. Soc.*, **126** (26), 8284–8294.

158 Palermo, V., Liscio, A., Gentilini, D., Nolde, F., Müllen, K., and Samorì, P. (2007) Scanning probe microscopy investigation of self-organized perylenetetracarboxdiimide nanostructures at surfaces: structural and electronic properties. *Small*, **3** (1), 161–167.

159 Han, J.J., Wang, W., and Li, A.D.Q. (2005) Folding and unfolding of chromophoric foldamers show unusual colorful single molecule spectral dynamics. *J. Am. Chem. Soc.*, **128** (3), 672–673.

160 Kauranen, M., Verbiest, T., Boutton, C., Teerenstra, M.N., Clays, K., Schouten, A.J., Nolte, R.J.M., and Persoons, A. (1995) Supramolecular second-order nonlinearity of polymers with orientationally correlated chromophores. *Science*, **270** (5238), 966–969.

161 Tornøe, C.W., Christensen, C., and Meldal, M. (2002) Peptidotriazoles on solid phase:[1,2,3]-triazoles by regiospecific copper(I)-catalyzed 1,3-dipolar cycloadditions of terminal alkynens to azides. *J. Org. Chem.*, **67**, 3057–3064.

162 Rostovtsev, V.V., Green, L.G., Fokin, V.V., and Sharpless, K.B. (2002) A stepwise Huisgen cycloaddition process: copper(I)-catalyzed regioselective "ligation" of azides and terminal alkynes. *Angew. Chem. Int. Ed. Engl.*, **41**, 2596–2599.

163 Schwartz, E., Kitto, H.J., de Gelder, R., Nolte, R.J.M., Rowan, A.E., and Cornelissen, J.J.L.M. (2007) Synthesis, characterisation and chiroptical properties of "click"able polyisocyanopeptides. *J. Mater. Chem.*, **17** (19), 1876–1884.

164 Kitto, H.J., Schwartz, E., Nijemeisland, M., Koepf, M., Cornelissen, J.J.L., Rowan, A.E., and Nolte, R.J.M. (2008) Post-modification of helical dipeptido polyisocyanides using the "click" reaction. *J. Mater. Chem.*, **18** (46), 5615–5624.

165 Schwartz, E., Koepf, M., Kitto, H.J., Espelt, M., Nebot-Carda, V.J., De Gelder, R., Nolte, R.J.M., Cornelissen, J.J.L.M., and Rowan, A.E. (2009) Water soluble azido polyisocyanopeptides as functional β-sheet mimics. *J. Polym. Sci. Part A: Polym. Chem.*, **47** (16), 4150–4164.

166 Le Gac, S., Schwartz, E., Koepf, M., Cornelissen, J.J.L.M., Rowan, A.E., and Nolte, R.J.M. (2010) Cysteine-containing

polyisocyanides as versatile nanoplatforms for chromophoric and bioscaffolding. *Chem. Eur. J.*, **16** (21), 6176–6186.
167 Amabilino, D.B., Ramos, E., Serrano, J.-L., and Veciana, J. (1998) Kinetic control of "unnatural" chiral induction in poly(isocyanide)s. *Adv. Mater.*, **10** (13), 1001–1005.
168 Tamura, K., Miyabe, T., Iida, H., and Yashima, E. (2011) Separation of enantiomers on diastereomeric right- and left-handed helical poly(phenyl isocyanide)s bearing l-alanine pendants immobilized on silica gel by HPLC. *Polym. Chem.*, **2** (1), 91–98.
169 Ramos, E., Bosch, J., Serrano, J.L., Sierra, T., and Veciana, J. (1996) Chiral promesogenic monomers inducing one-handed, helical conformations in synthetic polymers. *J. Am. Chem. Soc.*, **118** (19), 4703–4704.
170 Amabilino, D.B., Ramos, E., Serrano, J.-L., Sierra, T., and Veciana, J. (1998) Long-range chiral induction in chemical systems with helical organization. Promesogenic monomers in the formation of poly(isocyanide)s and in the organization of liquid crystals. *J. Am. Chem. Soc.*, **120** (36), 9126–9134.
171 Amabilino, D.B., Serrano, J.-L., and Veciana, J. (2005) Reversible and irreversible conformational changes in poly(isocyanide)s: a remote stereoelectronic effect. *Chem. Commun.*, (3), 322–324.
172 Amabilino, D.B., Ramos, E., Serrano, J.-L., Sierra, T., and Veciana, J. (2005) Chiral teleinduction in the polymerization of isocyanides. *Polymer*, **46** (5), 1507–1521.
173 Chen, J.P., Gao, J.P., and Wang, Z.Y. (1997) Long-distance chirality transfer in polymerization of isocyanides bearing a remote chiral group. *Polym. Int.*, **44** (1), 83–87.
174 Gomar-Nadal, E., Mugica, L., Vidal-Gancedo, J., Casado, J., Navarrete, J.T.L., Veciana, J., Rovira, C., and Amabilino, D.B. (2007) Synthesis and doping of a multifunctional tetrathiafulvalene-substituted poly(isocyanide). *Macromolecules*, **40** (21), 7521–7531.
175 Gomar-Nadal, E., Veciana, J., Rovira, C., and Amabilino, D.B. (2005) Chiral teleinduction in the formation of a macromolecular multistate chiroptical redox switch. *Adv. Mater.*, **17** (17), 2095–2098.
176 Hida, N., Takei, F., Onitsuka, K., Shiga, K., Asaoka, S., Iyoda, T., and Takahashi, S. (2003) Helical, chiral polyisocyanides bearing ferrocenyl groups as pendants: synthesis and properties. *Angew. Chem. Int. Ed. Engl.*, **42** (36), 4349–4352.
177 Vriezema, D.M., Hoogboom, J., Velonia, K., Takazawa, K., Christianen, P.C.M., Maan, J.C., Rowan, A.E., and Nolte, R.J.M. (2003) Vesicles and polymerized vesicles from thiophene-containing rod–coil block copolymers. *Angew. Chem. Int. Ed. Engl.*, **42** (7), 772–776.
178 van Dongen, S.F.M., Nallani, M., Cornelissen, J.J.L.M., Nolte, R.J.M., and van Hest, J.C.M. (2009) A three-enzyme cascade reaction through positional assembly of enzymes in a polymersome nanoreactor. *Chem. Eur. J.*, **15** (5), 1107–1114.

Index

a

ab initio calculations   24, 40, 41
Acantkella acuta   21
acetals   188, 189, 401
acetylenedicarboxylates
– nonclassical IMCRs   315, 316
– zwitterions   264–267, 269, 271, 274–277
achiral isocyanides   1
acid addition reactions   49–52
acid-mediated polymerization   25
acid-promoted condensations   355
α-acidic isocyanides   75–108
– dihydropyridone MCR   95–97
– 4,5-disubstituted oxazole MCR   83, 84
– 2-imidazoline MCR   91–95
– isocyanide reactivity   78–80
– multicomponent reactions   75–108
– nitropyrrole MCR   83, 84
– oxazole MCR and *in situ* Domino process   88–91
– synthesis   76–78
– 2,4,5-trisubstituted oxazole MCR   84–88
– 2,6,7-trisubstituted quinoxaline MCR   82, 83
– union of MCRs   93–95
– van Leusen imidazole MCR   81, 82
acrylonitriles   116
activated alkenes
– isocyanide reactivity   55–58
– α-isocyanoacetic acid derivatives   116, 117
– pyrrole synthesis   386–390
– zwitterions   279–281, 284–286, 288
activated alkynes   55–58, 390
activated azines   220–223
activated heterocumulenes   60, 61
activated isocyanides   299, 308–320
acyclic diaminocarbenes (ADC)   531–533, 537, 546

acyl halides   52–55
O-acyl isoimides   49–51
S-acyl isoimides   50, 51
acyl transfer reactions *see* Mumm rearrangements
acylating agents   129–133
2-acylbenzazepines   54
1-acyl-3,4-dihydroisoquinolines   54
α-acyloxyaminoamides   367
acylpiperazines   352
2-acylpyrrolines   54
acylthioamides   53, 54
2-adamantyl isocyanide   37
ADC *see* acyclic diaminocarbenes
AFM *see* atomic force microscopy
AIBN *see* azo-*bis*-isobutyronitrile
alcohols
– carbene coupling reactions   532–537
– chiral nonracemic isocyanides   2
– isocyanide reactivity   64–66
– zwitterions   273–275
aldazines   204
aldehyde-acids   362, 366–371
aldehydes
– amine surrogates   199
– carboxylic acid surrogates   180–184
– chiral nonracemic isocyanides   4, 5
– isocyanide reactivity   58–60
– α-isocyanoacetic acid derivatives   116, 121–126, 141, 142
– nonclassical IMCRs   305, 317
– zwitterions   275, 276, 283
aldol reactions   18, 121–126
alginate-coated gold electrodes   241, 242
aliphatic isocyanide reactivity   36–41, 43–46, 52
aliphatic nitriles   37–39

Isocyanide Chemistry: Applications in Synthesis and Material Science, First Edition. Edited by Valentine Nenajdenko.
© 2012 Wiley-VCH Verlag GmbH & Co. KGaA. Published 2012 by Wiley-VCH Verlag GmbH & Co. KGaA.

alkenes
- isocyanide reactivity   55–58
- pyrrole synthesis   386–390
- zwitterions   279–281, 284–286, 288
5-alkoxyoxazoles   80
alkyl isocyanoacetates   109, 110
alkyl 2-nitroethanoates   268
alkyl phenylglyoxylates   316
alkylamines   281
N-alkylated quinolines   220, 221
alkylation reactions   113–115, 351, 352
alkyne–azide cycloaddition reactions   442–444
alkynes
- isocyanide reactivity   55–58
- α-isocyanoacetic acid derivatives   119–121
- pyrrole synthesis   390, 401, 402
- zwitterions   263–282
allylation reactions   114
alternating peptoid–peptide chimeras   253
ambiguine H   22, 23
amides   50, 51, 271–273
amidines
- carbene coupling reactions   538, 539
- carboxylic acid surrogates   179, 184
- chiral nonracemic isocyanides   15
- molecular diversity   371–373
- nonclassical IMCRs   324
aminals   188, 189
amine surrogates   195–231
- activated azines   220–223
- enamines, masked imines and cyclic imines   223–227
- hydrazines   200–218
- hydroxylamines   196–200
- isocyanide reactivity   220–223
- miscellaneous surrogates   218–220
amines
- carbene coupling reactions   532–537
- chiral nonracemic isocyanides   2, 3
- isocyanide reactivity   64–66
- zwitterions   264
amino acids
- amine surrogates   203
- α-isocyanoacetic acid derivatives   111
- molecular diversity   362–365
- polyisocyanides   563
α-aminoaldehydes   436, 437
aminoboranes   184
aminocarbenes see carbenes
N-aminocarbonyl amides   218
amino carboxamides   182
aminocarbynes   532
2-aminofurans   451–453

amino isocyanides   11, 12
N-aminolactams   214
5-aminooxazoles   140, 141
2-amino-4H-pyrans   270
2-aminopyrazoles   466
aminopyrroles   278, 279
5-aminothiazoles   132
anthracenes   410, 411
anti-fouling activity   20, 21
antitubulin chalcones   471
Arbuzov reactions   10
arenesulfenyl chlorides   138
Armstrong benzodiazepine synthesis   433
aromatic nitro compounds   402–407
aromatic substitution reactions   355, 356
aryl transfer reactions see Smiles rearrangements
N-aryl amidation reactions   437, 438
N-aryl arylimines   64
N-aryl carboxamides   171, 172
aryl isocyanide reactivity   36, 39–41, 43–46, 52, 53, 65
aryl nitriles   39, 40
arylsulfenyl chlorides   55
arylsulfonylmethyl isocyanides   76–80
N-aryl thiocarboxamides   175, 176
arynes   283, 284, 316–318
Asinger reactions   362
atomic force microscopy (AFM)
- chiral nonracemic isocyanides   26
- polyisocyanides   558, 563, 567, 572
atropisomerism   23, 557
Axinella cannabina   18
axisonitrile-1   16
axisonitrile-3   18
aza-Diels–Alder reactions   89, 90
1-azadienes   269
6-azaindolines   89, 91
aza-β-lactams   212
azasugar imines   225, 226
aza-Wittig reactions   358, 359, 439, 440
azepine tetrazoles   170
azetidinones   186
azides
- amine surrogates   226
- chiral nonracemic isocyanides   11, 12
- polyisocyanides   570, 571
- tetrazole synthesis   471–473
2-azidophenyl isocyanide   540
azines
- amine surrogates   220–223
- nonclassical IMCRs   300, 304–306
azinium salts   300
azoalkenes   206

azo-*bis*-isobutyronitrile (AIBN) 14, 47
azolium salts 306
azomethines 59, 60, 310–312

b

Barton–Zard pyrrole synthesis 116, 117, 386, 390–410, 420–424, 458
base-promoted condensations 355
BBN *see* 9-borabicyclononanes
BCOD *see* bicyclo[2.2.2]octadiene
benzamides 284
benzazepines 342, 343
benzhydrylium ions 36
benzimidazoles
– IMCR examples and protocols 473
– molecular diversity 338, 340, 341, 345
benzimidazolium salts 307, 308
benzochromenes 274
1,5-benzodiazepine-2,4-diones 443–446
1,4-benzodiazepine-2,5-diones (BZD) 337–339, 345, 433–439
benzodiazepines 431–449
– 1,4-benzodiazepine scaffolds via IMCRs 433–443
– 1,5-benzodiazepine scaffolds via IMCRs 443–446
– fused-ring systems 440–443
– molecular diversity 337–340, 343–347, 349, 350
– two-ring systems 433–440
benzodiazepinones 353, 359, 437, 438, 445
benzofurans 473
benzofused heterocycles 175, 176
benzofused γ-lactams 366, 367
benzopiperazinones 176, 177, 357
benzoporphyrins 410
benzoquinones 317
1,4-benzothiazepines 368, 369
benzoxazepines 342, 343
benzoxazepinones 353
benzoxazinones 351, 353
benzoyl isothiocyanate 280, 281
benzyl isocyanide 38
BHT *see* butylated hydroxy toluene
BICA *see* β-bromo-α-isocyanoacrylates
bicyclic lactams 338, 339, 358, 364, 373
bicyclic tetrazoles 343, 344
bicyclo[2.2.2]octadiene (BCOD) 385, 386, 409, 410, 414–416, 419, 423, 424
bifunctional approach (BIFA) 361–374
– amino acids 362–365
– combined bifunctional and post-condensation modifications 373, 374
– cyclic imines 362, 365

– tethered aldehyde and keto acids 362, 366–371
– tethered heterocyclic amidines 371–373
Biginelli reactions 76
binucleation of mononuclear ligands 515, 516
biocatalytic desymmetrization 13
bioconjugates 240, 241, 258
biomimetic syntheses 20, 21
bipyrroles 396, 397
bis(aminofuryl)bicinchoninic amides 274, 275
Bischler–Napiralski synthesis 48
2,3-bis-iminoazetidines 59, 60
block copolymers
– chiral nonracemic isocyanides 26
– polyisocyanides 556, 557, 565, 574
BODIPY *see* boron–dipyrromethene
bond-forming efficiency 336
9-borabicyclononanes (BBN) 22, 23
boron–dipyrromethene (BODIPY) dyes 385, 386, 391, 420–424
boron trifluoride 186
borylative cyclization reactions 326
bovine serum albumin (BSA) 240, 241
bridging ligands
– carbenes 534, 535
– isocyanoarenes 494, 495, 514–517
2-bromobenzoic acids 437
β-bromo-α-isocyanoacrylates (BICA) 135, 136
Brønsted acids
– carboxylic acid surrogates 180, 181, 187
– molecular diversity 372
BSA *see* bovine serum albumin
BTPP *see* t-butyl-iminotris(pyrrolidino) phosphorane
Burgess reagent 2
butylated hydroxy toluene (BHT) 393
t-butyl-iminotris(pyrrolidino)phosphorane (BTPP) 401, 403, 404
BZD *see* 1,4-benzodiazepine-2,5-diones

c

carbamoylated dihydroisoquinolines 301
carbenes
– amine and alcohol coupling reactions 532–537
– dipole–isocyanide couplings 543, 544
– functionalized amine and alcohol couplings 537
– hydrazine and hydrazone couplings 537, 538
– imine and amidine couplings 538, 539

- intramolecular cyclizations 536, 537, 540–543
- α-isocyanoacetic acid derivatives 122, 128
- metal-bound isocyanides 531–550
- miscellaneous reactions 544–546
- polyisocyanides 554
- reactivity modes 531, 532
carbonic acid derivatives 163–165
carbonimidyl chlorides 52, 53
carbonylmetalate complexes 493–497, 502–505, 514, 515
carboxamides 368–370
carboxylation reactions 110
carboxylic acid surrogates 159–194
- carbonic acid derivatives 163–165
- cyanamide 179
- hydrazoic acid 167–171
- isocyanic acid derivatives 166, 167
- mineral and Lewis acids 180–189
- phenols and derivatives 171–179
- selenide and sulfide 165, 166
- silanol 165, 166
- thiocarboxylic acids 160–163
carboxylic acids
- chiral nonracemic isocyanides 2
- isocyanide reactivity 49–52
- zwitterions 274, 275
carvone 21
cationotropic 1,2-shifts 37, 38
cavitands 49, 50
CD see circular dichroism
chalcogenation reactions 321–323
chiral auxiliaries 3, 4
chiral nonracemic isocyanides 1–33
- amino and alcoholic functionalized 11–16
- amino isocyanides 11, 12
- applications 26
- azido isocyanides 11, 12
- carboxylic, sulfonyl and phosphonyl functionalized 4–10
- hydroxy isocyanides 12–16
- α-isocyano amides 7–9
- α-isocyano esters 4–7
- β-isocyano esters and amides 9
- isolation and natural sources 16, 17
- natural isocyanides 16–23
- phosphonylmethyl isocyanides 10
- polyisocyanide synthesis 23–26
- racemization in synthesis 4–6
- simple unfunctionalized isocyanides 1–4
- sulfonylmethyl isocyanides 10
- synthesis 1–23

chiral polyisocyanides 551–553
chiral racemic isocyanides 1
chiral stationary phases (CSP) 572
chlorination reactions 22, 53
chlorination–elimination strategy 226
α-chloroketones 173, 174
cholanic peptide–steroid hybrids 239, 240
chromenes 274
chromenylfurandicarboxylates 277
chromium complexes
- carbenes 540
- isocyanoarenes 511–516, 519–521
chromophore-functionalized polyisocyanides 563–570
cinchona alkaloid catalysts 117
circular dichroism (CD) spectroscopy
- chiral nonracemic isocyanides 24, 26
- polyisocyanides 557–559, 561–563, 565, 570, 573, 574
Claisen rearrangements 18, 77
click chemistry 361, 569–572
cobalt complexes
- α-isocyanoacetic acid derivatives 126
- isocyanoarenes 497–501, 522
collagenase-I 340, 341
compound chemical tractability 336
concanavalin A 237
conformation memory 560, 561
conformational constraints 96, 97
contracted porphyrins 414, 417–420
convergent approach to natural product mimics 256, 257
copper catalysts
- carboxylic acid surrogates 185
- α-isocyanoacetic acid derivatives 120, 121, 128, 140
- isocyanoarenes 517
- molecular diversity 360, 361
- nonclassical IMCRs 326
- polyisocyanides 569
corannuleneopyrroles 403, 404
core-modified porphyrins 414, 416, 418
Cotton effects 559, 573
crosslinking 242, 243
crown ethers 536
cryptands 246, 247, 255, 256
CSP see chiral stationary phases
CTV see cyclotriveratrylene
Curtius rearrangements 2, 3, 11, 20
CV see cyclic voltammetry
cyanamides 40, 179
α-cyano-divinylketones 118, 119
α-cyano-α-isocyanoacetates 137
N-(cyanomethyl)amides 85, 86

cyclic imines
- amine surrogates   223–227
- chiral nonracemic isocyanides   5, 6
- molecular diversity   362, 365
cyclic iminium derivatives   169
cyclic ketones   198
cyclic peptoids   254, 255, 358
cyclic thioimidic esters   282
cyclic voltammetry (CV)   555
cyclization reactions
- α-acidic isocyanides   80, 83–85
- amine surrogates   203, 216, 217
- benzodiazepine synthesis   432, 435, 436, 438–441
- carbenes   536, 537, 540–543
- carboxylic acid surrogates   160, 161, 166, 167, 177–179
- indole synthesis   474–477
- isocyanide reactivity   48, 53–56, 58–61, 65, 66
- α-isocyanoacetic acid derivatives   114, 115, 118, 119, 122, 129–132, 135, 136, 139–143
- molecular diversity   337–350, 355
- nonclassical IMCRs   314, 315, 321–323, 326
- pyrrole synthesis   385, 395, 396
- quinoline synthesis   477–479
- thiazole synthesis   464–466
cycloaddition reactions
- α-acidic isocyanides   78–82
- amine surrogates   198
- benzodiazepine synthesis   442–444
- chiral nonracemic isocyanides   11, 12
- isocyanide reactivity   35, 55–61
- α-isocyanoacetic acid derivatives   117, 118, 120, 121
- isoxazole synthesis   461, 464
- molecular diversity   357, 359–361, 371–373
- nonclassical IMCRs   303, 310–312, 318–320
- polyisocyanides   569
- pyrrole synthesis   385, 414–416
- zwitterions   264, 265, 280, 281, 285, 286
cyclobutenediones   277, 278
cyclobutyl isocyanide   38
cyclocondensation reactions   122
cycloelimination reactions   303
cyclopropanes   288, 289
cyclopropenimines   55
cyclo[8]pyrroles   385, 386, 419, 420
cyclosporines   6, 51, 144
cyclotriveratrylene (CTV)   238

cylindrical nanocapsules   49, 50
cysteine-based polyisocyanides   26

d
Davydov solitons   562
DBU *see* 1,8-diazabicyclo[5.4.0]undec-7-ene
DCC *see* N,N-dicyclohexylcarbodiimide
DDQ *see* 2,3-dichloro-5,6-dicyano-1,4-benzoquinone
Debye–Scherrer X-ray analyses   557
decarboxylation reactions   9
dehydration reactions
- α-acidic isocyanides   77, 78
- chiral nonracemic isocyanides   1, 2, 4, 9, 14, 15, 18
- α-isocyanoacetic acid derivatives   112
dendrimers   253, 254, 258
Dendrobinium nobile   54
density functional theory (DFT)   56, 512–514
2-deoxy-2-isocyano sugars   15
deoxypyrrololine   394
depsipeptides
- α-acidic isocyanides   96, 97
- α-isocyanoacetic acid derivatives   137
- multiple IMCRs   234, 251, 252
DFT *see* density functional theory
DHOP *see* dihydrooxazolopyridines
DHP *see* dihydropyridines
DIA *see* 2,6-diisocyanoazulene
gem-diactivated olefins   288
diarylglyoxal   205
3,4-diarylpyrroles   395
1,8-diazabicyclo[5.4.0]undec-7-ene (DBU)   389–406, 455, 456
diazaborolidines   310
diazadiborinines   309
DIB *see* 1,4-diisocyanobenzene
DIBA *see* 2,2′-diisocyano-6,6′-biazulenyl
2,3-dichloro-5,6-dicyano-1,4-benzoquinone (DDQ)
- multiple IMCRs   252, 253
- nonclassical IMCRs   301
- pyrrole synthesis   408, 409, 419
- quinoxaline synthesis   479
N,N-dicyclohexylcarbodiimide (DCC)   2, 51
Diels–Alder reactions
- α-acidic isocyanides   88–90
- chiral nonracemic isocyanides   20
- molecular diversity   360
- pyrrole synthesis   397–400, 410, 411
2-diethylaminoquinazolines   65, 66
dihalogeno isocyanides   52, 53

dihydroazaphenanthrene-fused
  benzodiazepinediones 440, 441
4,5-dihydro-3H-1,4-benzodiazepin-5-ones
  436
dihydroisoindoles 398, 399
3,4-dihydroisoquinolines 48
dihydroisoquinoxalinones 338, 339
dihydrooxazolopyridines (DHOP) 96, 97
dihydropyridines (DHP) 54, 307, 308
dihydropyridones 95–97
dihydroquinazoline–benzodiazepine
  tetracycles 442, 443
dihydroquinazolines 345–347, 359
diisocyanoarenes 514–521
2,6-diisocyanoazulene (DIA) 514–516,
  518–521
1,4-diisocyanobenzene (DIB) 515, 516, 519
2,2´-diisocyano-6,6´-biazulenyl (DIBA)
  515–519, 522
N,N-diisopropylethylamine (DIPEA) 571
diketopiperazines 338–340, 345, 352, 355,
  365
dimethyl acetylenedicarboxylate (DMAD)
– furan synthesis 451–453
– isocyanide reactivity 56
– nonclassical IMCRs 315, 316
– zwitterions 264–266, 276
dimethylamino isocyanoacrylates 135, 466,
  467
N-dimethylamino peptoids 202
β-dimethylamino-α-isocyanoacrylates 134
dimethyl-1,4-benzoquinone (DMBQ) 2
dimethyl 3-oxopentanedioate 269, 270
dimethyl sulfoxide (DMSO)
– isocyanide reactivity 42, 65
– polyisocyanides 560–562
– zwitterions 264
Dimroth-type rearrangements 276
dioxaphosphorinanes 10
DIPEA see N,N-diisopropylethylamine
dipeptides
– amine surrogates 226, 227
– multiple IMCRs 234, 257
– polyisocyanides 26
diphenyl phosphorylazide (DPPA) 131
diphenylketene 60
dipolar acid fluorides 306
dipolar compounds 263, 264
dipolarophiles 279, 280, 310–313
dipole–isocyanide couplings 543, 544
dipole moments 494
disaccharides 15
disubstituted imidazoles 467, 468
4,5-disubstituted oxazoles 83, 84, 459–461

2,5-disubstituted oxazoles 461
disubstituted pyrroles 453, 454
2,3-disubstituted quinolines 65, 66
disubstituted thiazoles 465
diterpenes 16, 18
dithiocarbamate 282
divinyl ketones 118, 119
DMAD see dimethyl acetylenedicarboxylate
DMBQ see dimethyl-1,4-benzoquinone
DMSO see dimethyl sulfoxide
Domino process 88–91, 321
double isocyanide insertions 318–320, 324,
  325
double MCRs 235, 244–246
DPPA see diphenyl phosphorylazide
drug discovery 335, 336, 374
dye-sensitized solar cells 423, 424, 464–467
dynamic kinetic resolution 7

e
electrocyclization reactions 167
electron paramagnetic resonance (EPR)
  spectroscopy 499, 500
electron spin resonance (ESR) spectroscopy
  39, 555
electronic structure 35, 36
electrophilic reactions
– chiral nonracemic isocyanides 22
– isocyanide reactivity 35, 36, 47–61
– α-isocyanoacetic acid derivatives 138, 139
– nonclassical IMCRs 317
eledoisin 203
elemental sulfur 281, 282
Ellman benzodiazepine synthesis 431, 432
enamines 223–227
enantiomerically pure isocyanides see chiral
  nonracemic isocyanides
encapsulation complexes 49, 50
endothiopeptides 161
7-epi-14-isocyano-isodauc-5-ene 21
epoxides 288, 320
EPR see electron paramagnetic resonance
ESR see electron spin resonance
(ethoxycarbonylmethyl)
  triphenylphosphonium bromide 270, 271
ethyl acetoacetate 203
ethyl isocyanoacetate 202, 203
eurystatin A 4, 5
expanded porphyrins 414, 417–420
exploratory power 336, 368, 369, 374

f
F-SPE see fluorous solid-phase extraction
ferrocenyl carbene complexes 533

ferrocenyl isocyanoarenes 509, 510, 514
ferrocenyl polyisocyanides 573, 574
ferrocenylphosphine complexes 125, 126, 128
file-enhancement strategies 336, 374
Fischer carbene complexes 545
five-coordinate isocyanidemetalates 502–504
flash vacuum pyrolysis (FVP) 3, 13
fluorescence spectroscopy 566, 569, 570
fluoro-benzaldehydes 286, 287
2-fluoro-5-nitrobenzoic acid 342, 343
N-fluoropyridinium fluorides 222, 471–473
fluorous solid-phase extraction (F-SPE) 438
formamides 77, 78, 112
formamidine salts 313
formylation reactions
– chiral nonracemic isocyanides 1, 2, 14, 15, 18
– α-isocyanoacetic acid derivatives 111, 112
– pyrrole synthesis 387, 388
four-coordinate isocyanidemetalates 497–502
fourfold MCRs 245, 246
free-radical *see* radical
Friedel–Crafts annulations 22
fullerenes 280, 281
functional dendrimers 253, 254
functional dyes 385, 386, 391, 420–424
functional polyisocyanides 570–575
functionalized benzodiazepinediones 437
furanone hydrazone pathway 276
furans
– IMCR examples and protocols 451–453
– α-isocyanoacetic acid derivatives 130
– zwitterions 275–279
furopyrrolones 89, 90
fused 3-amino imidazoles 372
fused-ring benzodiazepines 440–443
FVP *see* flash vacuum pyrolysis

g

gallium trichloride 288, 319, 320
GCNF *see* germ cell nuclear factors
gem-diactivated olefins 48, 49, 288
germ cell nuclear factors (GCNF) 432
glutamate analogues 360
glycoconjugates 236, 237
glycosyl isocyanides
– chiral nonracemic isocyanides 13–16
– isocyanide reactivity 41–43
glyoxaldehydes 349
glyoxylic acids 341

gold complexes
– carbenes 535–537, 543
– α-isocyanoacetic acid derivatives 125, 126, 128
– isocyanoarenes 517–522
– nonclassical IMCRs 325, 326
grafting on strategy 563, 564
Grignard reagents 62
Groebke–Bienaymé reactions 160, 180
Groebke–Blackburn reactions 252, 253

h

half-wave redox potentials 513, 514
halogenation reactions 22, 53, 387, 388
Hammett correlations 39, 40, 56
Hantzsch reactions 76
hapalindole alkaloids 17, 18, 21, 22
Heck coupling reactions 356, 357, 543
Heck-type cyclization–isomerizations 178
helical polymer configurations
– chiral nonracemic isocyanides 23–26
– polyisocyanides 551, 552, 556–561, 564
helix inversion barriers 552
heme 386, 393
hemicryptophanes 247
Henry reactions 392, 393
Hermann's catalyst 22
N-heteroaryl thiocarboxamides 175, 176
heteroarynes 283, 284
heterocumulenes 60, 61, 132
heterocyclic amidines 371–373
N-heterocyclic carbenes (NHC) 93, 94, 531, 532, 537, 540, 541, 546
heterocyclization reactions
– α-acidic isocyanides 83, 84
– isocyanide reactivity 54, 58–61, 65
– α-isocyanoacetic acid derivatives 118, 136
hexaphyrinogens 408, 409
high exploratory power 336, 368, 369, 374
highly oriented pyrolytic graphite (HOPG) 558
homoleptic isocyanide complexes 493–497, 510–514
homolytic bond breaking 39
HOPG *see* highly oriented pyrolytic graphite
Horner–Wadsworth–Emmons reactions 252, 387
horseradish peroxidase (HRP) 240–242, 575
Huisgen cycloaddition reactions
– chiral nonracemic isocyanides 11, 12
– molecular diversity 360, 361
– polyisocyanides 569

Hulme benzodiazepine synthesis   433
hybrid macrocycles   244–246, 255, 256
hybrid peptide–peptoid podands   237–240
hybrid peptide–steroid architectures
    238–240
hydantoins
– carboxylic acid surrogates   163, 164, 166,
    167, 184
– molecular diversity   349, 350
hydantonimides   249
hydration reactions   537
hydrazines
– amine surrogates   200–218
– carbene coupling reactions   537, 538
– carboxylic acid surrogates   169
hydrazino turns   207
hydrazinopeptides   203, 204, 206, 207,
    209–215, 250
hydrazo-Ugi reactions   200–218
hydrazoic acid   51, 52, 167–171
hydrazones   218, 537, 538
hydride ion transfer   57
hydrochloric acid addition reactions   52
hydrogels   241–243
hydrogen bonding
– amine surrogates   210, 215, 219
– polyisocyanides   561–563, 566, 575
hydrogen sulfide   165, 166
hydrogenation reactions   13
hydroselenide   160, 165, 166
hydroxides   64–66
hydroxy amides   181, 182
hydroxy benzaldehydes   286, 287
hydroxy heterocycles   175
hydroxy isocyanides   12–16
α-hydroxy-β-isocyano esters   10
hydroxylamines   196–200
α-hydroxylamino acetamides   199–200
α-hydroxylaminoamides   367
hydroxylaminomethyl tetrazoles   198
Hymeniacidon spp.   21

i

ibuprofen   3
IBX *see* 2-iodoxybenzoic acid
IEDDA *see* inverse electron demand
    Diels–Alder
IEP *see* iterative efficiency potential
imidate esters   284
imidates   324
imidazoisoquinolines   287
imidazoketopiperazines   370, 371
imidazoles
– α-acidic isocyanides   78, 79, 81, 82
– IMCR examples and protocols   466–469

– α-isocyanoacetic acid derivatives   132,
    133, 135, 138–140
– nonclassical IMCRs   314, 315
– polyisocyanides   26
imidazolidines   52, 186, 223, 224
imidazolidinetriones   350
imidazolines
– α-acidic isocyanides   78, 79, 91–95
– carboxylic acid surrogates   186
– α-isocyanoacetic acid derivatives   126–
    129, 134
– molecular diversity   340
– multiple IMCRs   251, 252
imidazolinones   115, 140, 141
imidazolium salts   306, 307
imidazolo[1,2-*a*]quinoxalines   252, 253
imidazopyridines   357
imidazoquinoxalinones   356
imidoyl anions   65, 66
imidoyl chlorides   185, 313–315
imine surrogates   195, 196, 223–227
imines
– carbene coupling reactions   538, 539
– isocyanide reactivity   58–60
– α-isocyanoacetic acid derivatives   126–129
– molecular diversity   336, 346, 347, 349,
    362, 365
– nonclassical IMCRs   302, 303
iminium intermediates
– amine surrogates   201, 223–227
– carboxylic acid surrogates   169
– nonclassical IMCRs   300, 307
– zwitterions   279, 280
iminoacyl species   532
iminoboranes   309–312
iminofurans   317
iminoindenones   284
2-iminoisoindolines   317
iminolactones   288, 315, 316
5-iminooxazolines   87, 88
iminooxetanes   58, 59
iminophorphoranes   470
iminopyrones   277, 278
in situ Domino process   88–91
in situ-generated hydrazoic acid   169
incident-photon-to-electron conversion
    efficiency (IPCE)   423
indazolinones   342, 343
Indinavir   352
indium triflate   188
indole-2-carboxylic esters   142
indoles
– carboxylic acid surrogates   178
– IMCR examples and protocols   473–477
– isocyanide reactivity   59

indolinones   357
indolizines   357
indoloketopiperazines   370, 371
infrared (IR) spectroscopy
– isocyanide reactivity   49, 50
– isocyanoarenes   497, 500, 505, 506, 513, 519–522
– polyisocyanides   560, 562, 563
insertion reactions   318–327
intramolecular cyclizations
– α-acidic isocyanides   80, 85
– amine surrogates   216
– benzodiazepine synthesis   435, 436, 438, 440, 441
– carbenes   536, 537, 540–543
– isocyanide reactivity   56
– α-isocyanoacetic acid derivatives   118, 119, 139
– molecular diversity   337
– nonclassical IMCRs   314, 322, 323
intramolecular cycloadditions   117, 118, 198
intramolecular elimination reactions   119
intramolecular hydrazo-Ugi reactions   212, 213
intramolecular hydrogen bonding   210, 215, 219
inverse electron demand Diels–Alder (IEDDA) reactions   360
iodocarbamoylation reactions   307, 308
2-iodoxybenzoic acid (IBX)   301, 302
ionic liquids   459–461
IPCE *see* incident-photon-to-electron conversion efficiency
IR *see* infrared
iron complexes
– carbenes   533, 541, 542
– chiral nonracemic isocyanides   13
– isocyanoarenes   501, 502
– quinoxaline synthesis   479
isocyanates   41–43, 278, 279
isocyanic acid derivatives   166, 167
isocyanide-cyanide rearrangements   35, 37–41
isocyanide dichlorides   52, 53
isocyanide reactivity   35–73
– acid reactions   49–52
– α-acidic isocyanides   78–80
– activated alkene and alkyne reactions   55–58
– activated heterocumulene reactions   60, 61
– amine surrogates   220–223
– carbonyl and imine reactions   58–60
– electronic structure   35, 36
– electrophilic reactions   35, 36, 47–61

– halogen and acyl halide reactions   52–55
– hydroxide, alcohol and amine reactions   64–66
– isocyanide–cyanide rearrangements   35, 37–41
– nonclassical IMCRs   299, 300, 308–320
– nucleophilic reactions   35, 36, 62–66
– organometallic reagents   46, 47, 62–64
– oxidation reactions   35, 41–45
– reduction reactions   35, 45–47
– sulfur and selenium reactions   42–45
isocyanide resins   312, 313
isocyanoacetamides   111, 112, 128, 129, 142, 143
isocyanoacetates   109, 110, 112, 115, 116, 391–407
α-isocyanoacetic acid derivatives   109–158
– acylating agents   129–133
– additional functional group chemistries   134–137
– alkylation reactions   113–115
– alkyne reactions   119–121
– allylation reactions   114
– carbonyl and imine reactions   116, 121–129
– imidazoline synthesis   126–129, 134
– Michael addition reactions   115–119, 122, 123, 135
– miscellaneous reactions   139–144
– multicomponent reactions   133, 134, 137, 141, 142
– organometallic reagents   121–124
– oxazole synthesis   118, 119, 129–132, 138, 142, 143
– oxazoline synthesis   122–126, 131
– pyrrole synthesis   116, 117, 119–121
– sulfur electrophile reactions   138, 139
– synthesis   109–113
– transition metal catalysts   124–127
isocyanoacetonitriles   112, 113
isocyanoamides   7–9, 40, 79–80
isocyanoarenes
– bridging ligands   494, 495, 514–517
– five-coordinate isocyanidemetalates   502–504
– four-coordinate isocyanidemetalates   497–502
– historical perspective   493–497
– homoleptic organometallic complexes   493–497, 510–514
– isocyanoazulenes   508–514, 518–521
– η^5-isocyanocyclopentadienide ligands   509, 510, 522
– low-valent complexes   497–508
– nonbenzenoid isocyanoarenes   508–521

- organometallic reagents 493–529
- redox-related complexes 497–502, 504–508
- self-assembled monolayers 517–521
- six-coordinate isocyanidemetalates 504–508
isocyanoazulenes 508–514, 518–521
10-isocyano-4-cadinene 19, 20
η^5-isocyanocyclopentadienide ligands 509, 510, 522
isocyano esters 4–7, 9, 77–80, 86
ent-2-(isocyano)trachyopsane 19–21
isoindoles 402–407
isoindolinones 356
isomeric porphyrins 414, 417–421
isoquinolines
- amine surrogates 221–223
- nonclassical IMCRs 301, 302, 304, 305, 317, 318
- zwitterions 287, 288
isoselenocyanates 45
isotactic polypropylene 552
isothiocyanates 42–45, 132
isotwistane dione 21
isoxazoles 361, 461, 464
isoxazolines 361
iterative efficiency potential (IEP) 336, 374

j
Joullie–Ugi reactions 225

k
Kennedy benzodiazepine synthesis 435
ketazines 204
ketenes 60, 129, 130
ketenimines
- isocyanide reactivity 61
- nonclassical IMCRs 313, 314
- zwitterions 267, 269, 271–274
keto-acids 362, 366–371
3-ketoamides 118, 119
2-keto-5-aminooxazoles 142, 143
ketoimidoyl halides 53
ketones
- amine surrogates 198, 199
- chiral nonracemic isocyanides 5
- isocyanide reactivity 58–60
- α-isocyanoacetic acid derivatives 116, 121–126, 142
- nonclassical IMCRs 305, 317
- zwitterions 283
ketopiperazines 338, 339, 343, 370, 371
kinetic isotope effect 38

kinetic resolution 7
Knoevenagel reactions
- amine surrogates 205
- α-isocyanoacetic acid derivatives 121, 122
- molecular diversity 355
Knorr pyrrole synthesis 385
Knorr-type pyrazolone syntheses 218

l
lactamization reactions 354
β-lactams
- chiral nonracemic isocyanides 17, 21
- molecular diversity 353, 354, 364, 368
γ-lactams 338, 366, 367
lactonization reactions 354
lanthanide catalysts 184
LC *see* liquid crystal
LDA *see* lithium diisopropylamide
levulinic acid 95
Lewis acids
- carboxylic acid surrogates 180, 181, 184–189
- molecular diversity 372
- nonclassical IMCRs 313, 314
- zwitterions 288, 289
library development 336, 339–343, 374
Liebeskind–Srogl cross-coupling reactions 93
Lindsey method 407, 412–414
liquid crystal (LC) systems 572–574
lithium diisopropylamide (LDA) 321, 479
living polymerizations 555, 556

m
MacDonald condensations 420
McFarland amidine synthesis 182–184
macrobicycles 246, 247, 255, 256
macrocycles
- α-acidic isocyanides 88
- carbenes 535
- molecular diversity 352, 361
- multiple IMCRs 243–247, 254–256
manganese complexes 502–504, 510, 537
Mannich reactions
- α-acidic isocyanides 76, 91, 92
- amine surrogates 205, 206
- carboxylic acid surrogates 169, 175
mannosides 237
Marcaccini benzodiazepine synthesis 436
masked imines 223–227
matrix metalloproteinases (MMP) 340, 341

merry-go-round mechanism   25, 554, 555, 575
metal-bound isocyanides
– amine and alcohol couplings   532–537
– carbenes   531–550
– dipole–isocyanide couplings   543, 544
– functionalized amine and alcohol couplings   537
– hydrazine and hydrazone couplings   537, 538
– imine and amidine couplings   538, 539
– intramolecular cyclizations   536, 537, 540–543
– miscellaneous reactions   544–546
– reactivity modes   531, 532
metal-to-ligand charge transfer (MLCT)   515, 516
N-methylated peptides   6
methyl $tert$-butyl ether (MTBE)   393
methyl $trans$-iminocrotonates   57, 58
N-methylmorpholine (NMM)   4, 7
– α-acidic isocyanides   78
– α-isocyanoacetic acid derivatives   110–112
MiB see multiple multicomponent macrocyclizations
Michael addition reactions
– α-acidic isocyanides   77
– benzodiazepine synthesis   437, 438
– chiral nonracemic isocyanides   21
– α-isocyanoacetic acid derivatives   115–119, 122, 123, 135
– pyrrole synthesis   392, 393
– zwitterions   271–273, 284, 286
microwave-assisted coupling reactions   143, 144, 346–348, 438, 439, 441–443
mineral acids   180–184
Mitsunobu reactions   352–354
MLCT see metal-to-ligand charge transfer
MMP see matrix metalloproteinases
molecular diversity   335–384
– alkylation reactions   351, 352
– amino acids   362–365
– aromatic substitutions   355, 356
– base- or acid-promoted condensations   355
– bifunctional approach   361–374
– combined bifunctional and post-condensation modifications   373, 374
– cyclic imines   362, 365
– cycloaddition reactions   357, 359–361, 371–373
– lactonizations and lactamizations   354
– Mitsunobu reactions   352–354

– nucleophilic additions and substitutions   351–356
– palladium-mediated reactions   356–358
– ring-closing metathesis   358, 373
– secondary reactions of Ugi products   350–361
– Staudinger–aza-Wittig reactions   358, 359
– tethered aldehyde and keto acids   362, 366–371
– tethered heterocyclic amidines   371–373
– TMSN$_3$-modified Ugi-4CC/one internal nucleophile   343, 344
– Ugi-4CC/one internal nucleophile   337–343
– Ugi-4CC/three internal nucleophiles   347, 348
– Ugi-4CC/two internal nucleophiles   344–347
– Ugi-5CC/one internal nucleophile   348–350
molecular weight distribution (MWD)   572
molybdenum complexes
– carbenes   540, 544, 545
– isocyanide reactivity   44
monamphilectine A   17, 21
MTBE see methyl $tert$-butyl ether
multicomponent polymerization   243
multicomponent reactions (MCR)
– α-acidic isocyanides   75–108
– benzimidazole synthesis   473
– benzodiazepine synthesis   431–449
– benzofuran synthesis   473
– chiral nonracemic isocyanides   1, 3, 4, 7, 9, 12–16, 21
– convergent approach to natural product mimics   256, 257
– dihydropyridones   95–97
– 4,5-disubstituted oxazoles   83, 84
– early examples   76
– formal isocyanide insertion reactions   320–327
– furan synthesis   451–453
– heterocycle synthesis   451–492
– hybrid peptide–peptoid podand synthesis   237–240
– imidazole synthesis   466–469
– 2-imidazolines   91–95
– indole synthesis   473–477
– isocyanide activation   299, 308–320
– isocyanide attack on activated species   299, 300–308
– isocyanide reactivity   35, 48, 49, 56–58, 67
– α-isocyanoacetic acid derivatives   133, 134, 137, 141, 142

- isocyanoarenes 496
- isoxazole synthesis 461, 464
- linear and branched scaffolds 248–254
- macrocyclizations 243–247, 254–256
- molecular diversity 335–384
- multicomponent polymerization 243
- multiple IMCRs 233–262
- multivalent glycoconjugate synthesis 236, 237
- nitropyrroles 83, 84
- nonclassical IMCRs 299–334
- one-pot multiple IMCRs 234–243
- oxadiazole synthesis 470, 471
- oxazole MCR and *in situ* Domino process 88–91
- oxazole synthesis 459–461
- protein modification and immobilization 240, 241, 258
- pyrazole synthesis 466, 470
- pyrrole synthesis 391, 453–459
- quinoline synthesis 477–479
- quinoxaline synthesis 479
- sequential IMCRs 248–257
- tetrazole synthesis 471–473
- thiazole synthesis 464–466
- triazole synthesis 470, 471
- 2,4,5-trisubstituted oxazoles 84–88
- 2,6,7-trisubstituted quinoxalines 82, 83
- type I processes 299–308
- type II processes 299, 308–320
- type III processes 299
- union of MCRs 93–95
- van Leusen imidazole MCR 81, 82
- zwitterions 263–265
- *see also* Passerini reactions; Ugi reactions
multiple isocyanide-based MCRs 233–262
- convergent approach to natural product mimics 256, 257
- hybrid peptide–peptoid podand synthesis 237–240
- linear and branched scaffolds 248–254
- macrocyclizations 243–247, 254–256
- multicomponent polymerization 243
- multivalent glycoconjugate synthesis 236, 237
- one-pot multiple IMCRs 234–243
- polysaccharide hydrogel assembly 241–243
- protein modification and immobilization 240, 241, 258
- sequential IMCRs 248–257
multiple multicomponent macrocyclizations (MiB) 243–247

multivalent glycoconjugates 236, 237
Mumm rearrangements
- amine surrogates 195, 198–200
- carboxylic acid surrogates 160, 161, 186
- molecular diversity 336, 363, 364, 367, 368
- multiple IMCRs 250, 251
muraymicyn D2 11
MWD *see* molecular weight distribution

n

nanocapsules 49, 50
nanoreactors 574, 575
naphthoporphyrins 398, 410
naproxen 3
natural isocyanides
- chiral nonracemic isocyanides 16–23
- isolation and natural sources 16, 17
- synthesis 17–23
Nef reactions
- α-isocyanoacetic acid derivatives 129, 135
- nonclassical IMCRs 313, 314
- pyrrole synthesis 392, 393
NHC *see* N-heterocyclic carbenes
nickel complexes
- carbenes 545, 546
- isocyanide reactivity 41, 42
- isocyanoarenes 522
- polyisocyanides 553, 554, 560, 563, 575
nicotinamides 274, 275
ningalin B 386, 395
niobium complexes 506–508
nitrilium intermediates
- carboxylic acid surrogates 160, 181, 186
- molecular diversity 336, 343, 373, 574
- nonclassical IMCRs 300, 302, 303, 306, 316, 317
nitroalkenes 389–396, 458
3-nitroindoles 404, 405
nitroketenes 401
nitrophenols 172–175
2-nitrophenyl isocyanide 540, 541
β-nitroporphyrins 405, 406
nitropyrroles 83, 84, 389, 390
nitroso compounds 320
NLO *see* nonlinear optical
NMM *see* N-methylmorpholine
NMR *see* nuclear magnetic resonance
NOESY *see* nuclear Overhauser effect spectroscopy
non-Evans *syn*-aldol reactions 18
nonbenzenoid isocyanoarenes 508–521

nonclassical isocyanide-based MCRs 299–334
– formal isocyanide insertion reactions 320–327
– isocyanide activation 299, 308–320
– isocyanide attack on activated species 299–308
– type I processes 299–308
– type II processes 299, 308–320
– type III processes 299
nonlinear optical (NLO) materials 567, 575
norbornenes 360
Norcardicin 363, 364
Norrish-type homolytic cleavages 22, 23
nuclear magnetic resonance (NMR) spectroscopy
– amine surrogates 209, 210
– carbenes 536
– chiral nonracemic isocyanides 24
– isocyanide reactivity 49, 50
– isocyanoarenes 494–496, 505, 506, 511, 512
– multiple IMCRs 236
– polyisocyanides 559–563
nuclear Overhauser effect spectroscopy (NOESY) 253
nucleophilic reactions
– amine surrogates 200
– carbenes 531, 532
– isocyanide reactivity 35, 36, 62–66
– α-isocyanoacetic acid derivatives 129, 135, 137
– molecular diversity 337–356
– zwitterions 264, 266–271
nucleophilicity parameters 36
Nutlins 93

o

OBO-esters 111
OFET *see* organic field effect transistors
oligomerization reactions
– isocyanide reactivity 35, 57, 58
– multiple IMCRs 243–247
– porphyrins 419–421
oligonucleotides 306
oligopyrroles 395, 396
one-pot multiple IMCRs 234–243
Ono method 407
OPV *see* organic photovoltaic
organic field effect transistors (OFET) 420, 423, 424
organic photovoltaic (OPV) devices 420, 423, 424, 545–547

organolithium reagents
– chiral nonracemic isocyanides 3, 4
– isocyanide reactivity 62, 63
– nonclassical IMCRs 309, 310, 321
organometallic reagents
– α-acidic isocyanides 77
– bridging isocyanoarene ligands 494, 495, 514–517
– carboxylic acid surrogates 185, 186
– chiral nonracemic isocyanides 3, 4, 14, 15
– five-coordinate isocyanidemetalates 502–504
– four-coordinate isocyanidemetalates 497–502
– historical perspective 493–497
– homoleptic isocyanide complexes 493–497, 510–514
– isocyanide reactivity 46, 47, 62–64
– α-isocyanoacetic acid derivatives 121–124
– isocyanoarenes 493–529
– isocyanoazulenes 508–514, 518–521
– η^5-isocyanocyclopentadienide ligands 509, 510, 522
– low-valent complexes 497–508
– nonbenzenoid isocyanoarenes 508–521
– redox-related complexes 497–502, 504–508
– self-assembled monolayers 517–521
– six-coordinate isocyanidemetalates 504–508
organosamarium reagents 64
organozinc reagents 64
oxadiazepins 269
oxadiazoles 470, 471
oxazepinones 367
oxazines 187
oxazoles
– α-acidic isocyanides 78–80, 83–91
– IMCR examples and protocols 459–461
– α-isocyanoacetic acid derivatives 118, 119, 129–132, 138, 142, 143
– nonclassical IMCRs 302
oxazolidines
– amine surrogates 223, 224
– nonclassical IMCRs 310–313, 320
oxazolines
– α-acidic isocyanides 78, 79, 87, 88
– chiral nonracemic isocyanides 13
– α-isocyanoacetic acid derivatives 122–126, 131
oxetanes 186, 187
oxidation reactions 35, 41–45

oxidation–substitution reactions 2
oximes 199
2-oxocarbapenems 364
oxodiazepines 359
oxyaminocarbene ligands 542, 543

p
palladium catalysts
– α-acidic isocyanides 77
– benzodiazepine synthesis 437, 440
– carbenes 534, 535, 537–539, 542–544
– carboxylic acid surrogates 178
– indole synthesis 474–476
– isocyanide reactivity 53, 64
– α-isocyanoacetic acid derivatives 114, 127, 128
– isocyanoarenes 522
– molecular diversity 356–358
– nonclassical IMCRs 324, 325
palladium–platinum catalysts 554–556, 560, 565, 572
PAMAM *see* Passerini–amine deprotection–acyl migration
parallel synthesis 336, 367
Parr–Knorr synthesis 385
Passerini reactions
– α-acidic isocyanides 75, 76, 94–97
– benzodiazepine synthesis 431, 432
– carbonic acid derivatives 163–165
– carboxylic acid surrogates 159–194
– chiral nonracemic isocyanides 3–5, 10, 12, 13, 16
– cyanamide 179
– hydrazoic acid 167–171
– isocyanic acid derivatives 166, 167
– α-isocyanoacetic acid derivatives 111, 133, 134, 137
– mineral and Lewis acids 180–189
– multiple IMCRs 233–235, 242, 243, 252, 253, 257
– nonclassical IMCRs 320
– oxazole synthesis 461
– phenols and derivatives 171–179
– selenide and sulfide 165, 166
– silanol 165, 166
– thiocarboxylic acids 160–163
– zwitterions 265
Passerini–amine deprotection–acyl migration (PAMAM) strategies 5, 8
PBG *see* porphobilinogen
PDI *see* polydispersity index
penicillin derivatives 362
Penicillium notatum 16
pentacenes 410, 411

pentafuranoses 14, 15
peptide-based hybrid macrocycles 244, 245
peptide nucleic acid (PNA) oligomers 248, 249
peptide–peptoid podands 237–240
peptide–steroid hybrids 238–240
peptidic tetrazoles 249, 250
peptidomimetics
– α-acidic isocyanides 97
– amine surrogates 203, 204, 206, 207, 209–215, 226, 227
– chiral nonracemic isocyanides 12, 13
– molecular diversity 352, 354, 358
– multiple IMCRs 240, 255–257
peptoid–peptide chimeras 253
peptoids
– amine surrogates 202
– hybrid peptide–peptoid podands 237–240
– molecular diversity 358
– multiple IMCRs 250, 251, 253–255
Perkow reactions 313, 314
perylene diimides 565–567
perylene-functionalized polyisocyanides 564–569, 575
phase-transfer catalysis 76
phenanthroline-Ru(III) complexes 414, 417
phenols 171–179
phenoxyiminoisobenzofurans 317, 318
phenylalanine methyl esters 202, 203
phosgenes 77, 78, 109–112
phosphazene super bases 130, 393, 403
phosphines
– carbenes 532, 533, 545, 546
– zwitterions 264
phosphonylmethyl isocyanides 10
photochemical cyclizations 479
phthalazinones 204, 205
Phyllidia varicosa 21
phytochromobolin 394, 395
Pictet–Spengler-type cyclizations 355
Plasmodium falciparum 367
platinum complexes 538, 539, 542, 543, 545, 546
PNA *see* peptide nucleic acid
podand architectures 237–240
polyacetylenes 552
poly(*tert*-butylisocyanide) 23
polychlorals 552
polydentate isocyanides 494–496
polydispersity index (PDI) 556, 566
polyguanidines 552
polyisocyanides 551–585
– applications 26
– chiral nonracemic isocyanides 23–26

- chiral polymers 551–553
- functional 570–575
- helical polymer configurations 551, 552, 555–561, 564
- monomeric isocyanides 553
- multiple IMCRs 238–240
- polyisocyanopeptides 561–563
- polymerization mechanism 553–556
- properties 24
- scaffolds for anchoring chromophores 563–570
- synthesis 25, 26
polyisocyanopeptides 561–563
polymer-supported IMCRs 312, 313, 339, 340, 366–371, 434–436
polymerization reactions
- chiral nonracemic isocyanides 25, 26
- isocyanide reactivity 35, 65
- multiple IMCRs 243
- polyisocyanides 553–556
polymersomes 574, 575
poly(methacrylate ester)s 552
polypropylene 552
polypyrroles 387, 391, 421, 422
polysaccharide networks 241–243
poly(triphenylmethyl methacrylate) 552
porphobilinogen (PBG) 386, 388, 393
porphyrin-fused phenanthroline-Ru(III) complexes 414, 417
porphyrinogens 408, 409
porphyrins
- cycloaddition reactions 414–416
- expanded, contracted and isomeric 414, 417–421
- functional dyes 420–424
- Lindsey method 407, 412–414
- oligomerization 419–421
- polyisocyanides 558, 559, 563–565, 575
- pyrroles 385–387, 391, 395, 396, 398, 407–423
- tetramerization of pyrroles 407–412
Povarov reactions 301
powder X-ray diffraction (PXRD) 563
prodigiosin 386
prolines 365
propiolyl chlorides 89
proteins
- covalent modification and immobilization 240, 241, 258
- multiple IMCRs 240, 241, 258
pseudo-peptidic imidazolines 251, 252
pump-probe spectroscopy 562
PXRD *see* powder X-ray diffraction
2*H*-pyran-3,4-dicarboxylates 282

pyran-2-ones 267
pyrazoles
- amine surrogates 205, 206, 215, 216
- IMCR examples and protocols 466, 470
pyrazol-3-ones 217, 218
pyrazolo[1,2-*a*]pyridazines 274
pyrazoloquinoxalinones 356
pyridazinones 205
pyridine *N*-oxide 42
pyridine-2,6-dicarboxyclic acid 240
pyridines
- nonclassical IMCRs 303, 304, 317, 318
- zwitterions 264, 288
pyridinium ylides 313, 314
pyridinyl–imidoyl adducts 181
pyrrenopyrroles 403, 404
pyrrole-2-carbaldehydes 395, 396
pyrrole-2-carboxylates 391–393, 400, 401
pyrroles 385–429
- α-acidic isocyanides 79, 80, 83, 84
- alkyne-based synthesis 390, 401, 402
- aromatic nitro compound-based synthesis 402–407
- IMCR examples and protocols 453–459
- isocyanoacetate-based synthesis 391–407
- α-isocyanoacetic acid derivatives 116, 117, 119–121
- isoindole synthesis 402–407
- nitroalkene-based synthesis 389–396
- nonclassical IMCRs 302, 303
- porphyrin synthesis 385–387, 391, 395, 396, 398, 407–424
- tetramerization 407–412
- TosMIC-based syntheses 386–390
- α,β-unsaturated sulfone-based synthesis 396–401
pyrrolizidines 118, 119
pyrrolo[1,2-a] [1,4]diazepines 369, 370
pyrroloketopiperazines 370, 371
pyrrolostatin 394
pyruvamides 53

q
quinazolines 338
quinolines 220–223, 477–479
quinoxalines 62, 82, 83, 479
quinoxalinones 341, 342, 347

r
radical inhibitors
- isocyanide reactivity 37–39
- pyrrole synthesis 393

radical reactions
- chiral nonracemic isocyanides   14, 15
- indole synthesis   474–477
- isocyanide reactivity   35, 38, 39, 47
random copolymerization   571
RCM *see* ring-closing metathesis
redox potentials   513, 514
redox-related complexes   497–502, 504–508
reduction reactions   35, 45–47
reductive amination reactions   22
reductive cleavage reactions   18
Reissert-type processes   300–307
resin-bound isocyanides   434–436
resonance light scattering (RLS)   565
retro Diels–Alder reaction   410, 420
retro-Michael reactions   89, 90
rhodium catalysts
   carbenes   545
- isocyanide reactivity   44, 45
- α-isocyanoacetic acid derivatives   122
ring-closing metathesis (RCM)
- carboxylic acid surrogates   177
- molecular diversity   358, 373
ring-opening reactions
- carboxylic acid surrogates   186–189
- isocyanide reactivity   58, 59
- nonclassical IMCRs   323, 324
- zwitterions   288, 289
Ritter reactions   21
RLS *see* resonance light scattering
ruthenium complexes
- carbenes   541
- indole synthesis   474–476
- isocyanoarenes   501
- pyrrole synthesis   414, 417
rylideneindolinones   356, 357

s

Saegusa deamination reactions   47
safety-catch linker isocyanides   338, 339
SAM *see* self-assembled monolayers
samarium catalysts   326, 327
samarium(II) iodide   38
sapphyrins   414, 417–420
Schöllkopf isocyanides   161–163, 170, 171
screening collections   335, 336
SEC *see* size-exclusion chromatography
selenium derivatives   321–323
seleno-carbamoylation reactions   307
selenocyanates   132, 167
selenoesters   132
seleno-oxidation reactions   45
self-assembled monolayers (SAM)   517–521
self-trapping states   562
semicarbazones   219, 220

semiconducting materials   567, 574, 575
sequential IMCRs   248–257
- amine surrogates   197, 210
- linear and branched scaffolds   248–254
- macrocycles   254–256
- natural product mimics   256, 257
sesquiterpenes   16, 18–20
Shaabani benzodiazepine synthesis   443–445
silaaziridines   323, 324
silanol   165, 166
siloxyamides   166
silver catalysts
- α-acidic isocyanides   92, 93
- isocyanide reactivity   48, 54
- α-isocyanoacetic acid derivatives   117, 126, 139, 140
- isocyanoarenes   517
silyl isocyanides   40, 41
single-crystal X-ray analyses   210, 369, 535
single reactant replacement (SRR) strategy   86–88
six-coordinate isocyanidemetalates   504–508
size-exclusion chromatography (SEC)   563
Smiles rearrangements
- carboxylic acid surrogates   160, 171–179
- multiple IMCRs   250, 251
- zwitterions   285, 286
sodium azide   198
sodium naphthalene/1,2-dimethoxyethane   46
solid-phase extraction (SPE)   341, 438
solid-phase synthesis
- benzodiazepines   431
- molecular diversity   366, 367
- multiple IMCRs   248, 249, 253
- nonclassical IMCRs   302
solid-supported IMCRs   312, 313, 339, 340, 366–371, 434–436
solution-phase synthesis   248–250, 253
Sonogashira cross-coupling reactions   356, 357, 539
Soret bands   564, 565
SPE *see* solid-phase extraction
spirocyclic iminolactones   277
γ-spiroiminolactams   278, 279
spiroisoindolines   178, 179
spiroisoquinolines   178, 179
spirostanic peptide–steroid hybrids   239, 240
SRR *see* single reactant replacement
Staudinger–aza-Wittig reactions   358, 359
steric factors   217
steroidal cages   247, 255
Straudinger reactions   439, 440

Strecker reactions 76
structure–reactivity relationships 39
styrylfuran derivatives 276, 277
5-substituted oxazoles 459
sugar-amino acid conjugates 15
sugar isocyanides 13–16
sugar polyisocyanides 26
sulfated glycoconjugates 236, 237
sulfenyl thiocyanates 138, 139
sulfonamides 218, 219
sulfones 396–401
sulfonylimines 317
N-sulfonylimines 129
sulfonylmethyl isocyanides 10
sulfur 281, 282
sulfur electrophile reactions 138, 139
sulfuration reactions 42–45
super-amphiphiles 574
supramolecular macrocycles 535
Suzuki coupling reactions
– benzodiazepine synthesis 438, 439
– carbenes 543
– isocyanide reactivity 53
– pyrrole synthesis 420
Suzuki–Miyaura cross-coupling reactions 538
syn–anti isomerism 559, 560
syndio conformation 24, 559, 560
synthetic hydrogels 241–243

t
Taft correlations 39, 40
tandem MCRs 326, 327
tantalum complexes 504–508
TAP *see* tetraanthra[2,3]porphyrins
TBAF *see* tetra-*n*-butylammonium fluoride
TBP *see* tetrabenzoporphyrin
TCAA *see* trichloroacetic anhydride
telaprevir 10
tellurium derivatives 321–323
tethered aldehyde and keto acids 362, 366–371
tethered heterocyclic amidines 371–373
tetraanthra[2,3]porphyrins (TAP) 413, 414
meso-tetraarylporphyrins 407, 410–414
tetrabenzoporphyrin (TBP) 385, 410–416, 423
tetra-*n*-butylammonium fluoride (TBAF) 77
tetracyclic scaffolds 345, 346
tetrahydroisoquinolines 301
tetrahydropyridazine-3,6-diones 212, 213
tetrahydropyridines 54
tetraisocyano-*meta*-cyclophane (TIMC) 518–520

1,1,3,3-tetramethylbutyl isocyanide (TMBI) 62, 63
tetranaphtho[2,3]porphyrins (TNP) 413, 414
tetraphenylporphyrin (TPP) 410–412, 423
tetrasubstituted imidazoles 468, 469
tetrasubstituted pyrroles 453, 458
tetrathiafulvalenes (TTF) 26, 573
tetrazoles
– amine surrogates 198, 200, 201
– carboxylic acid surrogates 167, 170, 171
– IMCR examples and protocols 471–473
– isocyanide reactivity 51–53, 55
– molecular diversity 343, 344
– multiple IMCRs 249, 250
tetrazolo[1,5-*a*][1,4]benzodiazepines 441–443
TFA *see* trifluoroacetic acid
TFAA *see* trifluoroacetic anhydride
TFT *see* thin-film transistors
1,4-thiazepines 368, 369
thiazoles
– α-acidic isocyanides 78, 79
– IMCR examples and protocols 464–466
– α-isocyanoacetic acid derivatives 137, 138
thiazolidines 223–225
thiazolines 78, 79
thin-film transistors (TFT) 567
thiocarbonic acid 164, 165
thiocarboxylic acids 160–163
thiocyanic acid 52
thioketones 124
thiophenols 48, 49, 175
threefold MCRs 235
TIMC *see* tetraisocyano-*meta*-cyclophane
tin complexes 502, 503
titanium catalysts 325, 326
titanium tetrachloride 184, 185, 187, 188
TMBI *see* 1,1,3,3-tetramethylbutyl isocyanide
TNP *see* tetranaphtho[2,3]porphyrins
Tolypocladium inflatum gams 51, 144
TON *see* turnover number
Torroba benzodiazepine synthesis 439, 440
N-tosylguanidines 314, 315
α-tosylpyrroles 389, 390
TPP *see* tetraphenylporphyrin
transition metal catalysts
– chiral nonracemic isocyanides 13, 25
– isocyanide reactivity 35, 41, 42, 44, 45
– α-isocyanoacetic acid derivatives 124–127
– isocyanoarenes 493–496
– nonclassical IMCRs 324–326

trapping agents
- carbonyl compounds 275–278, 283, 284
- CH-acids 266–271
- electron-deficient olefins 279–281
- imines 278–280
- isocyanides 287–289
- miscellaneous compounds 280–282
- NH-acids 271–273
- nonclassical IMCRs 313, 315–318
- OH-acids 273–275
- zwitterions 265–282
TREN see tris(2-aminoethyl)amine
triacid building blocks 238–240
triazadibenzoazulenones 441
triazines 52, 220
triazoles 361, 470, 471
tributyltin hydride 14, 15, 46, 47
trichloroacetic anhydride (TCAA) 305
trifluoroacetic acid (TFA)
- amine surrogates 204–208, 219
- molecular diversity 337, 341, 348
- multiple IMCRs 250, 257
- nonclassical IMCRs 313
- polyisocyanides 562, 563
trifluoroacetic anhydride (TFAA) 303–307
α-trifluoromethyl ketones 173, 174
triiminothietanes 61
triisocyanide building blocks 238–240
trimethylsilyl azide 471–473
9-triptycyl isocyanide 39
tris(2-aminoethyl)amine (TREN) 238–240
trisubstituted imidazoles 468, 469
2,4,5-trisubstituted oxazoles 84–88, 93, 461–463
trisubstituted pyrroles 453, 455–457
2,6,7-trisubstituted quinoxalines 82, 83
trisubstituted thiazoles 466
Trypanosoma brucei 367
TTF see tetrathiafulvalenes
tungsten complexes 514–516, 540
turnover number (TON) 537
two-component coupling reactions 35, 143, 144

u

UDC see Ugi/de-Boc/cyclize
Ugi reactions
- α-acidic isocyanides 75, 76, 94–97
- activated azines 220–223
- amine surrogates 195–231
- benzodiazepine synthesis 431–446
- bifunctional approach 361–374
- carbonic acid derivatives 163–165
- carboxylic acid surrogates 159–194
- chiral nonracemic isocyanides 4–6, 8, 11–13, 16
- cyanamide 179
- hydrazines 200–218
- hydrazoic acid 167–171
- hydroxylamines 196–200
- imine surrogates 195, 196, 223–227
- isocyanic acid derivatives 166, 167
- α-isocyanoacetic acid derivatives 111, 133, 134, 137
- mineral and Lewis acids 180–189
- molecular diversity 335–361
- multiple IMCRs 233–258
- nonclassical IMCRs 300–307, 320
- phenols and derivatives 171–179
- secondary reactions of Ugi products 350–361
- selenide and sulfide 165, 166
- silanol 165, 166
- tetrazole synthesis 471–473
- thiocarboxylic acids 160–163
- zwitterions 265, 285, 286
Ugi–Reissert reactions 301–304
Ugi/de-Boc/cyclize (UDC) strategy
- amine surrogates 204
- benzodiazepine synthesis 434, 435, 441
- carboxylic acid surrogates 163
- molecular diversity 337–350
- TMSN$_3$-modified Ugi-4CC/one internal nucleophile 343, 344
- Ugi-4CC/one internal nucleophile 337–343
- Ugi-4CC/three internal nucleophiles 347, 348
- Ugi-4CC/two internal nucleophiles 344–347
- Ugi-5CC/one internal nucleophile 348–350
Ullmann cyclizations 161, 177
ultraviolet (UV)-visible spectroscopy 515, 564–566, 570, 573, 574
union of MCRs 93–95
universal isocyanides 335
α,β-unsaturated aldehydes 276, 277
α,β-unsaturated ketones 116
α,β-unsaturated sulfones 396–401
ureas 349
ureido-linked disaccharides 15
urethanes 349
UV see ultraviolet

v

van Leusen reaction 81, 82, 386–390, 458, 466–471

vanadium complexes 504–508
variolins 53
vibrational circular dichroism (VCD) 558
Vilsmeier–Haack reactions 395, 396
vinigrol 47

w

Walborsky reagent 62, 63, 252
Wang resins 366, 367
Welwitindolinones 17, 21, 22
Wolff rearrangements 142, 143

x

X-ray diffraction (XRD)
– carbenes 535
– chiral nonracemic isocyanides 24
– molecular diversity 369
– multiple IMCRs 253
– polyisocyanides 560, 561, 563, 572
X-ray structural spectroscopy 499, 500, 513
xanthocillin 16

z

Zhu three-component reactions 244, 245
zinc bromide 52
zinc catalysts
– amine surrogates 199
– carboxylic acid surrogates 187
– oxadiazole and triazole synthesis 470, 471
– tetrazole synthesis 471–473
zinc porphyrins 565
zirconium catalysts 474–476
zwitterions 263–298
– carbonyl compounds as trapping agents 275–278, 283, 284
– CH-acids as trapping agents 266–271
– electron-deficient olefins as trapping agents 279–281
– imines as trapping agents 278–280
– isocyanide reactivity 47, 48, 56, 57
– isocyanide–alkyne reactions 263–282
– isocyanide–aryne reactions 283, 284
– isocyanide–electron-deficient olefin reactions 284–286
– isocyanides as trapping agents 287–289
– miscellaneous reactions 286, 287
– miscellaneous trapping agents 280–282
– NH-acids as trapping agents 271–273
– nonclassical IMCRs 313, 315–318
– OH-acids as trapping agents 273–275